Engineering of Glacial Deposits

Barry G. Clarke

CRC Press is an imprint of the
Taylor & Francis Group, an **informa** business

CRC Press
Taylor & Francis Group
6000 Broken Sound Parkway NW, Suite 300
Boca Raton, FL 33487-2742

First issued in paperback 2019

ISBN-13: 978-0-415-39865-7 (hbk)
ISBN-13: 978-0-367-86548-1 (pbk)

Library of Congress Cataloging–in–Publication Data

Names: Clarke, B. G. (Barry Goldsmith), 1950- author.
Title: Engineering of glacial deposits / Barry Clarke.
Description: Abingdon, Oxon ; New York, NY : Routledge is an imprint of the Taylor & Francis Group, an Informa Business, [2017] | Includes bibliographical references and index.
Identifiers: LCCN 2016047467| ISBN 9780415398657 (hbk : alk. paper) | ISBN 9781482265828 (ebk)
Subjects: LCSH: Soil mechanics. | Drift. | Geotechnical engineering.
Classification: LCC TA710 .C557 2017 | DDC 624.1/513--dc23
LC record available at https://lccn.loc.gov/2016047467

Visit the Taylor & Francis Web site at
http://www.taylorandfrancis.com

and the CRC Press Web site at
http://www.crcpress.com

Engineering of Glacial Deposits

Contents

Preface

The industrial revolutions of the nineteenth century saw the emergence of geotechnical engineering as a discipline as engineers undertook major earthworks to create the necessary alignments for the canal and rail networks. They relied on experience and observations to deal with a variety of ground conditions to create infrastructure that exists today. Towards the end of the nineteenth century, the science of soil mechanics developed to create a rational approach to geotechnical engineering, and in the twentieth century, codes were introduced to ensure geotechnical structures were fit for purpose, safe and economic. The nineteenth century also saw the emergence of the science of glacial geology with amateur and professional geologists debating the formation of glacial soils. This continues today but, as with engineering, with the support of increasingly sophisticated scientific tools. However, we still rely on the observational technique to verify the sophisticated analyses. This is especially the case with glacial soils which are complex, composite spatially variable soils that prove to be challenging not only in assessing their structure and properties but also in ensuring that economical, safe designs are constructed. Linking glacial geology, soil mechanics and geotechnical engineering is key to reducing risk when engineering glacial soils, which is the aim of this book.

This book sets out by exploring the development of engineering of glacial soils within the context of glacial geology, highlighting the fact that glacial soils do not conform to the assumption that soil can be assumed to be a gravitationally deposited homogeneous material that acts as a continuum. Indeed, the deposition of glacial soils and the impact on their properties are still not fully understood. The intrinsic link between the formation of glacial soils, their properties and the challenges of engineering these soils gives structure to this book, which uses case studies to highlight the behaviour of glacial soils in the natural and built environment.

Up to 30% of the world's land mass has been subject to periods of glaciation creating extensive deposits of glacial tills, glaciolacustrine clays, and glaciofluvial and glaciomarine soils. Studying their formation gives an insight into their composition, fabric and structure and how they may affect their behaviour. Appreciating the principles of glacial geology is critical in designing, implementing and interrogating a ground investigation that produces the parameters necessary to classify the soils and produce representative design parameters. This then leads onto the engineering characteristics of glacial soils which have proven to be difficult to assess because of the natural variation and variation introduced in the investigation process. Published data, principles of soil mechanics, constitutive models and statistical analyses are used to create a framework to interpret ground investigation data within the geological context. The engineering of glacial soils is set against the background of codes of practice, engineering principles and case studies to demonstrate how an appreciation of the formation of glacial soils and their characteristics can be used to reduce risk associated with earthworks and geotechnical structures.

Author

Barry G. Clarke is a Professor of Civil Engineering Geotechnics at the University of Leeds, UK and past president of the Institution of Civil Engineers.

Having been brought up in an area dominated by engineering within a glacial environment, it seemed inevitable that Clarke pursued a career as a geotechnical engineer. Clarke started as a student of civil engineering where he had the good fortune to be taught the elements of engineering geology by Professor Bill Dearman. After a spell in the ground investigation industry, he joined the Cambridge Soil Mechanics Group under the supervision of Professor Peter Wroth to study soil characterisation using *in situ* tests. This link between geology and soil mechanics has remained with him throughout his academic career as Professor of Geotechnical Engineering at Newcastle University and the University of Leeds and his professional career as a geotechnical engineer ending up as the 148th President of the Institution of Civil Engineers.

Chapter 1

Introduction

1.1 INTRODUCTION

In 1985, a failure took place during the construction of a UK motorway cutting (Arrowsmith et al., 1985), which was shown to be due to lenses of laminated clay embedded in the glacial till. The slope had been cut to a standard angle specified by the road authority, a typical angle for stiff clays. The laminated clay lenses proved to be nearly parallel to the slope. Ground investigations did indicate that the lenses were there but they were interpreted as pockets of laminated clay or horizontal layers of laminated clays, not unusual and not considered a risk. In 1993, a land fill design was based on the assumption that the underlying glacial till was impermeable. The arguments that a till can contain lenses of more permeable materials led to the proposal being rejected at a public inquiry (Gray, 1993). A dredging contractor claim that they did not expect boulders because the borehole logs did not indicate their presence was rejected on the grounds that glacial till can contain boulders. A piling contractor took as read that they would be constructing bored piles in stiff clay because that was the description given in the borehole logs. They chose to use smaller diameter piles than those considered at the design stage, which meant that they were longer and, importantly, extended below the depth of most of the boreholes. This proved to be an expensive mistake as the pile holes had to be cased to full depth to prevent water ingress and collapse of the sides of the boring when they encountered a layer of water-bearing sand. Hand excavation was proposed for a tunnelling scheme on the basis of the soil descriptions; the contractor ended up using explosives because the till was much stronger *in situ*. Bell and Culshaw (1991) consider the glacial till to be a problematic soil together with collapsible soils, quick sands, peat, expansive clays and frozen soils because of the variability in the composition. These examples highlight issues of misinterpreting borehole logs, the failure to appreciate the fabric and structure of glacial soils and poorly planned ground investigations.

A truly useful ground investigation would be staged to include preliminary boreholes to identify the ground profile and secondary boreholes to collect representative samples for geotechnical characterisation. Those samples are likely to be driven U100 samples or rotary cored samples. However, as pointed out by Taylor et al. (2011), neither of these sampling methods produces undisturbed samples; that is, it is often impossible to obtain quality representative samples of many glacial soils, which means that it is difficult to determine their characteristic stiffness and strength. Consequently, foundations are often overdesigned and inappropriate construction techniques may be used.

These examples highlight some of the challenges that glacial tills present and why conventional practice is often lacking, possibly inappropriate. This is because boulders, lenses and layers of water-bearing sands and gravels and laminated clays should be expected in glacial tills and glacial tills can be softer/stiffer than laboratory tests indicate. Other glacial soils such as glaciolacustrine, glaciofluvial and glaciomarine soils also have distinguishing features, which can be difficult to deal with. These include their anisotropic nature due to summer/winter deposition, variation in composition with distance from source, variation in composition associated with their source and speed of deposition and, possibly, very sensitive structure. Glacial soils are eroded by ice, transported by ice/water and deposited by ice/water, which leads to spatially variable soils in terms of their composition, fabric and structure. Failure to recognise this in glacial tills can have economic and, possibly, catastrophic consequences.

1.2 GLACIATION

It is estimated that at some time 30% of the world's land mass was covered by glaciers or ice sheets (Benn and Evans, 2010); a quarter of North America, one-third of Europe, and 60% of the United Kingdom were covered in glacial materials (Flint, 1971). A glacier is a slow moving mass or 'river' of ice formed by accumulation and compaction of snow falling on the upper reaches of a valley glacier or near the centre of ice sheets. About 22% of the Earth's land surface was covered by glaciers at the last ice age; currently, glaciers and ice sheets cover 9.6% of the terrestrial surface. Glacial ice is an important dynamic element of the earth system; for example, 25.7×10^6 km^3 of ice is found in the Antarctic Ice Sheet, equivalent to a rise of 61 m sea level; mountain glaciers are an important water resource. As glacial ice advances, it deforms and erodes the bedrock and underlying soil, including remnants of previous glaciations, and transports, homogenises and deposits glacial soils beneath the glacier, at the ice margins or remote from the ice margins. Most of the terrestrial glacial soils are remnants of the last glacial advance leaving extensive deposits of glacial drift (Table 1.1).

Table 1.1 Possible thicknesses of glacial soils

Continent	*Country*	*Area*	*Thickness (m)*
Europe	Germany	N Germany	58 ave
		Lunnendorf in Mecklenburg	470
		Heidelberg	397
	Sweden	Norrland	7
	Denmark	Denmark	50
	France	Grenoble	400
	Italy	Imola, Po Valley	800
	United Kingdom	East Anglia	143
		Isle of Man	175
		North Sea	920
North America	United States	Great Lakes	12
		Illinois	35
		Iowa	66
		Central Ohio	29 ave
		New Hampshire	10 ave
		Idaho	760
		Gulf of Alaska	5000
Antarctica		McMurdo Sound	>702
		Prjdz Bay	>480

Source: After Flint, R. F. *Glacial and Quaternary Geology*. Wiley, 1971; Hambrey, M. J. *Glacial Environments*. UBC Press, 1994.

Since the last glacial period, the ice has receded leaving ice sheets confined to Greenland and Antarctica and valley glaciers in the Alps, Himalayas, Andes and North Alaska (Table 1.2), with the ice sheets representing 96.6% of the current glaciated area.

There have been several ice ages (Table 1.3), the most recent being the Cenozoic Ice Age, 12,000 years ago. Remnants of earlier glacial periods exist in the form of lithified versions of glacial sediments and glaciomarine sediments. During the current period, there may have been 21 glacial cycles (Benn and Evans, 2010) comprising a cold period during which glaciers advance and warm periods when the glacier retreats. In the Northern Hemisphere, the most significant glaciers during the Cenozoic Ice Age were the Scandinavian, northern Asia and North American ice sheets while the Southern Hemisphere was dominated by the Antarctic ice sheet. In addition to the ice sheets, there are also mountain glaciers such as those found in the Alps, North Alaska, Andes and the Himalayas. The types of glaciers and the environment in which they move lead to a variety of landforms, which are a function of the mode of deposition with the glacial soils lying unconformably over the underlying bedrock or soils, which may be remnants of previous glaciations or deposits created during the current glacial period as the ice advances and retreats.

Table 1.2 Current areas of glaciation

Continent	*Region*	*Area (km²)*	*Total (km²)*
South America	Tierra del Fuego/Patagonia	21,200	25,908
	Argentina	1385	
	Chile	743	
	Bolivia	566	
	Peru	1780	
	Ecuador	120	
	Columbia	111	
	Venezuela	3	
North America	Mexico	11	2,002,500
	United States	75,283	
	Canada	200,806	
	Greenland	1,726,400	
Africa		10	10
Europe	Iceland	11,260	53,967
	Svalbard	36,612	
	Scandinavia	3174	
	Alps	2909	
	Pyrenees	12	
Asia	Commonwealth of Independent States	77,223	185,211
	Turkey/Iran/Afghanistan	4000	
	Pakistan/India	40,000	
	Nepal/Bhutan	7500	
	China	56,481	
	Indonesia	7	
Australasia	New Zealand	860	860
Antarctica	Sub-Antarctic islands	7000	13,593,310
	Antarctic continent	13,586,310	

Source: After WGMS: Global Glacier Change Bulletin No. 1 (2012–2013). ICSU(WDS)/IUGG(IACS)/UNEP/UNESCO/WMO, World Glacier Monitoring Service, Zurich, Switzerland, 2015: 230 p.

Table 1.3 Geological timescale highlighting the glacial phases and those that have taken place in the Quaternary Period

					Quaternary glacial phases						
Eon	*Era*	*Period*	*Epoch*	*Glacial phases*	*Alpine*	*N American*	*N European*	*United Kingdom*	*S American*	*Age (ka)*	*Inter/glacial*
Phanerozoic	Cenozoic	Quaternary	Holocene	Quaternary				Flandrian		0–12	Interglacial
			Pleistocene		Wurm	Wisconsin	Weichselian	Devensian	Merida	12–71	Glacial
					Riss-Würm	Sangamonian	Eemian	Ipswichian	Valdivia	115–130	Interglacial
					Riss	Illinoian	Saalian	Wolstonian	Santa Maria	130–200	Glacial
					Mindel-Riss	Pre-Illinoian	Holstein	Hoxnian		374–424	Interglacial
					Mindel		Elsterian	Anglian	Rio Llico	424–478	Glacial
					Günz-Mindel			Cromerian		478–563	Interglacial
					Günz		Elbe	Beestonian	Caracol	621–676	Glacial
		Neogene									
		Palaeogene									
	Mesozoic	Cretaceous									
		Jurassic									
		Triassic									
	Palaeozoic	Permian		Karoo							
		Carboniferous									
		Devonian									
		Silurian		Andean-Saharan							
		Ordovician									
		Cambrian									
Proterozoic	Neoproterozoic			Sturtian-Varangian							
	Mesoproterozoic										
	Palaeoproterozoic			Huronian							
Archaean											

1.3 ENGINEERING GLACIAL SOILS

Civil engineers are interested in the interaction of civil engineering structures with the ground. However, to fully understand the response, it is necessary to appreciate the science of the ground, which, in terms of glacial soils, has attracted the interest of geologists, physicists, mathematicians and sedimentologists. As with all geotechnical problems, it is necessary to understand the deposition and post-depositional processes, the current state of the soil and future environmental and loading changes. The methods of deformation, erosion, transportation and deposition of glacial soils, which are not fully understood, lead to the most diverse of any type of generic soil type. In terms of geotechnical engineering, this means the relationship between the depositional history and intrinsic properties is not fully understood. It is often difficult to determine the current state because of the difficulties in obtaining undisturbed and even representative samples. For example, the historical term for glacial tills, boulder clay, used by mining engineers, is a useful description since it emphasises that these soils can contain boulders and clays but it should be noted that boulder clay may contain neither boulders nor clay; the particle size distribution of glaciofluvial soils, mostly sands and gravels, varies with distance from the source; glaciolacustrine clays can be strongly anisotropic. This means that the generic term, glacial soil, indicates that it is a potential hazard because it does not indicate anything about the soil other than it may behave in an unexpected manner. In terms of future changes, it is not only the impact of loading/unloading and seasonal changes that have to be considered, but it is also necessary to consider the impact of climate change because of the design life of civil engineering structures. The most extreme example of this is that of nuclear waste repositories being affected by a future ice age because glacial action can have deep-seated impact. Climate change predictions now indicate that most of the built environment will be affected by environmental changes in their design life. Appreciating the consequence of intense, persistent rainfall events, rising groundwater levels and elevated temperatures are necessary. The impact on glacial soils is uncertain because of the spatial variability of these soils.

1.4 GLACIAL SOIL

Soil mechanics is traditionally developed in terms of fine-grained soils (clays, cohesive soils) and coarse-grained soils (sands, cohesionless soils), which undergo gravitational consolidation and swelling. Soils are described as normally consolidated, lightly over-consolidated or heavily over-consolidated depending on their gravitational stress history. This is based on the assumption that vertical and horizontal stresses are the principal stresses. This concept applies to glaciofluvial deposits but not to subglacial tills that undergo significant lateral deformation and intense shearing during their formation. Constitutive models to describe the response of soil to loading/unloading are often based on two extremes of behaviour defined by the particle size, governed by the hydraulic conductivity and expressed in terms of pore pressure development, that is, drained or undrained behaviour. Glacial soils are often described as intermediate or composite soils because they exhibit a range of responses. For example, a very stiff matrix-dominated till may be lightly over-consolidated, exhibiting very little change in pore pressure when loaded because it is so stiff.

Fine-grained soils are usually formed of clay minerals and silt size particles of rock. Glacial soils, which range in size from clay to boulders, are formed from source material, which may be soil or rock, by mechanical and fluvial processes. The shape and size of particles change but the composition does not. This means that the fine-grained components can be formed of clay minerals or particles of rock (rock flour) and the coarse-grained particles will be fragments of rock or agglomerates of soil particles. Post-depositional weathering may convert the fine-grained particles to clay minerals.

Glacial soils are sediments, which, in engineering terms, are described by their properties of particle size distribution, consistency and density. However, in terms of strength, some glacial tills can be described as rock. Glacial geologists have developed detailed descriptions covering clast fabric, clast shape, particle size distribution, composition and discontinuities. Clasts refer to large pebbles, cobbles and boulders and are useful because their orientation can be associated with the method of deposition. The clast shape can indicate the method of transport and the clast composition the possible source of material. Indeed, professional and amateur geologists in the nineteenth century spent considerable time collecting clasts to identify the source of the glacial material. Sedimentary structures can be used to distinguish between some glacial soils. This level of description requires extensive fieldwork studying exposures of glacial soils. Civil engineering projects rely on evidence from boreholes; it may not be possible to obtain large enough samples from the boreholes to produce a geological description of such diverse materials.

1.5 THE EVOLUTION OF GLACIAL GEOLOGY

The concept of glacial geology was first mooted in 1797 by a Swiss minister, Kuhn, according to Hambrey (1994) but it was not until the nineteenth century did the concept of widespread glaciation gain more support. However, there were then, as now, disputes within the scientific community. At that time, there were advocates, including eminent professors of geology, of the Great Flood who rejected the emerging glacial theory. To place this in context, the extent of the ice sheets of Greenland and Antarctic was yet to be discovered. So the debate about ice sheets could not be supported by visual evidence but rather by remnants of ice sheets.

In 1823, the novelist Goethe (Cameron, 1964) suggested an ice age based on the erratics found on the North German Plain. Lyell (1837) suggested that erratics were rafted by icebergs, which supported both the concept of an ice age and the Great Flood theory. A number of Swiss engineers and geologists (e.g. Charpentier, Veneto and Agassiz) developed the ice age theory, which was eventually accepted in the 1860s (e.g. Geikie, 1863; Jamieson, 1865). Four Quaternary ice ages were identified from terrestrial studies, and some evidence was found of earlier glacial events. In the 1950s, this proved to be an underestimate by the 1970s deep sea drilling programme, which showed that ice ages were more frequent (Table 1.3).

By the end of the nineteenth century, glaciation was accepted and the emphasis switched to the source of the glacial material. It was recognised that the pebbles, cobbles and boulders must be derived from bedrock. Therefore, the lithology of the clasts would indicate the source of the glacial material but not necessarily the route. Professional and amateur geologists observed exposures of glacial materials giving rise to much speculation as to how glacial material was deposited leading to public debate recorded in *Annals of British Geology*. However, from an engineering point of view, the nature of glacial materials was understood. Ansted (1888) commenting on river beds cut in glacial tills in northern England noted the valleys were covered with a coating of glacial tills completely masking the rock. He suggested that great caution is necessary when constructing works in valleys or on hillsides as glacial tills are so variable. He described it as *a very stiff tough clay, of dark brown or black colour, containing, distributed quite irregularly, a multitude of fragments of transported rock, generally angular. The dimensions of the stones vary from many cubic yards to fragments not larger than a nut or a pea. Occasionally, there is a large deposit almost entirely of clay, but not unfrequently the stones are very abundant. They are here and there in stratified beds, but more frequently if in large quantity they occupy lenticular hollows in the clay*. He noted that this material would be ideal for the construction of reservoirs. However, it was also known that *in any part of the deposit there may be running sand and silt, combined with loam, which, though firm when reached in the ground, and difficult to remove with the spade, runs at once to a thin-liquid mud on exposure to air and water*. These sands, silts and clay lenses were sometimes very local, extending for a few hundred metres but could also extend for a considerable distance and, importantly, be hydraulically linked to a source of water, which could be under pressure. Excavation disturbs this material to the extent that it stops all construction. Further, even if contained, these beds were recognised as confined aquifers. Those observations in 1888 are as relevant today as they were then.

Interest in the movement of ice started with valley glaciers in the eleventh century (according to Hambrey, 1994). It was recognised that they were 'rivers' of ice with a number of theories put forward to explain their movement. Scheuzer (1723) suggested the dilation theory in which rainfall collecting in the crevasses in a glacier froze and expands, thus causing the glacier to move forward; Altmann (1751) and Gruner (1760) stated that it was due to gravity with the ice sliding over the rock bed as a block. Bordier (1773) suggested that ice flowed as a viscous material and was dependent on the hydrostatic pressure. Tyndall (1873) introduced regelation theory in which ice under pressure melts and then refreezes allowing glaciers to slide. Chamberlin (1894) suggested that the ice moved in layers because of the evidence of shear planes within the ice and underlying sediment. In 1895, Deeley, a railway engineer with a passionate interest in geology, suggested that, despite the fact that the flow of viscous materials was well known, the principles had not been applied to glacial flow. Further, the theory of glacial flow developed separately from the experimental observations. The different theories of glacial movement stemmed from the fact that it was difficult to observe displacements within the ice and the underlying sediments.

By the early twentieth century, it was accepted that glaciers behave in a ductile manner with some basal slip if the temperature permitted regelation. However, a number of misconceptions developed including the concept that ice softened under pressure, which meant it was softer at greater depths, and therefore ice moved faster at depth; that is, the base of glacier moved faster than its surface, which contradicted field observations. Demorest (1941) promoted this theory, the extrusion theory, assuming that ice flows from the centre of the greatest accumulation in polar ice sheets with the flow rate depending on the pressure gradients. Nye (1951) using the work of Glen (1952) suggested that the extrusion theory was invalid because of the global imbalance of forces and proposed the view that glacial flow is a combination of basal sliding and ice deformation.

After accepting the concept of ice ages and glacial flow, the next major advance in glacial geology was to link the movement of the ice sheet or glacier with the sedimentary record. Three theories have been developed: the Coulomb friction model (Boulton, 1974), the friction model (Hallet, 1979) and the sandpaper model (Schweizer and Iken, 1992). Schweizer and Iken (1992) suggest that all three models are valid in different conditions; that is, there is no unified model. More recent studies have developed other constitutive models building on Boulton's hypothesis. These are explained in more detail in Chapter 2 because they affect a glacial soil and its properties. The current debate centres on the deposition of glacial soils, the landforms and the spatial variation in composition and structure.

1.6 THE PAST IS THE KEY TO THE PRESENT

Glacial soils can be divided into those that are deposited by ice and those deposited by water as a result of melting or sublimation. Glacial debris is derived from the base of a glacier and, in the case of valley glaciers, from the sides of the valley and rock falls above the glacier. A glacier moving over solid rock can erode the rock surface producing debris that is carried up into the glacier or transported at the base of the glacier undergoing further abrasion. A glacier moving over a soft sediment deforms that sediment, which can continue until the underlying sediment is completely homogenised. As a glacier advances, it eventually produces a homogenised deposit, which is subsequently deposited unconformably over the underlying rock or soil. Hence, there is a spectrum of subglacial deformation ranging from glaciotectonic deformed layers to homogenised material. Subglacial tills are also formed under stagnant or retreating ice due to melting or sublimation. The composition of these tills may be similar to those deposited by advancing ice, but their properties are different because the mode of deposition is different; the former undergoes shear as well as gravitational consolidation and the latter is gravitationally consolidated.

The hydraulic conditions at the base of a glacier are such that water can be flowing beneath the ice even when the ice is advancing, thus creating fluvial deposits, which can be deposited subglacially. Glaciofluvial deposits are also deposited beyond the margins of the ice whether they are advancing or retreating. A consequence of this and successive periods of glaciation means that previously deposited glacial deposits can be incorporated in subglacial tills and, depending on the hydraulic and temperature conditions at the base of a glacier and the amount of movement, retain some of their original structure. This is a reason for the pockets, lenses and layers of distinctly different soils within glacial deposits. Other examples of glacial soils that challenge engineering include the following:

- Glaciomarine deposits are those deposited in marine environment and include sea bed deposits on the continental shelf and terrestrial deposits that are no longer submerged because of isostatic uplift. Terrestrial glaciomarine deposits can be a very difficult soil to deal with because they are very sensitive as the pore fluid has changed with time.
- Subglacial tills lie unconformably over the underlying soil and rock masking the interface. Identifying that interface is very important but it is difficult at the scale of civil engineering projects.
- Glaciofluvial deposits lie on the surface of previously glaciated land and land that has never experienced glaciation. They are gravitationally deposited so could be studied using traditional theories of soil mechanics. Their composition is variable and depends on the method of transport, depositional environment and the distance from source.
- Glaciolacustrine clays are strongly anisotropic because of the mode of deposition.

1.7 GLACIAL SOILS ARE THE MOST VARIABLE OF ALL SOILS

The geological term glacial soil covers a wide range of particle size distribution, density, permeability, stiffness and strength. There are several types of glacial soils including subglacial and supraglacial tills, glaciofluvial soils, glaciolacustrine soils and glaciomarine soils. Glacial soils can be classified according to their depositional environment (terrestrial, fresh water or marine environments), position (ice margin, supraglacial or subglacial) or by process (deformation, lodgement, ice deposition or fluvial deposition). It should be noted that a soil description, which is based on a small sample, is only a description of that portion of stratum retrieved and may not be representative of the soil mass encountered during construction. This is relevant to glacial soils because of the spatial variability. It is very difficult to identify the type of glacial soil from borehole samples and even exposures because the difference between some of the glacial soils is subtle. Glacial soils can be deposited by ice or water, heavily over-consolidated or lightly over-consolidated, extensive or local. Indeed, while the term glacial soil does not help characterise the soil, it does highlight the fact that the soil composition, fabric, structures and properties are spatially variable.

Some glacial soils are very distinctive possibly because of the landform or because of the fabric. Lacustrine clays formed in pro-glacial lakes have a distinct anisotropic fabric associated with the depositional process. These extensive deposits of varved clays, colloquially known as bible leaf clays, comprise layers of clay/silt and sand/silt of varying thicknesses. Glaciofluvial deposits are predominantly sands and gravels, which can often be identified by the landform they create. Subglacial deposits are more difficult to classify, even identify. Their composition depends on the parent rock, which may be from various sources because of repeated glacial events, the hydraulic conditions and pressure at the base of the glacier, the distance transported and the degree of abrasion. Further, the most recent glacial event will incorporate deposits from the last glacial event and, depending on the process, many produce a very diverse range of structures, fabric and composition within a small zone. This can impact on the engineering properties of a soil within the zone of influence of a civil engineering project, leading to unintended consequences that can cause problems during construction, and over or under design. For example, excavated slopes at a safe angle can fail because of a pocket of water-bearing sand and gravel or a layer of weaker soil; piled foundations in glacial tills exceed their design capacity because the laboratory-measured strength is lower than the *in situ* values.

1.8 'WE KNOW MORE ABOUT THE STARS ABOVE US THAN THE SOILS BENEATH OUR FEET' (LEONARDO DA VINCI, c. 1600)

It is clear from this brief overview of glacial soils that they are challenging. There are conflicting views of their formation; they are the most variable of any type of soil; they may or may not conform to the basic principles of soil mechanics, which assumes gravitational consolidation. However, the most challenging aspect of working with these soils is the difficulty in obtaining representative samples. Glacial geologists study exposures at margins of ice sheets and glaciers, and geomorphologists study glaciated landforms. Thus, detailed knowledge of glacial soils formed at ice margins and the landforms created by glacial soils exists. A further challenge is the lack of understanding of how glacial tills are formed because it is difficult to study their creation as observations beneath glaciers are challenging. This is the subject of current debate among glacial geologists. The geotechnical engineer relies on representative samples from a spatially variable soil. The importance of studying exposures cannot be underestimated as they provide the best indication of the type of glacial soil, highlighting its variability and possibly its mode of erosion, transportation and deposition. However, it is known that the observed material may be a result of several glacial events, which means a simple classification is not valid. An exposure does give an indication of the lateral variation, an important feature of glacial soils. In many instances, there are no exposures to view, which means that remote sensing and sampling are required. Remote sensing has been used successfully to determine ice flow directions, which can provide evidence of fabric, and more recently the bedrock interface. A challenge is locating the interface between different types of glacial soil deposits and between glacial soils and the underlying soils and rock. The presence of erosion features means that the interface is not planar. Further, rafted rock can be incorporated in a glacial soil giving a false impression of rock-head. Glacial soils can be squeezed into fissures in the rock; there can be buried valleys filled with glacial soils.

Linear infrastructure projects may cross several glacial landforms requiring an informed approach to ground investigation. Superficial drift maps, produced by the British Geological

Survey for example, are a useful source of information as they demonstrate the potential variation in glacial soils, but they are not always available or available to the level of detail required for a civil engineering project. Therefore, it is important to complete a thorough desk study, which includes an assessment of the topography and geomorphology through the use of maps, archival material and walk-over surveys. Geological maps are useful and the borehole database held by the British Geological Survey is invaluable. Papers on regional characteristics including rock-head (e.g. Sissons, 1971; Menzies, 1981; Brabham and Goulty, 1988) and geotechnical properties (e.g. Clarke et al., 2008) allow greater confidence to be placed on the results from a site-specific ground investigation.

Investigations for structures either as part of infrastructure or as a stand-alone project require a different approach to those specified. Borehole databases in urban areas will be very important. However, the most useful activity of a desk study would be to investigate the construction of adjacent properties. Uncovering knowledge of excavations and geotechnical construction would be invaluable as it could highlight problems encountered. This point was raised by Ansted (1888) and is valid today as it was then. Yet failures still occur because of limited investigations.

Following the desk study it is necessary to undertake an intrusive investigation to develop a full understanding of the topographical, geological, geotechnical and hydrogeological models. Indeed, BS EN 1997-1:2004+A1:2013 (Eurocode 7) states that '*Knowledge of the ground conditions depends on the extent and quality of the geotechnical investigations. Such knowledge and the control of workmanship are usually more significant to fulfilling the fundamental requirements than is precision in the calculation models and partial factors*'.

BS EN 1997-1:2004+A1:2013 requires Class 1 samples for producing geotechnical design properties. All sampling techniques disturb the ground to some extent, and in some glacial soils, it is impossible to retain Class 1 samples (Taylor et al., 2011). The variation in particle size and strength means that care should be taken to select the most appropriate sampling technique. Therefore, a two-stage investigation is required; the first stage to establish the geological profile, and the second to collect representative samples. In the United Kingdom, thick-walled driven samplers are traditionally used for stiff clays, a category that includes some glacial soils. Pushed thin-walled samplers (pistons samplers) are used in soft clays. Rotary core samples are taken in weak rock.

Even if representative samples can be retrieved, any clasts present (Taylor et al., 2011) or discontinuities (McGown et al., 1977) could affect the strength of the samples. Thin-walled samplers are not robust enough to be pushed into many glacial soils either because they are too strong or because they contain too many stones. Rotary coring in stony clays can be difficult, especially if the matrix is weak.

The alternative is to use *in situ* tests. Penetrometer tests can be used but the strength and the presence of stones can affect the results or even stop penetration. Tests, such as pressuremeter tests, that need pockets to be formed suffer from the same issues as sampling.

A consequence of the challenge of sampling and testing glacial soils is that reliable values of strength and stiffness are difficult to obtain. BS EN 1997-1:2004+A1:2013 states that *reliable measurements of the stiffness of the ground are often very difficult to obtain from field and laboratory tests. In particular, owing to sample disturbance and other effects, measurements obtained from laboratory specimens often underestimate the in situ stiffness of the soil*, which is particularly relevant to these soils. This is one reason why construction in glacial soils is fraught with difficulty.

Boreholes provide limited information on the vertical soil profile. It is difficult to interpret between boreholes yet many glacial soils vary horizontally as well as vertically. Therefore, it is not unexpected to find 'running sand', inclusions of soft clay, boulders, stronger and denser materials and laminated clay lenses even though the ground investigation report has not identified them. This is especially the case in subglacial tills. Deformation during deposition and isostatic uplift following a glacial recession means that many subglacial tills contain discontinuities due to shear and unloading. This means that mass strength is often less than the intact strength. Further, the discontinuities may be aligned with the direction of ice flow, giving rise to construction problems depending on the relative alignment of the discontinuities with the project.

Isostatic uplift following the last ice age and the development of the drainage patterns within the glaciated areas led to numerous landslides, which are currently dormant. Engineering in this landscape and climate change can trigger these dormant landslides.

1.9 OBSERVATIONS

This brief overview has highlighted some of the issues that have to be addressed when constructing on, in or with glacial soils. The points to note include the following:

- The term glacial soil covers a wide range of soils that have been derived by ice in some way and been deposited by ice or water.
- The manner by which glacial soils are derived and deposited is not fully understood; there are a number of theories, depending on the conditions that existed during formation.
- A colloquial term for glacial till, one type of glacial soil, is boulder clay but it may contain neither boulders or nor clay.
- Some glacial soils are very dense and strong, but if disturbed, lose their strength.
- Some glacial soils are very dense and strong to an extent that it is difficult to obtain representative samples.
- The strength of glacial soils, particularly tills, is often underestimated leading to difficulties in construction and overdesign of foundations.
- Glacial tills can contain lenses of water-bearing sands and gravels or weak soils leading to local instability.
- Glacial soils may contain layers of sand and gravel, possibly hydraulically connected to a water source, and laminated clays.
- Some glacial soils contain sedimentary structures.
- Some glacial soils are so strong that they should be described as rock.
- Particle size distribution can vary from single size to bimodal to multimodal and this can vary over short distances or be extensive.
- Glacial soils lie unconformably over bedrock. The bedrock interface may not be planar to such an extent that buried valleys filled with glacial soils are possible.
- Glacial stratum can include glacial tills overlain by glaciofluvial soils but, because of several periods of glaciation, it is possible to find glaciofluvial soils beneath glacial tills.
- It is difficult to prove rock-head because glacial soils can contain boulders or even rafted rock.

For these reasons, it is important to fully understand the formation of glacial soils and the associated landforms, appreciate the variable nature of these soils, and understand the impact the formation has upon the geotechnical characteristics and how it affects construction in glacial soils and what can potentially go wrong. Glacial geologists, mathematicians and physicists have published many papers on glacial geology but have yet to reach a consensus or unified model on the processes that create glacial soils though their views are important when assessing glacial soils in construction. There are a number of books that summarise the formation of glacial soils (e.g. Hambrey, 1994; Ehlers and Gibbard, 1996; Benn and Evans, 2010; Bennett and Glasser, 2011; Eyles, 2013). These give a detailed account of the composition of glacial soils and the landforms they create but do not provide detail on their hydrogeological and geotechnical characteristics. It could be argued that classic soil mechanics theory and the practice of geotechnical engineering would be sufficient, but glacial soils are sufficiently different to warrant a study of their own. Hence, the structure of this book starts with an overview of relevant glacial geology. This is followed by a review of the characterisation of glacial soils using published and unpublished data to provide guidance on ground investigation, testing and interpretation. Case studies are used to highlight examples of design and construction.

Chapter 2

Glacial geology

2.1 INTRODUCTION

All civil engineering projects start with a desk study, which is normally followed by a ground investigation to develop geological, hydrogeological, geotechnical and topographical models to create the ground model for the site. These models are used to identify hazards, and to produce conceptual designs, the final design and the construction programme including any temporary works. The geological model provides information on the stratigraphy, including soil type and extent, and an indication of its classification. The hydrogeological model is an indication of the ground water level, the profile including perched water tables, aquifers and aquitards and how it responds with time. The topographical model provides information on landforms. Finally, the geotechnical model uses this information and the results of field and laboratory tests to produce design parameters and a risk assessment. This is the ideal situation but it is well known that many investigations are incomplete leading to delays, costs overruns and damage, possibly failure, during construction and in service. This is compounded in glaciated regions by the fact that the vertical and horizontal variation in glacial soils' composition and properties can lead to complex models that require more investment in the investigation and a better understanding of the formation of these soils than is normally specified when investigating gravitationally deposited soils.

The history of a glacial deposit and its impact on its geotechnical properties are difficult to assess. For example, a dense subglacial till may be described as over-consolidated because it is dense and it is assumed that the preconsolidation pressure was the weight of ice, yet laboratory tests may show that the till is lightly over-consolidated; laboratory tests on a dense subglacial till may show it to be normally consolidated but have similar properties to heavily over-consolidated gravitationally deposited clays.

The glacial process creates a variety of landforms, which can be an indication of the type of glacial soil; therefore, a study of the regional geomorphology should be incorporated in the ground investigation. The surface of a glacial soil profile is unlikely to reflect the bedrock interface. For example, the surface may be an undulating drumlin field created beneath a glacier, or hummocky moraines formed of glaciofluvial deposits yet the bedrock may be a former drainage system on an ice-eroded surface. Investigations in extensive plains underlain by glacial soils are likely to find that they are underlain by a very stiff glacial till that lies unconformably on the underlying bedrock. Post-glacial deposits formed following isostatic uplift and the formation of the current drainage system may lie unconformably over glacial deposits.

Glacial soils have been classified according to their mode of deposition, but the current view is that they should be classified by the modes of deposition and deformation. The composition, fabric and structures are spatially variable within any given deposit that leads to poorly sorted heterogeneous soils, which are difficult to classify on borehole samples alone. Descriptions of glacial soils can lead to misinterpretations when not placed in context.

The purpose of a ground investigation for a civil engineering project is different from an investigation as part of a scientific study as in the former case the investigation is to characterise the soils, identify the hazards and assess the design parameters and in the latter case it is to explain the formation of the soils and landforms. This leads to different classification and identification schemes. Further, different terms are used by geologists and engineers to describe the same thing. This can prove challenging, so studying the geological processes that create glacial soils from an engineering point of view is important when creating a ground model of a site.

Geological maps do provide information based on walk-over studies and previous investigations but are unlikely to give enough detail of glacial soils at the location of a specific project because they are so variable. They can be identified only from a detailed analysis of exposures, excavations, borehole samples and remote sensing. Glacial soils can be characterised in accordance with an engineering classification scheme for soils such as the European Soil Classification System (ESCS) or the American Unified Soil Classification System (USCS) or by a scientific lithofacies coding scheme. None of these schemes provide information on the history of the deposit, which is a crucial information for engineering investigations in glacial soils. Therefore, a further classification is needed, which could be based on a debris cascade system (Figure 2.1). It starts with the debris source, which influences the type (lithology) of particles, particle shape (morphology) and particle size distribution. The transport path includes deformation in the basal shear zone comprising the basal ice and underlying sediments, known as the substrate, movement within the ice and movement due to meltwater. Movement due to water or ice can be subglacial, englacial or supraglacial and further modifies the glacial debris by abrasion. The depositional and deformation processes influence the extent and thickness of the layer, create sedimentary structures and influence the particle size distribution, fabric and particle morphology. These in turn influence the geotechnical properties. These create primary and secondary deposits, as shown in Table 2.1, which lists the types of glacial soils and their relative position in the erosion, transport, deposition and deformation sequence. Glacial soils can be reworked by glaciofluvial action or further periods of glaciation, undergo further changes due to weathering and diagenesis and can be reworked by the formation of post-glacial drainage systems and landslides due to isostatic uplift.

Thus, in order to develop an appreciation of the engineering characteristics of glacial soils, it is necessary to understand glacier systems, including glacier dynamics; erosion, transport,

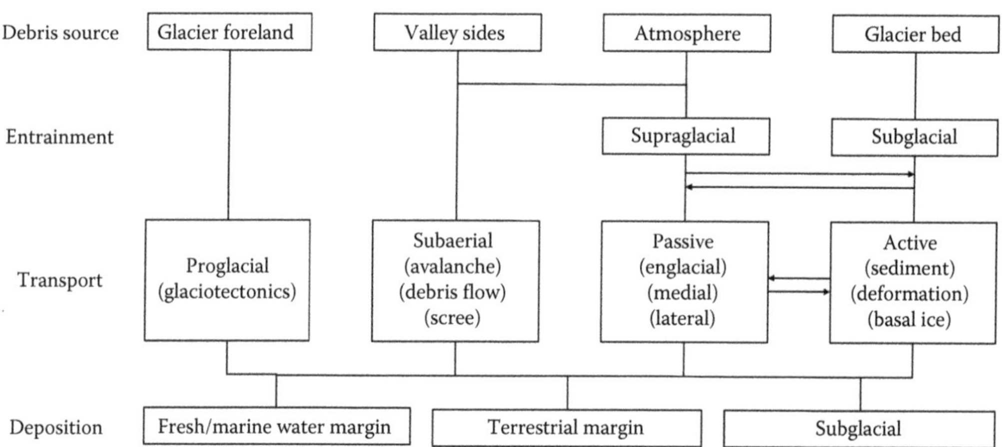

Figure 2.1 Debris cascade system relating the debris source, the relative position of the glacier and the modes of transport and deposition. (After Benn, D. and D. J. A. Evans. *Glaciers and Glaciation*. Routledge, London; 2010.)

Table 2.1 Classification of glacial soils and facies separated into primary and secondary deposits linked to the mode of transportation, deposition and deformation

Primary deposits (tills) (transported, deposited and deformed by ice)		*Secondary deposits*		
Deformed (subglacial traction tills)	*Gravitational (supraglacial and englacial melt-out tills)*	*Glaciofluvial deposits*	*Mass movement (transported by ice and gravity)*	*Sedimentation (transported by ice and deposited in water)*
Glaciotectonite Deformation till Comminution till Lodgement till	Melt-out till Sublimation till	Plane bed deposits Ripple cross-laminated Cross-bedded facies Gravel sheets Silt and mud drapes Hyper-concentrated flow deposits	Scree Debris fall deposits Gelifluction deposits Slide and slump deposits Debris flow deposits Turbidites	Cyclopels Cyclopsams Varves Dropstone mud Dropstone diamictons Undermelt diamicton Iceberg contact deposits Ice-keel turbate

Source: After Benn, D. and D. J. A. Evans. *Glaciers and Glaciation*. Routledge, London; 2010.

deposition and deformation of glacial debris; and landforms created by glaciation, deposition and deformation. There is extensive literature covering these aspects but they will be dealt with briefly here to highlight the impact they have on the engineering characteristics. Benn and Evans (2010), Hambrey (1994) and Bennett and Glasser (1996) amongst others provided a comprehensive view of the state of the art of glacial geology though the science of glacial geology continues to develop, importantly in the formation of glacial tills. Glacial geologists start by considering glacial dynamics, how they create glacial soils and landforms leading to a description of the various types of glacial soils. Their work is primarily based on field observations supported by experimental and theoretical studies. The approach used here is to describe glacial soils and landforms and then discuss glacial dynamics. This is consistent with an engineering approach to the classification of glacial soils, which is primarily based on samples from boreholes and trial pits. It focuses on factual information as there is much speculation over the formation of glacial soils. Landforms are important especially when dealing with infrastructure projects though many civil engineering projects are in urban areas where glaciated landforms may not be visible or even exist because of anthropogenic activity. The final section covers glacial dynamics that are very relevant to the history of the glacial deposits, especially glacial tills and, hence, their geotechnical characteristics.

2.2 GLACIAL SOILS

Glacial soils, or glaciogenic sediment (Dreimanis, 1989), are soils derived from glacier ice. The soils are formed of glacial debris, which is soil transported by ice and water. It may be derived from the sediments (bedrock or superficial deposits) underlying a glacier and, in the case of highland glaciers, from the sides of the valley. The glacial soils are deposited as glacial drift, a generic term for any glacially derived superficial deposits. Many glacial soils are diamictons, which are a wide range of non-sorted to poorly sorted soils or sediments (Flint, 1971). Glacial soils can range from clays/silts (muds) to boulders in various combinations depending on their source, mode of transport, deposition and deformation. Table 2.2, based on a classification of poorly sorted sedimentary rocks (Moncrief, 1989), is a non-genetic classification of lithified glacial soils showing how the clay, sand and gravel content affects the non-genetic classification of diamictite, lithified glacial soils. Glacial debris can

Table 2.2 Non-genetic classification of poorly sorted lithified sediments demonstrating the effect of the composition on the description of tills

		Increasing gravel content →						
		Gravel (>2 mm) in whole rock (%)						
		Trace	*<1*	*1–5*	*5–50*	*50–95*	*95–100*	% sand in matrix
Increasing silt/clay content ↑	*Silt/clay*	Mudstone	With dispersed clasts	Clast-poor muddy diamict	Clast-rich intermediate diamict	Muddy conglomerate	Conglomerate	<10%
	Composite soils	Sandy mudstone	Sandy mudstone with dispersed clasts	Clast-poor intermediate diamict	Clast-rich intermediate diamict	Conglomerate		10%–33%
		Muddy sandstone	Muddy sandstone with dispersed clasts	Clast-poor sandy diamict	Clast-rich sandy diamict			33%–66% 66%–90%
	Sand	Sandstone	With dispersed clasts	Gravelly sandstone		Sandy conglomerate		>90%

Source: After Moncrieff, 1989; Hambrey, M. J. *Glacial Environments*. UBC Press, 1994.

Table 2.3 Genetic classification of glacial soils based on the environment, position, process and composition

Environment	Subglacial, terrestrial, freshwater, marine
Position	Subglacial, ice margins (proximal), ice margins (distal)
Process	Deformation, subglacial traction, melt-out, sedimentation (fluvial, lake, marine)
Composition	Sands and gravels, clast-dominated till, matrix-dominated till, varved clays

be deposited to form glacial soil by ice, from ice, by water, and in fresh water or marine environments and can be deformed by ice. The fabric of the soil depends on the depositional environment and, in the case of glacial tills, subglacial deformation. While the source, transport and deposition (Figure 2.1) all affect the glacial soil, it is often not possible to relate the genetic classification (Table 2.3) to the debris cascade system, which means that it is difficult to classify glacial soils from descriptions alone, which has led to debate over the classification of glacial soils. For example, there are various definitions for till.

The International Union for Quaternary Research (Dreimanis, 1979) defines till as sediment that has been transported by or from glacier ice, with little or no sorting by water. Lawson (1981) suggests that this definition does not distinguish between primary deposits, those deposited by ice, and some secondary deposits, those glacial soils that have been reworked by some non-glacial process, an important issue when it comes to predicting geotechnical characteristics. A revised definition for till is a sediment deposited or deformed by glacial ice, without any further action; these are primary deposits. Table 2.1 lists those glacial tills that form primary deposits and this is the definition used here because subglacial tills are deposited in a glacial environment that results in a stress history very different from gravitationally consolidated soils, an issue when using classic theories developed for soil mechanics. These primary deposits include glaciotectonite, deformation till, comminution till and lodgement till, which are subject to shear and gravitational load when formed. van der Meer et al. (2003) suggested that all primary deposits are deformation tills because they all undergo deformation; Evans et al. (2006) refer to subglacial traction tills to cover tills deposited by ice and those that undergo deformation. These simple definitions remove the difficulties in distinguishing between the different types of till. It is helpful in that it means the unexpected should be expected. It is unhelpful as it does not provide information that may give further insight into the behaviour of till. The remaining glacial tills are gravitationally consolidated, which include melt-out till formed of debris deposited as ice melts, and sublimation till, formed as a result of vaporisation of ice that occurs in cold, arid regions such as Antarctica.

As a glacier melts, debris is transported by water to be deposited on, within, beneath or beyond the glacier creating glaciofluvial deposits that have similar geotechnical characteristics to other fluvial deposits as they are gravitationally consolidated. If a glacier terminates in water, sedimentation of glacial debris creates glaciolacustrine deposits when formed in a standing body of fresh water such as that found at ice margins; and glaciomarine deposits when a glacier terminates in the sea. The full list of glacial soils is given in Table 2.1 and their evolution in Figures 2.5 through 2.8.

Engineering soil descriptions are based on samples mostly taken from boreholes and trial pits. Descriptions are factual (Chapter 3) and do not refer to the geological type unless a geological investigation is undertaken. Describing soils for engineering purposes is different from describing sediment for geological purposes though there are common elements. Note that engineers refer to soils; geologists refer to sediments. This chapter focuses on geological descriptions of glacial soils highlighting the relevance to engineering. A glossary of terms is given in Appendix 2.

The generic term glacial soil covers a greater variety of soil types than other generic soil types. The term glacial soil is not very helpful in assessing the engineering characteristics of the soils but it is helpful in that it is a warning to expect a problem. Glacial soils can be distinguished by a combination of composition, fabric, structure, landform and land systems, as shown in Figure 2.2, a suggested hierarchical sediment classification (Walker, 1992) used by glacial geologists. A description of a glacial soil starts with identifying the extent of the soil with similar composition and formation, the facies. A stratigraphical model is built

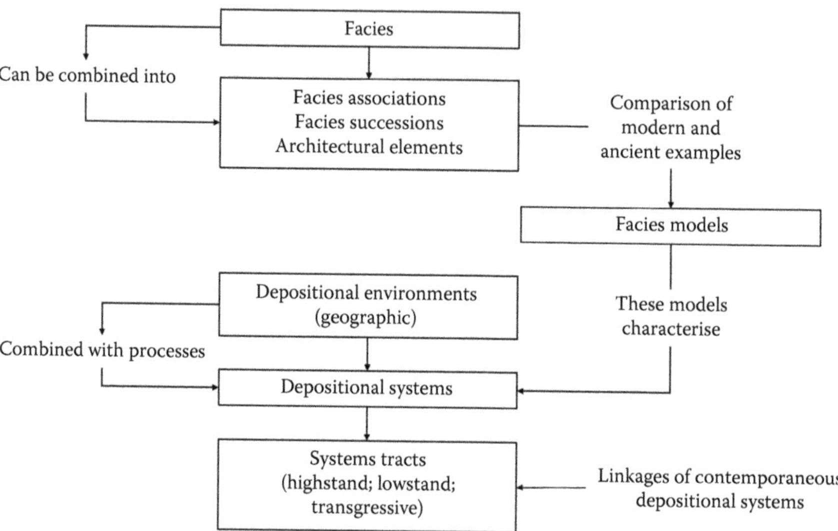

Figure 2.2 Sediment classification based on the facies, facies associations, depositional environments and land systems. (After Walker, R. G. Facies, *Facies Models: Response to Sea Level Change*. Geological Association of Canada, St. John's, Canada, 1992: 1–14.)

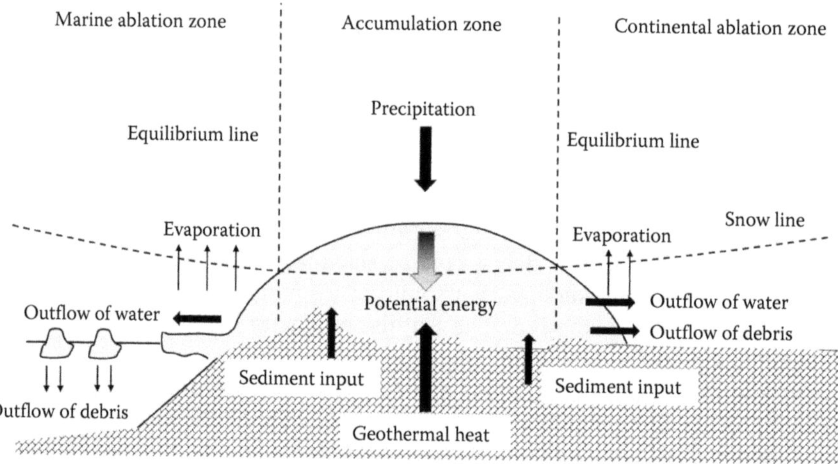

Figure 2.3 Cross section through an ice sheet showing the energy and mass input, the movement of material and the outlflow of debris and water. (After Brodzikowski, K. and A. J. Van Loon. *Development in Sedimentology*, 49; 1991: 688; Benn, D. and D. J. A. Evans. *Glaciers and Glaciation*. Routledge, London; 2010.)

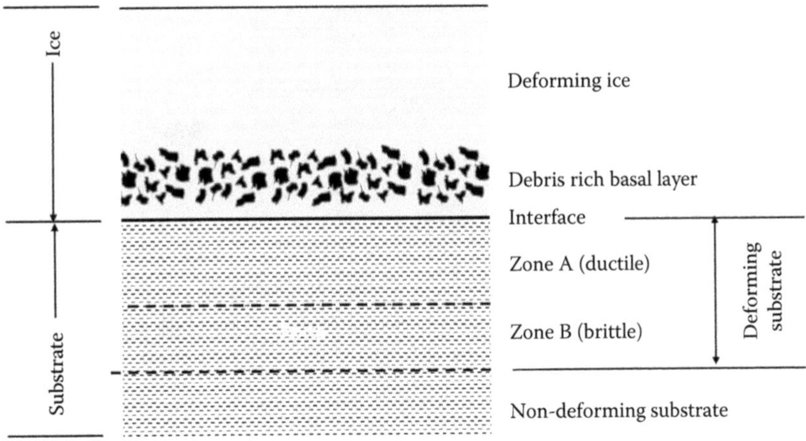

Figure 2.4 Terms used to describe the various zones within the ice and substrate.

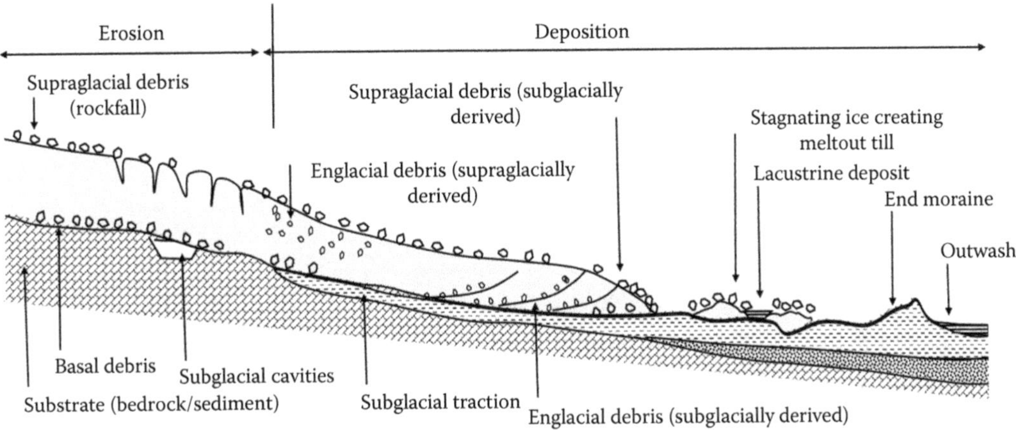

Figure 2.5 Sediment deposition associated with terrestrial ice margins as the ice recedes. (After Hambrey, M. J. *Glacial Environments*. UBC Press, London; 1994.)

from the facies associations based on the environment in which they were deposited. A full description of a glacial soil covers erosion, transport, deposition and deformation of the soil; its structure; and its position with respect to adjacent sediments and landforms. Thus, there are three aspects to consider: source of glacial debris; the modes of transport, deposition and deformation; and post-glacial activity. These will differ between ice sheets and highland glaciers.

Figure 2.3 shows a suggested model for subglacial erosion and deposition beneath an ice sheet. The ice sheet is created by precipitation falling within the accumulation zone and is lost through evaporation and melting in the ablation zone. These two zones are divided by the equilibrium line that separates the zone of energy and mass input from the zone of outflow of debris and water. The ice moves from the accumulation zone to the ablation zone by a combination of sliding, deformation of the ice and deformation of the underlying substrate. There are up to three zones (Figure 2.4) within the substrate: a stable non-deforming zone at depth, a slowly deforming zone and the upper layer that is rapidly deforming. The thickness of each zone depends on the temperature profile, the stiffness of the zone and

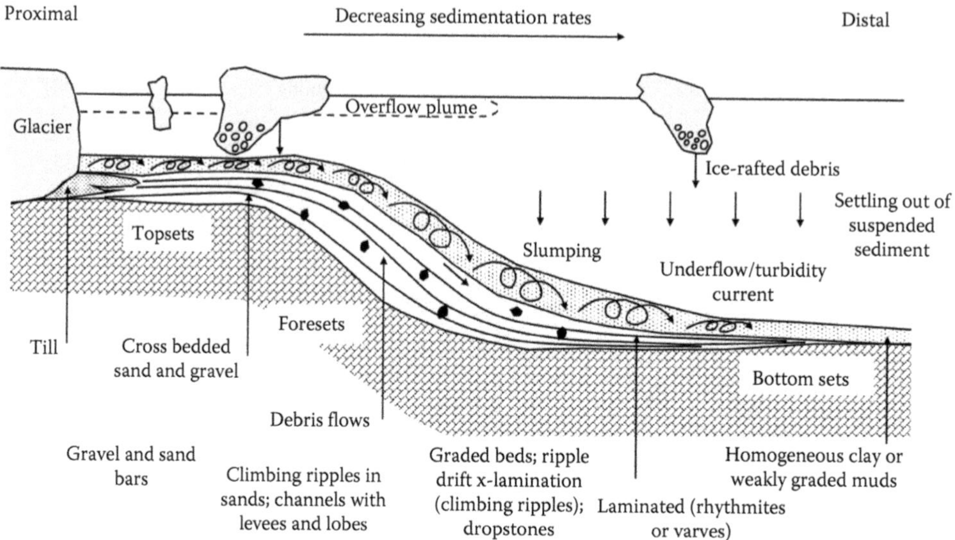

Figure 2.6 Sediment deposition associated with a freshwater environment. (After Bennett, M. R. and N. F. Glasser. *Glacial Geology: Ice Sheets and Landforms*. John Wiley, London; 1996: 364; Hambrey, M. J. *Glacial Environments*. UBC Press, London; 1994.)

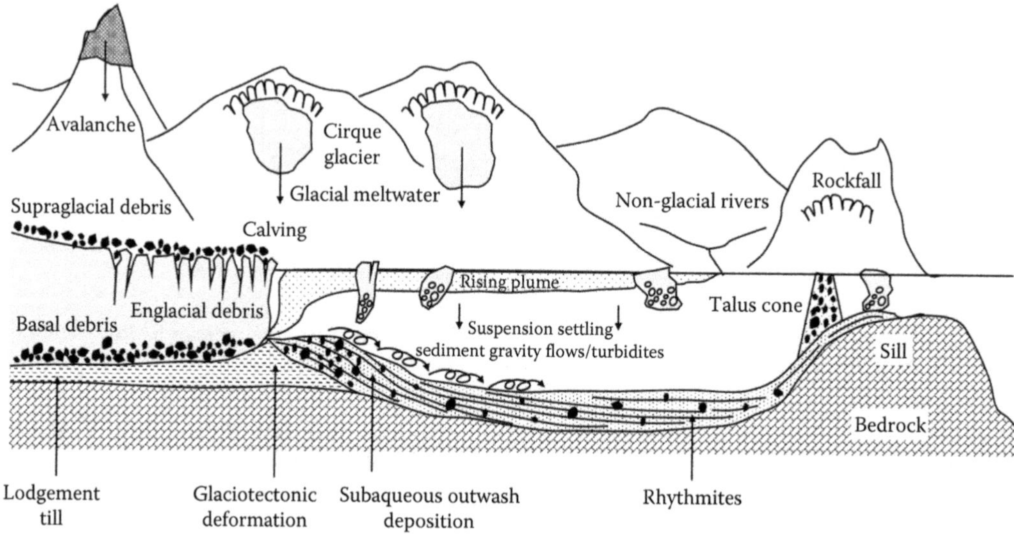

Figure 2.7 Sediment deposition associated with a glaciomarine environment within a fjord. (After Bennett, M. R. and N. F. Glasser. *Glacial Geology: Ice Sheets and Landforms*. John Wiley, London; 1996: 364; Hambrey, M. J. *Glacial Environments*. UBC Press, London; 1994.)

the pore pressure; as the pore pressure increases, the layer deforms more easily and the stiffness and strength reduce.

As the ice moves across the basal layer (Figure 2.5), which can be bedrock, remnants of a previous glaciation or gravitationally consolidated soil, the substrate is deformed and eroded. Some of the eroded material is moved up into the glacier to be carried along by the ice (englacial debris) and some remains as a debris-rich layer continuing to be deformed. In

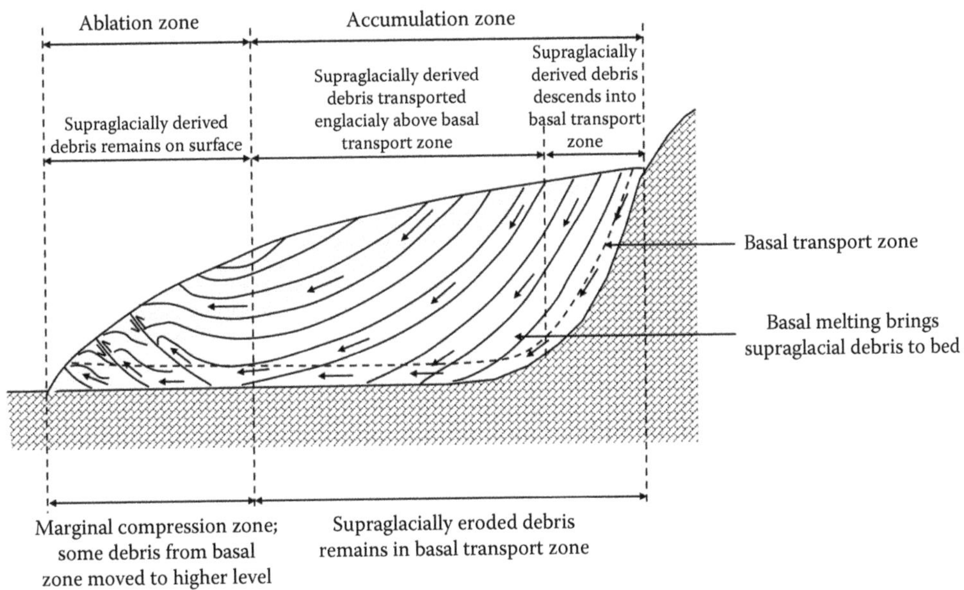

Figure 2.8 Glacial debris associated with highland glaciers showing the movement of supraglacial, englacial and subglacial debris within the glacier. (After Boulton, G. Glaciers and glaciation; In *Holmes' Principles of Physical Geology*, edited by Duff, P. M. D. and D. Duff, Taylor & Francis, 1993: 401–438.)

the compression zone, the debris is deposited as till and can undergo further deformation as the ice advances. Some of the subglacial debris may be eroded by water flowing through channels within the ice, which can lead to glaciofluvial deposits in the ice if the channels are closed for some reason. As the ice melts, the glacier retreats and englacial and supraglacial debris can be deposited as a till. The melting ice carries englacial and subglacial debris beyond the ice margin to form outwash deposits. As a glacier advances and retreats, deposits are continually reworked, leading to complex deposits beneath and beyond the ice. Deglaciation at the end of an ice age led to significant changes in sea level and isostatic uplift (10–100 m), which had a significant effect on the former glaciated landscape; glaciomarine deposits become terrestrial deposits; post-glacial drainage systems lie unconformably over historic landscapes, creating a stratigraphic sequence of non-conformable deposits.

If a glacier terminates in fresh water, the debris from the melting glacier is deposited on the bed of the pro-glacial lake (Figure 2.6). The composition of the soil changes with distance from the ice margin because the rate of sedimentation varies with the time of the year and the weight of the suspended particles within the meltwater. Sand and gravel are deposited near to the margin. In the summer, when the ice melt is the greatest, more debris is carried further into the fresh water lake. The coarser particles settle first. These are followed in winter by finer particles settling out of the lake creating a laminated deposit, glaciolacustrine clay. Icebergs breaking away from the glacier carry englacial debris, which can include particles of all sizes, some distance into the lake. As the iceberg melts, the englacial material falls onto the lake bed. These deposits are further complicated by the effects of currents within the lake creating turbites. Thus, pro-glacial deposits vary with depth and distance from the ice margin and seasonally. Glacial deposits in sea water, glaciomarine soils, are similar to those in fresh water but have additional debris derived from fresh water rivers (Figure 2.7).

The debris carried by highland glaciers includes eroded material from the substrate and sides of the valley, which can be carried as englacial or subglacial debris. Further debris is a result

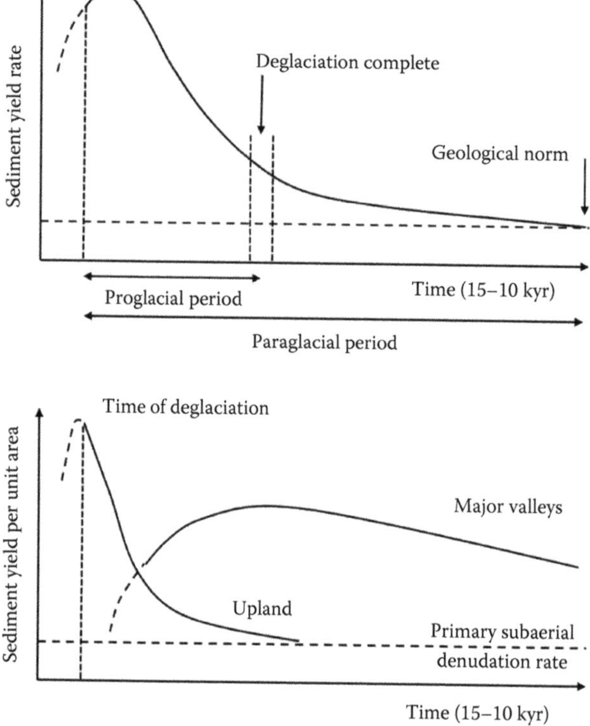

Figure 2.9 Sediment yield associated with period following deglaciation, the paraglacial period. (After Church, M., and J. M. Ryder. *Geological Society of America Bulletin* 83(10); 1972: 3059–3072. Church, M. and O. Slaymaker. Disequilibrium of Holocene sediment yield in glaciated British Columbia. *Nature*, 337(6206); 1989: 452–454; Ballantyne, C. K. and Benn, D. I. Paraglacial slope adjustment during recent deglaciation and its implications for slope evolution in formerly glaciated environments. In *Advances in Hillslope Processes*, edited by Anderson, M. G. and S. Brooks, volume 2, John Wiley, Chichester, 1996: 1173–1195.)

of rock falls and landslides in the valley sides above the glacier. This supraglacial debris can be transported as a result of basal melting (Figure 2.8) to create subglacial debris or englacial material or remains on the surface of the glacier to be deposited as melt-out till as the glacier retreats.

As a glacier retreats, the landscape moves towards a non-glacial equilibrium state, which is initially characterised by high rates of sediment yield (Figure 2.9), reworking of unconsolidated glacial sediments and slope failures, especially for highland glaciers as the ice support is removed leading to over-steepened slopes. This retreat is accompanied by isostatic rebound, leading to terrestrial glaciomarine soils and fluvial landscapes etched into the glacial landscape and continuing slope failures as the fluvial systems are created. A consequence of this paraglacial period, which can extend from a few hundred to a few thousand years, is the formation of extensive fluvially deposited sediment overlying glacial deposits, dormant landslides that may be triggered by subsequent engineering works and climate change and formation of very sensitive terrestrial glaciomarine clays.

2.2.1 Facies

Strictly, the term facies refers to a rock unit with specific characteristics that forms in a particular depositional environment but it has been used by glacial geologists to describe glacial

Table 2.4 Criteria used to describe lithofacies of glacial soils

Lithology	*Bedding* *Characteristics*	*Bedding* *Geometry*	*Sedimentary structures*	*Boundary relations*
Diamicton	Massive	Sheer	Grading (normal)	Sharp
Boulders	Weakly stratified	Discontinuous	Grading (reverse)	Gradational
Gravel	Well stratified	Lensoid	Grading (coarse tail)	Disconformable
Sand	Laminated	Draped	Cross-bedding (tabular)	Unconformable
Silts/clays	Rhythmic lamination	Prograding	Cross-bedding (trough)	
	Wispy stratification		Dropstones	
	Inclined stratification		Clast supported	
			Matrix supported	
			Clast concentrations (layers)	
			Clast concentrations (pockets)	
			Ripples	
			Scours	
			Load structures	

Source: After Hambrey, M. J. *Glacial Environments*. UBC Press, 1994.

sediments. For example, a lacustrine clay is a glacial facies formed by glacial debris deposited in a fresh water environment, a pro-glacial lake. Lithofacies is used to describe the petrological characteristics of a sediment with particular characteristics including colour, clast fabric, clast shape, particle size distribution, composition and sedimentary structures, as indicated in Table 2.4. A coding scheme (Table 2.5), used to describe sediments, is based on the dominant particle size and the particle size distribution. It was developed by Miall (1978) for braided stream deposits and modified to cover diamictons (poorly sorted sediments) by Eyles et al. (1983) and Benn and Evans (2010). This scheme describes a glacial soil but has to be used with other data to correctly interpret a glacial soil (Dreimanis, 1984; Karrow, 1984; Kemmis and Hallberg, 1984). The lithofacies classification can be compared to the European and US engineering soil classifications, which are also based on the dominant particle size and the particle size distribution. However, engineering soil classifications are extended to include consistency and density whereas a lithofacies scheme may be extended to the deposition processes, so genetic facies (Table 2.1), which imply a specific mode of deposition, are used for glacial environment (e.g. subaqueous flow deposits, Ghibaudo, 1992). A lithofacies description distinguishes between boulders (>256 mm), gravels (8–256 mm), granules (2–8 mm), sands (0.063–2 mm) and silts and clays (<0.063 mm). Engineering soil descriptions distinguish between very coarse soil (large boulders [>630 mm], boulders [200–630 mm], cobbles [63–200 mm]), coarse soil (gravel [2–63 mm], sand [0.063–2 mm]) and fine soils (silts and clays [<0.063 mm]). Thus, a description of the coarse soils (>2 mm) will be different depending on whether the glacial deposit is being described as an engineering soil or a glacial sediment.

Glacial sediments, particularly tills, can be described as clast or matrix dominated. Clasts refer to particles of rock (coarse and very coarse particles); matrix refers to the fine-grained sediment (silts and clays). Thus, a clast-dominated till will be a coarse-grained till containing some fine-grained particles; a matrix-dominated till will be a fine-grained till containing some coarse-grained particles. The former will have similar characteristics to a coarse-grained soil; the latter a fine-grained soil.

The orientation of the coarser clasts (gravels, cobbles and boulders) can be associated with the method of deposition; the clast shape can indicate the method of transport; and the clast type the possible source of material. Sedimentary structures, which are formed during deposition, can also be used to distinguish between some glacial soils. For example, glaciotectonite, a sediment that has not been transported but subject to shear, retains many

Table 2.5 Lithofacies coding scheme proposed by Miall (1985) and modified by Eyles et al. (1983) and Benn and Evans (2010) to be used with glacial soils

Particles	*Size (mm)*	*Coding scheme*	*Description*
Diamictons	<0.063 to >256	Dmm	Matrix supported, massive
		Dcm	Clast supported, massive
		Dcs	Clast supported, stratified
		Dms	Matrix supported, stratified
		Dml	Matrix supported, laminated
		---(c)	Evidence of current reworking
		---(r)	Evidence of re-sedimentation
		---(s)	Sheared
		---(p)	Includes clast pavements
Boulders	>256	Bms	Matrix supported, massive
		Bmg	Matrix supported, graded
		Bcm	Clast supported, massive
		Bcg	Clast supported, graded
		Dfo	Deltaic foresets
		BL	Boulder lag or pavement
Gravels	8–256	Gms	Matrix supported, massive
		Gm	Clast supported, massive
		Gsi	Matrix supported, imbricated
		Gmi	Clast supported, massive (imbricated)
		Gfo	Deltaic foresets
		Gh	Horizontally bedded
		Gt	Trough cross-bedded
		Gp	Planar cross-bedded
		Gfu	Upward fining (normal grading)
		Gcu	Upward coarsening (inverse grading)
		Go	Openwork gravels
		Gd	Deformed bedding
		Glg	Palimpsest (marine) or bed load lag
Granules	2–8	GRcl	Massive with clay laminae
		GRch	Massive and infilling channels
		GRh	Horizontally bedded
		GRm	Massive and homogeneous
		GRmb	Massive and pseudo-bedded
		GRmc	Massive with isolated outsize clasts
		GRmi	Massive with isolated, imbricated clasts
		GRo	Openwork structure
		GRruc	Repeating upward-coarsening cycles
		GRruf	Repeating upward-fining cycles
		GRt	Trough cross-bedded
		GRcu	Upward coarsening
		GRfu	Upward fining
		GRp	Cross-bedded
		GRfo	Deltaic foresets
Sands	0.063–2	St	Medium to very coarse and trough cross-bedded
		Sp	Medium to very coarse and planar cross-bedded
		Sr (A)	Ripple cross-laminated (Type A)
		Sr (B)	Ripple cross-laminated (Type B)

(Continued)

Table 2.5 (Continued) Lithofacies coding scheme proposed by Miall (1985) and modified by Eyles et al. (1983) and Benn and Evans (2010) to be used with glacial soils

Particles	*Size (mm)*	*Coding scheme*	*Description*
		Sr (S)	Ripple cross-laminated (Type S)
		Scr	Climbing ripples
		SSr	Starved ripples
		Sh	Very fine to very coarse and horizontally/plane bedded or low angle cross-laminated
		Sl	Horizontal or draped laminations
		Sfo	Deltaic foresets
		Sfl	Flaser bedded
		Se	Erosional scours with intraclasts and crudely cross-bedded
		Su	Fine to coarse with broad shallow scours and cross-stratification
		Sm	Massive
		Sc	Steeply dipping planar cross-bedding (non-deltaic foresets)
		Sd	Deformed bedding
		Suc	Upward coarsening
		Suf	Upward fining
		Srg	Graded cross-laminations
		SB	Bouma sequence
		Scps	Cyclopsams
		---(d)	With dropstones
		---(w)	With dewatering structures
Silts and clays	<0.063	Fl	Fine lamination often with minor fine sand and very small ripples
		Flv	Fine lamination with rhythmites or varves
		Fm	Massive
		Frg	Graded and climbing ripple cross-laminations
		Fcpl	Cyclopels
		Fp	Interclast or lens
		---(d)	With dropstones
		---(w)	With dewatering structures

of its original features and structures such as folds and faults due to the deformation. The alignment of clasts and sedimentary structures are best observed in excavations and natural exposures. It is unlikely that this level of detail can be observed in boreholes. Trial pits can provide some information on near-surface deposits, but given the depth of the zone of influence of civil engineering projects, boreholes will always be the prime source of samples. This means that it is unlikely that the structural features and very coarse particles will be retrieved, making it difficult to identify the type of glacial soil and, importantly, potential hazards associated with structure and composition.

2.2.2 Primary deposits

Primary deposits, those deposited by ice, are the subglacial deposits of glaciotectonite and subglacial traction till (deformation, lodgement and comminution tills) and melt-out till (supraglacial, englacial and sublimation) are summarised in Table 2.6. Deformation tills can be divided into Type A, ductile deformation, and Type B, brittle deformation. The prime

Table 2.6 Summary of characteristics of glacial tills

	Subglacial traction tills		*Gravitationally deposited*		
	Deformation till	*Lodgement till*	*Subglacial melt-out till*	*Supraglacial melt-out till*	*Sublimation till*
Deposition	Sediment detached from sedimentary base	Sediment deposited through pressure melting and mechanical processes	Sediment deposited due to melting of stagnant or slow moving ice	Sediment deposited due to melting of stagnant or slow moving ice	Sediment deposited due to melting of stagnant or slow moving ice
Landforms	No distinguishing landforms	Ground moraines, drumlins, flutes and other subglacial landforms	Ice marginal landforms	Ice marginal landforms	Ice marginal landforms
Basal contact	Variable basal contact	Distinct contact with base	Distinct contact with base; may have variable basal contact	Distinct contact with base; may have variable basal contact	Distinct contact with base; may have variable basal contact
Particle shape	Same as the characteristics of the deforming sediment with some basal debris; little modified	Clast characteristics typical of basal transport; rounded edges, spherical form, striated and faceted faces; large clasts have bullet appearance	Clast characteristics typical of basal transport; rounded edges, spherical form, striated and faceted faces; but less pronounced than lodgement till	Usually dominated by sediment of high-level transport but subglacial transported particles may be present; clast not normally striated or faceted	Clast typical of basal transport
Particle size	Diverse range of sizes reflecting those in the deforming sediment; rafts of original sediment	Particle size distribution is typical of basal transport; either bimodal or multimodal	Particle size distribution is typical of basal transport; either bimodal or multimodal; sediment sorting associated with dewatering and sediment flow visible	Particle size distribution is usually coarse and unimodal; some sorting if meltwater involved	Particle size distribution typical of basal debris transport; bimodal or multimodal
Particle fabric	Some particle fabric aligning with direction of shear	Strong particle fabrics with elongated particles aligned with the direction of ice flow	Fabric may be strong in the direction of the ice flow but not as obvious as lodgement till	Clast fabric unrelated to ice flow; poorly developed and spatially highly variable	Aligned with the direction of ice flow though not as much as lodgement till

(Continued)

Table 2.6 (Continued) Summary of characteristics of glacial tills

	Subglacial traction tills		*Gravitationally deposited*		
	Deformation till	*Lodgement till*	*Subglacial melt-out till*	*Supraglacial melt-out till*	*Sublimation till*
Particle packing	Very dense packing but can include pockets of low density due to dilatancy as sediment is sheared	Very dense packing	Medium dense packing	Loose-to-medium dense packing	Loose-to-medium dense packing
Particle lithology	Diverse range of particle composition but consistent with underlying sediments	Dominated by local rock types especially towards base	Clast lithology may show an inverse superposition	Variable clast lithology and can include erratics	Clast lithology may show an inverse superposition
Structure	Fold, thrust, fault structures present at low shear strains; homogenisation possible; rafts of non-deformed sediment	Massive structure less sediments with well-developed shear planes and foliations; boulder clusters or pavements may occur towards base; sub-horizontal jointing common with possible vertical and transverse joints	Usually massive but if subject to flow, may contain folds and flow structures; crude stratification sometimes present; no evidence of shearing; draping of sorted sediments over large clasts	Usually massive but if subject to flow, may contain folds and flow structures; crude stratification sometimes present; no evidence of shearing; draping of sorted sediments over large clasts	Usually massive but if subject to flow, may contain folds and flow structures; crude stratification sometimes present; no evidence of shearing; draping of sorted sediments over large clasts

Source: After Bennett and Glasser, 1996; Dreimanis, 1988; Elson, J. A. Comment on glacitectonite, deformation till, and comminution till. In *Genetic Classification of Glacigenic Deposits*, edited by Goldthwait, R. E. and Matsch, C. L., Balkema, Rotterdam, 1988: 85–88.

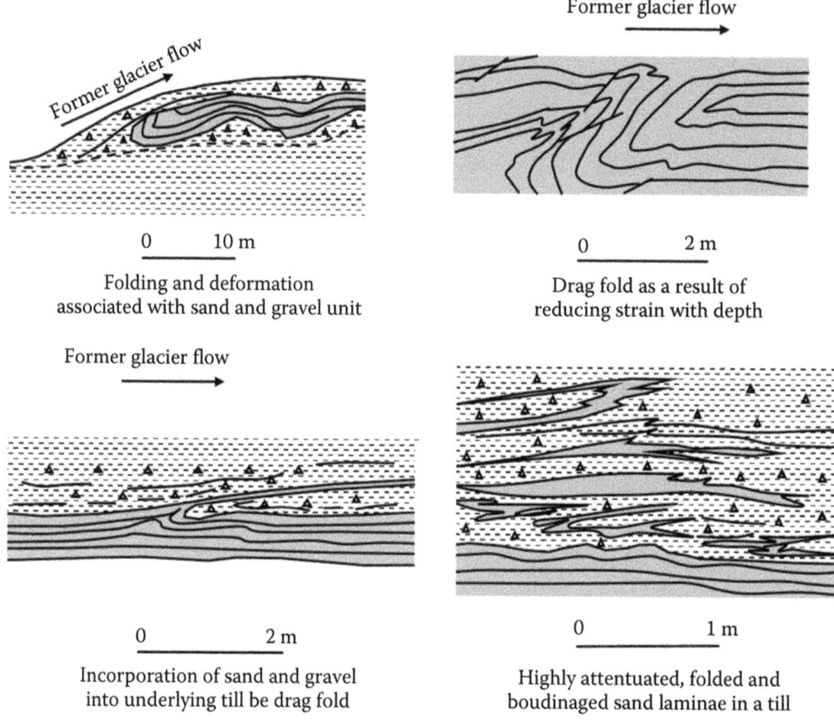

Figure 2.10 Schematic diagram of glaciotectonic processes associated with different levels of subglacial deformation. (After Hart, J. K. and G. S. Boulton. The interrelation of glaciotectonic and glaciodepositional processes within the glacial environment. *Quaternary Science Reviews* 10(4); 1991: 335–350.)

source of the deposits is the substrate, which is rock and superficial deposits possibly including remnants of previous glaciations. In highland glaciers, additional debris is collected from the sides of the valley and rock falls above the glacier. Glaciers sliding over the substrate initially deform the sediment to create glaciotectonite, which retains features of the original sediment and tectonic structures (Figure 2.10). Further movement creates a deformation till, which still contains features of the original sediment. Some of the eroded material is carried up into the ice; some remains at the base of the ice but increasingly the original structure is lost and particles are abraded until a completely homogenised soil is formed. The degree of homogenisation and structural deformation is a function of the temperature of the sliding zone, the distance transported, the amount of abrasion and the strength of the particles. Deformation tills can be formed of sediment from the substrate and assimilated debris from within the glacier. It is deposited when the imposed shear stress is less than the strength of the till. As the glacial debris is transported in the basal zone, an object may be encountered, which resists the movement of the glacial debris creating lodgement till. Lodgement till may be subject to deformation and erosion as the ice advances; therefore, the history of any glacial till is unknown. This means the composition of deformation and lodgement tills may be similar though it would be difficult, if not impossible, to identify this from borehole samples. This is why it may be better to consider the subglacial till as deformation till (van der Meer et al., 2002) or subglacial traction till (Evans et al., 2006).

These is one exception to this generic view and that is melt-out till, which is formed during a glacial recession as the glacial debris in the ice is lowered onto the underlying sediment.

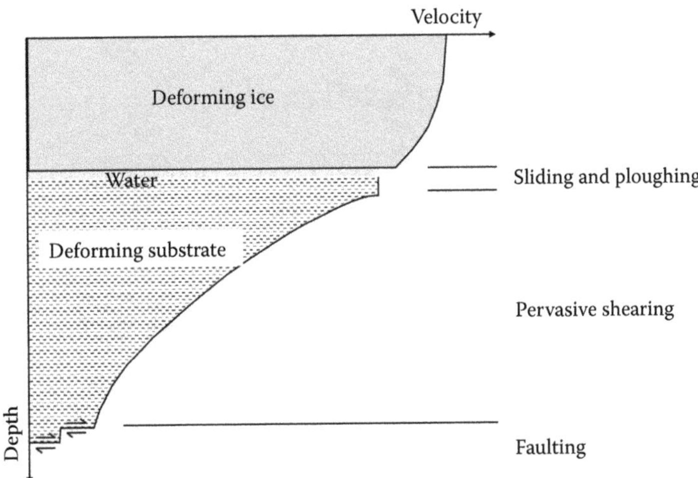

Figure 2.11 Various modes of movement including ice creep, sliding, ploughing, pervasive shear and faulting within the basal zone. (After Boulton, G. S. *Annals of Glaciology*, 22(1); 1996: 75–84; Boulton, G. S. *Journal of Glaciology*, 42(140); 1996: 43–62.)

Melt-out tills can be distinguished from other types of till because they are less dense not having undergone shear during deposition.

Evans et al. (2006) undertook a review of subglacial processes to show that current thinking suggests that deformation, flow, sliding, lodgement and ploughing coexist beneath temperate glaciers because the ice and bed are coupled (Figure 2.11) and these processes vary spatially and temporally. They also concluded that there were three distinct types of till: glaciotectonite, subglacial traction till and melt-out till. Subglacial traction till is a hybrid material that reflects the mosaics of deformation and sliding, warm- and cold-based conditions and hydraulic conditions. It includes deformation, lodgement and comminution tills and is defined as a sediment deposited by a glacier either sliding over and/or deforming a bed, the sediment having been released directly from the ice by pressure melting and/or liberated from the substrate and then disaggregated and completely or largely homogenised by shearing (Evans et al., 2006).

Van der Wateren et al. (2000) used the concept of progressive simple shear to reconstruct the deformation history of glacial sediments because it produces most of the characteristic asymmetric structures, in which the principal direction of finite extension is subparallel to the direction of shearing. The features (Figure 2.12) include discontinuities, folds, boudins and clast alignment with increasing strain. Sedimentary and deformation structures may completely disintegrate in the most intensely deformed sediments leading to complete homogenisation, although the typical shear zone fabric may still be identified in thin section.

Menzies et al. (2006), based on microstructural characteristics of subglacial tills (van der Meer, 1987, 1993, 1996, 1997; Menzies and Maltman, 1992; Tulaczyk, 1994, 1998, 2000; Menzies and van der Meer, 1998; Carr, 1999, 2000; Phillips and Auton, 2000; Van der Wateren et al., 2000; Khatwa and Tulazyk, 2001; Hiemstra and Rijsdijk, 2003; van der Meer et al., 2003; Lachniet et al., 2007), suggested that subglacial till formation is a result of structural rather than depositional processes and therefore a deformable bed (Alley et al., 1987; Engelhardt et al., 1990; Boulton, 1996b; Hart et al., 1999; van der Meer et al., 2003).

According to Boston et al. (2010), geochemical analysis used to detect major sources of ore within glacial deposits (e.g. Kettles and Shilts, 1989; Shilts and Smith, 1989; Klassen

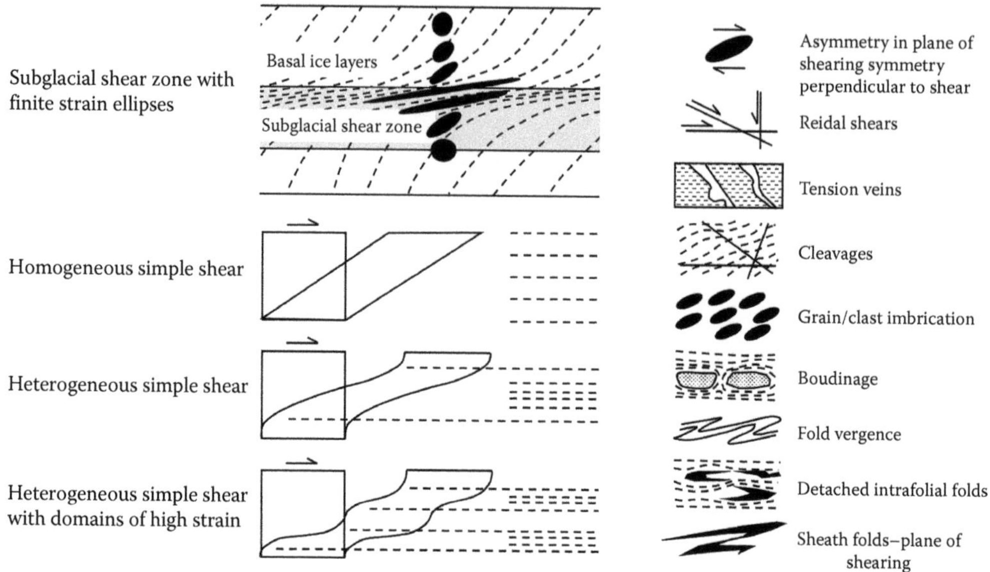

Figure 2.12 Effects of simple shear within the basal zone, the effect it has on the substrate structure and the structures produced by progressive simple shear. (After Van der Wateren, F. M., S. J. Kluiving, and L. R. Bartek. *Geological Society, London, Special Publications*, 176(1); 2000: 259–278.)

and Thompson, 1993; Shilts, 1993) can also be used to determine the variation in composition of glacial tills and therefore an indication of the source of sediment and direction of flow (e.g. Shilts et al., 1979; Saarnisto, 1990; Bölviken et al., 1990; Dyke and Morris, 1988; Klassen, 1999; Evans, 2007). They studied the tills on the Holderness Coast and observed that there were two distinct groups based upon their geochemical composition; a chalk/limestone clast-rich group derived from the local bedrock and a more diverse geotechnical signature indicative of distal sources such as NW England (Lake District) and Scotland. This suggested that repeated ice advances led to mixing of local and distal sources of debris. They also concluded that the traditional interpretation of a till sequence based on the lithology may not be supported by geochemical analysis.

2.2.2.1 Subglacial traction tills

Primary glacial deposits are difficult to classify, are subject to debate about their formation and have a complex history, and their composition, fabric and structure are spatially variable. This is why these soils are considered difficult soils for the construction industry. They include glaciotectonite, deformation, lodgement and comminution tills.

2.2.2.1.1 Glaciotectonite

Glaciotectonite is either subglacially deformed rock or superficial deposit that retains some of the original structure of the parent material (Banham, 1977; Pedersen, 1988; Benn and Evans, 2010). The tectonic features developed during deformation depend on whether brittle or ductile deformation takes place. Brittle deformation results in shear planes and faults; ductile deformation produces folds. Extensive shear strain can produce laminations, with distinctly different soils between the laminations. Pods of stiffer material can be generated because of the variation of stiffness in the basal zone. Further deformation leads to

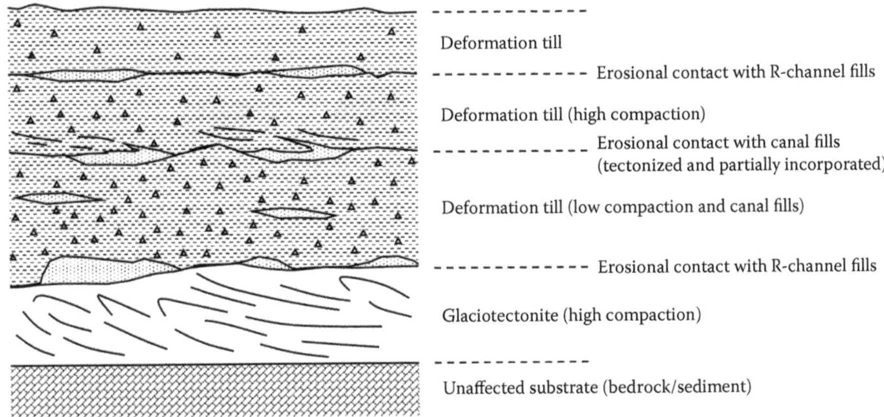

Figure 2.13 **Possible profile through the substrate showing the effect on the type of substrate varying from undisturbed substrate to completely homogenised till with channel fills within the till. (After Benn, D. and D. J. A. Evans. *Glaciers and Glaciation*. Routledge, London; 2010.)**

increasing disaggregation of the sediment and abrasion, thus breaking down the sediment until it is eventually completely homogenised. If ice is moving across glacial soils deposited under the previous advance or recession, a glaciotectonite will contain features of those glacial soils. Thus, it is possible for a glaciotectonite to appear to be similar to another type of glacial soil, making it extremely difficult to classify the soil. Further, a vertical profile through a till deposit may show a sequence ranging from undisturbed sediment at the base of the deposit to the deformation till at the top because of the variation in strain through the sequence (Figure 2.13). Hence, the distinction between deformation and glaciotectonite tills is not clear, suggesting that, for engineering purposes, they can all be considered subglacial traction tills.

2.2.2.1.2 Deformation till

Deformation till is completely disaggregated and possibly homogenised sediment by shearing in a subglacial deforming layer (Elson, 1961; Benn and Evans, 2010). Pre-existing structures are destroyed, but the lithology is retained. There is no distinct boundary between glaciotectonite and deformation till as it depends on the degree of strain. Further, a glacial stratum can be completely homogenised at the top and undisturbed at the bottom, as shown in Figure 2.13. This would help explain why tills can contain remnants of previous glaciations, e.g. the treacherous pockets of water-bearing sands and gravels and softer clays highlighted by Ansted (1888).

Deformation tills can vary in density and structure. The particle size distribution varies widely depending on the amount of strain and therefore abrasion. The deforming regime is a function of the *in situ* stresses and till density. Deformation tills can contain boulder pavements, a wide variety of microstructures and fabric, faults and folds, brittle and ductile features, rafts of intact rocks and boudins (Benn and Evans, 2010).

2.2.2.1.3 Lodgement till

Lodgement till results from the lodgement of glacial debris beneath a glacier by pressure melting or other mechanical processes (Chamberlin, 1985; Dreimanis, 1989) against a fragment of rock which, in glacial tills, can be a particle of any size (Figure 2.14) or when the

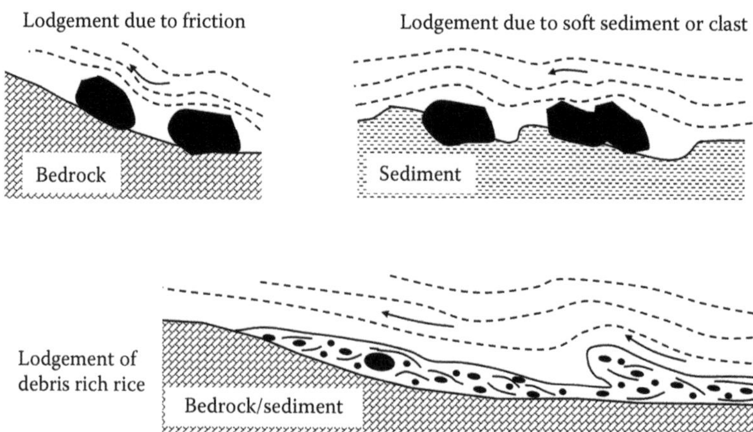

Figure 2.14 Lodgement due to frictional drag on a rigid substrate, obstacles on a soft substrate or debris-rich ice. (After Boulton, G. S. *Proceedings of the 6th Guelph Symposium on Geomorphology*, Vol. 1980, 1982: 1–31; Benn, D. and D. J. A. Evans. *Glaciers and Glaciation*. Routledge, London; 2010.)

frictional resistance between a clast and the underlying ground exceeds the frictional drag of the glacier. Lodgement tills are usually very dense with a low water content because of the combination of the pressure of the ice and shear. They are sometime described as over-consolidated though this may refer to the density rather than the geotechnical process of consolidation. They are often fissile with potential slip planes suggesting shear during deposition. These tills have bimodal or multimodal particle size distributions with distinct rock flour and gravel ranges. Cobbles and boulders are aligned with the direction of the ice flow. Lodgement tills and deformation tills may have a similar appearance (Eyles et al., 1994) and from an engineering point of view may not be different.

2.2.2.1.4 Comminution till

Comminution till is a particular type of deformation till, which is formed entirely of rock flour (clay-size particles) as a result of abrasion during deformation (Elson, 1988); that is, the clasts have been completely broken down. They tend to have a very high density and strength.

2.2.2.2 *Melt-out till*

Melt-out till is formed of glacial debris being deposited from stagnant or slow moving ice without further transport or deformation (Benn and Evans, 2010). As the ice melts, supraglacial and englacial debris are deposited at the base of the glacier. The source of heat can be solar or geothermal, creating supraglacial and subglacial melt-out tills. The clast content reflects high-level transport in which particles retain their angularity. It is generally poorly consolidated because it has not been subject to high pressures or shear. Therefore, it has a relatively low density compared to other tills. More importantly, this till has been subject to gravitational consolidation whereas the other tills are subject to shear as well as gravitational loads.

2.2.3 Secondary deposits

Secondary deposits are formed of glacial debris transported by water or deposited in water. These include glaciofluvial deposits, mass movements due to gravity, glaciolacustrine deposits in pro-glacial lakes and glaciomarine deposits in a marine environment. The source of these deposits include subglacial, englacial and supraglacial debris, which are released by the glacier as the ice melts or by erosion by water flowing through or beneath the ice.

2.2.3.1 Glaciofluvial deposits

Glaciofluvial deposits are terrestrial sediments deposited from flowing water either on (channels), within (tunnels), beneath (tunnels) or beyond the ice margin. The water and suspended sediment cause further abrasion of the suspended particles and sides of the channels/tunnels adding to the suspended load. Debris is carried as suspended load or bed load. The sediment can be deposited subglacially in tunnels in the ice or channels in the underlying sediment, along ice margins and beyond the ice margin. These deposits are often subject to further glacial or fluvial actions. It is difficult to determine the amount of sediment carried by meltwater streams but, given the scale of the deposits beyond the ice margin, it is substantial. Sedimentation beyond the ice margin follows the same process as conventional fluvial deposits. The discharge is seasonal and, because the water is cooler and therefore more dense and viscous than that in conventional fluvial processes, the settling rate of the suspended load is less, which means the suspended load is carried further. Close to the ice margins, the glaciofluvial deposits are coarse, poorly sorted clasts. The deposits become finer with distance from the ice margin as the velocity of the flow reduces. The pattern of flow also changes from a braided channel sequence to a single channel. Glaciofluvial deposits can be extensive either as valley in fill or outwash fans in lowland areas. Glaciofluvial deposits occur throughout glaciated regions and are a valuable source of sands and gravels.

2.2.3.2 Glacial sedimentation

Glaciers that terminate in water produce different sediments to the terrestrial deposits. Some 90% of the Antarctic ice sheet terminates in seawater and many highland glaciers terminate in fjords. A glacier that is grounded produces subglacial deposition. The glacier then acts as a source of debris as meltwater carries the glacial debris into the water; the deposition is controlled by sedimentation. This sedimentation process produces deposits, which are unique to glacial deposition and are significant for engineering projects. The deposits may have the appearance of glacial tills in some cases. Pro-glacial lakes may form at the ice margin creating a lacustrine environment. These lakes can form at the ice margin due to the damming action of a moraine or ice dam during the retreat of a melting glacier, or by meltwater trapped against an ice sheet due to isostatic depression of the crust around the ice. Glacial debris can be carried into the water suspended in meltwater or encased in ice. Other water-lain deposits are those deposited in a marine environment. Deposition in fresh water and that in sea water are different because the factors, temperature and density stratification, which affect the sedimentation process, are different. Figure 2.15 shows the seasonal temperature profile within lake water, which affects the seasonal deposition of glacial debris carried into a lake.

Sedimentation within a glacial lake includes deposition from meltwater debris falling from the ice margin, debris from icebergs and settling from suspension. The deposits may be reworked due to unstable submerged slopes, currents and wave action (Figures 2.6 and 2.7). During the last glaciation, enormous pro-glacial lakes were created leading to extensive glaciolacustrine deposits.

Debris-laden water entering a freshwater lake tends to flow to the bottom of the lake because of its relative density (Figure 2.16). Glacial meltwater is less dense than the seawater, so it tends to enter a marine environment as an overflow even allowing for the fact that

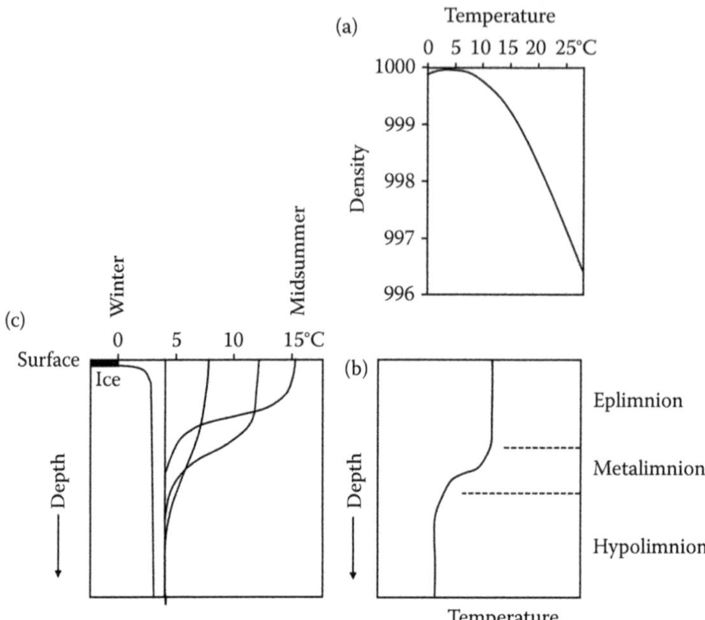

Figure 2.15 Temperature characteristics of lake water showing the (a) effect of temperature on density, (b) surface effects of wind and sun on the temperature profile and (c) the seasonal changes. (After Smith N. D. and G. M. Ashley. Proglacial lacustrine environment. In *Glacial Sedimentary Environments*, edited by Ashley, G. M, John Shaw, and N. D. Smith, No. 16. Society of Economic Paleontologists and Mineralogists; 1985.)

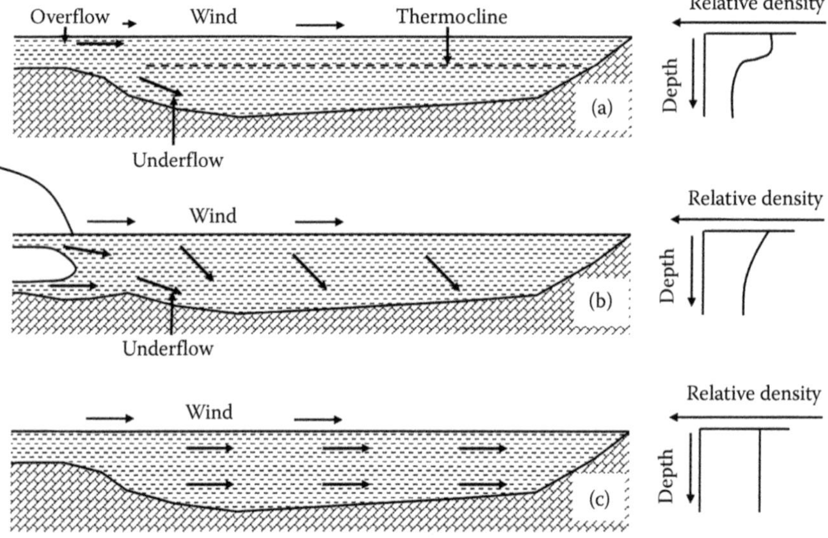

Figure 2.16 Density stratification in lakes showing (a) overflows and underflows in non-glacial lakes, (b) underflows in pro-glacial lakes and (c) no stratification. (After Smith N. D. and G. M. Ashley. Proglacial lacustrine environment. In *Glacial Sedimentary Environments*, edited by Ashley, G. M, John Shaw, and N. D. Smith, No. 16. Society of Economic Paleontologists and Mineralogists; 1985.)

the meltwater contains debris. This creates turbulence. Seasonal changes in water temperature and surface ice change the thermal profile. Waves, currents and tides also impact on the effect of the mixing of the meltwater and fresh or seawater.

Supraglacial, englacial and subglacial debris can be released directly from the ice margin into the water. Meltwater carries sediment into the water, and subglacial sediment can be pushed into the water. The dominant input depends on a number of factors. Glacial meltwater tends to dominate in temperate glaciers. Basal debris-rich ice is mostly deposited from icebergs from ice sheets. Supraglacial debris occurs in mountainous terrains, which enters the water through gravitational and glaciofluvial processes. The depositional environments include grounding line fans, moraine banks, grounding line wedges, deltas and distal environments.

Within lakes, two facies can be identified: basin margin and lake floor facies. The facies architecture in a glaciomarine environment is more complex. Glacial deposits visible today, whether terrestrial or subaqueous, were not necessarily deposited in those environments. For example, there is a debate over whether the glacial sediments exposed in the coastal cliffs at Norfolk, England are glaciomarine diamictons (Eyles et al., 1989) or subglacial deformed sediments (Hart and Boulton, 1991). From an engineering point of view, this is important as the properties will depend on whether the deposit was gravitationally deposited in water or subject to shear. Hart and Roberts (1994) suggested criteria to distinguish between these two types of sediments (Table 2.7).

2.2.3.2.1 Glaciolacustrine deposits

Figure 2.6 shows the transport and deposition processes in a pro-glacial lake in which glaciolacustrine deposits are formed. The deposition process is a function of the density profile within the lake, which is a function of the suspended load of the meltwater entering the lake and the temperature profile within the lake (Figure 2.15). The variation in surface temperature and density leads to seasonal changes, which affects the deposition of the glacial debris (Figure 2.16). Most sediments are formed as either deltaic sediments or lake bottom sediments formed as topsets, forests or bottom sets. Deltaic sediments are typically sand and gravel as they are nearest to the source of the sediment. Deltaic sediments include deltas, delta moraines, De Geer moraines, shorelines deltas, areas of debris slumping and sedimentation of fine sediment. Delta moraines are a product of glaciofluvial deposition in

Table 2.7 Criteria to distinguish the differences between glaciotectonite and glaciomarine sequences when they appear as terrestrial glacial deposits

	Glaciomarine	*Glaciotectonite*
Sedimentary units	Laterally continuous, on lapping relationships	Laterally discontinuous, tectonic boundaries
Basal boundary	Sedimentary	Decollement surface
Laminations	Graded	Non-graded
Shells	Common, *in situ*	Rare, not *in situ*
Folds	Gravitational flow folds, restricted to local areas, orientated downslope	Tectonic folds, deformation throughout, orientated in the direction of ice flow
Boudins	Rare	Common
Lone stones	Dropstones	Sinking clasts
Fabric	Variable, if present will reflect local slopes or flow directions	Variable, but may be well developed in the direction of shear

Source: After Hart, J. K. and D. H. Roberts. *Sedimentary Geology*, 91(1); 1994: 191–213.

front of a stationary ice margin creating shorelines (e.g. Glen Roy, Scotland). As the lake drains, the shoreline remains. Lake bottom sediments can be laminated couplets of silt and clay representing summer and winter deposition. They may contain glacial debris including dropstones released from icebergs. Three types of varved deposits exist; those in which the thicker clay layers are separated by thin silt layers; those in which the clay and silt layers are equal in thickness; and those which are deposited near to the ice margin forming thick silt layers separated by thin layers of clay. The sources of these materials include subglacial, englacial or supraglacial sediment and glacier melt streams.

2.2.3.2.2 Glaciomarine deposits

Glaciomarine sediments are deposited in fjords, on the continental shelf and deep sea environments. The deposition process is complex (Figure 2.17) because it is the interaction between the glacial, marine, biogenic environments and inputs from rivers and wind. In fjords, the sedimentation is influenced by tidal water, floating glaciers, river streams and slope and marine processes. Glaciomarine deposits on the continental shelf are influenced by grounded ice margins, ice shelves and open marine processes. Figure 2.7 shows the type and distribution of glaciomarine deposits in a fjord, highlighting the fact that glacial soils can range from diamictons, outwash, rhythmites, turbidites to bioturbidites. Rhythmites are a result of periodic sedimentation resulting in laminated deposits. Turbidites are a result of the suspension of fine particles creating a dense fluid, which is able to transport coarser particles.

2.3 GLACIAL DEPOSITIONAL LANDFORMS

The next descriptive level refers to depositional landforms, which include subglacial associations, ice marginal moraines, supraglacial associations, pro-glacial associations and glaciolacustrine and glaciomarine associations (Benn and Evans, 2010). Table 2.8 refers

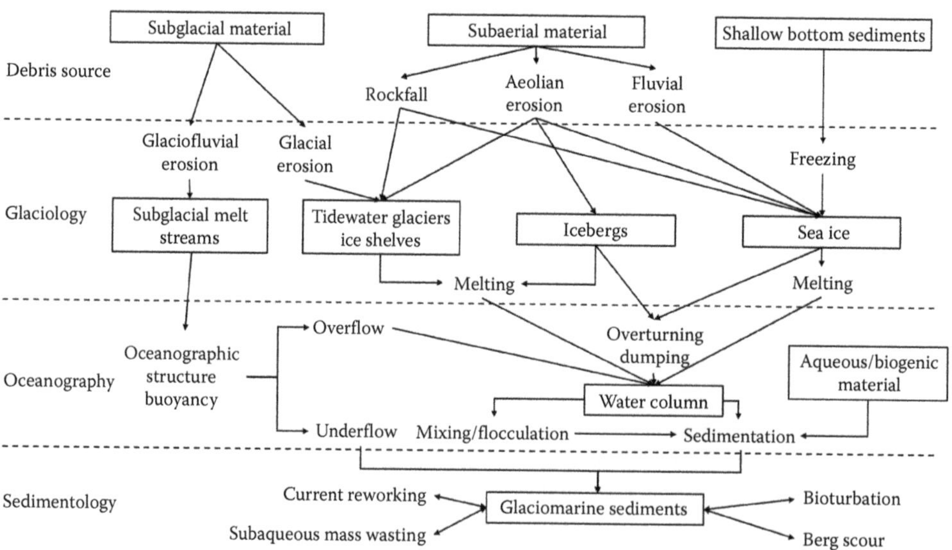

Figure 2.17 Flow chart showing the complex processes in the glaciomarine environment showing the source, glacial processes and depositional environment. (After Dowdeswell, J. A. *Progress in Physical Geography* 11(1); 1987: 52–90.)

Table 2.8 Landforms in terrestrial glacial depositional environment highlighting their location relative to the glacier and direction of ice flow and their scale

Location relative to glacier	*Alignment with ice flow*	*Landform*	*Scale (m)*							
			0.01	*0.1*	*1*	*10*	*100*	*1k*	*10k*	*100k*
Supraglacial (during accumulation)	Parallel	Lateral moraine				←	—	—	—	→
		Medial moraine				←	—	—	—	→
	Transverse	Thrust moraine			←	—	—	→		
		Rockfall			←	—	—	→		
	Non-orienteered	Dirt cone		←	→					
		Erratic			↔					
		Crevasse filling		←	—	—	→			
Supraglacial (during deposition)	Parallel	Moraine dump			←	—	→			
	Non-orienteered	Hummocky moraine				←	—	—	—	→
		Erratic			←	→				
Subglacial	Parallel	Drumlin				←	—	—	→	
		Drumlinoid ridge					←	—	→	
		Fluted moraine				←	—	→		
		Crag and tail ridge				←	—	—	→	
	Transverse	De Geer moraine				←	—	→		
		Rogen moraine				←	—	→		
	Non-orienteered	Till plain						←	—	→
		Gentle hill			←	—	—	—	→	
		Hummocky ground moraine						←	→	
		Cover moraine						←	—	→
Ice marginal	Transverse	Terminal moraine			←	—	—	—	—	→
		Recessional moraine			←	—	—	—	→	
		Annual moraine		←	—	→				
		Push moraine			←	—	—	—	→	
	Non-orienteered	Hummocky moraine			←	—	—	→		
		Rockfall		←	—	—	—	→		
		Slump		←	—	—	—	→		
		Debris flow		←	—	—	—	→		

Source: After Hambrey, M. J. *Glacial Environments*. UBC Press, 1994.

to the landforms (moraines, mounds and ridges of glacial tills and drumlins) formed at the sides, front and beneath a moving glacier and at the ice margins. Table 2.9 refers to the landforms created by meltwater, which can occur beneath a glacier, at the ice margins and beyond the ice. There are many types of landforms associated with glacial erosion, deposition and deformation, which provide an insight into the surface glacial soils

Table 2.9 Landforms created by subglacial meltwater erosion highlighting their location relative to the glacier, the process by which they are formed and their scale

Location relative to glacier	*Process*	*Landform*	*Scale (m)*							
			0.01	*0.1*	*1*	*10*	*100*	*1k*	*10k*	*100k*
Subglacial	Erosion by subglacial water	Tunnel valley							←	→
		Subglacial gorge				←				→
		Nye (bedrock) channel			←			→		
		Sediment channel				←				→
		Glacial meltwater chute			←	→				
		Glacial meltwater pothole		←		→				
		Sichelwannen		←	→					
	Deposition subglacial channels	Esker				←				→
		Nye channel fill			←	→				
		Moulin kame			←	→				
Ice marginal	Stream erosion	Meltwater channel					←			→
	Ice contact deposition from meltwater	Kame field					←		→	
		Kame plateau					←		→	
		Kame terrace				←		→		
		Kame delta				←				→
		Crevasse fillings		←	→					
Pro-glacial	Meltwater erosion	Scabland topography							←	→
	Meltwater deposition	Outwash plain (sandar)					←			→
		Valley train					←			→
		Outwash fan				←			→	
		Pitted plain					←	→		
		Outwash delta complex				←				→
		Kettle hole			←			→		

Source: After Hambrey, M. J. *Glacial Environments*. UBC Press, 1994.

present in the area. This fact means that a study of the regional geomorphology can help in identifying the glacial soils from which it may be possible to infer geotechnical characteristics. The landforms occur at different scales and different types of landform can be adjacent or overlapping. However, a study of the geomorphology alone does not provide information on the geological profile, particularly the stratum and bedrock interface. For example, it cannot be assumed that bedding planes are parallel to the surface or an interface identified in one borehole can be connected to a similar interface in an adjacent borehole.

Table 2.10 Classification of terrestrial glacial landforms

Deposition	*Ice marginal*	*Subglacial*
Glacial	Glaciotectonic moraines Dump moraines Ablation moraines	Flutes Megaflutes Drumlins Rogens Mega-scale glacial lineation Geometrical ridge networks
Glaciofluvial	Outwash fans Outwash plains Kame terraces Kames Kame and kettle topography	Eskers Braided eskers

Source: After Bennett, M. R. and N. F. Glasser. *Glacial Geology: Ice Sheets and Landforms*. John Wiley, London; 1996: 364.

Glaciofluvial deposits lie unconformably on glacial tills, which may include tectonic structures and lenses of distinctly different materials contributing to a complex geological profile. Infrastructure projects may cross many glacial landforms making a geomorphological study an essential component of any ground investigation. In urban areas, the geomorphological profile may not be obvious, though it could be there and be significant. It is possible to classify the terrestrial landforms according to their mode and location of deposition (Table 2.10). The landforms may provide some indication of the glacial history; for example, transverse moraines, if formed during glacial recession, will be low-density non-sorted diamictons. The glacial soils forming these landforms may provide a further indication of the glacial history. A drumlin field is aligned with the direction of ice flow and, more importantly, raises the possibility of fissures aligned with the longitudinal axis of the drumlin, a problem for excavations in line with that axis. Some landforms are associated with highland glaciers so would not be expected in areas of plains of glacial soils. Table 2.11 summarises the principal landforms created by glacial deposition and their relation to the glacial environment. Table 2.12 lists the landforms typical of glaciofluvial soils. A summary of the major landforms is given here to demonstrate the complexity and diversity of these structures. The formation of glaciofluvial landforms is understood but recognising them can be challenging. The formation of subglacial landforms is still a matter of debate. In conclusion, geomorphological and geological studies of a region provide a valuable insight into the type of glacial material and therefore the geotechnical characteristics.

2.3.1 Subglacial landforms formed by ice

The surface of subglacial tills is generally smooth but not necessarily level. The exception is the surface of melt-out tills, deposited during ice recession, which can be variable, possibly hummocky, because no shearing taking places during deposition. The surface of subglacial and englacial tills may be overlain by glaciofluvial deposits, which can in some cases further smooth out the surface of the till or create distinct landforms. Subglacial ice-generated landforms are longitudinal (drumlins, flutes and megaflutes) and transverse (Rogen moraines) accumulations of glacial debris. They are related to substrate morphology, sediment, stress level, and direction of ice flow. The distinction between the longitudinal deposits is a function of the elongation ratio (Figure 2.18).

Table 2.11 Landforms of ice marginal deposits and their relation to the glacial motion

Moraines	*Description*
Seasonal push	• Low sediment ridges transverse to the direction of the ice flow • Position of the ice margin and warm-based ice in a maritime environment • Winter ablation is less than winter ice velocity at the snout • Spacing is usually a function of summer ablation • Number of ridges provides an estimate of rate of retreat
Composite push	• Large multi-crested ridges transverse to the ice flow • Position of the ice margin; suggests possible surging behaviour or strong ice compression at the ice margin either due to thermal variation at the snout or due the presence of frontal tectonics
Thrust	• When ice covered, they consist of single or multi-crested ridges transverse to ice flow • Tectonic structure; thrust and share zones within the ice • Not related to ice margin
Dump	• Steep-sided ridges with well-developed scree-like bedding within them • Affected by withdrawal of lateral ice support • Commonly found as lateral moraines around the margins of warm-based glaciers • Can be found as frontal moraines, especially if the ice is cold based • Cross-valley asymmetry in moraine size indicates patterns of debris supply in glacier basin • May contain distinct stratification which could be seasonal
Ablation	• Variable, ranging from well-defined ridges to belts of mounds, ridges and enclosed hollows • Morphological form may be strong and organised when buried ice persists and may reflect structure of thrust and shear planes within the ice • Related to ice margin • Origin includes high supraglacial debris; high englacial debris content due to a mixed basal thermal regime and freezing on of abundant debris; strong compressive thrusting at the ice margin transferring basal debris to the ice surface
Hummocky	• Mounds, ridges and enclosed hollows with an irregular plan from distribution composed in part of supraglacial till • Ice marginal areas in which the surface cover of debris has prevented ablation • Possibly, a result of high supraglacial debris content; high englacial debris content due to a mixed basal thermal regime and freezing on of abundant debris; due to strong compressive thrusting at ice margin
Flutes	• Low linear sediment ridges formed in the lee of boulders or bedrock obstacles • Local ice flow directions and warm-based or thin ice
Megaflutes	• Linear sediment ridges • Local ice flow directions and warm-based or thin ice
Drumlins	• Smooth oval-shaped or elliptical hills composed of glacial sediments • Possibly with other superimposed landforms • Local ice flow direction • Subglacial deformation and warm-based ice
Rogens	• Streamlined ridges of glacial sediment transverse to direction of ice flow • Ridge may be lunate form and drumlinised • Subglacial deformation and warm-based ice
Mega-scale glacial lineation	• Board low ridges of glacial sediment only recognised from satellite imagery • Smaller landforms superimposed on surface • Regional ice flow patterns • Subglacial deformation and warm-based ice

Source: After Bennett, M. R. and N. F. Glasser. *Glacial Geology: Ice Sheets and Landforms*. John Wiley, London; 1996: 364.

Table 2.12 Landforms of glaciofluvial deposits and their relation to the glacial motion

Landform	Description
Outwash fans	• Low angled fan-shaped accumulations of sand and gravel • Braided surface and fan apex at meltwater portal • May contain kame and kettle topography • Stationary ice margin with relatively high meltwater/sediment discharge
Outwash plain	• Flat surface of sand and gravel formed by braided river systems • Retreating ice margin with relatively high meltwater/sediment discharge • Position of the ice margin; suggests possible surging behaviour or strong ice compression at the ice margin either due to thermal variation at the snout or due to the presence of frontal tectonics
Kames	• Irregular collection of mounds and ridges often with enclosed kettle holes or depressions • Areas of outwash deposition in which melt-out of buried ice occurred after the surface had been abandoned by the melt streams
Kame terraces	• Valley side terraces with outer edged which possess a concentration of kettle holes of belts of kame and kettle topography • Position of ice margin
Eskers	• Steep-sided sinuous ridges of variable extent and size • Location of discharge routes
Braided eskers	• Multi-pole series of steep crested sinuous ridges which form a bifurcating pattern • Glaciofluvial sediment on the surface of a glacier • Possible high magnitude flow events

Source: After Bennett, M. R. and N. F. Glasser. *Glacial Geology: Ice Sheets and Landforms*. John Wiley, London; 1996: 364.

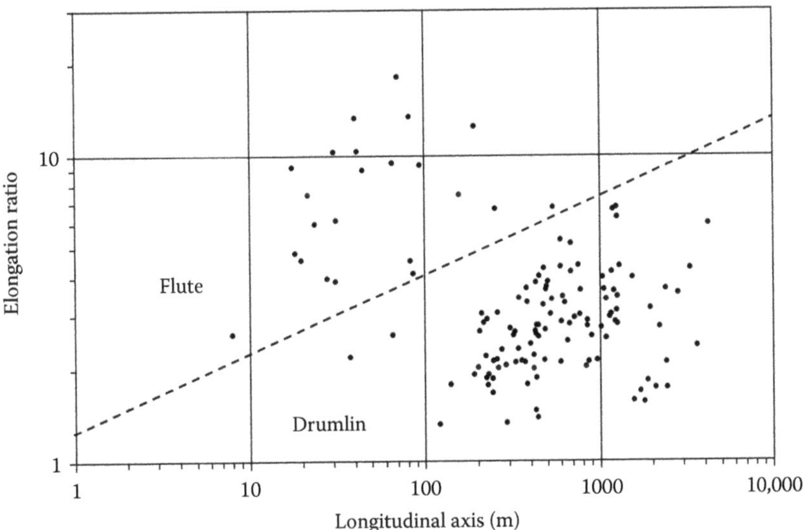

Figure 2.18 Relation between length and elongation of landforms showing the distinction between flutes and drumlins from Norway and Scotland. (After Rose, J. *Quaternary Newsletter*, 53(9); 1987.)

2.3.1.1 Drumlins

Drumlins can occur on their own or in fields of several thousands. Their height varies from 10 to 50 m and their length from 50 m to 20 km. They are typically smooth oval or elliptical shaped with the major to minor axis varying between 1.5 and 4.1 (Hambrey,

1994). They can be formed of lodgement till, bedrock, mixtures of glacial soils and sands and gravels. There are several suggestions as to how drumlins are formed including products of subglacial deformation, subglacial lodgement, fluvial infills, remnants of subglacial floods, or products of melting of debris-rich ice. Boulton (1987) suggested that subglacial deformation (Figure 2.19) is the most likely with drumlins forming around some obstacle as the ice erodes the softer material adjacent to the obstacle and deforms the obstacle. Figure 2.20 shows possible drumlin formation because of changes in the bedrock surface, which leads to the subglacial deforming till rising over the bedrock obstruction to create a drumlin. Figure 2.21 is the suggestion that drumlin formation is a consequence of a fold generated in the lower deforming zone and around glaciofluvial deposits formed during the last ice retreat. A deforming basal layer moving across pro-glacial meltwater streams, which have deposited gravel (Figure 2.22), can also lead to drumlins. These four models, based on excavations, depend on the characteristics of the surface, the ice and the deforming

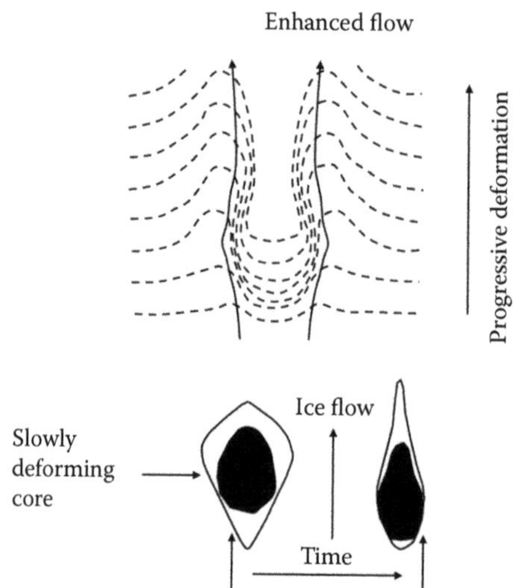

Figure 2.19 Drumlins formed by subglacial deformation as a result of an obstacle in the path of the ice. (After Boulton, G. S. *Drumlin Symposium*, 1987: 25–80.)

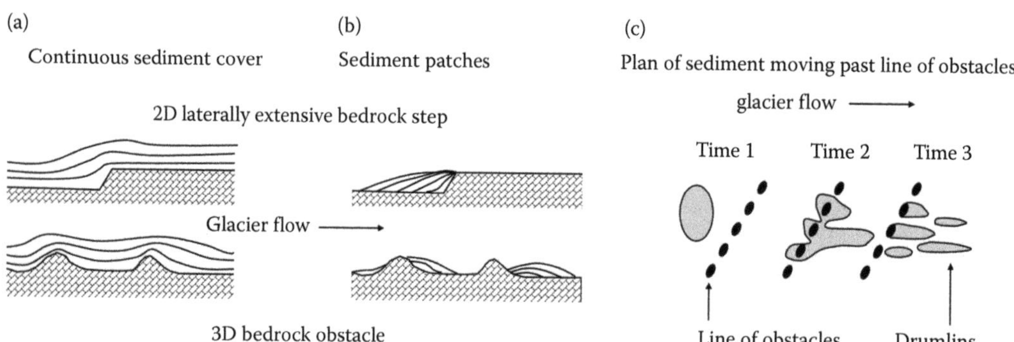

Figure 2.20 Drumlins formed by changes in the bedrock surface. (After Boulton, G. S. *Drumlin Symposium*, 1987: 25–80.)

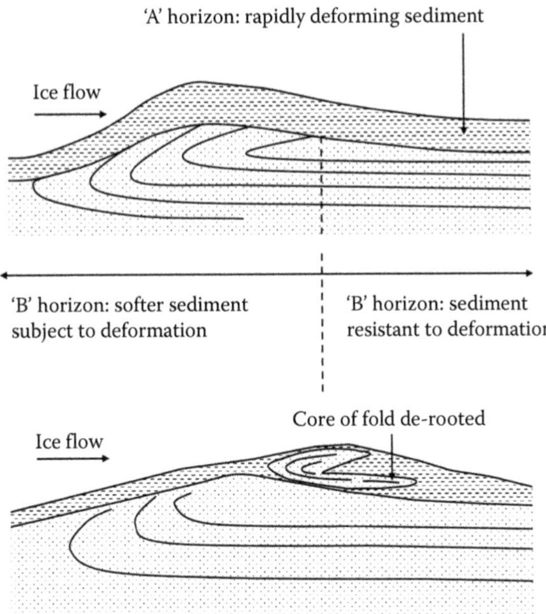

Figure 2.21 Formation of a drumlin as a result of a fold forming in the deforming substrate. (After Boulton, G. S. *Drumlin Symposium*, 1987: 25–80.)

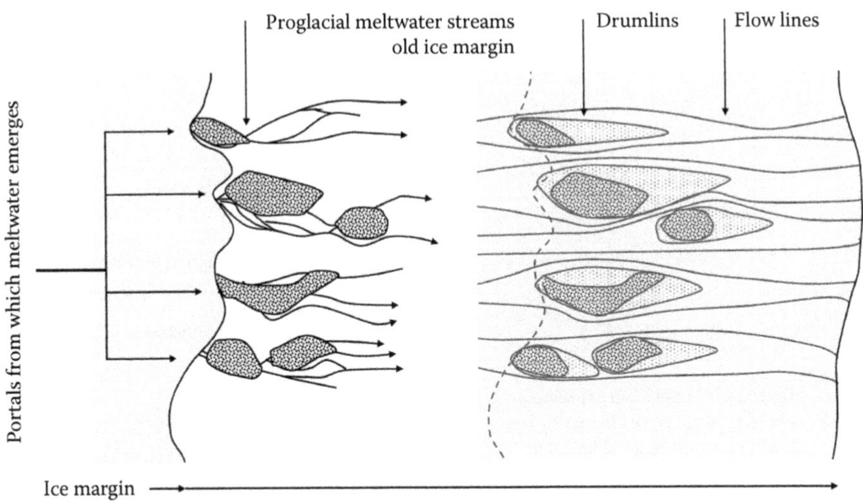

Figure 2.22 Formation of drumlins as the ice moves over pro-glacial meltwater stream deposits. (After Boulton, G. S. *Drumlin Symposium*, 1987: 25–80.)

substrate. It is difficult to prove these models because observations beneath glaciers are difficult.

An alternative view (Shaw, 1983) is that drumlins are large scours created by subglacial floods subsequently filled with glacial infill. The evidence is based on similarities with turbulent underflows and the fact that the infills are stratified. Thus, there are two opposing views, which reinforce the difficulty in linking the depositional history to the geotechnical properties. Boulton et al. (2001) and others suggest that the sediment deformation process is a much more likely cause of drumlins.

Drumlin fields can be extensive such as those found in Europe (United Kingdom, Switzerland, Poland, Estonia, Ireland, Latvia, Sweden, Germany, Denmark, Finland and Greenland), North America (State of New York, Eastern Massachusetts, New Hampshire, Michigan, Minnesota and Wisconsin) and Canada (Ontario, Alberta, Nova Scotia).

2.3.1.2 Flutes

Flutes look like a ploughed surface of ridges (Figure 2.23) typically less than 3 m wide and less than 3 m high. They usually start with a boulder or collection of boulders or bedrock obstacle, which leads to a linear ridge of lodgement till. Flutes are aligned with the direction of ice flow. Megaflutes are taller and broader than flutes.

2.3.1.3 Rogens (ribbed moraines)

Large, coherent fields of ribbed moraines occur in central Canada and Scandinavia (Hättestrand, 1997; Hättestrand and Kleman, 1999), northern and central Ireland (Knight and Marshall McCabe, 1997; McCabe et al., 1999; Clark and Meehan, 2001; Smith et al., 2005; Dunlop and Clark, 2006; Greenwood and Clark, 2008, 2009a,b), and the United Kingdom (Bradwell et al., 2008; Finlayson and Bradwell, 2008; Van Landeghem et al., 2009; Hughes et al., 2010, 2014). Rogens are irregular transverse moraines that are typically 10–20 m high, 50–100 m wide and 1–2 km long. They are composed of clast-rich poorly sorted sediments laid down by

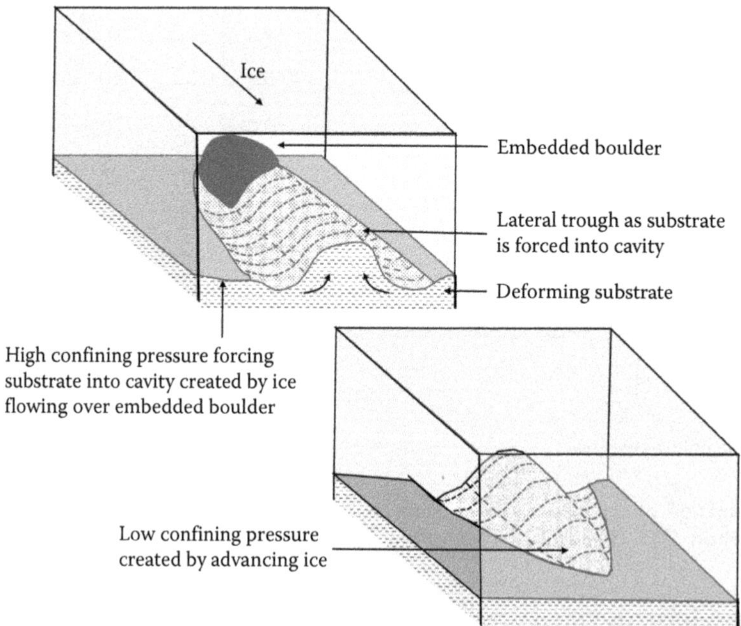

Figure 2.23 Formation of flutes due to embedded boulders creating a migrating cavity within the glacier, which is filled with till or outwash deposits due to the confining pressure acting on the substrate. (After Benn, D. I. Fluted moraine formation and till genesis below a temperate valley glacier: Slettmarkbreen, Jotunheimen, southern Norway. *Sedimentology*, 41(2); 1994: 279–292.)

water or as a result of the deforming substrate. Large boulders are often found on top of the moraines. Their origin is attributed to marginal moraines, subglacial moraines formed in the transitional zone between warm and cold ice, infilling of crevasses from supraglacial debris, filling of substrate crevasses by subglacial debris, and folding of debris-rich layers.

2.3.1.4 *Erratics*

Erratics are pieces of rock, which are not native to the area found on the surface. They have been transported by ice or water and deposited on the surface. They generally refer to boulders but can include any coarse-grained particle.

2.3.2 Subglacial landforms formed by water

Subglacial landforms are created by meltwater flowing in channels beneath a glacier creating ridges of glaciofluvial debris known as eskers. They are a function of the hydraulic potential within a glacier and the slope of the glacier. Eskers can be single ridges or braided ridges, which vary in length from a few hundred metres to a few hundred kilometres. The smaller eskers are typically 1–2 m high and 40–50 m wide; the long eskers are typically 400–700 m wide and 40–50 m high. They bear no relationship to the topography as they were formed by the glacier. There are four types of eskers according to Warren and Ashley (1994): tunnel fills, ice channel fills, segmented tunnel fills and beaded eskers. Single eskers form when the channel is blocked depositing the glacial debris held in suspension until then. The channels can exist as supraglacial, englacial or subglacial channels. An esker formed from a supraglacial channel or englacial channel is deposited when the ice melts. Beaded eskers are generally found at ice margins and formed as the ice retreats. Braided eskers are attributed to catastrophic subglacial floods or by lowering of englacial and supraglacial channels. Many eskers are aligned with the direction of the ice flow. Tunnel fills are effectively remnants of pressurised pipes or non-pressurised channels at the bed of a glacier.

Eskers comprise a diverse range of poorly sorted well-rounded sands and gravels, which have not travelled very far. They can contain sedimentary structures such as graded bedding, cross-bedding, slump folds, faults, laminations, load structures and climbing ripples reflecting the complex hydraulic flow in open channels and pipes. Slumping may occur at the margins as the ice walls melt. They are not stable as meltwater erosion can remove them as a glacier retreats.

2.3.3 Ice margin moraines

Ice marginal moraines are glaciotectonic moraines, dump moraines or ablation moraines formed by ice pushing, englacial and pro-glacial thrusting, rock fall or debris flow, ice dumping, ice melt-out or subglacial melt-out. From a geological point of view, ice marginal moraines are difficult to identify. However, the deposition process for the different types of moraines leads to similar geotechnical characteristics of the poorly sorted gravitationally consolidated material.

2.3.3.1 *Push moraines*

Push moraines form as the ice advances and bulldozes the pro-glacial sediment (Figure 2.24). They are typically composed of subglacial tills though may contain outwash sediments and other debris. Fines may be washed out as a result of meltwater flowing through the moraine. They may build up as seasonal advances and retreats of the ice front

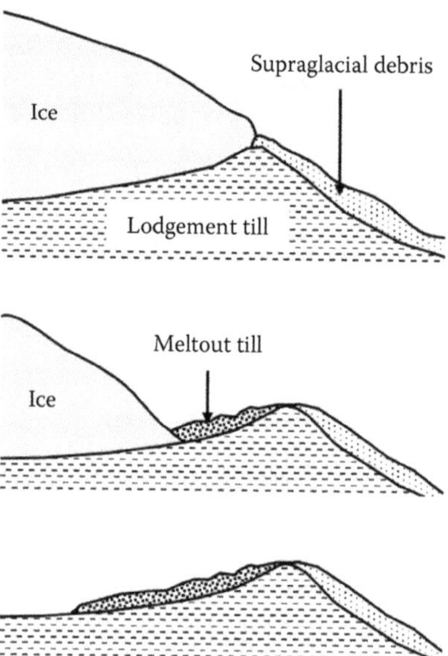

Figure 2.24 Composition of a typical seasonal push moraine. (After Bennett, M. R. and N. F. Glasser. *Glacial Geology: Ice Sheets and Landforms*. John Wiley, London; 1996: 364.)

or be more complex and large with a sustained advance. As a glacier advances, the ice may override a push moraine incorporating the debris into the subglacial debris (Figure 2.25). Very large, push moraines may result in the deformation of the underlying bedrock leading to rafts of rock embedded in glacial tills (e.g. Sidestrand, Norfolk). Englacial thrusts occur with complex basal thermal regimes (Figure 2.26).

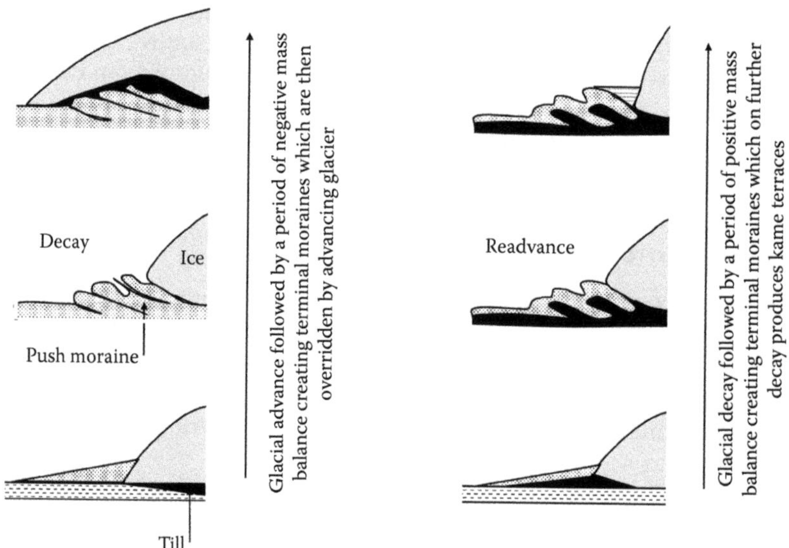

Figure 2.25 Push moraines within a glacial cycle showing how push moraines, outwash fans and kame terraces are created as the glacier advances and retreats. (After Boulton, G. S. Push-moraines and glacier-contact fans in marine and terrestrial environments. *Sedimentology*, 33(5); 1986: 677–698; Bennett, M.R. and N.F. Glasser. *Glacial Geology: Ice Sheets and Landforms*. John Wiley, London; 1996: 364.)

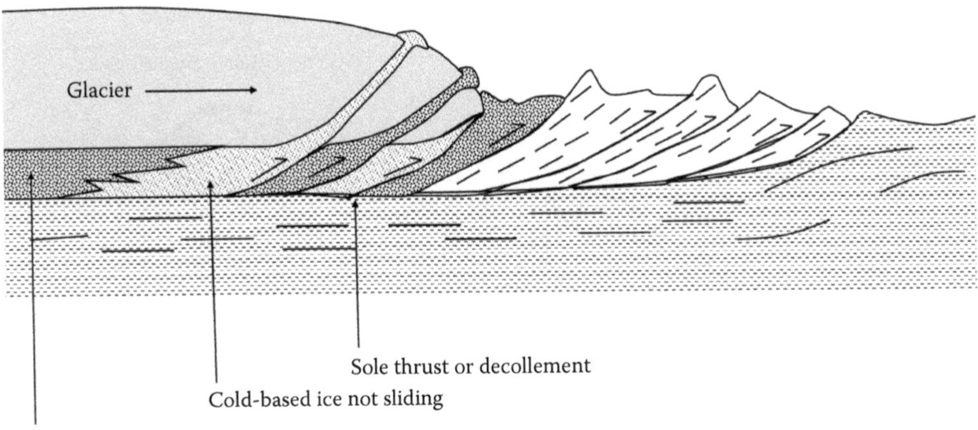

Figure 2.26 Moraine formation due to glacial thrusting at a polythermal boundary. (After Hambrey, M. J. *Glacial Environments*. UBC Press, London; 1994.)

2.3.3.2 *Dump moraines*

Debris delivered to a stationary steep ice margin forms a dump moraine. The size of the moraine depends on the debris content within the ice, the rate of retreat of the ice margin, and the speed of the flow of ice delivering debris to the front. If these conditions are not sufficient, then the debris is spread over a larger area rather than forming moraines. The conditions are more likely to be met at the sides of a valley glacier (Figure 2.27). The debris is variable, but there is some evidence of fabric and bedding.

2.3.3.3 *Ablation moraines*

Ablation moraines are formed of supraglacial debris that remains as a glacier retreats. The debris surfaces on the ice due to upward flowing ice bringing debris to the surface. If the extent of the debris is extensive, it gives rise to hummocky moraine. This is a simple explanation of a complex process (Figure 2.28).

2.3.4 Glaciofluvial ice marginal landforms

Meltwater emerging from a glacier carries debris to form outwash fans, kames and kame terraces. The debris spreads out in front of the ice margin and backs over the ice; therefore, the topology of the glaciofluvial landforms depends on their location with respect to the ice margin, the presence of buried ice and the amount of transported sediment. Braided river systems develop downstream of the ice margin creating an outwash fan as the glacial debris is deposited (Figure 2.29). While these river systems form at the ice margins, they can contain buried ice, which on melting leads to kettle holes, water-filled pits that are gradually filled with further glacial debris possibly leading to conical lenses of distinctly different materials from the surrounding outwash fan. If the outwash fan crosses an extensive area of ice as the ice melts, it creates a hummocky surface to the rear of the outwash fan known as kame and kettle (hollows) topography. Kames, consisting of well-sorted deposits of sand and gravel, are formed at the ice margin creating either isolated hummocks or broad flat-elevated areas (Figure 2.30). The velocity of meltwater reduces rapidly as it emerges from a glacier

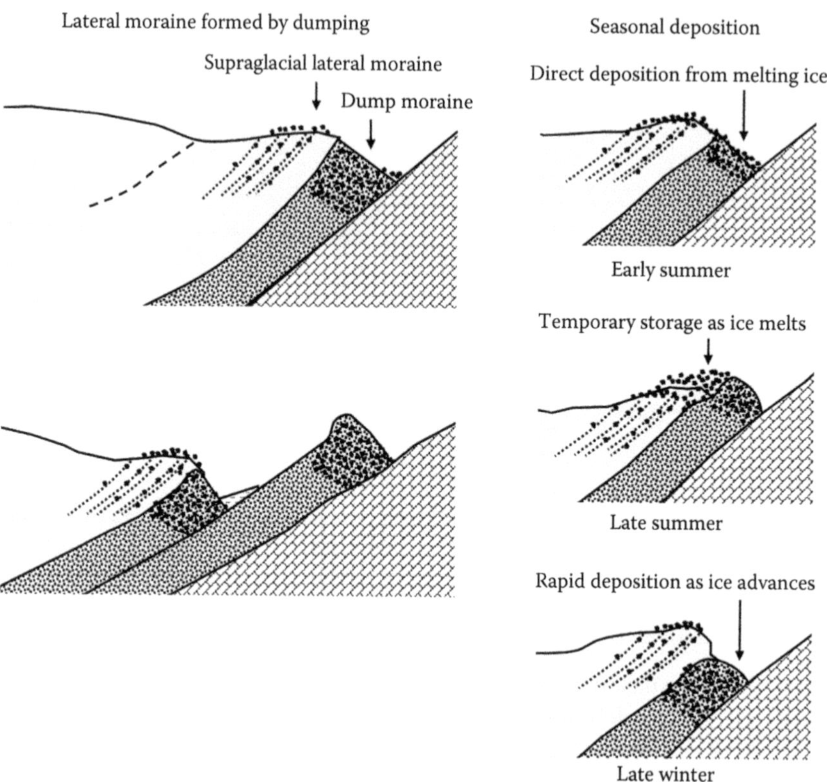

Figure 2.27 Formation of lateral moraines as supraglacial debris is dumped at the edge of a glacier seasonally and as ice retreats. (After Small, R. J. Englacial and supraglacial sediment: Transport and deposition. *Glacio-Fluvial Sediment Transfer: An Alpine Perspective*. John Wiley and Sons, New York, 1987: 111–145; Bennett, M. R. and N. F. Glasser. *Glacial Geology: Ice Sheets and Landforms*. John Wiley, London; 1996: 364.)

resulting in coarser materials being deposited near to the outlet and finer material being carried further afield (cf. pipe discharge into a lagoon). As the ice retreats, the meltwater may be diverted along the ice margin creating a kame terrace. Kames can vary from a few hundred metres to over a kilometre in length. Kame terraces form parallel to the direction of ice flow from streams running along the sides of a glacier.

2.4 GLACIAL LAND SYSTEMS

A land system is a means of combining the sediment, landforms and landscapes to explain the characteristics of the glacial soils (Eyles, 1983). A land system starts with the land elements (e.g. drumlins) to create the land facets (e.g. drumlin field), which collectively form the land system, a three-dimensional holistic approach to basin-scale patterns of glacial deposition. This three-dimensional approach enables likely subsurface conditions to be predicted which is of benefit to civil engineering projects, particularly infrastructure projects. The development of glacial land systems can be traced back to Boulton and Paul (1976), Boulton and Eyles (1979) and Eyles (1983). Sequences of glacial deposits are bounded by unconformable boundaries not too dissimilar to hydrocarbon reservoirs. The connection is the use of sequence stratigraphy developed by petroleum geologists for locating such reservoirs.

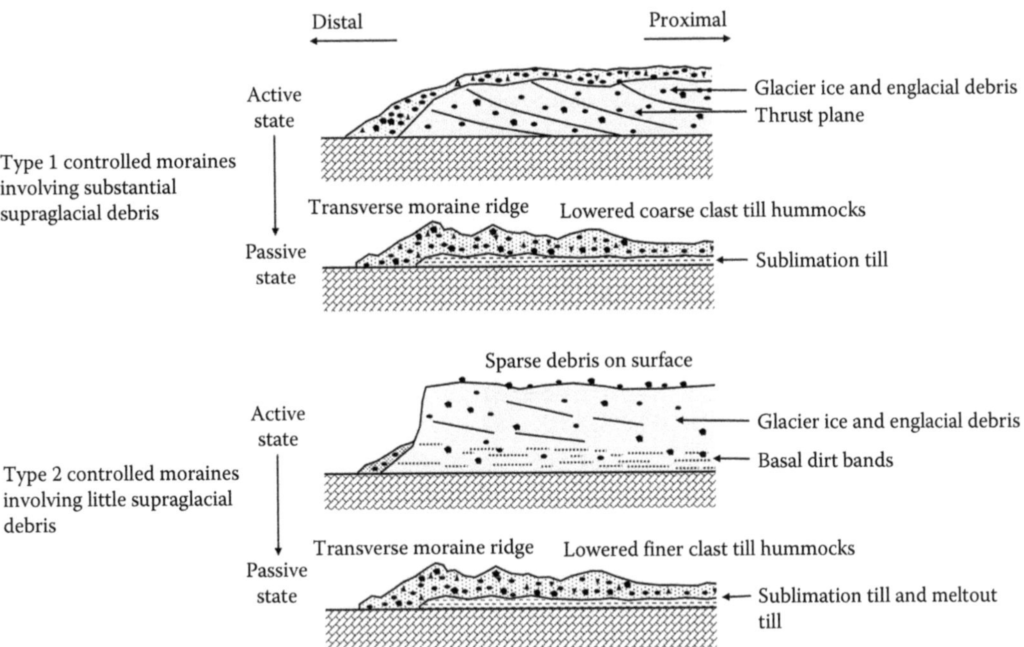

Figure 2.28 Formation of ablation moraines from supraglacial and englacial debris. (After Benn, D. and D. J. A. Evans. *Glaciers and Glaciation*. Routledge, London; 2010.)

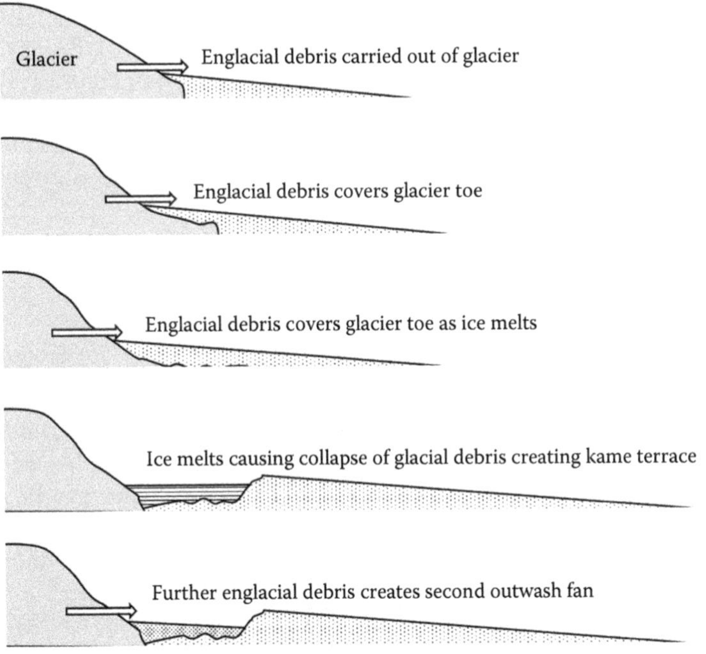

Figure 2.29 Formation and morphology of outwash fans. (After Bennett, M. R. and N. F. Glasser. *Glacial Geology: Ice Sheets and Landforms*. John Wiley, London; 1996: 364.)

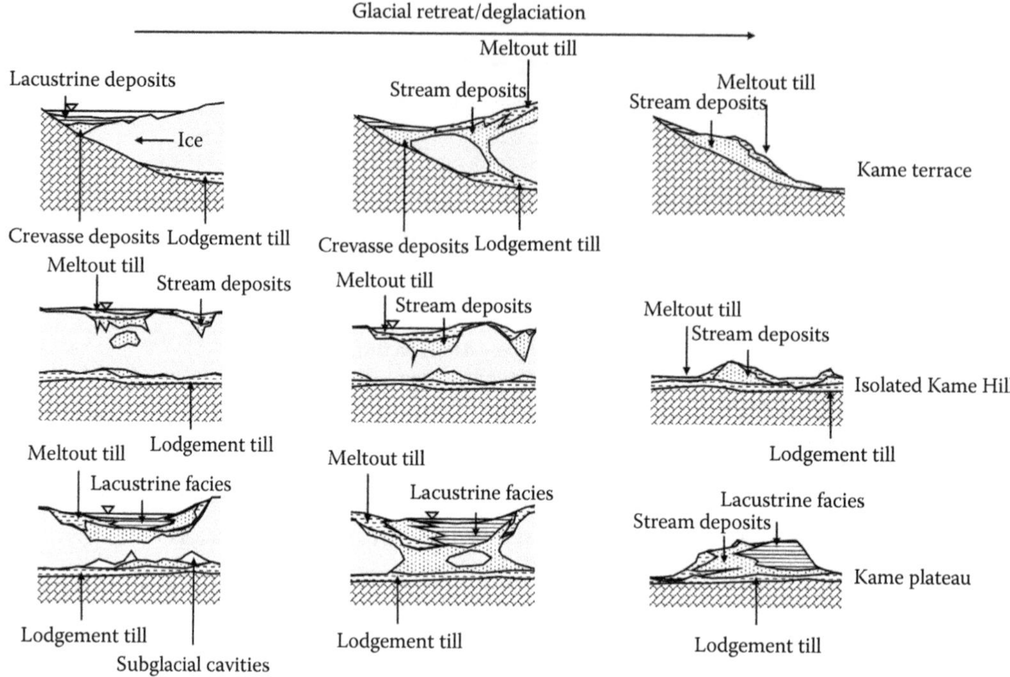

Figure 2.30 Formation of kames and kame terraces during glacial retreat and deglaciation highlighting the complex composition of deglacial landforms. (After Brodzikowski, K. and A. J. Van Loon. *Development in Sedimentology*, 49; 1991: 688.)

Sequence stratigraphy has been applied to glacial deposition by Boulton (1990), Eyles and Eyles (1992), Martini and Brookfield (1995), Benn and Evans (2010), Evans (2014) and others. The glacial land systems of the United Kingdom are shown in Figure 2.31 using the terrain to highlight likely glacial deposits. The large-scale distribution of depositional systems and land systems can be subglacial, ice marginal and supraglacial, subaqueous and glaciated valleys (Benn and Evans, 2010). A summary of glacial land systems is given in Table 2.13, which links the erosional landform with sediments.

2.5 GLACIAL DYNAMICS

Engineering properties of glacial soils depend on their source, methods of erosion, transportation, deposition and deformation, in the case of glacial tills, subsequent pro-glacial processes. Classic soil mechanics is based on an assumption that fine and coarse grained soils are gravitationally consolided and the fine-grained soils are composed of clay minerals. Most glacial soils are composite soils, the fine-grained content may be rock flour and primary deposits are not gravitationally consolidated. Therefore, an appreciation of glacial dynamics is important as it helps explain some of the anomalies encountered in ground investigations and interpretation of the data.

It was not until the late nineteenth century that the concept of an ice sheet laying down vast tracts of glacial soils was realised though the presence of these soils was known through exposures, excavations and the mining industry. Glacial soils can be observed at the ice margins, so it was possible to develop an understanding of their composition and, in the case

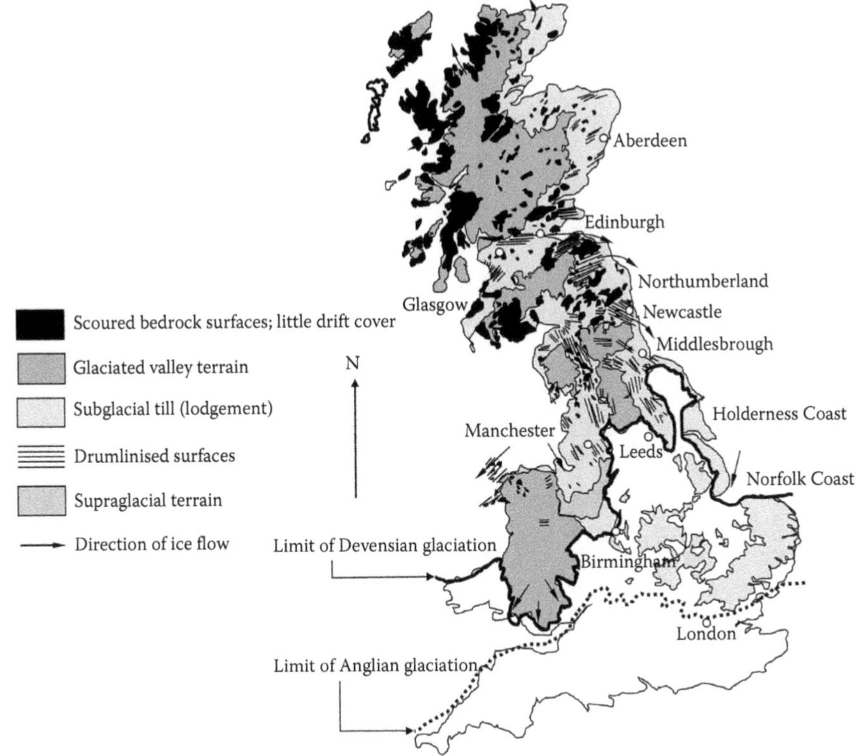

Figure 2.31 Glacial land systems in the United Kingdom. (After Eyles, N. and W. R. Dearman. *Bulletin of the International Association of Engineering Geology – Bulletin de l'Association Internationale de Géologie de l'Ingénieur,* 24(1); 1981: 173–184.)

of secondary deposits, their mode of deposition, by investigating excavations in the deposits. This was particularly the case in the Alps where much of the early ideas of glaciation were developed. The relationship between the ice and glacial soils is much more difficult to assess and remains today an area of debate.

A glacier can be separated into an accumulation zone (Figure 2.32) in which snow and ice accumulates and an ablation zone in which glacial soils are deposited. The snow and ice creates the glacier, which moves until it is lost through melting, evaporation or pieces of ice breaking away if the margin terminates in water. An advancing glacier is one in which accumulation exceeds ablation; a retreating glacier is the reverse. The thermal regime affects glacial erosion (Figure 2.33). Erosion is limited if the ice is cold based and there is no basal melting. If the base of a glacier starts to melt, whether due to pressure or temperature, the glacier can slide across the substrate and the meltwater will erode the substrate. As a glacier moves, it may erode the underlying sediments to create glacial debris, which is transported by ice and water and deposited in a variety of ways. Boulton et al. (2001) suggests that a large proportion of the forward movement of a glacier is due to deformation within the underlying sediment creating a coupled system (Boulton and Jones, 1979; Clark, 1994) such that the ice flow and substrate deformation interact governing the production and distribution of tills and subglacial landforms.

The glacial cycle and debris cycles are interlinked systems. The movement of a glacier is a combination of sliding, ice deformation and deformation of the glacier bed (Figure 2.34). The resistance to movement is balanced by the weight of snow and ice. Extensive deposits of subglacial

Table 2.13 Glacial erosional landforms and the underlying sediments

Land systems associated with erosion	Selective linear erosion	Landform	Deep glacial troughs separated by areas with little or no evidence of glacial erosion
		Sediment	Periglacial landforms and sediments may exist
		Glacial conditions	Warm based in low lying areas with thick ice cover; cold-based ice where ice is thin overlying upland areas
	Landscapes of little of no erosion	Landform	Periglacial landforms
		Sediment	Areas of periglacial sediments exist
		Glacial conditions	Evidence of cold-based ice where erosion is mostly ineffective, thin ice and ice divides where ice velocity is low
	Areas of areal scour	Landform	Rock surfaces show signs of glacial abrasion and plucking; landforms include roches moutonnees, whalebacks, glacial troughs, rock basins and striated surfaces
		Sediment	Periglacial sediments may exist
		Glacial conditions	Associated with warm-based ice or ice in thermal equilibrium
	Landforms of local glaciation	Landform	Valley glaciers and cirque glaciers cut through sand cirques roches moutonnees, whalebacks, glacial troughs, rock basins and striated surfaces exist
		Sediment	Periglacial sediments may exist
		Glacial conditions	Associated with warm-based ice
Land systems associated with deposition and deformation	Supraglacial landform-sediment assemblage	Landform	Areas of kames and hummocky moraines formed of sand, gravel and supraglacial till
		Sediment	High englacial and supraglacial debris due to compressive flow and englacial thrusting, high englacial debris
		Glacial conditions	Transition from a warm base to a cold base ice; belt of cold-based ice at margin
	Subglacial landform-sediment assemblage	Landform	Megaflutes, drumlins, rogens and mega-scale glacial lineation exist; subglacial eskers may be present
		Sediment	Subtraction till including deformation and lodgement till and possible interbedded sands and gravels
		Glacial conditions	Associated with warm base ice and areas of deformable sediment
	Maritime ice marginal landform-sediment assemblage	Landform	Seasonal push moraines, local areas of hummocky moraine, dump moraines and, in mountainous areas, hummocky moraine and ablation moraines; eskers.
		Sediment	Complex sediments including sub traction till, melt-out till, glaciofluvial sediments
		Glacial conditions	Warm-based ice margins in maritime climates in retreat.

(*Continued*)

Table 2.13 (Continued) Glacial erosional landforms and the underlying sediments

	Continental ice marginal landform-sediment assemblage	Landform	Limited ice margin moraines, ice marginal landforms and outwash fans; irregular spread of melt-out till
		Sediment	
		Glacial conditions	Associated with warm-based ice in continental climates with possible cold-based ice margins
	Surging landform-sediment assemblage	Landform	Distinctive landform assemblage including large composite push moraines, crevasse squeezed ridges, hummocky moraine and kames
		Sediment	
		Glacial conditions	
	Cold-based or poplar ice landform-sediment assemblage	Landform	Dump and ablation moraines due to englacial and supraglacial debris; hummocky moraine; no glaciofluvial deposits below ice
		Sediment	Sediments predominantly basal and supraglacial melt-out till; no glaciofluvial processes; little or no reworking of debris
		Glacial conditions	Cold-based ice
	Glaciofluvial-landform assemblage	Landform	Large outwash fans, meltwater channels, kame terraces, kames, eskers and braded eskers are common
		Sediment	Primarily glaciofluvial sands and gravels
		Glacial conditions	Develop along major meltwater routes within ice sheets with a stationary or slow moving ice margin; meltwater discharge reworks most of the glacial sediment
	Glaciolacustrine assemblage	Landform	Meltwater channels, lake shorelines, De Greer moraines, ice contact outwash fans, ice contact deltas, delta moraines, fluvial deltas, eskers and moraine banks
		Sediment	Sands and gravels, laminated silts and fine sands
		Glacial conditions	Pro-glacial and ice dam lakes
	Glaciomarine-fjord assemblage	Landform	Subaqueous push moraines, outwash fans, moraine banks, ice contact deltas and eskers may be present; lateral moraines and meltwater channels may exist above the water surface
		Sediment	Sands and gravels, laminated silts
		Glacial conditions	Glacier terminates in a marine environment in a fjord
	Glaciomarine continental shelf landform-sediment assemblage	Landform	Large moraine banks, delta moraines and, possibly, fans may be present
		Sediment	Sands and gravels, laminated silts and fine sands and diamictons deposited form ice raft debris or sediment gravity flows
		Glacial conditions	Ice sheet terminates on the continental shelf

Source: After Bennett, M. R. and N. F. Glasser. *Glacial Geology: Ice Sheets and Landforms*. John Wiley, London; 1996: 364.

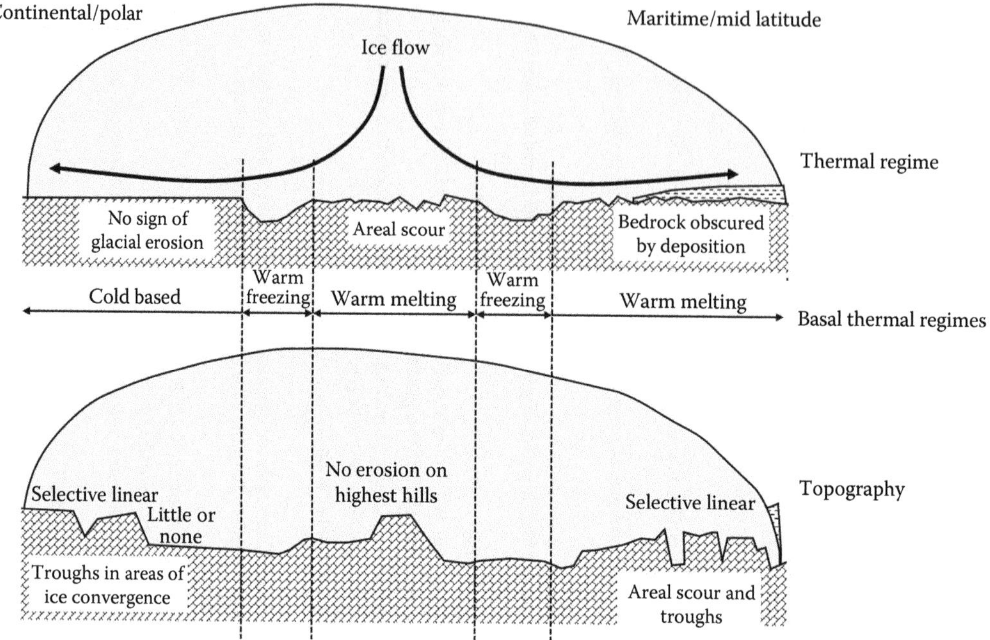

Figure 2.32 Patterns of erosion and deposition within an ice sheet related to the basal thermal regimes and topography. (After Chorley, R. J., S. A. Schumm, and D. E. Sugden. *Geomorphology.* Methuen, New York, 1984.)

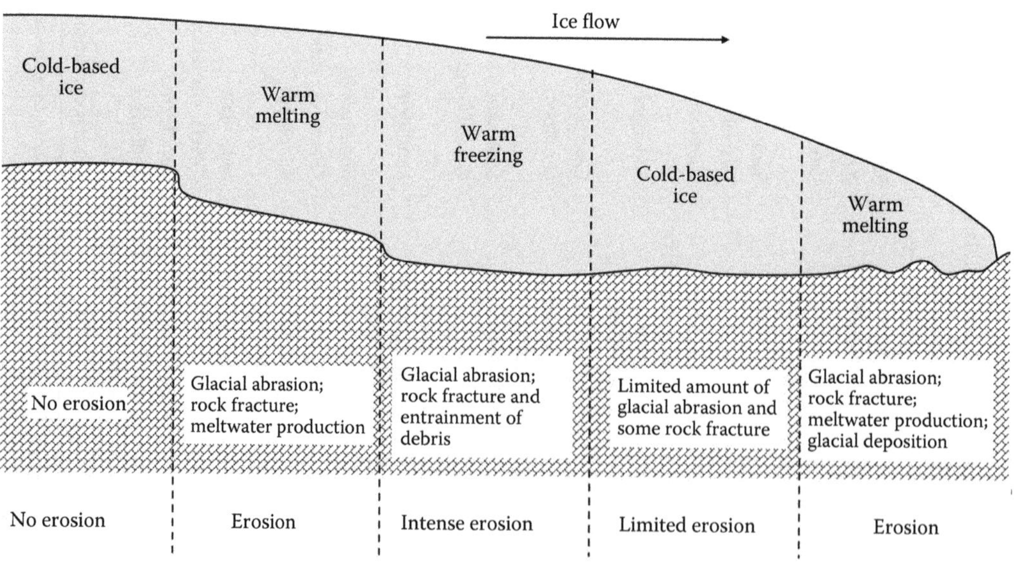

Figure 2.33 Thermal regime within a glacier and its impact on erosion. (After Bennett, M. R. and N. F. Glasser. *Glacial Geology: Ice Sheets and Landforms.* John Wiley, London; 1996: 364.)

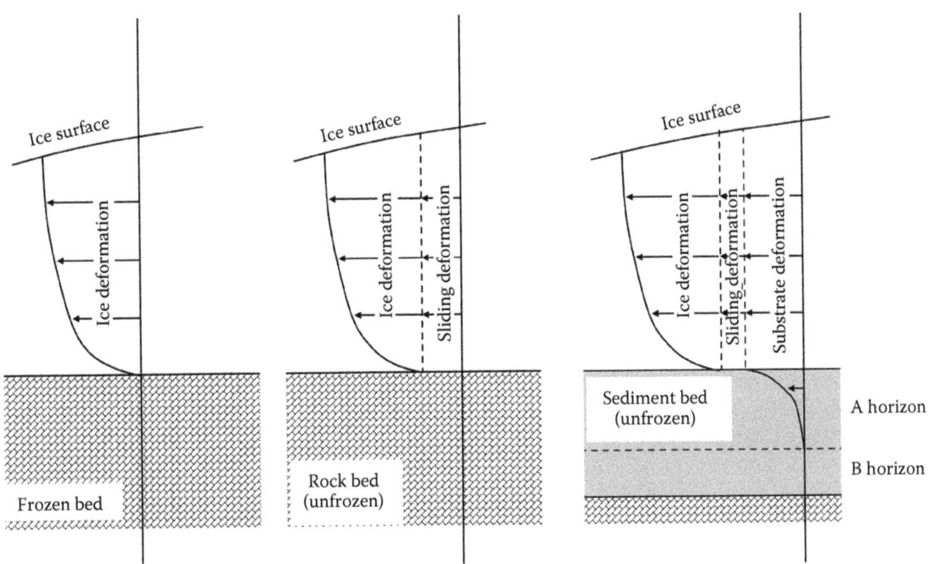

Figure 2.34 Relative displacement profile through a glacier and the underlying substrate showing ice deformation only, ice deformation and basal sliding, and ice deformation, sliding and substrate deformation. (After Boulton, G. S. *Journal of Glaciology*, 42(140); 1996: 43–62.)

material are the legacy of the last ice age with further glacial materials beyond the ice margin formed of glaciofluvial or glaciomarine deposits or remnants of earlier ice ages. In marine environments, it is possible to find glacial debris a considerable distance from the ice margin because glacial debris is transported by icebergs and deposited on the sea bed as the ice melts.

The form and structures of glaciers are beyond the scope of this book. Readers are referred to Benn and Evans (2010), for example, for a detailed account of glaciers and glaciation. Glaciers can be divided into those controlled by the topography (highland) and those that

Table 2.14 Classification of glaciers according to their size and their shape and relationship to the topography

	Area (km²)								
Glacier type	*0*	*1*	*10*	*100*	*1k*	*10k*	*100k*	*1m*	*10m*
Ice sheets									
Ice sheet							←	—	→
Ice cap			←	—	—	→			
Ice shelf				←	—	—	→		
Ice stream				←	—	→			
Ice tongue				←	→				
Highland									
Highland ice field				←	—	→			
Valley glacier			←	—	—	→			
Piedmont glacier			←	—	→				
Cirque glacier		←	→						
Hanging glacier	←	→							
Rejuvenated glacier	←	—	→						

Source: After Hambrey, M. J. *Glacial Environments*. UBC Press, 1994.

are not (ice sheets) (Table 2.14). Glaciers controlled by the topography are important in mountainous areas. The legacy of the ice sheets has a greater effect on civil engineering and offshore engineering because of their scale; it affects the urban environment, national infrastructure and engineering on the continental shelf. While the focus here is on terrestrial glacial soils or sediments, understanding glacier dynamics—that is, ice erosion, transportation, deposition and deformation; advance and retreat of glaciers, deglaciation; isostatic uplift and changes in sea level—is also important because of the effects of glacier movement on the engineering properties of glacial soils.

According to Boulton et al. (2001), there are two dominant modes of deformation within a substrate due to the shear forces exerted by an overriding glacier:

- Shear deformation is a maximum immediately beneath the glacier sole. The net strain increases upwards towards a decollement surface that generally represents the former location of the glacier sole.
- Shear deformation increases downwards towards an underlying decollement.

Glacial movement erodes/deforms the bed, but this will depend on the temperature and geology of the bed (Figure 2.34). Figure 2.34a shows a glacier sitting on a frozen bed of rock or soil. The ice deforms, but there is no movement at the base of the ice, which means that there is erosion of the bed. Figure 2.34b shows the glacier sliding over an unfrozen bed. Figure 2.34c shows that the glacial movement is a function of sliding across the substrate that is undergoing deformation.

Therefore, the resistance to flow depends on the temperature and pressure at the base of the glacier, the interface friction between the ice and the bed, which is a function of the bed roughness and the strength of the underlying sediment.

It is extremely difficult to observe what happens beneath glaciers, so many of the views are based on theoretical models. Some field work has been carried out to observe subglacial deformation (e.g. Boulton, 1979; Boulton and Hindmarsh, 1987; Blake et al., 1992; Iverson et al., 1995). Seismic sounding has been used to infer deformations beneath Ice Stream B in Western Antarctica (Alley et al., 1986, 1987; Blankenship et al., 1986, 1987; Alley, 1989a,b). There have been a number of studies to monitor subglacial behaviour (Hodge, 1976; Fischer and Clarke, 1994; Iverson et al., 1995, Hooke et al., 1997; Engelhardt and Kamb, 1998; Murray and Porter, 2001; Fischer and Clarke, 2001; Martinez et al., 2004; Hart et al., 2006; Hart and Martinez, 2006; Hart et al., 2009), which have shown that the motion at the base of a glacier and the drainage of water beneath the ice are strongly interdependent.

2.5.1 Glacier movement due to substrate deformation

Glaciers move because the rate of precipitation exceeds the rate of ablation. This movement gives rise to permanent deformation of the ice and substrate, erosion of the substrate and deposition of glacial debris. A glacier will slide over a substrate until the yield stress of the substrate is exceeded. At that point, the substrate starts to deform and erosion can occur. The critical shear stress, τ^*, is defined by the Coulomb equation:

$$\tau^* = c' + \sigma' \tan\varphi'$$

where σ' is the effective pressure and φ' the angle of friction.

Once the critical shear stress is exceeded, the substrate is assumed to deform as a viscous material such that the rate of strain, $\dot{\varepsilon}$, is

$$\dot{\varepsilon} = \frac{K(\tau - \tau^*)^a}{\sigma'^b}$$

where K, a and b are constants and τ the current shear stress. The critical shear stress will depend on the pore water pressure within the substrate and, since this fluctuates due to dilation as the substrate shears, the critical shear stress is spatially variable creating a phenomenon known as stick–slip.

2.5.2 Sliding

Glaciers can slide across a substrate, but the resistance to sliding is controlled by adhesion, roughness of the interface and the debris held within the ice, particularly in the basal zone. If the glacier/substrate interface is at or below the pressure melting point, the ice adheres to the substrate preventing or restricting the rate of sliding.

As ice slides across the interface, it encounters obstacles at various scales giving rise to regelation sliding and enhanced creep. Regelation occurs when the ice encounters an obstacle; the interface pressure increases, melting the ice, which allows the ice to slide over the obstacle. As it passes the obstacle, the pressure drops refreezing the ice. The strain rate of ice depends on the shear stress, and as the shear stress increases when the ice encounters an obstacle, the strain rate increases. This is known as enhanced creep. Regelation dominates for smaller obstacles; enhanced creep for larger particles. Water is necessary for a glacier bed to slide to lubricate the interface.

2.5.3 Friction and sliding

Debris within the basal layer of the glacier impacts on the movement of a glacier because of frictional drag and erosion. Studies of glacier beds, subglacial landforms and sediments together with field and laboratory experiments and theoretical models have been used to develop an understanding of the interaction between the ice and embedded debris and the underlying bed, which leads to erosion of the bed and abrasion of the debris. Boulton (1974) and Hallet (1979) showed that the shear stress and the strength of the material influence the erosion, transport and deposition of the subglacial shear zone. This may seem obvious from an engineering point of view but it proved to be a transformational way of thinking when it was first proposed as it allowed glacial dynamics to be linked to the creation of tills.

Three models have been proposed: the Coulomb model (Boulton, 1974), friction model (Hallett, 1979) and the sandpaper model (Schweizer and Iken, 1992).

In the Coulomb model, it is assumed that the friction between the substrate and the ice is proportional to the vertical effective stress. The constant of proportionality is the angle of friction. This model assumes that the ice and substrate are rigid; it ignores deformation of the ice and substrate. The Coulomb friction model assumes that the shear force, F, is a function of the weight of ice less the pore pressure acting over the area of contact.

$$F = (\gamma h - u_w)A \tan\varphi'$$

where γ is the ice unit weight, h the ice thickness, u_w the water pressure in cavities at the base of a particle, A the contact area and φ the interface friction. This model assumes that a column of ice above the particle is the weight acting on the particle. It does not take into account arching in the ice or the increase in density of ice due to the embedded particles. Nor does it take into account the fact that ice provides some buoyancy and the ice can deform.

The Hallett model assumes that the contact forces are independent of ice thickness as the ice deforms around subglacial particles. The contact forces are the drag force due to the ice and the effective weight of the particle due to buoyancy because of the ice. Ice flows towards the bed due to geothermal heat and sliding friction, regelation and vertical straining of the ice. This model is more realistic when the basal debris is less than 50% by volume. The Hallet model assumes that the contact force is the buoyant weight of the particle. Thus, the shear force, F, is

$$F = \mu_n \left(f 4\pi \tan\varphi' \frac{R^3}{(R_{ax}^2 + R^2)} \right) \tag{2.1}$$

where u_n is the ice velocity normal to the bed, f a correction factor, R the particle radius and R_{ax} the critical radius controlling regelation and enhanced creep. This leads to frictional drag.

The sandpaper model assumes that the density of debris-rich basal ice is high such that the ice does not flow around the particles. The ice is in contact with the substrate as the basal zone deforms. In this case, the shear force is proportional to the buoyant weight of the particles taking into account water-filled cavities that tend to reduce drag. The sandpaper model takes into account the pressure in the water-filled cavities:

$$F = (p - su_w)A \tan\varphi' \tag{2.2}$$

where p is the ice overburden pressure and s the proportion of the bed occupied by cavities.

Schweizer and Iken (1992) suggested that the Coulomb model applies to transient conditions, a rigid base or ice-free base; the Hallet model to ice with a low density of particles; and the sandpaper model to ice with a high density of particles. These three simple models do not fully explain the complex processes taking place within the interface zone because this zone is a combination of rock particles, water and ice continually undergoing change as the ice moves and melts in cycles, the debris slides, rolls, deforms and abrades and the water pressure at the interface and in underlying bed fluctuates. The size of rock particles and the particle size distribution vary due to erosion and abrasion; the percentage of ice and water is continually changing; the density of the basal ice varies as particles are removed by meltwater or become lodged to the bed.

Desai et al. (2010) used the disturbed state concept (DSC) – a constitutive model that allows for elastic, plastic and creep deformations; rate dependence; microstructural changes leading to softening; and critical conditions – to model a deforming bed. The parameters are determined from triaxial and creep tests. The DSC predicts that failure occurs within the whole specimen unlike the Mohr–Coulomb model, which predicts distinct failure planes (Figure 2.35) and infinite strains. Desai et al. (2010) claims that it is more realistic because it predicts deformation of the whole mass, which has been observed in the field.

2.5.4 Erosion

Erosion can be due to abrasion caused by debris-rich ice gradually wearing away the substrate; plucking of blocks of rock from the substrate; and meltwater erosion. Figure 2.36 shows the forces acting on a basal particle resting on the bed. As the particle moves because the drag force exceeds the shear resistance, erosion takes place and the particle is transported

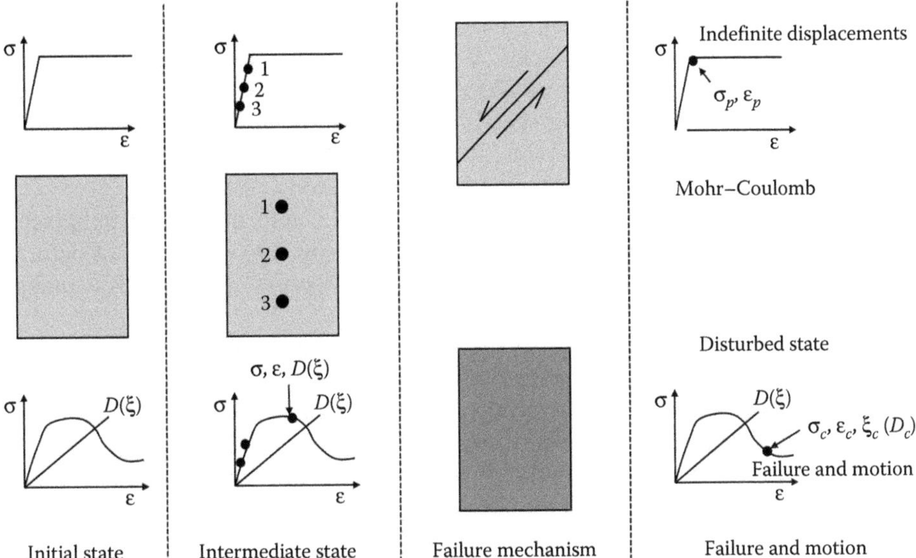

Figure 2.35 Comparison between the linear elastic perfectly plastic model and the disturbed state concept showing how the Mohr–Coulomb model leads to shear surfaces and indefinite displacements and the Disturbed State Model leads to failure of the mass and creep deformation. (After Desai, C. S., S. Sane, and J. Jenson. *International Journal of Geomechanics*, 11(6); 2010.)

by the ice. As soon as the drag force is less than the shear resistance, the particle stops moving, thus depositing the particle.

It is difficult, possibly impossible, to study detailed erosion within the basal zone of a glacier. The theories of erosion can be studied in the laboratory using ring shear apparatus (Iverson et al., 2015). Iverson et al. (1997) and Larsen et al. (2006) used a ring shear apparatus similar to that used in geotechnical engineering. Iverson and Petersen (2011) developed the Iowa State Sliding Simulator (ISSS), which simulates the effect of a glacier by dragging a ring of ice (0.9 m O.D., 0.25 wide, ~0.22 m tall) at the pressure melting temperature over either a hard or a soft substrate. The slip across a hard substrate accounts for most of a glacier‘s surface velocity (Cuffey and Paterson, 2010), rates of bedrock abrasion (Hallet, 1979)

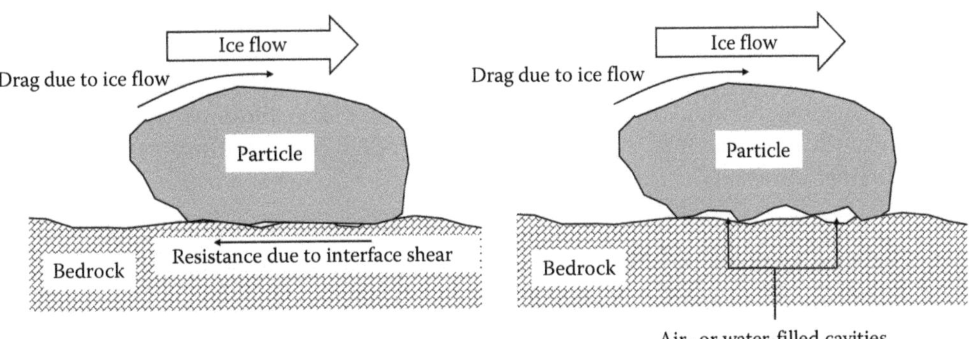

Figure 2.36 Forces on a particle on a glacier bed to show the effect of ice, interface shear and water-filled cavities on erosion. (After Benn, D. and D. J. A. Evans. *Glaciers and Glaciation*. Routledge, London; 2010.)

and quarrying (Iverson, 2012) depending on sliding speed. The drag should increase with sliding speed (Weertman, 1957; Kamb, 1970; Nye, 1970). However, it is known that the ice separates from the lee of bed obstacles (Walder and Hallet, 1979; Hallet and Anderson, 1980; Iken et al., 1983; Anderson et al., 2004; Hooyer et al., 2012), which means that the drag should decrease with increasing speed because of cavities formed downstream of an obstacle within the bed. Zoet and Iverson (2015) were able to demonstrate this using a sinusoidal bed with the ISSS confirming limited field observations.

Experiments on till deformation showed that elastic deformation of subglacial tills accounts for little glacier motion but could replicate in situ measurements of deformation (e.g. Blake et al., 1994; Iverson et al., 2003, 2007; Harrison et al., 2004; Kavanaugh and Clarke, 2006).

Iverson et al. (1996) used the ISSS to study grain size evolution as a function of shear strain in a deforming bed using mudstone particles initially 2.0–3.3 mm in diameter. The rate of displacement (320 m/year) and effective normal stress (84 kPa) were constant. They observed that the grain size distribution became fractal with progressive crushing and abrasion of grains achieving a fractal dimension of about 2.85 with sufficient shearing. The fractal dimension, m, for N, the number of particles at size d, is

$$N = N_o\left(\frac{d}{d_o}\right)^{-m} \tag{2.3}$$

where N_o is the number of particles at a reference dimension, d_o. Hooke and Iverson (1995) predicted that m would lie between 2.8 and 3 for this situation. Local normal stresses measured with load cells fluctuated between 50 and 250 kPa during the initial phases of shear, but fine sediment produced by crushing of grains caused homogenisation of intergranular stresses that promoted abrasion over crushing, with a steady fractal dimension larger than that from crushing alone (~2.6; Biegel et al., 1989).

Hooyer and Iverson (2000) showed using layered soil that mixing and diffusion took place with sufficient strain though the dominant process depended on local strains and the topology of the interface between the layers.

2.5.5 Deposition

Subglacial deposition from debris-rich ice includes lodgement and melt-out. Figure 2.14 shows the lodgement processes, which occur when the shearing resistance exceeds the frictional drag. Boulton's model predicts lodgement below thick ice; Hallet's model predicts lodgement where basal melting rates are high.

Debris that is transported in the basal shear zone is subject to further abrasion and fracture producing bimodal or polymodal particle size distributions; this is known as active transport. Englacial and supraglacial debris undergo passive transport as they do not undergo further abrasion.

2.6 SUBGLACIAL DEFORMATION

Primary glacial deposits are formed at the base of a glacier through a process of deformation and deposition. Experimental evidence of subglacial deformation (Boulton et al., 2001) suggests clast rotation in the direction of shear and reverse rotations, possibly due to consolidation (Blake et al., 1992; Iverson et al., 1995), irregular, slip–stick motion (e.g. Fischer and Clarke, 1997) and the hydraulic geometry of the system plays a key role in determining effective pressures and the location of decollement. There is also considerable evidence that subglacial deformation can extend to some depth.

The ability to monitor subglacial processes has led to the conclusion that glacial tills are a result of (a) deformation (glaciotectonite) or (b) a combination of deposition and deformation (subglacial traction till) or (c) deposition alone (melt-out till).

Recent studies of subglacial processes include an assessment of the temporal and spatial changes (e.g. Engelhardt and Kamb, 1998; Boulton et al., 2001; Fischer and Clarke, 2001; Murray and Porter, 2001) and integration of the processes (e.g. van der Meer, 1993; Hart and Rose, 2001; Evans et al., 2006; Menzies et al., 2006; Piotrowski et al., 2006; Hart, 2007). This includes *in situ* subglacial experiments (e.g. Hart and Rose, 2001; Hart et al., 2009).

It is now accepted that till undergoes deformation (at low effective stress) and lodgement and ploughing (at high effective stress) (e.g. Brown et al., 1987; Hart and Boulton, 1991; van der Meer et al., 2003; Evans et al., 2006). The effective stress can vary laterally creating areas of low effective stress and areas of high effective stress that change with time (e.g. Alley, 1993; MacAyeal et al., 1995; Piotrowski et al., 2004; Stokes et al., 2007). Distinct till fabric is created (e.g. Andrews, 1971; Dowdeswell and Sharp, 1986; Hart, 1994; Benn, 1995; Carr and Rose, 2003). The local strength of tills depends on the constraints to the deforming layer; a constrained layer has a high strength (Benn, 1995; Hart, 2006) and a thicker, less constrained layer will have a low strength (Dowdeswell and Sharp, 1986; Hicock et al., 1996; Hart et al., 2004). There is debate over clast behaviour in a deforming zone, which impacts on the fabric strength. Some assume that the clasts rotate in a Newtonian fluid to form the fabric (Jeffery, 1922) or a model in which the clasts act as passive markers within the shear plane (March, 1932) (Figure 2.37). Hooyer and Iverson (2000), Thomason and Iverson (2006), Hooyer et al. (2008), Iverson et al. (2008) and Shumway and Iverson (2009) carried out ring shear tests to study the effect on particles using five different tills and a linear-viscous putty. Hooyer and Iverson (2000) observed that particles in putty are consistent with the theory of Jeffery (1922); that is, particles will rotate indefinitely in a shearing, linear-viscous, laminar fluid, slowing their rotations when their orientations are near to that

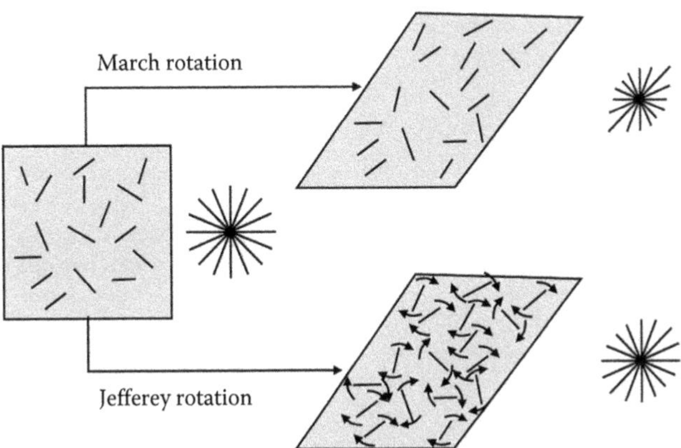

Figure 2.37 Models of particle orientation in a deforming medium in which (a) the particles rotate passively such that the deformation ellipsoid reflects the deformed shape and (b) particles are continuously subject to rotational forces such that the alignment of the deformation modulus is less clear. (Adapted from March, A. *Zeitschrift für Kristallographie-Crystalline Materials*, 81(1–6); 1932: 285–297; Jeffery, G. B. Proceedings of the *Royal Society of London A: Mathematical, Physical and Engineering Sciences*, 102(715); 1922: 161–179 (The Royal Society).)

of the shear plane but rotating through it. However, in till, particles also rotate towards the shear plane but with sufficient strain attain a steady orientation parallel to the shearing direction; they generally do not rotate through the plane of shear because fine matrix particles can slip across surfaces of rotating clasts (Hooyer and Iverson, 2000), and Riedel shears develop in the shear zone at orientations that differ from that of the macroscopic shear plane (Thomason and Iverson, 2006). This means that particles stop rotating and are held at a steady-state orientation. Therefore, with sufficient strain, shear deformation of till with initially near-random particle orientations results in strongly clustered fabrics, with particles' long axes oriented parallel to the shearing direction and plunging mildly up glacier. Hooyer and Iverson (2000) and Iverson et al. (2008) shows that the deformation rate and effective pressure on fabric development were small, so cumulative shear strain is the dominant independent variable that causes fabric evolution.

Alley (1989a,b) and Hart (1995) suggested that the basal motion of a glacier over a sedimentary bed (Figure 2.38) can be due to sliding between ice and bed, ploughing of clasts through the upper layer of the bed, pervasive deformation of the bed or shearing across discrete planes in the bed, which depend on the degree of coupling at the ice–bed interface. Strong coupling between the glacier and the underlying sediment can be due to a high density of clasts at the bed surface, which can inhibit sliding motion. This means that pervasive bed deformation is more likely to occur, especially if high pore water pressures cause the sediment yield strength to drop below a critical shear stress that can be supported by the ice–bed interface (Boulton and Hindmarsh, 1987; Alley, 1989a,b). Complete decoupling of ice and sediment can occur if a layer of highly pressurised water develops at the interface increasing the tendency for the glacier to slide over the bed (Iken and Bindschadler, 1986; Cuffey and Alley, 1996). Incomplete coupling creates a transitional state between sliding and pervasive bed deformation known as 'ploughing', in which clasts that protrude across the ice–bed interface are dragged through the upper layer of the sediment. This ploughing process, assisted by local elevated pore pressures developed in front of clasts, leads to a local reduction in strength and therefore local deformation (Brown et al., 1987;

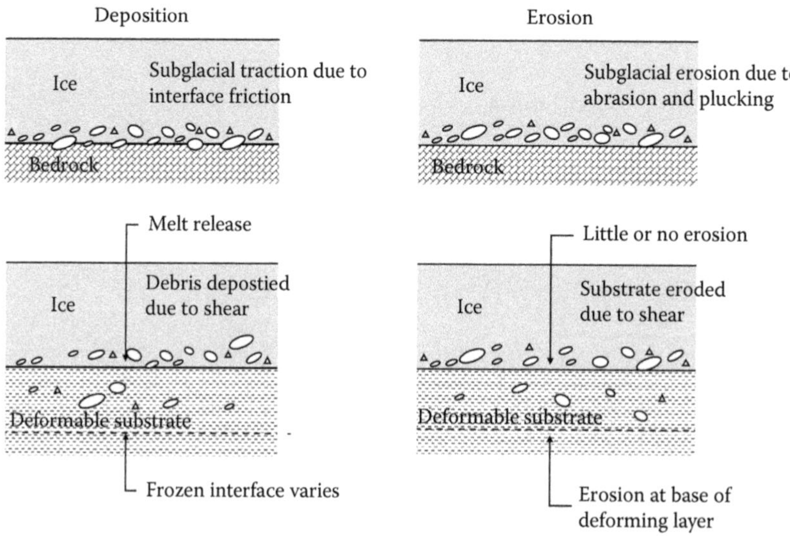

Figure 2.38 Conceptual profile through a basal layer showing the erosion and deposition processes of a glacier moving over (a) a rock and (b) a soil. (After Hart, J. K. *Progress in Physical Geography,* 19(2); 1995: 173–191; Benn, D. and D. J. A. Evans. *Glaciers and Glaciation.* Routledge, London; 2010.)

Alley, 1989a,b). Subsole deformation may be the primary mechanism to sustain fast slow of ice (e.g. Clarke et al., 1984; Alley et al., 1986; Clarke, 1987; Sharp, 1988; Engelhardt et al., 1990; Humphrey et al., 1993).

Historically, tills have been classified according to their perceived mode of deposition (Jaap et al., 2003; Evans et al., 2006), but it is now suggested that tills may be part of a deforming glacier bed; therefore, a till can be considered as a tectonic deposit if the till possesses deformational characteristics. Jaap et al. (2003) suggested that a subglacial till containing deformation features could be described as a 'tectomict'. Till continues to be the term used by glacial geologists and, for consistency, is used here.

2.6.1 Hydraulic conditions

A key component of a glacial system, which has an impact on glacial deposits, is the hydraulic conditions that exist beneath the glacier formed of meltwater and the pore fluid. Water transports glacial debris to create glaciofluvial deposits, meltwater can lubricate the base of a glacier and a rise in pore water pressure weaken the subglacial sediments. The water in the basal zone, whether it is a pore fluid in substrate, a result of pressure melting of ice or meltwater flows through tunnels and channels within a glacier's basal zone, affects the movement of ice and the stress history of the glacial deposits. Meltwater flowing in the basal zone facilitates erosion of the sediment, and the suspended and bed loads form glacial debris.

Water is unlikely to flow through rocks of low permeability but will flow through connected cave systems that exist in some rocks such as limestone. Water will flow through coarse-grained soils but not fine-grained soils. However, the pore pressure in fine-grained soils can change due to the weight of ice. An increase in pore pressure will weaken the soil layer allowing it to deform more easily. There is evidence of diurnal cycles of water pressure within soils removing fines (Boulton and Hindmarsh, 1987; Hubbard et al., 1995).

Meltwater not only transports sediment but in subglacial conduits erodes the underlying rock and sediment. Erosion is through abrasion, cavitation, hydraulic pressure, particle entrainment and chemical action. Subglacial and englacial flow can carry suspended sediment, and debris can move along the bed through sliding, rolling and saltation.

Subglacial drainage controls the rheology and strength of glacier beds and the glacial motion (Benn and Evans, 1996). Subglacial drainage includes the following:

- Bulk movement of pore water and soil particles within the deforming sediment (Clarke, 1987)
- Movement of pore fluid (Boulton and Jones, 1979; Murray and Dowsedell, 1992; Boulton et al., 1994)
- Pipe flow (Smart 1986; Boulton et al., 1994)
- Dendritic channel networks at the ice–substrate interface in the ice, in the bed or within tunnel valleys (Röthlisberger, 1972; Shreve, 1972; Nye, 1973; Boulton and Hindmarsh, 1987)
- Linked cavity systems (Walder and Hallet, 1979; Hallet and Anderson, 1980; Kamb, 1987; Sharp et al., 1989)
- Braided canal networks formed of wide, shallow channels between the ice and deformable bed (Clark and Walder, 1994; Walder and Fowler, 1994)
- Thin films of water at the ice–bed interface (Weertman, 1972; Hallet, 1979)

Benn and Evans (2010) summarised the effect on an increase in pore pressure and the effect on till deformation:

Low pore water pressure: The glacier is coupled to the till (Boulton and Hindmarsh, 1987) mobilising the strength of the very dense till, which resists deformation. The glacier does move with some brittle shearing taking place in the till (Boulton and Hindmarsh, 1987; Benn and Evans, 1996).

Medium pore water pressure: Reduction in bed strength to an extent that ductile deformation takes place (Alley, 1989a,b).

High pore water pressure: Decoupling of the glacier from the till causing the glacier to slide over the till (Iverson et al., 1995; Fischer and Clarke, 1997; Boulton et al., 2001).

The effect of subglacial water pressures on deformation and sliding is shown in Figure 2.39.

Hart et al. (2009) using wireless probes inserted into glacier ice and the underlying substrate in Briksdalsbreen, Norway confirmed that this was the case for that glacier. They noticed that in the summer water pressures increased leading to basal sliding and in winter deformation took place because of the drop in pore pressures. The wireless probes acted as clasts because of their size (16 cm by 5.5 cm), which allowed Hart et al. (2009) to predict clast behaviour. They found that the clasts underwent continuous rotation with the reduction in dip related to the glacial velocity confirming weak fabric in the shear zone. They made further observations relating the till characteristics with the water pressures and season, as shown in Table 2.15. They concluded that low water pressures were associated with high stress variability, a consequence of velocity-driven stick–slip events directly transmitted through the grain structure of a relatively strong till, leading to brittle deformation. Intermediate pressures are associated with intermediate stress variability, leading to friction associated with deformation. High water pressures are associated with low stress variability and ductile deformation.

2.6.2 Deformation

Bedrock is likely to abrade whereas superficial deposits beneath a glacier will deform. Once that deformation starts, the sediment is classed as a subglacial till. Most of the theories

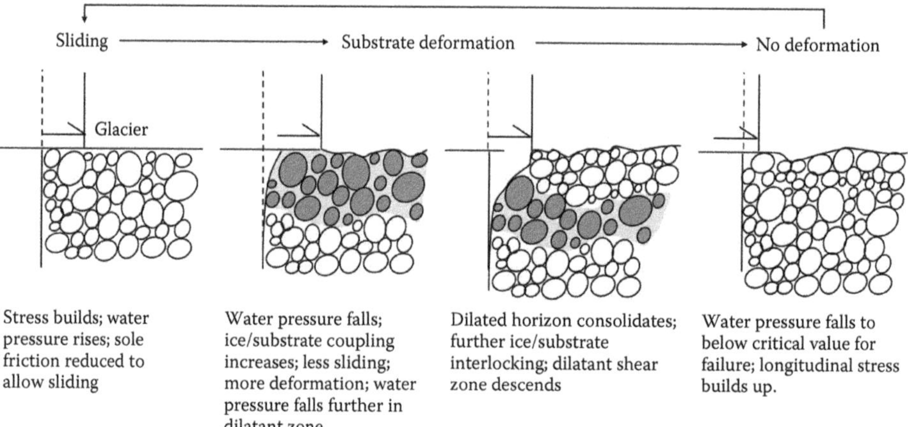

Figure 2.39 Diagram showing how subglacial water pressure affects glacier sliding and substrate deformation. (After Bennett, M. R. Ice streams as the arteries of an ice sheet: Their mechanics, stability and significance. *Earth-Science Reviews*, 61(3); 2003: 309–339; Boulton, G. S., K. E. Dobbie, and S. Zatsepin. Sediment deformation beneath glaciers and its coupling to the subglacial hydraulic system. *Quaternary International*, 86(1); 2001: 3–28.)

Table 2.15 The response of an artificial clast (wireless probe) in the substrate beneath an advancing glacier

Property	*Water pressure*		
	Low	*Intermediate*	*High*
Rheology	Elastic	Viscous	Viscous
Case stress variability	High	Intermediate	Low
Clast temperature	Moderate	High	Low
Clast tilt	Slow changes in dip	a-axis rotation	Dip oscillations
Water content	Low	Saturated	Saturated
Till strength	Strong	Dilation strengthening	Dilation strengthening
Deformation	Brittle	Ductile	Ductile
Season	Autumn/winter	Late summer	Autumn/spring/early summer

Source: After Hart et al., 2010.

assume that deformation of the basal deposits will take place only if it is unfrozen and the shear stress exceeds a critical shear stress defined by the Mohr–Coulomb criterion. This is not consistent with the fact that ice deforms. Once the applied shear stress exceeds the critical shear stress, the strain rate is defined as

$$\dot{\epsilon} = K\frac{(\tau - \tau^{*})^{a}}{\sigma_{v}^{\prime\, b}} \tag{2.4}$$

where τ is the current shear stress; σ_v' the normal effective vertical stress; and K, a and b material constants. This concept shows that the deforming layer is confined to the surface of the substrate because the critical shear stress increases with depth; the strain rate is greatest at the surface of the bed where the effective vertical stress is a minimum; and strain rates increase as the pore pressure increases resulting in a reduction in critical shear stress and effective vertical stress.

This basic concept is modified to take into account dilatancy, sediment grain size, thermal processes, spatial variations in bed strength and decoupling of the glacier from the bed. If the basal zone is dense, then deformation is accompanied by dilation, resulting in a thin low-density layer of till (Benn, 1995). However, there is a rapid transition from the less dense, dilatant till and the underlying dense till. Alley (1991) suggested that dilatancy cannot be sustained below a critical strain rate. Iverson et al. (1998) suggested that the net strain in the subglacial system may be the outcome of a large number of individual strain events related to fluctuations in the drainage system.

Sliding movement at the ice–substrate interface is governed by a Coulomb friction law. When the critical water pressure for failure is achieved, sliding can occur and the sliding rate will be water pressure dependent. As water pressures fall, the substrate will consolidate but the rate will depend on the composition of the substrate.

The dominant grain size influences the mobilised shear strength. The strength and permeability of matrix-dominated tills are generally less than the strength and permeability of clast-dominated tills. The consequence of a lower permeability means that pore pressures can develop reducing the mobilised strength further. The implication is that matrix-dominated subglacial tills are likely to deform more than clast-dominated tills. A frozen bed will not deform but pressure melting of ice could occur, which will affect the substrate allowing it to deform.

All of the models assume that the basal zone is a continuum but there is evidence for a layer of pressurised water developed at the interface, which reduces the interface friction to the extent the glacier slides over the bed reducing the deformation of the bed.

2.6.3 Local deformation

A fundamental assumption made routinely in geotechnical practice is that soils are 'cross-anisotropic'; that is, the stresses in the horizontal plane are equal. This can be true for gravitationally deposited soils and is often assumed to apply to subglacial tills with the weight of ice being responsible for the preconsolidation pressure. Since subglacial tills undergo deformation, it is more likely that the horizontal stress varies. Further, the vertical and horizontal stresses may not be the principal stresses because of shear taking place during formation. A number of models for subglacial tills have been developed. Feeser (1988) proposed a model for subglacial stresses where the principal axis is rotated due to shearing under an advancing glacier but behind the ice front, the principal stress is vertical due to the weight of the ice sheet and that this vertical loading causes consolidation. Boulton and Dobbie (1993) suggested a model that was based on one-dimensional consolidation under a melting ice sheet. One-dimensional consolidation would create isotropic horizontal stresses, which is not the case according to Gareau et al. (2004). Further, the vertical distribution of preconsolidation pressures within till profiles varies and is not consistent with the concept that they are due to the weight of ice (Sauer et al., 1993). The reason for this is that the weight of ice leads to an increase in pore pressure and therefore a reduction in preconsolidation pressures.

Geophysical data from beneath glaciers in Antarctica, Canada and Sweden have shown a relationship between subglacial sediment deformation, pore water pressure, ice velocity, and strain at the ice–bed interface (Clarke, 1987; Alley, 1993; Anandakrishnan and Alley, 1994; Blake et al., 1994; Harbor et al., 1997; Hooke et al., 1997; Fischer et al., 1999), which suggests stick–slip behaviour (Alley, 1993; Anandakrishnan and Alley, 1994) and an oscillation between quasi-steady slow and fast modes of basal ice flow (Beeman et al., 1988; Bahr and Rundle, 1996). This behaviour gives rise to subglacial structures and landforms at the ice–bed interface (Piotrowski and Kraus, 1997; Piotrowski and Tulaczyk, 1999; Knight, 2000) because it creates a stick–slip mechanism. Figure 2.40 shows the changes between different aspects (subglacial shear stress, pore water pressure, basal ice velocity and hydraulic gradients in the pore water and at the ice–bed interface) of the subglacial environment over a single stick–slip cycle.

The stick phase is related to the presence of high friction (high strength) asperities or sticky spots at the ice–bed interface (Alley, 1993; Anandakrishnan and Alley, 1994), and spatial differences in the distribution of free basal water (Fischer et al., 1999). The stick phase terminates when water pressure over a single sticky spot exceeds the frictional resistance of overlying ice to sliding (Baumberger et al., 1994; Bahr and Rundle, 1996; Fischer et al., 1999).

Slip is associated with subglacial cavities, which causes ice–bed uncoupling when they become connected. During stick phases, subglacial stress is accommodated by brittle fracture and internal deformation of basal ice layers (Harbor et al., 1997). Slip at the ice–bed interface is due to ice–bed uncoupling and fast ice flow supported by a low-resistance meltwater layer of variable thickness (Iverson et al., 1995; Hooke et al., 1997; Fischer et al., 1999), which can be a result of meltwater redistribution in a closed hydraulic system. This depends on subglacial topography and hydraulic gradients. At the moment of uncoupling, the shear stress decreases dramatically because of the presence of a low-friction meltwater layer, and substrate conditions become isotropic over the entire uncoupled portion of the ice mass (Dolgoushin and Osipova, 1973; Fowler and Johnson, 1995). Peak ice sliding rates

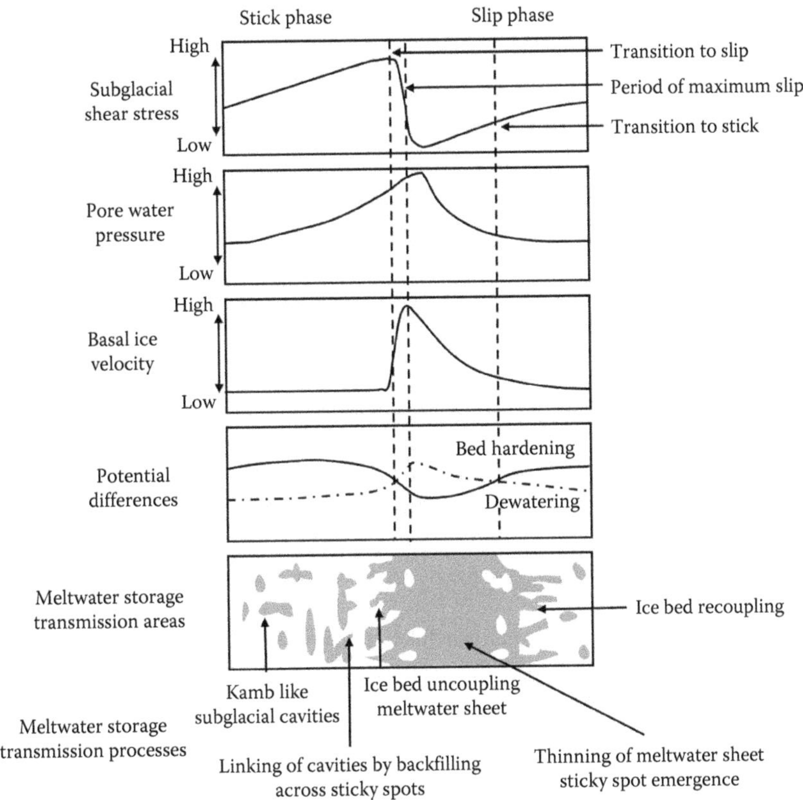

Figure 2.40 Conceptual variation in shear stress, pore water pressure, ice velocity, hydraulic gradient between the pore water and interface water, the meltwater storage areas and the processes during a single stick–slip cycle in the subglacial environment. (After Anandakrishnan and Alley, 1994; Fischer, U. H., G. K. C. Clarke, and H. Blatter. Evidence for temporally varying 'sticky spots' at the base of Trapridge Glacier, Yukon Territory, Canada. *Journal of Glaciology*, 45(150); 1999: 352–360.)

do not always coincide with peaks in recorded pore water pressure (Iverson et al., 1995; Hooke et al., 1997), suggesting that the meltwater migrates between isolated meltwater-filled cavities, which become progressively linked over time – evidence for stick–slip ice flow (Fischer and Clarke, 1997). Meltwater movement decreases over time as the meltwater reservoir is depleted (Kamb, 1987), which leads to ice–bed recoupling as the water layer thins. A difference in potential between the subsurface and interface creates an upward pore water flow, which can initially soften the upper layers of the substrate increasing deformation, possibly erosion (Boulton, 1975). With time, the deformation will stop as the substrate consolidates (Boulton, 1975).

Till modelled as a linear elastic soil (e.g. Kamb, 1991; Iverson et al., 1998; Tulaczyk et al., 2000) means that the shearing resistance is insensitive to the strain rate. However, at glacial rates of strain, it is expected that there would be a relationship between shearing resistance and strain rate and field observations suggest that there can be stick–slip motion (Wiens et al., 2008; Winberry et al., 2009), which is associated with till weakening with increasing shearing rate. The actual motion is more complicated because it can be associated with local variations in pore pressure due to dilation (e.g. Damsgaard et al., 2013) leading to an increase in shearing resistance (e.g. Clarke, 1987; Iverson et al., 1998) and as the excess pore pressure dissipates a reduction in shear resistance (Moore and Iverson, 2002). This may account of subglacial landforms such as drumlins (e.g. Piotrowski, 1987).

Field measurements in the soft beds of glaciers indicate that when effective stress is falling or low, the tendency is for basal motion to be focused at or very near the substrate surface (Fischer and Clarke, 1997; Engelhardt and Kamb, 1998; Iverson et al., 1999, 2007; Boulton et al., 2001; Truffer and Harrison, 2006), which results in clasts being dragged across the substrate surface (ploughing) causing the till to yield (Brown et al., 1987; Alley, 1989a,b; Iverson, 1999; Tulaczyk, 1999). This was confirmed experimentally by Thomason and Iverson (2008) and Iverson (2010).

Figure 2.40 suggests how this stick–slip cycle of sliding and sediment deformation might occur, and how distributed cumulative strain might occur as a consequence of failure at progressively lower depths in the till using the mechanism suggested by Iverson et al. (1998). Figures 2.11 and 2.41 (Alley, 1989a,b) show how deformation changes from sliding and ploughing on the ice–bed contact, through pervasive shearing to discrete shearing at the base of the deforming bed. Figure 2.41 shows that increasing water content leads to increasing displacement and an increase in the thickness of the deformed zone, which is consistent with a reduction in strength (assuming the deforming bed is saturated); and a reduction in clay content leads to a reduction in displacement and a reduction in the thickness of the deformed zone, consistent with an increase in strength because of the increase in granular content.

The spatial variability of a deforming bed can be expressed in terms of 'H' (eroding substrate), 'Q' (mix of erosion and deformation) and 'M' (deforming substrate) classes (Figure 2.42), which represent the spatial variation in composition, water content, shear strength and applied shear stress levels interacting with variations in thickness and velocity. This spatial variation leads to an undulated surface, which will be the start of landforms, either flutes or drumlins (Rose and Letzer, 1977; Boulton, 1987; Menzies, 1987, 1989; Rose, 1987, 1989a,b; Hindmarsh, 1998, 1999; Menzies and Shilts, 2002; Kjaer et al., 2003).

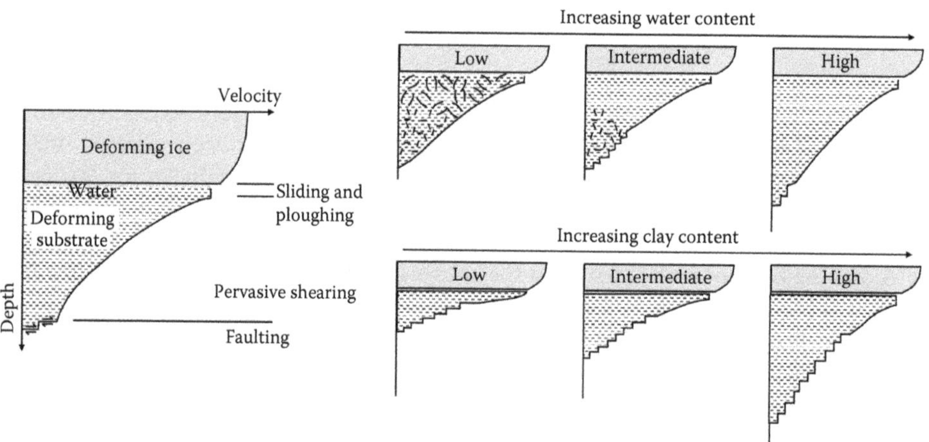

Figure 2.41 Effect of water content and clay content on the deformation of the substrate. An increase in water or clay content increases the depth of deformation and possible switch from brittle to ductile behaviour. (After Alley, R. B. *Journal of Glaciology*, 35(119); 1989: 108–118; Alley, R. B. *Journal of Glaciology*, 35(119); 1989: 119–129; Menzies, J. *Sedimentary Geology*, 62(2); 1989: 125–150; van der Meer, J. J. M., J. Menzies, and J. Rose. *Quaternary Science Reviews*, 22(15); 2003: 1659–1685.)

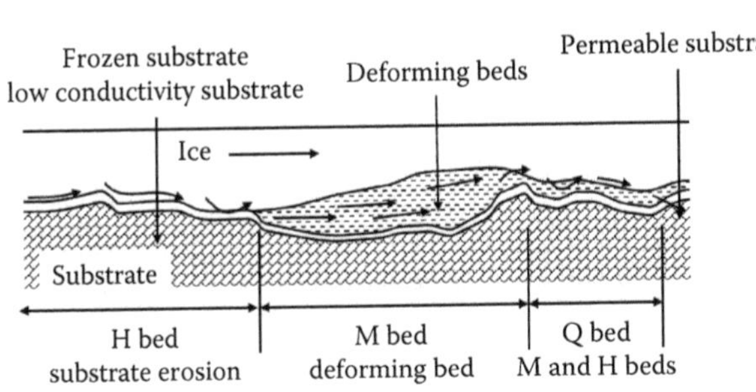

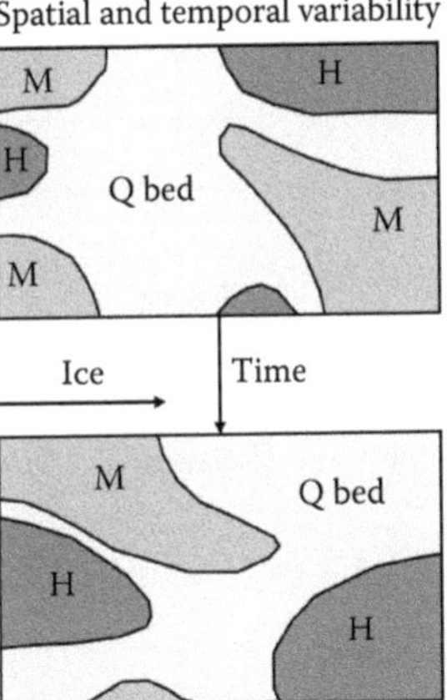

Figure 2.42 Temporal and spatial variability leading to substrate erosion (H bed), substrate deformation (M bed) and a mix of substrate erosion and deformation (Q bed). (After Alley, R. B. Water-pressure coupling of sliding and bed deformation: II. Velocity-depth profiles. *Journal of Glaciology*, 35(119); 1989b: 119–129; Menzies, J. *Sedimentary Geology*, 62(2); 1989: 125–150; van der Meer, J. J. M., J. Menzies, and J. Rose. *Quaternary Science Reviews*, 22(15); 2003: 1659–1685.)

Figure 2.42 shows the conceptual variation in the different types of deforming substrate, which, through time, will change (Truffer et al., 2001). The mobility of particles in a substrate means that the composition of the whole deforming substrate will change if the deforming substrate encompasses the full thickness of the till substrate. Changes in water content of tills beneath ice means that the thickness of the deforming layer will also change. These figures represent moments in time, so deformation will cease only following deglaciation. These changes will be recorded in the microstructures.

A study of the microscopic behaviour highlights a range of microfabrics and microstructures within the plasma and S-matrix (organisation of plasma, skeleton grains and voids) of glacial sediments (Table 2.16), showing that even with a fully homogenised till, tectonic features still exist within the microstructure.

Therefore,

- All subglacial tills are former deforming glacier beds with the exception of melt-out tills, which are gravitationally deposited through a period of deglaciation.
- The intensity of deformation is influenced by glacier velocity, water content and clay content.
- A combination of spatial changes in water and clay content results in a strongly diversified deforming bed, continuously changing its configuration over space and time.

This explains the close associations of tills of markedly different composition and without apparent mixing, geochemical anomalies, the development of fissility in till, the development of deformation macrostructures such as shears, folds and fractures, the development of deformation microstructures in till including birefringent plasmic fabrics and marble-bed configuration and the development of internal and lower boundaries of till beds, including till wedges.

Deformation of subglacial traction tills causes preferred orientations of particles and micro- and macro-fabric features. Field studies have resulted in at least four hypotheses for fabric development resulting from subglacial shear of till:

Table 2.16 Possible microstructures found within the plasma and S-matrix of glacial sediments

	Soil skeleton				
Plasma					
		S-matrix			
Plasma microfabric	*Plasma microfabric/ S-matrix*	*Ductile*	*Brittle*	*Polyphase (ductile/brittle)*	*Pore water influenced or induced*
Masepic	Skelsepic	Strain caps and shadows	Faulted domains	Multiple diamicton domains	Cutans
Lattisepic		Fold structures	Discrete shear lines	Comet structure	Water escape structures
Omnisepic		Layering and foliation	Shear zones	Sill and dyke structure	Silt caps
Unistrial		Necking structures	Reverse fault	Tiled units of laminated clays and silts	Polygonal structures
Insepic		Rotational structure	Kink bands		Silt and clay coatings
Kinking		Secondary foliation	Crushed grains		
Banded		Crenulation foliation			

Source: After Menzies, J. *Geological Society, London, Special Publications*, 176(1); 2000: 245–257, Brewer, 1976; van der Meer, J. J. M. *Quaternary Science Reviews*, 12(7); 1993: 553–587; Jaap et al., 2003.

- Weak fabrics parallel to the shearing direction (Dowdeswell and Sharp, 1986; Hicock, 1992; Hart, 1994; Clark 1997)
- Transverse fabrics (Glen et al., 1957; Carr and Rose, 2003)
- Variable fabric strength dependent on till porosity, water content, or layer thickness (Dowdeswell et al., 1985; Hart, 1994; Evans et al., 2006)
- With sufficient strain it results in strong flow-parallel fabrics (Benn, 1995; Benn and Evans, 1996, 2010)

This overview of local deformation of subglacial tills explains the effect shear has upon the fabric of the tills and, therefore, a reason for the differences between the geotechnical behaviour of subglacial tills and gravitationally deposited soils of similar densities.

2.7 OBSERVATIONS

A review of the history of glacial geology has emphasised the debate that has taken place over the last 150 years highlighting the complex nature of glacial soils. The development of our understanding of the subsurface is heavily influenced by field observations, but this has proved difficult for glacial soils because of the extent of ice cover preventing access to the process of glacial erosion, deposition and deformation. Over the years, the debate has shifted from ice as a means of creating a geological environment, through ice as an erosive medium to the formation of glacial tills. Improvements in instrumentation and numerical methods have created a better understanding of the glacial geological processes and therefore a better understanding of the geotechnical characteristics of glacial soils. The current thinking suggests the following:

- The origin of glacial soils is erosion products from the substrate, which can be rock or superficial deposits including remnants of previous glaciations and, in the case of highland glaciers, erosion products from valley sides or rockfalls from above the glacier.
- Glacial soils can be deformation products, deposits transported and deposited by ice, or deposits transported by ice and water and deposited in a terrestrial, freshwater or marine environment.
- The primary products are those soils that are either deformed (glaciotectonite) or transported and deposited by ice either through subglacial traction with, possibly, further deformation or through deglaciation (meltwater tills).
- The secondary products are those soils which are deposited by or in water (glaciofluvial soils, glaciolacustrine clays and glaciomarine deposits).
- The composition, fabric and structure of glacial soils, both primary and secondary deposits, are spatially variable because of the erosion, transport, deposition and deformation processes.
- Secondary products and meltwater tills are deposits that are gravitationally consolidated but are not isotopic as they are influenced by the deposition process.
- Primary products are subject to shear during deformation and deposition, so they are truly anisotropic.
- It can be difficult to classify a glacial soil from the description alone.
- Landforms can be a useful indicator of the underlying glacial soils.
- The formation of tills and their associated landforms is not fully understood.

This review explains why glacial soils are considered difficult soils from a geotechnical point of view, but the work of glacial geologists has made a substantial contribution to our understanding of what to expect when investigating glacial soils and therefore the design of ground investigations, and how glacial soils may perform when subject to a change in environment.

Chapter 3

Ground investigation in glacial soils

3.1 INTRODUCTION

A ground investigation is a critical part of the design and construction process because it addresses the inherent risk associated with the ground. The hazards include the spatial variation in the design parameters of strength, stiffness and permeability of the soils; and the groundwater conditions. The principles of a ground investigation are set out in various codes; for example, BS 5930:1999; BS EN 1997-2:2007; and publications such as Clayton et al. (1995) and SISG (1993). The primary objectives of a ground investigation are to assess whether a site is suitable, to identify hazards, to produce design parameters, to plan the construction process and to assess the impact of the construction on the ground, adjacent structures and the environment. Glacial soils are also a valuable source of construction materials: deposits of sands and gravels, clays for bricks, clay for landfill liners and suitable materials for embankments. This is especially important when considering linear infrastructure projects where cut and fill techniques and excavations are routine.

This chapter focuses on the ground investigation in glacial areas highlighting the issues to be addressed.

There are six stages to an ideal ground investigation: desk study, site reconnaissance, preliminary exploratory boreholes and trial pits, main investigation including sampling and field and laboratory testing, factual reporting and interpretive reporting. While the objectives of a ground investigation are universal, techniques vary from country to country. There are international and national standards for most, but not all, tests. This chapter focuses on the aspects of glacial soils that have to be considered when planning an investigation and specifying tests.

A review of the formation of glacial soils suggests the following:

- The composition, fabric and structure of glacial soils are spatially variable because of spatial and temporal variations during their formation.
- Glacial soils are composite soils, and all glacial soils can contain a diverse range of particle sizes including very coarse particles.
- Glacial soils can be divided into primary deposits (tills) and secondary deposits (soils deposited by water in a terrestrial environment, and soils deposited in fresh water and marine environments).
- Primary deposits can be divided in glaciotectonite, subglacial traction till and melt-out till.
- Secondary deposits include glaciofluvial, glaciolacustrine and glaciomarine deposits.
- Isostatic uplift and the creation of the current drainage system led to reworking of glacial soils due to mass movement, fluvial processes and weathering.
- Glacial soils can lie unconformably on underlying bedrock and superficial deposits.

- Land systems and landforms are indicators of the type of glacial soils.
- Fluvial sediments can lie unconformably on glacial soils.

This knowledge can be used to ensure a ground investigation is planned to reduce the risk using a strategy to produce the geological and hydrological models with some confidence leading to representative values of geotechnical characteristics.

There are six stages to a ground investigation:

- Desk study to develop knowledge of the site including the topography, geology, potential hazards, groundwater regime and subsurface structures
- Site reconnaissance to view exposures and confirm findings of the desk study
- Preliminary investigation (Stage A) to identify the geological model from boreholes, *in situ*, laboratory and geophysical tests
- Main investigation (Stage B) to identify the ground model, including the geotechnical characteristics, hydrogeological model and potential hazards for construction, design and operation of the ground-related aspects of the civil engineering project, from boreholes, *in situ*, laboratory and geophysical tests
- Factual report covering the results of the desk study, site reconnaissance, exploratory investigations, laboratory tests and field tests
- Interpretative report covering the hazards that will affect the design, construction and operation of the civil engineering project; the design parameters of strength, stiffness and permeability; and the groundwater profile

3.2 DESIGN OF A GROUND INVESTIGATION

The stages of a ground investigation (Figure 3.1) and what is expected at each stage is well documented (e.g. BS EN 1997-2:2007). Here, the focus is on aspects that are particular to glaciated areas based on the points discussed in Chapter 2. The primary objectives of a ground investigation for a civil engineering project are to assess the suitability of the site; provide information to be able to produce a safe, economic and sustainable design that meets the needs of the users; to assess the consequences of the construction on the environment, and adjacent properties; and to identify hazards that could affect the design, construction and operation of the project. In order to achieve these objectives, an assessment of the regional geology, geomorphology, topography, hydrogeology and geotechnical characteristics are required, as well as a detailed assessment of the ground conditions to the particular project. The regional assessment is particularly important in areas of glacial soils since a glacier creates landforms, which gives some indication of the likely types of the glacial soils in the area (see Table 2.6). Further, depending on the landforms, it may provide helpful information on the hydrogeological conditions. This applies to both infrastructure and building projects. In the case of infrastructure projects, the regional assessment is essential because the project will be crossing an extensive glaciated region. It is also important for building projects because it provides information on what may be expected at the site because the site will be in a glaciated region. For example, a construction project in Glasgow may be in a drumlin field, which has characteristics described in Section 3.2. Therefore, exploratory boreholes will be positioned to locate the features expected.

Time and cost pressures often impact on the quality of an investigation to the detriment of the project. Indeed, a poorly planned and executed ground investigation is a hazard that can lead to delays and additional costs. Failures of excavations in glacial soils, overdesigned pile foundations, inadequate excavation equipment and failure to detect permeable layers

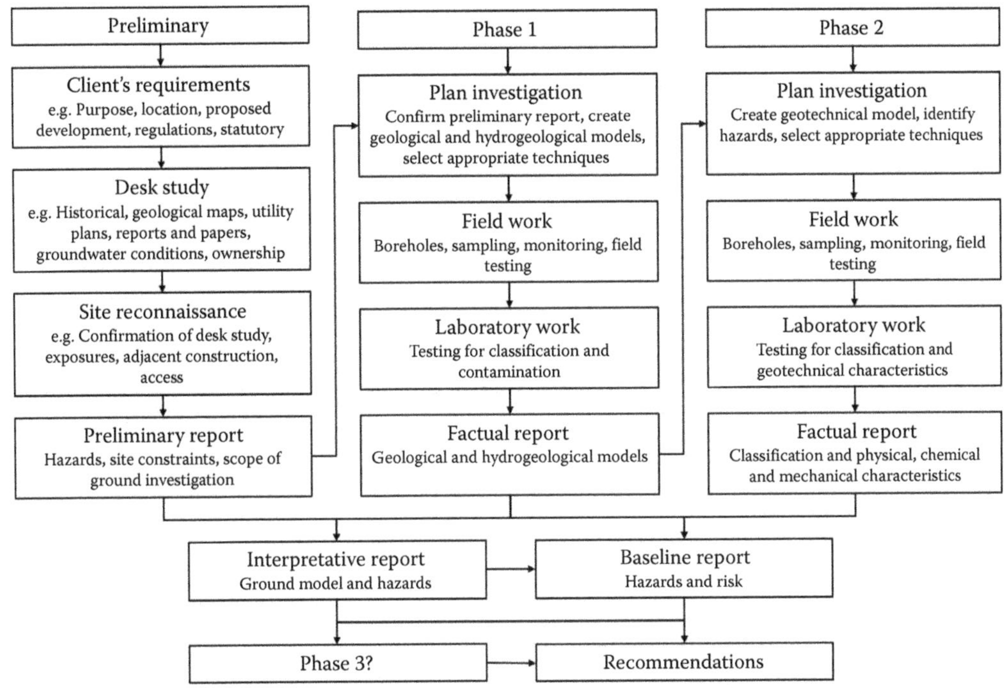

Figure 3.1 Stages of a ground investigation highlighting the technical aspects.

described in Chapter 1 are examples of consequences of inadequate investigations. Many ground investigations focus on environmental issues because of concerns of contamination, yet the same care is not necessarily paid to the geological, hydrogeological and geotechnical characterisation. This is short-sighted; there is enough evidence to show that an inadequate ground investigation adds to the cost of a project possibly some years after the construction is complete.

3.3 DESK STUDY

A ground investigation starts with a desk study, which includes studies of topographical, historical and geological maps, aerial photographs, geological memoirs and historical evidence of ground movement (BS5930:1999). The topographical, geological and engineering geology maps provide an indication of landforms, the generic geological profile and potential hazards (e.g. BGS, 2015). This is particularly important for some types of glacial soils, which can be intrinsically linked to the landform. The history of glacial soils, that is, the erosion, transport, deposition and deformation of a glacial deposit, and its impact on its geotechnical properties are difficult to assess from a desk study because of the nature and diversity of glacial soils, which makes it difficult to produce generic design parameters at this stage. However, an understanding of the formation of glacial soils and the landforms created provide a useful guide to what may be expected.

Geological maps are unlikely to give much detail of glacial soils because they are so variable and can only be identified from a combination of a detailed analysis of exposures, excavations, borehole samples and remote sensing. Glacial soils can be described using an engineering classification scheme for soils such as the European Soil Classification System

(ESCS) or Unified Soil Classification System (USCS) or with a scientific lithofacies coding scheme. None of these schemes provide information on the history of the deposit, which is a crucial information for engineering investigations in glacial soils. Therefore, a further classification is needed, which could be based on the debris cascade system (Figure 2.1).

3.4 SITE RECONNAISSANCE

The desk study should be followed by a site visit, which aims to confirm the findings of the desk study. A site reconnaissance is an opportunity, indeed an essential requirement, to observe regional landforms to identify any obvious glacial features and record exposures of glacial soils. Exposures of glacial soils are extremely valuable as they provide a cross section that is not available from exploratory holes and, given the spatial variability of glacial soils, an opportunity to assess the composition, fabric and structure of the soils. Local knowledge of previous construction projects from consultants, contractors and local authorities should be collected. Given the scale of a glaciated terrain, the site reconnaissance should not be constrained by the project boundaries. Indeed, lessons can be learnt of the nature of glacial deposits from visits to quarries, river banks, coastal cliffs or construction projects, that is, anywhere where natural or anthropogenic excavations have taken place.

3.5 PRELIMINARY INVESTIGATION

Given the diversity of glacial soils, it is recommended, indeed essential, to follow a desk study with a preliminary investigation and a more detailed investigation. The preliminary investigation includes a series of exploratory holes to establish more details of the geological profile to help plan the main investigation.

A review of the geological maps produces a generic geological profile, which can be used to produce preliminary designs based on published values of strength, stiffness and permeability. However, the final design must be based on characteristic values derived for that particular site. This is especially important for glacial soils as they are spatially variable, both vertically and horizontally.

A geological model starts with geological maps, topographical maps, aerial photographs and a walk-over survey. In the glaciated terrain, it should be an aim to produce an overview of the likely types of glacial soils from the geomorphological features and the geological maps. It should be noted that exposures in the region provide an indication of the type of glacial soil, but it does not mean that the engineering soil type (as opposed to the geological sediment) noted in the exposure will be found at the site of the project. For example, an adjacent exposure may show a subglacial till. It is likely that the site of interest will be underlain by a subglacial till because these deposits are extensive. However, the engineering characteristics of the soils (e.g. matrix-dominated till containing lenses of sands and gravels and laminated clays) in the exposure may be different at the site because of the spatial variation of glacial soils. Trial pits and trenches at this stage would be a useful addition to help plan the main investigation and should be considered an essential part of a preliminary investigation.

The extent of a ground investigation depends on the character and variability of the ground, the type of project and the results of the desk study. In the case of glacial soils, it is prudent to assume that the soils will be variable irrespective of the size of the project. The depth and extent of the exploratory work will depend on the type of project, but in glacial soils, it is anticipated that the geological profile will have an impact on the design of the investigation; it is not sufficient to specify borehole depth and location based on the project requirements alone. For example, proving rock head in glacial tills can be difficult because the bedrock may have been subject to glacial erosion, leading to an irregular surface and the misinterpretation of boulder beds and rafted rock as bedrock. So it is prudent to specify a greater distance to drill to prove rock than is normal in the non-glaciated terrain. Identifying whether any sand or gravel encountered within a glacial till is a pocket or lens and, if a lens, the extent of that lens is important, especially if it is an aquifer. Identifying weaker layers or lenses within a dense till is important as they can lead to slope failures and excessive local settlement.

These examples show why a preliminary investigation is important and flexibility is required in the main investigation because the features may not be uncovered in the preliminary investigation. A preliminary investigation is essential in glacial soils to determine the most appropriate sampling and testing regime in the main investigation, which depends on the particle size and particle size distribution.

3.6 THE MAIN INVESTIGATION

The spacing of exploratory boreholes, trial pits and test profiles depend on the category of the project and complexity of the ground conditions. For example, BS5930:1999 suggests 10–30 m for structures, a minimum of three locations for structures with a small plan area. Structures involving major geotechnical works (e.g. retaining structures, dams, tunnels, excavations and deep foundations) require a greater understanding of the geology to reduce risk and delays. Given the spatial variation in glacial soils compared to that for gravitationally consolidated soils, it is likely that the number of boreholes, samples and *in situ* and laboratory tests will be greater in order to develop the ground model and select the design parameters.

BS EN 1997-2:2007 recommends that boreholes should be spaced at 15–40 m apart for high rise and industrial structures; 20–200 m for linear structures such as roads, retaining walls, tunnels and pipelines; 25–75 m for weirs and dams at a number of sections; and for specialist foundations for bridges, machinery for example, two to six boreholes per foundation. It is prudent when working in a glaciated terrain to err on the cautious side. The depth of exploration extends beyond the zone of influence of the structure and, in particular, beyond any layers of weak or compressible soils.

BS5930:1999 suggests that rock head should be proved to at least 3 m and this should be in more than one borehole to assess whether it is a boulder or bedrock. However, the presence of rafted rock and undulating rock head that could be dissected by valleys filled with glacial soils means that 3 m may be insufficient. Encountering rock in only one borehole does not necessarily mean that a boulder is encountered; it could be evidence of an irregular bedrock surface.

BS5930:1999 suggests that the depth of exploration should be at least one and a half times the width of the loaded area. For shallow foundations, this means the area of an individual footing or the plan area of the structure if the contact stress is significant or the foundations are close together or it is raft foundation. The desk study and the first stage of the exploratory work should provide sufficient information to carry out a conceptual design. This allows the depth of exploration to be linked to a possible design solution. However, it must be noted that in glacial soils, foundations can be overdesigned because of the difficulties of determining characteristic strengths; therefore, the type of foundation may change following the ground investigation. This means that the depth of the exploration should be extended in places. Table 3.1, a summary of the extent of exploratory work based on BS

Table 3.1 Recommended depth of exploration

Structure	*Recommended depth of exploration*	*Comments*
High rise structures and civil engineering projects	6 m or 3× breadth of the foundation whichever is the greatest	Deeper boreholes may be required to locate bedrock surface if within zone of influence; possible weaker and water-bearing layers within zone of influence, if a piled solution is likely
Raft foundations and structures with several foundations that interact at depth (interaction is likely if the foundations are less than B apart where B is the width of the foundation)	1.5× minimum width of the structure	Deeper boreholes may be required to locate bedrock surface if within zone of influence; possible weaker and water-bearing layers within zone of influence, if a piled solution is likely
Embankments	6 m or between 0.8h and 1.2h whichever is the larger (where h is the maximum height of the embankment)	Need to locate possible aquifers in matrix-dominated tills
Cuttings	2 m or 0.4h whichever is the larger (where h is the maximum depth of the cutting)	Need to locate the bedrock surface if it is irregular and within the cutting
Roads and airfields	At least 2 m below the formation level	
Trenches and pipelines	2 m or 1.5× breadth of the trench below the invert level whichever is the greatest	Need to be aware of potential hard spots due to embedded boulders
Small tunnels and caverns	Between the width and twice the width below the base of the excavation	Possibility of encountering water-bearing lenses and layers
Excavations	Where the piezometric surface and the groundwater tables are below the excavation base, either 0.4h or $(t + 2)$m whichever is the largest (where t is the embedded length of the support and h is the excavation depth)	Deeper borehole maybe required to locate aquifers below the base of the excavation
	Where the piezometric surface and the groundwater tables are above the excavation base, $(H + 2)$m or $(t + 2)$m whichever is the largest (where H is the height of the groundwater level above the excavation base and t is the embedded length of the support)	Lens of permeable soils may be misinterpreted as aquifers
	If no stratum of low permeability is encountered, then the boreholes should be increased to $(t + 5)$m	In glacial tills layers of permeable material may exist
Cut-off walls	At least 2 m below the surface of the stratum impermeable to groundwater	May need deeper boreholes to locate permeable layers in matrix-dominated tills
Piles	5 m and 3D_F and b_g (where D_F is the pile base diameter and b_g is the smaller side of the rectangle circumscribing the group of piles forming the foundation at the level of the pile base)	Deeper boreholes may be required to locate bedrock surface if within zone of influence; possible weaker and water-bearing layers within zone of influence

Source: After BS EN 1997-2:2007. *Eurocode 7: Geotechnical Design – Part 2: Ground Investigation and Testing (Incorporating Corrigendum 2010)*. British Standards Institution, London.

EN 1997-2:2007, gives more detailed recommendations for depth of exploration, which are, with reference to the lowest point of the foundation, structural element or excavation. Of course, at the time of the investigation, these may not be known because the design will depend on the ground conditions. This is another reason to carry out a two-stage ground investigation and why some boreholes should extend beyond the zone of influence.

3.6.1 Field work

BS5930:1999 suggests that the methods of ground investigation will be influenced by the character of the site, the availability of the equipment and personnel and the cost of the methods. In glacial soils, it is also a function of the particle size, particle size distribution and the lithology of the glacial soils. The prime purpose of a ground investigation for a civil engineering project is to identify hazards and to produce characteristic design values. Field work includes trial pits, trenches, boreholes, sampling, *in situ* tests and geophysical tests from which the ground model is developed. Most useful design parameters for civil engineering projects will be derived from *in situ* and laboratory tests, so appropriate exploratory techniques should be selected for the types of soils likely to be encountered, the depth of exploration and the design parameters required. In the United Kingdom, boreholes are normally drilled using light percussion or rotary rigs, the choice depending on the ground conditions and the depth of exploration. Light percussion rigs can be used in all glacial soils, but the composition of the soils means that it can be difficult to obtain quality samples necessary for design characteristics. The alternative, rotary rigs, can improve the quality of a borehole and samples, but clasts can have a significant effect on the quality of a sample and *in situ* test. Therefore, a borehole is designed to take samples or carry out *in situ* tests. Table 3.2 is a summary of the recommendations of BS5930:1999 for coarse-grained soils, fine-grained soils and matrix-dominated soils; all of which can be found in glacial soils. Tills can either be matrix-dominated or clast-dominated tills and both could contain gravels, cobbles or boulders. Drilling techniques for coarse-grained soils or clays containing gravels and cobbles should always be considered. It is not possible to obtain Class 1 samples or even Class 2 samples, that is, samples suitable for assessing geotechnical characteristics, from many glacial soils. However, in matrix-dominated tills, it is possible to recover samples that can be used to describe the lithology and fabric of the till and carry out tests to determine strength and stiffness. The value of those results is discussed in Chapter 5. Penetration tests are also used in tills, but again, the quality of the results depends on the composition of the till. It is possible to create a borehole in which an *in situ* testing device is inserted, but the quality of the results will be affected by the composition of the till.

Boreholes in secondary deposits are less challenging since the composition of the glacial soils are typically fine grained (lacustrine deposits) or coarse grained (sands and gravels), though cobbles and boulders should be expected. Hence, in lacustrine deposits, it should be possible to obtain Class 1 samples using thin-walled samplers from the base of boreholes drilled using light percussion or rotary rigs. In other secondary deposits, it will be possible to obtain disturbed samples and carry out appropriate *in situ* tests from boreholes drilled using light percussion or rotary rigs.

The choice of drilling method, sampling techniques and *in situ* tests for the main investigation will depend on the results of the preliminary ground investigation.

3.6.1.1 Field investigation

Trial pits and trenches are very useful in glaciated terrains as they allow an exposure of glacial soils to be observed, something that is not possible from boreholes. They also help

Table 3.2 Selection of exploration methods in the United Kingdom

Soil type	*Bh diameter*	*Drilling method*	*Support*	*Drilling aid*	*Sample quality*	*In situ tests*
Coarse-grained soils containing boulders, cobbles or gravel	250 mm (gravel) 450 mm (cobbles) Boulders	Light cable percussion Rotary Rotary	Casing	Dry	Class 4 open tube	SPT (cone) SPT (cone) (?)
Sand		Light cable percussion Rotary	Casing	Added water to maintain stability	Class 5 open tube Class 2/3 piston sampling	SPT (split barrel) Pressuremeter Permeability
Silt		Light cable percussion Rotary			Class 2/3 open tube	SPT (split barrel)
Soft to firm clays		Light cable percussion Rotary		Dry	Class 2/3 open tube Class I piston sampling	Vane
Dense clays		Light cable percussion Rotary		Dry	Class 2/3 open tube Class I thin-walled sampling	SPT (split barrel)
Matrix-dominated soils				Dry	Class 4/5 open tube	SPT (split barrel) Pressuremeter

Source: After BS 5930:1999+A2:2010. *Code of Practice for Site Investigation*. British Standards Institution, London.

confirm the likely type of glacial soil in the upper layers allowing the engineering descriptions from the borehole samples to be placed in context. For health and safety reasons, the trenches and pits must be no deeper than 1.2 m if unsupported. The pits and trenches must not be located where they may affect the future structure.

Light percussion rigs are in common use in the United Kingdom for historical reasons. They have proved successful in obtaining samples from many types of soils. The borehole is advanced by repeatedly dropping a clay cutter or shell onto the base of the hole. The soil is removed, thus advancing the borehole. In appropriate soils, a borehole is drilled dry and without casing. There are soils, such as stiff clays, that can stand unsupported. Otherwise, casing is used to line the hole preventing collapse. Holes can be drilled up to 60 m in suitable soils and weak rock. Boreholes are typically 150 or 200 mm diameter though in soils containing cobbles and deep boreholes, the diameter may increase to 300 mm. Continuous flight augurs with a hollow stem can be used in matrix-dominated tills, if the clast content is limited, and lacustrine soils. Rotary drilling, developed for drilling in rock, can be used in some soils. The drill bit is either driven by a downhole motor or from the surface using a drill string. The cuttings are flushed to the surface using air, foam, water or mud flush. Holes can be advanced using a drill bit or core barrel. A core barrel brings a Class 2 sample to the surface, so is more useful in ground investigations. Conventional or wireline, double or triple core barrels fitted with diamond or tungsten-tipped core bits are used. Rotary coring works best in fine-grained glacial soils, which contain little coarse material or coarse material embedded in a strong matrix. Wash boring can be used in fine-grained soils and sands. The soil is broken up by water pressure and is flushed to the surface. It is not used in gravels, which may discount its use in clast-dominated tills and secondary deposits other than lacustrine deposits.

3.6.1.2 Sampling

BS5930:1999 suggests that where suitable information is available it is unnecessary to determine the character and structure of the strata. It can be assumed that this does not apply to glacial soils because of their spatial variability. Therefore, samples of sufficient quality to describe the geological features are required, and, of particular importance, the lithology and fabric of the soils, Table 3.3 summarises types of samples that can be obtained from soils. This table suggests that none of these sample types are suitable for composite soils, such as glacial soils, because of the coarse particle content. In practice, samples are required so representative geotechnical characteristics of these composite soils may be assessed even on poorer quality samples. This could explain why, *in situ*, composite soils can often be stiffer and stronger than expected.

Table 3.4 is a summary of the class of sample that can be used in glacial soils and what can be expected of the sample. This is based on BS5930:1999 and BS EN ISO 22475-1:2006 description of sampling by drilling, sampling with samplers and block sampling. The quality of a sample is linked to the laboratory tests (e.g. Class 1 samples are required for assessing design parameters). The sampling methods are divided into three categories: Type A samples of quality 1–5, Type B samples of quality 3–5 and Type C for sample quality 5 only. Class 1 and 2 samples are required for geotechnical design parameters as they retain the same water content and porosity as *in situ*. Samples of quality 3 and 4 can provide useful geological information and can be used to classify a soil if the fabric is retained. Samples of quality 5 indicate only the lithology of the soil; no information is provided on fabric as that is completely destroyed during drilling.

It is only possible to obtain Class 1 samples from completely homogenised tills and lacustrine deposits. However, with careful sampling, it should be possible to obtain Class 2

Table 3.3 Sampling in soils

Type of sampler[a]	Preferred sample dimensions		Technique used	Applications and limitations		Sampling category[a]	Achievable quality class[a]
	Diameter (mm)	Length (mm)		Unsuitable for	Recommended for use in		
Thin-walled (OS-T/W)	70–120	250–1000	Static or dynamic driving	Gravel, loose sand below water surface, firm cohesive soils, soils including coarse particles	Cohesive or organic soils of soft or stiff consistency	A	1
					(medium) Dense sand below water surface	B (A)	3 (2)
					Cohesive or organic soils of stiff consistency	A	2 (1)
Thick-walled (OS-TK/W)	>100	250–1000	Dynamic driving	Gravel, sand below water surface, pasty and firm cohesive or organic soils, soils including coarse particles	Cohesive or organic soils of soft to stiff consistency, and including coarse particles	B (A)	3 (2)
Thin-walled (PS-T/W)	50–100	600–800	Static driving	Gravel, very loose and dense sands, semi-firm and firm cohesive or organic soils, soils including coarse particles	Cohesive or organic soils of pasty or stiff consistency, and sensitive soils	A	1
					Sand above groundwater	B	3
Thick-walled (PS-TK/W)	50–100	600–1000	Static driving	Gravel, sand below water surface, pasty and firm cohesive or organic soils, soils including coarse particles	Cohesive or organic soils of soft to stiff consistency, and sensitive soils	B (A)	2 (1)
Cylinder (LS)	250	350	Static rotating	Sand	Clay, silt	A	1
Cylinder (S-SPT)	35	450	Dynamic driving	Coarse gravel, blocks	Sand, silt, clays	B	4
Window	44–98	1500–3000	Static or dynamic driving	Sand, gravel	Silt, clay	C	5

Source: After BS EN ISO 22475-1:2006. *Geotechnical Investigation and Testing. Sampling Methods and Groundwater Measurements. Technical Principles for Execution*. British Standards Institution, London.

Note: OS-T/W, open-tube samplers, thin-walled; OS-TK/W, open-tube samplers, thick-walled; PS-T/W, piston samplers, thin-walled; PS-TK/W, piston samplers, thick-walled; LS, large sampler; S-SPT, SPT (standard penetration test) sampler.

[a] The sampling categories and achievable quality classes given in parentheses can only be achieved in particularly favourable soil conditions, which shall be explained in such cases.

Table 3.4 Examples of sampling methods with respect to the sampling category in glacial soils

	Quality				
Property	*1*	2	3	4	5
Sequence of layers	√	√	√	√	√
Stratum boundaries (broad)	√	√	√	√	
Stratum boundaries (fine)	√	√			
Consistency limits	√	√	√	√	
Particle size	√	√	√	√	
Water content	√	√	√		
Density	√	√			
Permeability	√	√			
Stiffness	√				
Strength	√				
Sample category according to BS EN ISO 22475-1:2006			A	B	C

Source: After BS EN ISO 22475-1:2006. *Geotechnical Investigation and Testing. Sampling Methods and Groundwater Measurements. Technical Principles for Execution.* British Standards Institution, London.

samples from matrix-dominated tills and class 4 samples from clast-dominated tills and secondary deposits. It is likely that the strength and stiffness of matrix-dominated soils will be underestimated because of sample disturbance (Class 2 samples). This can lead to the overdesign of foundations and inappropriate excavation techniques. However, while tests on subglacial tills may underestimate the mechanical properties because of sample disturbance, the size of specimen has to be sufficient to take into account discontinuities since the fabric of these soils influences the mechanical properties. *In situ* tests will be used in clast-dominated tills and coarse-grained secondary deposits because of the difficulty in obtaining anything other than Class 3 samples.

It is very difficult to identify the type of glacial soil from borehole samples because, as Figure 3.2 shows, a borehole may penetrate a lens or layer of sand and gravel, but without further investigation it is not known whether it is a lens or a layer or, if a layer, whether it is inclined or horizontal. Samples of glacial till, no matter the type, may have a similar composition yet be formed in different way. Samples of glaciofluvial soils are possibly easier to identify, but it may be difficult from borehole samples to distinguish them from post-glacial fluvial deposits. The fabric of glacial soils influences the geotechnical characteristics, yet the fabric may not be easily observed in borehole samples. The spacing and orientation of discontinuities in subglacial tills will be difficult to assess. Samples of rock may help distinguish between bedrock, boulders derived from that bedrock and boulders transported to that area.

A correctly designed ground investigation will produce sufficient specimens and test results to produce the geological profile, classification of the soil types and characteristic geotechnical properties and to identify hazards. Given the spatial variability of the composition, the fabric and structure of glacial soils, the difficulty in retrieving representative samples and the impact clasts have on the quality of *in situ* and laboratory tests, it is prudent to specify more boreholes, samples and *in situ* tests in glacial soils than would be expected in gravitationally consolidated soils, which are often less variable.

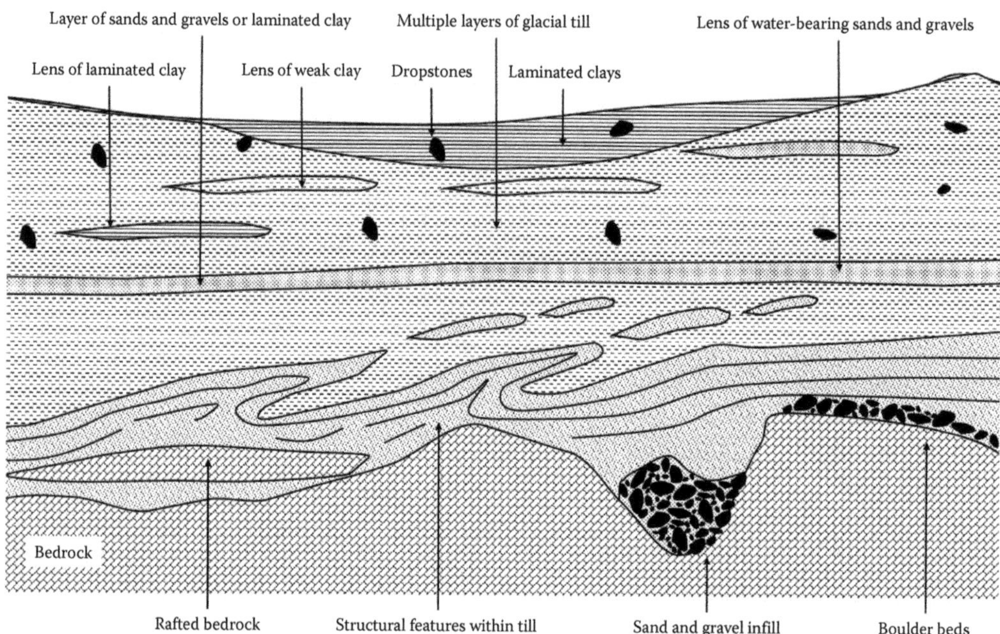

Figure 3.2 Relation between the ground conditions and the borehole highlighting the challenge of creating a 3D image of glacial tills because of structural features associated with deformation, difficulty in identifying bedrock due to rafted rock and boulder beds, lens and layers of weaker clays/water-bearing sands and gravels, dropstones.

BS5930:1999 suggests samples every 1.5 m and when the stratum changes. It would be prudent to take samples more frequently, especially if the preliminary investigations show the soils to be variable to obtain sufficient samples to describe the soil profile and obtain enough representative samples to assess the geotechnical characteristics. Table 3.5 can be used as a guide to determine what types of samples are required to take account of the composition of glacial soils. For example, consistency limits are based on the fine-grained content of matrix-dominated tills; therefore, account has to be taken of the coarse-grained content including clast content when determining the minimum quantity of sample. Tests for strength and stiffness on matrix-dominated tills are likely to be on 100 mm diameter specimens because of composition and fabric. All glacial soils can contain gravels though this is more likely in tills and glaciofluvial soils.

It is a normal practice for the operators of drilling equipment to make notes of the strata encountered using samples obtained from the drilling process while the borehole is advanced. This is a useful source of information, which is often used to identify stratum boundaries. Table 3.6 shows the category of samples for a variety of drilling methods. It shows that rotary dry core drilling with single, double or triple-tube core barrels may be used to obtain samples for geotechnical characterisation from matrix-dominated tills and lacustrine deposits, though triple-tube core barrels are the best. However, it must be noted, in the case of matrix-dominated tills, that this depends on the strength of the matrix and the presence of clasts. If the matrix is too soft, the fine-grained material may be washed away when drilling through clasts. It also suggests that percussive drilling in matrix-dominated tills with particles less than a third of the diameter of the clay cutter and lacustrine deposits can provide samples for geotechnical characterisation though it would be usual to use a

Table 3.5 Quality of samples needed for identification, classification and geotechnical characteristics

Glacial soil	*Soil type*	*Suitability depends on*	*Sampling method* A	B	C
Fully homogenised till; lacustrine clays	Clay	Stiffness or strength sensitivity	PS-PU OS-T/W-PU OS-T/W-PE[a] OS-TK/W-PE[a] CS-DT, CS-TT LS, S-TP, S-BB	OS-T/W-PE OS-TK/W-PE CS-ST HSAS AS[a]	AS
Fully homogenised till; lacustrine clays	Silt	Stiffness or strength sensitivity; groundwater surface	PS OS-T/W-PU OS-TK/W-PE[a] LS, S-TP	CS-DT, CS-TT OS-TK/W-PE HSAS	AS CS-ST
Glaciofluvial sands	Sand	Sizes of the particles; density; groundwater surface	S-TP OS-T/W-PU[a]	OS-TK/W-PE CS-DT, CS-TT HSAS	AS CS-ST
Glaciofluvial gravels	Gravel	Size of the particles; density; groundwater surface	S-TP	OS-TK/W-PE[a] HSAS	AS CS-ST
Matrix-dominated tills		Stiffness or strength sensitivity; % of clasts	CS-DT, CS-TT OS-TK/W-PE	OS-TK/W-PE HSAS	AS CS-ST
Clast-dominated tills		Size of the particles; density; groundwater surface; % of fines	S-TP	OS-TK/W-PE HSAS	AS CS-ST
Glaciofluvial sands and gravels		Size of the particles; density; groundwater surface	S-TP	OS-TK/W-PE HSAS	AS CS-ST

Source: After BS EN ISO 22475-1:2006. *Geotechnical Investigation and Testing. Sampling Methods and Groundwater Measurements. Technical Principles for Execution.* British Standards Institution, London.

Note: OS-T/W-PU, open-tube samplers, thin-walled/pushed; OS-T/W-PE, open-tube samplers, thin-walled/percussion; OS-TK/W-PE, open-tube samplers, thick-walled/percussion; PS, piston samplers; PS-PU, piston samplers, pushed; LS, large samplers; CS-ST, rotary core drilling, single tube; CS-DT, CS-TT, rotary core drilling, double or triple tube; AS, auguring; HSAS, hollow stem auguring; S-TP, sampling from trial pit; S-BB, sampling from borehole bottom.

[a] Can be used only in favourable conditions.

separate sampler. None of the methods can provide quality samples of coarse-grained secondary deposits and clast-dominated tills.

There are a number of points of good practice highlighted in BS EN ISO 22475-1:2006. The inside of the sampling tube or liner has to be clean and smooth. If casing is used with percussive drilling, the percussion process must cease when it is within 0.25 m or five times the borehole diameter of the sampling depth. In the case of rotary drilling, the casing can be lowered to the bottom of the borehole except in sensitive clays where it must stop 2.5 times the borehole diameter above the sampling depth. The bottom of a borehole must be cleaned before the sample is taken. Table 3.5 shows that only thin-walled samplers can be used to obtain samples of sufficient quality to characterise soils. The table also shows that these samplers can be used only in fine-grained soils. This means that it is only possible to obtain samples of glacial soils of sufficient quality if they are completely homogenised tills or lacustrine deposits. The only sampler recommended for matrix-dominated tills is a dynamically driven thick-walled sampler (e.g. U100), but depending on the amount of clasts, it may be possible to obtain samples for geotechnical characterisation. None of the samplers are suitable for secondary glacial soils unless they are lacustrine deposits or pure sands.

It is possible to cut block samples from trial pits provided there is sufficient cohesion to retain the intact sample. Therefore, it should be possible to obtain Class I samples of

Table 3.6 Sampling by drilling in soils

Drilling method				Equipment		Application and limitations[a]			
Soil cutting technique	Use of flushing medium	Extraction of sample by	Designation	Tool	Typical borehole diameter (mm)	Unsuitable	Preferred	Sampling category[b]	Quality
Rotary drilling	No	Drilling tool	Rotary dry core drilling[f]	Single-tube core barrel	100–200	Coarse gravel, cobbles, boulders	Clay, silt, fine sand, silt	B(A)	4 (2–3)
				Hollow stem auger	100–300		Clay, silt, sand, organic soils	B(A)	3 (1–2)
	Yes	Drilling tool	Rotary dry core drilling	Single-tube core barrel	100–200	Non-cohesive soils	Clay, clayey and cemented composite soils, boulders	B(A)	4 (2–3)
				Double-tube core barrel[c]				B(A)	3 (1–2)
				Triple-tube core barrel[c]				A	1
	Yes	Drilling tool	Rotary dry core drilling	Double/triple-tube core barrel with extended inner tube	100–200	Gravel, cobbles, boulders	Clay, silt	A	2 (1)
	No	Drilling tool	Auger drilling	Drill rods with shell or flight auger; hollow stem auger	100–2,000	Boulders larger than $D_e/3$[a]	All soils above water surface, all cohesive soils below water surface	B	4 (3)
	Yes	Reverse flow of flushing medium	Reverse circulation drilling	Drill rods with hollow chisel	150–1,300		All soils	C (B)	5 (4)
	No	Drilling tool	Auger drilling with light equipment	Shell auger or spiral flight auger	40–80	Coarse gravel with a particle size larger than $D_e/3$[a], dense soils, cohesion-less soils beneath groundwater surface	Clay to medium gravel above water surface; cohesive soils below water surface	C	5
Hammer driving	No	Drilling tool	Percussive core drilling	Percussion clay cutter with cutting edge inside; also with sleeve (or hollow stem auger)[d]	80–200	Soils with a particle size larger than $D_e/3$[a] laminated soil, e.g. varve	Clay, silt and soils with a particle size up to $D_e/3$[a]	Cohesive: A Non-cohesive soil: B (A)	2 (1) 3 (2)
	No	Drilling tool	Percussive core drilling	Percussive clay cutter with cutting edge outside[d]	150–300	Soils with a particle size larger than $D_e/3$[a]	Gravel and soils with a particle size up to $D_e/3$[a]	B	4
	No	Drilling tool	Small diameter hammer driving	Hammer driving linkage with tube sampler	30–80	Soils with a particle size larger than $D_e/2$	Soils with a particle size up to $D_e/5$	C[e]	5

(Continued)

Table 3.6 (Continued) Sampling by drilling in soils

Drilling method				*Equipment*		*Application and limitations*[a]			
Soil cutting technique	*Use of flushing medium*	*Extraction of sample by*	*Designation*	*Tool*	*Typical borehole diameter (mm)*	*Unsuitable*	*Preferred*	*Sampling category*[b]	*Quality*
Rotary hammer driving	Yes	Drilling tool	Rotary percussive drilling	Single- or double-tube core barrel	100–200	Composite and pure sands with a particle size larger than 2 mm, gravel, firm and stiff clays	Clay, silt, fine sand	Cohesive soil: A Non-cohesive soil: B	2 (1) 4 (3)
Vibration drilling with an optional slow rotation	No (only for lowering casing)	Drilling tool	Resonance	Thick-wall sampler or single-tube core barrel with optional plastic lining tube	80–200			Cohesive soil: B Non-cohesive soil: C	4 5
Percussion	No	Drilling tool	Cable percussion drilling	Cable with percussion shell auger	150–500	Gravel above water surface, silt, sand and gravel below water surface	Clay and silt above water surface, clay below water surface	C (B)	4 (3)
	No	Drilling tool	Cable percussion drilling	Cable with valve auger	100–1,000	Recovery above water surface	Gravel and sand in water	C (B)	5 (4)
Pneumatic/ continuous thrust	No	Drilling tool	Small diameter pneumatic/ continuous thrust drilling	Pneumatic/ continuous thrust linkage, with tube sampler	30–80	Dense and coarse-grained soils	Clay, silt, fine sand	C[e]	5
Grabbing	No	Drilling tool	Grab drilling	Cable with grab	400–1,500	Firm, cohesive soils, boulders of size larger than $D_e/2$	Gravel, boulders of size less than $D_e/2$, cobbles below water surface	C[e]	5

Source: After BS EN ISO 22475-1:2006. *Geotechnical Investigation and Testing. Sampling Methods and Groundwater Measurements. Technical Principles for Execution.* British Standards Institution, London.

[a] D_e is the internal diameter of the sampling tool.
[b] The sampling categories and quality classes given in parentheses can only be achieved in particularly favourable ground conditions, which shall be explained in such cases.
[c] Conventional or wireline core barrel.
[d] Using the hammer driving technique, the drilling tool will be driven by a special driving tool. Using the percussion technique, the drilling tool will be driven by its repetitive lifting and falling.
[e] Sampling category B is sometimes possible in light cohesive soils.
[f] Rotary dry core drilling is commonly used if the observation of the groundwater surface is the most important aim of the ground investigation.

matrix-dominated tills and lacustrine deposits. However, trial pits are useful in all glacial soils as it is possible to produce a geological description of the soil, thus classifying the glacial soil.

3.6.1.3 Groundwater profile

Failures during construction because of groundwater are not uncommon, but developing the hydrogeological model is challenging. BS EN ISO 1997-2:2007 states that assessing the groundwater conditions is critical, but it is often difficult to obtain meaningful information from a routine investigation, especially when investigating fine-grained soils. In that case, the time taken to reach equilibrium conditions exceeds the time of the investigation and, if there is no means of monitoring groundwater levels in the long term, the groundwater pressures will have to be estimated. In that case, a worst-case scenario might be to consider hydrostatic pressure with a phreatic surface at or near ground level. This might apply to matrix-dominated tills, but these tills also contain pockets and lenses of water-bearing sands and gravels. These are a particular problem if encountered during excavations or in open holes for piling, especially if they are connected to a source of water. Therefore, locating these lenses and establishing continuity are essential. If these pockets are encountered, a water strike will be noted. The water level may rise rapidly up the borehole, but this should not be read as a measure of groundwater pressure since it may be a confined layer, that is, an aquifer. Further, if it does rise up the borehole, it should not be considered a measure of the groundwater pressure because it may be a confined pocket.

As a matter of routine, any water strikes in exploratory boreholes should be noted and the standing level recorded sometime later. In clast-dominated glacial soils, this will provide an indication of groundwater pressures, but not seasonal pressures. Therefore, it is necessary to install piezometers and monitor them through a full seasonal cycle. In matrix-dominated tills, there may be no water strikes during drilling but that does not mean no groundwater pressure. Therefore, piezometers are essential. A key issue in glaciolacustrine clays is that the drilling process can smear the sides of the borehole altering the rate of inflow as the hydraulic conductivity of glaciolacustrine clays is highly anisotropic; the horizontal conductivity far exceeds the vertical conductivity. Therefore, as groundwater conditions are critical,

- An investigation should be designed to measure water pressure at several depths to identify the groundwater profile to determine phreatic surfaces, aquifers and aquitards.
- Seasonal changes in the groundwater profile should be determined.

These are relevant to construction and design. The alternative is to assume the worst-case credible conditions.

3.6.2 Field tests

Field tests can be carried out in all glacial soils and there are advantages to using field tests in glacial soils:

- They can be used in those soils that are difficult to sample such as clast-dominated tills and glaciofluvial sands and gravels.
- They can be used where sampling disturbance can affect the test results such as permeability assessment of glaciolacustrine clays.

- Some field tests test larger volume of soils, which may be relevant in composite soils where clast size can have an impact on the results of laboratory tests.
- More frequent tests and possibly a near continuous record can be obtained, which is useful in such spatially glacial soils.
- Tests can be used to identify zones for representative sampling.

There are disadvantages of using field tests in glacial soils:

- In many tests, the soil type has to be inferred from borehole samples, and given the spatial variability of glacial soils it means that the interpretation of field tests may be incorrect if they depend on knowledge of the soil being tested unless a specimen of the soil tested (as with the standard penetration test [SPT]) can be retrieved.
- Test results are dependent on the *in situ* permeability as it cannot be assumed that tests are fully drained or fully undrained, which may be more relevant with composite soils than coarse- or fine-grained soils.
- The fact that many glacial soils are truly anisotropic, or at least cross anisotropic, means that the test results depend on the direction of loading in relation to the *in situ* stress regime.
- There may be some disturbance to the soil before a test is carried out due to the formation of the test pocket.

Normally, only one field test is carried out on a volume of soil unlike laboratory tests where several tests on a sample may be possible. As with planning borehole locations, boreholes in the preliminary investigation should be used to position the field tests to maximise the information. Field tests include destructive tests (e.g. penetrometers, pressuremeter tests and other tests) in which the soil fabric is destroyed during testing, non-destructive tests (e.g. geophysical tests) and tests to assess groundwater. The confidence that *in situ*, intrusive tests can be used in glacial soils and the parameters that can be derived from the results are listed in Table 3.7. It shows that at least one form of these tests can be used to determine the geotechnical characteristics, and in some glacial soils, this may be the only means of obtaining relevant information. The results of *in situ* tests and their applicability to glacial soils are given in Table 3.8.

3.6.2.1 Penetration tests

The first field test was the penetrometer test, which can be used to produce a profile of ground resistance either from frequent tests or semi-continuous records. Penetrometers are either hammered or pushed into the ground and at least one form of penetrometer can be used in the diverse range of glacial soils.

3.6.2.1.1 Standard penetration tests

The standard penetrometer test (BS EN ISO 22476-3, BS EN 1997) is either a thick-walled sampling tube driven into matrix-dominated soils or a cone driven into clast-dominated soils though the latter is no longer recommended. It is used to measure the relative density of coarse-grained soils from which an estimate of the angle of friction of the soil can be made provided the test is carried out according to the specification and the relevant correlation is used. It is also used as a means of measuring the strength index of matrix-dominated soils, but note that the blow count may be affected by any clasts encountered during driving. This may explain the typical scatter in N_{60} profiles (Figure 4.25).

Table 3.7 In situ tests, their relevance and suitability in glacial soils

				Soil parameters											Soil type				Till	
Group	*Device*	*Soil type*	*Profile*	u	Φ'	c_u	I_D	m_v	c_v	K	G_o	σ_h	*OCR*	σ–ε	*Gravel*	*Sand*	*Silt*	*Clay*	*Matrix*	*Clast*
Penetro-meters	DPT	C	B	–	C	C	C	–	–	–	C	–	C	–	B	A	B	B	A	B
	Mechanical	B	A/B	–	C	C	B	C			C	C	C	–	C	A	A	A	A	B
	CPT	B	A	–	C	B	A/B	C	–	–	B	B/C	B	–	C	A	A	A	B	C
	CPTU	A	A	A	B	B	A/B	B	A/B	B	B	B/C	B	C	–	A	A	A	B	C
	Seismic	A	A	A	B	A/B	A/B	B	A/B	B	A	B	B	B	–	A	A	A	B	C
	SPT	A	B		C	C	B	–	–	–	C	–	C	–	A	A	A	A	A	C
	Resistivity	B	B	–	B	C	A	C	–	–	–	–	–	–	A	A	A	A	B	C
Pressure-meters	PBP	B	B		C	B	C	B	C	–	B	C	C	A	B	B	A	B	B	C
	SBP	B	B	A	B	B	B	B	A	B	A	A/B	B	A/B	B	B	A	B	C	–
	FDP	B	B	–	C	B	C	C	C	–	A	C	C	C	B	B	A	A	C	–
Others	Vane	B	C	–	–	A	–	–	–	–	–	–	B/C	B	–	–	A	B	–	–
	Plate	C	–	–	C	B	B	B	C	C	A	C	B	B	B	A	A	A	A	A
	DMT	B	A	–		B	B	B	–	–	–	B	–	–	C	B	A	A	B	–

Source: After Lunne, T., P. K. Robertson, and J. J. M. Powell. *Cone Penetration Testing in Geotechnical Practice*, Chapman and Hall, London, 1997.

Note: A, high; B, moderate; C, low; –, not applicable.

Table 3.8 Field test results of geotechnical standards

		Glacial soil applicability			
		Primary deposits		*Secondary deposits*	
Field test	*Test results*	*Matrix-dominated tills*	*Clast-dominated tills*	*Glaciofluvial deposits*	*Lacustrine deposits*
Cone penetration test (CPT)	Cone penetration resistance (q_c) Local unit side friction (*fs*) Friction ratio (R_f)	Depends on stone content and strength of till	Unlikely because of gravel content	Only in sand deposits with a limited amount of gravel	Possible
Piezocone (CPTU)	Corrected cone resistance (q_t) Local unit side friction (f_s) Measured pore pressure (u)	Depends on stone content and strength of till	Unlikely because of gravel content	Of no additional value unless there is a significant fines content	Able to identify the layers
Dynamic probing	Number of blows N_{10} for the following tests: DPL, DPM, DPH Number of blows (N_{10}) or (N_{20}) for the DPSH test	Possible but results will be strongly influenced by coarse particles; may have to limit the blow count	Limited possibility especially if very dense; may have to limit blow count	Limited possibility especially if dense gravel	Possible
Standard penetration test (SPT)	Number of blows N Energy correction E_r Soil description	Possible but results will be strongly influenced by coarse particles; may have to limit the blow count	Limited possibility especially if very dense; may have to limit blow count	Limited possibility especially if dense gravel	Possible but other tests may be more appropriate
Ménard pressuremeter test (MPM)	Pressuremeter modulus (E_M) Creep pressure (p_f) Limit pressure (p_{LM}) Expansion curve	Results depend on the quality of the test pocket which is affected in the same way as sampling	May be difficult to create a suitable test pocket because of the stone content	May be difficult to create a suitable test pocket because of the stone content	Possible
Prebored pressuremeter test	Dilatometer modulus (E_{FDT}) Deformation curve	Results depend on the quality of the test pocket which is affected in the same way as sampling	May be difficult to create a suitable test pocket because of the stone content	May be difficult to create a suitable test pocket because of the stone content	Possible
Self-bored pressuremeter test	Expansion curve	Only possible in homogenised tills	Not possible	Only possible in sands	Possible
Full displacement pressuremeter test	Expansion curve	Only possible in homogenised tills; depends on strength	Not possible	Only possible in sands	Possible

(*Continued*)

Table 3.8 (Continued) Field test results of geotechnical standards

		Glacial soil applicability			
		Primary deposits		*Secondary deposits*	
Field test	*Test results*	*Matrix-dominated tills*	*Clast-dominated tills*	*Glaciofluvial deposits*	*Lacustrine deposits*
Field vane test	Undrained shear strength (uncorrected) (c_{fv}) Remoulded undrained shear strength (c_{rv}) Torque rotation curve	Not possible	Not possible	Not possible	Possible
Plate loading test	Ultimate contact pressure (p_u)	Possible	Possible but results may be invalid if clasts are too large	Possible but results may be invalid if very coarse particles are present	Possible
Flat dilatometer test (DMT)	Corrected lift off pressure (p_0) Corrected expansion pressure (p_1) at 1.1 mm expansion Dilatometer modulus E_{DMT} Material index (I_{DMT}) Horizontal stress index (K_{DMT})	Only possible in homogenised tills; depends on strength; results could be affected by large gravel	Not possible	Only possible in sands	Possible

Source: After BS EN 1997-2:2007. *Eurocode 7: Geotechnical Design – Part 2: Ground Investigation and Testing (Incorporating Corrigendum 2010)*. British Standards Institution, London.

The test is carried out in a borehole, typically at 1.5 m intervals of depth or when there is a change in strata, particularly important when testing tills that contain tectonic features. More frequent testing in glacial soils is recommended to take account of the spatial variability. A standard 50 mm outside diameter 'split-spoon' penetrometer is driven into the soil using repeated blows of a 63.5 kg weight falling through 760 mm. The N_{60} value is the number of blows required to achieve a penetration of 300 mm, after an initial seating drive of 150 mm. This value, corrected for standard hammer energy and overburden pressure, is used with empirical correlations, to estimate the stiffness and strength of soils. While there is an international procedure for the test that is carried out from the base of a borehole, the results are dependent on the quality of the drilling, especially in clast-dominated and coarse-grained soils, where the soil can be loosened or compacted. Hence, N_{60} values from such soils need to be treated with caution if used for design parameters. Powell and Clayton (2012) suggest that a small diameter, uncased, carefully drilled borehole full of water at all times reduces the disturbance of silts, sands and gravels.

Design parameters from N_{60} are based on empirical correlations. If these are used as generic correlations rather than site-specific correlations, then it is important to ensure that a standard procedure has been followed and the ground conditions are similar to those for which the correlations were developed. The latter is dealt with through extensive publications of results. The International Reference Test Procedure (IRTP, 1999) suggests that N_{60} values should be corrected to 60% of the free-fall energy, N_{60}, which is current British practice. However, given the variations in equipment and procedure, it is prudent to treat N_{60} values with caution and not to use them as the sole design input. N_{60} values are dependent on the effective angle of friction (granular soils), the relative density (granular soils), effective stress level (granular soils), grain size (coarse granular soils and silty granular soils), undrained shear strength (cohesive soils), cementing (weak rocks, granular soils) and jointing (weak rocks). This means that the soil type must be known before a correlation can be applied. Most N_{60} correlations are based on sands, yet many glacial soils are composite soils, which implies that the correlations may not be correct. The standard correlations and interpretations used in granular soil are for sands. Correlations obtained for sands cannot be assumed to apply to gravels or coarse-grained soils with a percentage of fine-grained particles, that is, matrix-dominated tills. For example, N_{60} should be reduced by 55:60 for fine sands and increased by 65:60 for coarse sands (BS EN ISO 22476-3:2005+A1:2011).

A correction factor for N_{60} in sands for the effect of the overburden pressure is given in Table 3.9. N_{60} in clays is a function of the undrained strength; therefore, the N_{60} value is not corrected for overburden pressure in clays. Design methods based on N_{60} may not state whether a corrected value is used, so care must be taken when using such methods.

Table 3.9 Correction factor for N_{60} in sands for the effect of the overburden pressure

Type of consolidation	*Density index, I_D*	*Correction factor, C_N*
Normally consolidated	40–60	$\frac{200}{100+\sigma'_v}$
	60–80	$\frac{300}{200+\sigma'_v}$
Over-consolidated	–	$\frac{170}{70+\sigma'_v}$

Source: After BS EN ISO 22476-3:2005+A1:2011. *Geotechnical Investigation and Testing. Field Testing. Standard Penetration Test.* British Standards Institution, London.

Table 3.10 Correlation between the density index, I_D, and N_{60}

	Very Loose	*Loose*	*Medium*	*Dense*	*Very dense*
I_D	0–15	15–35	35–65	65–85	85–100
N_{60}	0–3	3–8	8–25	25–42	42–58

Source: After BS EN 1997-2:2007. Eurocode 7: *Geotechnical Design – Part 2: Ground Investigation and Testing (Incorporating Corrigendum 2010)*. British Standards Institution, London.

An example of a relationship between the blow count (N_{60}), density index, I_D ($=(e_{max} - e)/(e_{max} - e_{min})$) and the effective vertical stress, σ_v', is

$$\frac{N_{60}}{I_D^2} = a + b\sigma_v' \tag{3.1}$$

The parameters a and b for normally consolidated sands are nearly constant for $0.35 < I_D < 0.85$ and $50 \text{ kPa} < \sigma_v' < 250 \text{ kPa}$. The b factor is increased to allow for over-consolidation by $[(1 + 2K_o)/(1 + 2K_{NC})]$, where values of a and b are given by Skempton (1986).

Table 3.10 gives a relationship between N_{60} and I_D for normally consolidated natural sand deposits; Table 3.11 gives a relationship between N_{60} and φ'.

3.6.2.1.2 Dynamic probing

The dynamic probe is a low cost, simple, rapid *in situ* test used to obtain profiles of the number of blows every 10–20 cm of a standard weight falling a standard height to drive the cone a certain distance. It is used to explore near-surface deposits as the depth is limited because of the energy used to drive the probe into the ground. It can be used in areas of restricted access because it is light and portable. There are five types of probes (Table 3.12) in use depending on the strength of the soil. The results are affected by gravels and cobbles, so apart from profiles of lacustrine deposits, profiles of dynamic probing test (DPT) in glacial soils are likely to produce scattered profiles of blow count. In the United Kingdom, this probe is used as a profiling tool to provide preliminary information but not as a test to produce geotechnical characteristics. This is not the case in other countries where correlations with geotechnical properties have been developed.

The results are recorded as the number of blows needed to drive the probe 10 cm (N_{10}) or 20 cm (N_{20}). The blow count can be converted into unit cone resistance, r_d, or dynamic

Table 3.11 Correlation between the density index, I_D, and the angle of friction, φ', for silica sands

Density index, I_D	*Fine*		*Medium*		*Coarse*	
%	*Uniform*	*Well graded*	*Uniform*	*Well graded*	*Uniform*	*Well graded*
40	34	36	36	38	38	41
60	36	38	38	41	41	43
80	39	41	41	43	43	44
100	42	43	43	44	44	46

Source: After BS EN 1997-2:2007. *Eurocode 7: Geotechnical Design – Part 2: Ground Investigation and Testing (Incorporating Corrigendum 2010)*. British Standards Institution, London.

Table 3.12 Types of dynamic probes

Factor	*Test specification*				
	DPL	*DPM*	*DPH*	*DPSH-A*	*DPSH-B*
Hammer mass (kg)	10 ± 0.1	30 ± 0.3	50 ± 0.5	63.5 ± 0.5	63.5 ± 0.5
Mass of anvil and rod guide (kg)	0.5 ± 0.01	0.5 ± 0.01	0.5 ± 0.01	0.5 ± 0.01	0.75 ± 0.02
Rebound (max) (%)	6	18	18	18	30
Rod length (m)	1 ± 0.1%	1 ± 0.1%	1 ± 0.1%	1 ± 0.1%	1 ± 0.1%
Mass of rod (max) (kg)	3	6	6	8	8
Rod eccentricity (max) (mm)	0.2	0.2	0.2	0.2	0.2
Rod OD (mm)	22 ± 0.2	32 ± 0.2	32 ± 0.2	32 ± 0.2	35 ± 0.2
Rod ID (mm)	6 ± 0.2	9 ± 0.2	9 ± 0.2	9 ± 0.2	9 ± 0.2
Cone apex angle	90°	90°	90°	90°	90°
Cone area (mm²)	10	15	15	16	20
Cone diameter (new) (mm)	35.7 ± 0.3	43.7 ± 0.3	43.7 ± 0.3	45.0 ± 0.3	50.0 ± 1.0
Cone diameter (worn) (mm)	34	42	42	43	49
Mantle length (mm)	35.7 ± 1	43.7 ± 1	43.7 ± 1	90 ± 2	50.5 ± 2
Blow count interval	10	10	10	20	20
Standard range of blows	3–50	3–50	3–50	5–100	5–100
Specific work/blow (kJ/m)	50	98	167	194	238

Source: After BS EN ISO 22476-2:2005+A1:2011. *Geotechnical Investigation and Testing. Field Testing. Dynamic Probing.* British Standards Institution, London.

cone resistance, q_d, to normalise the results allowing comparisons between different probes to be made:

$$q_d = \frac{M}{M + M'} r_d \tag{3.2}$$

$$r_d = \frac{Mgh}{Ae} \tag{3.3}$$

where M is the mass of the hammer in kilograms, h is the height of fall of the hammer in metres, A is the projected area of the cone in m^2, e is the average penetration in metres per blow ($0.1/N_{10}$ from DPL, DPM15, DPM and DPH, and $0.2/N_{20}$ from DPSH), and M' is the total mass of the extension rods, the anvil and the guiding rods in kilograms.

Since the probe is driven from the top of the rods, it is likely that the driving rods will be forced to bend, thus increasing the number of blows needed. Experience has shown that torque readings in excess of 200 Nm generally mean that the driving rods have been forced off-line and it is suggested that tests should be terminated when a torque reading reaches 120 Nm. BS EN ISO 22476-2:2005 suggests that the results depend on the density, the grain structure, the grain size distribution, the grain shape and grain roughness, the mineral type, the degree of cementation and the strain condition for coarse-grained soils and density and rod friction for fine-grained soils. Blow counts below groundwater level are lower than those above water level in coarse-grained soils. In coarse-grained soils,

- The penetration resistance increases linearly with increasing density index of the soil.
- Angular soils possess a higher penetration resistance than soils with round and smooth particles.

Table 3.13 Examples of density index, I_D, from the DPT for different values of uniformity coefficient, C_U, ($3 < N_{10} < 50$)

Soil type	C_U	Groundwater	DPT	
			Light	Heavy
Poorly graded sand	<3	Above	$I_D = 0.15 + 0.26 \log N_{10}$	$I_D = 0.10 + 0.435 \log N_{10}$
Poorly graded sand	<3	Below	$I_D = 0.21 + 0.23 \log N_{10}$	$I_D = 0.23 + 0.380 \log N_{10}$
Well-graded sand and gravel	>6	Above	–	$I_D = -0.14 + 0.550 \log N_{10}$

Source: After BS EN 1997-2:2007. *Eurocode 7: Geotechnical Design – Part 2: Ground Investigation and Testing (Incorporating Corrigendum 2010)*. British Standards Institution, London.

- Cobbles and boulders can significantly increase the penetration resistance.
- Particle size distribution (uniformity coefficient and grading curve) influences the penetration resistance.
- Penetration resistance is considerably increased by cementation.
- Penetration resistance increases when there are thin layers with embedded cobbles; locally occurring peaks of penetration resistance do not represent a measure of the bearing capacity of the whole layer.
- The fluctuations are greater in soils with mixed grain sizes (e.g. glacial soils) owing to the higher proportion of coarse grains.

Table 3.13 gives examples of density index, I_D, and the blow count and Table 3.14 angle of friction. The stress-dependent oedometer modulus, E_{oed}, can be found from DPT results using

$$E_{oed} = 100w_1\left(\frac{\sigma'_v + 0.5\Delta\sigma'_v}{100}\right)^{w_2} \tag{3.4}$$

where w_1 is a stiffness coefficient; w_2 is a stiffness exponent; σ'_v is the effective vertical stress at the base of the foundation or at any depth below it due to overburden of the soil; $\Delta\sigma'_v$ is the increase in effective vertical stress caused by the structure at the base of the foundation or at any depth below it; I_p is the plasticity index; and w_L is the liquid limit. For sands with a uniformity coefficient $C_U \le 3$, $w_2 = 0.5$; for clays of low plasticity ($I_p \le 10$; $w_L \le 35$), $w_2 = 0.6$. Values for the stiffness coefficient (w_1) can be derived from DPT using Table 3.15.

Table 3.14 Correlation of angle of friction of coarse soil with density index and uniformity coefficient

Soil type	Grading	Range of ID		Angle of friction
Slightly fine-grained sand; sand; sand and gravel	Poorly graded ($C_U < 6$)	15–35	Loose	30
		35–65	Medium dense	32.5
		>65	Dense	35
Sand; sand and gravel; gravel	Well graded ($6 < C_U < 15$)	15–35	Loose	30
		35–65	Medium dense	34
		>65	Dense	38

Source: After BS EN 1997-2:2007. *Eurocode 7: Geotechnical Design – Part 2: Ground Investigation and Testing (Incorporating Corrigendum 2010)*. British Standards Institution, London.

Table 3.15 Stiffness coefficient, w_l, from DPT tests used to determine the oedometer modulus, E_{oed}

Soil type	*Groundwater*	*DPT*	
		Light	*Heavy*
Poorly graded sand $C_U < 3$	Above	$w_l = 71 + 24 \log N_{10}$ $4 < N_{10} < 50$	$w_l = 161 + 249 \log N_{10}$ $3 < N_{10} < 10$
Low plasticity stiff clays $0.75 < I_c < 1.30$	Above	$w_l = 30 + 4 \log N_{10}$ $6 < N_{10} < 19$	$w_l = 50 + 6 \log N_{10}$ $3 < N_{10} < 13$

Source: After BS EN 1997-2:2007. *Eurocode 7: Geotechnical Design – Part 2: Ground Investigation and Testing (Incorporating Corrigendum 2010)*. British Standards Institution, London.

3.6.2.1.3 The cone penetration test

The more advanced and more versatile penetrometer is the static cone penetrometer, which is a cone pushed at 20 ± 5 mm/s into the ground to rapidly give a semi-continuous profile of resistance, which can be used to classify the ground and determine a wide range of geotechnical parameters. The test is covered by IRTP (1999) and described in BS EN ISO 22476-1. More details are given by Lunne et al. (1997) and Meigh (2013). It can be pushed from the surface in suitable soils, but in those soils containing very coarse particles, for example, it may be necessary to operate the cone in conjunction with a drilling rig.

The diameter of the standard 60° cone is 35.7 mm (cross-sectional area of 10 cm^2) and the area of the friction sleeve is 150 cm^2. The cone is fitted with sensors to measure, for example, tip resistance, side friction resistance and pore pressure. Specialist cones such as acoustic, resistivity, pressuremeter and environmental cones do exist. As with all penetrometers if the results are to have any value, the equipment and procedure must comply with the specification; a summary is presented in Table 3.16. The piezocone (CPTU) is particularly useful in lacustrine deposits as they may indicate the thickness of the varves. Electric cones are more susceptible to damage, hence the need to be aware of the ground conditions prior to the test. Clasts will tend to deflect the cone but if the cone is fitted with an inclinometer, recommended for profiles in excess of 15 m, a correction can be made for depth. It can be used in all glacial soils provided there are a limited number of larger particles and the soil is not too dense. In both cases, it will not be possible to push the probe into the soil without damaging the probe, again emphasising the importance of a preliminary investigation to determine the site-specific stratigraphy. Static cone tests can be carried out from the base of predrilled holes, which can be useful when there are different layers of glacial soils, some of which may stop the cone because of the particle size or density.

The pore pressure may be measured at the cone tip, behind the cone shoulder or above the friction sleeve though it is normal to measure it just behind the cone shoulder as it is less likely to be damaged and relatively easy to saturate, very important when using the CPTU as a profiling tool in clays. The CPTU can be used to determine the coefficient of consolidation by carrying out a pore pressure dissipation test.

The total force acting on the cone tip divided by the projected area of the cone gives the cone resistance, q_c, and if pore pressure is measured, the corrected cone resistance ($q_t = q_c + u(1 - a)$), where a is the area ratio and u the pore pressure immediately behind the cone. The total force acting on the friction sleeve divided by its surface area gives the sleeve friction resistance, f_s. The friction ratio, R_f, is the ratio of the sleeve friction resistance to the cone resistance. A depth correction is also applied if the cone deviates from the vertical.

Table 3.16 Application classes for electrical cone and piezocone penetration testing

Class	*Soil type*	*Test type*	*Measured parameter*	*Allowable minimum accuracy*[a]	*Maximum length between measurements*	*Use*	
						soil[b]	*Interpretation/ evaluation*[c]
1	Soft to very soft soil deposits; not suitable for mixed bedded soil profiles with soft to dense layers	TE2	Cone resistance	35 kPa or 5%	20 mm	A	G, H
			Sleeve friction	5 kPa or 10%			
			Pore pressure	10 kPa or 2%			
			Inclination	2°			
			Penetration length	0.1 m or 1%			
2	Mixed bedded soil profiles with soft to dense layers	TE1	Cone resistance	100 kPa or 5%	20 mm	A	G, H*
		TE2	Sleeve friction	15 kPa or 15%		B	G, H
			Pore pressure[d]	25 kPa or 3%		C	G, H
			Inclination	2°		D	G, H
			Penetration length	0.1 m or 1%			
3	Mixed bedded soil profiles with soft to dense soils	TE1	Cone resistance	200 kPa or 5%	50 mm	A	G
		TE2	Sleeve friction	25 kPa or 15%		B	G, H*
			Pore pressure[d]	50 kPa or 5%		C	G, H
			Inclination	5°		D	G, H
			Penetration length	0.2 m or 2%			
4	Mixed bedded soil profiles with soft to very stiff or loose to dense layers	TE1	Cone resistance	500 kPa or 5%	50 mm	A	G*
			Sleeve friction	50 kPa or 20%		B	G*
			Penetration length	0.2 m or 2%		C	G*
						D	G*

Source: After BS EN ISO 22476-1:2012. *Geotechnical Investigation and Testing. Field Testing. Electrical Cone and Piezocone Penetration Test.* British Standards Institution, London.

[a] The allowable minimum accuracy of the measured parameter is the larger value of the two quoted. The relative accuracy applies to the measured value and not the measuring range.

[b] According to ISO 14688-2: A: homogeneously bedded soils with very soft to stiff clays and silts (typically $q_c < 3$ MPa); B: mixed bedded soils with soft to stiff clays (typically $q_c \leq 3$ MPa) and medium dense sands (typically 5 MPa $\leq q_c <$ 10 MPa); C: mixed bedded soils with stiff clays (typically 1.5 MPa $\leq q_c <$ 3 MPa) and very dense sands (typically $q_c > 20$ MPa); D: very stiff to hard clays (typically $q_c \geq 3$ MPa) and very dense coarse soils ($q_c \geq 20$ MPa).

[c] G: profiling and material identification with low associated uncertainty level; G*: indicative profiling and material identification with high associated uncertainty level; H: interpretation in terms of design with low associated uncertainty level; H*: indicative interpretation in terms of design with high associated uncertainty level.

[d] Pore pressure can only be measured if TE2 is used. TE1, cone resistance and sleeve friction; TE2, cone resistance, sleeve friction and pore pressure.

Table 3.17 Deriving φ′ and *E*′ from CPT tests in quartz and feldspar sands

I_D (%)	q_c (MPa)	φ′	E′ (MPa)
Very loose	0–2.5	29–32	<10
Loose	2.5–5	32–35	10–20
Medium dense	5–10	35–37	20–30
Dense	10–20	37–40	30–60
Very dense	>20	40–42	60–90

Source: After Bergdahl, U., E. Ottosson, and B. Stigson Malmborg. *Plattgrundläggning*. Stockholm: Svensk Byggtjänst. ISBN91-7332-662-3, 1993.

Note: Angle of friction – values are given for sands; reduce by 3° for silty sands and increase by 2° for gravels; Drained modulus – likely to be lower in silty sands and higher in gravels.

Soil parameters are best assessed from site-specific correlations though there are published generic correlations (e.g. Table 3.17). The oedometer modulus can be determined from Equation 3.4 using Table 3.18.

Empirical relations between strength and stiffness and the cone resistance exist. For example,

$$c_u = \frac{q_c - \sigma_v}{N_k} \tag{3.5}$$

$$E_{oed} = \alpha q_c \tag{3.6}$$

where N_k is a cone factor and α is a coefficient given in Table 3.19.

There are many published profiles of penetration resistance as static cones are used in glacial soils to characterise the deposits. There are a number of papers that demonstrate appropriate use of cones (e.g. Baker and Gardener, 1989; Dobie, 1989; Hird and Springman, 2006), and other examples are given in the chapters covering geotechnical design. Baker and Gardener (1989) reported (Figure 3.3) profiles of piezocone, temperature, conductivity and seismic cones to detect thin sandy horizons in a clay glacial till in Northern England. Dobie (1989) found that cone penetration test (CPT) and SPT tests (Figure 3.4) provided more consistent results of the undrained shear strength of a matrix-dominated till than those from undrained triaxial tests on 102 mm specimens. He used an N_k factor of 18 but found a representative range of 15–22 (Figure 3.5). Hird and Springman (2006) undertook an investigation in a deep deposit of glacial lacustrine clay using piezocones with cross-sectional

Table 3.18 Stiffness coefficient, w_l, from CPT tests used to determine the oedometer modulus, E_{oed}

Soil type	Groundwater	w_l
Poorly graded sand $C_U < 3$	Above	$w_l = 113 + 167 \log q_c$ 5 MPa < q_c < 30 MPa
Well-graded sand $C_U < 6$	Above	$w_l = -13 + 463 \log q_c$ 5 MPa < q_c < 30 MPa
Low plasticity stiff clays $0.75 < I_c < 1.30$	Above	$w_l = 50 + 15.2\, q_c$ 0.6 MPa < q_c < 3.5 MPa

Source: After BS EN 1997-2:2007. *Eurocode 7: Geotechnical Design – Part 2: Ground Investigation and Testing (Incorporating Corrigendum 2010)*. British Standards Institution, London.

Table 3.19 Empirical coefficient, α, used to determine the oedometer modulus, E_{oed}

Soil	*Cone resistance*	$E_{oed} = \alpha\ q_c$
Low plasticity clay	$q_c < 0.7$ MPa	$3 < \alpha < 8$
	$0.7 < q_c < 2$ MPa	$2 < \alpha < 5$
	$q_c > 2$ MPa	$1 < \alpha < 2.5$
Low plasticity silt	$q_c < 2$ MPa	$3 < \alpha < 6$
	$q_c > 2$ MPa	$1 < \alpha < 2$
Very plastic clay	$q_c < 2$ MPa	$2 < \alpha < 6$
Very plastic silt	$q_c > 2$ MPa	$1 < \alpha < 2$
Sands	$q_c < 5$ MPa	$\alpha = 2$
	$q_c > 10$ MPa	$\alpha = 1.5$

Source: After BS EN 1997-2:2007. *Eurocode 7: Geotechnical Design – Part 2: Ground Investigation and Testing (Incorporating Corrigendum 2010).* British Standards Institution, London.

areas of 5 and 10 cm². They found that a 5 cm² piezocone was better at detecting the thin silty layers, as thin as 2–4 mm, than a 10 cm². There were no significant differences between the magnitudes of the cone resistance and excess pore pressure recorded in the clay between the two cones. Pore pressure dissipation test results were variable, but in a region where no silt layers were detected, similar results were obtained with piezocones of each size. They did highlight the need to check for hard layers within the glacial soils, which could damage a 5 cm² cone by using a 10 cm² first.

3.6.2.2 *Pressuremeter tests*

The aim of a pressuremeter test, a test in which a cylindrical flexible membrane is inflated within the soil, is to obtain the stiffness, and in weaker materials the strength, of the ground, by measuring the relationship between the applied radial pressure and the resulting deformation. There are three categories of pressuremeters (Table 3.20), which are based on the concept of an expanding cylindrical membrane. Pressuremeters can be inserted in a predrilled borehole (Menard and prebored pressuremeters), self-drilled (self-boring pressuremeters [SBPs]) or push-in (full displacement pressuremeters). The expansion of the probe can be pressure or displacement controlled, and the expansion can be measured with volume or displacement transducers. Pressuremeter tests can be used directly in design (e.g. Menard pressuremeter guidelines) or to produce the mechanical characteristics of a soil. More details are given by Clarke (1994), Mair and Wood (1987) and Baguelin et al. (1978).

The original pressuremeter, the Menard pressuremeter, is lowered down a predrilled borehole. The quality of the results depends on the quality of the borehole. Normally, a special pocket is drilled ahead of the casing about 10% larger diameter than the probe. Ideally, the predrilled pocket should be uniform, but any clast will affect the quality of the pocket (cf. sampling). Clasts affect the test because they affect the diameter of the test pocket and influence the expansion of the membrane. Further, the method of measuring the expansion of the membrane is affected by the position of any clasts. It is assumed that the membrane expands as a right circular cylinder, thus volume and displacement measuring devices would, ideally, give the same results. However, clasts mean that this may not be the case. A transducer positioned next to a clast will give a different result to one positioned next to the matrix.

The second group of pressuremeters are those that are drilled into the ground. It is unlikely to be able to drill these into clast-dominated tills or glaciofluvial deposits. They can be used in lacustrine deposits and some matrix-dominated tills with low clast content. Note that the drilling process does not break up clasts, so unless they can pass through the drill string, they

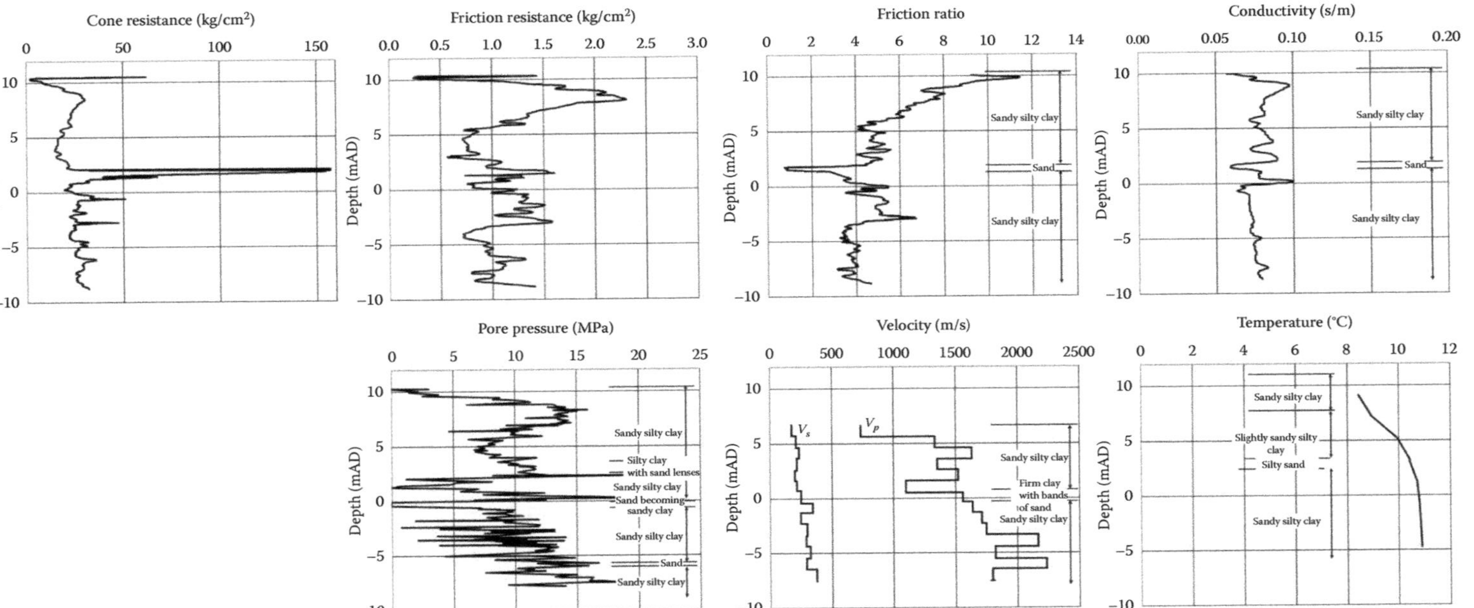

Figure 3.3 Use of piezocone, temperature cone and seismic cones to detect thin sandy horizons in a clay glacial till in Northern England. (After Baker, P. J. and Gardener, R. Penetration testing in glacial till. In *Penetration Testing in the UK, Geotechnology Conference*, Birmingham, United Kingdom, 1989: 223–226.)

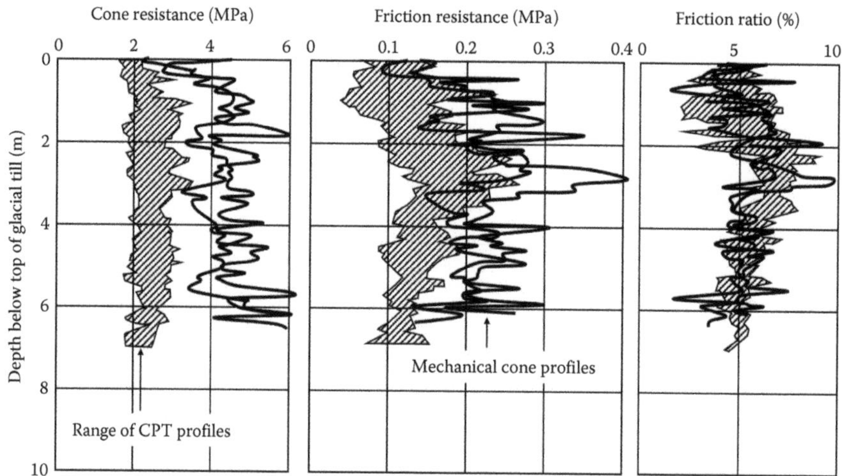

Figure 3.4 Comparison between CPT and mechanical cone profiles in a matrix-dominated till. (After Dobie, M. J. The use of cone penetration tests in glacial till. In Penetration Testing in the UK, Geotechnology Conference, Birmingham, United Kingdom. 1989: 212–222.)

will not be removed. Typically, this means that the SBP cannot drill through tills containing anything larger than medium gravel and only occasional pieces of gravel. In lacustrine deposits, this pressuremeter can be drilled continuously from the top of the deposit carrying out tests every metre. It would be unlikely to do this in matrix-dominated tills unless they were completely homogenised. In matrix-dominated tills, it is better to drill the pressuremeter from the base of a borehole; that is, the borehole is advanced between test positions with percussive or rotary drilling techniques.

The third type of pressuremeter, the full displacement pressuremeter, is a cone, so it can be used in soils in which it is possible to push a cone, that is, lacustrine deposits, glaciofluvial sands and matrix-dominated tills with a limited amount of gravel and nothing larger than gravel. The information from the expansion phase depends on the presence of clasts in relation to the measuring system.

The pressuremeter test involves the expansion of the flexible membrane either at a constant rate of expansion (strain-controlled test) or in pressure increments (stress-controlled

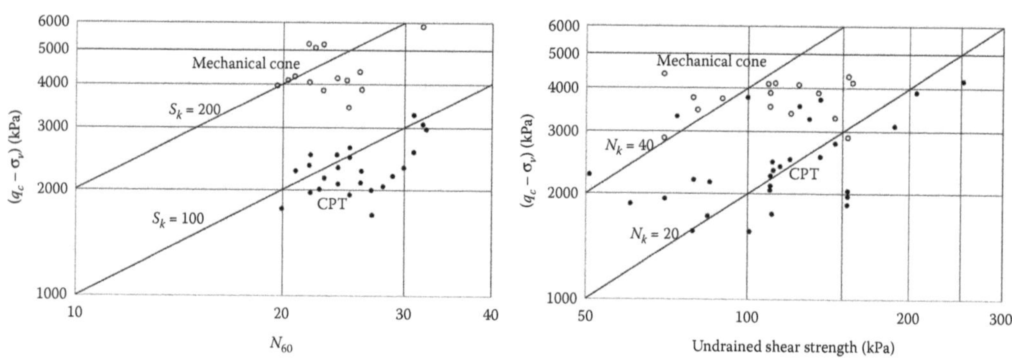

Figure 3.5 Relationships between mechanical cone and CPT resistances and N_{60} and undrained shear strength from tests on 100-mm samples of matrix-dominated till. (After Dobie, 1989.)

Table 3.20 Categories and applications of pressuremeters

Type	*Pressure capacity (MPa)*	*Strain capacity*	*Diameter (mm)*	*Length/diameter expanding section*	*Measuring system*
Menard (MPM) – soil	4	53%–55%	74	6.5	Surface volume
Menard (MPM) – rock	20	53%–55%	74	6.5	Surface volume
Prebored (PBP) – soil	2.5–10	12%–55%	66–88	6.1–7.4	Surface displacement/ radial displacement
Prebored (PBP) – rock	20–100	25%–55%	66–108	6.1–10.5	Surface displacement/ radial displacement
Self-bored (SBP)	4.4–20	10%–15%	84	4.8	Radial displacement
Pushed-in (FDP)	2.5–3.5		66–89	4–5	Surface displacement/ radial displacement

Soil	*State*	*Pressuremeter*	σ_h	c_u	φ'	G_i	G_{ur}	p_l	c_v
Clay	Soft	PBP		B			A	BE	B
		SBP	A	A	B	A	A	A	A
		FDP	CE	B			A	BE	A
	Stiff	PBP	C	B			A	BE	B
		SBP	A	A	B	A	A	A	A
		FDP	CE	B			A	BE	A
Sand	Loose	PBP			CE		A	CE	
		SBP	B		A	A	A	A	
		FDP			CE		A	CE	
	Dense	PBP			CE		A	CE	
		SBP	C		A	A	A	A	
		FDP			CE		A	BE	
Gravel	Loose	PBP			CE		C	CE	
		SBP			N	N	N	N	
		FDP			N	N	N	N	
Matrix-dominated till		PBP		B			A	BE	B
		SBP	B	A	B	B	A	A	A
		FDP	CE	B			A	BE	A
Clast-dominated till		PBP			CE		C	CE	
		SBP	C		N	N	N	N	
		FDP			N	N	N	N	
Glaciolacustrine clay		PBP		B			A	BE	B
		SBP	A	A	B	A	A	A	A
		FDP	CE	B			A	BE	A

Source: After Clarke, B. G. *Pressuremeters in Geotechnical Design*. CRC Press, 1994.

Note: A – excellent; B – good; C – possible; N – not possible; E – empirical.

test). The applied pressure and displacement are measured to produce a stress strain curve (Figure 3.6), which can be interpreted to produce the shear modulus and strength of the soil (undrained strength of matrix-dominated tills and lacustrine deposits; angle of friction in clast-dominated tills and glaciofluvial deposits).

Figure 3.6 shows that the ground response to the three types of pressuremeter is different. SBP tests should represent the ideal situation as a test starts at the horizontal total stress if the probe is installed correctly. The borehole wall is unloaded before a prebored pressuremeter is used and full displacement pressuremeters displace the soil during installation;

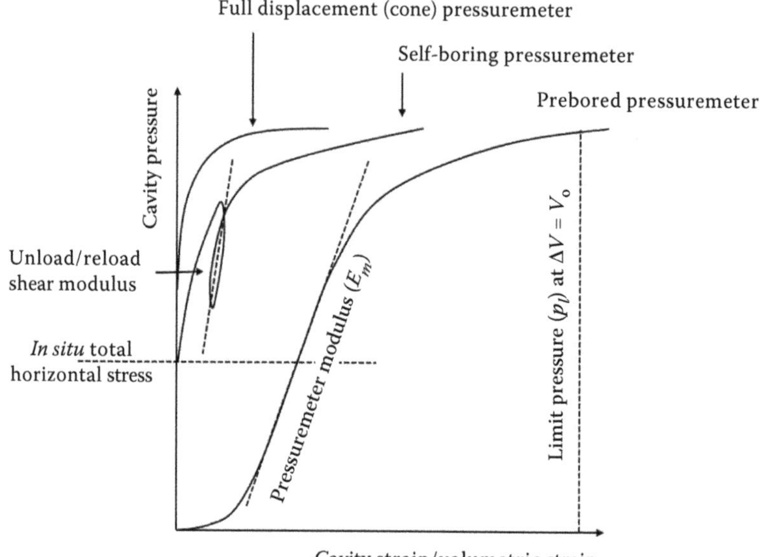

Figure 3.6 Comparison between the test curves from prebored, self-bored and full displacement pressuremeters showing the key parameters of total horizontal stress, shear modulus, pressuremeter modulus and limit pressure.

therefore, a test starts at a higher stress than the *in situ* horizontal stress. The increase in radial stress in all three probes should take the soil to failure though in some of the denser glacial soils this may not be possible. Unload/reload cycles are a useful means of assessing the stiffness of the ground and are independent of probe type. Soil properties can be determined from first principles because the boundary conditions are defined though the prebored pressuremeter (the Menard pressuremeter) was developed to design foundations directly, so the interpretation is semi-empirical. The test has to follow a standard procedure (BS EN ISO 22476-4:2012) to produce the pressuremeter modulus (E_m) and limit pressure (p_l). These two parameters are used in conjunction with design tables and curves to directly produce design of foundations and retaining walls, for example. Analysis of pressuremeter tests has led to numerous studies to derive the stress–strain behaviour of soil to give total horizontal stress, shear modulus and strength (expressed as undrained shear strength, angle of friction or limit pressure). This type of analysis provides intrinsic properties that are used in semi-empirical design methods or numerical studies.

3.6.2.3 *Other intrusive tests*

The third category of *in situ* intrusive tests includes the vane test, the flat dilatometer test and plate loading test.

3.6.2.3.1 *The vane test*

Vane tests (BS EN 1997-2:2007) are used in fine-grained soils (lacustrine deposits) in which it is possible to push the vane into the soil and rotate it to obtain the undrained shear strength. It may be possible to use it in matrix-dominated tills provided they are not too stiff and contain a very little coarse-grained material. A cruciform vane mounted on a solid rod is pushed into the soil, a torque applied to the vane and the rotation and torque measured.

Vane tests can be carried out from the surface or the base of a borehole. The field vane has four rectangular blades and a height-to-diameter ratio (H/D) of two. In the United Kingdom, the test is only considered suitable for clays with strengths less than 75 kPa. The test is generally not suitable for composite soils including matrix-dominated tills and glaciolacustrine clays. The vane test is routinely used to determine the 'undisturbed' peak undrained shear strength and the remoulded undrained shear strength to give an assessment of a soil's sensitivity. It is assumed that the penetration of the vane causes negligible disturbance, that no drainage occurs during shear and that the soil fails on a cylindrical shear surface, whose diameter is equal to the width of the vane blades. The results of a vane shear test may be influenced by many factors, namely,

- Type of soil, especially when a permeable fabric exists or stones are present
- Strength anisotropy
- Disturbance due to insertion of the vane
- Rate of rotation (strain rate)
- Time lapse between insertion of the vane and the beginning of the test
- Progressive failure of the soil around the vane

3.6.2.3.2 Marchetti dilatometer test

The Marchetti dilatometer test (DMT) (BS EN 1997-2:2007; ASTM D6635:2015) can generate profiles of horizontal stress, stiffness and strength of soils relatively quickly. A 250-mm-long, 94-mm-wide and 14-mm-thick blade with a tip angle of 16° is pushed into the ground. It has a flat, 60-mm diameter steel membrane mounted flush on one side, which is used to load the soil, thus obtaining the response of the soil to load. A test is carried out every 0.2 m. Gas pressure is applied to the membrane and the pressures required to bring the membrane flush with the blade and to move it a further 1.1 mm are recorded. The gas pressure is then reduced, and the pressure when the membrane is once again flush with the blade is recorded. These three pressures, corrected for membrane stiffness, are converted to a material index, I_D, the horizontal stress index, K_D, and the dilatometer modulus, E_D, which, through empirical correlations, are related to soil type, shear strength, over-consolidation ratio, stiffness and density.

The DMT is suitable for use in sands, silts and clays, where the grains are small compared to the membrane diameter (60 mm), with a very wide range of strengths, from extremely soft clay to a stiff soil.

3.6.2.3.3 Plate testing

Plate bearing tests (ASTM D1194-72; BS EN ISO 22476-13) can be used at ground level or at the base of an excavation or borehole to determine the strength and stiffness of a soil. The test consists of loading a 300 mm (or larger) diameter rigid metal plate bedded onto the soil in increments of about one-fifth of the design load, holding each increment until the rate of settlement is reduced to an acceptable level – 0.004 mm/min over 60 min (Clayton et al., 1995). The test is terminated when the soil fails or when the contact pressure is twice the design bearing pressure.

Results are presented as time–settlement curves for each applied load and a load–settlement curve for the entire test. A minimum of three tests are required to take account of any variability, though for tests on glacial tills more are required because of their spatial variability. The plate diameter should be at least six times the maximum soil particle size though the stiffness obtained from a test only represents the stiffness of the soil within the zone of influence of the plate. The stiffness, E, is

$$E = \frac{\pi q B}{4} \frac{(1 - \upsilon^2)}{\rho} \tag{3.7}$$

where q is the applied pressure, B is the plate width, ρ is the settlement under the applied pressure and ν is Poisson's ratio. Poisson's ratio will normally be between 0.1 and 0.3 for coarse-grained soils.

3.6.2.4 Geophysical tests

Geophysical tests have potential to determine the soil profile and to detect lines of unconformity (e.g. glacial till/bedrock interface), boulders and other subsurface anomalies and small-strain stiffness. Specialists are needed to choose the most appropriate method, design the array and interpret the results. According to Reynolds (2012), geophysical testing should reduce risk by detecting buried utilities, locating voids and other key features on a site; should reduce uncertainty when used in conjunction with boreholes and *in situ* tests. The primary benefits of geophysical surveying (Reynolds, 1996) include the following:

- Rapid areal coverage (hectares per day)
- Fine spatial resolution (<1 m)
- Volumetric sampling rather than spot measurements
- Non-invasive and environmentally benign nature
- Time-lapse measurements
- Quantitative rather than qualitative data

An assessment of the quality of geophysical techniques and the physical property upon which they depend is given in Table 3.21. Surface geophysics includes potential field methods, electrical methods, electromagnetic methods and seismic methods. Continuous surface wave testing, downhole testing and cross-hole testing can be used to determine the seismic shear wave, velocity from which the shear modulus can be calculated.

Abbiss (1981) described the use of shear wave refraction and surface wave methods at three sites including Cowden. The dynamic moduli G_{dyn} is given by

$$G_{dyn} = \rho V^2 \tag{3.8}$$

where ρ is the mass density and V is the shear wave velocity. These dynamic low-strain values can be corrected to values, G, for longer times and higher strains using

$$G = G_{dyn} \left(\frac{T_o}{T} \right)^{\frac{2}{\pi Q}} \tag{3.9}$$

where T_o represents the time of the dynamic measurement and T the time of interest. Q relates to the damping factor allowing comparison with other *in situ* tests. Abbiss (1983) was able to predict with some accuracy the settlement of plate tests on the Cowden till.

Ku and Mayne (2012) proposed that the K_o profile could be estimated from the small-strain stiffness anisotropy ratio in soils using

$$K_o = (1 - \sin\varphi') \left(\alpha \frac{\sigma_{atm}}{\sigma'_{vo}} \left(\frac{G_{0HH}}{G_{0VH}} \right)^{\beta} + 1 \right)^{\sin\varphi'} \tag{3.10}$$

Table 3.21 Dependent properties for geophysical methods and their applications with reference to glacial soils

Geophysical method		*Physical property*	*Regional studies*	*Depth to bedrock*	*Stratigraphy*	*Lithology*	*Fractured zones*	*Fault displacement*	*Buried channels*	*Natural cavities*	*Groundwater*
Potential field	Gravity	Density	4	1	0	0	0	2	2	4	1
	Magnetics	Susceptibility	4	0	0	0	0	2	1	0	0
Electrical	Resistivity (sounding)	Resistivity	2	4	3	3	2	2	3	2	4
	Resistivity (tomography)	Resistivity	2	3	2	2	4	3	3	3	4
	Induced polarisation	Resistivity; capacitance	2	2	2	3	1	1	2	0	3
	Self-potential	Potential difference	0	0	0	0	2	2	1	1	4
Electromagnetic	FDEM	Conductivity; inductance	4	2	2	2	4	2	3	4	4
	TDEM	Conductivity; inductance	4	2	2	2	3	2	3	1	3
	VLF	Conductivity; inductance	2	0	0	0	1	1	1	2	3
	GPR	Dielectric permittivity; conductivity	0	2	3	1	2	3	2	3	2
Seismic	Refraction	Elastic modulus; density	4	4	3	2	3	4	4	1	2
	Surface wave profiling	Elastic modulus; density	0	3	4	3	4	3	3	2	2
	Reflection	Elastic modulus; density	4	2	2	2	1	2	1	2	2

Source: After Reynolds, J. M. *An Introduction to Applied and Environmental Geophysics*. John Wiley & Sons, Chichester, UK; 2011.

Note: 0, not suitable; 1, limited use; 2, can be used but there are limitations; 3, excellent potential; 4, techniques well-developed and excellent approach.

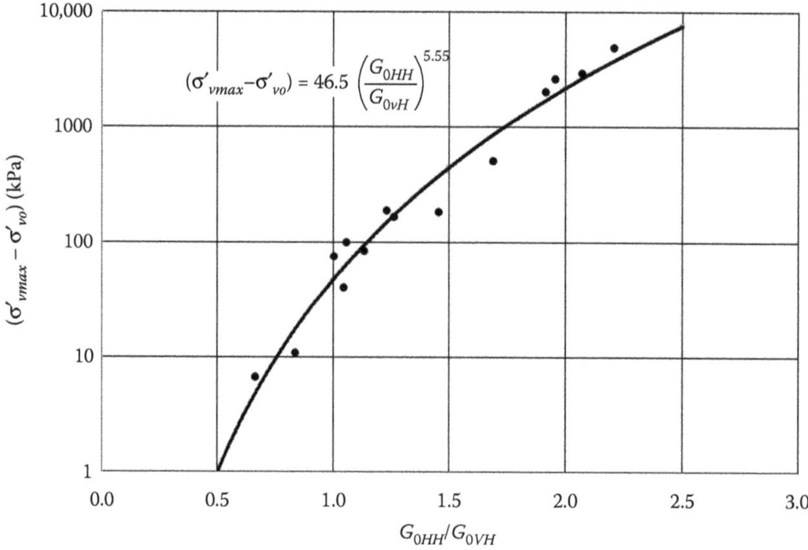

Figure 3.7 Variation in the difference between the preconsolidation stress and the current effective vertical stress and the small-strain stiffness ratio based on a number of soils including glacial till. (After Ku, T. and P. W. Mayne. *Journal of Geotechnical and Geoenvironmental Engineering,* 139(5); 2012: 775–787.)

where σ_{atm} is the atmospheric pressure, σ'_{vo} the vertical effective stress, G_{0HH} the small-strain stiffness in the horizontal plane, G_{0VH} the small-strain stiffness in the vertical plane and α and β soil constants taken from Figure 3.7, results from various sites including the Cowden till site in East Yorkshire. They compared their predicted K_o with results from a variety of field and laboratory tests including SBPs, total stress cells (TSCs), triaxial tests, instrumented consolidometers and suction measurements. The data showed a strong relationship between the K_o predicted from small-strain stiffness measured by geophysical tests and that determined from the other tests (Figure 3.8).

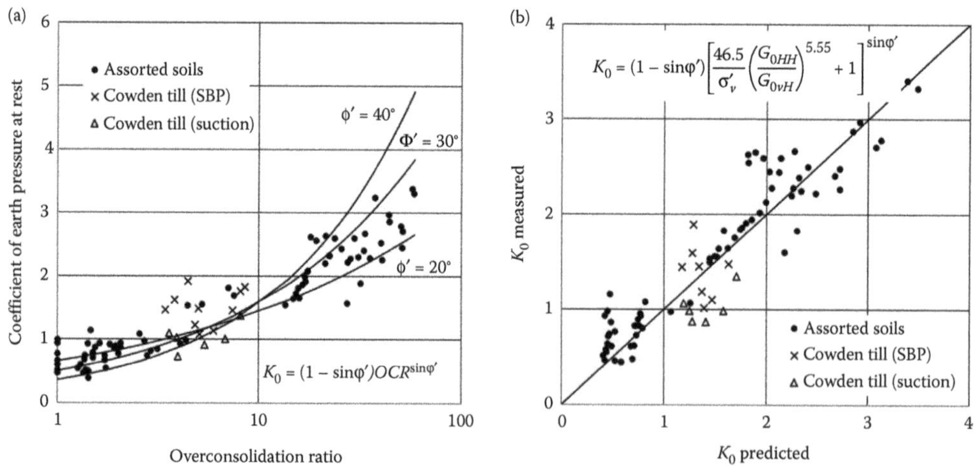

Figure 3.8 Comparisons between K_0 measured from a variety of tests including glacial till and those predicted from (a) the angle of friction and (b) small-strain geophysical tests. (After Ku, T. and P. W. Mayne. *Journal of Geotechnical and Geoenvironmental Engineering,* 139(5); 2012: 775–787.)

Geophysical sensors can also be used in boreholes (Table 3.22) either as (a) single downhole techniques to produce a vertical log of a measured parameter, or (b) cross-hole tomography using two boreholes with sources in one borehole and sensors in the other. Cross-hole tomography use similar techniques to surface geophysics, that is, seismic, electrical or radar methods. The spacing between boreholes should not normally be more than 10 times the minimum dimension of the target being sought. Borehole logging can have a very high vertical resolution but may have a very limited penetration beyond the wall of the borehole.

Donohue et al. (2012) used electromagnetic conductivity mapping, electrical resistivity tomography, seismic refraction and multichannel analysis of surface waves to investigate glaciomarine deposits in Scandinavia and North America to map their occurrence and extent. These results were compared to geotechnical data from laboratory and *in situ* tests. They found that electrical resistivity tomography and electromagnetics were able to delineate the zone of quick clay; seismic refraction was able to assess the sediment distribution and to indicate the presence of shallow bedrock; the multichannel analysis of surface waves highlighted differences between the intact stiffness of quick and unleached clay. They suggested that intrusive exploratory work was still required but could be reduced.

Gibson et al. (2014) used geomorphological mapping with ERT to identify the main stratigraphic and hydrostratigraphic units of Bull Island. ERT data allowed the depth to bedrock and the delineation of the spatial distribution of the hydrostratigraphic units to be estimated.

Sarala et al. (2015) undertook geomorphological mapping based on an aerial light detection and ranging (LiDAR)-derived digital elevation model, field observations, ground penetrating radar measurements and test pit surveys over 370 km^2, with the LiDAR data having a pixel size of 2 m × 2 m and vertical resolution of 0.3 m. The geomorphology of the area consists of large till-covered hills, ground moraine plains, glaciofluvial sand and gravel deposits composed of esker systems and related delta and outwash formations, followed by pro-glacial glaciolacustrine and post-glacial lacustrine and fluvial sand/silt deposits. The benefits of LiDAR data compared to traditional aerial-photo-based interpretation were more detailed identification of surface deposits and more precise edging of the morphologies.

BTS (2005) undertook a useful review of geophysical methods that had potential to be used to detect subsurface anomalies relevant to the application of closed-face tunnelling machines (Table 3.23). They included microgravity survey, magnetic survey, electrical resistivity imaging, electromagnetic traversing (conductivity survey), very-low-frequency (VLF) radio survey, ground probing radar, cross-hole seismic survey, surface refraction survey, in-tunnel seismic reflection survey, infrared (IR) thermography, marine seismic reflection and marine side-scan sonar. Table 3.23 briefly describes the methods available and their advantages and disadvantages.

3.6.2.5 Remote sensing

Remote sensing is a wide spectrum of techniques based on optical, IR and radar imaging, from orbiting satellites, aircraft, drones, vehicles and fixed platforms, which is increasingly being used in ground investigation. For example, Christensen et al. (2015) used airborne electromagnetic (AEM) to supplement a geotechnical investigation for a highway project in Norway. Heterogeneous glacial geology and variable bedrock led to the development of an automated algorithm to extract depth to bedrock by combining borehole data with AEM data. They were able to reduce the number of boreholes but not remove them altogether. In particular, they were able to locate shallow bedrock, steep or anomalous bedrock topography, and to estimate the spatial variability of depth at earlier phases of investigation.

Table 3.22 Appropriate borehole geophysical methods for geotechnical applications in glacial soils

Method	*Depth to bedrock*	*Stratigraphy*	*Lithology*	*Buried channels*	*Natural cavities*	*Groundwater*
General	4	4	4	3	2	4
Cross-hole seismic	0	2	3	3	3	0
Cross-hole GPR	0	3	3	3	4	0
Magnetometer	0	0	0	0	0	0

Method	*Bed boundaries*	*Bed thickness*	*Bed type*	*Porosity*	*Density*	*Permeable zones*	*Borehole fluid*	*Formation fluid*	*Fluid movement*	*Direction of dip*	*Fracture/ joints*	*Casing*	*Dia- meter*	*Type of borehole*
Self-potential	√	√	√					√						O,W
Long and short normal and lateral resistivity	√	√	√	√				√			√	√	√	O,W
Natural gamma	√	√	√									√		A
Gamma-gamma	√	√	√	√	√	√					√			A
Spectral gamma	√	√	√											A
Neutron	√	√	√	√		√								A
Fluid conductivity						√	√		√		√			L, O,W
Fluid temperature						√			√		√			L, O,W
Flow meter						√			√		√			L, O,W
Dip meter	√									√	√	√		O,W
Sonic (velocity)	√	√	√	√							√			L, O,W
Calliper	√	√				√					√	√	√	A
Televiewer	√	√				√					√			O,W

Source: After Reynolds, J. M. In *ICE Manual of Geotechnical Engineering*, edited by Burland, J., Chapman, T., Skinner, H., and Brown, M., Thomas Telford Ltd, London; 2012: 601–618, Chapter 45.

Note: 0, not applicable; 1, limited use; 2, used but not best approach; 3, excellent but still to be developed; 4, excellent and developed; L, lined; O, open hole; W, water filled hole; A, all types of borehole.

Table 3.23 Advantages and disadvantages of geophysical tests for closed-face tunnelling and ground stability

Method	*Description*	*Advantages*	*Disadvantages*
Microgravity	Measures slight differences in the gravitational field associated with local variations in mass density in the shallow subsurface. Implemented by taking readings from a sensitive gravity meter at a series of grid points of known elevation across the survey area. Data processing includes the removal of all known gravity effects from the data set, such as nearby basements or tunnels.	Well-established and accepted method for detecting subsurface voids. Viability of the technique to discern a particular size and depth of anomaly can be modelled and assessed before geophysics field work starts. Requires direct access to the area to be surveyed but is not intrusive. A typical minimum grid spacing is about 2 m by 2 m.	Resolution decreases with depth. Width of survey zone must extend well beyond footprint of feature of interest. Rather slow and laborious in the field. Can be adversely affected by disturbances such as traffic. Requires a topographic survey at each survey grid point. Requires full details of other subsurface features such as tunnels, basements and cellars, sewers etc. to 'model out' the effects of these objects.
Magnetic survey	Depends on there being a difference in the magnetic properties of the feature of interest and the surrounding ground. The technique will not identify a zone of locally loosened ground that is sitting within similar but more densely packed material.	Well-established method for investigating the subsurface. Relatively rapid data acquisition, requiring a walk-over survey only.	In an urban setting, the method is highly vulnerable to interference from buried cables and a variety of ferrous and other magnetic objects. Unlikely to identify variations in packing density of natural ground.
Electrical resistivity imaging	Measures variations in the electrical resistance of the ground, which depends on soil type and water content. Implemented by placing a line of electrodes along the ground surface.	Well-established method for imaging the shallow subsurface. Data acquisition and processing are relatively rapid.	Requires direct access to area to be imaged, and insertion of pencil sized metal electrodes into the surface. In an urban setting, at-surface surveys are highly vulnerable to interference from buried cables etc. Survey results are presented as two-dimensional images but may be affected by three-dimensional effects. Depth of survey penetration is a function of the length of survey spread. May be difficult to achieve adequate spread length in urban settings. Confident interpretation of presence or absence of relevant ground anomalies may be difficult.

(*Continued*)

Table 3.23 (Continued) Advantages and disadvantages of geophysical tests for closed-face tunnelling and ground stability

Method	*Description*	*Advantages*	*Disadvantages*
Electromagnetic traversing (conductivity)	Well-established family of methods in which electromagnetic waves are generated from a portable source and are 'listened' to at a nearby receiver. Variations in the received signal indicate variations in the electrical conductivity of the ground and surroundings.	Well-established method for investigating the shallow subsurface. Devices are carried by the operative and survey requires walk-over only. Data acquisition and processing are rapid.	In an urban setting at-surface surveys are highly vulnerable to interference from buried cables etc. Survey penetration is inadequate for deep anomalies.
Very-low-frequency (VLF) radio survey	Passive system, akin to conductivity surveying but which instead monitors localised changes in signals from VLF radio station broadcasts. Changes arise due to local variations in the conductivity of the ground and other surroundings.	Well-established method for investigating the shallow subsurface. Rapid data acquisition in the field requiring a walk-over survey only.	In urban settings the method is highly vulnerable to interference from buried cables etc. and adjacent structures. Generally shallow depth of penetration.
Ground probing radar (GPR)	Established principles of radar (radio range finding) to image the shallow subsurface.	Well-established method for investigating the shallow subsurface. Rapid data acquisition in the field. Requires direct access to the area to be surveyed but is not intrusive.	Very poor penetration of GPR signal through clay. Presence of even a thin band of clay at a site will render the technique 'blind' beyond the clay layer. In an urban setting, the method is likely to pick up numerous false-positive results associated with shallow benign features. Interpretation of results is relatively time-consuming and requires an experienced operator. Must compromise between resolution and penetration. Low-frequency GPR devices of less than about 100 MHz give better penetration while high-frequency devices of more than about 1 GHz give better resolution. Survey penetration is inadequate for deep anomalies.

(*Continued*)

Table 3.23 (Continued) Advantages and disadvantages of geophysical tests for closed-face tunnelling and ground stability

Method	*Description*	*Advantages*	*Disadvantages*
Cross-hole seismic survey	Two types of cross-hole seismic survey: parallel cross-hole surveys in which the source and receiver are kept at matching elevations and a plot of the variation of seismic properties with depth is obtained; and cross-hole tomographic surveying, in which a two-dimensional image of the variation of seismic properties in the inter-borehole plane is obtained.	Possibility to infer presence of cavities from parallel seismic amplitude (not velocity) surveys, subject to quality control achieved in grouting of borehole liner, and suitable repeatability of seismic source used.	Only limited coverage is achievable. Data are obtained only for zone between boreholes. Low success rate in determining the presence of cavities using velocity-based cross-hole seismic methods. Highly intrusive technique requiring installation of boreholes at each survey area. Requires relatively low ambient levels of ground-borne vibration. In soils, probe holes must be lined using grouted-in-place plastic casings.
Surface refraction survey	Potential means of investigating variations in stratigraphic levels between boreholes (e.g. undulating strata) and variations in the depth to rock, provided that the seismic velocity properties of the ground increase with depth.	Well-established method. Provides stratigraphic information between boreholes including depth to rock.	Needs a lengthy corridor of 100 m or more to spread sensors along ground surface. This may be difficult to achieve in urban settings. Does not work if seismic wave propagation velocity (in effect, soil stiffness) decreases with depth. Low velocity 'sandwich' layers cannot be identified. Localised features of limited horizontal extent (e.g. voids) cannot be imaged. Requires relatively low ambient levels of ground-borne vibration. Maximum achievable depth of penetration is a function of geology and, in part, apparatus used.

Source: After BTS. *Closed-Face Tunnelling Machines and Ground Stability*. Thomas Telford, London, 2005.

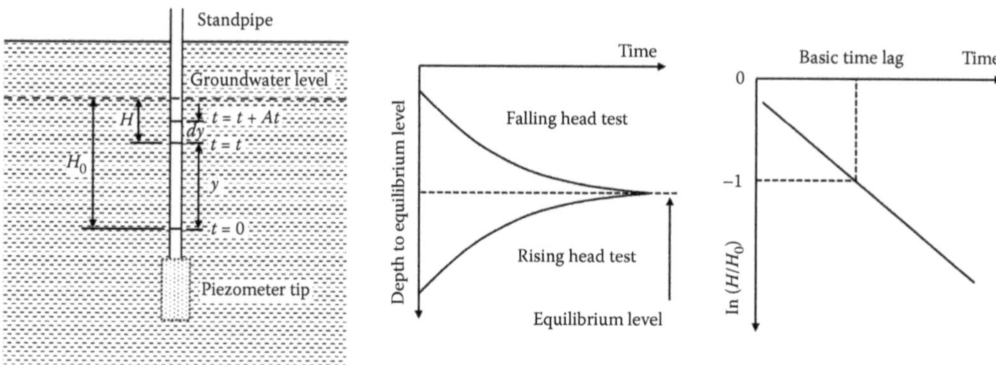

Figure 3.9 Use of standpipes to determine the *in situ* hydraulic conductivity by rising/falling head tests using Hvorslev's method showing the terms used to determine the basic time lag.

3.6.2.6 *Groundwater testing*

The mass permeability of a soil, especially in spatial variable glacial soils, fissured matrix-dominated tills and anisotropic soils such as glaciolacustrine clays is much more important than the intrinsic permeability determined from laboratory tests. Tests can be carried out in open boreholes, using piezometers, or in sections of a borehole sealed by inflatable packers. Rising and falling head tests are used in permeable coarse-grained soils and may possibly be used in fissured fine-grained soils or interbedded soils (BS 5930:1999). They rely on water flowing in or out of sealed borehole section. Figure 3.9 shows a typical test result and the terms used to calculate the hydraulic conductivity of the soil using Hvorslev's basic time lag method.

The time lag, T, is

$$T = \frac{A}{Fk} \tag{3.11}$$

where A is the cross sectional area of the borehole, k is the hydraulic conductivity of the soil and F is the shape factor of the test section. For a cylindrical piezometer or standpipe sand pocket, or a cased borehole, of length L and diameter D, the shape factor, F, is

$$F = \frac{2\pi L}{\ln[L/D + \sqrt{1 + (L/D)^2}]} \tag{3.12}$$

The time for a test to be carried out increases as the mass permeability reduces to such an extent that consolidation of fine-grained soils can become an issue.

3.6.3 Laboratory tests

There are two groups of laboratory tests to consider: classification tests and tests to produce geotechnical design characteristics. Table 3.24 is a summary of the results obtained from laboratory tests and the quality of those results with respect to glacial soils taking into account the composition and fabric of the soils assuming that the best quality samples are available. The classification tests used in conjunction with field tests, sample description and drillers' logs produce the geological model and identify the main strata that can be used to

Table 3.24 Summary of the results obtained from laboratory tests and their relevance to glacial soils taking into account the effects of sampling and specimen preparation

		Glacial soil applicability			
		Primary deposits		*Secondary deposits*	
Laboratory test	*Test results*	*Matrix-dominated tills*	*Clast-dominated tills*	*Glaciofluvial deposits*	*Lacustrine deposits*
Water content	• w	Relevant if Class 1 or 2 samples can be retrieved	Only relevant if significant fines content and Class 1 or 2 samples can be retrieved	Not relevant	Relevant
Bulk mass density	• ρ	Relevant if Class 1 or 2 samples can be retrieved	Not relevant	Not relevant	Relevant
Particle mass density	• ρ_s	Assume 2.65–2.72	Assume 2.65–2.72	Assume 2.65–2.72	Assume 2.65–2.72
Particle size distribution	• Grain size distribution curve	Important to assess likely behaviour	Relevant but only if Class 1, 2 or 3 samples recovered	Relevant but only if Class 1, 2 or 3 samples recovered	Not necessary
Consistency limits	• Plastic and liquid limit values (w_P), (w_L)	Relevant for matrix	Not relevant	Not relevant	Relevant
Density index	• e_{max}, e_{min} and I_D	Not relevant	Useful	Useful	Not relevant
Organic content	• Loss on ignition (C_{OM})	Not usually relevant	Not usually relevant	Not usually relevant	Not usually relevant
Carbonate content	• Carbonate content (C_{CaCO_3}				
Sulfate content	• Sulfate content ($C_{SO_4^{2}}$) or ($C_{SO_3^{2}}$)				
Chlorite content	• Chlorite content (C_{Cl})				
pH	• pH				
Compressibility oedometer	• Compressibility curve (different options) • Consolidation curves (different options) • Secondary compression curve (creep curve) • E_{oed} (stress interval) and σ'_p or C_s, C_c, σ'_p • C_α	Only on fully homogenised tills	Not relevant	Not relevant	Relevant

(*Continued*)

Table 3.24 (Continued) Summary of the results obtained from laboratory tests and their relevance to glacial soils taking into account the effects of sampling and specimen preparation

		Glacial soil applicability			
		Primary deposits		*Secondary deposits*	
Laboratory test	*Test results*	*Matrix-dominated tills*	*Clast-dominated tills*	*Glaciofluvial deposits*	*Lacustrine deposits*
Laboratory vane	• Strength index (c_u)	Unlikely to be possible because of strength	Not relevant	Not relevant	Relevant but allow for effect of varves
Unconsolidated undrained compression	• Undrained shear strength (c_u)	Only if 100 mm diameter Class 1 or 2 samples can be retrieved; tests on 100 mm samples to allow for composition and fabric	Not relevant	Not relevant	Tests on three 38-mm specimens from single 100-mm sample
Consolidated triaxial compression (drained and undrained with pore pressure measurements)	• Stress–strain curve(s) and pore pressure curve • Stress paths • Mohr circles • c', φ' or c_u • Variations of c_u with σ'_c • Deformation parameter(s) (E') or (E_u)	Only if 100-mm diameter Class 1 or 2 samples can be retrieved; tests on 100-mm samples to allow for composition and fabric; tests on three samples	Specimens have to be reconstituted to *in situ* density	Specimens have to be reconstituted to *in situ* density	Tests on three 38-mm specimens from single 100-mm sample
Consolidated direct shear box	• Stress displacement curve • *t s* diagram • c', φ' • Residual parameters	Possible	Specimens have to be reconstituted to *in situ* density	Specimens have to be reconstituted to *in situ* density	Possible
California bearing ratio	• CBR index (I_{CBR})				
Permeability (soil)	• Coefficient of permeability (k): • Directly from permeameter or triaxial test • Indirectly from oedometer test	Oedometer tests possible on homogenised till	Large permeameter	Large permeameter	Useful to assess anisotropy

assign properties from a combination of published information, regional databases and field and laboratory test results. Laboratory tests will be carried out on Class 1, 2 and possibly 3 samples. The quality of the sample has an impact on the classification, especially if the recovered sample is not representative of the layer. Ideally, Class 1 samples should be recovered but this is unlikely in glacial soils because of clasts, the fabric and strength of the soils. These all affect the quality of a sample. BS EN 1997-2:2007 suggests that there are five types of samples: undisturbed, disturbed, re-compacted, remoulded and reconstituted. Chapters 4, 5 and 6 provide case studies of geotechnical applications where these types of samples may be used successfully. For example, reconstituted specimens of matrix-dominated tills can be used to determine effective strength parameters (Clarke et al., 1998); compacted specimens can be used to determine the properties of glacial soils as engineered fills. It is likely that it will only be possible to obtain disturbed samples in coarse-grained soils such as clast-dominated tills and glaciofluvial soils and undisturbed samples in fine-grained soils such as fully homogenised matrix-dominated tills and glaciolacustrine clays. Undisturbed samples can permit an assessment of the fabric (clasts, varves and fissures), which may affect the performance of the soil. It is likely that most samples of glacial soils will be disturbed to some extent.

Once the geological model is complete, representative samples can be tested to create the geotechnical model. Tests are carried out in accordance with the relevant standards. Not all tests are covered here. The focus is on those tests that are used to classify glacial soils to produce the geological model from which tests can be carried out on representative samples to create the geotechnical model. Chemical tests and tests to determine organic content are excluded from this chapter.

3.6.3.1 *Classification tests*

These tests (Table 3.25) are used to classify, identify and describe the soil. In Europe, the standards BS EN ISO 14688-1 and 14688-2 apply; in North America ASTM D2487 – 11, ASTM D4318 – 10e1, ASTM D6913 – 04(2009)e1. Table 3.26 is a suggested number of classification tests for each stratum, but it should be noted that in glacial soils there may be more than one soil type in each stratum. For example, a glacial till may contain lenses or layers of laminated clays or sands and gravels. Therefore, the number of tests needed to classify a stratum may be greater than that shown in Table 3.26. Table 3.27 is the recommended mass of soil required to assess the classification and compaction properties of soils.

It is important to measure the water content of a soil because that is related to the strength of a clay soil. However, the water content is of value only if it represents the water content of the *in situ* soil. So, the minimum sample quality is Class 3. Further, the water content of a matrix-dominated till may vary within a sample as it will be the average of the water content of the matrix and the water content of the clasts, which may be different.

Bulk density is a classification and a characterisation parameter. It is used to determine the soil as an action as well as to provide information on the strength of the soil; an increase in density means an increase in strength. The bulk density of glacial soils has possibly the greatest range of all soils as it can vary from a loose glaciofluvial sand to a very dense fully homogenised till. Within a stratum, the density may vary especially in deformation tills containing remnants of secondary deposits. Bulk density is obtained from Class l or 2 samples, which limits it to matrix-dominated tills and lacustrine deposits. Further, in tills, the bulk density may be the density of the sample and not a representative density because of the issues of obtaining representative samples of a highly variable soil.

Particle densities of glacial soils are typical of those of other inorganic soils, that is, in the range of 2.65–2.72. However, within glacial soils, there can be a range of particle densities

Table 3.25 Laboratory tests for soil classification

Parameter	*Type of soil*							
	Fine-grained soil (clay)			*Fine-grained soil (silt)*			*Sandy, gravelly soil*	
	Undisturbed	*Disturbed*	*Remoulded*	*Undisturbed*	*Disturbed*	*Remoulded*	*Undisturbed*	*Remoulded*
Geological description and soil classification	√	√	√	√	√	√	√	√
Water content	√	(√)	(√)	√	(√)	(√)	(√)	(√)
Bulk density	√	(√)		√	(√)			
Minimum and maximum density				(√)	(√)	(√)	√	√
Consistency limits	√	√	√	√	√	√		
Particle size distribution	√	√	√	√	√	√	√	√
Strength index (c_u)	√			(√)				
Permeability	√			√	(√)	(√)	(√)	(√)
Sensitivity	√							

Source: After BS EN 1997-2:2007. *Eurocode 7: Geotechnical Design – Part 2: Ground Investigation and Testing (Incorporating Corrigendum 2010)*. British Standards Institution, London.

Note: √, recommended; (√), may be possible but not necessarily representative.

Table 3.26 Suggested minimum number of samples to be tested in one soil stratum but more samples will be necessary in glacial soils because of their spatial variability

	Comparable experience	
Classification test	No	Yes
Particle size distribution	4–6	2–4
Water content	All samples of Quality Class 1–3	
Strength index	All samples of Quality Class 1	
Consistency limits (Consistency limits)	3–5	1–3
Bulk density	Every element test	
Density index	As appropriate	
Particle density	2	1

Source: After BS EN 1997-2:2007. *Eurocode 7: Geotechnical Design – Part 2: Ground Investigation and Testing (Incorporating Corrigendum 2010).* British Standards Institution, London.

Table 3.27 Mass of soil for tests on disturbed samples for classification tests and tests on engineered fill

				Minimum mass of prepared test specimens			
Test		*Initial mass required*		*Fine-grained soils*	*Sand*	*Matrix- and clast-dominated till and gravels*	
Water content		Twice specimen mass		30 g	100 g	D = 2–10 mm	D > 10 mm 500 g min
Grain size	Sieve	2× MMS				MMS	
	Hydrometer	250 g		50 g	100 g		
	Pipette	100 g		12 g	30 g		
Consistency limits		500 g		300 g (D < 0.4 mm)			
Density index		8 kg		Depends on soil behaviour during test			
Dispersibility		400 g					
Compaction		S	NS				
	Proctor mould	25 kg	10 kg				
	CBR mould	80 kg	50 kg				
CBR		6 kg					
Permeability	100 mm permeameter	4 kg					
	75 mm permeameter	3 kg					
	50 mm permeameter	500 g					
	38 mm permeameter	250 g					

Source: After BS EN 1997-2:2007. *Eurocode 7: Geotechnical Design – Part 2: Ground Investigation and Testing (Incorporating Corrigendum 2010).* British Standards Institution, London.

Note: D, maximum particle size of 10% or more of dry mass; NS, soil particles not susceptible to crushing; S, soil particles susceptible to crushing.

from that of intact rock forming gravel to boulders to that of rock flour. Rock flour depends on the source rock and can typically be ground-up quartz and feldspar or clay minerals.

Particle size distribution of glacial soils is important but only if the distribution is linked to the fabric. For example, the particle size distribution of a varved clay could show a bimodal distribution of clay and silt whereas the fabric shows it to be distinct layers of clay and silt. Particle size distribution of most glacial soils may not be representative of the soil mass because larger particles are not sampled.

Consistency limits have been used to identify glacial soils as explained in Chapter 4; however, only the fine-grained component of a soil is tested. This means that the consistency limit of the matrix of a matrix-dominated till can be determined. The consistency limit of a varved clay is the average of the layers not the consistency limit of each layer. Therefore, any relationship to other soil properties, for example strength, must be treated with caution unless they are site specific.

The undrained shear strength of clay is both a classification test and a characteristic parameter. It is used for matrix-dominated tills and lacustrine deposits. BS EN 1997-2:2007 suggests that vane and fall cone tests can be used but only in lacustrine deposits since matrix-dominated tills are too stiff. The difficulties of sampling these tills mean that the shear strength may be an underestimate of the *in situ* strength and may not be representative of the stratum, especially if it is a subglacial traction till.

3.6.3.2 Geotechnical characteristics

Geotechnical characteristics include tests to measure strength, stiffness and permeability (Table 3.28), and the laboratory tests needed to produce those characteristics are given in Table 3.29. The composite nature and the effects of formation on the fabric of glacial soils

Table 3.28 Geotechnical characteristics of glacial tills highlighting the relative values of the characteristics and the fabric that could affect the values

				Relative comparison of geotechnical properties			
Till	*Class*	*Fabric*	*OCR*	*Density*	*Compressibility*	*Permeability*	*Anisotropy*
Deformation	G W Mg Mc	Deformed basal sediments or bedrock	1	5	3	5-8	
Lodgement	G W Mg Mc	Interlaying of glaciofluvial, joints, fissures, contortions; preferred clast orientation	2–5	4–7 5–8 6–8 6–8	1 2 2 3	5–6 2–3 4–5 2	7
Melt-out	G W Mg Mc	Occasional interlaying with glaciofluvial; clast orienteered with englacial state	1–2	2–4 2–6 2–6 2–7	2–4 3–5 3–6 4–7	7–9 4–5 5–8 3–4	3–5

Source: After McGown, A. and E. Derbyshire. *Quarterly Journal of Engineering Geology and Hydrogeology*, 10(4); 1977: 389–410.

Note: G, granular till; W, well-graded till; Mg, matrix till (granular); Mc, matrix till (cohesive).

Table 3.29 Laboratory tests for geotechnical characteristics

	Type of soil				
Parameter	*Gravel*	*Sand*	*Silt*	*NC clay*	*OC clay*
m_v			Oedometer	Oedometer	Oedometer
E, G	Triaxial	Triaxial	Triaxial	Triaxial	Triaxial
c', φ'	Triaxial shear box	Triaxial shear box	Triaxial shear box	Triaxial shear box	Triaxial shear box
c'_r, φ'_r			Ring shear	Ring shear	Ring shear
c_u			Triaxial	Triaxial	Triaxial
c_v			Oedometer triaxial	Oedometer triaxial	Oedometer triaxial
K	Triaxial constant head	Triaxial constant head	Triaxial constant head	Triaxial constant head	Triaxial constant head

Source: After BS EN 1997-2:2007. *Eurocode 7: Geotechnical Design – Part 2: Ground Investigation and Testing (Incorporating Corrigendum 2010).* British Standards Institution, London.

mean that tests should be carried out on specimens that are representative of the soil. This means that more samples are required and the specimens have to be large enough to be representative. A combination of Tables 3.30 and 3.31 can be used to determine the size of sample and whether the results will be representative of the stratum. For example, it should be possible to use three 38-mm specimens to determine the effective strength of fully homogenised till if there is no visible fabric, but if there is visible fabric or the sample contains clasts, then tests will be carried out on three 100-mm specimens, that is, three samples

Table 3.30 Mass of soil for tests on undisturbed samples for geotechnical characteristics

	Specimen dimensions		
Type of test	*Diameter (mm)*	*Height (mm)*	*Minimum mass required (g)*
Oedometer	50	20	90
	75	20	200
	100	20	350
Triaxial	35	70	150
	38	76	200
	50	100	450
	70	140	1,200
	150	300	12,000
Shear box	60 × 60	20	150
	100 × 100	20	450
	300 × 300	150	30,000
Density	D < 5.6 mm		125
	D < 8 mm		300
	D < 10 mm		500
	D > 10 mm		1,400

Source: After BS EN 1997-2:2007. *Eurocode 7: Geotechnical Design – Part 2: Ground Investigation and Testing (Incorporating Corrigendum 2010).* British Standards Institution, London.

Note: D, largest particle size.

Table 3.31 Maximum particle size for laboratory tests for geotechnical characteristics

Test	*Maximum size of particle*
Oedometer	H/5
Direct shear	H/10
Triaxial test	D/5
Permeability	D/12

Source: After BS EN 1997-2:2007. *Eurocode 7: Geotechnical Design – Part 2: Ground Investigation and Testing (Incorporating Corrigendum 2010).* British Standards Institution, London.

Note: H, height of specimen; D, diameter of specimen.

from a stratum rather than one. This means that the number of representative samples will be three times that recommended in the standards.

3.6.3.2.1 Strength tests

Tests can be carried out on Class 1 samples of fine-grained soils to obtain the total (undrained) and effective strength parameters of the soil. These include triaxial (ASTM D2850 – 15, ASTM D4767 – 11, ASTM D7181 – 11), shear box (ASTM D3080/D3080M – 11) and ring shear (ASTM D6467 – 13) tests. Class 1 samples means that it is only possible to measure the strength of glaciolacustrine deposits or fully homogenised tills. In practice, however, the shear strength of matrix-dominated tills is determined from inferior quality samples because of the effect of clasts and fabric on the samples. Indeed, it is prudent to test as large as specimen as possible to obtain a more representative strength. In the 1970s, Anderson (1974) proposed to test three vertically adjacent specimens of a glacial till to give an average value of undrained strength. Another procedure was to load a single specimen under one confining pressure until it was about to fail, then increase the confining pressure and continue loading in three stages until near failure. This test, a multistage undrained triaxial test, overcame the problem of relying on one result, but it introduced the concept of an undrained angle of friction, which is now considered unsafe. Given that tills are remoulded due to glacial action during deposition, Clarke et al. (1998) have suggested that it should be possible to test reconstituted specimens provided coarse material is removed and the specimens are consolidated to the *in situ* density. This means preparing a specimen in a rigid wall chamber by applying pressures in excess of those found in standard laboratories.

Undrained triaxial tests are used to classify glacial clays and produce characteristic values of undrained shear strength. Consolidated drained and undrained triaxial tests are used to obtain effective strength parameters. It is normal to test three specimens at three different confining pressures, but for the reasons of difficulty in sampling and the effect of clasts and fabric, tests should be carried out on three adjacent specimens or representative samples. However, it is important to check that those three specimens have similar classification properties. Any differences will affect the interpretation. Table 3.32 gives the recommended number of soil specimens to be tested to determine the representative total and effective strength parameters from a triaxial shear test. Note that this will require three times as many samples of matrix-dominated tills compared to samples of pure clays and silts.

Shear box tests are carried out on sands and gravels reconstituted to the field density determined from a field test such as the SPT. Table 3.33 gives the recommended number of soil specimens to be tested to determine the representative effective strength parameters from a direct shear test. It is likely that 300-mm shear box tests will be necessary when testing glacial soils because of the clast content.

Table 3.32 Recommended number of soil specimens to be tested to determine the representative total and effective strength parameters from a triaxial shear test[a]

Variability in the strength envelope derived from a minimum of three tests	*Comparable experience*		
	None[b]	*Medium*[c]	*Extensive*[d]
Consolidated drained and undrained tests with pore pressure measurement for effective strength parameters			
Coefficient of correlation > 0.95	4	3	2
0.95 < coefficient of correlation < 0.98	3	2	1
Coefficient of correlation < 0.98	2	1	1
Undrained tests for total strength parameters			
$c_{umax}/c_{umin} > 2$	6	4	3
$1.25 < c_{umax}/c_{umin} < 2$	4	3	2
$c_{umax}/c_{umin} < 1.25$	3	2	1

Source: After BS EN 1997-2:2007. *Eurocode 7: Geotechnical Design – Part 2: Ground Investigation and Testing (Incorporating Corrigendum 2010).* British Standards Institution, London.

[a] Tests on fine-grained soils will usually be based on three subsamples from the same depth; tests on matrix-dominated soils will be usually be based on three separate adjacent samples from one borehole or three representative samples from the stratum.
[b] Results of previous investigations unavailable and no regional database of results.
[c] Results of previous investigations unavailable but there is a regional database of results.
[d] Results of previous investigations available and there is a regional database of results.

3.6.3.2.2 Stiffness

The stiffness of a fine-grained soil can be determined from an oedometer test including constant stress and constant strain tests. The aim is to assess the swelling and compression characteristics of the soil. Specimens are constrained laterally in a cell with vertical drainage. Typically, the cell is 75 mm diameter and 19 mm high. It is used with fine-grained soils with no gravel content so is restricted to lacustrine clays and fully homogenised tills. Large oedometers do exist (e.g. the 250-mm-diameter Rowe Cell), which can be used to test matrix-dominated tills containing some gravel. However, sampling is restricted to trial pits because of the size of specimen required.

An oedometer test is used to determine the coefficient of compressibility (m_v), the compression and swelling indices (C_c and C_s) and preconsolidation pressure of fine-grained soils. It must be noted that there is an upper limit to the pressure in a standard oedometer test of 1,600 kPa, which is equivalent to 160 m of ice; that is, measuring preconsolidation pressures

Table 3.33 Recommended number of soil specimens to be tested to determine the representative effective strength parameters from a direct shear test

Variability in the strength envelope derived from a minimum of three tests	*Comparable experience*		
	None[a]	*Medium*[b]	*Extensive*[c]
Coefficient of correlation > 0.95	4	3	2
0.95 < coefficient of correlation < 0.98	3	2	2
Coefficient of correlation < 0.98	2	2	1

Source: After BS EN 1997-2:2007. *Eurocode 7: Geotechnical Design – Part 2: Ground Investigation and Testing (Incorporating Corrigendum 2010).* British Standards Institution, London.

[a] Results of previous investigations unavailable and no regional database of results.
[b] Results of previous investigations unavailable but there is a regional database of results.
[c] Results of previous investigations available and there is a regional database of results.

Table 3.34 Recommended number of soil specimens to be tested to determine the representative coefficient of compressibility

Range of values of coefficient of compressibility	*Comparable experience*		
	None[a]	*Medium*[b]	*Extensive*[c]
$m_V > 50\%$	4	3	2
$20\% < m_V < 50\%$	3	2	2
$m_V < 20\%$	2	2	1

Source: After BS EN 1997-2:2007. *Eurocode 7: Geotechnical Design – Part 2: Ground Investigation and Testing (Incorporating Corrigendum 2010)*. British Standards Institution, London.

[a] Results of previous investigations unavailable and no regional database of results.
[b] Results of previous investigations unavailable but there is a regional database of results.
[c] Results of previous investigations available and there is a regional database of results.

of glacial tills may not be feasible if the till was deposited in drained conditions. Of course, this is assuming that the preconsolidation pressure has any meaning when applied to a soil that was subject to shear and compression during deposition. Table 3.34 is the recommended number of soil specimens to be tested to determine the representative coefficient of compressibility. Given that it is likely that tests will be carried out only on fine-grained glacial soils, these number of tests are acceptable.

The triaxial test is more appropriate method of determining the deformation moduli of a soil provided an undisturbed specimen can be obtained. The test procedure is similar to a consolidated undrained or drained triaxial test with measurements of local displacement taken across the middle of the specimen to avoid the constraints of the bottom and top platens. To have any value, these tests should be carried out only on Class 1 samples, which means that it is unlikely that they will be carried out on tills or coarse-grained glacial soils. However, such is the value of the results, especially if undertaking any kind of numerical analysis it is worth considering these tests. Therefore, it is important to ensure that the best quality samples are taken.

3.6.3.2.3 Hydraulic conductivity

Characteristic values of permeability can be found from *in situ* tests and laboratory tests on undisturbed and reconstituted specimens. *In situ* tests provide information on the mass permeability by measuring in flow or out flow from a length of a borehole or by observing the phreatic surface created by lowering the water level in a borehole. There are time-consuming and expensive tests; hence, they would not be used in the majority of civil engineering projects. They are used when the hydrogeological model is a critical aspect of a design, for example, waste containment facilities. Laboratory tests to measure permeability include constant and falling head tests in a permeameter and constant flow tests in a triaxial cell. Coarse-grained soils are normally compacted into a permeameter to the *in situ* density; Class 1 or 2 samples of fine-grained soils can be tested in a permeameter using a falling head, but it is preferable to use a triaxial specimen, which means that the specimen can be consolidated to the required density/pressure before applying an appropriate hydraulic gradient across the specimen. The derived permeability depends on the density, degree of saturation, pore fluid and hydraulic gradient. Therefore, to have any value, *in situ* conditions should be replicated in a test unless it is known that a different hydraulic gradient will exist in future. Table 3.35 is the recommended number of soil specimens to be tested to determine the representative coefficient of permeability.

Table 3.35 Recommended number of soil specimens to be tested to determine the representative coefficient of permeability

Ratio of maximum to minimum coefficient of permeability	*Comparable experience*		
	None[a]	*Medium*[b]	*Extensive*[c]
$k_{max}/k_{min} > 100$	5	4	3
$10 < k_{max}/k_{min} < 100$	5	3	2
$k_{max}/k_{min} < 10$	3	2	1

Source: After BS EN 1997-2:2007. *Eurocode 7: Geotechnical Design – Part 2: Ground Investigation and Testing (Incorporating Corrigendum 2010).* British Standards Institution, London.

[a] Results of previous investigations unavailable and no regional database of results.
[b] Results of previous investigations unavailable but there is a regional database of results.
[c] Results of previous investigations available and there is a regional database of results.

3.7 THE REPORT

There are three reports that can be produced: the factual, interpretative and baseline reports. Table 3.36 lists the information expected in the reports. The factual report covers the desk study, field work and laboratory tests, which presents all relevant topographical, geomorphological, geological, geotechnical and hydrogeological factual data. The interpretative report uses those data to provide an assessment of the geological, geotechnical and hydrological models. The baseline report, first produced in the tunnelling industry in the United States, provides an assessment of risk based on the interpretative report and assessment of the category of the structure. While such reports may not be in common use, they are valuable as they provide an indication of the level and type of risk, which can help allocate responsibility for that risk. The interpretative report, sometimes combined with the factual report, uses the information in the factual report to produce the models and parameters required for design. The geological model is based on geological maps and memoirs, the borehole logs and the classification test results. In this way, a 3D image of the soil profile can be developed for the site. However, it must be stressed that interpretation of glacial tills is challenging as explained in Section 3.2. It is possible to have a stratum of glacial till with characteristic values of strength and stiffness containing lenses and layers of soil with distinctly different characteristic values. Further, it may not be possible to classify the soil layers in accordance with the geological classification of glacial soils. Indeed, it is necessary to use a combination of information to produce an interpretation of a site's geology.

3.8 OBSERVATIONS

A ground investigation for a civil engineering project is designed to reduce risk, which means that it has to take into account the performance of the project throughout its life as well as the method of construction. In order to reduce risk, the investigation has to produce an appropriate ground model that covers the topography, geomorphology, geological and hydrogeological conditions and the geotechnical characteristics.

It is understood that the ground is a risk, though this does not mean that an adequate ground investigation is undertaken. Failures due to inadequate ground investigation are well known and it is estimated that it costs the construction industry. For example, according to Littlejohn et al. (1991), 37% of all industrial building projects overran due to unforeseen ground conditions; a significant number of roads and bridges were subject to remedial

Table 3.36 Information within the factual, interpretative and baseline reports

Factual	• The purpose and scope of the geotechnical investigation including a description of the site, the planned structure and the stage of the planning • Geotechnical category of the structure • The names of all consultants and subcontractors • The dates between which field and laboratory investigations were performed • A description of the site including an assessment of the topography, geology, hydrogeology, sites of scientific or historical interest, environmental issues, historical use • Field reconnaissance of the site and the surrounding area noting particularly (a) evidence of groundwater; (b) behaviour of neighbouring structures; (c) exposures in quarries and borrow areas; (d) areas of instability; (e) any exposures of mining activity at the site and in the neighbourhood; (f) difficulties during excavation; (g) history of the site; (h) geology of the site, including faulting; (i) survey data with plans showing the structure and the location of all investigation points; (j) information from aerial photographs; (k) local experience in the area; (l) information about the seismicity of the area • A description of the field work including the borehole locations and levels, the borehole logs, water strikes and monitoring • The results of the field investigations and laboratory tests
Interpretative	• A review of the results of the site and laboratory investigations and all other information • A description of the geometry of the strata (the geological model) • A description of the hydrogeological conditions (the hydrogeological model) • Detailed descriptions of all strata including their physical properties and their deformation, strength and drainage characteristics (the geotechnical model) • An assessment of the quality of the results taking into account the groundwater table, ground type, drilling method, sampling method, transport, handling and specimen preparation • Comments on irregularities such as cavities and zones of discontinuous material • Identification of hazards relevant to design, construction and operation of the project • Tabulation and graphical presentation of the results of field investigation and laboratory testing in cross sections of the ground showing the relevant strata and their boundaries including the groundwater table in relation to the requirements of the project • The values of the geotechnical parameters for each stratum • A review of the derived values of geotechnical parameters based on the results of the ground investigation and published data
Baseline	• Contractual statements describing the geotechnical conditions • A specification for the ground covering the geological, geotechnical and hydrogeological conditions in the context of the project

works as a result of ground conditions; cost overruns on major infrastructure projects; and 25% of the cost of construction projects is attributed to ground-related problems (Geo Impuls, 2015).

There are publications, guidelines and specifications that cover ground investigation, so it should be possible to design a ground investigation that reduces the risk of failure, cost overrun and delays due to ground conditions. However, a review of the formation of glacial soils and an assessment of current practice suggest that many of the recommendations in the standards and guidelines have to be adapted for glacial soils. The following should be considered when undertaking a ground investigation in glacial soils:

- A study of the regional land system will provide an indication of the type of landforms and therefore an indication of the likely glacial soils to be encountered.

- Glacial soils are composite soils, which can include particles ranging from clay to boulders.
- The composition, fabric and structure of glacial soils are spatially variable; therefore, more samples and *in situ* tests are needed to classify and characterise the soils compared to the number recommended in codes of practice.
- The composition and fabric of glacial soils mean that larger samples have to be tested if representative design parameters are going to be determined. This requires more samples than recommended in a number of specifications for laboratory tests.
- The composite nature of glacial soils means that it is often difficult to retrieve undisturbed samples.
- The ground investigation should include a preliminary investigation to establish the geological model, thus allowing the main investigation to focus on the geotechnical and hydrogeological characteristics.
- The hydrogeology of glacial soils is complex, so groundwater pressures should be assessed at various depths and over time so that the groundwater profile can be established taking into account seasonal changes.
- A baseline report should be produced to assess the risk based on the category of structure to be built.

This review of codes of practice, specifications and guidelines has highlighted the fact that the principles are relevant when investigating glacial soils but the details may be inappropriate for glacial soils because of their spatial variability in composition, fabric, classification and geotechnical characteristics.

Chapter 4

Characterisation of glacial soils

4.1 INTRODUCTION

Chapter 2 highlights the challenges glacial geologists have had over the years in gaining consensus as to the formation of glacial soils, particularly tills. Their research has made a significant contribution to an appreciation of the history of the various types of glacial soils, a key factor when assessing the performance of glacial soils. It has highlighted anomalies between the classical approach to soil mechanics and the actual behaviour, which may influence the selection of geotechnical parameters.

Chapter 3 highlights the ground investigation process and its impact on establishing the engineering aspects of glacial soils. Reference is made to the difficulty in obtaining representative samples and enough samples, of identifying the type of soil and of interpreting *in situ* tests.

This chapter covers the description, classification and engineering characteristics of glacial soils emphasising the relevance of standard tests, techniques used to enhance those tests to produce useful data and data on glacial soils. Engineering soils are characterised by their physical, chemical and mechanical properties assessed using national and international standards. Physical properties, that is, the classification properties, together with the soil description are used to produce the geological profile and identify potential hazards. This provides information on which samples have to be tested to obtain representative design parameters for the various strata. The chapter starts with an overview of the issues identified in Chapters 2 and 3 to highlight the points that have to be addressed to assess the geotechnical characteristics of glacial soils. This is followed by an overview of the procedures used to describe the soil. The limitations of the classification tests and the impact they have on establishing the geological profile are discussed using examples to provide a database to validate future work. The section on geotechnical characteristics covers the tests used to assess the deformation, strength and conductivity of saturated and partially saturated glacial soils providing examples to show the key characteristics of those soils.

4.2 THE CHALLENGES OF ASSESSING PROPERTIES OF GLACIAL SOILS

Glacial soils are deposited by ice or by water or in water creating, in some instances, a unique landform, which helps identify the possible soil type. A review of the formation of glacial soils suggests that they can be divided into primary and secondary deposits for engineering purposes, as shown in Table 2.1. The formation of secondary deposits can generally be observed, especially those that are deposited beyond the ice margins. Secondary deposits are formed in a similar manner to alluvial soils, so their stress history can be evaluated.

However, the composition and fabric of a given deposit can vary by a significant amount because of the erosion, transport and deposition processes. Primary deposits are unlike any other soils in that they are subject to shear as well as compression during deposition. There is much speculation as to the formation of primary deposits because they are formed below an ice sheet, which can be tens or hundreds of metres thick. This creates very dense deposits that cannot be simply replicated through standard consolidation procedures in a laboratory environment. Glacial soils are composite soils that can contain boulders, cobbles, gravels, sands, silts and clays in various proportions and, as composite soils, they are difficult to sample and test. The stone content and fabric of glacial soils means that tests will be carried out on larger samples; hence, more samples are needed. Descriptions are based on the deposition process, the dominant or principal particle fraction and the engineering behaviour. Tests are related to the dominant particle type, but the response may be governed by the minor fractions. Therefore, when planning an investigation, it is necessary to

- Obtain sufficient samples for soil description and classification
- Obtain sufficient (possibly more than suggested in standards) representative samples to determine the geotechnical characteristics
- Understand the effect of sampling, testing and interpretation on the geotechnical characteristics
- Validate the results against published data

4.3 DESCRIPTION

There are a number of systems used to describe/classify a soil: geological, agricultural and engineering. The first assists with an appreciation of the formation of the soil, while the others are used in the application of soils either for horticultural purposes or to create engineering structures. Agricultural soils are typically topsoil and subsoil and engineering soils are those below the topsoil. Generally, agricultural soils and engineering soils are different because of the organic content with the exception of organic soils such as peat and the subsoil at the interface between topsoil and the underlying inorganic soils. The classifications systems are different because their purpose is different. Examples of the classification system used by glacial geologists in Chapter 2 are not the same as the engineering classifications described in this chapter. In general, this is not an issue because the classifications are being used for different purposes. However, in the case of glacial soils, the geological classification is of use to the engineer as it provides additional information when interpreting the behaviour of these soils.

Figure 4.1 is an example of the process used to describe soils, which includes preliminary borehole logs produced by site operatives, the logs produced from sample descriptions by geologists, engineering geologists and geotechnical engineers, and the full description based on test samples. Figure 4.1 is based on the dominant fraction not the engineering behaviour; that is, descriptions of soils based on engineering behaviour can be different from those based on classification. A summary of the field descriptions used to describe inorganic soils is given in Table 4.1. The descriptions are modified by results of field and laboratory tests.

Glacial soils are natural soils that are unlikely to have little or no organic content. Therefore, the description, according to Table 4.1, will be based on the very-coarse-, coarse- and fine-grained content. Borehole samples will not contain particles greater than the borehole diameter (e.g. 200 mm), so most descriptions of borehole samples will be based on the coarse- and fine-grained particle content only.

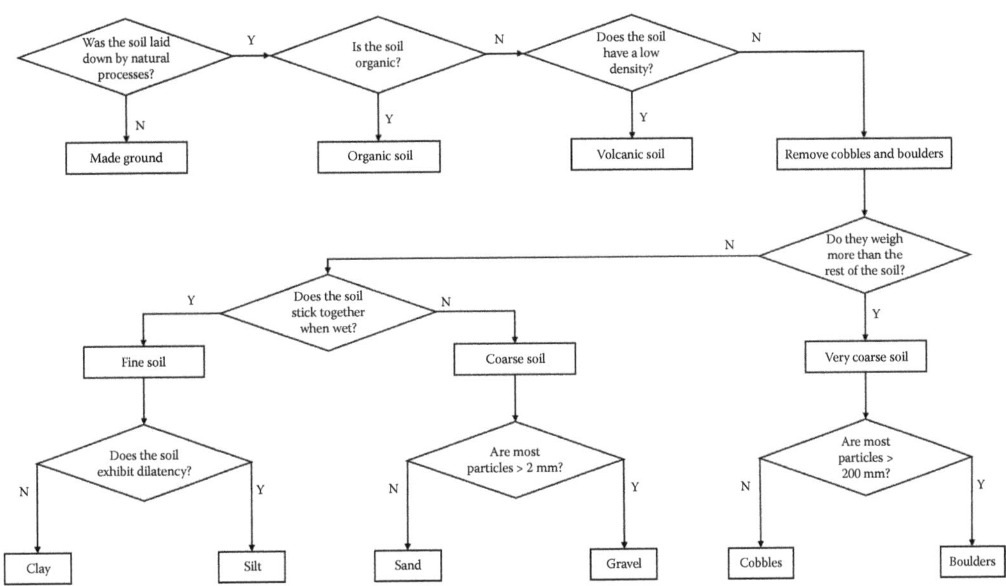

Figure 4.1 Process to identify and describe soils. (After BS EN ISO 14688-1:2002+A1:2013. *Geotechnical Investigation and Testing – Identification and Classification of Soil – Part 1: Identification and Description.* British Standards Institution, London.)

Coarse-grained soils are separated into sands and gravels; that is, cobbles and boulders are not included in the classification but should be included in the description if they are present in the site operatives' logs. Fine-grained soils are divided into low, medium and high plasticity soils and into clays and silts.

A simple field test based on the engineering behaviour (Table 4.2) can be used to separate out coarse-grained and fine-grained soil behaviour. Coarse-grained glaciofluvial soils and clast-dominated tills will not stick together; glaciolacustrine clays and matrix-dominated tills will. This field description is based on the engineering behaviour, particle size, discontinuities, bedding, colour, dominant soil type and relative density. At this stage, it is possible to consider further tests to give more detailed descriptions.

The mechanical and hydraulic behaviour of a soil depends on the particle fractions indicated in Table 4.3. Soils consist of the principal fraction, which is modified according to the percentage by weight of other fractions. For example, a sandy clay will show characteristics of fine-grained soil modified by the presence of coarse-grained soil. This is important in glacial soils as their engineering behaviour will be dominated by the fine fraction.

Table 4.4 lists the subdivisions of particle sizes found in glacial soils and the symbols used to denote those soils. Glacial soils are composite soils that contain principal and minor fractions, which are used to qualify the principal fractions. For example, a sandy CLAY (saCl) would be a soil formed of at least 40% clay (the principal fraction) and sand. If the principal fraction of a soil formed of coarse particles (sand and gravel) and fine-grained particles is more than 60% by weight of coarse particles, it will be, according to the standard, a coarse-grained soil even if it exhibits a dry strength or plasticity. The dry strength is determined by a simple field test in which the soil is dried and its resistance to disintegration assessed.

1. *Low dry strength*: Soil disintegrates under a light pressure.
2. *Medium strength*: Soil disintegrates under a medium pressure.
3. *High dry strength*: Can only be broken with force.

Table 4.1 Field identification and description of soils

Soil group	*Principal soil type*	*Particle size (mm)*	*Visual identification*	*Relative density/ consistency*	*Discontinuities*	*Bedding*	*Colour*	*Composite soils*	*Particle shape*	*Principal soil type*
Very coarse soils	Boulders	Large boulder	Only seen in pits and exposures Difficult to recover from boreholes	Qualitative description of packing by inspection	Spacing of features such as fissures, shears, partings, isolated beds or laminae, desiccation cracks, rootlets etc. Fissured – soil breaks into blocks along unpolished discontinuities Shears – soil breaks into blocks along polished discontinuities Very widely spaced Widely spaced Medium spaced Closely spaced Very closely spaced Extremely closely spaced	Thickness of beds in a accordance with geological definition Interbedded or interlaminated Alternating layers of different types Prequalified by thickness term if in equal proportions Otherwise thickness of and spacing between subordinate layers Very thickly Thickly Medium Thinly Very thinly Thickly laminated Thinly laminated	Lightness – light, dark Chroma – pinkish, reddish, yellowish, orangish, brownish, greenish, bluish, greyish Hue – pink, red, yellow, orange, brown, green, blue, white, grey, black		Very angular Angular Subangular Subrounded Rounded Well rounded Cubic Flat Elongate	Boulders
		Boulder								
	Cobbles	Cobble								Cobbles
Coarse-grained soils	Gravel	Coarse	Easily visible to naked eye; particles shape can be described; grading can be described	Very loose (0–4) Loose (4–10) Medium dense (10–30) Dense (30–50) Very dense (>50)				Slightly sandy (<5%) Sandy (5%–20%) Very sandy (>20%) Sand and gravel (50%)		Gravel
		Medium								
		Fine								
	Sand	Coarse	Visible to naked eye; no cohesion when dry; grading can be described							Sand
		Medium								
		Fine								
Fine-grained soils	Silt	Coarse	Only coarse silt visible with hand lens; exhibits little plasticity and marked dilatancy; slightly granular or silky to touch; disintegrates in water; possess cohesion but can be powdered easily between figures	Very soft Soft Firm Stiff Very stiff Hard				Slightly sandy (<35%) Sandy (35%–65%) Very sandy (>65%) Silty clay Clayey silt		Silt
		Medium								
		Fine								
	Clay		Dry lumps can be broken but not powdered between the fingers; disintegrate under water but more slowly than silt; smooth to the touch, exhibits plasticity but no dilatancy; stick to the fingers and dries slowly; shrinks appreciably on drying usually showing cracks							Clay

Source: After BS 5930:1999+A2:2010. *Code of Practice for Site Investigation*. British Standards Institution, London.

Table 4.2 Criteria used to identify those soils that behave as coarse-grained soils and those that behave as fine-grained soils

Criterion	*Soil group*	*Quantity*	*Groups of similar properties*			*Further subdivision*
Wet soil does not stick together	Very coarse	Most particles >200 mm	Bo	xBo		Requires special consideration
				boCo	coBo	
		Most particles >63 mm	Co	saCo, grCo	sagrCo	
	Coarse	Most particles >2 mm	Gr		cosaGr	Particles size grading
				soGr		Shape of grading curve
				saGr, grSa	sasiGr, grsiSa	Relative density Permeability
		Most particles >0.063 mm	Sa	siGr, clGr orSa	siSa, clSa, saclGr	(Mineralogy) (Particle shape)
Wet soils stick together	Fine	Low plasticity dilatant	Si	saSi	sagrSi saclSi	Plasticity Water content
				clSi, siCl		Strength, sensitivity
		Plastic non-dilatant	Cl		sagrCl	Compressibility, stiffness (Clay mineralogy)

Source: After BS EN ISO 14688-1:2002A1:2013. *Geotechnical Investigation and Testing – Identification and Classification of Soil – Part 1: Identification and Description.* British Standards Institution, London.

Note: The secondary fractions as adjectives shall be placed with the term describing the principal fraction in the order of their relevance, as shown in the following examples: sandy gravel (saGr), coarse sandy fine gravel (csaFGr), medium sandy silt (msaSi), fine gravely coarse sand (fgrCSa), silty fine sand (siFSa), fine gravelly, coarse sandy silt (fgrcsaSi) and medium sandy clay (msaCl). If coarse secondary fractions are present in a particularly small or particularly large proportion, the term 'slightly' or 'very' shall precede the qualifying term. Bo, boulders; Co, cobbles; Gr, gravel; Sa, sand; Si, silt; Cl, clay.

A soil with 10% silt, 30% sand and 60% gravel will be described as slightly silty, sandy gravel whereas a soil with 45% silt, 30% sand and 30% gravel will be described as a gravelly, sandy silt. A clay content of 15% is critical because experience suggests that it represents a boundary between different types of hydraulic behaviour. A soil with 15% or more clay content will behave, in engineering terms, as a fine-grained soil. This means that

Table 4.3 Use of particle size fractions in soil descriptions

Fraction	*Content of fraction in wt% of material <63 mm*	*Content of fraction in wt% of material <0.063 mm*	*Name of soil*	
			Modifying term	*Main term*
Gravel	20–40 >40		Gravelly	Gravel
Sand	20–40 >40		Sandy	Sand
Silt and clay	5–15	<20 >20	Slightly silty Slightly clayey	
	15–40	<20 >20	Silty Clayey	
	>40	<10 10–20 20–40 >40	Clayey Silty	Silt Silt Clay Clay

Source: After BS EN ISO 14688-2:2004+A1: 2013. *Geotechnical Investigation and Testing – Identification and Classification of Soil – Part 2: Principles for a Classification.* British Standards Institution, London.

Table 4.4 Particle size fractions found in glacial soils

Soil fractions	*Sub-fractions*	*Symbols*	*Particle sizes*
Very coarse soil	Large boulder	LBo	>630
	Boulder	Bo	200–630
	Cobble	Co	63–200
Coarse soil – gravel	Gravel	Gr	2–63
	Coarse gravel	CGr	20–63
	Medium gravel	MGr	6.3–20
	Fine gravel	FGr	2–6.3
Coarse soil – sand	Sand	Sa	0.063–2
	Coarse sand	CSa	0.63–2
	Medium sand	MSa	0.2–0.63
	Fine sand	FSa	0.063–0.2
Fine soil – silt	Silt	Si	0.002–0.063
	Coarse silt	CSi	0.02–0.063
	Medium silt	MSi	0.0063–0.02
	Fine silt	FSi	0.002–0.0063
Fine soil – clay	Clay	Cl	<0.002

Source: After BS EN ISO 14688-2:2004+A1: 2013. *Geotechnical Investigation and Testing – Identification and Classification of Soil – Part 2: Principles for a Classification.* British Standards Institution, London.

a soil with 20% clay, 30% sand and 40% gravel may be described as a clayey sand and gravel but its behaviour will be that of a very gravelly, very sandy clay. This is important when distinguishing between matrix- and clast-dominated tills. A matrix-dominated till will have at least 30% clay content; a clast-dominated till less than 15% (McGown and Derbyshire, 1977).

It is possible for glacial tills to be described as coarse-grained according to Table 4.2 but behave as fine grained; that is, it is not necessarily the dominant or principal fraction by weight that influences the behaviour. This discrepancy could lead to an inappropriate initial assessment of the engineering behaviour of a glacial till and possibly inappropriate tests to determine the mechanical characteristics. Therefore, it is prudent to use Table 4.1 to produce an initial assessment of the engineering behaviour of a glacial soil and a combination of Tables 4.1 and 4.2 to decide on the classification tests. For example, a glacial soil that sticks together but contains less than 40% by weight of clay and silt will be a coarse-grained soil according to Table 4.2, but should be assessed using classification tests for both coarse- and fine-grained soils. This means that glacial soils may not be classified, from an engineering point of view, from their principal fraction but by their engineering behaviour. Hence, a glacial soil that behaves as a fine-grained soil should be classified as a fine-grained soil because that will provide information on its engineering behaviour. The fact that it is classified as a fine-grained soil does raise concerns when it comes to construction since the coarser particles will affect the excavation process, stability of excavations and compaction. Further, whereas a fine-grained soil may be stable during construction, composite soils that appear to be fine grained may not because they are prone to collapse during construction as the timescale is of a similar order of magnitude as the rate of pore pressure dissipation.

Figure 4.2 shows ranges of particle size distributions found among UK glacial soils. Note that this may not be the true range since very coarse particles (Table 4.5) may not

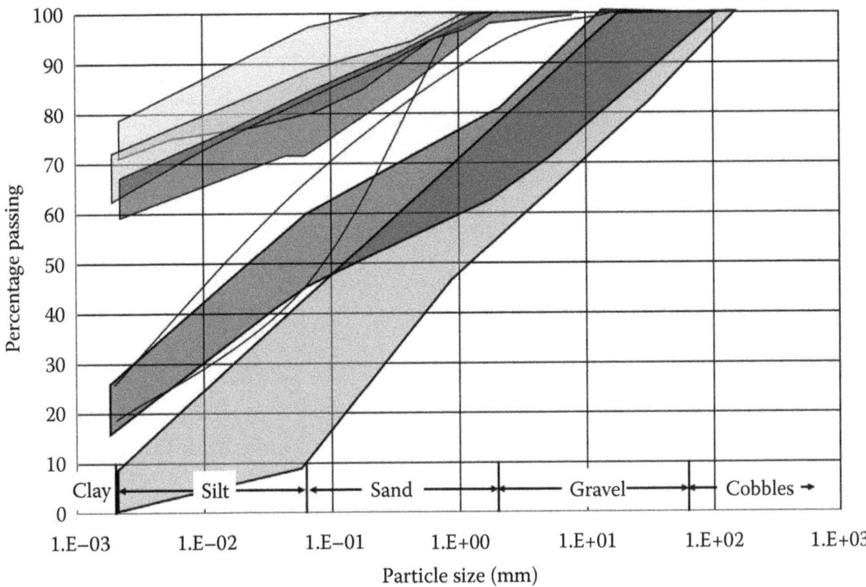

Figure 4.2 Particle size distributions of glacial soils from across the United Kingdom. (After Trenter, N. A. *Engineering in Glacial Tills*. CIRIA, London, 1999.)

be recovered from boreholes. Therefore, a description based on borehole samples and laboratory tests may not refer to very coarse particles including cobbles and boulders because they are not in the samples. Site operatives' descriptions may show their presence but not always, especially when they are randomly distributed throughout the soil mass. It is prudent to assume that very coarse particles, that is, cobbles and boulders, do exist in glacial tills and coarse-grained glaciofluvial deposits. Very coarse particles can also be found in glaciolacustrine deposits (Figure 2.6) and in glaciomarine deposits as dropstones (Figure 2.7).

The fine-grained particles found in unweathered glacial soils include silt and clay size particles but, as they are products of erosion, they are not necessarily clay minerals; it depends on the source material. The process of erosion and transport reduces the particle sizes but does not change their composition. Therefore, the fine-grained fraction of glacial soils may include rock particles and clay minerals (Table 4.6) depending on the source.

Table 4.5 Classification of very coarse soil

Fraction	*% by mass*	*Term*
Boulders	<5	Low boulder content
	5–20	Medium boulder content
	>20	High boulder content
Cobbles	<10	Low cobble content
	10–20	Medium cobble content
	>20	High cobble content

Source: After BS EN ISO 14688-2:2004+A1:2013. *Geotechnical Investigation and Testing – Identification and Classification of Soil – Part 2: Principles for a Classification*. British Standards Institution, London.

Table 4.6 Examples of the composition of glacial tills highlighting the presence of rock flour in some tills

	Chicago tills		*NE England tills*	
Mineral	*Blodgett till*	*Deerfield till*	*Upper till*	*Lower till*
Illite	57	53	14–29	22–36
Dolomite	15	21		
Chlorite	10	8		
Variscite	7	9		
Calcite	6	5		
Kaolinite	5	4	54–63	67–70
Quartz			45–52	53

Discontinuities form in glacial tills (Boulton, 1975; Kirkaldie and Talbot, 1992) due to

- Reduction in vertical stress due to melting of the ice surcharge
- Horizontal tensile stress due to isostatic rebound
- Induced failure due to shear stresses
- Contraction during thawing (deglaciation)
- Shrinkage associated with drying

The discontinuities can be horizontal (reduction in overburden and shearing) and vertical (isostatic unloading, freeze/thaw and shrinkage). The discontinuities affect the mass geotechnical engineering characteristics of fine-grained tills. It should be assumed that matrix-dominated tills contain discontinuities (Allred, 2000). Discontinuities in matrix-dominated tills reduce the stability of slopes due to more rapid build-up of pore pressure due to infiltration, seepage into excavations greater than predicted from laboratory tests, water loss from reservoirs underlain by glacial till, contamination of groundwater level due to landfill construction in glacial tills and increased rate of foundation settlement. Discontinuities can be described by terms used to describe rock discontinuities (BS EN 14689, 2003), the most relevant terms being those that apply to sedimentary rocks. Tables 4.7 and 4.8 list the terms used to describe discontinuity spacing and apertures. These discontinuities can have a significant effect on the mechanical and hydraulic behaviour of the till, so these should be recorded (dip and azimuth). Glacial soils also exhibit sedimentary structures, which are most easily observed in excavations and exposures. Figures 2.10 and 2.12 show examples

Table 4.7 Terms used to describe discontinuity spacing

Term	*Spacing (mm)*
Very wide	>2000
Wide	600–2000
Medium	200–600
Close	60–200
Very close	20–60
Extremely close	<20

Source: After BS EN ISO 14689-1:2003. *Geotechnical Investigation and Testing – Identification and Classification of Rock – Part 1: Identification and Classification.* British Standards Institution, London.

Table 4.8 Terms used to describe discontinuity aperture

Description	*Aperture size*
Very tight	<0.1 mm
Tight	0.1–0.25 mm
Partly open	0.25–0.5 mm
Open	0.5–2.5 mm
Moderately wide	2.5–10 mm
Wide	1–10 cm
Very wide	10–100 cm
Extremely wide	>1 m

Source: After BS EN ISO 14689-1:2003. *Geotechnical Investigation and Testing – Identification and Classification of Rock – Part 1: Identification and Classification.* British Standards Institution, London.

of structures associated with subglacial deformation and Tables 2.4 and 2.5 list the types of bedding and how they are used in a lithofacies coding scheme. Interbedding is frequently found, especially in deformation tills and glaciolacustrine deposits. Structural terms include bedded, interbedded, laminated, folded, massive and graded. Terms used to describe bedding thickness are listed in Table 4.9. For example, a glaciolacustrine clay may be referred to as thinly laminated, and a glacial till may include tight fissures.

Since the descriptions of engineering soils are of samples taken from boreholes, it is likely that any bedding noted will be thin to thinly laminated; and discontinuities close to extremely close. This does not mean that more widely spaced discontinuities do not exist, which can be an issue, since they may govern the behaviour of the soil. A consequence of the sample size is that it is not possible to describe the size of block bounded by discontinuities unless the spacing is very close. It may be possible to describe the roughness of the discontinuities at the small (several millimetres) and medium scale (several centimetres) using the terms rough or smooth; planar, stepped or undulating as shown in Figure 4.3. The discontinuities can be described as very tight to extremely wide (Table 4.8) though discontinuities in borehole samples will vary between very tight and open because of the scale.

The shape of gravel and very coarser particles is described in terms of their angularity (Table 4.10). Ideally, the azimuth of the particles should be recorded though this may not be feasible for samples taken from boreholes unless care is taken to preserve the orientation of

Table 4.9 Terms used to describe bedding thickness

Term	*Spacing (mm)*
Very thick	>2,000
Thick	600–2,000
Medium	200–600
Thin	60–200
Very thin	20–60
Thickly laminated	6–20
Thinly laminated	<6

Source: After BS EN ISO 14689-1:2003. *Geotechnical Investigation and Testing – Identification and Classification of Rock – Part 1: Identification and Classification.* British Standards Institution, London.

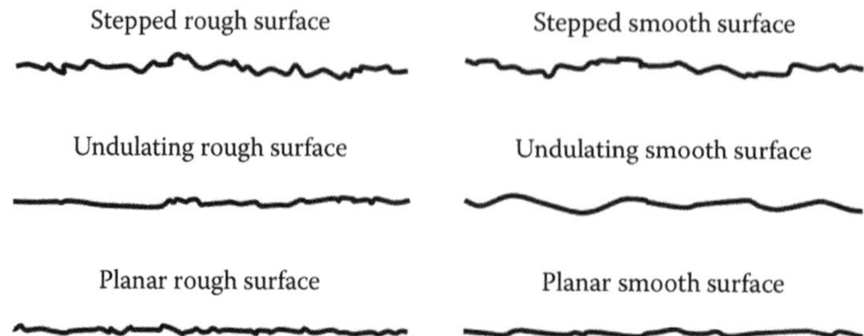

Figure 4.3 Surfaces of discontinuities. (After BS EN ISO 14689-1:2003. *Geotechnical Investigation and Testing – Identification and Classification of Rock – Part 1: Identification and Classification.* British Standards Institution, London.)

the sample. The mineral composition of the particles should be recorded. Rock fragments (gravel, pebbles, cobbles and boulders) are an indication of the source of the glacial soil though a glacial soil may contain rock fragments from several locations because of periods of glacial advance and recession.

The soil structure, also known as fabric, refers to the orientation and distribution of particles in the soil. At a microscopic level, there are two extremes of soil structure: flocculated as observed in marine clays and dispersed such as those found in subglacial tills (Figure 4.4). The microstructure affects the permeability; for example, a flocculated structure will be more permeable than a dispersed structure and a dispersed structure will be more anisotropic than a flocculated structure. Fabric can also refer to structural features such as discontinuities that affect the mechanical properties of the soil. Rowe (1972) suggested that soil fabric refers to the size, shape and arrangement of the solid particles and associated voids and structure is the element of fabric that deals with a particular size range and can include discontinuities. Derbyshire et al. (1985) suggested that the fabric and structure of tills are due to the erosional, depositional and post-depositional processes (Figure 4.5). This is a simpler version of the geological description of fabric (Tables 2.2, 2.4 and 2.5). Table 4.11 is a summary of the relevant geotechnical characteristics of subglacial and supraglacial tills

Table 4.10 Terms used to describe particle shape

Parameter	*Description*
Angularity/roundness	Very angular
	Angular
	Subangular
	Subrounded
	Rounded
	Well rounded
Form	Cubic
	Flat
	Elongate
Surface texture	Rough
	Smooth

Source: After BS EN ISO 14688-1:2002+A1:2013. *Geotechnical Investigation and Testing – Identification and Classification of Soil – Part 1: Identification and Description.* British Standards Institution, London.

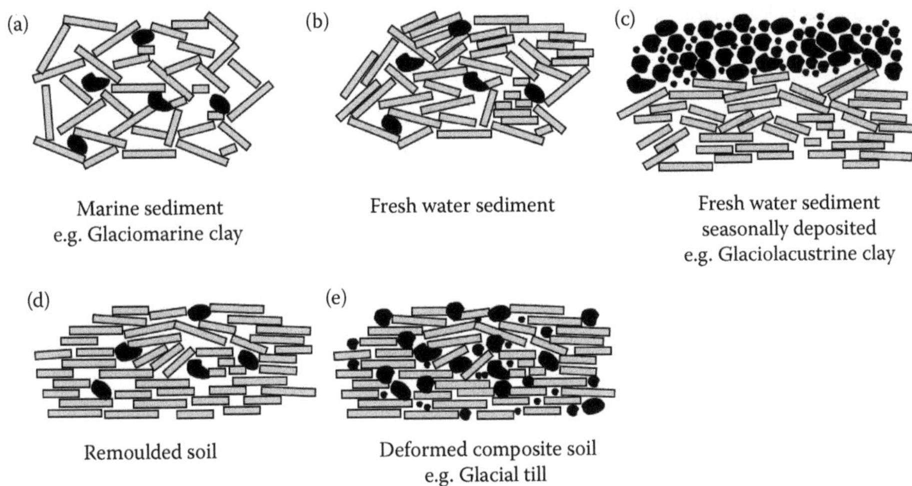

Figure 4.4 Conceptual models of the microstructure of soils including glacial soils showing the effect of the deposition processes on the microstructure.

in relation to their fabric, highlighting the impact deposition has upon the properties. For example, the densities of melt-out tills (2–7) are less than those of lodgement tills (6–8) even though they may have the same composition. Melt-out tills are deposited when the ice melts; the lodgement till undergoes shear and compression during deposition.

Thus, the complete description based on field observations and samples covers the particle sizes, discontinuities and engineering behaviour. It is also possible, using local knowledge and geological maps and memoirs, to suggest the geological origin of the soil, but given the reservations expressed in Chapter 3, any reference to geological origin has to be treated with caution. As a consequence, a glacial soil may be simply described as till or glaciofluvial deposit. It is important that those making use of the description fully realise the limitations and contradictions of the descriptions as the *in situ* behaviour may be different from that inferred from those descriptions.

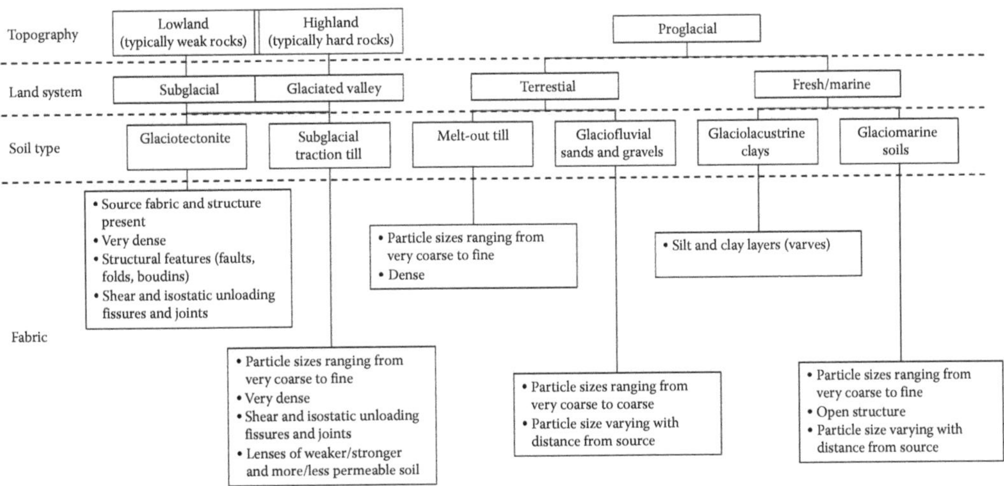

Figure 4.5 General description of fabric type in glacial soils.

Table 4.11 Geotechnical characteristics of glacial tills

				Relative comparison of geotechnical properties[a]			
Till	*Class*	*Fabric*	*OCR*	*Density*	*Compressibility*	*Permeability*	*Anisotropy*
Deformation	G W Mg Mc	Deformed basal sediments or bedrock	1	5	3	5–8	
Lodgement	G W Mg Mc	Interlaying of glaciofluvial, joints, fissures, contortions; preferred clast orientation	2–5	4–7 5–8 6–8 6–8	1 2 2 3	5–6 2–3 4–5 2	7
Melt-out	G W Mg Mc	Occasional interlaying with glaciofluvial; clast orienteered with englacial state	1–2	2–4 2–6 2–6 2–7	2–4 3–5 3–6 4–7	7–9 4–5 5–8 3–4	3–5

Source: After McGown, A. and E. Derbyshire. *Quarterly Journal of Engineering Geology and Hydrogeology*, 10(4); 1977: 389–410.

Note: G, granular till; W, well-graded till; Mg, matrix till (granular); Mc, matrix till (cohesive).

[a] 1 is low; 9 is high.

4.4 CLASSIFICATION

Table 4.12 is a guide to the parameters used to classify glacial soils. It refers to the type of glacial soil and sample and the possible tests that could be carried out depending on the quality of the sample, highlighting the fact that some glacial soils, which include fine- and coarse-grained particles, will be subject to a range of classification tests. The table allows for the fact that it is not always possible to obtain Class 1 or 2 samples; it may only be possible to retrieve disturbed samples.

The European principles of soil classification for engineering purposes (Table 4.13) are based on particle size distribution, plasticity, organic content and genesis. The classification starts with the principal fraction, which is organic, fine-grained or coarse-grained soil. Coarse-grained soils are separated into sands and gravels; that is, cobbles and boulders are not included in the classification but should be included in the description. Fine-grained soils are divided into low, medium and high plasticity soils and into clays and silts. Glacial soils can be fine-, coarse-, very-coarse-grained or composite soils; glacial soils are not organic. The classification of coarse-grained soils depends on the particle size distribution; and for fine-grained soils, consistency limits. Many glacial soils are composite soils and can contain fine-grained or coarse-grained particles or both, which means that they be classified using the full spectrum of classification tests.

The European classification is one of several engineering classification schemes. Table 4.14 is the American USCS classification scheme. These two schemes are different, so care must be taken when using classification data to derive engineering properties. For example, a soil may be described either as a sandy clay or as a clayey sand, a subjective description yet critical when it comes to interpreting the engineering behaviour. A soil may be described as thinly laminated silt and clay (i.e. a glaciolacustrine deposit), yet the consistency limits may show it to be a silt. If 50% or more by weight passes through the 0.063-mm sieve (ESCS),

Table 4.12 Relevance of sample type to physical characteristics to classify glacial soils

Parameter	Type of soil							
	Matrix-dominated soil			Clast-dominated soil			Sands and gravels	
	Undisturbed[a]	*Disturbed*[b]	*Remoulded*[c]	*Undisturbed*[a]	*Disturbed*[b]	*Remoulded*[c]	*Disturbed*[b]	*Remoulded*[c]
Particle description	√	√	√	√	√	√	√	√
Fabric	√	√[d]		√	√[d]		√[d]	
Geological description	√	√[d]		√	√[d]		√[d]	
Soil classification	√	√[d]		√	√[d]		√[d]	
Water content	√			√				
Bulk density	√			√				
Minimum and maximum density					√	√	√	√
Consistency limits	√	√	√					
Particle size distribution	√	√	√	√	√	√	√	√
Undrained shear strength	√							
Permeability	√			√			√[e]	
Sensitivity	√							

[a] Class 1 or 2 samples which may only be possible in glacial soils largely composed of fine grained soils.
[b] Class 2, 3 or 4 samples? May contain some features of the fabric.
[c] Class 4 or 5 samples which have retained all the particles but no features of the fabric.
[d] This depends on the amount of disturbance.
[e] Provided the *in situ* density is known and there is no effect of the fabric.

Table 4.13 European Soil Classification System (ESCS)

Criteria for allocation of symbols and names to individual soil groups based on laboratory testing[a]				*Soil classification*	
				Symbol	*Group name*[b]
Coarse-grained soils (more than 50% retained on the 0.063-mm sieve)	Gravel (more than 50% retained on the 2-mm sieve)	Pure gravel (less than 5% of fine grains[e,f])	$C_u \geq 15; 1 \leq C_c \leq 3$[c]	GrW	Well-graded gravel[d]
			$6 < C_u < 15; C_c < 1$[c]	GrM	Medium-graded gravel[d]
			$C_u < 6; C_c < 1$[c]	GrP	Poorly graded gravel[d]
		Gravel with fine grains (more than 15% of fine grains[e,f])	Fine grains are classified as siL, sil or siH	siGR	Silty gravel[d]
			Fine grains are classified as clL, cll or clH	clGR	Clayey gravel[d]
	Sand (50% or more passing through the 2-mm sieve)	Pure sand (less than 15% of fine grains[e,f])	$C_u \geq 15; 1 \leq C_c \leq 3$[c]	SaW	Well-graded sand[g]
			$6 < C_u < 15; C_c < 1$[c]	SaM	Medium-graded sand[g]
			$C_u < 6$ i/ili $C_c < 1$[c]	SaP	Poorly graded sand[g]
		Sand with small grains (more than 15% fine grains[f,h])	Fine grains are classified as siL, sil ili siH	siSa	Silty sand[g]
			Fine grains are classified as clL, cll ili clH	clSa	Clayey sand[g]
Fine-grained soils (50% or more passing through the 0.063-mm sieve[i])	Liquid limit less than 35%	Inorganic[f]	At or above A line	ClL	Low plasticity clay[i]
			Below A line	SiL	Low plasticity silt[i]
	Liquid limit from 35%–50%	Inorganic[f]	At or above A line	Cll	Medium plasticity clay[i]
			Below A line	Sil	Medium plasticity silt[i]
	Liquid limit greater than 50%	Inorganic[f]	At or above A line	ClH	High plasticity clay[i]
			Below A line	SiH	High plasticity silt[i]
Organic soil		Primary organic matter, dark colour and organic odour		Or	Organic soil

Source: After BS EN ISO 14688-2:2004+A1: 2013. *Geotechnical Investigation and Testing – Identification and Classification of Soil – Part 2: Principles for a Classification.* British Standards Institution, London.

a Based on materials passing through the 63-mm sieve.

b If soil samples *in situ* contain pieces or blocks or both, the name of the soil group must be extended with 'with pieces' or 'with blocks' or 'with pieces and blocks'.

c $C_u = D_{60}/D_{10}$; $C_c = (D_{30})2/(D_{10} \times D_{60})$.

d If the soil contains ≥15% of sand, then the mark 'sa' should be added in front of the group symbol, while the wording 'sandy' should be added in front of the group name.

e Depending on grading and plasticity, sands with 5%–15% of fine grains are marked as follows: siGrW, silty well-graded sand; siGrM, silty medium-graded gravel; siGrW, silty poorly graded gravel; clGrW, clayey well-graded gravel; clGrM, clayey medium-graded gravel; clGrW, clayey poorly graded gravel.

f If fine grains contain organic matter, then the mark 'or' should be added in small letters in front of the group name symbol, while the wording 'organic' should be added in front of the group name.

g If the soil contains ≥15% of gravel, then the mark 'gr' should be added in small letters in front of the group name symbol, while the wording 'gravelly' should be added in front of the group name.

h Depending on grading and plasticity, sands with 5%–15% of fine grains get the following marks: siSaW, silty well-graded sand; siSaM, silty medium-graded sand; siSaW, silty poorly graded sand; clSaW, clayey well-graded sand; clSaM, clayey medium-graded sand; clSaW, clayey poorly graded sand.

i If the soil contains ≥15% of coarse-grained material, then the mark 'sa' or 'gr' should be added in small letters in front of the group name symbol, while the wording 'sandy' or 'gravelly' should be added, depending on which of the two materials is dominant.

Table 4.14 American Soil Classification System (USCS)

Criteria for allocation of symbols and names to individual soil groups based on laboratory testing[a]				*Soil classification*	
				Symbol	Group name[b]
Coarse-grained soils (more than 50% remains on sieve No. 200–0.075 mm)	Gravel (more than 50% retained on the sieve No. 4–4.75 mm)	Pure gravel (less than 5% of fine grains[e])	$C_u \geq 4$ and $1 \leq C_c \leq 3$[c]	GW	Well-graded gravel[d]
			$C_u < 4$ and/or $1 > C_c > 3$[c]	GP	Poorly graded gravel[d]
		Gravel with fine grains (more than 12% of fine grains[e])	Fine grains are classified as ML or MH	GM	Silty gravel[d,f,g]
			Fine grains are classified as CL or CH	GC	Clayey gravel[d,f,g]
	Sand (50% or more grains passing through sieve No. 4–4.75 mm)	Pure sand (less than 5% of fine particles[i])	$C_u \geq 6$ and $1 \leq C_c \leq 3$[c]	SW	Well-graded sand[h]
			$C_u < 6$ and/or $1 > C_c > 3$[c]	SP	Poorly graded sand[h]
		Sand with fine grains (more than 12% of fine grains[i])	Fine grains are classified as ML or MH	SM	Silty sand[f,g,h]
			Fine grains are classified as CL or CH	SC	Clayey sand[f,g,h]
Fine-grained soils (50% or more passing through sieve No. 200–0.075 mm)	Silt and clay (liquid limit less than 50%)	Inorganic	IP > 7 and at or above A-line[j]	CL	Clay[k,l,m]
			IP < 4 or below A-line[j]	ML	Silt[k,l,m]
		Organic	(Liquid limit – drying in oven)/(Liquid limit – without drying in oven) <0.75	OL	Organic clay[k,l,m,n] Organic silt[k,l,m,n]
	Silt and clay (liquid limit in excess of 50%)	Inorganic	IP at and above A-line	CH	Fat clay[k,l,m]
			IP below the A-line	MH	Elastic silt[k,l,m]
		Organic	(Liquid limit – drying in oven)/(Liquid limit – without drying in oven) <0.75	OH	Organic clay[k,l,m,p]
				MH	Organic silt[k,l,m,q]
Highly organic soil		Primary organic matter, dark in colour, with organic odour		PT	Peat

Source: After ASTM D2487 – 11. *Standard Practice for Classification of Soils for* Engineering *Purposes (Unified Soil Classification System)*. ASTM International, West Conshohocken, PA.

[a] Based on materials passing through the sieve of 3 in, 75 mm.
[b] If soil samples *in situ* contain pieces or blocks or both, the name of the soil group must be extended with 'with pieces' or 'with blocks' or 'with pieces and blocks'.
[c] $C_u = D_{60}/D_{10}$; $C_c = (D_{30})2/(D_{10} \times D_{60})$.
[d] If soil contains ≥15% of sand, the name of the soil group must be extended with 'with sand'.
[e] Gravels with 5%–12% fine grains get double symbols: GW-GM, well-graded gravel with silt; GW-GC, well-graded gravel with clay; GP-GM, poorly graded gravel with silt; GP-GC, poorly graded gravel with clay.
[f] If fine grains are classified as CL-ML, then double symbols GC-GM or SC-SM should be used.
[g] If fine grains are organic, the name of the soil group should be extended by adding 'with organic fine grains'.
[h] If soil contains ≥15% of gravel, the name of the soil group should be extended by adding 'with gravel'.
[i] Sand with 5%–12% of fine grains get double symbols: SW-SM, well-graded sand with silt; SW-SC, well-graded sand with clay; SP-SM, poorly graded sand with silt; SP-SC, poorly graded sand with clay.
[j] If a pair of values (wL, IP) in the plasticity diagram is situated within the hatched area (4 < IP < 7), the soil is designated as CL-ML, as silty clay.
[k] If the soil contains 15%–30% of material above the sieve No. 200–0.075 mm, the name of the soil group should be extended by adding 'with sand' or 'with gravel', depending on which of these two materials is dominant.
[l] If the soil contains ≥30% of material above the sieve No. 200–0.075 mm, and if the sand is dominant, the name of the soil group should be extended by adding 'sandy'.
[m] If the soil contains ≥30% of material above the sieve No. 200–0.075 mm, and if the gravel is dominant, the name of the soil group should be extended by adding 'gravelly'.
[n] IP ≥ 4 and at or above the A-line.
[o] IP < 4 or below the A-line.
[p] IP at or above the A-line.
[q] IP below the A-line.

then a soil is classified as a fine-grained soil. In the USCS scheme, it is 0.075-mm sieve. There are many empirical correlations in geotechnical engineering that, usefully, provide an initial assessment of geotechnical characteristics, a framework to establish validity of test results highlighting any anomalies and parameters that may not be obtained from the ground investigation. These include correlations between geotechnical characteristics and classification data. Therefore, it is important to appreciate the standard used to classify a soil when applying a published empirical relationship.

Many glacial tills are bimodal, which means that they could be classed as fine or coarse grained depending on the percentage of fine or coarse grains. However, the engineering behaviour of a coarse-grained soil will be dominated by the fine-grained content even if it is less than 50% but above 15%. This means a matrix-dominated till with more than 50% by weight of coarse particles would be considered a coarse-grained soil even though it exhibits a medium to high dry strength. This is not consistent with the engineering behaviour of these tills, which are often described as clays. However, the standard does suggest that, for composite soils, the dry strength should be an indication of whether a soil is fine grained or not, because if it does have a dry strength, it will behave as a fine-grained soil. In practice, this means that it may be possible to excavate without support but collapse could occur shortly after excavation; the bearing capacity will exceed that derived from the undrained shear strength because of the rate of dissipation of excess pore pressure.

Classification tests are normally carried out on samples retrieved from boreholes and trial pits, and, for the reasons given in Section 3.1, may not be representative of the soil being sampled. It is impossible to retrieve representative samples from soils containing a significant number of boulders and cobbles because it is impossible to retrieve boulders and large cobbles from boreholes. Samples may not include lenses of clay, sands and gravels, which can be present in some glacial soils because none of the boreholes penetrate the lenses (Figure 4.6). This means that care must be taken when interpreting borehole logs to ensure that the soil description is truly representative of the stratum and not just the samples. The description and classification process not only helps identify strata but is also used to

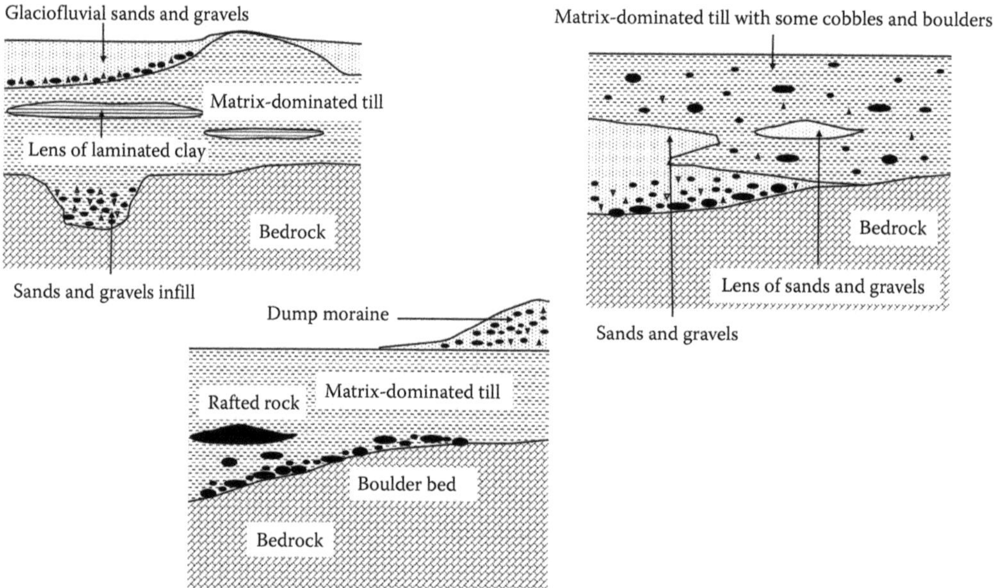

Figure 4.6 Examples of features found in glacial soils that can lead to errors in creating the ground model without additional ground investigation to establish the extent of these features.

identify which tests have to be carried out to determine the design parameters. For example, a soil described as a glaciolacustrine deposit will be known to be anisotropic so that tests for permeability have to determine both vertical and horizontal hydraulic conductivities. The classification parameters of relative density and strength index give an indication of the mechanical characteristics, but they are normally derived from a range of tests specifically designed to determine mechanical characteristics.

Glaciofluvial deposits (excluding glaciolacustrine clays) and clast-dominated tills are likely to be coarse grained, so classification for engineering purposes will be based on the particle size distribution. Matrix-dominated tills, excluding those containing a significant amount of boulders and cobbles, are classified according to their consistency limits and particle size distribution. Given that many types of glacial soils contain significant amount of coarse particles, it is possible to use the grading curve to describe the soil. For example, a glacial till may be described as a gravelly, sandy clay based on the particle size distribution and strength index.

Engineering soils are further classified according to their relative density (if dominated by coarse-grained particles) or strength index (if dominated by fine-grained particles). Unlike the ESCS and USCS classifications in which dominance refers to particles by weight, engineering dominance refers to behaviour. Field tests (e.g. SPT) are used to determine the relative density of coarse-grained glacial soils, which means that the soils can be described as very loose to very dense using an empirical relationship between the field test results and relative density. Thus, a full classification of glaciofluvial soils (excluding glaciolacustrine clays) and clast-dominated tills will be based on the relative density, particle size distribution and principal fraction with secondary descriptions based on the percentage by weight of other fractions.

It is possible to use particle size distribution to distinguish between silt and clay, but consistency limits are found adequate for that purpose for soils in which the fine-grained particles dominate the engineering behaviour, that is, matrix-dominated tills and glaciolacustrine deposits. If a glacial soil contains sufficient quantity of fine-grained material, then the strength index (undrained shear strength) is used to classify those soils. The distinction between a fine-grained and coarse-grained soil in terms of classification is if a soil contains more than 40% by weight of silt and/or clay (Table 4.3). However, Gens and Hight (1979) and Stephenson et al. (1988) suggested that soils containing more than 15% by weight of fine-grained material will behave as a fine-grained soil; that is, pore pressures will develop if the soil is loaded relatively quickly compared to the construction period. Thus, a composite soil may be classed as a coarse-grained soil, but its engineering behaviour is that of a fine-grained soil. Some matrix-dominated tills may exhibit a strength index (undrained shear strength) in excess of 300 kPa, which means that they are classified as weak rock according to BS EN ISO 14688 (2013) though it is likely to be described as a soil. This can be an issue when choosing appropriate excavation techniques.

Sensitivity is also used to classify fine-grained soils though many matrix-dominated tills are insensitive as the depositional process effectively remoulds the till. However, some glacial soils, such as quick clays, are extremely sensitive to the extent that they may turn from a solid to a semi-liquid when subject to load. These clays include Leda Clay and Champlain Sea Clay from Canada, and the quick clays in Norway, Russian, Sweden, Finland and the United States. The clay size particles were deposited in a marine environment creating a strongly bonded marine deposit in which the bonds were created by negative sodium cations. While the clay deposit was strong *in situ*, it had a very open structure. Following glacial recession, the land mass rose converting the marine soil to a terrestrial soil. Overtime, the sodium salts were washed away by rainwater leaving a weakened structure, which is highly susceptible to collapse.

4.4.1 Water content

The water content (w) of saturated soils is an indirect measure of the void ratio ($e = wG_s$). An increase in water content, therefore void ratio, leads to a reduction in density, stiffness and strength and an increase in permeability. It is used with consistency limits to develop empirical correlations with geotechnical characteristics for fine-grained soils. Water content, the mass of water expressed as a percentage of the dry mass, is based on 30 g, 300 g or 3 kg of an undisturbed sample depending on whether it is a fine-grained, a fine-to-coarse-grained or a coarse-grained soil though it is difficult to recover undisturbed samples of coarse-grained soils, so it is unusual to assess the water content of those soils. Therefore, water content measurements usually relate to those glacial soils that are considered fine grained, which includes laminated clays and matrix-dominated tills. This means that at least 300 g of soil is required for most glacial soils. The water content is expressed as the water content of the whole mass not the matrix. This has implications when using empirical correlations described in Section 4.4.3 developed for fine-grained soils.

4.4.2 Particle size distribution

Particle size distribution helps with identifying the type of glacial soil and the stratum. It is used for selecting fill materials for embankment construction, liners for impermeable barriers, road sub-base materials, drainage filters, ground stabilisation and modification techniques and aggregate sources. Glacial soils have the most diverse range of particle size distribution of any soil, yet within any category of glacial soil, it is possible to make some observations that help identify the type of soil and how it will behave as shown in Figure 4.7a, a ternary diagram highlighting the relationship between particle size distributions and description. McGown and Derbyshire (1977) suggested a description of tills based on the percentage of fines (Table 4.15). Gens and Hight (1979) suggested that soils containing more than 15% fines will have characteristics of fine-grained soils. Stephenson et al. (1988) suggested that there is a difference in hydraulic behaviour depending on whether the fines criteria are more or less than 15%–20%. Barnes (1988) suggested that more than 40% granular content led to a significant reduction in dry density. Winter et al. (1998), using gravel greater than 20 mm with sand and clay mixes, suggested that 45% gravel content led to a significant reduction in dry density of compacted soils.

Soils are divided into very-coarse-, coarse- and fine-grained soils, but glacial soils can exist in any one of these three categories or more than one category. For example, glaciofluvial soils can be very-coarse- or coarse-grained soils or mixture of these soils; glaciolacustrine clays are fine-grained soils; and glacial tills can fall into any category or be a mixture of two or three of these categories.

Table 4.4 lists the sizes of particles and the symbols used to describe the particles. This shows that, in routine ground investigations, samples are unlikely to include coarse particles

Table 4.15 Relation between percentage of fines and type of till

Dominant soil fraction	*Nature of dominant fraction*	*% of fines*	*Textural description*
Clasts	Coarse	0–15	Granular (G)
Non-dominant fraction		15–45	Well graded (W)
Matrix	Coarse	45–70	Granular matrix (Mg)
	Fine	70–100	Cohesive matrix (Mc)

Source: After McGown, A. and E. Derbyshire. *Quarterly Journal of Engineering Geology and Hydrogeology*, 10(4); 1977: 389–410.

greater than all but the smallest cobbles since it is impossible to sample the larger particles from boreholes. Therefore, the full classification of glacial soils containing very coarse particles (Table 4.5) is likely to be based on samples from excavations and exposures. It is important to gather information on large particles as they can impact on construction processes (e.g. piling, dredging and earthworks).

There are several tests used to determine the particle size distribution depending on the range and maximum particle sizes, the stability of the soil grains and the presence of fine-grained soils. The sample size depends on the maximum particle size. Clean (no fines content) sands and gravels can be assessed by dry sieving. Sands and gravels containing fines are washed first to remove the fines. The retained sample is dry sieved using the total weight of soil (including the fines) to assess the distribution. Fine-grained soils with some sand are pretreated to break down the soil and remove the sand before carrying out a sedimentation test.

The particle size distribution of engineering soils is usually expressed as the percentage by weight of particles between certain sizes plotted on a logarithmic scale (Figure 4.2). An alternative way to present the particle size distribution is a ternary diagram (Figure 4.7b), which shows the composition of various glacial soils.

BS EN ISO 17892-4 describes the procedures to determine the particle size distribution. Samples with less than 10% of particles by weight smaller than 0.063 mm do not require a sedimentation test; samples with less than 10% larger than 0.063 mm do not require a sieve analysis. Table 4.16 are the recommended masses required for sieving.

Soils can be described as even-graded soils (Table 4.17) in which the particles are of a similar size (e.g. even-graded sand), multi-graded soil (e.g. glacial tills) and gap-graded soils (e.g. bimodal glacial tills). A multi-graded soil is the densest of these soils since the voids between the largest particles are filled with smaller particles. Table 4.18 shows a theoretical relationship between particle size and surface area based on spherical particles.

The theoretical maximum density can be calculated from these data and the Fuller curve:

$$P = \sqrt{\left(\frac{D}{D_{max}}\right)} \times 100 \tag{4.1}$$

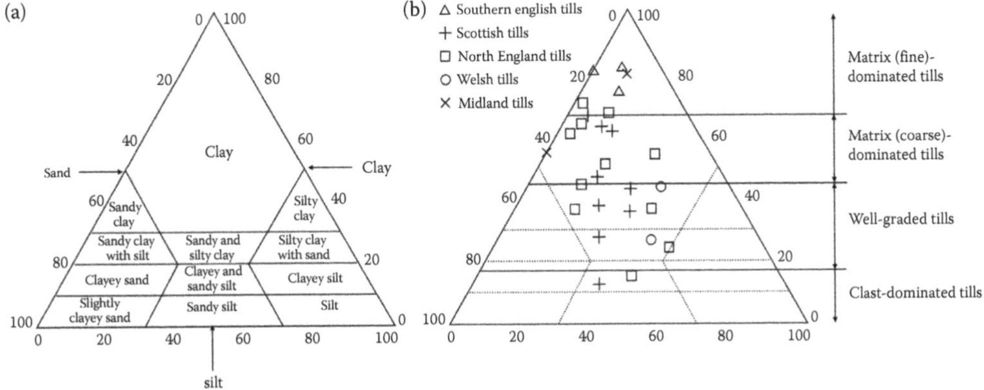

Figure 4.7 Ternary diagrams showing (a) the range of soils composition and their relation to engineering descriptions and (b) the composition of a number of UK tills. (After Trenter, N. A. *Engineering in Glacial Tills. CIRIA*, London, 1999.)

Table 4.16 Recommended minimum mass of soil for sieve analysis

Particle diameter (mm)	*Recommended minimum mass required for sieving*
<2	100 g
2	100 g
6.3	300 g
10	500 g
20	2 kg
37.5	15 kg
63	70 kg
100	150 kg
150	500 kg
200	1000 kg

Source: After BS EN ISO 17892-4:2014. *Geotechnical Investigation and Testing. Laboratory Testing of Soil. Part 4. Determination of Particle Size Distribution.* British Standards Institution, London; Head, K. H. *Manual of Soil Laboratory Testing, Vol. 1, Soil Classification and Compaction Tests.* Pentech, London, 1984.

Table 4.17 Shape of grading curve

Description	C_u	C_c
Multi-graded	>15	$1 < C_c < 3$
Medium graded	6–15	<1
Even graded	<6	<1
Gap graded	Usually high	Any (usually <0.5)

Source: After BS EN ISO 14688-2:2004+A1: 2013. *Geotechnical Investigation and Testing – Identification and Classification of Soil – Part 2: Principles for a Classification.* British Standards Institution, London.

Table 4.18 Theoretical assessment of the number of particles per gram and the surface area demonstrating the distribution of particle sizes and the impact on the behaviour

Soil category	*Particle size (mm)*	*Approximate mass of particle (g)*	*Approximate number of particles per gram*	*Approximate surface area (mm²/g)*	*Approximate surface area (m²/g)*
Cobble (largest from SI)	75	590	1.7/kg	30	
Coarse sand	1	0.0014	720	2300	
Fine sand	0.1	1.4×10^{-6}	7.2×10^{6}	23000	0.023
Medium silt	0.01	1.4×10^{-9}	7.2×10^{8}	23×10^{5}	0.23
Clay	0.001	1.4×10^{-12}	7.2×10^{11}	2.3×10^{6}	2.3

Source: After Head, K. H. *Manual of Soil Laboratory Testing, Vol. 1, Soil Classification and Compaction Tests.* Pentech, London, 1984.

where P is the percentage by weight of particles smaller than diameter D and D_{max} is the maximum particle size. The theoretical density of composite soils varies between 1.92 and 2.30 Mg/m^3 for a water content of 30%–10% assuming a particle density of 2.65 Mg/m^3. These water contents are typical of tills as are the theoretical densities suggesting that these soils are very dense.

It is necessary with multimodal and bimodal glacial soils to carry out a stage process to assess the distribution because of the restriction to the weight that can be retained on each nest of sieves. The sample is passed through a set of coarse sieves. It is then riffled to obtain a subsample for the smaller sieves. The riffling process is repeated until the distribution of the whole sample is obtained.

This is feasible for sands and gravels, but if a glacial soil contains cobbles and boulders, then it is necessary to obtain a much larger sample and remove particles greater than 75 mm by hand. Figure 4.8 shows the process for coarse-grained soils (e.g. glaciofluvial soils) containing very-coarse-grained particles and coarse-grained soils containing fines (e.g. clast-dominated tills).

Matrix-dominated tills are mixtures of fine- and coarse-grained particles, which means that the distribution is assessed using wet sieving and sedimentation procedures (if the percentages of clay and silt are required). The soil is first pretreated with a dispersing agent to break down the fine-grained component and ensure that it does not adhere to the coarse-grained particles. The sample is then washed through a set of sieves to remove the fine-grained particles. The fine-grained particles are subject to a sedimentation test.

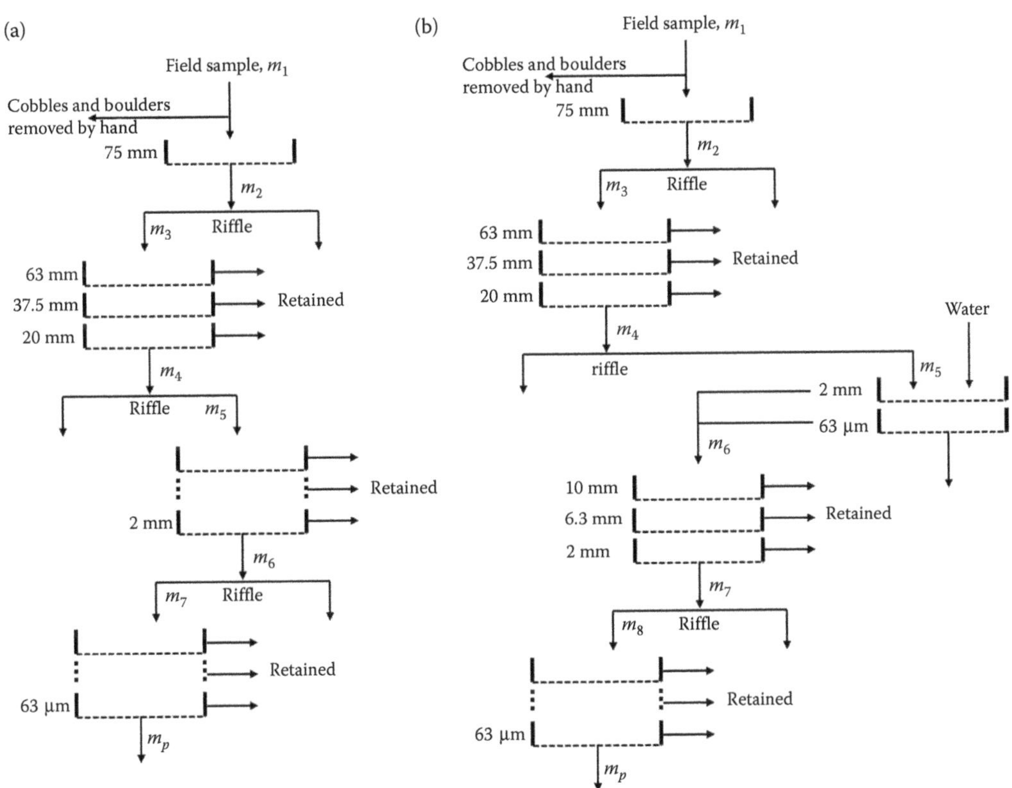

Figure 4.8 (a) The dry sieving process to assess the particle size distribution of samples of coarse-grained soils (e.g. glaciofluvial soils) containing very coarse and coarse particles and (b) the wet sieving process for coarse-grained soils containing fines (e.g. clast-dominated tills).

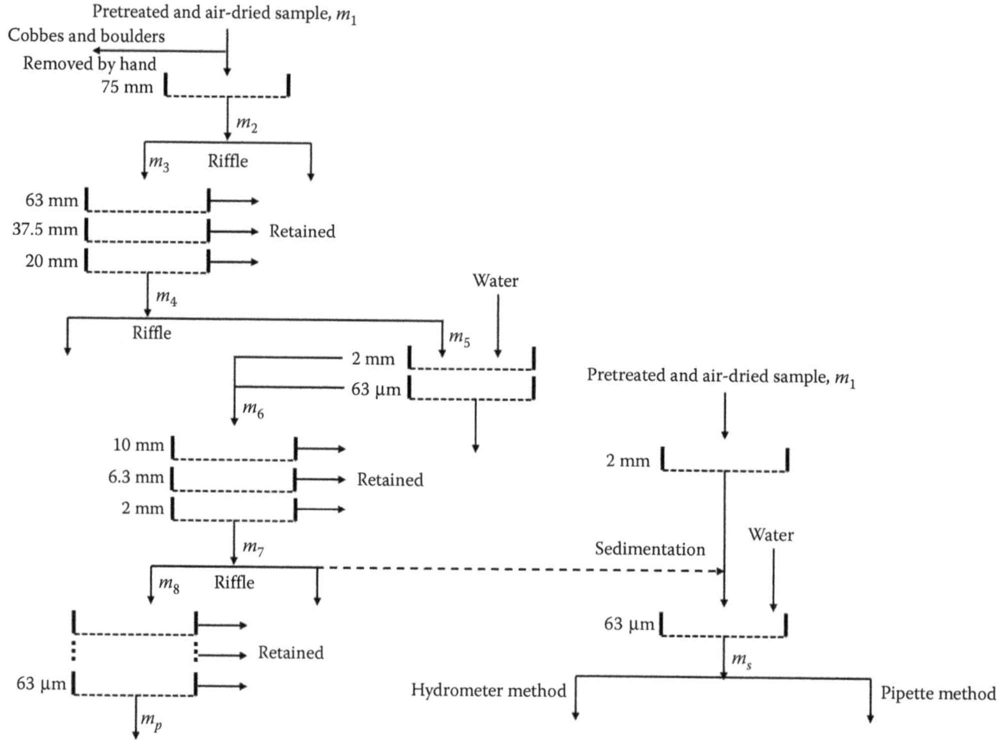

Figure 4.9 Wet sieving process to assess the particle size distribution of samples containing coarse and fine particles (e.g. matrix-dominated glacial till) and fine particles (e.g. glaciolacustrine clays).

The procedure for a matrix-dominated till containing cobbles is shown in Figure 4.9, which includes soaking the soil in dispersant solution, removing very coarse material by hand, wet sieving the sample through the coarse set of sieves, possibly riffling the sample if there is sufficient gravel present, dry sieving of the sand particles and sedimentation test, if necessary, on the fine-grained particles.

Thus, there are three cases to consider when dealing with composite soils: coarse-grained soil (dry sieve), fine-grained soil (sedimentation) and composite soils (wet sieve and sedimentation). If the soil contains particles no greater than 2 mm, then the size fraction obtained from the sedimentation test is expressed as a percentage of the total sample used for the test. If the soil is a clast-dominated till or glaciofluvial deposit, then a larger sample is used, which means that it is necessary to riffle the sample a number of times to obtain the distribution of the sample finer than the 2 mm. A portion of the sample may be lost in the pretreatment. A correction is applied to correct for that loss. The loss due to pretreatment is given by

$$\text{Loss due to pretreatment} = \frac{m_3 - m}{m_3} \times 100\% \tag{4.2}$$

where m is the dry mass after pretreatment and m_3 is the mass of riffled soil. If the percentage is less than 1%, no further correction is necessary and the corrected percentages are the ratio of the dry mass passing the 2-mm sieve to the dry mass of the original sample. If the pretreatment loss exceeds 1%, the mass removed by pretreatment is

$$\text{Mass removed by pretreatment} = \frac{m_3 - m}{m_3} \times m_2 \tag{4.3}$$

where m_2 is the dry mass less than 2 mm. The mass, m_0, of the whole sample after pretreatment is

$$m_0 = m_1 - \frac{m_3 - m}{m_3} \times m_2 \tag{4.4}$$

The corrected mass passing the 2-mm sieve is m_4, where

$$m_4 = m_2 - \frac{m_3 - m}{m_3} \times m_2 \tag{4.5}$$

4.4.3 Consistency limits

Fine-grained soils can exist in four phases (Figure 4.10): solid, semi-solid, plastic and liquid states. These states are separated by the plastic (I_P), liquid (I_L) and shrinkage (I_S) limits. The plastic and liquid limits are known as consistency or Atterberg limits and the difference between the consistency limits is the plasticity index (*PI*). Particles smaller than 425 µm are used to assess the consistency limits, which are related to particle size and mineral composition.

Skempton (1953) showed that the plasticity index is related to the clay fraction (<2 µm) such that for a given clay, its activity is given by

$$\text{Activity} = \frac{PI}{\text{clay fraction}} \tag{4.6}$$

The water content and consistency limits are an indication of the current state of the clay expressed as the liquidity index (*LI*):

$$LI = \frac{w - I_P}{PI} \tag{4.7}$$

Phase	Solid	Semi solid	Plastic	Liquid	Suspension
Limits		Shrinkage limit	Plastic limit	Liquid limit	
Shrinkage	Constant volume	← Volume decreasing / Water content decreasing			
Condition	Hard to stiff	Workable	Sticky	Slurry	Water held suspension
Shear strength	Shear strength decreasing →				Negligible

Figure 4.10 Phases of fine-grained soil and their relationship to water content.

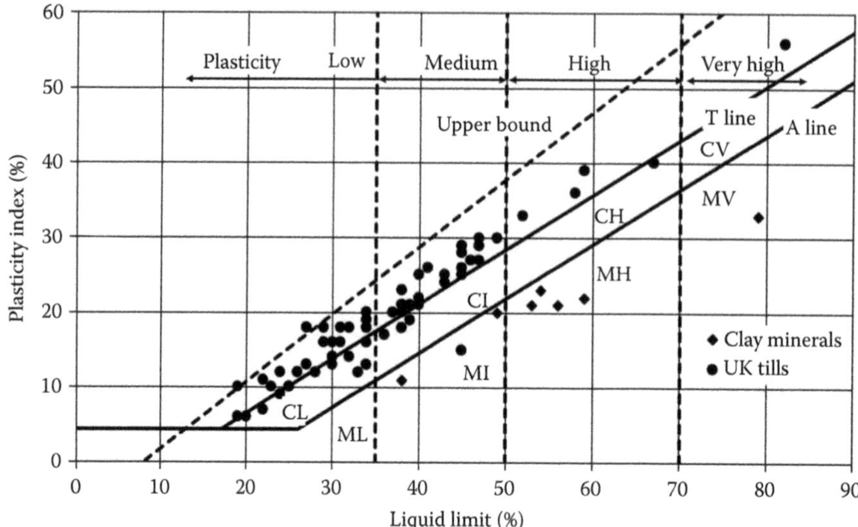

Figure 4.11 Consistency limits for UK matrix-dominated tills and clay minerals. (After Trenter, N. A. *Engineering in Glacial Tills*. CIRIA, London, 1999; Clarke, B. G., D. B. Hughes and S. Hashemi. *Géotechnique*, 58(1); 2008: 67–76.)

The limits are based on the fraction of fine-grained soil; the water content on the whole mass. In fine-grained soils such as glaciolacustrine clays and completely homogenised tills, the liquidity index applies to the whole mass. In composite soil, which includes coarse-grained particles, the liquidity index is based on the water content expressed in terms of the total mass not the mass of the fine-grained fraction. If the water content is expressed in terms of the weight of the fine-grained particles, then the liquidity index will be greater. For example, consider two matrix-dominated tills with 20% and 80% particles less than 425 μm. Assume the liquid limit of the matrix is 50%, the plastic limit is 15% and the water content of the total sample is 18%. The liquidity index of both samples is 0.09, but the liquidity index of the matrix is 2.14 and 0.21 assuming that the coarse-grained particles do not contain any water.

The consistency limits are used to classify a soil using Figure 4.11. The A-line, based on experimental evidence, separates clays from silts. There is also an experimental upper line, the U-line. The soils are divided into low, medium, high, very high and extremely high plasticity. The consistency limits of many glacial soils lie astride the T-line (Boulton and Paul, 1976). The T-line lies in the clay zone, yet many glacial tills compromise erosion products that are silt sized. Erosion products are not necessarily clay minerals (Table 4.6), but most glacial soils do contain some clay minerals, the quantity depending on their source rock and the degree of weathering. Therefore, care has to be taken when applying correlations of geotechnical characteristics with consistency limits based on sedimented soils. However, the consistency limits are a useful means of identifying the stratum though the variation in water content and consistency limits in a matrix-dominated till can be significant.

The character of glacial sediments is a function of the lithology and geochemical properties, the nature and distance of sediment transport and the mode of sediment deposition. This means that source terrains composed of carbonate rocks produce carbonate-rich glacial deposits. The consistency limits reflect the source rock. Subglacial tills can be formed of the underlying bedrock though glacial advances and recessions can complicate that simple view as the source material may have travelled much further. Glacial tills in England south

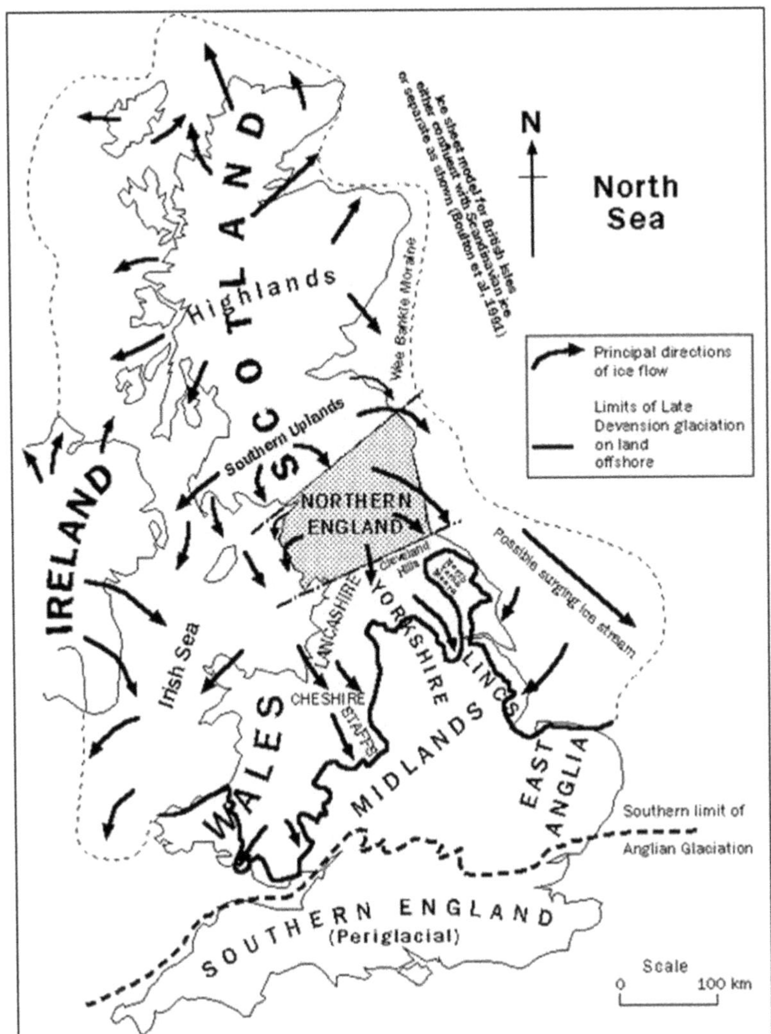

Figure 4.12 Distribution of glacial tills in the United Kingdom and the relevance to the underlying solid geology. (After Eyles, N. and W. R. Dearman. *Bulletin of the International Association of Engineering Geology-Bulletin de l'Association Internationale de Géologie de l'Ingénieur*, 24(1); 1981: 173–184.)

of the Tees Exe line (Figure 4.12) are predominantly derived from chalk, shales, clays and mudstones, which lead to matrix of low to medium plasticity clay (CL to CI); to the north of that line, the matrix-dominated tills are more likely to be low plasticity because the source is dominated by igneous and metamorphic rocks (Trenter, 1999). However, Clarke et al. (2008) has shown that a sequence of glacial soils within a region can vary from low to medium plasticity.

Given that glacial tills are multimodal distribution of particles and the consistency limits are based on particles finer than 425 μm, then the results represent only the fine-grained fraction of matrix-dominated tills. The consistency limits vary with the clay content and liquid and plastic limits (Figure 4.13) such that a reduction in clay content leads to a reduction in the limits and plasticity index consistent with data of glacial soils. Further, Trenter (1999) showed that if the clay fraction is less than 40% the soil will be described as a low plasticity

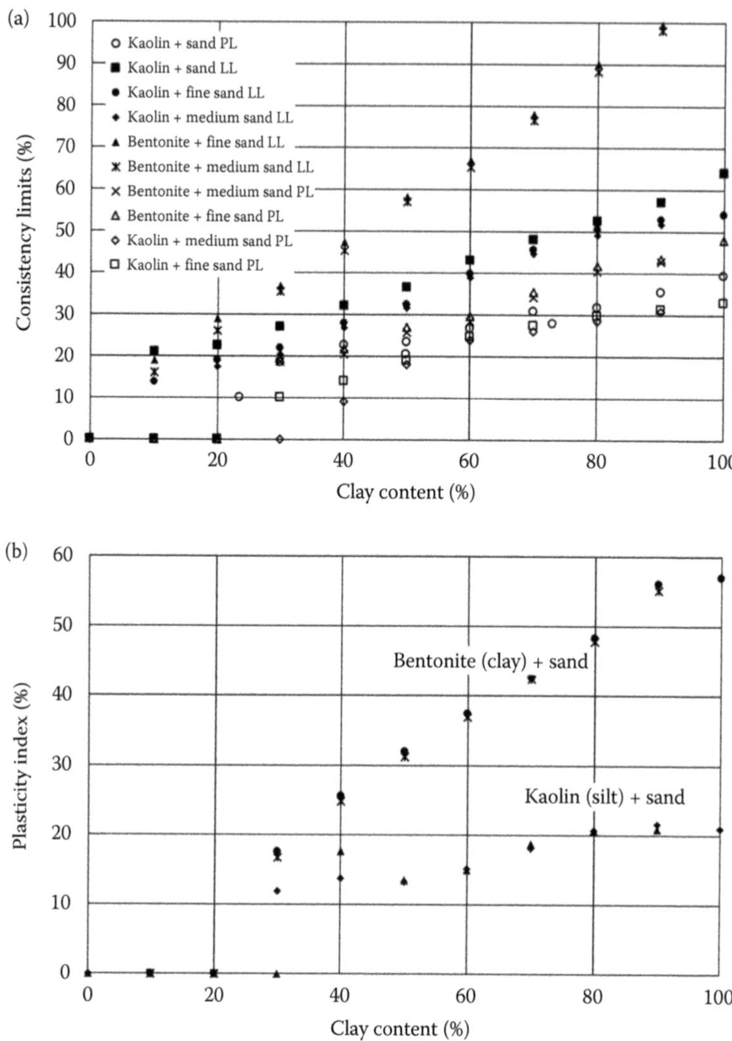

Figure 4.13 Effect of clay fraction on the consistency limits showing an increase in clay content increases (a) the liquid and plastic limits and (b) the plasticity index.

clay and lie about the T-line. Clarke et al. (2008) showed that glaciolacustrine deposits from the NE of England also lay astride the T-line. They developed a regional database of glacial tills; lower, upper and upper weathered tills to show (Figure 4.14) the increase in clay content either due to the deposition process or weathering increased the plasticity index such that the glacial soils moved from low plasticity clays at the bottom of the geological sequence to high plasticity clays at the top of the sequence.

Glacial soils weather by oxidation, hydration, leaching and mechanical disintegration (Eyles and Sladen, 1981; Sladen and Wrigley, 1983). This can be explained by the chemical composition of the source rock. For example, Madgett and Catt (1978) suggest that oxidation of pyrite can create the necessary conditions for solution of carbonate. Sladen and Wrigley (1983) suggest that weathering of tills leads to increased silt and clay content by disintegration, increased clay content and formation of clay minerals due to chemical weathering. They also observed that weathered tills had a higher water content than the

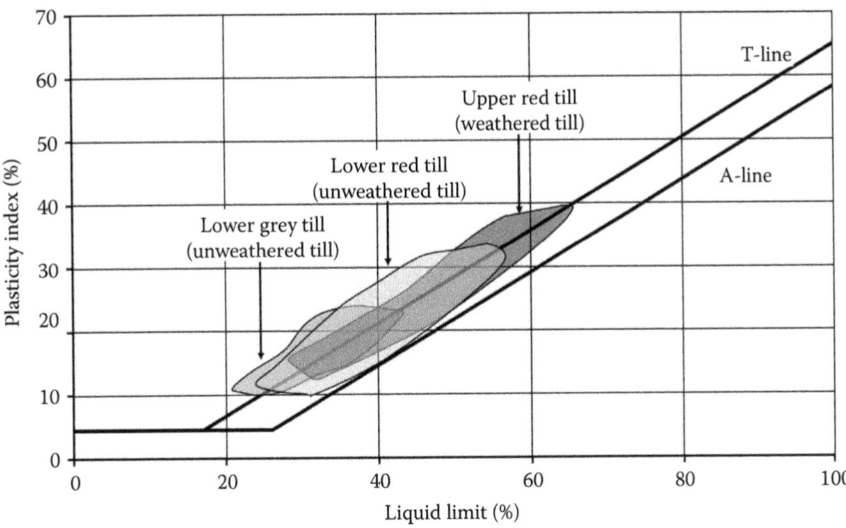

Figure 4.14 Consistency limits for glacial soils in the North East of England showing that they cluster around the T-line and that weathering appears to increase the liquid limit and the plasticity index. (After Clarke, B. G. et al. *Géotechnique*, 58(1); 2008: 67–76.)

unweathered till. Robertson et al. (1994) showed a number of differences between weathered and unweathered till in the NE England (Table 4.19). Mckinlay et al. (1974) showed for tills of the Central Lowlands of Scotland that the percentage of fine particles increased from 18% (unweathered) to 40% (weathered), a reduction in bulk density (2.26–2.05 Mg/m^3) and an increase in water content (12.3%–20.5%). They also showed that the plasticity index of weathered tills was greater than that of the unweathered till, which is similar to the observations of the tills of NE England (Bell and Forster, 1991; Clarke et al., 2008). The top of a till stratum, if near the surface, is likely to be weathered. However, there is some debate as to the extent of the weathering zone as some people suggest that the upper layers of till are a separate till whereas others suggest that it is weathered till. It should be expected that the top of a till will be weathered; therefore, it is important to test representative samples at different depths to distinguish between the weathered and unweathered till even if there are no visible distinguishing features.

Figure 4.10 shows the relation between consistency limits and the soil phases. The consistency limits are often used to identify geotechnical characteristics of silts and clays using empirical correlations, which relate the *in situ* water content to the liquid, plastic and shrinkage limits. This is expressed in terms of the liquidity index (Equation 4.7) or relative consistency (Equation 4.8). Wroth and Wood (1978) showed that the shear strength of remoulded clay at the liquid limit was about 1.7 kPa and at the plastic limit about 170 kPa, which can be used to estimate the shear strength of remoulded clays.

Table 4.19 Effect of weathering on properties of glacial tills in NE England

Description	*Water content (%)*	*Plasticity index (%)*	*Liquidity index (%)*	*Dry density (Mg/m³)*	*Undrained shear strength (kPa)*
Weathered upper till	17 (11–30)	23 (8–36)	0.3 (−0.45 to 0.8)	1.75 (1.50–1.96)	150 (30–375)
Unweathered upper till	14 (9–34)	20 (9–39)	0 (−0.65 to 0.65)	1.83 (1.62–1.93)	180 (50–410)

Source: After Robertson, T. L., B. G. Clarke and D. B. Hughes. *Ground Engineering*, 27(10); 1994: 29–34.

Table 4.20 Consistency index of silts and clays

Term	*Consistency index*
Very soft	<0.25
Soft	0.25–0.50
Firm	0.50–0.75
Stiff	0.75–1.00
Very stiff	>1.00

Source: After BS EN ISO 14688-2:2004+A1: 2013. *Geotechnical Investigation and Testing – Identification and Classification of Soil – Part 2: Principles for a Classification.* British Standards Institution, London.

The consistency index, I_c, used to describe fine-grained soils is expressed as

$$I_c = \frac{(I_L - w)}{PI} \tag{4.8}$$

Since the water content is related to the shear strength, these terms are often interpreted as a proxy for strength; a glacial soil with a water content less than its plastic limit is a very stiff soil, which is likely to have a very high shear strength. BS EN 14688-1 (2013) describes a field test to determine the consistency (Table 4.20). A very soft fine-grained soil is described as very soft if it exudes between fingers when squeezed; soft if it can be moulded with little pressure; firm if it cannot be moulded but can be rolled into 3-mm threads without breaking or crumbling; stiff if those 3-mm threads break; and very stiff if it crumbles under pressure. These qualitative descriptions reflect the water content of the fine-grained component.

It is likely that a sample of glaciolacustrine clays will comprise particles less than 425 μm, but these soils are not a mixture of clays and silts but distinct layers of clays and silts. Therefore, the liquidity index will not represent the two layers but a mix of the two layers. This may not be an issue when considering macro behaviour, but the behaviour of these soils can often be dominated by the fabric and its relation to the direction of loading. It would be very difficult to separate the silt and clay layers and determine the water contents of each layer but the effect of layer thickness on the average liquidity index can be assessed by considering the consistency limits of each layer. In this case, the layers are separated and the properties of each type of layer are determined separately.

The third limit is the shrinkage limit, the water content at which there is no further reduction in volume even if the water content reduces. If the water content falls below the shrinkage limit, cracking can occur, which affects the mass permeability of the soils. Glacial clays are used for impermeable barriers such as clay cores of earth dams, clay liners for rock fill dams and impermeable barriers for land fill sites; these soils often have a natural water content less than the I_P, so the shrinkage limit is an important factor when considering these soils.

4.4.4 Density

The density of undisturbed samples from boreholes is simply the weight of the sample divided by its volume. On occasion, block samples will be retrieved from an excavation. In that case, the density of a subsample may be assessed using the water displacement method. The density (1.90–2.30 Mg/m³) of subglacial tills is typically greater than that for coarse- or fine-grained soils because of the particle size distribution and the method of deposition.

4.4.5 Density index

The density index (I_D) or relative density is a term used for coarse-grained soils, which expresses the *in situ* void ratio, e, in terms of the maximum (e_{max}) and minimum (e_{min}) void ratio:

$$I_D = \frac{e_{max} - e}{e_{max} - e_{min}} \tag{4.9}$$

In practice, for the purposes of soil description, this is based on *in situ* penetration tests as it is not feasible to measure the *in situ* void ratio directly because of the difficulty in obtaining Class 1 samples. The maximum density of coarse-grained particles can be found from a dynamic compaction test. The minimum density can be found by carefully pouring the soil into a chamber full of water.

The penetration resistance depends on grain size as well as density and overburden pressure. Meyerhof (1957) proposed that the N_{60} blow count, N_{60}, was related to the overburden pressure, σ'_v, expressed in kPa and density index by

$$N_{60} = \left(17 + 24\frac{\sigma'_v}{98}\right) I_D^2 \tag{4.10}$$

Skempton (1986) suggested a more generic relationship where the constants a and b were site specific to take into account grain size.

$$N_{60} = \left(a + b\frac{\sigma'_v}{98}\right) I_D^2 \tag{4.11}$$

Thus, the normalised blow count, that is, the blow count at an effective overburden pressure of 98 kPa, is

$$\frac{N_{60}}{I_D^2} = a + b \tag{4.12}$$

which provides a relationship between relative density and SPT N_{60}. Cubinovski and Ishihara (2001) undertook a series of tests on high-quality undisturbed samples obtained by ground freezing to determine a comparison between relative density and blow count. Figure 4.15 shows the variation of void ratio range with grain size, showing that the range increases as the particle size decreases. Cubrinovski and Ishihara (2001) suggested that relative density was related to the normalised value of SPT N_{60} by the void ratio range (Figure 4.16), which led to the proposal:

$$I_D = \left[\frac{N_{60}(e_{\max} - e_{\min})^{1.7}}{9}\left(\frac{98}{\sigma'_v}\right)^{0.5}\right]^{0.5} \tag{4.13}$$

Thus, it is possible to estimate the relative density or strength index from SPT N_{60} and use Table 4.21 to refine the soil classification. This is based on the assumption that the soil tested is coarse grained and does not contain very coarse-grained particles, which will affect SPT N_{60}.

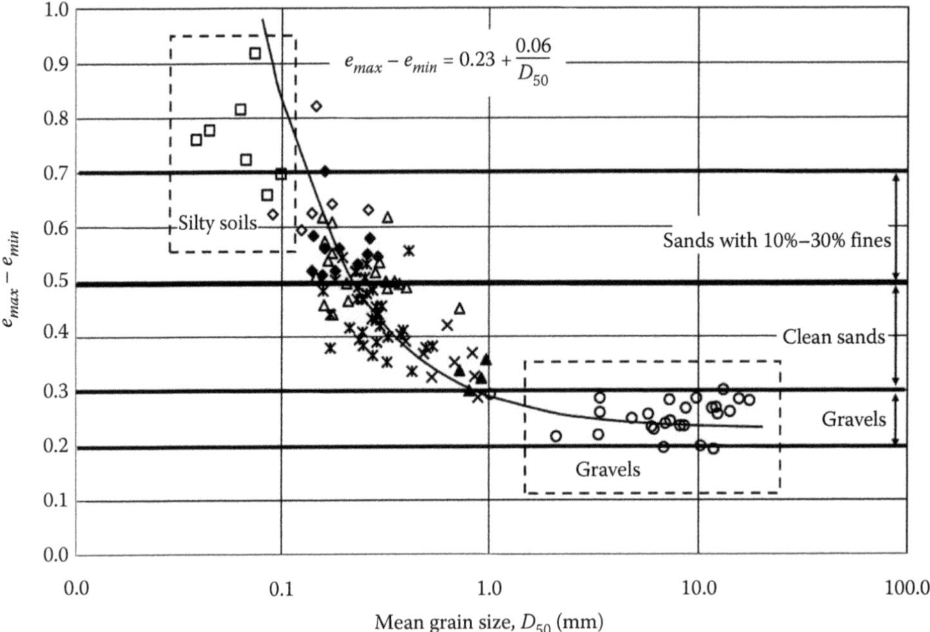

Figure 4.15 Variation of void ratio range with mean grain size. (After Cubrinovski, M. and K. Ishihara. Correlation between penetration resistance and relative density of sandy soils. Istanbul, Turkey: *15th International Conference on Soil Mechanics and Geotechnical Engineering*, University of Canterbury. 2001: 393–396.)

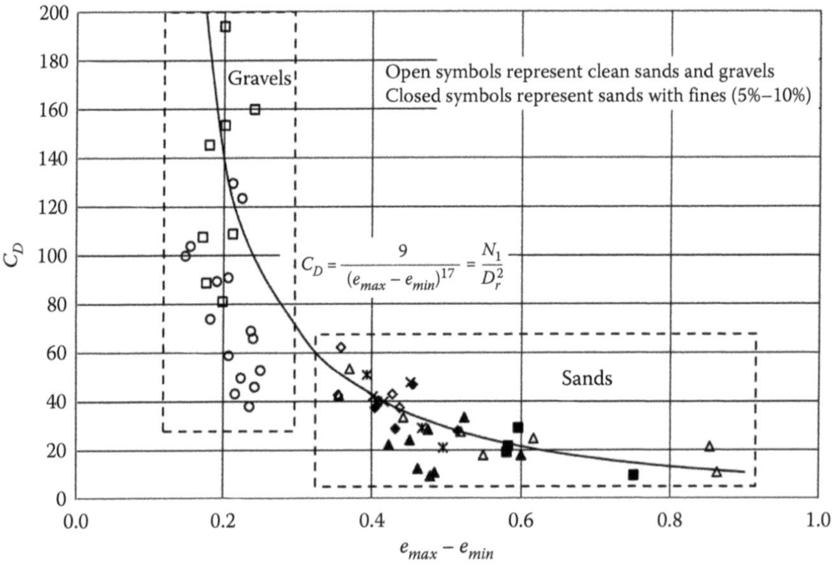

Figure 4.16 Relation between relative density, SPT blow count and void ratio range. (After Cubrinovski, M. and K. Ishihara. Correlation between penetration resistance and relative density of sandy soils. Istanbul, Turkey: *15th International Conference on Soil Mechanics and Geotechnical Engineering*, University of Canterbury. 2001: 393–396.)

Table 4.21 Relation between density description and density index

Term	*Density index (I_D%)*
Very loose	0–15
Loose	15–35
Medium dense	35–65
Dense	65–85
Very dense	85–100

Source: After BS EN ISO 14688-2:2004+A1: 2013. *Geotechnical Investigation and Testing – Identification and Classification of Soil – Part 2: Principles for a Classification.* British Standards Institution, London.

Table 4.22 Relation between strength description and strength index

Term	*Strength index (undrained shear strength) (kPa)*
Extremely low	>10
Very low	10–20
Low	20–40
Medium	40–75
High	75–150
Very high	150–300
Extremely high[a]	>300

Source: After BS EN ISO 14688-2:2004+A1: 2013. *Geotechnical Investigation and Testing – Identification and Classification of Soil – Part 2: Principles for a Classification.* British Standards Institution, London.

[a] These may also be considered as weak rocks.

4.4.6 Strength index

It is common to measure the undrained strength of matrix-dominated tills and glaciolacustrine deposits and use it in geotechnical design but the undrained strength is actually a strength index used to classify fine-grained soils. The strength index is not an intrinsic property of a soil since it depends on the quality of sample, type of test, the sample fabric, the water content and the test procedure. The strength of a fine-grained soil is often measured directly in unconsolidated undrained triaxial tests but is also estimated from penetration tests using empirical correlations, which can be enhanced by using site-specific correlations. The relations between the undrained shear strength and the descriptive term for strength index are listed in Table 4.22.

4.5 GEOTECHNICAL CHARACTERISTICS

Glacial soils include those (a) deposited in water by sedimentation, (b) those deposited in water but can be affected by other processes such as currents and (c) those deposited by ice either accompanied by shearing or (d) by melting of ice. The deposition processes for (a) and (d) are the classic deposition processes used in soil mechanics theory and can be described by the variation in void ratio with consolidation pressure in which the principal stresses are vertical and horizontal. There is an element of shearing in the deposition of the other glacial soils being most pronounced for subglacial tills. The thickness of ice, the pressure and temperature in the basal zone, the movement of the ice and the drainage conditions at the base of the glacier affect the deposition of the till. Boulton (1975) proposed three possible mechanisms: a glacier sliding over a frozen bed, a glacier moving with a deforming bed, in which there is no relative movement between the glacier and the bed, and a glacier sliding over a deforming bed (Figure 2.34). Boulton (1975) further suggested that different drainage conditions would exist in the basal zone. Drained conditions exist when the pore pressure can dissipate because of the permeability of the underlying soils and rocks or because of water channels formed within the basal zone. Undrained conditions exist when there are no water channels and the permeability of the underlying soils and rock is low. Even if the permeability is high, undrained conditions can exist because of the distance to the ice margins. The implication is that a subglacial till can be shown to be lightly over-consolidated, yet be very dense; the density is a result of the particle size distribution and vertical and shear stresses.

Figure 4.17 shows possible stress paths to compare the processes of glaciation, deglaciation and isostatic uplift for subglacial tills with sedimentation and erosion. These stress paths are a simple representation of what might occur. Over time, the till is loaded and unloaded due to the advance and retreat of a glacier, fluctuating pore pressures in the basal zone due to stick/slip phenomenon and thermal variations, fluctuating groundwater levels post-glaciation and post-glacial deposition and erosion. This results in a very complex stress history superimposed upon the major stress changes.

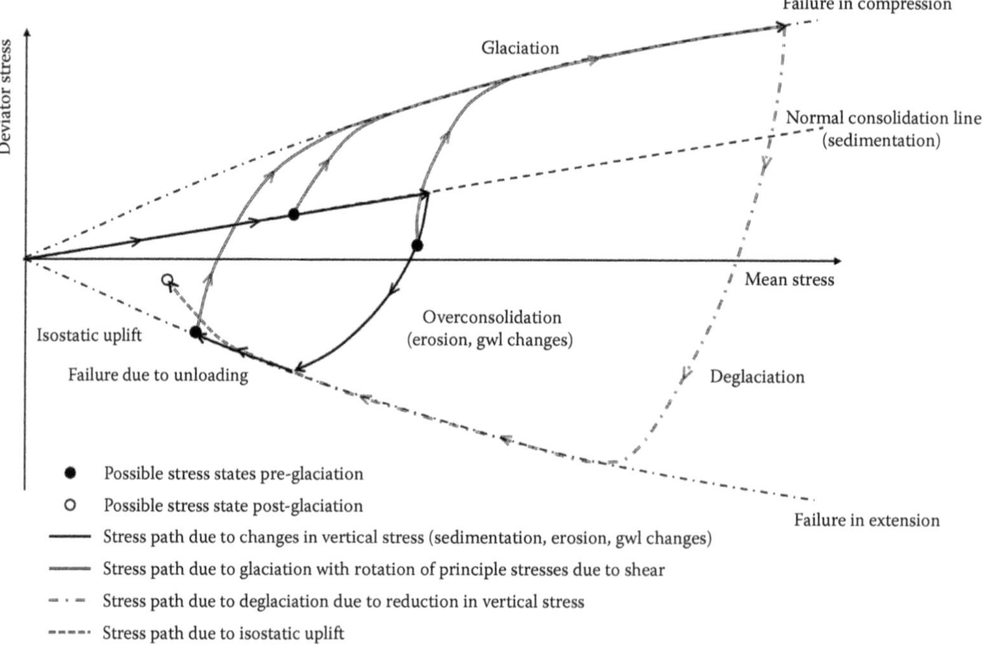

Figure 4.17 Effect of glaciation, deglaciation and isostatic uplift on the stress state within subglacial tills assuming fully drained conditions compared to the stress changes associated with sedimentation and erosion for dense soils.

During glaciation, the soils undergo shear, which means that the vertical stress is no longer a principal stress; that is, there is a rotation of the principal stress axes. During deglaciation, the till is no longer subject to shear and undergoes a reduction in vertical stress leading to failure in extension creating a fissured material. The till undergoes further stress changes due to creep as a result of isostatic uplift, which could lead to a reduction in horizontal stress. These paths have assumed fully drained conditions. It is often assumed that the maximum vertical stress acting on the till due to the weight of the ice is the maximum stress the till was subject to. If that were the case, it would be possible to estimate the thickness of ice from the preconsolidation pressure, a procedure based on the assumption that a soil is gravitationally consolidated. This is not possible because of the effects of shear and isostatic uplift.

However, the simple model of gravitational consolidation in which the vertical stress representing the weight of ice is a principal stress helps explain another anomaly; some subglacial tills are lightly over-consolidated. Two extremes can exist: an undrained condition in which the pore pressure increases with the thickness of ice, which means that there is no change in effective vertical stress; and a drained condition in which there is no change in pore pressure. The former case leads to a normally consolidated till because unloading as the ice melts means that the pore pressure will reduce; thus, the effective vertical stress remains constant. The latter leads to a heavily over-consolidated soil. It is likely that the conditions exist somewhere between these two extremes.

The stress paths in Figure 4.17 are for fully drained conditions. Fully undrained conditions would result in very little change in deviator stress compared to the weight of ice. The process of deformation accompanied by particle breakage leads to an increase in density and hence strength.

This simple model does not take into account the complexity of the local conditions at the base of a glacier. Piotrowski (1987) suggested that these could vary such that the stress history of a glacial till layer can vary vertically and horizontally. This could account for the scatter in strength index, density and water content with depth often associated with glacial soils.

These models demonstrate why subglacial tills can be very dense yet appear to be lightly over-consolidated. It shows why it is impossible to make any definitive statement about the stress history of a subglacial till. It also explains why it is not possible to determine the preconsolidation pressure using the Casagrande method even if the theoretical past total stress (based on ice thickness) can be achieved with the laboratory equipment. The simple assessment of the complex stress changes that take place during deposition of glacial tills explains why tills can be dense and lightly over-consolidated. It also explains why constitutive models developed for soils may not be so relevant.

The description and classification of the soils are used to identify the geological profile. Tests on representative samples from each of the stratum in that profile are used to determine the characteristics for design, which include the deformation, strength and time-dependent characteristics. There are numerous ways to determine these characteristics including those derived empirically from field tests and directly from field and laboratory tests. A routine investigation is likely to use tests set out in standards such as the European and American standards, but these do not necessarily provide the parameters used in design, particularly if the design is based on numerical methods. For example, Table 3.24 lists the laboratory tests set out in BS EN 1977-2 (2007) showing which are relevant to glacial soils. Compare that to Table 4.23, which includes the design parameters used in commercially available software. It may be necessary to interpret tests in a different way to that specified or carry out different tests or use published data to obtain further parameters needed for design.

Table 4.23 Relevance of sample type to mechanical properties to produce design parameters for glacial soils

Parameter		*Type of soil*		
		Matrix-dominated soil	*Clast-dominated soil*	*Sands and gravels*
Stiffness	Oedometer	√		
	Triaxial	√	√[a]	√[a]
	Electric cone	√	√[a]	√[a]
	Pressuremeter	√	√[a]	√[a]
Effective strength	Triaxial	√	√[a]	√[a]
	Shear box			√
Undrained shear strength	Triaxial	√		
	Electric cone	√		
	Pressuremeter	√		
Residual shear strength	Ring shear	√		
Bulk density		√	√[a]	
Permeability	Oedometer	√		
	Falling head	√	√[a]	
	Constant head			√
Coefficient of consolidation	Oedometer	√	√[a]	

Source: After BS EN 1997-2:2007. *Eurocode 7: Geotechnical Design – Part 2: Ground Investigation and Testing (Incorporating Corrigendum 2010)*. British Standards Institution, London.

[a] Depends on the quality of the sample and the maximum particle size.

4.5.1 *In situ* stresses

An important parameter in numerical studies is the horizontal earth pressure usually expressed in terms of K_o, the coefficient of earth pressure at rest. There are a number of empirical methods to estimate K_o such as that given by

$$K_o = K_{nc}(OCR)^n \tag{4.14}$$

where K_{nc} is the coefficient of earth pressure at rest for normally consolidated clay, assumed to be $(1 - \sin \varphi')$, OCR the over-consolidation ratio and n an empirical factor often assumed to be $\sin \varphi'$. OCR is normally based on the preconsolidation pressure measured in oedometer tests, but it is not certain that is possible for glacial tills because of the pressures involved. Further, as explained in Section 4.5, it is not certain what is meant by over-consolidation in tills. Preconsolidation pressures of basal tills are normally less than the expected based on the thickness of ice because of the thermal and hydrogeological conditions in the basal layers (Boulton, 1975). For example, Edil and Mickelson (1995) showed that OCR for tills in SE Wisconsin ranged between 2 and 31.

An alternative method, first proposed by Skempton and Sowa (1963), is to use suction measurements on Class 1 samples to determine the *in situ* effective horizontal stress assuming the soil behaves as an isotropic material. Doran et al. (2000) suggested that it was important to take into account cross-anisotropy. They showed from tests on Belfast Upper Boulder Clay that a cross-anisotropic consolidation approach gives reasonable predictions of suction pressures. Figure 4.18 shows the effect of the assumption of isotropic and anisotropic elasticity based on the following equations:

$$\frac{p'_k}{\sigma'_{vO}} = \frac{1+2K_0}{3} - \frac{J}{3G^*}(1-K_o) \quad \text{for cross-anisotropic behaviour} \tag{4.15}$$

$$\frac{p'_k}{\sigma'_{vO}} = \frac{1+2K_0}{3} \quad \text{for isotropic behaviour} \tag{4.16}$$

where p'_k is the measured effective stress in the specimen, σ'_{vO} is the *in situ* vertical effective stress, J is a coupling parameter linking mean stress with shear strain and deviator stress with volumetric strain when a soil is subjected to loading or unloading (Graham and Houlsby, 1983) and G^* is the anisotropic shear modulus. (J/G^*) is typically (−0.25). These reduce to

$$\frac{p'_k}{\sigma'_{vO}} = 0.58 + 0.42K_0 \quad \text{for cross-anisotropic elasticity} \tag{4.17}$$

$$\frac{p'_k}{\sigma'_{vO}} = 0.33 + 0.67K_0 \quad \text{for isotropic elasticity} \tag{4.18}$$

This compares to the proposal of Skempton (1961):

$$\frac{p'_k}{\sigma'_{vO}} = K_0 - A_s(K_0 - 1) \tag{4.19}$$

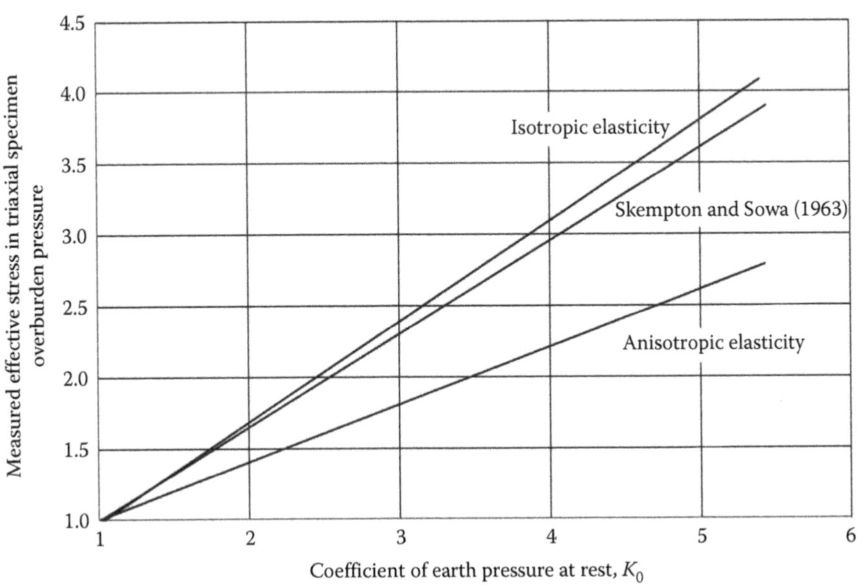

Figure 4.18 Influence the method has upon estimated values of the coefficient of earth pressure. (After Doran, I. G. et al. *Géotechnique*, 50(2); 2000: 189–196.)

where A_s is the pore pressure coefficient during sampling. For a typical value of A_s of 0.3, this reduces to

$$\frac{p'_k}{\sigma'_{vO}} = 0.30 + 0.70K_0 \tag{4.20}$$

This method may be applicable to glaciolacustrine clays and fully homogenised glacial tills but may prove difficult in many matrix-dominated basal tills because of the challenge of obtaining Class 1 samples.

It is also possible to measure the total horizontal stress directly in the field using pressuremeter tests but only in fine-grained soils or sands, not composite soils or gravel.

4.5.2 Strength

The behaviour of soils subject to loading or unloading depends on the rate, direction and type of loading and the response of a soil depends on its density, particle size distribution, particle type and fabric. In fully undrained conditions, it is assumed, for practical purposes, that there are no volume changes; that is, any change in load causes a change in pore pressure if the soil is fully saturated. This is the theoretical condition assumed for clays. It means that the water and soil particles are incompressible. The density of tills is such that the soil skeleton is so stiff that some of the load is taken by the skeleton, which means that, during undrained loading, the pore pressure will be less than the applied load (see Figure 4.31). Therefore, if unconsolidated undrained tests are carried out on a till at three different confining pressures, a low angle of friction and high cohesion are obtained (see Figure 4.19). This is an unsafe result. It is better to interpret unconsolidated undrained tests as fully undrained tests, quoting the average deviator stress at failure.

The other extreme is fully drained conditions in which pore pressure does not change. The difference between these undrained and drained conditions depends on the speed of loading. For example, all soils will behave in an undrained mode during transient loading conditions imposed by an earthquake, and all soils will behave as drained if the loading rate is slow enough to allow pore pressures to dissipate but it does depend on the coefficient of hydraulic conductivity. The timescale between fully undrained and fully drained conditions can vary by a factor of 10^9 (Head, 1988b). It is often assumed that construction in clays takes place in undrained conditions and construction in sands takes place in drained conditions. This assumption is reasonable for glaciofluvial sands and gravels and possibly glaciolacustrine clays but does not necessarily apply to glacial tills. Glacial tills, as composite soils, will exhibit both pore pressure changes and volume changes during construction depending on their fabric, density and particle size distribution. Fully homogenised tills are more likely to behave in an undrained manner, but it is safer to assume that there will be some drainage during construction, especially if discontinuities are present. This means that stability of temporary excavations should not be assumed.

Determining the strength of coarse-grained soils, that is, glaciofluvial soils, clast-dominated tills and glaciomarine deposits, is difficult because of the difficulty in recovering undisturbed samples or knowing the *in situ* density to prepare reconstituted samples. *In situ* tests using empirical correlations are possibly the most appropriate (see Section 4.5.2.1) but note the effect of coarse particles on the results.

The strength of a soil depends on many factors including the imposed stress changes, known as the stress path. For example, the strength of a soil in extension is different from that in compression. Tests carried out in routine ground investigations include compression tests (e.g. triaxial tests), shear tests (e.g. direct shear tests), penetration tests (e.g. SPT) and expanding cavity tests (e.g. pressuremeter test). The stress path followed is different for different tests, which mean that a comparison between results is not helpful. The stress paths imposed by the construction process and the subsequent operation of a structure also do not relate to those followed in the laboratory and field tests. This is one reason why many design methods include an empirical correction factor. The increasing use of numerical methods requires a more sophisticated approach to geotechnical investigation and interpretation so that appropriate constitutive models with the correct input parameters are used. Numerical methods provide a powerful means of undertaking scenario analyses to identify the most critical serviceability and ultimate limit states.

It is often stated that ground investigations are inadequate, which can lead either to overdesign because conservative parameters are assumed, or possibly failure because of unsafe assumptions. Therefore, it is important to ensure in any ground investigation to specify the correct test and the correct number of tests to give the appropriate design parameters. Table 4.24 gives examples of tests that could be carried out for various geotechnical structures. The effective strength of a soil is defined by the effective strength parameters (c', φ'), which are a function of the soil density and particle size distribution. Note that c', φ' are convenient ways of expressing the strength of a soil. It assumes that the failure envelope is a straight line and independent of the stress path to failure. The implication is that triaxial tests on three samples consolidated to three different pressures will fail such that the tangent to the three Mohr's circles at failure is defined by c', φ'. Saturated soils, which are not cemented, do not exhibit cohesion, which appears to exist in triaxial tests at the typical pressures used if a linear Mohr–Coulomb criterion is assumed. This is because the failure line is actually non-linear. Tests at very low confining pressures will show no cohesion; tests at very high pressures will show cohesion and low angle of friction.

Tests on a fully saturated soil, assuming a linear failure envelope, will produce c', φ'. Tests on fissured soils or soils containing a range of particle sizes may produce high values or even negative values of cohesion because of the influence of discontinuities/coarse particles on the failure mechanism of each specimen. The effect of fabric on sampling and testing and the interpretation of the test, which is a simple curve fitting routine, suggest that scatter is

Table 4.24 Applications of laboratory tests to geotechnical problems involving glacial soils

Geotechnical structure	*Critical period*	*Type of analysis*	*Parameters*	*Type of test*
Foundation capacity	End of construction	Total stress	c_u	UU
	Long term	Effective stress	c', φ'	CU, CD
Foundation settlement	End of construction	Elastic	E_u	
	Long term	Consolidation	m_v	Oedometer
		Numerical analysis	G	CU, CD with local strain measurements
Earth retaining structures	End of construction	Total stress	c_u	UU
	Long term	Effective stress	c', φ'	CU, CD
Embankment fill stability	During Construction	Effective stress	c', φ'	CU, CD
Embankment settlement	End of construction	Elastic	E_u	
	Long term	Consolidation	m_v	Oedometer
		Numerical analysis	G	CU, CD with local strain measurements

Note: UU, undrained triaxial test; CU, consolidated undrained triaxial test with pore pressure measurements; CD, consolidated drained triaxial test.

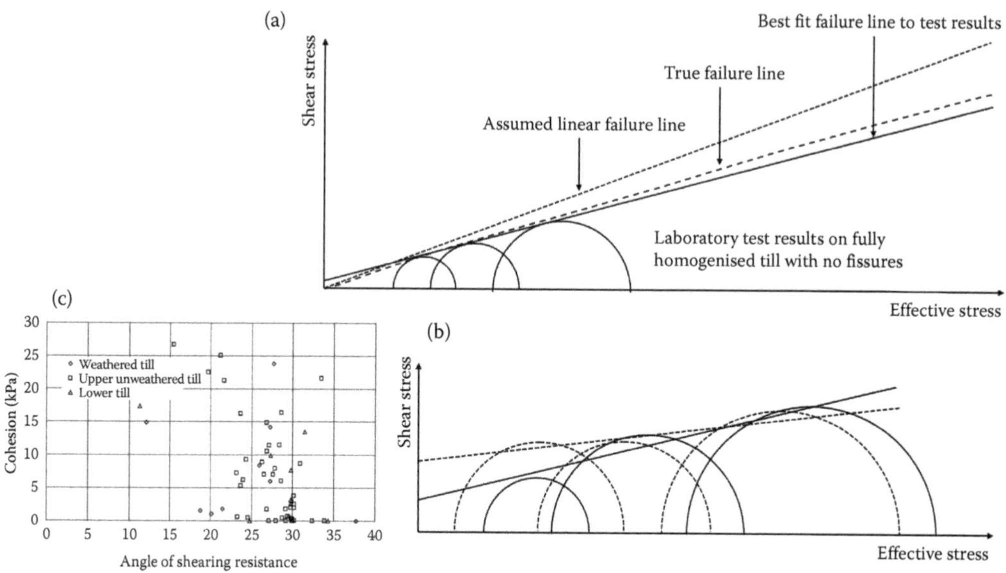

Figure 4.19 Effects of (a) the assumed failure model and (b) the fabric and composition of glacial tills have upon the interpreted effective strength parameters from consolidated tests on those tills showing how c' can be overestimated and φ' can be underestimated producing (c) scatter in standard test results.

inevitable. Figure 4.19c shows a range of values of cohesion and angle of friction from commercial tests on glacial tills and how they can be affected by the method of interpreting tests and the fabric and composition of the soil. Thus, there is no relationship between c' and φ' (Figure 4.19c).

For these reasons, values of cohesion and angle of friction should be treated with caution. There are three ways to obtain more representative values:

1. Plot the stress parameters (t, s') at failure of all tests on specimens from the same stratum. This may mean that more tests are required to be significant. In an ideal situation, the results would lie on a line, possibly curved. The failure line in this case is defined by

$$t' = a' + s' \tan\alpha' \tag{4.21}$$

where $t' = (\sigma_v' - \sigma_h'/2)$; $s' = (\sigma_v' + \sigma_h')/2$; a' and α' are constants that represent the cohesion and angle of friction such that $\sin\alpha' = \tan\varphi'$ and $c' = (a'/\cot\varphi')$.
2. Compare the results with published data. Terzaghi et al. (1996) suggested a relation between the angle of shearing resistance and the plasticity index, which shows that an increase in clay content reduces the angle. The range of plasticity indices in Figure 4.11 suggests that the angle of friction of glacial tills would vary between 35 and 25 (Figure 4.20). Data from various sources suggest that the results of triaxial tests on a glacial till do lie about this line but the scatter in the data suggests caution.
3. Clarke et al. (1997a), Atkinson et al. (1985) and Lewin and Powell (1985) suggested that tests can be carried out on the reconstituted till at the same density as the natural till (Figure 4.21). This can apply to glacial sands and tills that contain very coarse

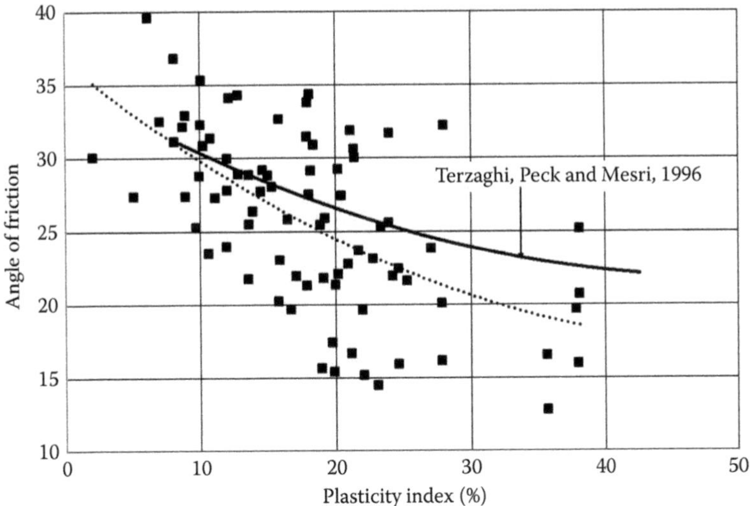

Figure 4.20 Variation of angle of friction with the plasticity index for UK glacial tills compared to the suggested of Terzaghi et al. (1996) showing the scatter in the results due to fabric, composition and interpretation. (After Trenter, N. A. *Engineering in Glacial Tills*. CIRIA, London, 1999.)

particles. In this way, the effects of fabric and large coarse particles are removed leading to more consistent results. It is unlikely that the samples can be consolidated to the *in situ* density in standard triaxial test equipment because of the pressure needed to achieve the required density. Samples may have to be prepared in a consolidation cell and transferred to the triaxial cell. Reconstituting the soil reduces the scatter in the data as it removes the effect large particles can have on the failure mechanism and the impact of discontinuities, which, as Figure 4.22 shows, can reduce the strength. The properties of matrix-dominated tills depend on the fines content based

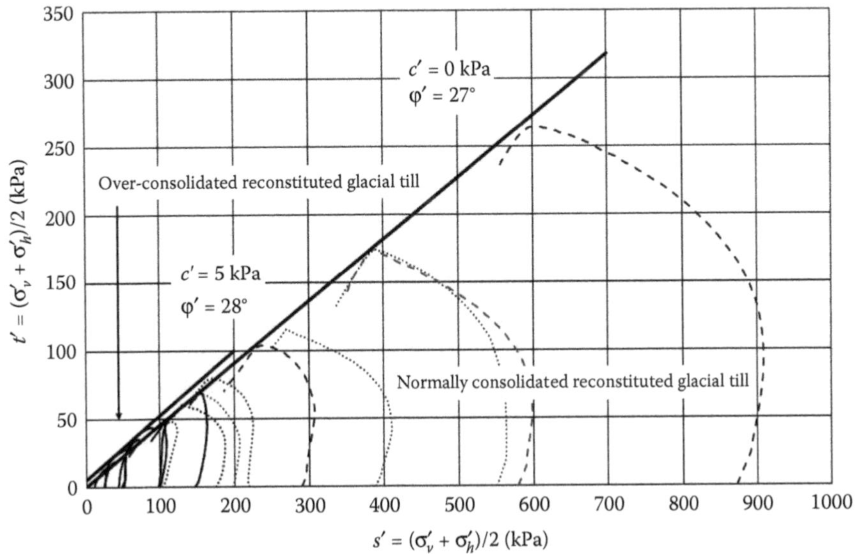

Figure 4.21 Effect of reconstitution on the failure line of glacial soils.

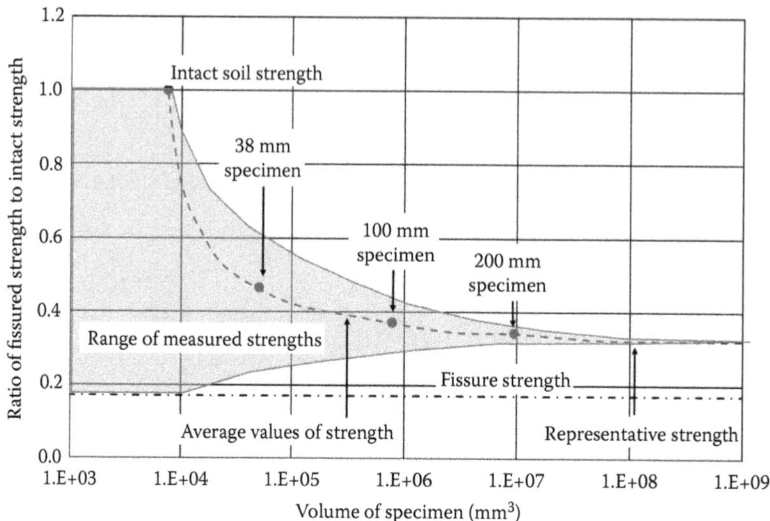

Figure 4.22 Effect of specimen size on strength of fissured tills highlighting the need to test larger specimens. (After McGown, A., A. M. Radwan, and A. W. A. Gabr. Laboratory testing of fissured and laminated clays. In *Proceeding of the 9th International Conference on Soil Mechanics and Foundation Engineering*, Tokyo, VI, 1977: 205–210.)

on studies of the effect of clasts on strength (e.g. Gens and Hight, 1979) but the clasts affect the quality of the sample and therefore, the properties. Clarke et al. (1997a) used reconstituted samples in which samples were consolidated one dimensionally to achieve similar densities to those found *in situ*. They compared the effective strength of the reconstituted samples with those from tests on routine samples to show that the strength of the reconstituted samples formed the lower bound to the tests on the 'undisturbed' samples. Atkinson et al. (1985) undertook tests on reconstituted and remoulded Cowden Till; Lewin and Powell (1985) carried out tests on thin-walled push samples of the same till. They observed that the stress paths were similar for all tests giving an angle of friction of 27.5°. Therefore, tests on the reconstituted matrix-dominated till appears to produce a failure envelope, which is equal to the average from tests on the undisturbed till or a lower bound to those tests.

4.5.2.1 Field tests

In situ vane tests can be carried out in some glacial soils provided they are not too strong or contain coarse-grained particles that will impact on the vane. There are three sizes of vane (100 mm × 200 mm; 40 mm × 80 mm; 33 mm × 66 mm), which are used in different soils (Table 4.25); the stronger the soil the smaller the vane. A correction factor (μ) (BS ENV 1997-2, 2007) is applied to the vane shear strength (c_{uv}) to obtain the undrained shear strength. In soft clays, the correction factor is related to the liquid limit (Figure 4.23a):

$$c_u = \mu c_{uv} \tag{4.22}$$

This can be reduced to 0.3 in fissured clays. Figure 4.23b shows a correction factor for depth for over-consolidated clays based on the plasticity index. The degree of over-consolidation is expressed in terms of (c_u/σ'_v). These correlations are based on the work by Aas et al. (1986) and Hansbo (1957).

Table 4.25 Application classes for the field vane tests

Application class	*Application*	*Suggested vane*	*Test*[d]	*Allowable minimum accuracy*[a]		*Maximum rotation between measurements*	*Suggested use*	
							Soil[b]	*Interpretation*[c]
1	Special purposes (e.g. very soft soils)		FVTa	Torque Rotation angle Depth of test	0.5 Nm 1° 0.1 m	1°	A to D	H
2	Soft to very soft soils	10–20 kPa, 70–120 mm Very soft clays <10 kPa, 70–150 mm	FVTa FVTb	Torque Apparent rotation angle Depth of test	1 Nm 5° 0.1 m	2°	A to D	H, H*
3	Stiffer soils	Stiff clays and silts 50–100 kPa, 50–70 mm Medium stiff clays and silts 20–50 kPa, 75 mm	FVTc	Torque Apparent rotation angle Depth of test	2 Nm 10° 0.2 m	5°	A to D	H*
4	Very stiff soils	Very stiff clays and silts 100–300 kPa, 50–70 mm	FVTd	Torque Depth of Test	5 Nm 0.2 m	Not relevant	C to D	H*

Source: After BS EN ISO 22476-9:2014. *Ground Investigation and Testing. Field Testing. Part 9. Field Vane Test.* British Standards Institution, London.

[a] The allowable minimum accuracy of the measured parameter is the larger value of the two quoted. The relative accuracy applies to the measured value and not the measuring range.

[b] A: homogeneously bedded soils (typically $Cu < 2$ MPa); B: clays, silts and sands (typically 2 MPa $\leq Cu < 4$ MPa); C: clays and silts (typically 4 MPa $\leq Cu \leq 10$ MPa); D: clays and silts (typically $Cu > 10$ MPa).

[c] H: interpretation in terms of engineering parameters with associated low uncertainty level; H*: interpretation in terms of engineering parameters with associated high uncertainty level.

[d] FVT: field vane test; a: continuous downhole measurement of torque versus rotation; b: continuous uphole measurement of torque versus rotation; c: continuous uphole measurement of torque versus rotation; d: uphole measurement of maximum torque.

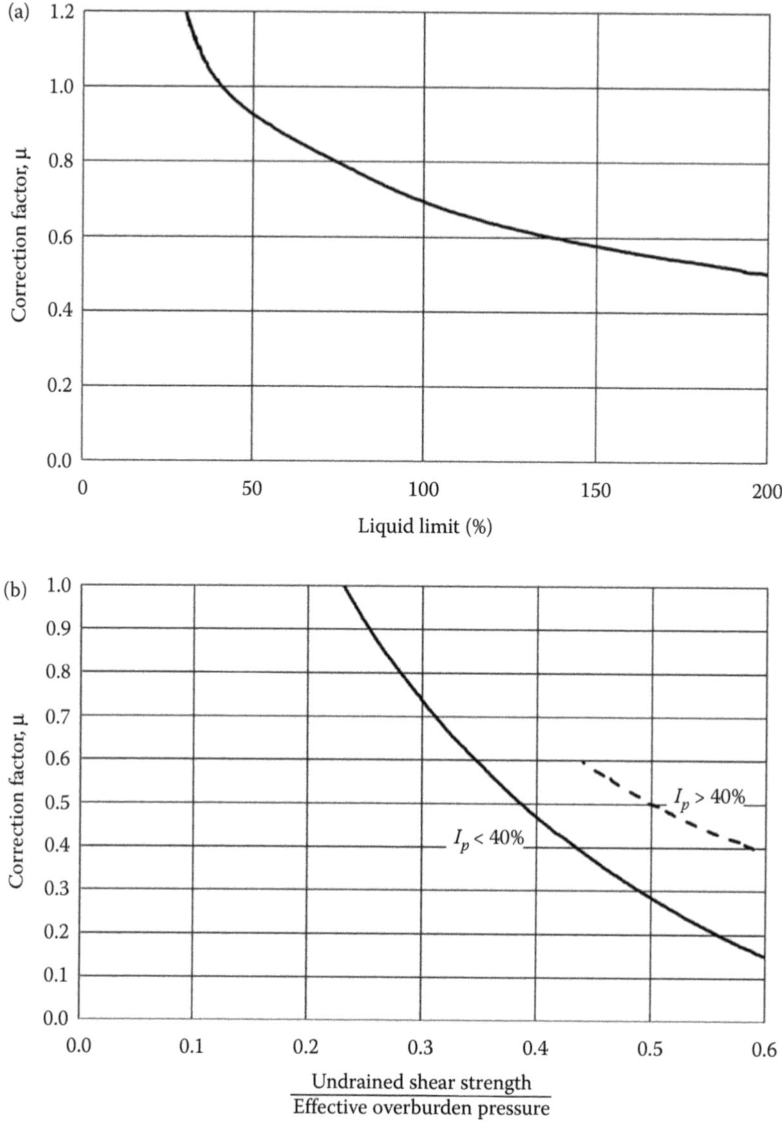

Figure 4.23 Examples of factors used to correct the vane strength to obtain the undrained shear strength based on (a) the liquid limit and (b) the plasticity index for over-consolidated clays. (After ENV 1997-2:2006; BS EN 1997-2:2007. *Eurocode 7: Geotechnical Design – Part 2: Ground Investigation and Testing (Incorporating Corrigendum 2010)*. British Standards Institution, London.)

It is not uncommon to specify SPT tests in tills. This may be prudent given the difficulty of obtaining representative samples. Stroud and Butler (1975) proposed a relationship between undrained shear strength and SPT N_{60} of the form shown in Figure 4.24, which shows that the factor is a function of the plasticity index. However, it may be necessary to develop site-specific correlations because of the effect of fabric, particle size and sampling quality on the results (Figure 4.25), which shows the profile of undrained shear strength, an estimate of the undrained shear strength based on SPT results assuming an average plasticity index of 25% and an empirical relationship with over-consolidation ratio. This highlights the difficulties of classifying matrix-dominated tills and selecting a design profile.

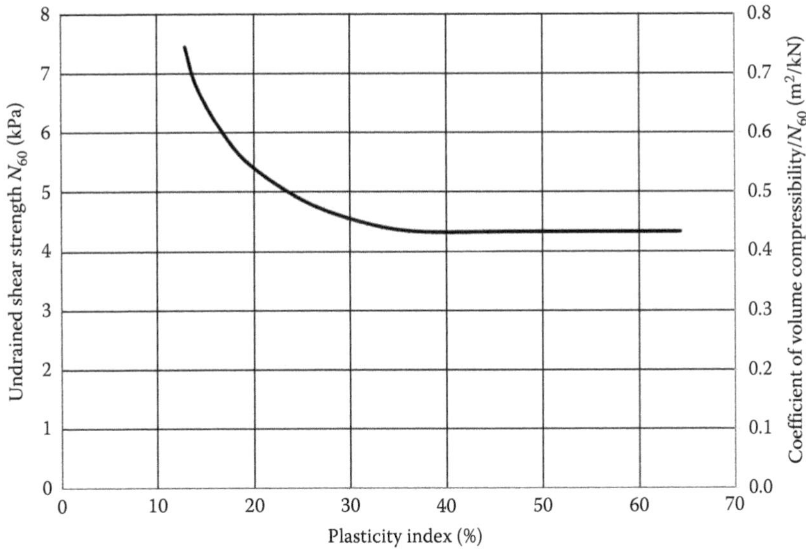

Figure 4.24 Relationship between undrained shear strength and coefficient of volume compressibility and SPTN as a function of the plasticity index. (After Stroud, M. A. and F. G. Butler. The standard penetration test and the engineering properties of glacial materials. In *Symposium on Engineering Properties of Glacial Materials*, Midland Geotechnical Society, 1975.)

4.5.2.2 Direct shear test

The direct shear test is a test in which the upper half of the soil is sheared against the lower half of the specimen. The shear force and horizontal and vertical displacements are measured. Shear box samples can be 60, 100 or 300 mm in plan. It is possible to carry out multi-reversal tests to determine the residual shear strength though the ring shear may be more

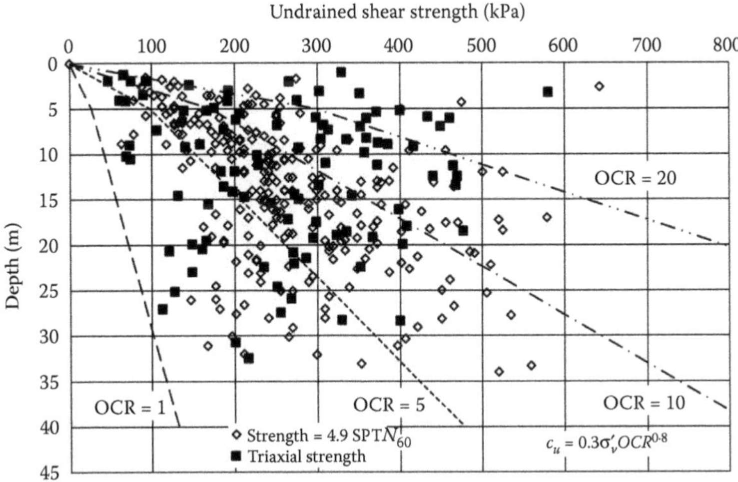

Figure 4.25 Comparison of the variation in undrained shear strength with depth based on an empirical relationship with OCR, the relationship in Figure 4.24 and strengths determined from undrained triaxial tests highlighting the difficulty of selecting a design profile because of the effects of composition and fabric on a matrix-dominated till.

appropriate because the movement is continuous. It is usual to carry out three tests at three different normal forces to obtain the variation in shear force. This is expressed in terms of shear stress and normal stress. The results are assumed to lie on a straight line:

$$\tau_f = c' + \sigma_n' \tan\varphi' \tag{4.23}$$

where τ_f is the shear stress at failure, σ_n' the normal stress, c' the cohesion and φ' the angle for friction. This is routinely used in geotechnical engineering as a means of assessing the strength of coarse-grained soils. Equation 4.23 applies to drained tests. Tests on coarse-grained soil will be drained, and this is the most common laboratory method used to determine the effective strength parameters provided the specimens are prepared at the *in situ* density. Tests on composite soils such as glacial tills may exhibit drained behaviour, but unless the rate of loading is slow enough, it should not be assumed that the parameters are effective strength parameters.

Undrained tests on clays may also show an apparent angle of friction because the specimens may be partially saturated or the specimen may be partially consolidated and, in the case of very stiff soils such as glacial tills, because some of the normal load is taken by the very stiff soil skeleton. In these cases, the shear stress will increase as the normal force increases. This may be interpreted to give an apparent cohesion and angle of friction. It is unsafe to use these parameters in design. If undrained tests are carried out on a fine-grained soil, then the average shear stress should be quoted. A consequence of partial consolidation is that the strength index of a composite soil can be overestimated.

Large shear box tests are useful when testing composite soils as it is possible to include particles up to 37.5 mm in the specimen. Small shear box tests on laminated clays are useful as they can assess strength parallel and perpendicular to the laminations. Shear box tests on coarse-grained soils are useful because it is possible to assess the effective shear strength parameters at various densities.

There are a number of limitations:

- The failure plane is predetermined.
- The stress distribution on the failure plane is not uniform.
- There is no control over drainage.
- The displacement is limited.
- The area of the failure surface reduces with displacement.

Given the difficulty of obtaining Class 1 samples of coarse-grained soils, the direct shear test may be the only suitable test for coarse-grained soils containing gravel, provided tests are carried out at the *in situ* density.

4.5.2.3 Triaxial test

The triaxial test is the most common laboratory test used to determine the undrained shear strength of clays and the effective strength parameters of clays and sands. Figure 4.26 shows the definitions of failure used in triaxial testing of soil, which are the peak deviator stress $(\sigma_1 - \sigma_3)$, maximum principal stress ratio (σ_1'/σ_3'), limiting strain, critical state and residual strength where σ_1 is the principal axial total stress and σ_3 is the principal radial total stress. The definition of strength leads to different values, an example of which is shown in Figure 4.27 where the peak deviator stress and peak stress ratio are used to plot the strength with depth for a glacial till. There is a trend that shows that the variation in strength at the peak stress ratio is less than the peak deviator stress despite the scatter in the data. Note that the deviator stress is the same whether it is expressed in terms of total or effective stress since the pore pressure is isotropic at any point in the soil. The principal stress ratio, expressed in the terms of effective stress in undrained tests, produces better correlations with other parameters (Head, 1988b). Figure 4.28 shows the typical response of compression tests on loose and dense soils. Note that dense soils exhibit peak strength; loose soils and dense soils of the same composition reach a constant volume at large strains, which is the critical state. All soils contract initially when loaded; dense soils subsequently dilate. In loose soils, the limiting strain, usually 20%, is defined as failure. The critical state strength is an intrinsic property of the soil as it is independent of the initial density. The angle of friction is made up of two components: the critical state angle of friction and a variable component, which is the dilatant component and depends on the initial density.

Three tests are normally carried out on three specimens from one sample consolidated to three different confining pressures. The peak deviator stress is plotted against the confining effective stress in the form of Mohr's circles (Figure 4.29) to produce the failure envelope. The failure envelope is curved, but in practice, it is usually expressed in terms of cohesion and angle of friction as a linear failure line known as the Mohr–Coulomb failure (Terzaghi, 1936). If an undrained test is carried out at three different confining pressures but the specimens are not consolidated, then three different Mohr's circles are obtained, which should be the same diameter if the specimens have the same composition, fabric and density and are fully saturated. In practice, the circles are different diameter particularly for tests on glacial tills, as shown in Figure 4.30. These tests are sometimes interpreted to give an undrained cohesion and undrained angle of friction. This is unsafe and should not be considered. The reason for this apparent increase in strength is due to fabric, soil stiffness, partial saturation and larger particles.

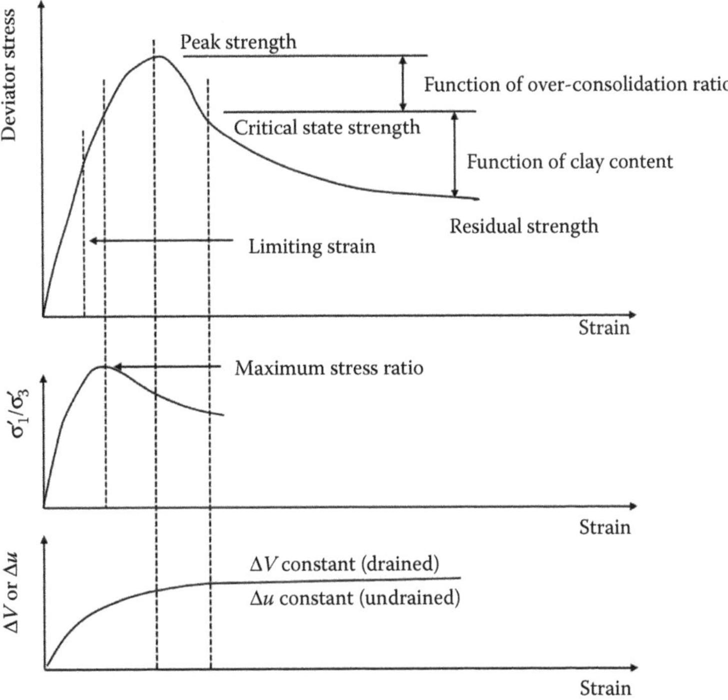

Figure 4.26 Failure criteria for soils.

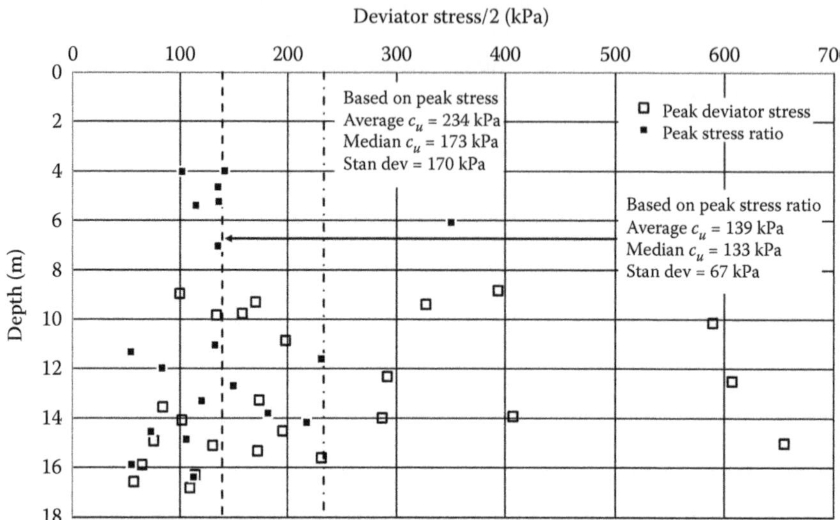

Figure 4.27 Comparison between the strength based on the peak deviator stress and that based on the peak stress ratio showing the reduction in standard deviation and the reduction between the mean and median strength. (After Marsland, A. *Quarterly Journal of Engineering Geology and Hydrogeology*, 10(1); 1977: 1–26.)

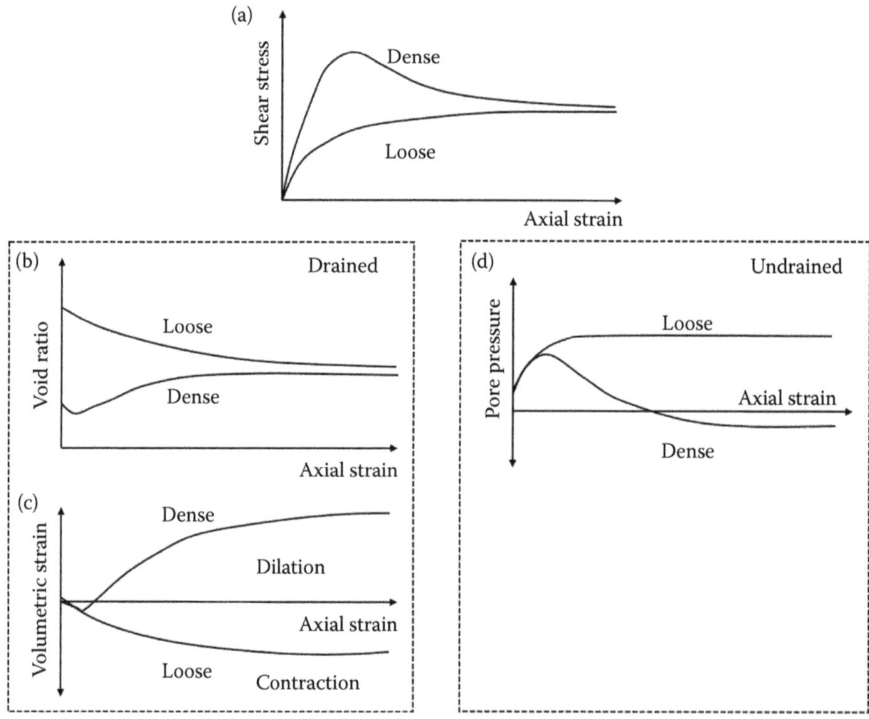

Figure 4.28 Shear characteristics of soils showing (a) deviator stress against strain and the changes during loading of (b) the volume, (c) the pore pressure and (d) the voids ratio using the concept of dense and loose soils.

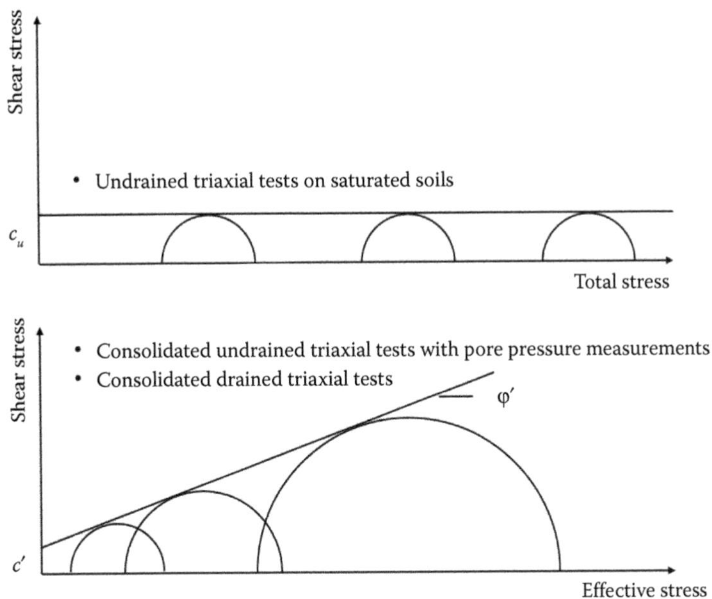

Figure 4.29 Interpreting triaxial tests on soils showing the difference between undrained and effective strength tests.

There are three types of triaxial tests: quick undrained to obtain the strength index of clay; consolidated undrained test with pore pressure measurements to determine the effective strength parameters of clays; and drained tests on clays and sands to determine effective strength parameters. These are standard tests. There are variations, which include stress path and anisotropic consolidation tests as listed in Table 4.26.

Figure 4.28 shows the volume changes that take place during triaxial tests. While dilatant behaviour is associated with over-consolidated soils, it is actually a consequence of density of packing of the soil particles. Figure 4.26 shows the effect of over-consolidation (density), level of strain and definition of failure; Figure 4.29 shows the total and effective stress circles for quick undrained, consolidated undrained and drained triaxial tests. Glacial soils can

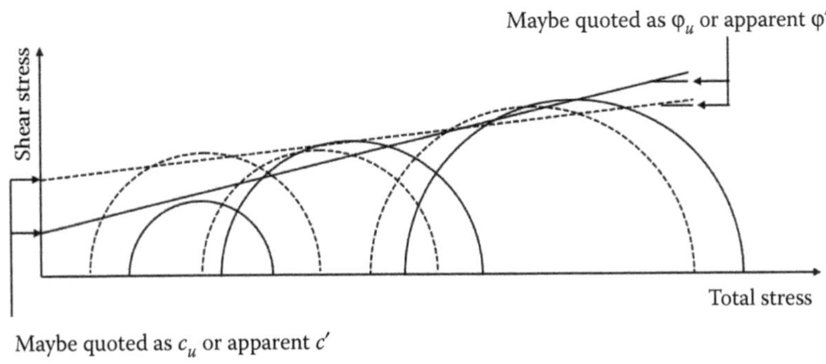

Figure 4.30 Mohr's circles for typical undrained triaxial tests on matrix-dominated glacial tills showing the effect of composition, fabric and partial saturation on the results potentially leading to overdesign or unsafe design due to incorrect interpretation.

Table 4.26 Types of triaxial tests

Type of test	*Abbreviation*	*Consolidation*	*Drainage*	*Rate of strain*	*Parameters*
Unconsolidated (quick) undrained compression test	UU	No	No	Failure in 10 min	c_u
Isotropically consolidated undrained compression test with pore pressure measurements	CU	Isotropic	No	Steady state pore pressure maintained	c', φ'
Isotropically consolidated drained compression test with volume change measurements	CD	Isotropic	Yes	Failure in 10 min	c', φ'
Anisotropic consolidated undrained compression test with pore pressure measurements	CAUC	Anisotropic (K_o)	No	Steady state pore pressure maintained	c', φ'
Anisotropic consolidated undrained extension test with pore pressure measurements	CAUE	Anisotropic (K_o)	No	Steady state pore pressure maintained	c', φ'

exhibit these types of behaviour depending on their density. So it is possible for a normally consolidated subglacial till to appear to behave as a heavily over-consolidated soil.

Ideally, tests should be carried out on saturated samples, or fully drained tests should be carried out if the samples are taken from below the groundwater level, allowing for the fact that the groundwater level can rise. The processes of sampling, transport, storage and preparation can lead to loss of pore water resulting in partial saturation. In order to ensure saturation, a back pressure is applied in stages and the increase in pore pressure is measured at each stage. The sample is consolidated between each stage so that the pore pressure parameter, *B*, can be measured. Theoretically, the increase in pore pressure should equal the increase in confining pressure. The pore pressure parameter, *B*, is the ratio of increase in pore pressure to the increase in confining pressure. Figure 4.31 shows typical values of

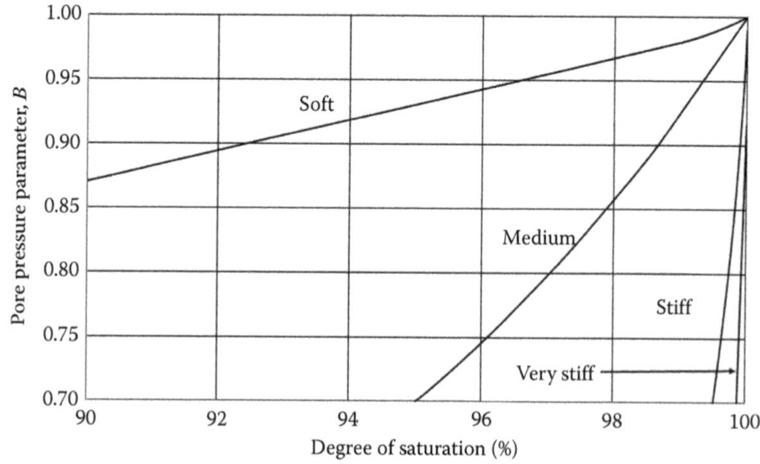

Figure 4.31 Variation in the pore pressure coefficient, *B*, with stiffness and degree of saturation showing that very stiff glacial clays may exhibit very little change in pore pressure when loaded unless fully saturated. (After Black, D. K. and K. L. Lee. Saturating laboratory samples by back pressure. *Journal of Soil Mechanics & Foundations Div* 99(SM1); 1973: 75–93.)

B for different degrees of saturation and soil stiffness. This figure shows that, in practice, it is impossible to achieve a theoretical value of 1 for B because of the stiffness of the soil skeleton. This is particularly the case for very stiff tills.

The triaxial test was developed to test cylindrical samples of soil. This allowed the radial and vertical principal stresses, the pore pressure and the rate of loading to be varied. The test is widely used to determine the strength index of clays (undrained shear strength) and effective strength of all soils. Tests can be carried out on soils in their natural, remoulded or reconstituted state. A key advantage of this test over the shear box test is that samples are easier to prepare, the boundaries are controlled and shear failure mode is not predetermined.

Typically, tests are carried out on three specimens at three different total confining pressures. The confining pressure is held constant in a standard test and the axial stress increased to a maximum displacement of 20% though a test may be terminated earlier if the soil has obviously failed. Tests include uniaxial compression and triaxial compression though it is now more usual to specify triaxial compression tests. Drained tests on all soils and undrained tests on clays can be specified.

Test results will be affected by the quality of the sample, the fabric of the soil and its composition. Tests are normally carried out on fine-grained soils as it is possible to retrieve Class 1 samples. Tests can be carried out on coarse-grained soils provided the maximum particle size is less than 3.35 mm for 38-mm specimens or 37.5 mm diameter for 150-mm specimens. This means it is inevitable that, unless the soil is a sandy clay or clay, most tests on composite soils will be on 100-mm samples because that is a typical field sample diameter retrieved from boreholes.

It is entirely feasible with fully homogenised tills and glaciolacustrine clays to carry out tests on 38 mm specimens taken from one U100 sample. However, the effects of fabric are less pronounced in small specimens (Figure 4.22) but the effects of random stone content may be more pronounced. There are three ways to deal with this: tests on reconstituted soil thus destroying the fabric and removing stones; carry out a multistage test on a single sample; or carry out tests on three samples consolidated to three different confining pressures. If a multistage test (Figure 4.32) is going to be used, then it has to be carefully monitored so that the pressure increases are correctly carried out.

Anderson (1974) suggested that many tills exhibit ductile behaviour so tests can show an increase in deviator stress up to 20% strain. This led to the proposal that the confining pressure should be increased at 18%, 22% and 22%. Thus, a failure criterion is used, hence, the need to monitor the test. Failure can be

- The development of a slip surface particularly in brittle soils
- Approaching the peak deviator stress
- A predetermined strain for more ductile soils
- A peak stress ratio
- Peak pore pressure that coincides with the peak deviator stress

A better method is to use a stress path plot, in which, the radius $(\sigma_1 - \sigma_2)/2$ of the Mohr's circle is plotted against the centre of the circle $(\sigma_1 + \sigma_2)/2$ during a test, as shown in Figure 4.33. It is possible to observe the stress path approaching the failure envelope and therefore to stop loading, increase the confining pressure and allow the specimen to consolidate before increasing the axial stress further.

The alternative is to test several samples from the same stratum at different confining pressures. Normally, three adjacent samples would be tested to give a failure envelope. The effect of fabric and stone content means that the failure envelope may not be representative of the strength of the stratum. The best way to present the data is to produce a stress path

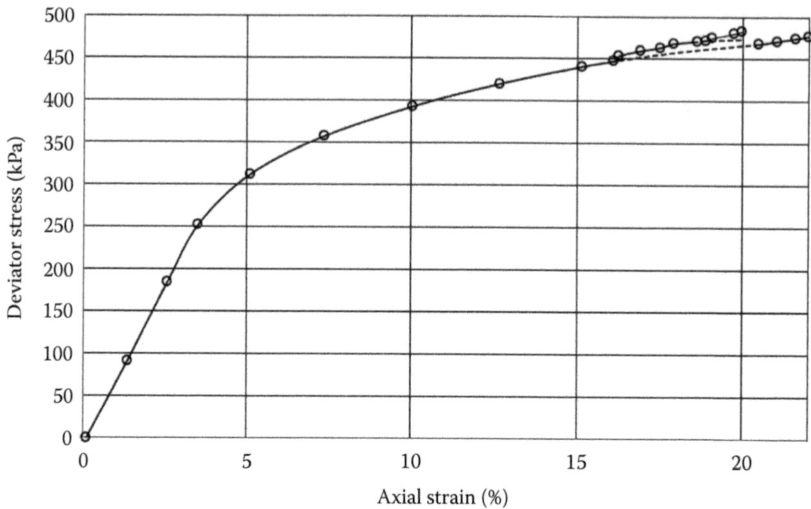

Figure 4.32 Multistage triaxial test showing the points pre-failure at which the confining pressure is increased. (After Anderson, W. F. The use of multi-stage triaxial tests to find the undrained strength parameters of stony boulder clay. In *Proceedings of Institution of Civil Engineers*, 57(2); 1974: 367–372.)

plot for all tests from one stratum and use that plot to determine a representative strength of the stratum. It does mean that more samples are required.

4.5.2.4 CBR test

Glacial soils are often used as a source of the engineered fill for embankments and sub-base materials. The Californian Bearing Ratio test was developed to test sub-bases and subgrades as part of an empirical design procedure. A 50.8 mm diameter plunger is pushed at 0.05 mm/min into the soil held in a standard container. Tests can be carried out on

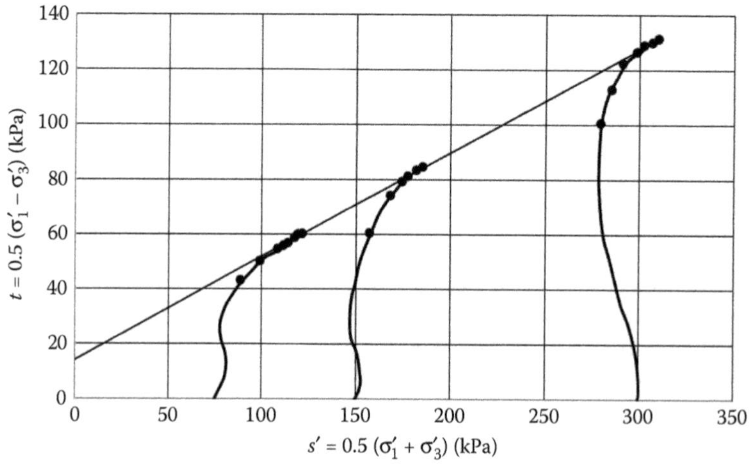

Figure 4.33 Use of stress paths in consolidated undrained multistage triaxial test with pore pressure measurements to determine the failure envelope.

as-received, as-compacted or soaked samples with compacted material soils being prepared using the same effort as *in situ* density tests. The penetration resistance is expressed as a percentage of the force required to push the penetrometer 2.5 and 5 mm into a compacted limestone (the standard load). Particles greater than 20 mm must be removed so that the test is applicable only to fine-grained soils and soils containing nothing greater than medium gravel. Tests can be carried out at the *in situ* water content or a range of water contents to assess the variation of CBR with density and water content. The final design depends on a standard test procedure; hence, it is important to follow that procedure. The test will be used when glacial soils, particularly tills and glaciofluvial soils, are used to form a sub-base or subgrade. The issue of particle size has to be addressed.

4.5.2.5 Undrained shear strength

Undrained shear strength is a strength index used to classify fine-grained soil in accordance with Table 4.22 . It is also used in geotechnical design as explained in Chapters 5 and 6. The ratio between undisturbed and remoulded undrained shear strength is known as sensitivity, which varies from low (<8), medium (8–30) to high (>30). Soils with a sensitivity >50 are quick clays. Most glacial tills have a low sensitivity because of the remoulding that took place during deposition.

Figure 4.34 shows the variation of shear strength with depth for a glacial till at Cowden, NE England where results from plate tests and triaxial tests on 100-mm samples give similar results but less than those from pressuremeter tests, whereas Figure 4.35 shows that plate tests give lower values in a glaciolacustrine clay. These figures highlight the scatter commonly observed when testing glacial tills and the fact that the shear strength is a function of test procedure. The scatter in the results is common in fissured tills and those containing

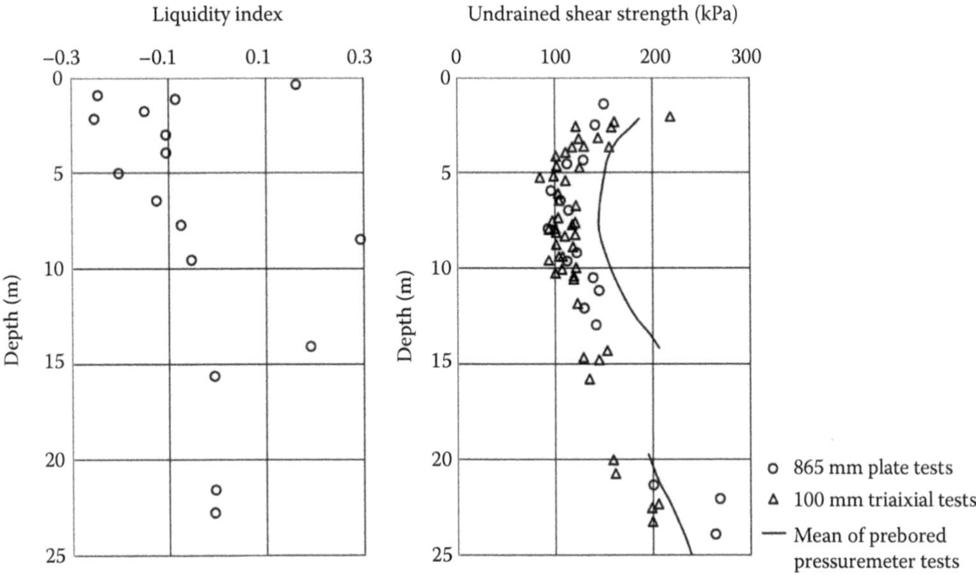

Figure 4.34 Variation in shear strength with depth at Cowden, NE England showing the effect of the type of test on the measured strength. (After Marsland, A. and J. J. M. Powell. Field and laboratory investigations of the clay tills at the building research establishment test site at Cowden, Holderness. In *Proceedings of the International Conference on Construction in Glacial Tills and Boulder Clays*, Edinburgh, 1985: 147–168.)

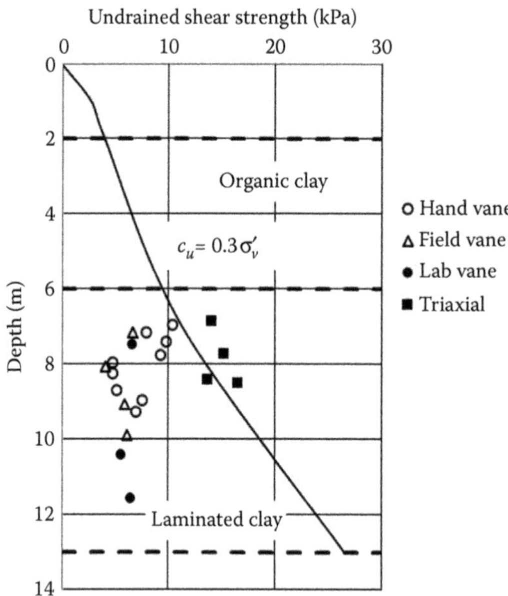

Figure 4.35 Variation in undrained shear strength of the glaciolacustrine clay Athlone with depth. (After Long, M. Sample disturbance effects on medium plasticity clay/silt. In *Proceedings of the Institution of Civil Engineers-Geotechnical Engineering*, 159(2); 2006.)

gravel. McGown et al. (1977) suggested that the ratio of fissured strength to intact strength varies with the volume of the specimen (Figure 4.22) to such an extent that the undrained strength should be determined from tests on 100-mm samples. Since it is often impossible to obtain 38-mm diameter samples from matrix-dominated tills, it may be possible to test only 100-mm diameter samples.

Sampling disturbance is clearly an issue for glacial tills because of their density, fabric and particle size distribution; and for glaciofluvial deposits and glaciomarine sands and gravels because of their composition and fabric. La Rochelle and Lefebvre (1971), Lacasse et al. (1985) and Long (2006) have shown that it is also an issue for glaciolacustrine and glaciomarine clays. These composite soils are subject to both densification and destructuring during sampling, which can lead to an overestimate of the strength and stiffness. Long (2006) carried out anisotropically consolidated triaxial tests and oedometer tests on samples of Athlone clay, a glacial lake deposit shown in Figure 4.35. The specimens were taken from 100-mm fixed piston, MOSTAP continuous and Sherbrooke block samplers. Table 4.27 is a summary of the soil properties and properties of other composite soils and marine clays used in the investigation into sample disturbance. The CUAC tests showed that $\left(c_u/\sigma'_v = 0.3\right)$, which was expected for these normally consolidated soils, the *in situ* vane tests gave much lower values of strength, suggesting that the installation disturbance was sufficient to destruct the soil. Hence, *in situ* vane tests are not recommended in these varved clays. The effort of sampling disturbance can be judged by the volumetric strain (Kleven et al., 1986) or void ratio change (Lunne et al., 1997) required to reinstate *in situ* conditions. Figure 4.36 shows how the sampling affected the volume and void ratio. Table 4.28 lists the average parameters obtained from tests on specimens from the three samplers. Figure 4.37 shows how the sampling quality affected the geotechnical characteristics. Long (2006) concluded that sampling disturbance causes densification and destructurisation, which increases the

Table 4.27 Summary of the properties of soils considered by Long (2006) investigating sample disturbance effects on medium plasticity clay/silt

	Soil type	*% clay*	*% silt*	I_p *(%)*	*Sensitivity*	*Permeability (m/s)*	*Reference*
Soils that can show dilative behaviour	Athlone	35	60	18	4–10	$k_h = 1$ to 4×10^{-9} $k_h/k_v \approx 1.5$	Long and O'Riordan (2001) Long and Gudjonsson (2004) Long (2000)
	Athlone C	25	65	12	4–10	$k_h = 1$ to 4×10^{-9} $k_h/k_v \approx 1.5$	Long and O'Riordan (2001) Long (2000)
	Bothkennar laminated	15–30	60–75	20–40	9.5–15	$k_h = 2 \times 10^{-9}$ $k_h/k_v \approx 5$	Long (2003) Nash et al. (1992)
	Eidsvold	37–48	47–58	13–19	2–5		Karlsrud (1995)
	Kvenild	42	58	6–10	70		Seierstad (2000)
	Drammen	40	60	10–20	5–7		Lunne et al. (1997)
Soils that can show contractive behaviour	Bothkennar bedded	19–30	65–76	30–50	8	$k_h = 1.6 \times 10^{-9}$ $k_h/k_v \approx 1.2$	Hight et al. (1992) Long (2000)
	Onsoy	60	40	30–50	8	$k_h = 1 \times 10^{-9}$	Lacasse et al. (1985) Lunne et al. (1997) Lunne et al. (2003)
	Loiuse-ville	75–80	20–25	40	>20	$k_h = 1.4 \times 10^{-9}$ $k_h/k_v \approx 1.2$	La Rochelle and Lefebvre (1971) Hight and Leroueil (2003) Lefebvre and Poulin (1979)
	Araike	58	42	50–70	25		Tanaka et al. (1996)

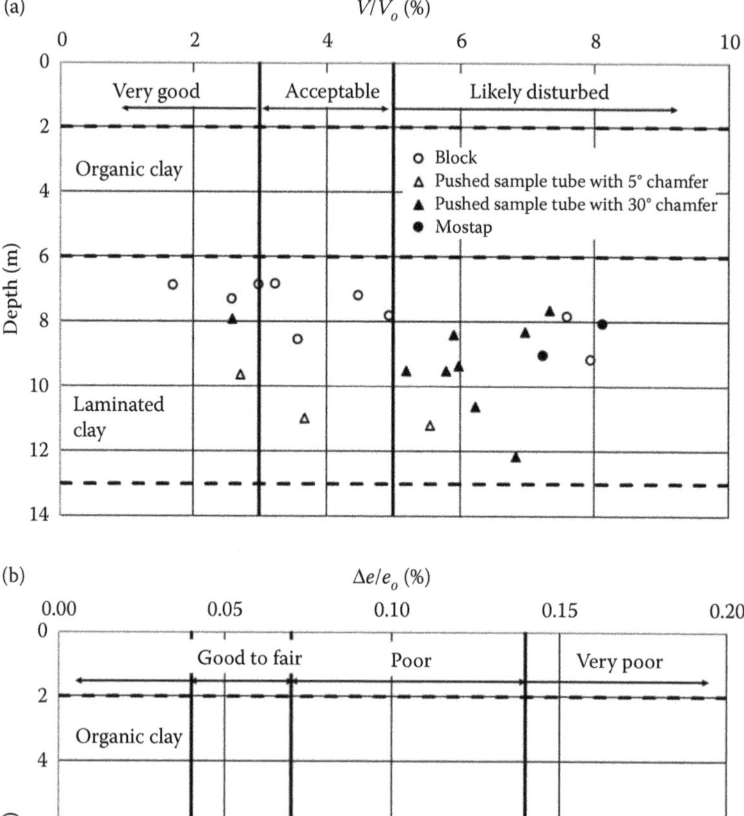

Figure 4.36 Assessment of sample quality based on the (a) volumetric changes (Kelven et al., 1986) or (b) void ratio changes (Lunne et al., 1997) that take place during consolidation to the *in situ* stresses showing that typical U100-driven samples are unlikely to obtain quality samples. (After Long, M. Sample disturbance effects on medium plasticity clay/silt. In *Proceedings of the Institution of Civil Engineers-Geotechnical Engineering*, 159(2); 2006.)

Table 4.28 Average parameters measured in CUAC triaxial tests

Parameter	*30° tube*	*5° tube*	*MOSTAP*	*Block*
Strain at peak deviator stress (%)	5.8	6.8	6.6	1.7
Secant stiffness at 0.1% (MPa)	213	243	257	96
c_u/σ'_v	0.43	0.50	0.65	0.39
$A_{f1.0}$	1.0	0.4	−0.04	0.9

Source: After Long, M. Sample disturbance effects on medium plasticity clay/silt. In *Proceedings of the Institution of Civil Engineers-Geotechnical Engineering*, 159(2); 2006.

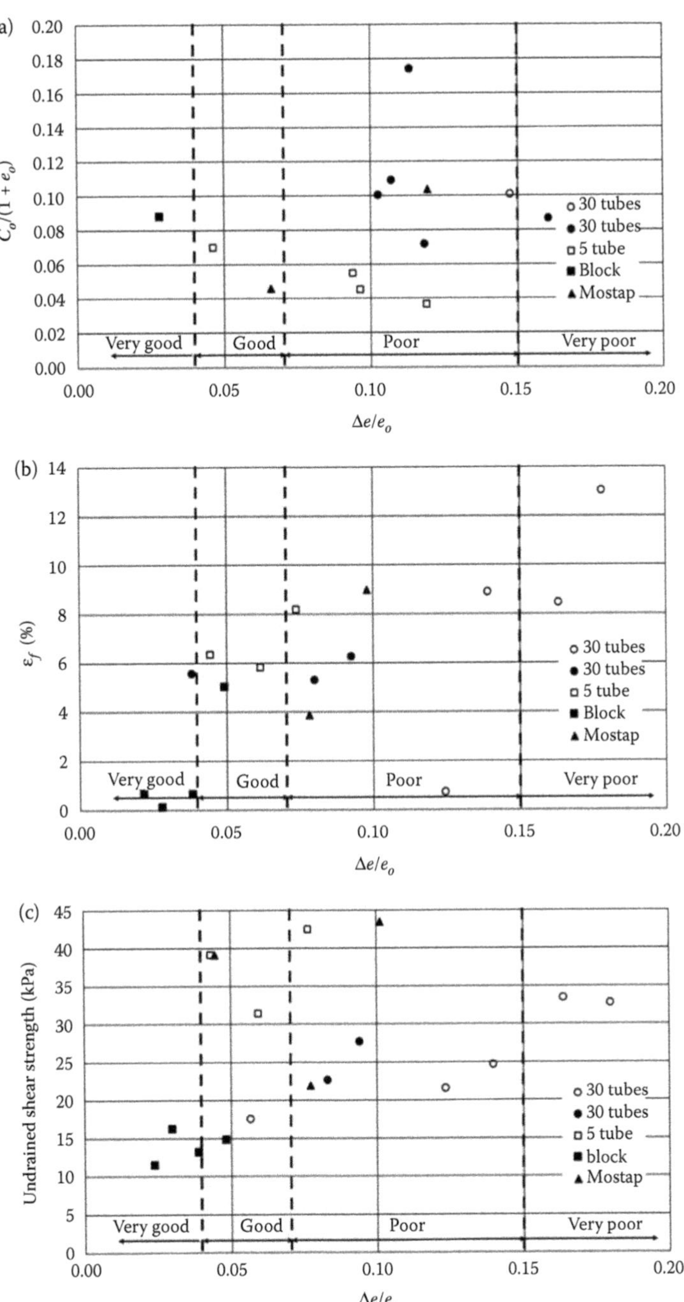

Figure 4.37 Effect of sample disturbance on (a) undrained shear strength, (b) strain to failure and (c) stiffness (showing quality criteria of Lunne et al., 1997). (After Long, M. Sample disturbance effects on medium plasticity clay/silt. In *Proceedings of the Institution of Civil Engineers-Geotechnical Engineering*, 159(2); 2006.)

small-strain stiffness, undrained strength and strain to peak stress, reduces the pore pressure parameter, A_f; and leads to post-failure dilatancy.

Anderson (1974) suggested that triaxial tests on three 38-mm specimens from a single U100 sample of the glacial till tested at different soil pressures rarely gave a unique Mohr–Coulomb failure envelope. This was attributed to disturbance in sampling and subsampling because of gravel but could also be due to discontinuities. Testing three representative U100 samples from a single stratum at different pressures also proved to be unsuitable because of sample disturbance since they did not produce a unique failure envelope. It was noted that the deviator stress in these tests rarely reached a peak value even at 20% axial strain. This suggests ductile behaviour.

4.5.2.6 Effective strength

The effective strength parameters are not intrinsic properties of soil as they depend on density, water content, confining pressure, rate and direction of loading and drainage conditions. The effective strength is expressed in terms of cohesion and angle of friction and the angle of friction can be the peak, post-peak, critical state or residual values depending on the amount of strain.

The impact of composition and fabric on strength led some authors to consider testing reconstituted glacial tills to reduce the difficulty in obtaining representative values of effective strength parameters because subglacial tills are remoulded during deposition.

Skempton and Bishop (1954), Skempton and Brown (1961), Bishop and Vaughan (1962), Vaughan et al. (1975) and Vaughan et al. (1978) all showed that the remoulded strength was satisfactory. However, Vaughan et al. (1978) showed that a small change in water content could have a significant effect on strength, a point noted by Millmore and McNicol (1983) on tests on glacial tills (Figure 4.38). This may also contribute to the scatter in results from routine ground investigations in glacial tills because of the variation in water content due to the slip/stick mode of deposition. Clarke et al. (1997a) suggested that tests on reconstituted soils, provided they are consolidated to the *in situ* density, will produce a failure line that forms the lower bound to tests on undisturbed specimens (Figure 4.39).

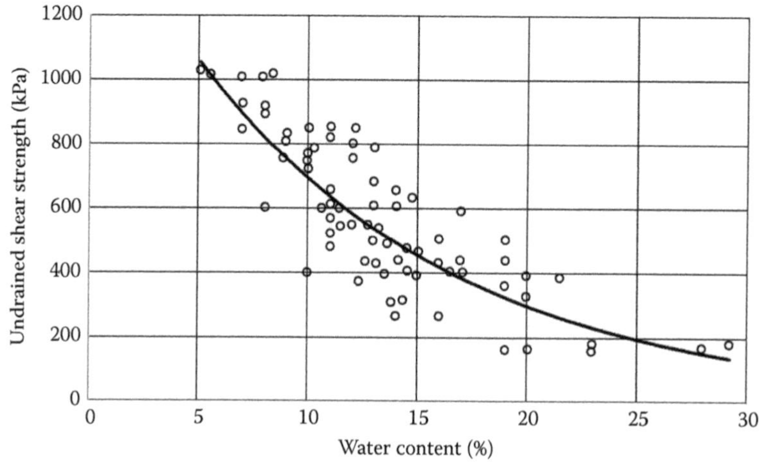

Figure 4.38 Variation in undrained shear strength with water content for remoulded matrix-dominated tills from Kielder, Northern England. (After Millmore, J. P. and R. McNicol. Geotechnical aspects of the Kielder Dam. In *Proceedings of the Institution of Civil Engineers*, 74(4); 1983: 805–836.)

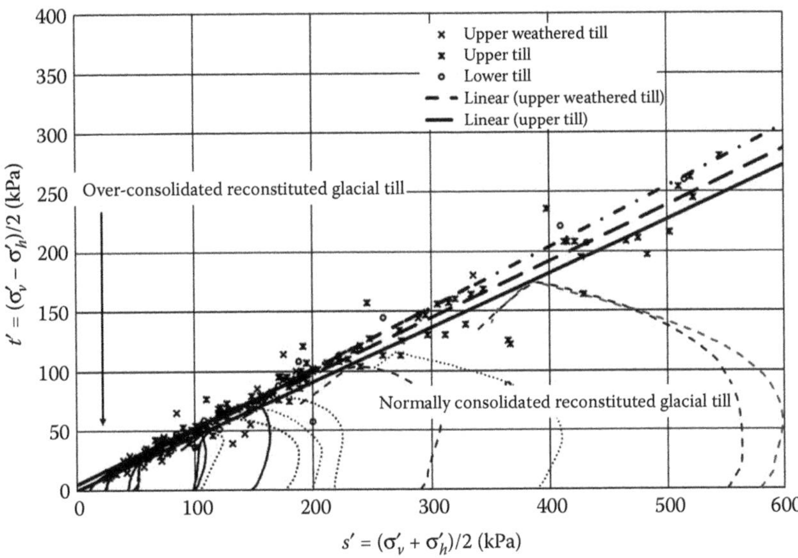

Figure 4.39 Comparison between the failure envelope from tests on undisturbed and reconstituted samples of glacial till showing that the tests on reconstituted samples produce a lower bound to the field specimens provided the density is similar.

Skempton and Brown (1961) carried out tests on reconstituted and intact specimens to show that they gave similar results for compression tests for both increasing axial stress and reducing radial stress (Table 4.29).

Jung et al. (2012) reported a comparison of triaxial and oedometer tests on undisturbed and reconstituted Chicago clay of the Deerfield stratum (Chicago clays) to show that the strengths and stiffness were not similar, concluding that it is inappropriate to use reconstituted specimens for design parameters. However, though they modelled the assumed stress history of the clay, they did not achieve the *in situ* density. The undisturbed samples were stronger and stiffer than the reconstituted soil, raising the question as to whether stress history or density is more important.

Table 4.29 Results of triaxial tests on reconstituted and undisturbed glacial till

	Consistency limits						*Effective strength parameters*		
Depth (m)	I_L	I_P	*PI*	*Water content (%)*	*Bulk density (kg/m³)*	*Clay fraction*	*Cohesion (kPa)*	*Angle of friction*	*Undisturbed (U) remoulded (R)*
0.9	23	11	12	11.9	2.24		8.1	34	R
1.8	28	13	15	12	2.23	14	5.7	32	U
2.7	26	12	14	15.1	2.16	15	9.6	32	U
7.6	25	12	13	11.2	2.26	15	3.8	34	R
12.8	28	15	13	14.4	2.15	18	9.1	29	R
3.7	26	12	14	9.6	2.26	17	12.9	33	U
3.0	27	14	13	11.1	2.24	25	12.5	30	R
2.7				10.4	2.26		7.7	32	R

Source: After Skempton, A.W. and J. D. Brown. *Geotechnique*, 11(4); 1961: 280–293.

Given the stiffness of subglacial tills, sensitivity of strength to water content and the difficulty of establishing the stress history, it would appear that tests on reconstituted matrix-dominated tills should provide a reasonable assessment of the characteristic effective strength, provided the specimens are reconstituted at the *in situ* water content and consolidated to the *in situ* density. The values obtained are likely to a lower bound to the *in situ* values and do not include the effect of fabric. The main advantage of this approach is to remove the gravel, which, *in situ*, will have little effect on mass strength because of their random distribution and quantity but will influence the strength of laboratory specimens. Trenter (1999) listed four disadvantages to using remoulded or reconstituted samples to determine the strength of subglacial tills:

1. The cost of preparing samples.
2. Ensuring that the water content was correct and possibly carrying out tests at different water contents to determine the sensitivity of strength to water content.
3. A decision has to be made on the largest particle size. For example, Atkinson et al. (1985) removed all gravel. Gens and Hight (1979) measured the total and effective strength of reconstituted and remoulded samples with varying gravel content up to 12%. They showed that the critical parameter was the water content provided there are no particles >2 mm.
4. Any cementation is lost.

McGown (1975) investigated the effect of fines content and dry density on reconstituted samples of till. They showed, as expected, that the angle of friction reduces as the dry density reduces (Figure 4.40) and there is an optimum fines content for a maximum angle of friction. This is probably coincident with maximum density, a result of particle size distribution.

Subglacial tills have a low plasticity; therefore, according to Lupini et al. (1981), these tills should exhibit turbulent shear with little reduction in strength. This is not the case with glaciolacustrine deposits. Trenter (1999) presented data from a number of sites to produce a design curve for the residual strength of subglacial tills based on Lupini et al. (1981) (Figure 4.41).

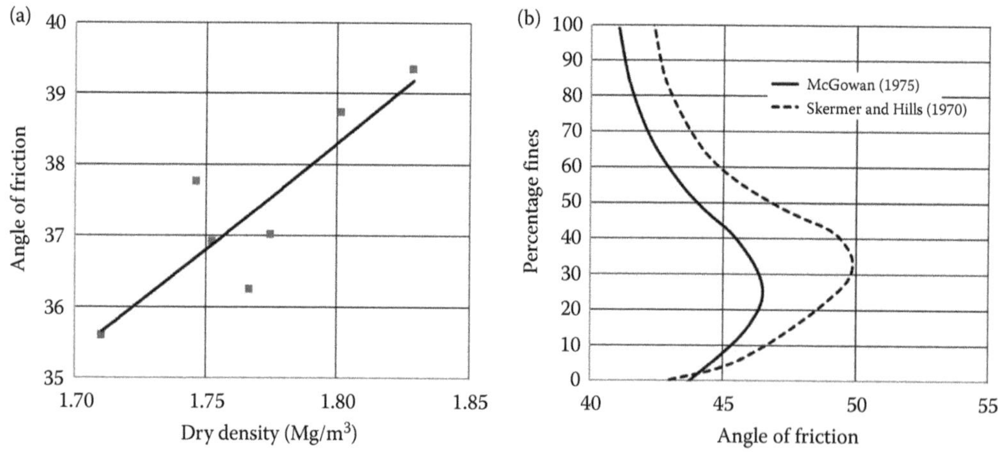

Figure 4.40 Increasing the fines content of glacial till increases (a) the dry density leading to an increase in the angle of friction and (b) there is a maximum value of angle of friction between 20% and 40% fines content consistent with the observations of composite soils.

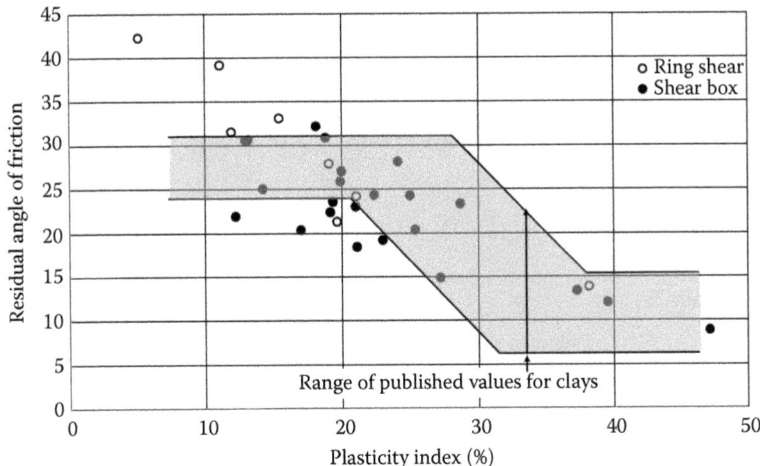

Figure 4.41 Variation of residual angle of friction with the plasticity index for glacial tills showing a similar behaviour to other clays. (After Trenter, N. A. *Engineering in Glacial Tills*. CIRIA, London, 1999.)

The stress path in a standard triaxial test is assumed to be that beneath the centre of a foundation (Figure 4.42) where the vertical principal stress increases and the horizontal principal stress remains constant. The vertical stress may not be the principal stress in matrix-dominated tills at the time of deposition, but it is assumed that with time it is a reasonable assumption that it is today. In practice, because of the homogeneous nature of the till and the other factors that affect the strength, this may not be an issue.

Stress path testing is not covered here but the use of stress paths to explain soil behaviour is a powerful tool to understand changes that will occur when soil is subject to loading or unloading. Stress paths can be plotted in terms of q, p' (known as the Cambridge method) or t, s' (known as the MIT method) where

$$s = \frac{\sigma_1 + \sigma_2}{2} \tag{4.24}$$

$$t = \frac{\sigma_1 - \sigma_2}{2} \tag{4.25}$$

$$s' = \frac{\sigma_1' + \sigma_2'}{2} \tag{4.26}$$

where σ_1, σ_2 are the principal total stresses. Figure 4.43 shows the effective stress path for drained and undrained tests and the effect of over-consolidation on the stress path. The failure envelope is given by

$$t = a' + s' \tan\alpha' \tag{4.27}$$

where ($a' = c' \cot \varphi'$) and ($\tan \alpha' = \sin \varphi'$).

Triaxial tests on glaciolacustrine clays produce a characteristic strength, which is useful for foundation design but may not be appropriate for slope design as they are strongly

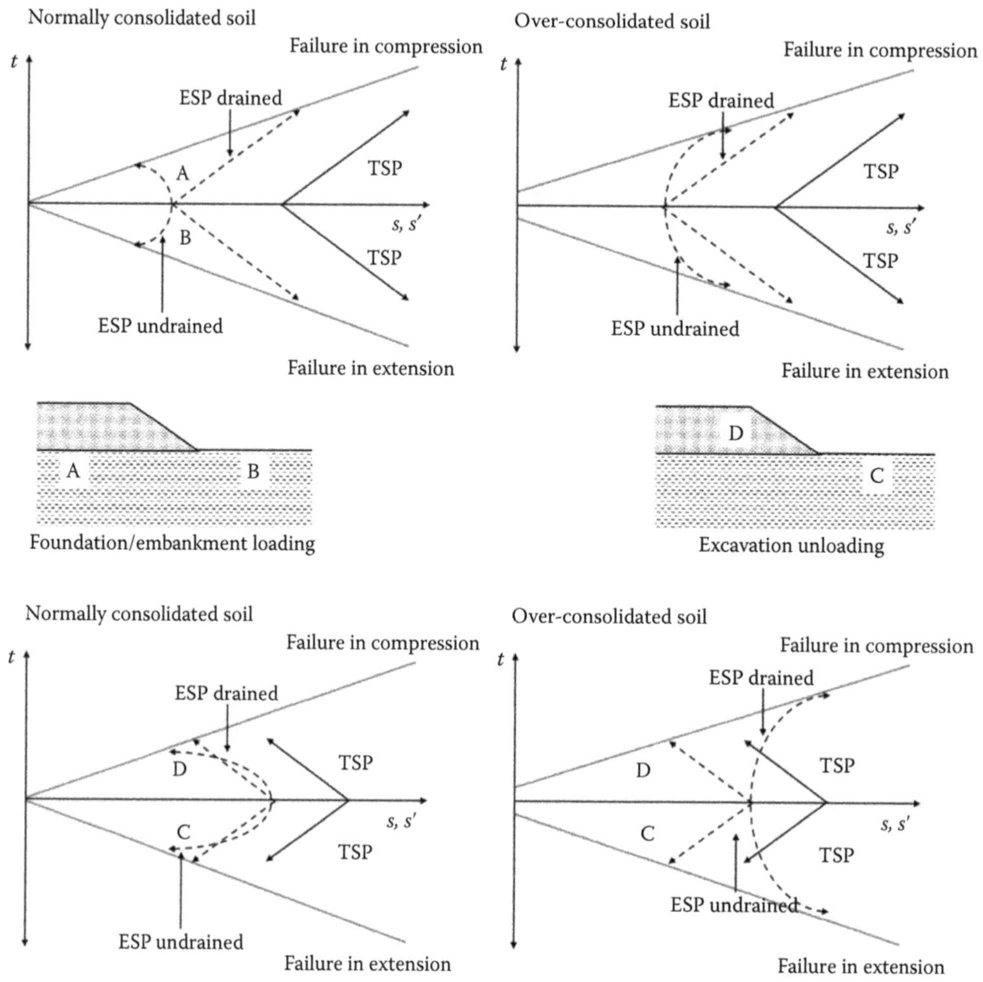

Figure 4.42 Comparison between drained and undrained stress paths triaxial tests on normally consolidated and heavily over-consolidated clays to model the stress changes (a) beneath and adjacent to a foundation or embankment and (b) in the side slope and base of an excavation.

anisotropic. The silt laminations, if they exist, 'reinforce' the clay when subject to a compression test but not when a specimen is subject to direct shear, the path followed in a slip surface parallel to the laminations.

Chegini and Trenter (1996) presented the results of a detailed investigation for a nuclear facility in SW Scotland. The investigation included 100 boreholes for *in situ* testing and sampling to determine the geological and geotechnical characteristics in some detail. Tests were carried out on reconstituted and undisturbed samples of a glacial till, a matrix-dominated basal till derived from Permo-Triassic sandstone. The results were typical for these tills – scatter in results, little correlation between parameters – due to the fabric and composition of the till. They observed that the SPT N_{60} relationship with undrained shear strength varied between 6N and less than 1N. Figure 4.43 shows results of CD and CU triaxial tests on 100-mm samples that lie about a failure line defined by an angle of friction of 31°. Tests on the reconstituted till also failed on the same line. They concluded that the variable soil composition meant that no useful relationship would be found between SPT N_{60} and

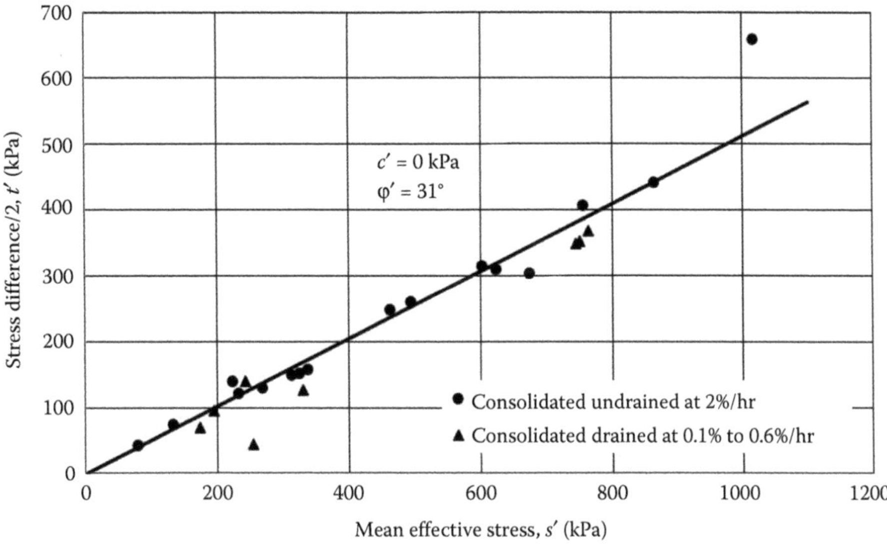

Figure 4.43 Effective strength of glacial tills from Chapelcross, United Kingdom showing the advantage of plotting the peak deviator stress from all tests on a stress path to obtain a representative strength. (After Chegini, A. and N. A. Trenter. The shear strength and deformation behaviour of a glacial till. In *Proceedings of Conference on Advances in Site Investigation Practice*, London, 1996.)

undrained shear strength and the angle of friction was clearly defined and could be obtained from consolidated undrained and drained triaxial tests on the reconstituted till.

Finno and Chung (1992) reported a detailed assessment of Chicago glacial clays, which are typical of the soils in the Great Lakes area. These tills are low to medium plasticity supraglacial and subglacial clay tills. Otto (1942) suggested that there are six distinct till sheets deposited during the Wisconsin period in the Chicago area – Valparaiso, Tinley, Park Ridge, Deerfield, Blodgett and Highland Park – based on their water content and strength index. Finno and Chung (1992) undertook consolidation, shear and stress path tests on 71-mm diameter piston and Shelby tube samples. The tills were predominantly formed of illite (51%–57%) and dolomite (15%–21%) and smaller percentages of chlorite, variscite, calcite and kaolinite. Figure 4.44 shows the variation of consistency limits and c_u/σ'_v with depth. Triaxial and compression tests were carried out on 71-mm diameter Shelby tube samples. Gravel was present but the particles were typically less than 10 mm in diameter and less than 5% by weight. Anisotropic consolidation was used to prepare samples up to an over-consolidation ratio of four. Figure 4.45 shows the variation in shear strength normalised by the effective vertical stress with over-consolidation ratio for extension and compression tests. The site-specific relationships are as follows:

$$\left(\frac{c_u}{\sigma'_v}\right)_{TXC} = 0.46(0.90 - w)OCR^{0.9} \tag{4.28}$$

$$\left(\frac{c_u}{\sigma'_v}\right)_{TXE} = 0.31(0.90 - w)OCR \tag{4.29}$$

where w is the natural water content.

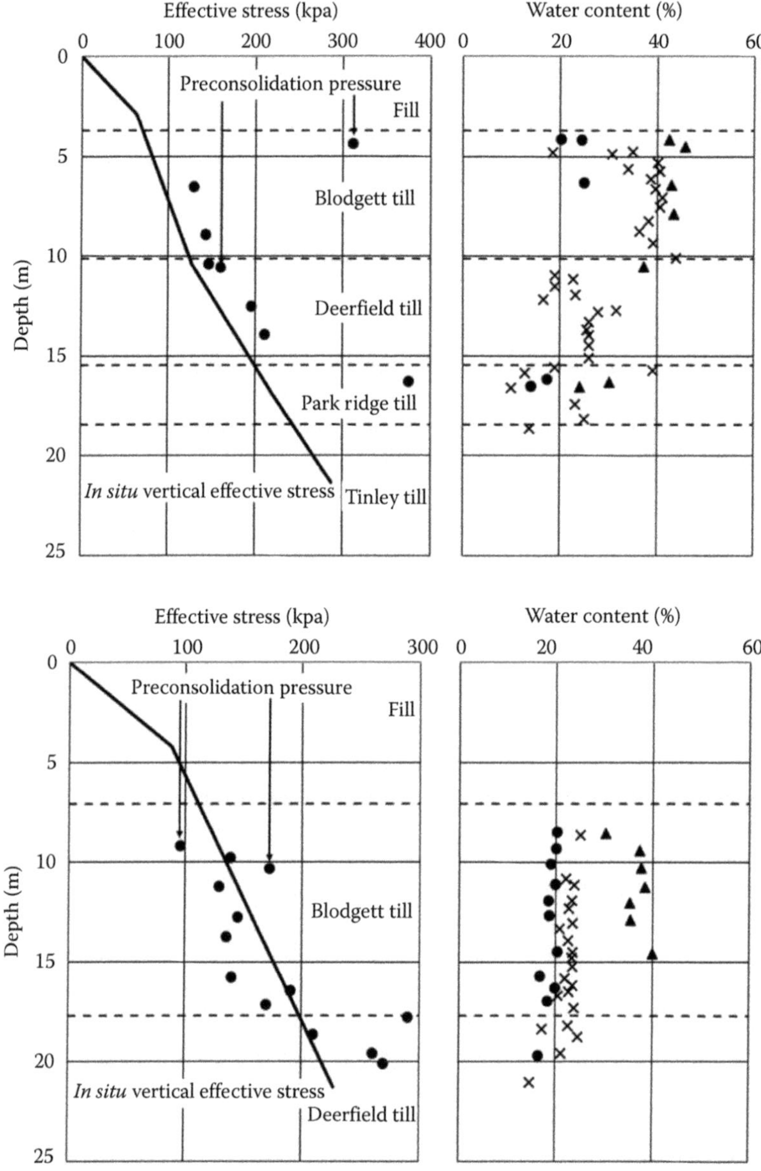

Figure 4.44 Profiles of consistency limits, water content, *in situ* vertical effective stress and preconsolidation pressure for Blodgett Till (supraglacial till) and Deerfield Till (subglacial till), which are part of the Chicago glacial clays sequence. (After Finno, R. J. and C.-K. Chung. *Journal of Geotechnical Engineering*, 118(10); 1992: 1607–1625.)

Figure 4.46 shows a comparison between shear strength results from *in situ* and laboratory tests together with the predicted strength from the laboratory tests confirming that scatter in results can mask the differences between the sets of the results. The stress path tests, however, did provide a means of determining a more consistent profile though not necessarily helpful given the differences between the best fit to the tests on undisturbed specimens and field tests.

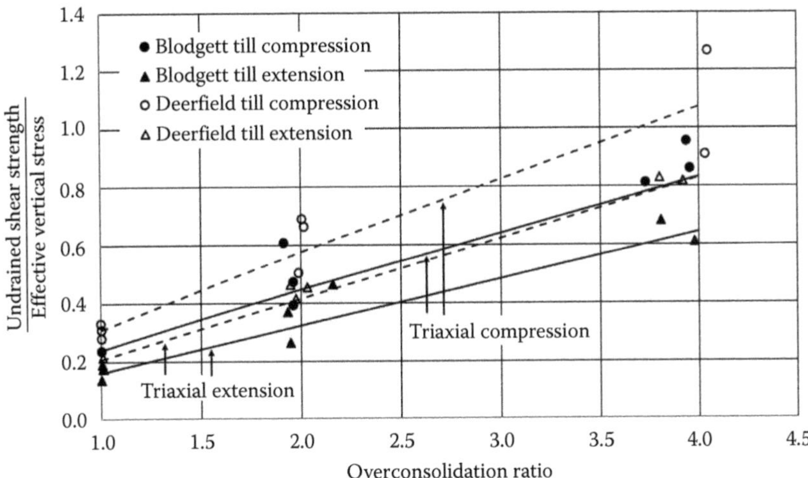

Figure 4.45 Variation in normalised undrained shear strength from triaxial compression and extension tests on specimens of Blodgett Till (supraglacial till) and Deerfield Till (subglacial till), which are part of the Chicago glacial clays sequence, with over-consolidation ratio. (After Finno, R. J. and C.-K. Chung. *Journal of Geotechnical Engineering*, 118(10); 1992: 1607–1625.)

The angle of friction was found to be between 28.3° and 34.6° and the post-peak value between 27.6° and 32.3°.

Long and Mentiki (2007) produced a summary of the characteristics of Dublin Boulder Clay using an extensive number of quality samples from major construction projects. They took block samples using 300- and 350-mm cubical, thin-walled samplers with 20° or 45° angled cutting edges with 9-mm-thick walls and wireline triple tube rotary coring with a

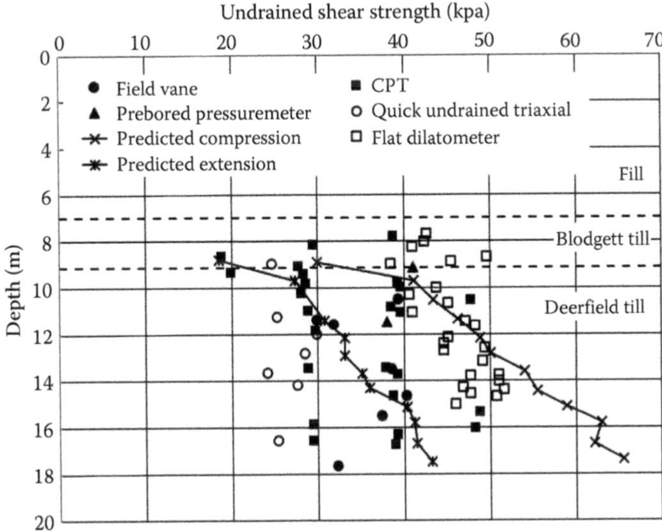

Figure 4.46 Comparison between undrained shear strength derived from field and laboratory tests and those predicted from the best fit to the triaxial test results shown in Figure 4.45 using the *in situ* water content. (After Finno, R. J. and C.-K. Chung. *Journal of Geotechnical Engineering*, 118(10); 1992: 1607–1625.)

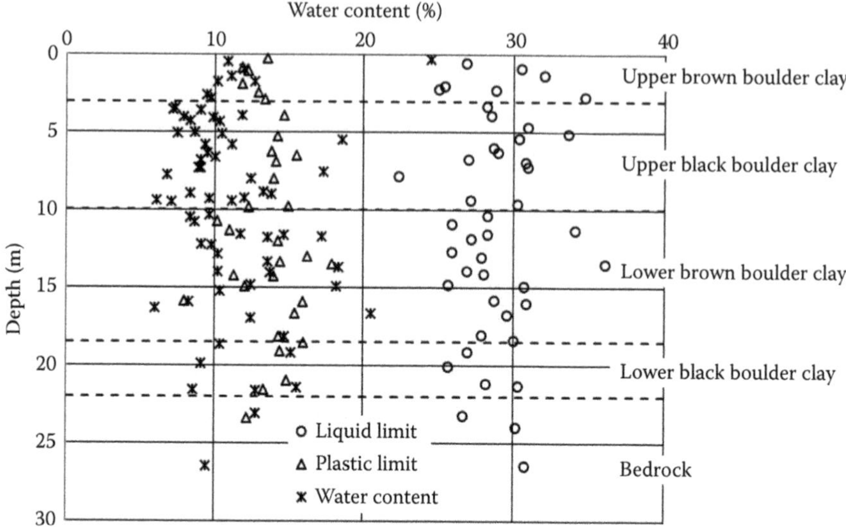

Figure 4.47 Classification data for Dublin Boulder Clay at the Dublin Port Tunnel site. (After Long, M. and C. O. Mentiki. *Geotechnique*, 57(7); 2007: 595–611.)

polymer flush. Dublin Boulder Clay is a lodgement till that is divided into four layers: Upper Brown Boulder Clay, a 2- to 3-m-thick weathered till; Upper Black Dublin Boulder Clay, 4–12 m thick; Lower Brown Boulder Clay, 5–9 m thick; and Lower Black Dublin Brown Boulder Clay, possibly a boulder pavement that is mostly less than 2 m thick. The stone content increases with depth. Figure 4.47 shows the classification data and groundwater profile and Table 4.30 is a summary of the properties of the four tills that are predominantly formed of clay minerals (76%). Table 4.31 summarises the average undrained shear strength of the four layers. The peak angle of friction was 44° and the post-peak 36° similar to the critical state angle (Lehane and Faulkner, 1998), and the failure envelope is curved with negligible cohesion.

Bell (2002) presented an extensive overview of the geotechnical characteristics of exposures of a glacial till along the east coast of England. Table 4.32 summarises the description

Table 4.30 Properties of Dublin Boulder Clay

Property	*Upper brown boulder clay*	*Upper black boulder clay*	*Lower brown boulder clay*	*Lower black boulder Clay*
Water content (%)	13.1	9.7 (11 ± 3)	11.5	11.3
Bulk density (Mg/m³)	2.23	2.34	2.28	2.28
Liquid limit (%)	29.3	28.3 (25 ± 4)	30.0	29.5
Plastic limit (%)	15.9	15.1	14.9	17.8
Clay content (%)	11.7	14.8 (15 ± 5)	17.8	17.5
Silt content (%)	17.0	24.7	28.3	30.5
Sand content (%)	25.0	24.7	25.7	34.0
Gravel content (%)	46.3	35.9 (30 ± 5)	28.0	35.5

Source: After Long, M. and C. O. Mentiki. *Geotechnique*, 57(7); 2007: 595–611.

Note: Values in brackets were reported by Lehane and Simpson (2000), which represent the most commonly found glacial till.

Table 4.31 Average values of $(c_u/(N_{60}))$ and (c_u/σ'_v) for Dublin Boulder Clay

Test	*Parameter*	*Upper brown*	*Upper black*	*Lower brown*	*Lower black*
SPT	N_{60}	19	53	53	68
CIUC	c_u		287	297	240
	c_u/σ'_v		1.93	2.11	0.98
	$c_u/(N_{60})$		5.4	5.6	3.5
CAUC	c_u	84	373	520	
	c_u/σ'_v	2.25	3.23	2.58	
	$c_u/(N_{60})$	4.4	7.0	9.8	
CAUE	c_u	21	87	129	
	c_u/σ'_v	0.46	0.87	0.75	

Source: After Long, M. and C. O. Mentiki. *Geotechnique*, 57(7); 2007: 595–611.

of the tills and their properties. Typical values of strength and compressibility are given in Table 4.32, suggesting that the tills are not heavily over-consolidated confirming the view of complex pore pressure regime during deposition. Triaxial and shear box tests were used to determine the sensitivity, total and effective strength parameters. The results confirm observations of others that these tills are insensitive, which, together with observations of Anderson (1974) and McGowan (1975), suggests that many UK tills can be considered to be insensitive, supporting the view that tests on the reconstituted till at the same density as the natural till will provide a consistent value of *in situ* strength. Values of residual strength and angles of friction from total stresses are quoted, but these should be treated with caution as the level of strain (24%) in the shear box tests was insufficient to achieve residual conditions and the triaxial test specimens may be partially saturated.

Discontinuities also affect the strength of a soil. Terzaghi et al. (1996) suggested that the shear strength reduced with time as discontinuities opened up on excavation, water softened the soil adjacent to the discontinuities and further discontinuities form due to the softening of the soil adjacent to existing discontinuities. This can take time. For example, Terzaghi et al. (1996) and Duncan and Dunlop (1968) showed that some engineered slopes fail 20–80 years after construction. Aldred (2000) attributed the soil softening of fractured glacial till to the softening of the soil adjacent to discontinuities, which can open due to stress relief caused by excavation. However, Skempton and Brown (1961) in analysing the Selset landslide (NE England) suggested that cohesion of glacial tills does not reduce with time unlike stiff fissured over-consolidated clays. This may explain why many natural till slopes stand at 45°. Lo (1970) proposed a relationship between sample size and strength of the soil:

$$c_u = c_{um} + (c_{u0} - c_{um})e^{-\alpha(A-A_0)^\beta} \quad \text{for } A > A_0 \tag{4.30}$$

where c_u is the undrained shear strength of the specimen, c_{um} the mass strength, c_{u0} the intact strength, A the area of the failure plane and A_0 the area of the failure plane for an intact sample. α and β are constants derived from unconsolidated undrained triaxial tests. This could be used to assess whether an excavation in glacial till is likely to fail in the long term. It requires a detailed description of the till to establish the characteristics of the discontinuities and sufficient samples to determine the relationship between discontinuities and strength.

Table 4.32 Geotechnical properties of glacial tills in Teesside, Holderness and Cromer, East Coast of England

Area	Till		Water content (%)	Plastic limit (%)	Liquid limit (%)	Plasticity index (%)	Liquidity index	Consistency index	Activity	Intact c_u	Remoulded c_u	Sensitivity	Cohesion	Angle of friction
North Norfolk	Hunstanton	Max	18.6	23	40	23	0.07	0.97	1.00	184	164	1.22	18	34
		Min	16.8	15	34	15	−0.19	0.89	0.75	152	128	1.18	8	26
		Mean	17.6	18	37	20	−0.02	0.92	0.85	158	134	1.19	12	29
	Chalky boulder clay	Max	25.2	21	45	26	0.48	0.85	0.50	120	94	1.49	16	28
		Min	22.4	18	32	14	0.15	0.50	0.40	104	70	1.28	7	21
		Mean	23.6	20	37	18	0.32	0.68	0.45	110	81	1.34	11	24
	Contorted drift	Max	18.9	18	29	13	0.33	0.86	0.80	180	168	1.67	20	33
		Min	13.2	9	19	8	0.07	0.72	0.65	124	76	1.08	6	27
		Mean	15.6	14	25	11	0.16	0.78	0.75	160	136	1.23	11	30
	Cromer till	Max	15.8	20	40	24	−0.16	1.16	0.95	224	188	1.19	19	32
		Min	11.9	14	27	13	−0.18	0.98	0.65	154	140	1.10	12	26
		Mean	13.2	17	35	19	−0.17	1.09	0.80	176	156	1.13	14	29
Holderness	Hessle till	Max	26.6	26	53	32	0.07	1.15	2.10	138	116	1.31	8	24
		Min	18.5	20	38	17	−0.02	0.79	0.06	96	74	1.10	5	13
		Mean	22.6	22	47	25	0.04	0.97	1.24	106	96	1.19	7	25
	Withernsea	Max	19.3	21	39	20	−0.28	1.02	1.21	172	148	1.18	19	34
		Min	12.3	15	22	12	−0.10	0.83	0.72	140	122	1.15	5	16
		Mean	16.9	18	34	17	−0.16	0.99	0.93	160	136	1.16	9	25
	Skipsea	Max	18.2	19	36	18	−0.29	1.29	0.67	194	168	1.15	21	36
		Min	13.5	14	20	9	−0.04	0.98	0.51	182	154	1.08	10	24
		Mean	15.5	16	30	14	−0.19	1.11	0.56	186	164	1.13	12	30
	Basement	Max	20.4	23	42	22	−0.16	1.08	0.59	212	168	1.27	17	36
		Min	15.6	16	28	12	−0.03	0.98	0.53	163	140	1.19	6	20
		Mean	17	20	36	19	−0.13	1.01	0.55	186	156	1.21	9	29
Teesside	Upper boulder clay	Max	21	20	49	34	0.23	1.77	1.50					
		Min	5	11	22	10	−0.46	0.42	0.57					
		Mean	14	15	33	19	−0.04	1.23	0.97					
	Lower boulder clay	Max	17	16	38	23	0.32	1.92	1.22					
		Min	10	9	27	13	−0.31	0.71	0.63					
		Mean	13	13	31	18	−0.03	1.13	0.81					

Source: After Bell, F. G. *Engineering Geology*, 63(1); 2002: 49–68.

4.5.2.7 *Unsaturated strength*

Since matrix-dominated tills are abundant and are considered to be relatively impermeable when remoulded, they are used as liners and capping layers for landfills. The performance of these partially saturated layers is critical to prevent contamination. Partially saturated tills have also been used to construct embankments over the last 200 years in the United Kingdom. The compaction procedures have changed over that time from 'dig and dump' to properly engineered fills, which means that these embankments are prone to changes in pore pressure, which is an increasing problem due to climate change. Therefore, an understanding of unsaturated matrix-dominated tills is necessary.

Fredlund et al. (1995) undertook tests on a compacted glacial till to show that the shear strength of a partially saturated till could be predicted. Figure 4.48 shows the typical variation of shear strength and degree of saturation with matric suction. The shear strength, τ_f, of unsaturated soils is given by (Fredlund et al., 1978)

$$\tau_f = c' + (\sigma_n - u_a)\tan\varphi' + (u_a - u_w)\tan\varphi^b \tag{4.31}$$

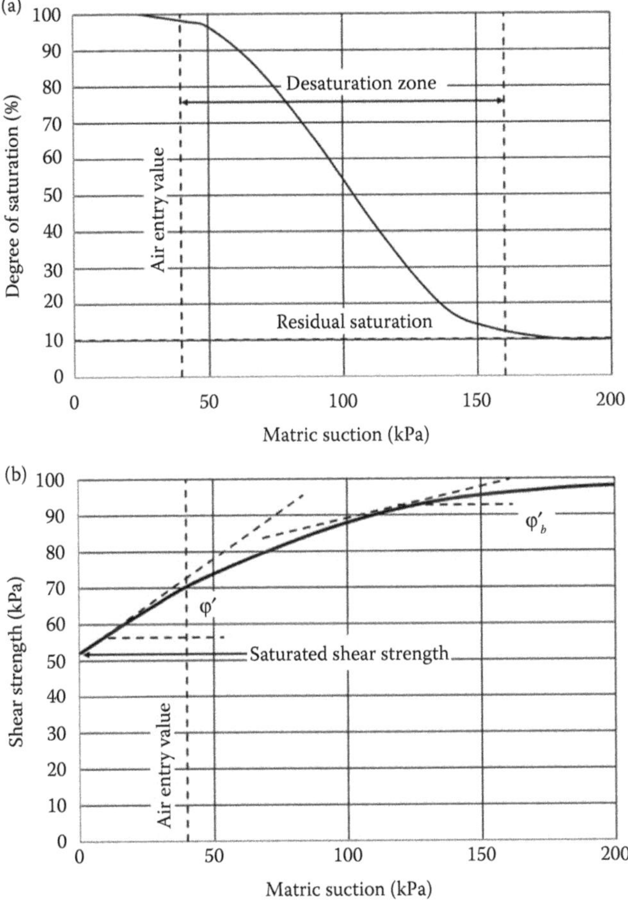

Figure 4.48 Relationship between (a) degree of saturation and matric suction and (b) shear strength and matric suction. (After Fredlund, D. G. et al. Predicting the shear strength function for unsaturated soils using the soil-water characteristic curve. In *First International Conference on Unsaturated Soils*, Paris, France, 1995: 6–8.)

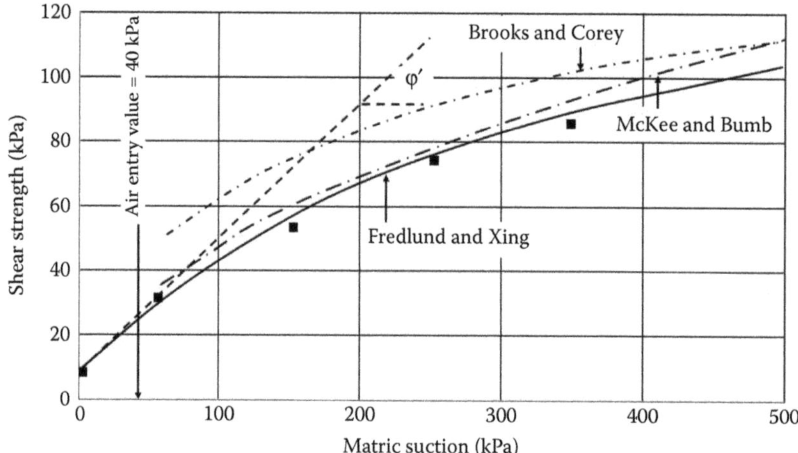

Figure 4.49 Comparison of experimental and predicted variations of shear strength with matric suction (Fredlund, D. G. et al. Predicting the shear strength function for unsaturated soils using the soil-water characteristic curve. In *First International Conference on Unsaturated Soils*, Paris, France, 1995: 6–8.)

where $(\sigma_n - u_a)$ is the net normal stress, $(u_a - u_w)$ the matric suction and φ^b the angle of shearing resistance to matric suction.

Fredlund and Xing (1994) developed a rigorous solution for the soil water characteristic curve given by

$$S = \left[1 - \frac{\ln(1 + (\psi/\psi_r))}{\ln(1 + (1{,}000{,}000/\psi_r))}\right]\left[\frac{1}{\ln(2.72 + (\psi/a)^n)}\right]^m \quad (4.32)$$

where S is the degree of saturation, ψ the soil suction, ψ_r the suction corresponding to the residual water content and a the air entry value. This leads to a shear strength prediction:

$$\tau = c' + (\sigma - u_a)\tan\varphi' + \tan\varphi' \int_0^{\psi} \left[\left(\frac{S - S_r}{1 - S_r}\right)\right] d(u_a - u_w) \quad (4.33)$$

which is compared with the experimental results on the glacial till in Figure 4.49.

4.5.3 Compressibility and deformation

While strength is important, deformation parameters are increasingly more useful as methods of analysis have improved to such an extent that some confidence can be placed on predicted deformations. However, as mentioned in Eurocode 7, if the output is going to be relevant, then the quality of sampling and testing has to be of the highest standard. This means Class 1 samples with local strain measurements. The difficulty of obtaining such samples, obtaining representative samples and the cost of carrying out local strain measurements means that these tests are often restricted to major projects.

Deformation and compressibility characteristics can be determined in a variety of ways ranging from empirical correlations with results of field tests to local strain measurements in triaxial tests. The characteristics are particularly susceptible to soil disturbance so only Class 1 samples are appropriate for laboratory tests. This means that laboratory assessments of deformation characteristics can be made only on samples of glacial clays. There are three methods: oedometer, Rowe cell and triaxial tests.

4.5.3.1 One-dimensional consolidation tests

The Rowe cell and oedometer tests are one-dimensional consolidation tests. The oedometer test is described in BS EN ISO 17892-5:2014. The specimen has to be at least 35 mm diameter and 12 mm high with a diameter to height ratio of not less than 25. In the United Kingdom, specimens are usually 75 mm diameter, 19 mm high. BS EN 1997-2:2007 suggests that the maximum particle size should be H/5, which means that specimens with particles exceeding 4 mm would be unacceptable. Rowe (1972) suggests that the fabric of a soil will affect the results, which means a large representative sample is needed (cf. triaxial test specimens for glacial soils). At depth, the earth pressures are great enough to close the soil discontinuities, which means that the stiffness will increase with depth but on excavation will reduce due to the reduction in vertical stress and opening of discontinuities. This means that slopes would be more unstable because of the reduction in strength and foundations settle more quickly because of the increase in permeability. This means that oedometer tests should be restricted to glaciolacustrine clays with laminations less than 4 mm thick and matrix-dominated tills with particles less than 4 mm and no visible structural features.

There has to be seven stages of loading, doubling the stress at each stage up to a maximum stress of five times the maximum stress likely to be experienced *in situ*. Given the importance of stress history, it is useful to determine the preconsolidation pressure. This is feasible for tests on glaciolacustrine clays, but it depends on the drainage environment during deposition as to whether it is possible in matrix-dominated tills (see Section 2.6.1). It is recommended that at least two unload/reload cycles be carried out to reduce effects of sample disturbance and system compliance. The test results can be used to show the variation of effective stress with void ratio, which can be used to determine the compression stiffness index (S_c) and the compression index (C_c) from the linear portion of the compression curve (post-yield) and the swelling stiffness index (S_s) and swelling index (C_s) from the swelling curve, the preconsolidation pressure ($\sigma'_{v\max}$) and, for each load increment, the coefficient of volume compressibility (m_v), the oedometer modulus (E_{oed}), the coefficient of consolidation (c_v) and the coefficient of secondary compression (c_a).

The oedometer test is used to estimate the preconsolidation pressure (Figure 4.50), which, in the case of subglacial tills, is often assumed to be due to the weight of ice.

This method is valid for gravitationally consolidated clays such as glaciolacustrine clays, but may not be relevant to subglacial tills because

- It is assumed that the normal consolidation line on the ($e\ v \log_{10} \sigma'_v$) plot is linear. It is likely to be concave, especially at the stresses imposed on the tills during deposition, because there is a minimum void ratio.
- It is assumed that the soil is gravitationally consolidated whereas subglacial tills undergo shear. This means that the vertical stress is no longer a principal stress; the maximum principal stress will exceed the stress due to the weight of the ice because of the rotation of the principal axes.
- It is assumed that the swelling line is 'elastic'; that is, on reloading the change in void ratio follows the swelling line allowing for hysteresis. The stress changes that take place during deglaciation lead to failure in extension (Figure 4.17).

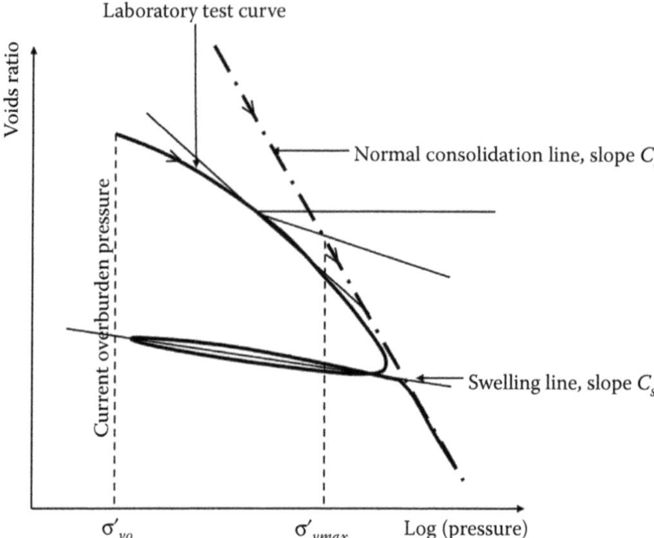

Figure 4.50 Variation in void ratio with effective stress showing the compression and swelling indices and the Cassagrande method of assessing the preconsolidation pressure.

- Isostatic uplift following deglaciation led to further stress changes.
- The void ratio is less than that predicted from the weight of ice because of the shear and increase in principal stress.
- Reloading an intact specimen of a matrix-dominated subglacial till will produce a compression curve that is similar to that for a gravitationally consolidated soil. If the effects of shearing and isostatic uplift are ignored, then the applied loads will have to exceed the weight of ice (e.g. the Antarctic Ice Sheet is estimated to be up to 2,500 m thick; equivalent to 25,000 kPa acting on the subglacial till).
- The stress changes that occur during deglaciation are limited by failure in extension such that

$$\frac{\sigma'_h}{\sigma'_v} < K_{omax} = \frac{1+\sin\varphi'}{1-\sin\varphi'} \tag{4.34}$$

Based on a typical value of angle of friction for glacial soil, K_{omax} is 3. Hence, further stress changes during deglaciation are accompanied by a reduction in horizontal and vertical stress as the soil fails in extension.

There will be a relationship between void ratio and vertical effective stress, which will have similar characteristics to those assumed in the Cassagrande construction but this will not be the same relationship that the till has undergone. Reloading the till will indicate a lightly over-consolidated soil because the increase in pressure necessary to re-establish a normal consolidation line (Figure 4.17) is much less than that required to create the stress history of the till. The Cassagrande construction is useful because it helps identify a change in behaviour, the yield stress, where irreversible strains take place.

These points are based on the assumption that the compression and swelling of a subglacial till takes place in drained conditions. There is evidence that undrained conditions exist at the base of a glacier. In that case, the preconsolidation pressure will be similar to the current *in situ* stress, but the effects of shearing and isostatic uplift still apply.

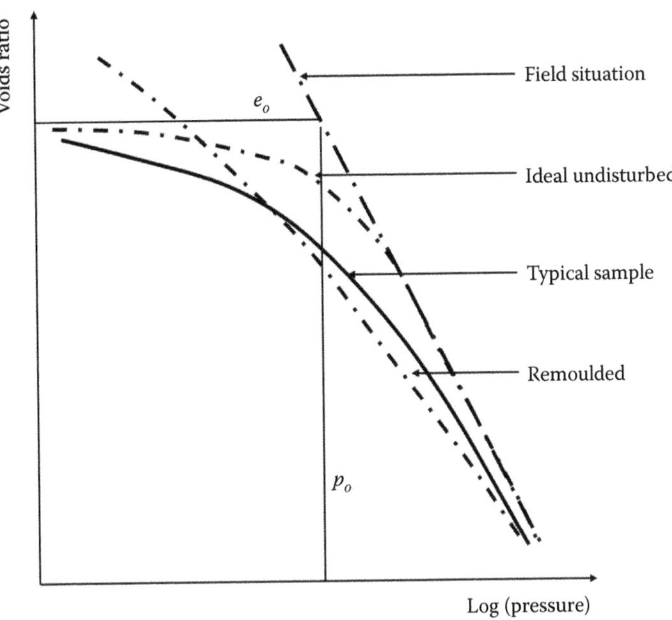

Figure 4.51 Effect of disturbance on the consolidation properties of normally consolidated soil. (After Head, K. H. *Manual of Soil Laboratory Testing, Vol. 2, Permeability, Shear Strength and Compressibility Tests*. Pentech, London, 1988a.)

The standard odometer test is used to determine the compressibility characteristics of fine-grained soils, that is, the coefficient of compressibility characteristic, a_v, coefficient of volume compressibility, m_v, coefficient of consolidation, c_v and the time factor, T_v. The sampling process affects results because sampling reduces the stress acting on the sample creating a state of over-consolidation that means the reloading curve is different from the geological reloading curve (Figure 4.51). Normally and lightly over-consolidated soils are particularly sensitive to disturbance; over-consolidated soils are sensitive to stress relief (Figure 4.52). While this test is the standard test for clays, it tends to underestimate the time of the settlement as the fabric *in situ* dominates the behaviour.

It is the dissipation of pore pressure that controls the consolidation process. The coefficient of volume compressibility varies from 0.1 m^2/MN for glaciofluvial clays to below the 0.05 m^2/MN for very stiff glacial tills. The shape of the time settlement of a curve for a single loading increment depends on the percentage of silt. The test curve for a single loading increment is used to produce the coefficient of consolidation. It was developed for clays where the start of the consolidation can be easily identified. As the silt content increases (Figure 4.53), this becomes more difficult because the theoretical start of consolidation cannot be determined. Head (1988a,b) suggested corrections to select appropriate points for silty clays and silts, that is, glacial fine-grained soils.

4.5.3.2 *Triaxial consolidation tests*

Tests are carried out on saturated specimens. The confining pressure is applied in undrained conditions so that there is a build-up of water pressure. Once this is constant, the drainage valve is opened and the pore pressure and volume changes are measured with time until steady-state conditions are reached. This can be repeated at different confining pressures to

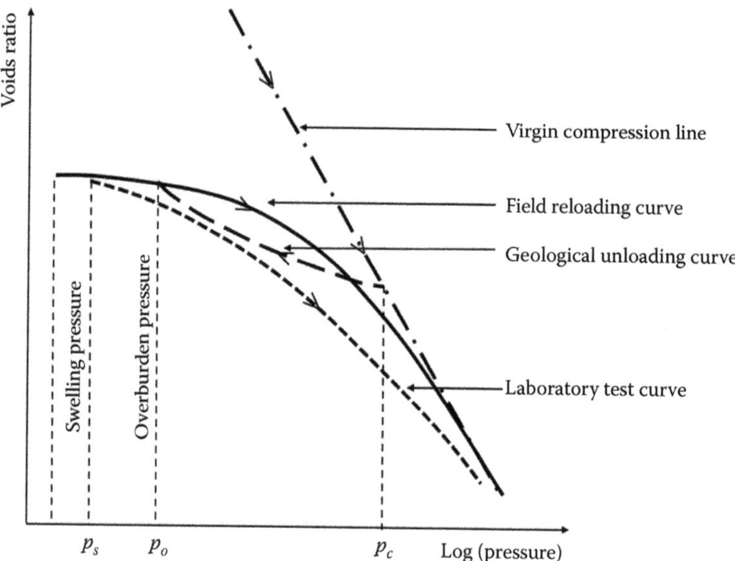

Figure 4.52 Effect of over-consolidation on the laboratory compression curve. (After Head, K. H. *Manual of Soil Laboratory Testing, Vol. 2, Permeability, Shear Strength and Compressibility Tests*. Pentech, London, 1988a.)

obtain the variation in compression characteristics with effective stress. The height of the specimen, H, at the end of each stage assuming the specimen is isotropic is given by

$$H = H_o\left(1 - \frac{1}{3}\frac{\Delta V}{V_o}\right) \quad (4.35)$$

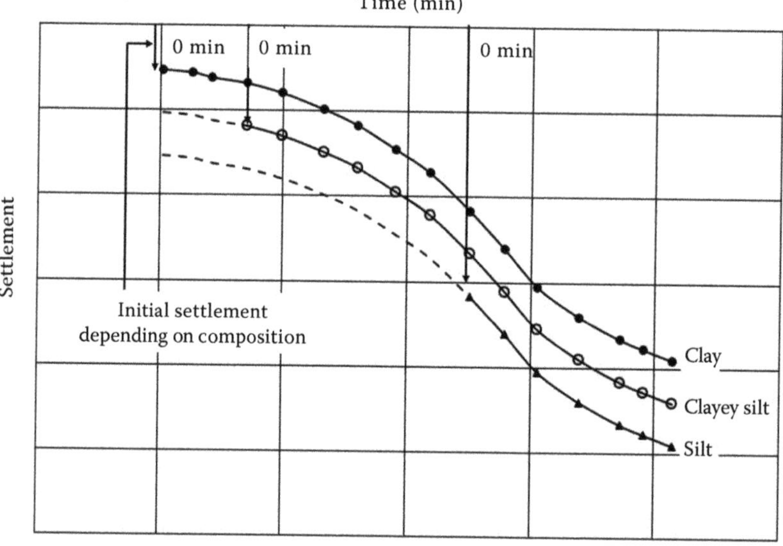

Figure 4.53 Shape of the time settlement curve during a consolidation test depends on the percentage of silt making it difficult to assess the start and end of primary consolidation. (After Head, K. H. *Manual of Soil Laboratory Testing, Vol. 2, Permeability, Shear Strength and Compressibility Tests*. Pentech, London, 1988a.)

where H_o and V_o are the height and volume at the start of consolidation and ΔV is the change in volume during consolidation. The void ratio, e, is

$$e = e_s - (1 + e_s)\frac{\Delta V}{V_o} \tag{4.36}$$

where e_s is the void ratio at the start of the consolidation stage. The coefficient of volume compressibility, m_{vi}, for isotropic consolidation is given by

$$m_{vi} = \frac{\delta e}{\delta \sigma'} \frac{1000}{1 + e_1} m^2/kN \tag{4.37}$$

where δe is the change in the void ratio during the stage and $\delta\sigma'$ the change in effective stress. Experience has shown that the coefficient of volume compressibility from isotropic compression is about 1.5 times the value from 1D compression.

4.5.3.3 Stiffness

Marsland (1975, 1977) and Marsland and Powell (1985, 1991) compared profiles of shear modulus from plate, pressuremeter and triaxial tests on a matrix-dominated till (Figure 4.54). These comparisons suggest that there is a unique value of stiffness for a soil and soil is linear elastic. It is known that soils are non-linear, which means that the strain range over which the stiffness is determined has to be stated or the stiffness degradation curve presented. Further, the stiffness is test dependent because the loading path and the direction of loading vary between tests. For example, a plate test is a model foundation, which

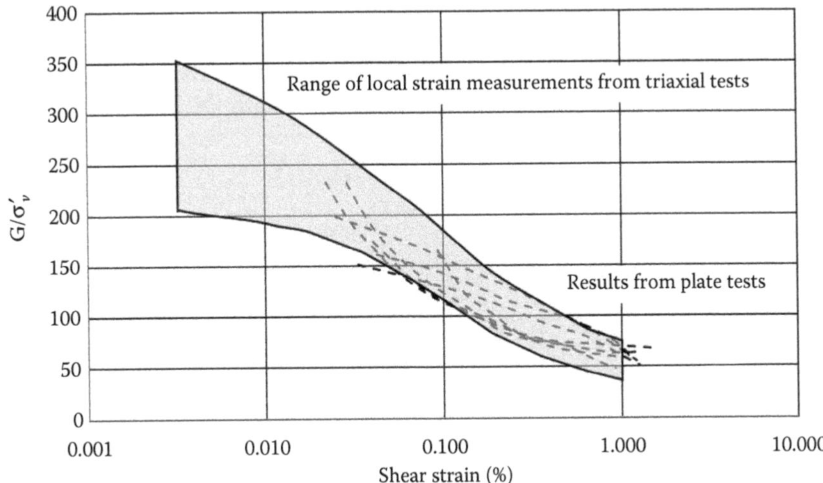

Figure 4.54 Profiles of shear modulus from plate, pressuremeter and triaxial tests on matrix-dominated tills at Cowden and Redcar, NE England highlighting the effect of test type on the values and the need to provide more details of the stress and strain levels if the results are to be of use. (After Marsland, A. *In-situ* and laboratory tests on Boulder clay at Redcar. In *Midland Soil Mechanics and Foundation Engineering Society Symposium on Engineering Behaviour of Glacial Materials*, Birmingham, 1975: 7–17; Marsland, A. and J. J. M. Powell. Field and laboratory investigations of the clay tills at the test bed site at the Building Research Establishment, Garston, Hertfordshire. *Geological Society, London, Engineering Geology Special Publications*, 7(1); 1991: 229–238.)

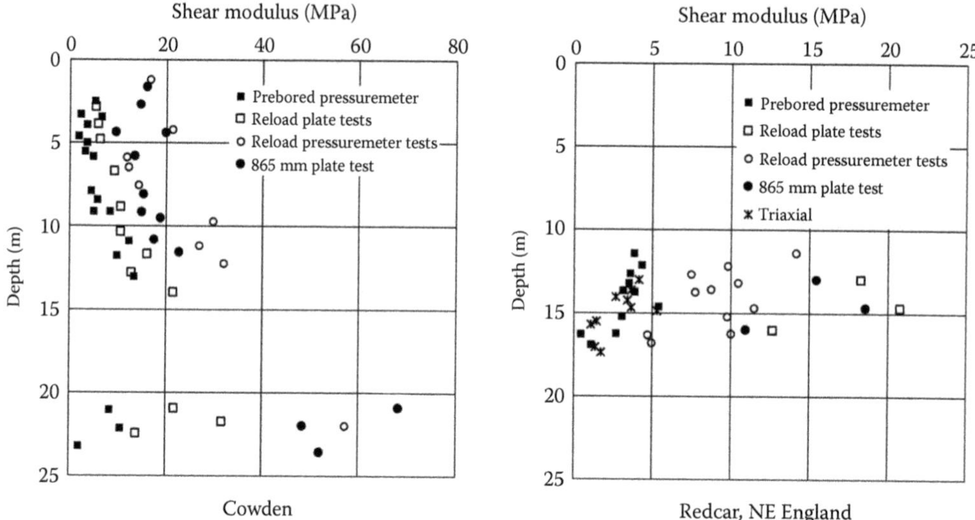

Figure 4.55 Variation in secant stiffness with shear strain from plate bearing and triaxial tests on glacial tills. (After Hird, C. C. et al. Investigations of the stiffness of a glacial clay till. In *Proceeding of the 10th European Conference on Soil Mechanics and Foundation Engineering*, Florence, 1991: 107–110.)

means that the mobilised stiffness is a function of the horizontal and vertical stiffness profile beneath the plate whereas the pressuremeter loads the soil horizontally. Therefore, there is no obvious relationship between results from different tests.

Hird et al. (1991) carried out tests on samples of the highest quality of Cowden Till and compared the results to *in situ* instrumented plate tests to find that the range of stiffness, expressed as deformation degradation curves, from both test methods was similar (Figure 4.55). Atkinson and Little (1988) and Chegini and Trenter (1996) presented results on the effect of OCR on shear modulus from tests on undisturbed and reconstituted samples (Figure 4.56) to show that the stiffness increased with OCR and the stiffness of the

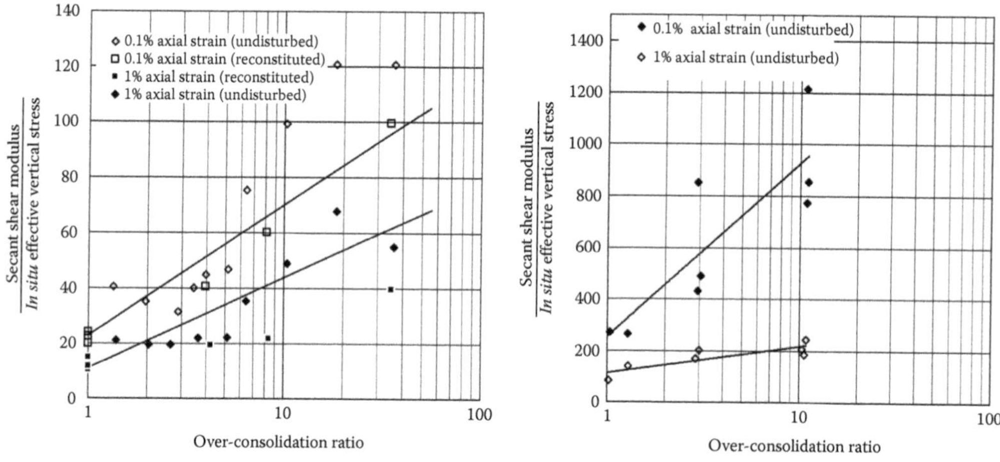

Figure 4.56 Variation of shear modulus with over-consolidation ratio for St. Albans till and Chapelcross till. (After Atkinson, J. H. and J. A. Little. *Canadian Geotechnical Journal*, 25(3); 1988: 428–439; Chegini, A. and N. A. Trenter. The shear strength and deformation behaviour of a glacial till. In *Proceedings of Conference on Advances in Site Investigation Practice*, London, 1996.)

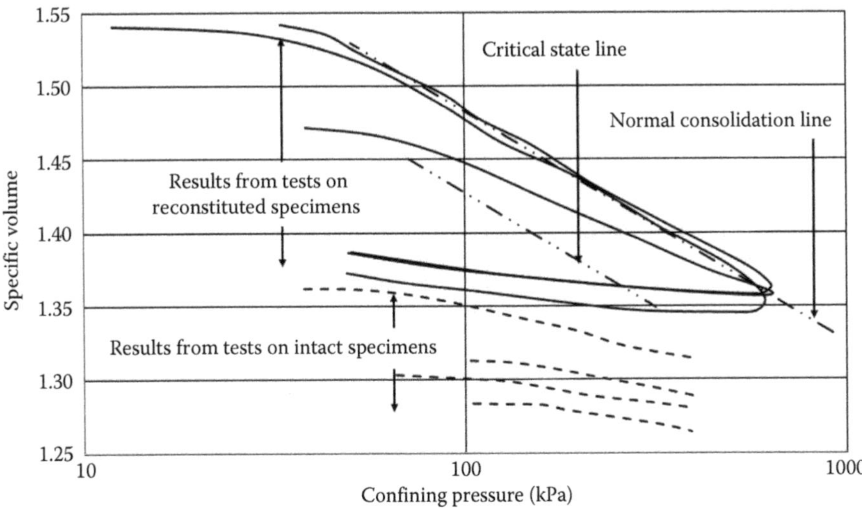

Figure 4.57 Isotropic consolidation of 100-mm specimens of the intact matrix-dominated till and 38-mm specimens of the reconstituted till showed that the reconstituted till exhibited classic normal and over-consolidation curves whereas the intact till appeared to be over-consolidated at a much lower void ratio. (After Chegini, A. and N. A. Trenter. The shear strength and deformation behaviour of a glacial till. In *Proceedings of Conference on Advances in Site Investigation Practice*, London, 1996.)

undisturbed till tended to be greater than the stiffness of reconstituted till though no information is given on density.

Oedometer tests on the intact and reconstituted till showed that the reconstituted till exhibited classic normal and over-consolidation curves whereas the intact till appeared to be over-consolidated at a much lower void ratio (Figure 4.57), suggesting that the reconstituted specimens were not consolidated to the *in situ* density. Therefore, any measurement of stiffness has to be made on Class 1 specimens consolidated to the *in situ* stress or on reconstituted specimens at the *in situ* density and then consolidated to the *in situ* stress. Replicating the stress history to create representative specimens may not be feasible because it is unknown.

Small-strain stiffness (Figure 4.58) was determined from *in situ* seismic surveys, bender element triaxial tests and resonant column tests on a till from Chapelcross (Chegini and Trenter, 1996). This demonstrates the difficulty of measuring a representative stiffness for a soil and a diverse range for any given test procedure. The results from the torsional resonant column tests appear to be consistent with the results of the crosshole geophysics results at small strains and the average of the degradation curves. These were used to produce the design curve (Figure 4.59), which compares favourably with the Ramberg–Osgood model.

Long and Menkiti (2007) reported the results of geophysical tests and triaxial tests (Figure 4.60) to show that stiffness of the tills increased with depth, stiffness in triaxial compression and extension are similar, there was no significant difference between the stiffness of block samples and the rotary cored samples and the projected small-strain stiffness from triaxial test results was similar to that derived from the *in situ* MASW testing. A comparison with the results of pressuremeter tests suggested that the tills were strongly anisotropic. They also showed that the stiffness of Dublin tills was five times that of the Cowden till, reinforcing the fact that till may be a convenient term to describe the type of geological deposit but different tills have different properties. Further, they reported stiffness back analysed using finite elements to be significantly greater than those measured in the laboratory. This was attributed to sample disturbance and anisotropic behaviour.

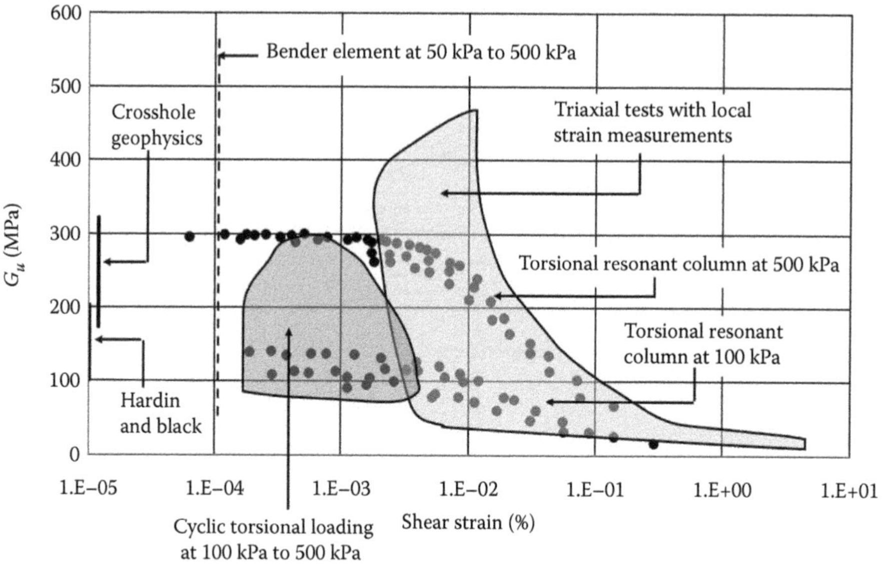

Figure 4.58 Comparison between different measurements of small-strain stiffness of glacial till at Chapelcross, Northern England. (After Chegini, A. and N. A. Trenter. The shear strength and deformation behaviour of a glacial till. In *Proceedings of Conference on Advances in Site Investigation Practice*, London, 1996.)

Cho and Finno (2009) reported stress path tests on the Deerfield stratum (Chicago clays). They observed that specimens K_o consolidated had similar propagation velocities using bender elements as the *in situ* shear wave velocity from seismic cones. The Chicago clays were non-linear over the strain level of 0.002%. The drained stress path tests showed that the stiffness depended on the direction of loading (Figure 4.61), which contradicts Long and Menkiti's (2007) conclusion.

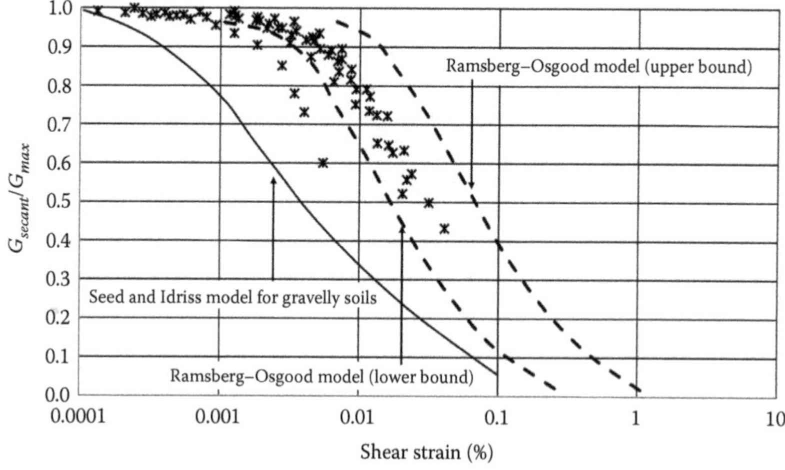

Figure 4.59 Design curve for degradation of undrained secant shear modulus with shear strain from torsion resonant column tests on glacial till from Chapelcross, Northern England. (After Chegini, A. and N. A. Trenter. The shear strength and deformation behaviour of a glacial till. In *Proceedings of Conference on Advances in Site Investigation Practice*, London, 1996.)

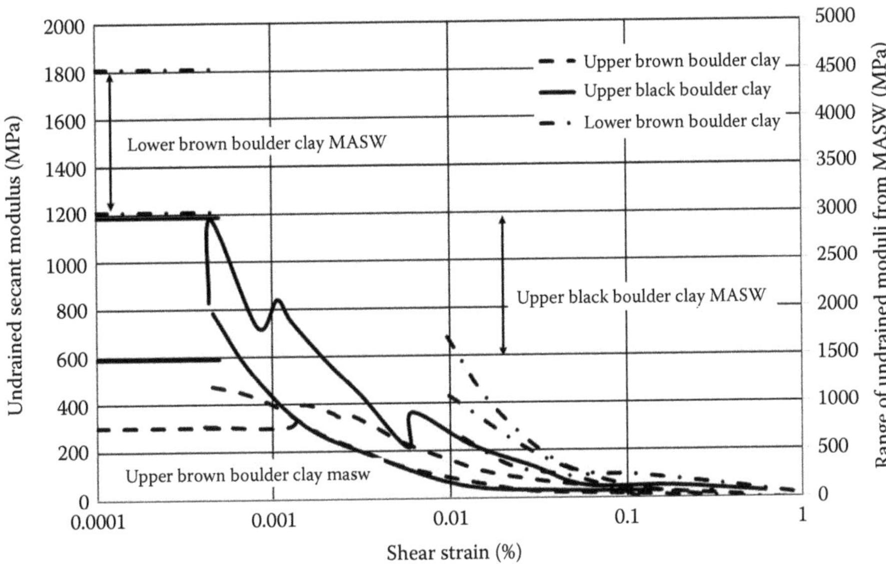

Figure 4.60 Small-strain stiffness of Dublin Boulder Clay measured by geophysical and triaxial tests with local strain measurements. (After Long, M. and C. O. Mentiki. *Geotechnique*, 57(7); 2007: 595–611.)

Thus, it is possible to assess the module decay curve from triaxial tests on Class 1 samples using local strain measurements. Fabric and composition will affect the stiffness, so tests on reconstituted specimens prepared at the *in situ* density are necessary if the fabric and composition are significant.

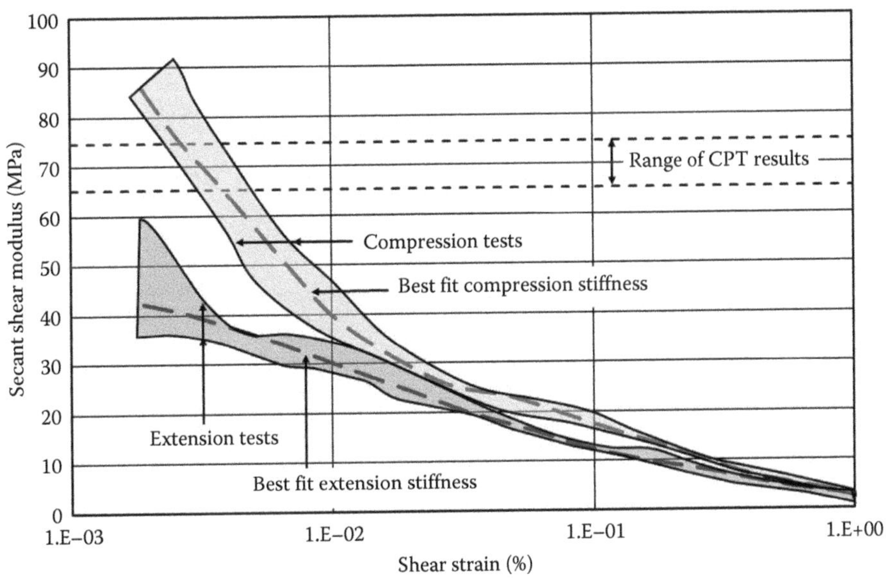

Figure 4.61 Small-strain stiffness degradation compression and extension curves for Chicago clay from local strain measurements in triaxial tests on block specimens and *in situ* measurements using a seismic cone. (After Cho, W. and R. J. Finno. *Journal of Geotechnical and Geoenvironmental Engineering* 136(1); 2009: 178–188.)

4.5.3.4 Partially saturated soils

Volume changes associated with partially saturated soils can be defined by four parameters from oedometer, pressure plate and shrinkage tests (Ho et al., 1992). These are a_t and a_{ts} for compressibility and swelling with respect to the net normal stress $(\sigma_n - u_a)$ and a_m and a_{ms} for the matric suction $(u_a - u_w)$. Figure 4.62 shows the conceptual volume change response for partially saturated soils. Ho et al. (1992) undertook oedometer tests according to ASTM D4546, the standard for tests on unsaturated specimens; pressure plate tests to ASTM D2325-68 (replaced by ASTM D6836 – 02, 2008); and shrinkage tests ASTM D427 (replaced by ASTM D4943 – 08). The volume change relationships for a glacial till compacted dry of optimum and wet of optimum are given in Figures 4.63 and 4.64, showing that standard tests can be used on matrix-dominated glacial tills to obtain the compression and swelling characteristics of partially saturated soils.

4.5.4 Conductivity

Groundwater conditions are critical in all geotechnical projects; hence, knowledge of the hydraulic conductivity, groundwater level and groundwater profile are essential. Yet, many ground investigations do not provide this information. One reason for this is the failure to install instruments such as standpipes or pore pressure transducers, to monitor the groundwater pressure over a period of time to establish equilibrium conditions and to determine the *in situ* hydraulic conductivity. Even then, the groundwater level can fluctuate seasonally and with rainfall intensity, and the hydraulic conductivity can change due to excavation and/or loading. This means that in geotechnical design the worst case is to have no knowledge of the groundwater conditions and hydraulic conductivity and the best case where the hydraulic conductivity of the intact and mass of soil are known and the groundwater levels and profile are known over several seasons. Even with the best case, there is no guarantee that the groundwater conditions will not exceed the observations, especially as rainfall events will become more intense and frequent due to climate change.

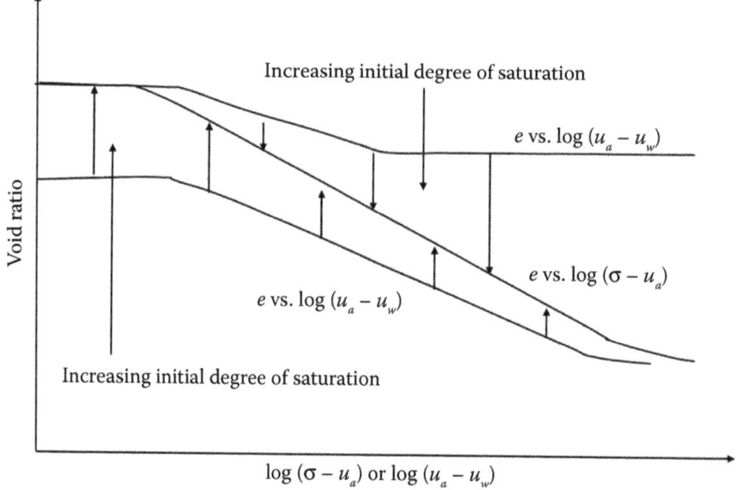

Figure 4.62 Variation of void ratio with effective stress for different degrees of saturation. (After Ho, D. Y. F., D. G. Fredlund, and H. Rahardjo. *Canadian Geotechnical Journal*, 29(2); 1992: 195–207.)

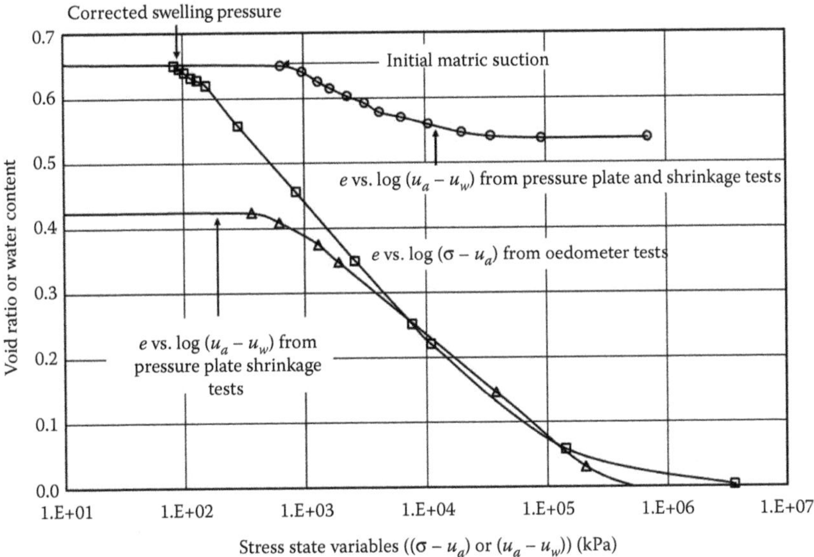

Figure 4.63 Volume change relationships for glacial till compacted dry of optimum. (After Ho, D. Y. F., D. G. Fredlund, and H. Rahardjo. *Canadian Geotechnical Journal*, 29(2); 1992: 195–207.)

Any excavation or structure below ground surface is affected by groundwater, which can flow and increase pressure. Examples include the following:

- Water flowing into an excavation requiring the groundwater level to be temporarily lowered or water pumped out of the excavation
- Design of an impermeable wall around an excavation to prevent water entering the excavation and instability due to liquefaction at the base of the excavation

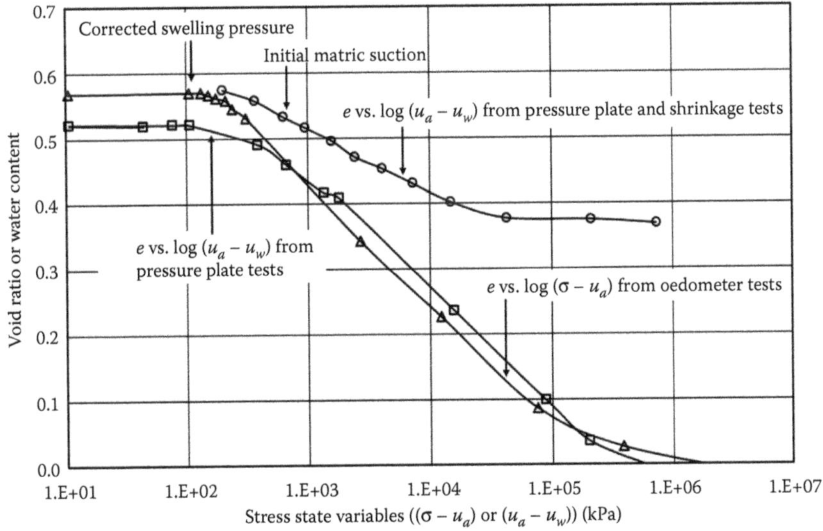

Figure 4.64 Volume change relationships for glacial till compacted at optimum water content. (After Ho, D. Y. F., D. G. Fredlund, and H. Rahardjo. *Canadian Geotechnical Journal*, 29(2); 1992: 195–207.)

- Reduction of effective stress resulting in a reduction in strength due to rising groundwater level leading to increased pressures on retaining structures and reduced stability of slopes
- Erosion of soil within earth dams
- Water pressures on landfill liners
- The rate of settlement of foundations and the movement of retaining walls and the settlement of embankments due to consolidation

Glacial deposits can be a major groundwater source (e.g. Stephenson et al., 1988) and are susceptible to contamination due to agriculture and waste disposal.

Table 4.33 is a summary of the characteristics of glacial soils relevant to hydraulic conductivity.

Glacial soils can be grouped into those that transmit water readily and those that do not. The erosion, transport and depositional processes control the grain size distribution and fabric of glacial soils. Examples relevant to hydraulic conductivity include the following:

- Matrix-dominated tills contain discontinuities that vary in width with depth, which means that the mass conductivity is greater than the intact conductivity. As a till is excavated, the *in situ* stress reduces opening up the discontinuities increasing the

Table 4.33 Typical properties of glacial soils

	Diamicton		*Sand and gravel*		
Characteristic	*Till*	*Pro-glacial and supraglacial deposits*	*Ice contact stratified deposits*	*Pitted outwash*	*Outwash*
Sorting	Poor	Poor	Moderate	Good	Excellent
Stratification	None or poor	None or poor	Locally collapsed	Locally collapsed	Well developed
Surface from	Flat hummocky or stream lined	Hummocky	Hummocky or ridges	Gently sloping with depressions	Gently sloping
Site of deposition	Beneath ice	Ice margin	Ice margin or beneath ice	In front of or on ice margin in valley, apron or plain	In front of margin in valley, apron or plain
Grain size	Sandy to clayey, uniform	Sandy to clayey, variable	Variably usually coarse	Variable gravel near source, sand further away	Uniform gravel near source, sand further away
Rounding of clasts	Angular	Moderate	Variable	Fairly well rounded	Well rounded
Compaction	Fairly compact, often over-consolidated	Loose, usually not over-consolidated	Loose, granular	Loose, granular	Loose, granular
Jointing	Often well developed, vertical	Often closely spaced, fissile	Not common	Not common	Not common
Lateral continuity	Well developed	Poorly developed	Poorly developed	Well developed	Well developed
Aquifer potential	Poor	Poor to fair	Fair	Good	Excellent

Source: After Stephenson, D. A., A. H. Fleming, and D. M. Mickelson. Glacial deposits. In *Hydrogeology*, edited by Back, W., J. S. Rosensheim and P. R. Seaber, Geological Society of America, Boulder, CO, 1988: 301–314.

hydraulic conductivity. Conversely, loading the till will reduce the mass conductivity though it may have little effect on the intact conductivity.
- Glaciolacustrine clays are strongly anisotropic.
- The particle size distribution of glaciofluvial soils reduces with distance from the source.
- The conductivities of clast-dominated and melt tills are highly variable because of the variability in composition.
- Hydraulic conductivity is also stress dependent because it depends on the void ratio, pore entry distribution and the discontinuity apertures. This means, for a given till, the conductivity will decrease with depth and excavation will increase the conductivity as the discontinuities open up due to stress relief.
- Glacial soils can contain lenses of coarse- or fine-grained soils, which have significantly different conductivity to the surrounding soil.

The hydraulic conductivity of a soil depends on the test procedure as well as the size of specimen tested. Therefore, it is difficult to assess whether the conductivity of glacial soils represents the mass or intact conductivity without knowledge of the scale of the sample as well as the test procedure. In Table 4.34, typical values of hydraulic conductivity of glacial deposits are given. The intact conductivity is usually assessed on small specimens or reconstituted samples. This is referred to as the primary hydraulic conductivity. The secondary hydraulic conductivity is a function of the post-depositional processes such as discontinuities and weathering. Figure 4.65 shows that there is a strong relation between grain size distribution and primary hydraulic conductivity, which is strongly influenced by the amount of clay. Figure 4.65 shows that a clay or silt content of 15%–20% marks a threshold between soils of low permeability and those of medium to high permeability. It is also noted that two tills with the same particle size distribution can have values of conductivity differing by two orders of magnitude, a reflection of the mode of deposition affecting the density of the deposit.

Table 4.35 shows the difference between primary (laboratory tests on intact soils) and secondary (field or mass) conductivity of glacial deposits. In order to obtain the mass conductivity, it is necessary to carry out *in situ* tests unless it can be demonstrated that an intact sample is large enough to be representative of the soil. However, as Figure 4.66 shows, field measurements of hydraulic conductivity are affected by the position of the piezometer or by the size of the test pocket in relation to the discontinuities.

Flow of water in soils is laminar and can be expressed in terms of Darcy's law in which the rate of flow is proportional to the hydraulic gradient. The hydraulic conductivity of a soil is not an intrinsic property as it depends on the particle size distribution, particle shape and texture, mineralogical composition, voids ratio, degree of saturation, soil fabric, nature of pore fluid and temperature. The *in situ* fabric can have a significant effect making laboratory test results of little value.

Table 4.34 Ranges of hydraulic conductivity of glacial soils

Glacial soil	*Unweathered (m/day)*	*Weathered (m/day)*	*Fractured (m/day)*
Basal till	10^{-2} to 10^{-6}	10^{-1} to 10^{-4}	1 to 10^{-4}
Supraglacial till	1 to 10^{-4}	1 to 10^{-4}	1 to 10^{-4}
Lacustrine clays	10^{-4} to 10^{-8}	n.a.	10^{-3} to 10^{-4}
Loess	1 to 10^{-4}	10^{-2} to 10^{-5}	n.a.
Outwash	10^{2} to 10^{-2}	n.a.	n.a.

Source: After Stephenson, D. A., A. H. Fleming, and D. M. Mickelson. Glacial deposits. In *Hydrogeology*, edited by Back, W., J. S. Rosensheim and P. R. Seaber, Geological Society of America, Boulder, CO, 1988: 301–314.

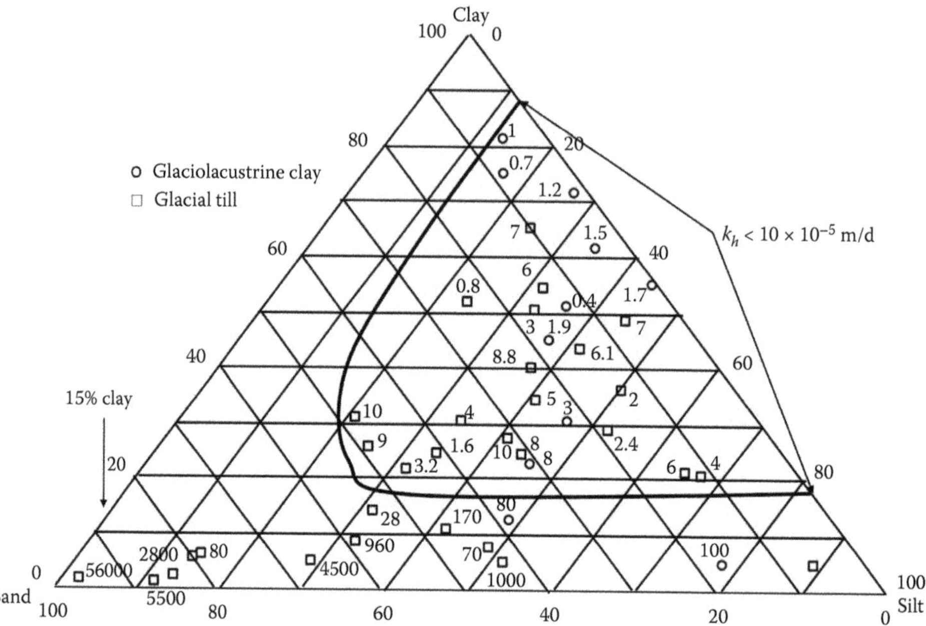

Figure 4.65 Values for the coefficient of permeability for glaciolacustrine clays and glacial tills showing the relationship to composition. (After Stephenson, D. A., A. H. Fleming, and D. M. Mickelson. Glacial deposits. In *Hydrogeology*, edited by Back, W., J. S. Rosensheim and P. R. Seaber, Geological Society of America, Boulder, CO, 1988: 301–314.)

Flow through soil is a function of the piezometric head, which is given by

$$p = h\gamma_w \tag{4.38}$$

where p is the pressure, h the head of water and γ_w the unit weight of water.

Table 4.35 Comparison between field and laboratory values of hydraulic conductivity for glacial soils where the laboratory values are possibly intrinsic (primary) values and the field values include the effects of discontinuities.

Glacial soil	*Hydraulic conductivity (m/d)*		*Reference*
	Laboratory	*Field*	
Clayey till	3×10^{-5}	1×10^{-2}	Gordon and Huebner (1983)
Clayey till	3×10^{-5}	3×10^{-5}	
Lacustrine clay	7×10^{-6}	3×10^{-5}	
Clayey till	10^{-5} to 10^{-6}	10^{-1} to 10^{-2}	Sharp (1984)
Basal till	10^{-5} to 10^{-6}	10^{-1} to 10^{-2}	Herzog and Morse (1986)
Ablation till	10^{-3} to 10^{-4}	10^{-1} to 10^{-2}	
Clay-loam till	10^{-5} to 10^{-6}	10^{-4}	Grisak and Cherry (1976)
Lacustrine clay	10^{-7}	10^{-4}	
Sandy clay till	10^{-4} to 10^{-5}	10^{-2} to 10^{-4}	Hendry (1982)
Clayey till	10^{-5}	10^{-4} to 10^{-5}	Prudic (1982)
Clay-loam till/lacustrine sediments	10^{-5} to 10^{-6}	10^{-2} to 10^{-4}	Grisak et al. (1976)

Source: After Stephenson, D. A., A. H. Fleming, and D. M. Mickelson. Glacial deposits. In *Vol 02 Hydrology: The Geological Society of America*, edited by Back, W., J. S. Rosenshein, and P. R. Seaber, The Geology of North America; 1988: Chapter 5.

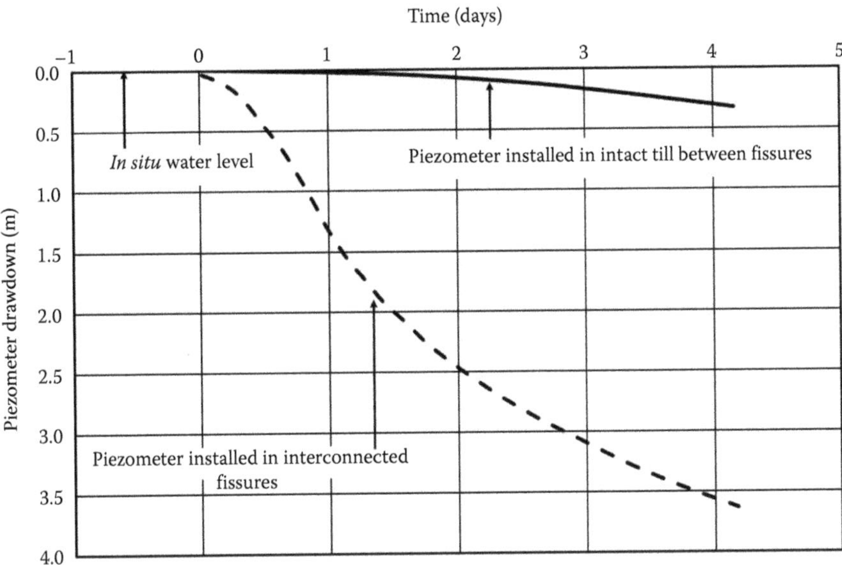

Figure 4.66 Effect of piezometer position on field measurements of hydraulic conductivity showing the impact of discontinuities on the drawdown curve. (After Stephenson, D. A., A. H. Fleming, and D. M. Mickelson. Glacial deposits. In *Hydrogeology*, edited by Back, W., J. S. Rosensheim and P. R. Seaber, Geological Society of America, Boulder, CO, 1988: 301–314.)

It is the difference in potential that causes the pore fluid to flow. The hydraulic gradient, i, is the difference in head $(h_1 - h_2)$ over the length (L); the difference is measured as follows:

$$i = \frac{h_1 - h_2}{L} \tag{4.39}$$

The quantity of flow per minute is proportional to the hydraulic gradient, and that constant of proportionality is the coefficient of permeability or hydraulic conductivity. This is the basis of laboratory tests to measure the permeability of an element of soil in which a difference in hydraulic head is applied across the specimen. There are two configurations: constant head and falling head. In a constant head test, for example, permeameter tests on sands and triaxial tests on sands and clays, a constant head is maintained across the specimen. In the falling head used in the permeameter test on clays, the head drops as the test proceeds. The falling head test is also used *in situ* to determine the mass permeability of a soil by observing the rising or falling head in a standpipe.

Figure 4.67 shows typical values of conductivity of soils and how they can be determined and Figure 4.68 the relation between particle size and hydraulic conductivity. It is very difficult to obtain undisturbed samples of coarse-grained soils or to determine the *in situ* density in order to reconstitute laboratory specimens correctly, which means that any laboratory measurement of hydraulic conductivity of these soils has to be treated with caution. *In situ* tests will be more reliable.

It is possible to use empirical correlations such as the Hazen (1892) formula based on particle size and the Kozeny–Carman formula (Carman, 1939), which also takes into account particle shape, porosity and grading. The preferred formula for coarse-grained soil is

$$k = \frac{\gamma_w}{C\eta_w S^2}\frac{e^3}{1+e} \tag{4.40}$$

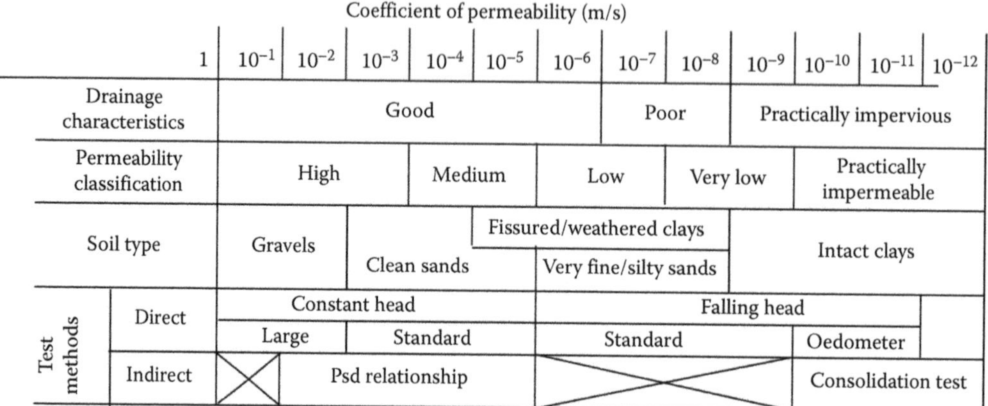

Figure 4.67 Permeability and drainage characteristics of soils. (After Head, K. H. *Manual of Soil Laboratory Testing, Vol. 2, Permeability, Shear Strength and Compressibility Tests*. Pentech, London, 1988a.)

where S is a shape factor, which depends on the surface area of a unit volume of soil

$$S = \frac{6}{\sqrt{(d_1 d_2)}} \tag{4.41}$$

where d_1 and d_2 are the range of particle sizes, C is a shape factor which varies from 5 for spherical particles to 7 for angular grains, e is the void ratio and η_w is the viscosity of the pore fluid. The angularity of the particles is examined during a particle size distribution test by visual inspection.

A constant head permeameter can be used to assess the hydraulic conductivity of coarse-grained particles. The cell diameter has to be at least 12 times the largest particle size, so for sands a cell diameter of 75 mm is acceptable and medium gravel 114 mm diameter.

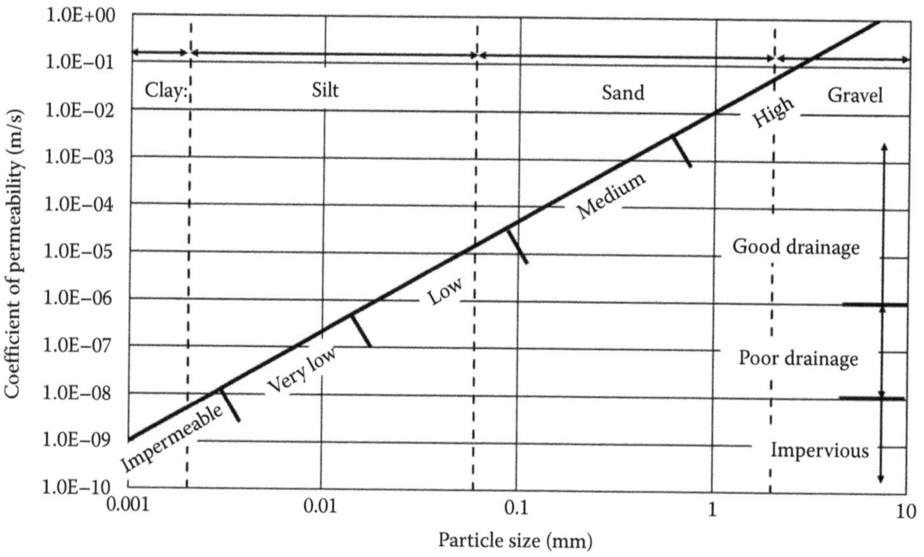

Figure 4.68 Permeability classification related to particle size.

Much larger cells are required for coarse-grained glacial soils because of the size of particles, which suggests that *in situ* tests are the only feasible way of determining hydraulic conductivity of these soils. The hydraulic conductivity of the soil mass will be little affected by very coarse particles, so a judgement on whether laboratory tests are feasible is based on the principal fraction.

Provided undisturbed samples of glacial clays can be obtained, it is possible to determine the hydraulic conductivity using falling head tests in a falling head permeameter, consolidation cell or triaxial cell, or constant head tests in a triaxial cell. In all tests, the concerns of selecting a representative specimen of the undisturbed till are significant. Further, the fabric, whether it is laminations or fissures, will influence the value to such an extent that the hydraulic conductivity will neither be that of intact soil or the soil mass.

Trenter (1999) presented results of tests on a number of tills showing the variation of coefficient of permeability with specific volume (Figure 4.69). These results highlight the challenge of obtaining a characteristic value of conductivity. Little (1984) compared the permeability of undisturbed and reconstituted samples to show that the conductivity of undisturbed samples was much greater than that of the reconstituted samples, suggesting that fabric had a dominant effect, a point noted by Keller et al. (1986) and Mckay et al. (1993) though it is not certain that the specimens were reconstituted to the *in situ* density. Sims et al. (1996) suggested that fissures within matrix-dominated tills have a significant effect on the permeability but the effect reduces with depth because the density of fissuring decreases with depth and the *in situ* stress increases closing the fissures. They suggested that if the effective stress exceeded 120 kPa (i.e. about 12 m), then the *in situ* conductivities of the intact and fissured clay are similar. According to Trenter (1999), many studies of English tills were carried out as part of major infrastructure investment in the 1960s, 70s and 80s. Generally, the coefficient of consolidation was low and variable. For example, Bishop and Vaughan (1962), Vaughan et al. (1975) and Millmore and McNicol (1983) reported on the coefficient of consolidation of glacial clays used in the construction of Selset, Cow Green and Kielder Dams in the North of England; Anderson (1974) and Hossain and McKinley

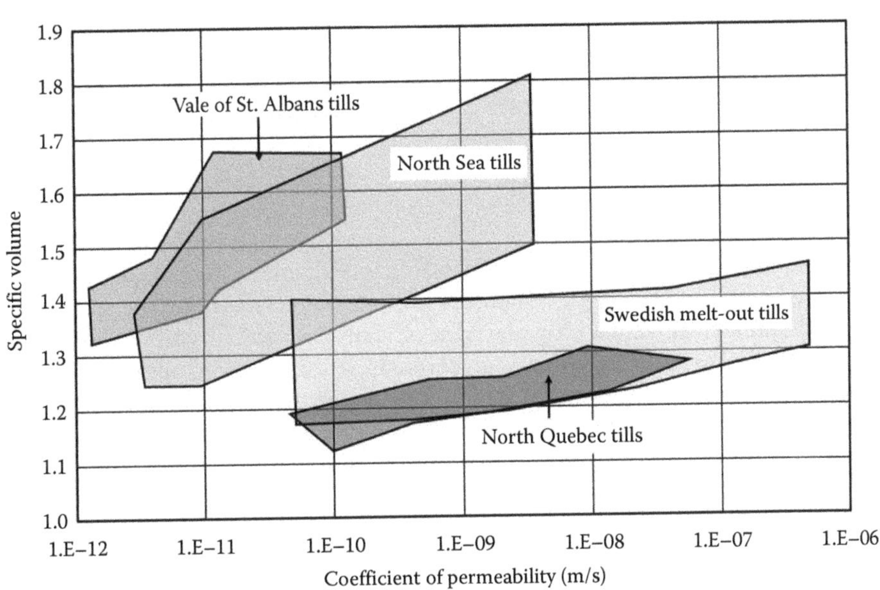

Figure 4.69 Variation of specific volume with permeability for a number of tills. (After Trenter, N. A. *Engineering in Glacial Tills*. CIRIA, London, 1999.)

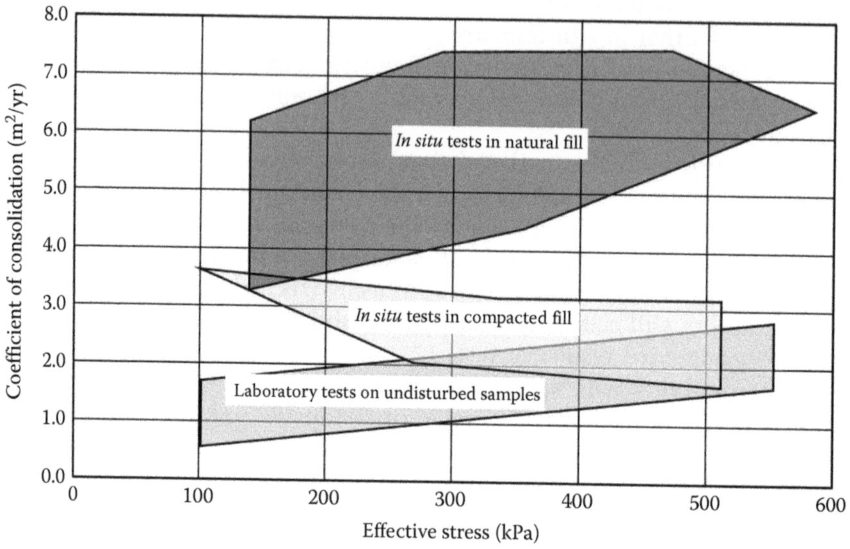

Figure 4.70 Variation of coefficient of consolidation with effective stress from laboratory tests on intact till, *in situ* tests on natural and compacted till at Selset reservoir, N England showing the effect of discontinuities on the coefficient when comparing the laboratory and field tests, and the effect of compaction on reducing the coefficient. (After Bishop, A. W. and P. R. Vaughan. *Proceedings of the Institution of Civil Engineers*, 21(2); 1962: 305–346.)

(1991) tested West Scotland tills. Figure 4.70 reinforces the fact that the properties of tills vary, so it is not possible to quote typical values for tills.

Empirical relationships with particle size distribution do not apply to clays because of the effect of pore size distribution. This has led to empirical relationships with pore size distribution. Tanaka (2003) studied a number of marine soils and came to the conclusion that most pore size distributions of natural clays (Figure 4.71) are similar and can be expressed in terms of a characteristic value D_{p50}, the entrance pore size diameter for 50% of the cumulative pore volume. They showed that there is a correlation with k:

$$k \infty n D_{p50}^2 \tag{4.42}$$

where n is the porosity.

Probabilistic capillary models relating hydraulic conductivity to pore entrance size distribution have been developed by Marshall (1958) and Garcia-Bengochea et al. (1979) and updated by Watabe et al. (2006) for glacial tills. The pore size distribution or pore entry diameter can be measured by intrusion (e.g. mercury), gas expansion, optical, tomography or imbibition tests. Typically, in geotechnical engineering, intrusion methods are used. The capillary models are based on Poiseuille's equation for laminar flow through a cylindrical cavity:

$$k_{cap} = \frac{\gamma_w d_{cap}^2}{32\mu} \tag{4.43}$$

where k_{cap} is the hydraulic conductivity of the capillary, d_{cap} the diameter of the capillary and μ the viscosity of the pore fluid. Garcia-Bengochea et al. (1979) showed that the hydraulic conductivity of the soil is given by

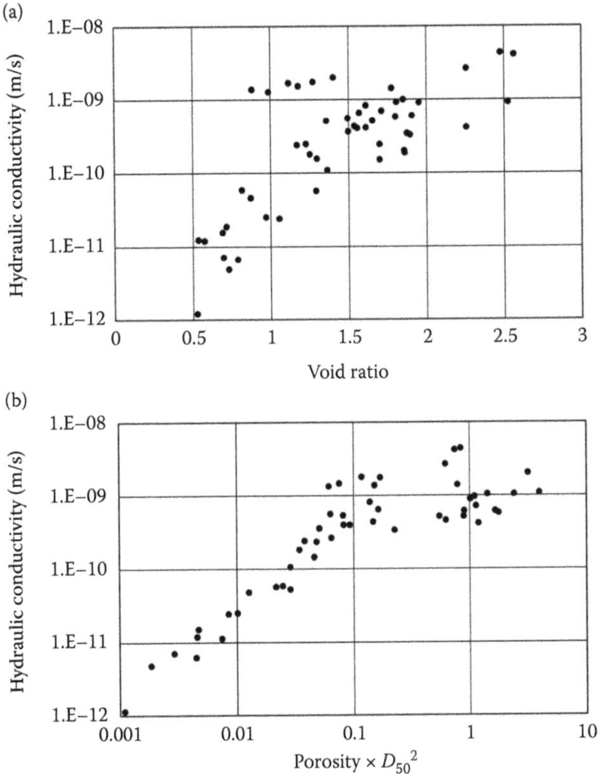

Figure 4.71 Relationship between (a) hydraulic conductivity and void ratio for a wide variety of marine soils including glacial soils and the relationship between (b) hydraulic conductivity, porosity and pore entry size. (After Tanaka, H. Pore size distribution and hydraulic conductivity characteristics of marine clays. In *2nd International Symposium on Contaminated Sediments*, 2003: 151–157.)

$$k = \frac{\gamma_w n}{32\mu} \sum_{i_i=1}^{m} (d_{i_{i1}}^2 f(d_{i_1})) \tag{4.44}$$

where n is the porosity, d_{i1} is a pore entrance diameter and $f(d_{i1})$ is the volumetric probability of having a pore entrance diameter of d_i in the section under consideration. Watabe et al. (2006) developed this further when investigating the capillary model for a compacted glacial till. They found (Figure 4.72) that a general capillary model with a typical pore connection of 3 as opposed to 1 gives a better fit to the data. They concluded that the compaction process influences the hydraulic conductivity. However, Kilfeather et al. (2008) undertook tests on low-porosity clast-dominated Irish tills using image analysis to estimate pore size distribution to find that the fabric of the till has a much greater effect on hydraulic conductivity than the pore size distribution, suggesting that field description and *in situ* measurements of conductivity are essential.

Discontinuities increase the hydraulic conductivity and reduce the shear strength and stiffness. For example, the hydraulic conductivity, k, of a fractured till can be expressed as

$$k = k_{intact} + k_{fractures} = \frac{kg}{\eta} + \frac{b^3 g}{12\eta L} \tag{4.45}$$

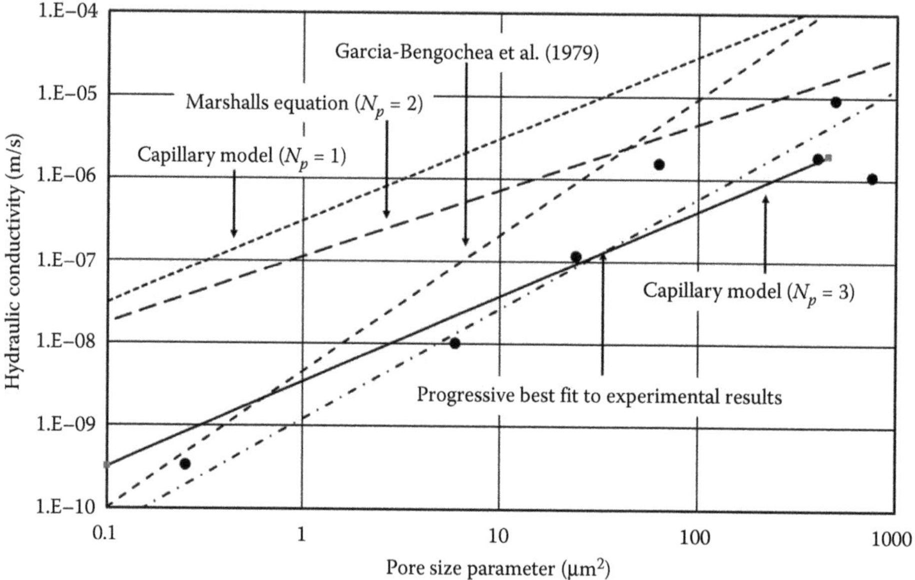

Figure 4.72 Relationships between saturated hydraulic conductivity and pore size parameter using the capillary model with (N_p = 1), Marshall's model with (N_p=2) and the general capillary model with (N_p= 3) compared with results of tests on three glacial clays from Northern Quebec. (After Watabe, Y., J-P. LeBihan, and S. Leroueil. *Géotechnique*, 56(4); 2006: 273–284.)

where k_{intact} is the conductivity of the matrix between the discontinuities, that is, the intact conductivity, $k_{fractures}$ the conductivity of the discontinuities, η is the kinematic viscosity, *b* the aperture width and *L* the distance between fractures. This equation has little practical use because of the difficulty in measuring discontinuity spacing and aperture but it does demonstrate the impact discontinuities have upon the mass conductivity of the till. The mass conductivity of a till can be several orders of magnitude greater than the conductivity of the intact till (e.g. Grisak et al., 1976; Hendry, 1982; Keller et al., 1986; Mckay et al., 1993).

It is possible to estimate the discontinuity spacing/apertures from a back analysis of groundwater discharge. Table 4.36 shows the effect of discontinuity spacing/aperture on *k* and c_v. Garga (1988) suggested that a better estimate of the coefficient of consolidation, c_{vfield}, is

$$c_{vfield} = \frac{k_{field}}{\gamma_w m_{vlab}} \tag{4.46}$$

to take into account mass conductivity. k_{field} is the field hydraulic conductivity and m_{vlab} is the coefficient volume compressibility from oedometer tests.

4.5.4.1 *Triaxial permeability test*

It may be possible to retrieve representative samples, especially of glacial soils that are highly fissured or intact. In that case, the hydraulic conductivity can be measured in the laboratory using triaxial tests. They have the advantage that larger samples can be tested thus taking fabric into account, and drainage can be vertical or radial thus assessing anisotropy. Various loading and drainage conditions can be applied including isotropic and anisotropic consolidation; constant head and falling head and constant flow. Constant flow tests in triaxial cells on glacial clays are possible. The flow rate is recorded until steady-state conditions are reached. At that point, the hydraulic conductivity, k, is given by

$$k = \frac{q}{Ai} \tag{4.47}$$

where A is the cross-sectional area of the specimen, q the rate of flow of water and i the hydraulic gradient across the specimen.

The average effective stress, σ, for isotropic consolidation is

$$\sigma' = \sigma - \frac{1}{3}(u_2 - 2u_1) \tag{4.48}$$

where u_1 and u_2 are the pore pressures at the top and bottom of the specimen to take into account the variation in void ratio across the specimen.

The coefficient of consolidation, c_{vi}, is given by

$$c_{vi} = \frac{0.199H^2}{t_{50}} \text{m}^2/\text{year} \tag{4.49}$$

where H is the mean height in mm and t_{50} the time in minutes for 50% consolidation. This is not the same as the value obtained from an odometer test. Head (1988b) suggests that

$$c_{vi} = \frac{1}{1 - B(1 - A)(1 - K_o)} c_v \tag{4.50}$$

where A and B are the pore pressure coefficients and K_o is the earth pressure at rest.

It is also possible to determine the effect of radial drainage, which may be relevant for glaciolacustrine clays. The horizontal coefficient of consolidation is either

Table 4.36 **Effect of fracture spacing on properties of glacial tills**

Fracture aperture (cm)	*Fracture spacing (cm)*	*Hydraulic conductivity (cm/s)*	*Fracture porosity*	*Coefficient of consolidation (cm²/s)*	t_{90} *(days)*
0	∞	1.0×10^{-8}	0	1.0×10^{-3}	2.4×10^{-4}
0.0005	5	4.2×10^{-7}	2.0×10^{-4}	4.3×10^{-2}	5.7×10^{1}
	20	1.1×10^{-7}	5.0×10^{-5}	1.1×10^{-2}	2.2×10^{2}
	100	3.0×10^{-8}	1.0×10^{-5}	3.1×10^{-3}	7.9×10^{2}
0.0025	5	5.1×10^{-5}	1.0×10^{-3}	5.2×10^{-0}	4.7×10^{-1}
	20	1.3×10^{-5}	2.5×10^{-4}	1.3×10^{-0}	1.9×10^{0}
	100	2.5×10^{-6}	5.0×10^{-4}	2.6×10^{-1}	9.4×10^{0}
0.0075	5	1.4×10^{-3}	3.0×10^{-3}	1.4×10^{2}	1.8×10^{-2}
	20	3.4×10^{-4}	7.5×10^{-1}	3.5×10^{1}	7.0×10^{-2}
	100	6.9×10^{-5}	1.0×10^{-1}	7.0×10^{-8}	1.0×10^{-1}

Source: After Allred, B. J. *Ohio Journal of Science*, 100(3/4); 2000:63–72.

$$c_h = \frac{0.026D^2}{t_{100}} \text{m}^2/\text{year based on the square-root time method} \tag{4.51}$$

$$c_h = \frac{0.023D^2}{t_{50}} \text{m}^2/\text{year based on the log time method} \tag{4.52}$$

When applying a difference in hydraulic potential across a specimen, it can lead to consolidation of the specimen. Therefore, for a given hydraulic conductivity, there is a maximum hydraulic gradient.

4.5.4.2 Hydraulic conductivity

Benson and Trast (1994) created a database of 67 compacted clay liners throughout the United States, and Benson and Trast (1995) studied 13 of those clay liners including glacial tills and glaciolacustrine clays, in detail. The hydraulic conductivity of the soils compacted at different densities are shown in Figure 4.73. At that time, US Regulations required a hydraulic conductivity no greater than 10^{-9} m/s for clay liners. The samples were compacted using modified, standard and reduced Proctor methods and then subject to falling head tests in a flexible wall permeameter. The cell pressure was 20 kPa and the influent and effluent pressures 5 and 18 kPa, respectively. It shows that the coefficient of permeability reduces with the degree of saturation.

Vaughan (1994) highlighted the fact that hydraulic conductivity varies with *in situ* effective stress even though it is often assumed to be a constant. Clarke and Chen (1997) carried out constant flow tests on reconstituted glacial tills from NE England to determine their intrinsic hydraulic conductivity. The equipment, described by Araruna et al. (1995), has the advantages that it measures the movement of pore fluid through a specimen at low hydraulic gradients in about 12 h for low permeability clays. Tests were carried out on reinstituted

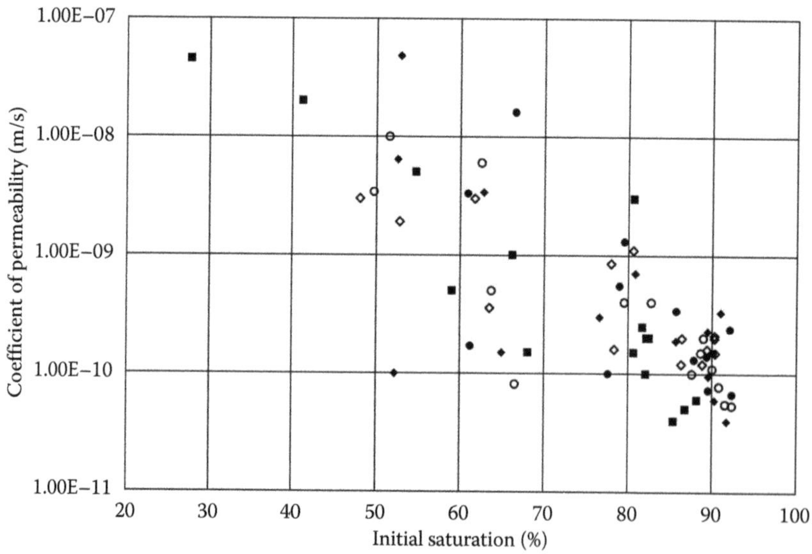

Figure 4.73 Variation in hydraulic conductivity with initial degree of saturation for five glacial soils compacted to reduced, standard and modified Proctor effort. (Data from Benson, C. H. and J. M. Trast. *Clays and Clay Minerals*, 43(6); 1995: 669–681.)

Table 4.37 Classification properties of tills of North East England

		Upper till		*Lower till*	
Property		*Lab*	*Field*	*Lab*	*Field*
Natural w	%	15.9	9–31	20.6	9–23
c_u	kPa		50–410[b]		65–410[b]
LL	%	57.0		31.1	
PL	%	25.6		17.1	
PI	%	31.4		14.0	
Clay fraction	%	38.6		22.6	
Silt fraction	%	44.5		34.2	
Sand fraction	%	13.5		31.6	
Gravel fraction	%	2.1		8.5	
Kaolinite	%		54–63[a]		67–70[a]
Illite	%		14–29[a]		22–36[a]
Quartz	%		45–52[a]		53[a]
Activity	%	0.81	0.65–0.70	0.62	0.59–0.67
Specific gravity		2.69		2.65	
Density	Mg/m^3		1.62–1.93[b]		1.76–2.00[b]
c'	kPa	4.7	0–25[c]	2.3	0–15[c]
ϕ'		22.5	27–35[c]	26.3	32–37[c]

Source: After Clarke, B. G. and C.-C. Chen. Intrinsic properties of permeability. In *Proceedings of the International Conference on Soil Mechanics and Foundation Engineering*, Vol. I, AA Balkema, 1997: 259–262; Thabet, 1973; Robertson, T. L. et al. *Ground Engineering*, 27(10); 1994: 29–34; Eyles and Sladen, 1981.

[a] Thabet, K. M. A. Geotechnical Properties and Sedimentation Characteristics of Tills in S.E. Northumberland, PhD Thesis, Newcastle University, UK, 1973.

[b] Robertson, T. L., B. G. Clarke and D. B. Hughes. Geotechnical properties of Northumberland Till. *Ground Engineering*, 27(10); 1994: 29–34.

[c] Eyles, N. and J. A. Sladen. *Quarterly Journal of Engineering Geology and Hydrogeology* 14.2, 1981: 129–14.

tills; the properties are listed in Table 4.37. Figure 4.74 shows that the hydraulic conductivity decreases with void ratio and Figure 4.75 shows that it decreases as the confining effective pressure increases. Figure 4.75 also shows that weathering reduces the permeability of a till. The measured values show that these tills are acceptable for barriers provided they are remoulded to remove discontinuities.

4.6 SELECTION OF GEOTECHNICAL CHARACTERISTICS

The increasing use of numerical methods to study soil behaviour has led to numerous constitutive models including those incorporated in commercial software such as linear elastic, linear elastic perfectly plastic, hardening, small-strain stiffness, soft soil and modified Cam Clay to explain stress–strain behaviour. Models incorporating drainage conditions and degree of saturation allow coupled deformation/seepage analysis to be undertaken. Constitutive models fall into four categories: linear and non-linear elastic perfectly plastic model, critical state framework, hyperbolic model and visco-elastic/plastic model. All of the models assume a continuum, which means that the fabric is not taken into account. Further, many of the models assume that soils are gravitationally consolidated. Soil is anisotropic,

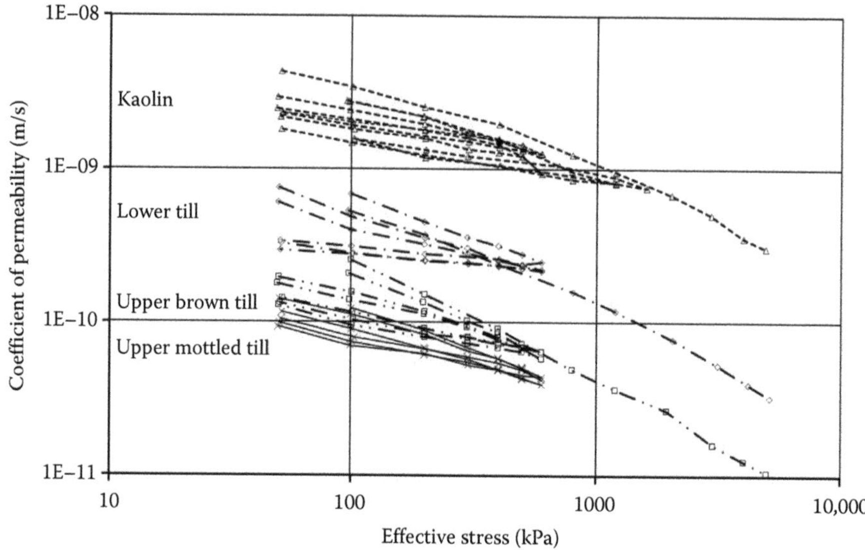

Figure 4.74 Variation of hydraulic conductivity with void ratio for reconstituted glacial tills in the NE England and kaolin showing the reduction permeability with void ratio. (After Clarke, B. G. and C.-C. Chen. Intrinsic properties of permeability. In *Proceedings of the International Conference on Soil Mechanics and Foundation Engineering*, Vol. 1, AA Balkema, 1997: 259–262.)

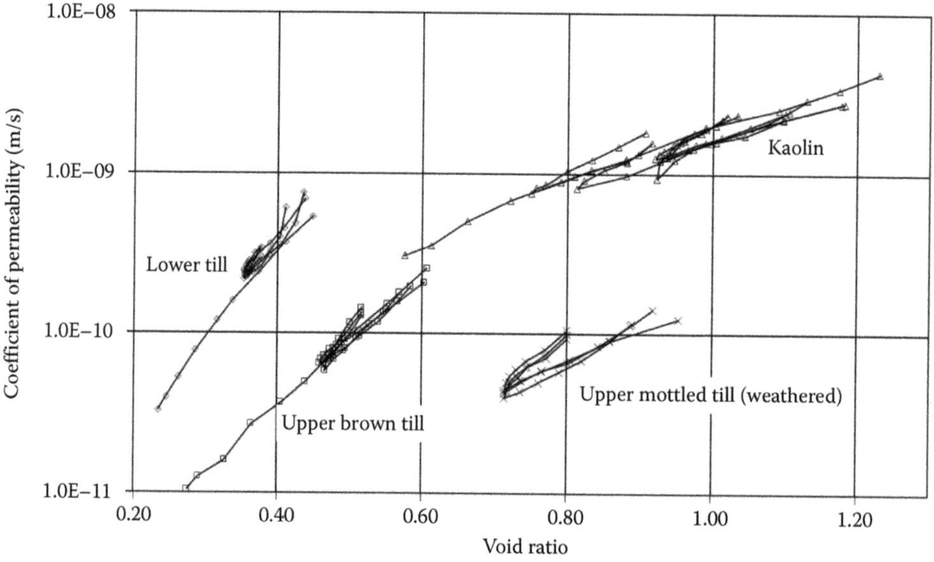

Figure 4.75 Variation in hydraulic conductivity with effective stress for reconstituted glacial tills in the NE England compared with kaolin showing the effect of an increase in silt content in the tills leading to an increase in permeability. (After Clarke, B. G. and C.-C. Chen. Intrinsic properties of permeability. In *Proceedings of the International Conference on Soil Mechanics and Foundation Engineering*, Vol. 1, AA Balkema, 1997: 259–262.)

non-linear, time dependent and load dependent. Modifications to constitutive models can take these into account. Hence, the selection of the most appropriate soil model is important to take into account the anticipated behaviour and what is to be investigated. For example, a slope stability analysis may be undertaken using a limit equilibrium approach with a perfectly plastic model such as Mohr–Coulomb providing a means of analysing the stability of an excavation. If a homogenous soil is analysed, then it is reasonable to assume a circular failure mechanism, which can be simply identified by investigating a number of circular failure mechanisms to determine the most critical. In heterogonous glacial soils, a circular failure surface may be an unsafe assumption because the failure could be governed by the fabric, stratum or internal weaker lenses of soil. In that case, a non-circular analysis can be carried out, provided the discontinuities have been clearly identified. The alternative is to carry out a numerical analysis such as a finite element or finite difference analysis as the failure surface is not predetermined. It is still important to model the geological profile but as the failure mechanism is not specified it is possible to observe the development of the likely mechanism. However, the actual failure mechanism cannot be predicted because a continuum is assumed.

Once a model is selected, it is necessary to determine the correct parameters. A perfectly plastic model based on the Mohr–Coulomb model uses the cohesion and angle of friction, that is, parameters that can be obtained from routine ground investigations. Parameters not routinely determined are required for other models. An initial analysis can be carried out using a linear elastic perfectly plastic model, which requires five parameters as indicated in Table 4.23. A routine investigation will provide only two of those parameters, which means that the others have to be assumed. There is a lack of published values of those parameters for glacial soils and, given the variability of these soils, it is unlikely that typical parameters can be published. Therefore, a combination of theoretical, empirical and semi-empirical relationships is used. This introduces risk, which can be assessed by undertaking a sensitivity analysis.

The second model is the hardening model, which incorporates stress-dependent stiffness only available from triaxial and pressuremeter tests with measurements of local strain.

In order to generate the initial stress condition, an assumption is made about the stress history. This is normally an estimate based on the geological history and the interpretation of oedometer tests on clays. The stress history is likely to be unknown for glacial tills and may not be inferred from laboratory tests because of the unknown pressures at the time of glaciation and the stress changes that take place during glaciation and isostatic uplift.

More sophisticated models take into account small-strain stiffness, anisotropy, ductility and creep. It is important to appreciate that numerical modelling will not give a precise answer but helps understand the response of the soil to change in loading or environmental conditions.

4.6.1 Frameworks

Given the challenges of obtaining representative samples of sufficient quality to be able to determine the geotechnical characteristics and produce the most credible design parameters from such variable results, it is useful to investigate the use of the frameworks. There are numerous correlations between field measurements and soil properties based on experimental observations (e.g. SPTN and undrained shear strength of clay) and theoretical studies (e.g. shear strength related to liquidity index for normally consolidated clays) but they may not apply to glacial soils. The scatter in results due to composition, fabric and stress conditions at the time of deposition means that selecting a credible value is difficult. The risk can be reduced by selecting the worst credible value leading to overdesign. An average value can lead, at times, to an unsafe design because of local weaknesses. Therefore, a framework to support the selection of the correct parameters would be helpful.

The first to consider is the T-line, the relationship between the plasticity index and the liquid limit used to classify silts and clays, which defines glacial soils even if the tills are predominantly eroded rock rather than clay minerals. This is referred to in Section 4.4. It is recognised that UK glacial clays lie about the T-line (Boulton and Paul, 1976). Baranski (2004) shows that Polish tills lie about a line ($PI = 0.66(I_L - 7)$), which is above the T-line, and Figure 4.19 shows that Chicago clays lie about the T-line. It is also observed that the increase in clay content associated with glaciolacustrine deposits and weathered matrix-dominated tills means that the liquid limit increases but remains about the T-line. Thus, the T-line can be used to make an assessment of a soil to appreciate whether it is a glacial soil and if it is weathered.

It is not certain that glacial soils can be considered to be over-consolidated from the ice surcharge because of the complex pore pressure regime in the basal zone. Further, if the soils are over-consolidated due to the ice pressure, then it is unlikely that the preconsolidation pressure could be measured in routine laboratory tests because of the pressure involved. However, the strength of clays is a function of density; it may be possible to reconstitute the glacial till to remove the effects of fabric and stone content and measure the intrinsic properties.

Terzaghi (1941) proposed a sedimentation compression curve (SCC) and showed that the natural compression of clays was similar but not the same as those for reconstituted clays. Skempton and Jones (1944) compared the SCC for natural soils with clays prepared from slurries and observed that the undisturbed samples are closer to SCC. In 1971, Skempton (1970) showed that the position of the SCC depends on the liquid limit. Burland (1990) used the intact compressibility and strength of reconstituted clay to provide a framework for the properties of natural undisturbed samples clay. He introduced the void index, I_v, given by

$$I_v = \frac{e - e^*_{100}}{e^*_{100} - e^*_{1,000}} = \frac{e - e^*_{100}}{C^*_c} \tag{4.53}$$

where e^*_{100} is the void ratio at a pressure of 100 kPa during one-dimensional compression, e^*_{1000} is the void ratio at a pressure of 1000 kPa and C^*_c is the intrinsic compression index (Figure 4.76). Experimentally the void index is a unique line given by

$$I_v = 2.45 - 1.285 \log \sigma'_v + 0.015(\log \sigma'_v)^3 \tag{4.54}$$

Clarke et al. (1998) and Baranski (2004) confirmed that glacial clays followed that unique line (Figure 4.77).

Liu and Carter (1999) showed that there is a correlation between the deformability of intact and reconstituted clays defined by the structural compression factor, A_v and the structure index, S_v.

$$e = e^* + S_v \frac{\sigma'_{vy}}{\sigma'_v} \ln \sigma'_{vy} - C^*_c \ln \sigma'_v \tag{4.55}$$

Clarke et al. (1998), Baranski (2004), Skempton and Bishop (1954), Atkinson and Little (1988), Gens and Hight (1979) and Coop et al. (1995) all found that the strength of the reconstituted till was similar to the strength of the natural till; that is, the sensitivity is very low. The implication is that tests on reconstituted samples prepared to the same density as the natural till can be used to obtain the strength of a till.

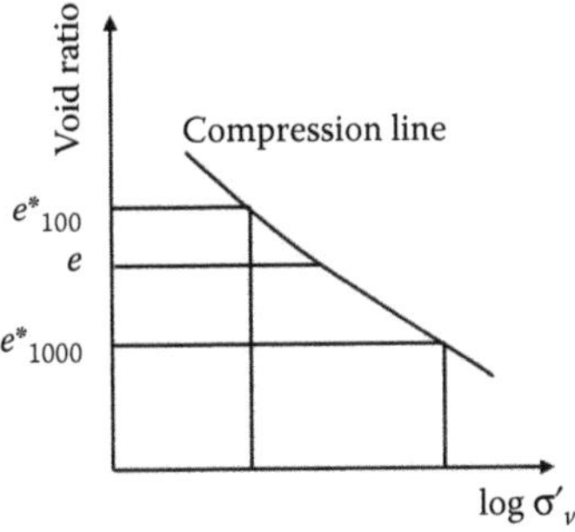

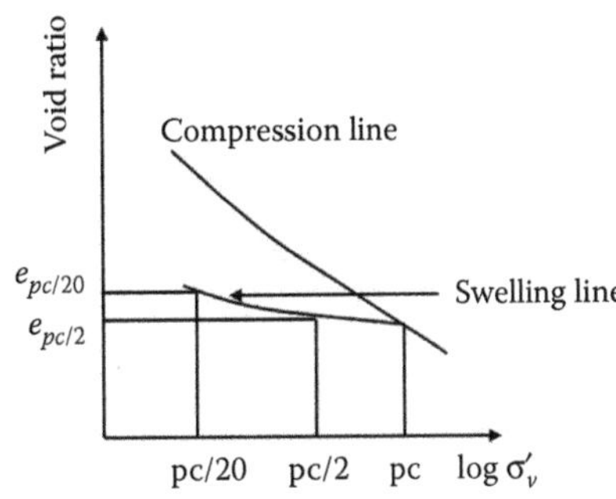

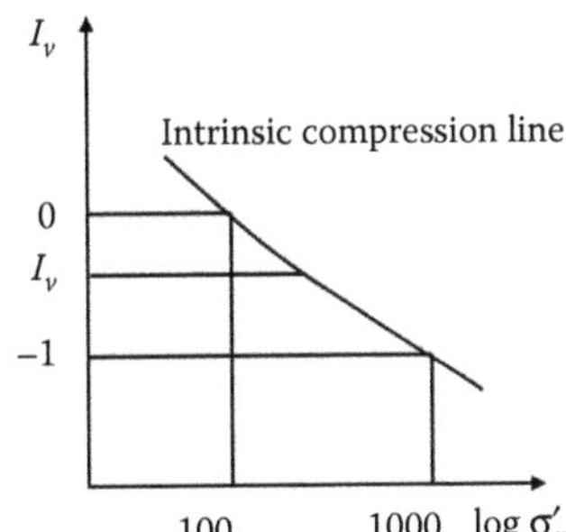

Intrinsic compression line

$$I_v = \frac{e - e^*_{100}}{e^*_{100} - e^*_{1000}} = \frac{e - e^*_{100}}{C^*_c}$$

Intrinsic swelling line

$$I_{vs} = (e - e^*_{pc/2})\,\frac{\log \sigma'_v - \log \sigma'_{pc/2}}{e^*_{pc/20} - e^*_{pc/2}}$$

Figure 4.76 Definition of the (a) void index and the (b) intrinsic compression line. (After Burland, J. B. On the compressibility and shear strength of natural clays. *Géotechnique*, 40(3); 1990: 329–378.) Definition of the intrinsic swelling line. (After Clarke, B. G. et al. A framework for characterization of glacial tills. In *Proceedings of the International Conference on Soil Mechanics and Foundation Engineering*, Vol. 1, AA Balkema, 1997: 263–266.)

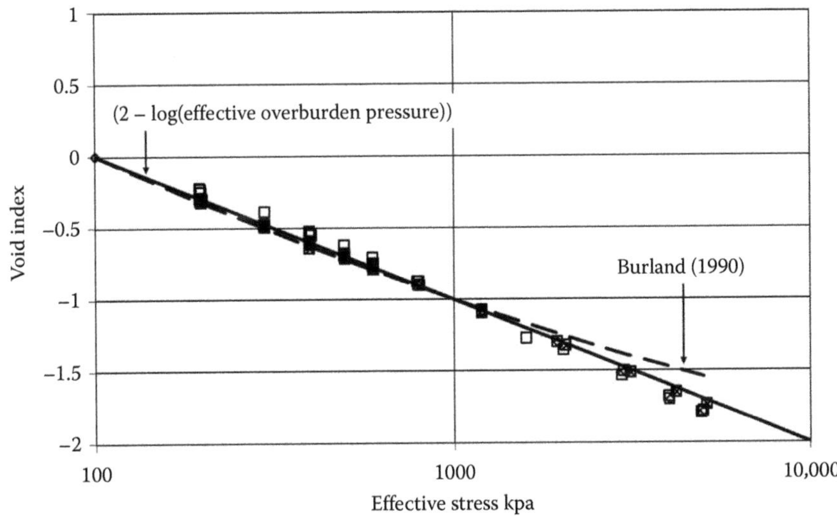

Figure 4.77 Variation in void index with effective stress derived from consolidation tests on upper weathered till, upper till and lower till from NE England. (After Clarke, B. G. et al. *Quarterly Journal of Engineering Geology and Hydrogeology*, 31(3); 1998: 235–246.)

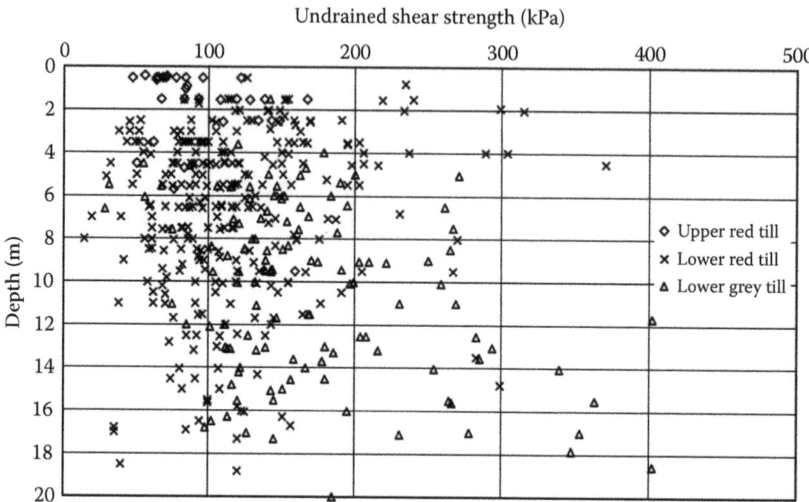

Figure 4.78 Profile of undrained shear strength from a site in NE England shows the typical scatter in the results of undrained triaxial tests on two tills with the top layer of the upper till being weathered. (After Clarke, B. G. et al. *Géotechnique*, 58(1); 2008: 67–76.)

Clarke et al. (1997a) used the concept of the intrinsic swelling line to develop a relationship to predict the undrained strength. A profile of undrained shear strength from a site in NE England shows the typical scatter in the results (Figure 4.78) for tills. Figure 4.25 shows the undrained shear strength taken from triaxial tests on U100 samples and the results derived from SPTN results using the Stroud and Butler correlation ($c_u = 4.9N$). Also shown on this plot are the values of OCR based on the relationship:

$$\frac{c_u}{\sigma'_v} = 0.23\ OCR^{0.8} \tag{4.56}$$

which shows that it is difficult to establish a design profile for undrained shear strength as OCR is unknown. Clarke et al. (1997a) used the concept of the intrinsic swelling line to show the relationship shown in Figure 4.79.

The intrinsic swelling line is not unique as it depends on the preconsolidation pressure. However, it falls in a narrow band around the line as shown in Figure 4.79:

$$I_{vs} = -0.242 + 0.738 \log OCR + 0.196(\log OCR)^2 - 0.028(\log OCR)^3 \tag{4.57}$$

The concept of the intrinsic void index can be used to develop relationships between geotechnical characteristics. An example is shown on Figure 4.80, which is means of assessing the undrained shear strength from the intrinsic swelling index, clay activity and undrained shear strength for normalised consolidated clay defined by

$$\frac{c_u}{\sigma'_v} = \left(\frac{c_u}{\sigma'_v}\right)_{nc} \frac{10^{I_v}}{\text{activity}} \tag{4.58}$$

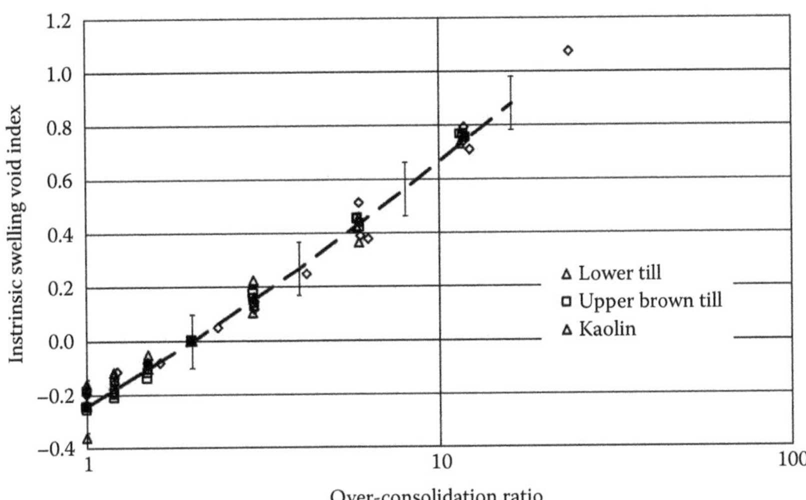

Figure 4.79 Intrinsic swelling line, for example for reconstituted glacial tills, is an average of the normalised swelling lines as they depend on the preconsolidation pressure. (After Clarke, B. G. and C.-C. Chen. Intrinsic properties of permeability. In *Proceedings of the International Conference on Soil Mechanics and Foundation Engineering*, Vol. 1, AA Balkema, 1997: 259–262.)

Clarke and Chen (1997) extended the concept to intrinsic conductivity (Figure 4.81) defined by

$$k_n = \frac{\log(k) - \log(k^*_{100})}{\log(k^*_{1000}) - \log(k^*_{100})} \tag{4.59}$$

where k_n is the intrinsic hydraulic conductivity of normally consolidated clay, for normally consolidated clay, k^*_{100} is the coefficient of permeability at an effective stress of 100 kPa and $k^*_{1,000}$ at an effective stress of 1000 kPa; and for over-consolidated clay,

$$k_s = \frac{\log(k) - \log(k^*_{pc/2})}{\log(k^*_{1pc/20}) - \log(k^*_{pc/2})} \tag{4.60}$$

where k_s is the intrinsic hydraulic conductivity of over-consolidated clay, $k^*_{pc/2}$ and $k^*_{pc/20}$ are the values of k at effective vertical pressures at OCR of 2 and 20, respectively.
Figure 4.82a shows the intrinsic line for normally consolidated clay which is a unique line given by

$$k_n = -0.016 + 0.836I_v - 0.126I_v^2 + 0.014I_v^3 \tag{4.61}$$

Figure 4.82b shows the band defining an intrinsic line for over-consolidated clay. This is not a unique line as it varies with the preconsolidation stress about a line:

$$k_s = 0.013 + 1.270I_v - 0.980I_v^2 + 0.850I_v^3 \tag{4.62}$$

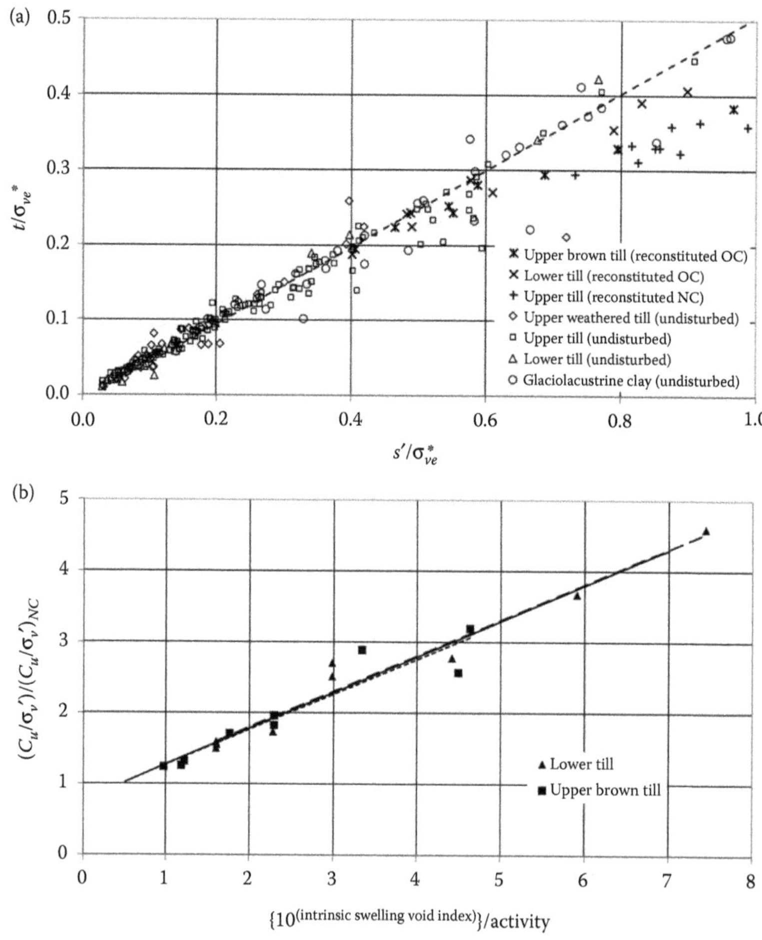

Figure 4.80 Use of the intrinsic void index to normalise the peak shear strength from tests on undisturbed glacial tills and glaciolacustrine clay and reconstituted tills showing a unique (a) failure line in terms of effective stress and (b) relationship between the undrained shear strength of overconsolidated and normally consolidated reconstituted tills.

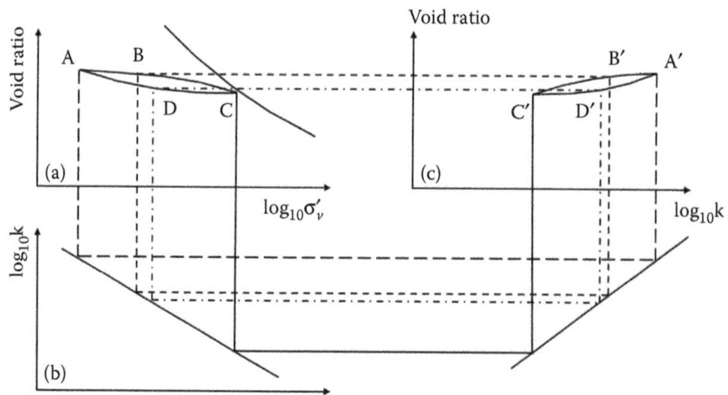

Figure 4.81 Stress state–permeability relationship showing (a) compression curve; (b) variation of permeability with stress; and (c) variation of permeability with void ratio.

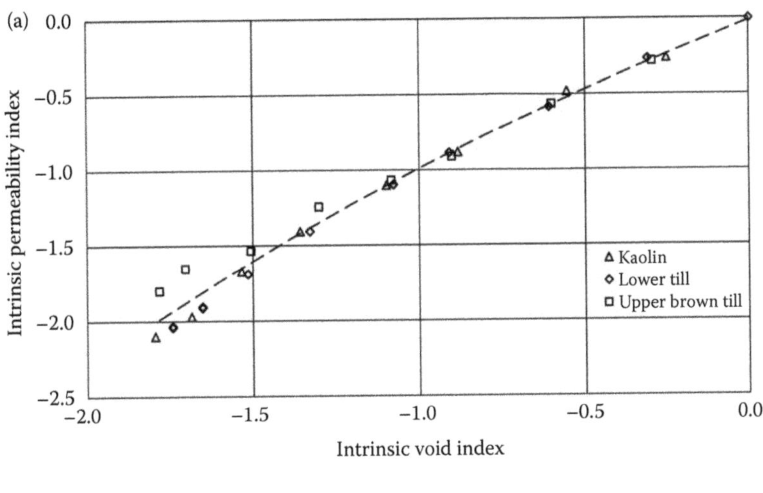

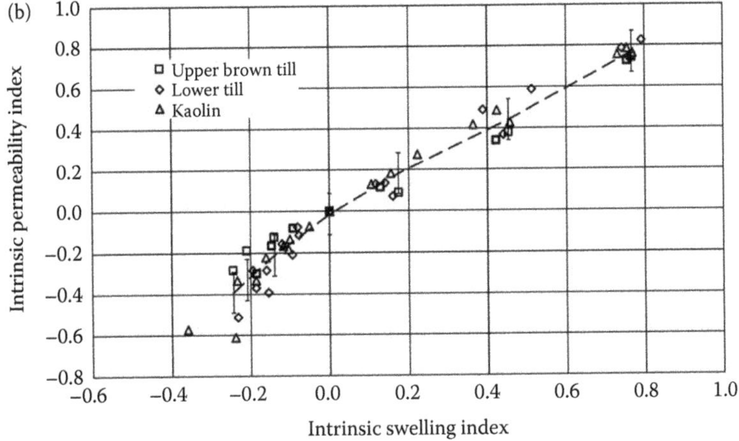

Figure 4.82 Intrinsic permeability index for (a) normally and (b) over-consolidated glacial tills. (After Clarke, B. G. and C.-C. Chen. Intrinsic properties of permeability. In *Proceedings of the International Conference on Soil Mechanics and Foundation Engineering*, Vol. 1, AA Balkema, 1997: 259–262.)

4.6.2 Databases

Databases of typical values of geotechnical design parameters are essential to validate results from ground investigations and provide data should the ground investigation have not been designed to produce the information required. Clarke et al. (2008) extended this concept using the relational database developed by Hashemi et al. (2006), which is a collection of data from numerous investigations in NE England undertaken as part of the programme to exploit coal reserves through open cast mining.

The regional database of the physical properties of glacial tills has been interrogated to produce characteristic design values and baseline construction values. Figure 4.83 highlights the difficulty in assigning characteristic values to this spatially variable soil and identifying the differences between the three tills. The consistency limits of the three tills lie about the T-line (Figure 4.14), and the average values are different as they actually reduce with depth. Figure 4.84 shows the variation in average water content, consistency limits and density with depth, reducing the scatter in the data.

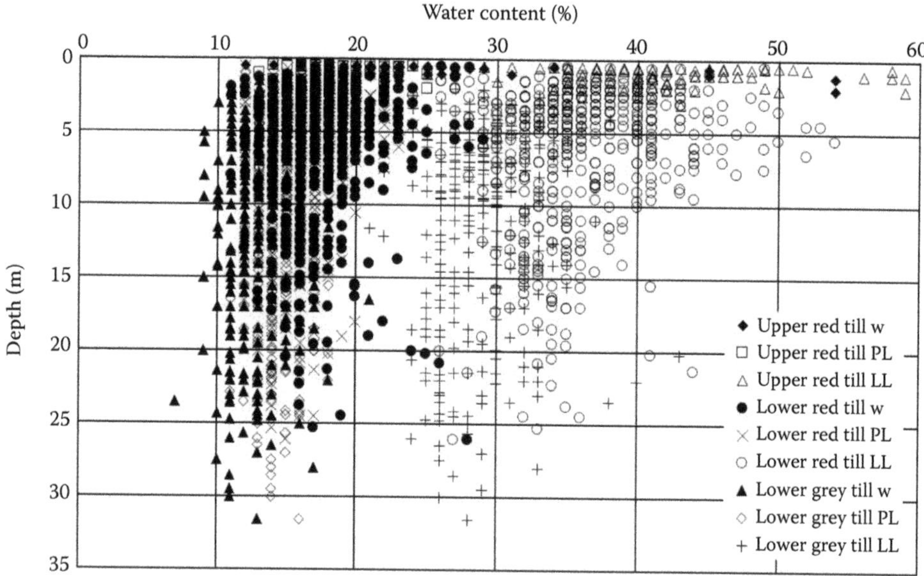

Figure 4.83 Profile of consistency limits for a site in NE England highlighting the difficulty in assigning characteristic values to this spatially variable soil. (After Clarke, B. G. et al. *Géotechnique*, 58(1); 2008: 67–76.)

Figure 4.85 shows the variation in undrained shear strength with depth taking the average value at each depth.

This regional database represents a 'global' situation; therefore, a cautious mean represents the characteristic value (Frank et al., 2004), with the probability of the worst case occurring being less than 5%. Frank et al. (2004) suggest that if there is no significant trend in the data, the characteristic value X_k is given by

$$X_k = X_{\text{mean}}(1 \pm k_n V_x) \tag{4.63}$$

where X_{mean} is the mean value of the parameter. V_x, the coefficient of variation, is equal to the standard deviation divided by the mean value if there is no trend to the data and there is no *a priori* knowledge. $X_k > X_{\text{mean}}$ is used when the cautious mean is a high value, for example, when calculating active pressures on retaining structures; $X_k < X_{\text{mean}}$ is used when the cautious mean is a low value, for example, when calculating bearing capacity. Clarke et al. (2008) give values of V_x as *a priori* known value when characteristic values are being derived from a new investigation in the region. k_n is a statistical coefficient that takes into account the number of samples, the volume of the ground, the type of results and the level of confidence. Schneider (1999) suggested that the characteristic value for a new investigation is simply the mean less half the standard deviation, since the regional coefficient of variation is likely to be unknown, which leads to an over-cautious estimate of the mean. The characteristic values for design may be different from baseline values for construction, since baseline values are established for contractual reasons to limit claims arising from unforeseen circumstances (Essex, 1997). The values are based on an assessment of the subsurface conditions to produce rational limits to the likely worst-case values. Given that these govern construction processes, then the local low (or high) value could be considered. Eurocode 7 (2004) suggests that this is represented by the 5% fractile values with $\left(k_n = 1.64\sqrt{1/n+1}\right)$.

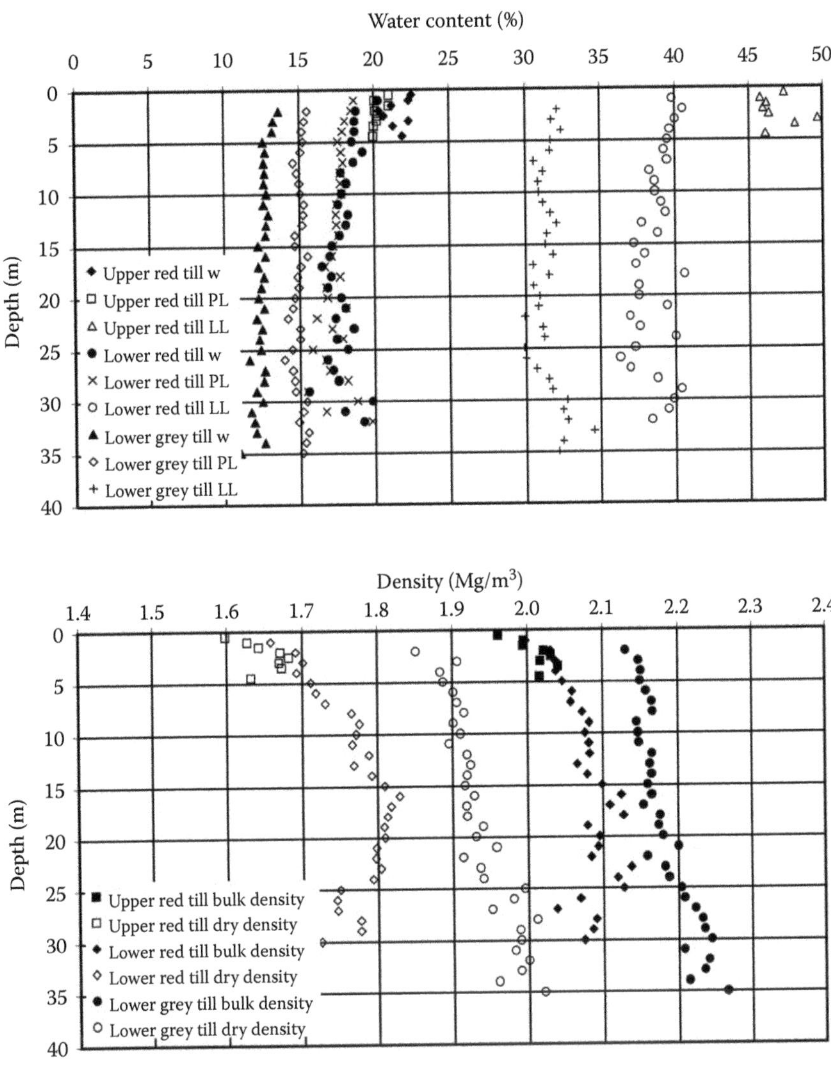

Figure 4.84 Variation in average water content, consistency limits and density with depth showing consistent profiles. (After Clarke, B. G. et al. *Géotechnique*, 58(1); 2008: 67–76.)

4.7 OBSERVATIONS

A review of characteristics of glacial soils has highlighted a number of points that should be considered when characterising glacial soils:

- Glacial soils can contain boulders, cobbles, gravel, sands, silts and clays in various proportions.
- Many glacial soils are composite soils containing a mix of very coarse, coarse and fine garden particles.
- Glacial soils are anisotropic because of depositional processes and subsequent unloading in the case of subglacial tills.

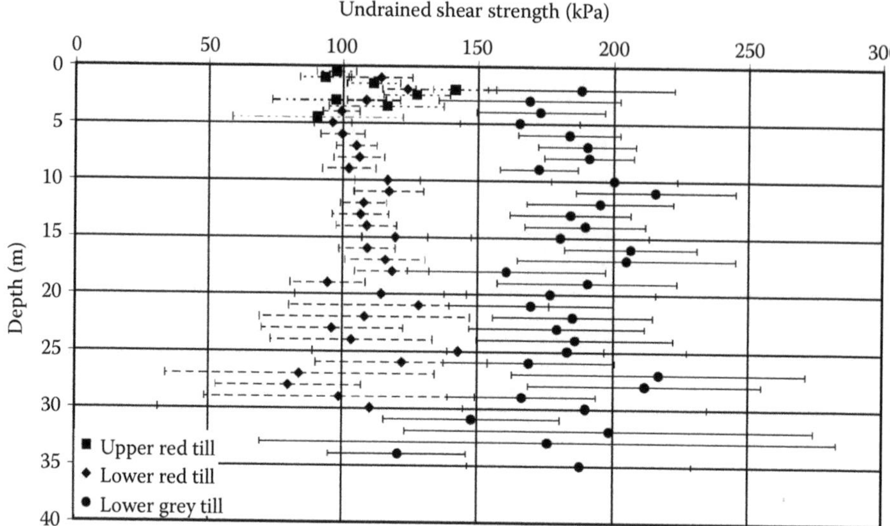

Figure 4.85 Variation in undrained shear strength with depth showing the mean profile and range. (After Clarke, B. G. et al. *Géotechnique*, 58(1); 2008: 67–76.)

- While typical values of geotechnical characteristics are published, the evidence is that there are no typical values as glacial soils are so variable because of their composition, fabric and local stress history during formation.
- The consistency limits of glacial clays tend to lie astride the T-line.
- Boulders, cobbles and gravel have a significance on the quality of a sample and test results but, *in situ*, may have little effect on the behaviour of the glacial soils. It depends on whether they are randomly distributed throughout the soil or not. However, they do impact on construction.
- It is likely that laboratory tests to determine geotechnical characteristics will be based on reconstituted glaciofluvial soils, clast-dominated tills and, possibly, matrix-dominated tills. It is important to carry out the tests at the *in situ* density.
- *Glacial tills*
    - There are essentially two types of glacial soils that have been transported by ice, which may have the same composition but their density and fabric are different. They are subglacial tills and melt-out tills with subglacial tills being denser because they subject to shear during deposition.
    - Subglacial tills contain discontinuities as a result of deposition and post-depositional processes. These discontinuities have a significant effect on the strength, stiffness and conductivity characteristics, which means that tests should be carried out on as large as samples as possible.
    - Soil descriptions are based on the principal fraction, but engineering behaviour is a function of the dominant fraction. Hence, a matrix-dominated till can have as little as 15% fine-grained particles to behave as a fine-grained soil. This is because the hydraulic conductivity of the till changes if it contains about 15%–20% by weight fine-grained particles.
    - Discontinuities open up during excavation, so the properties change.

- The effective strength of tills should be based on tests on a number of samples rather than specimens from one sample to overcome the difficulty of obtaining representative samples.
- Matrix-dominated tills are remoulded during deposition so can be reconstituted to the in situ density to give the intact properties of the till, thus removing the effects of fabric and coarse-grained composition. This produces a lower bound value, which may have to be adjusted to take into account the fabric.
- *In situ* and laboratory test results should not be expected to give the same results. Site-specific correlations need to be developed to relate *in situ* penetration test results with geotechnical characteristics.
- Unweathered matrix-dominated tills can contain fine-grained rock particles and/or clay minerals depending on the source rock.
- The clay mineral content of a weathered matrix-dominated till will be greater than the clay mineral content of an unweathered matrix-dominated till.

- *Glaciolacustrine clays*
  - These are highly anisotropic, which can be significant depending on the directions of loading and hydraulic potential.
  - The strength of these clays should be assessed by loading perpendicular to the laminations and by shear parallel to the laminations.
  - The hydraulic conductivity should be assessed both perpendicular and parallel to the laminations.
- *Glaciofluvial soils*
  - These are predominantly coarse and very coarse grained with the particle size distribution varying with distance from the source.
  - It is unlikely that Class 1 or 2 samples will be retrieved unless special measures are taken (e.g. freezing). Hence, geotechnical characteristics are derived from *in situ* tests and laboratory tests on reconstituted soils.

Chapter 5

Earthworks

Slopes, cuttings, embankments and tunnels

5.1 INTRODUCTION

This chapter covers earthworks, that is, design and construction involving major volumes of glacial soils. It covers the stability of slopes, tunnels and embankments and stabilised ground. The principles of stability for mass movements of soil also apply to the overall stability of geotechnical structures covered in Chapter 6.

Geotechnical design is based on codes, good practice and experience, especially when working with glacial soils because the composition, fabric and structure of the soils and their relation to the underlying bedrock and the overlying, more recently deposited soils have an influence on the behaviour of a glacial soil stratum. A study of the methods of erosion, transportation and deposition suggests that excavating glacial soils is hazardous because of the uncertainty of dealing with such a variable soil and the possibility of catastrophic failure should the hazards not be investigated and mitigation measures applied. Slope failures and major surface and subsurface excavations can have a detrimental effect on adjacent structures, infrastructure and utilities. Therefore, design has to take into account the behaviour of a significant volume of Earth and, given the spatial variability of glacial soils, requires a realistic assessment of the mass characteristics and local characteristics of glacial soils. For example, matrix-dominated tills can be fissured and contain weaker layers or lenses, which means that it is possible for the predicted mass stability to be acceptable but local instability could trigger failure. Excavation reduces the overburden pressure causing any discontinuities to open up, thus reducing the mass strength; weaker layers are local zones of lower strength soil. It is difficult to assess stability because of the difficulty in obtaining representative samples of sufficient size to determine the mass characteristics and to undertake sufficient exploratory techniques to uncover the spatial variability of a glacial soil to take into account the influence of the fabric, structure and composition on the soil behaviour.

5.2 OVERALL STABILITY

Overall stability covers mass movement of the ground leading to damage or loss of serviceability of a structure and neighbouring structures, roads or services. Natural processes that cause excessive movement include landslides, infiltration, rising/falling groundwater levels, freeze/thaw, seismic activity, erosion, collapse of underground cavities, wave action and vegetation or its removal. Construction processes, cuttings, embankments or structures on or near a particular site can also cause excessive ground movement. The forces causing instability include the effective weight of soil and external loads (e.g. buildings); the restoring force is a mainly a function of the soil strength, which in turn depends on the effective weight of soil and the effective strength of the soil. The mobilised strength can change if the pore pressure changes. This is a particular challenge because of the difficulty in establishing the seasonal variations in groundwater conditions and future variations due to climate change. Therefore, it is necessary to assess the groundwater conditions thoroughly and to take a worst credible estimate of likely groundwater conditions throughout the life of the structure providing adequate drainage, maintenance of those drains and measures to adapt the structure in the future to cope with potential changes. The alternative is to assume the worst possible conditions and overdesign. Therefore, an investigation must include an assessment of the following:

- The groundwater level, including perched water levels and seasonal changes
- The effects of the regional hydrogeological conditions and site-specific conditions including effects of infiltration
- The mass permeability and its variation vertically and horizontally
- The effect of excavation/construction on the permeability, especially if the soil is fissured as excavation can increase the permeability of the soil

Overall stability of slopes can be improved by:

- Regrading the slope
- Vegetation to reduce infiltration and increase the strength of the upper layers
- Drainage to prevent water flowing onto the slope, remove water falling onto the slope and lower the groundwater level within the slope
- Concrete cover to prevent infiltration and local erosion and anchored if overall stability is an issue
- Soil nails or anchors with appropriate facing depending on the steepness of the slope
- Retaining structures including gravity walls such as gabions to restrain the toe and embedded walls to extend the slip surface
- Grouting to fill voids and fractures to reduce mass permeability

Overall stability of geotechnical structures such as foundations and retaining walls is improved by extending the depth of the foundations or walls if the excessive movements are a result of poor ground conditions or instability if built on a slope; ground improvement if poor ground conditions; or infilling of voids and fractures.

5.2.1 Stability of slopes

There are four types of slopes to consider: natural slopes, natural slopes with toe erosion (e.g. coastal cliffs and banks of rivers), cuttings and embankments. Potential failure modes include falls and slides, shown in Table 5.1. Falls are normally associated with rock slopes as they are governed by joints and bedding planes, but they also occur in stiff matrix-dominated tills, especially if there is over-steepening due to erosion and the till is fissured. Slides include translational and rotational slides. Slides can lead to flows especially in glacial soils high upon valley sides and saturated tills with a low fines content, an issue that crippled the Scottish road network in 2004 (Winter et al., 2013).

Stability calculations must take into account the topographical, geological, hydrogeological and geotechnical conditions, which, in glacial soils, includes

- The spatial variation in stratum thickness and layers of weaker or water-bearing soils
- The variation in composition vertically and horizontally
- The effect of fabric and structure on the strength and permeability of the soil and the effect of excavation on the fabric

Table 5.1 Types of failure that can occur in glacial soils

Type of failure	*Description*	*Glacial soils*	*Mechanism*
Circular slide	Movement of a block of soil along a curved failure surface	Homogenised glacial tills Clast-dominated tills Glaciofluvial soils	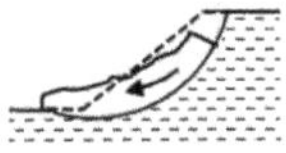
Non-circular slide	Movement of a block of soil along a non-circular surface	Matrix-dominated tills containing weaker lenses or layers Glaciolacustrine clays	
Translational slide	Movement of a shallow mass of soil along a surface approximately parallel to the surface	Thin layers of glacial soils overlying bedrock Poorly compacted edge of embankments	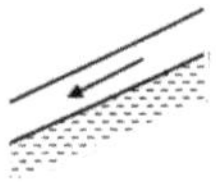
Compound slide	Combination of rotational and translational slides	Commonly found in failed slopes in glacial soils	
Flow slide	Translational slide in saturated soil due to an increase in water pressure causing the soil to flow as a viscous fluid possibly considerable distances	Clast-dominated tills in mountainous regions; glaciofluvial soils	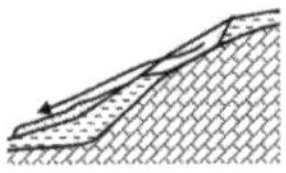
Debris slide	Translational slide of debris triggered by rainfall or surface water creating a mantle on the slope	Clast-dominated tills in mountainous regions	
Slab slide	Translational slide in which the sliding mass remains intact	Mantle of glacial soils overlying more dense soil/rock	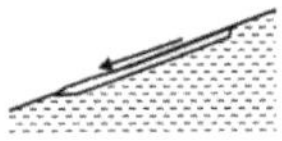
Block slide	Translational slide of block	Matrix-dominated tills with discontinuities	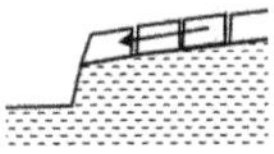
Progressive failure	Failure surface develops in brittle soils due to loss of strength post-peak progressively transferring the load along the failure surface		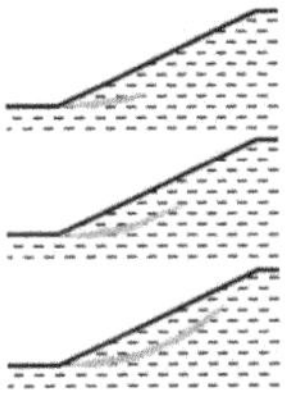
Scour	Water flowing across the surface (e.g. run-off, water course) leading to gullies		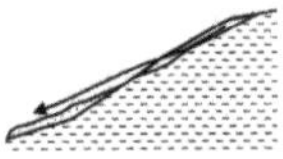
Internal erosion	Seepage of groundwater along a preferential flow path causing loss of fines or slumping	Sand layers within matrix-dominated tills	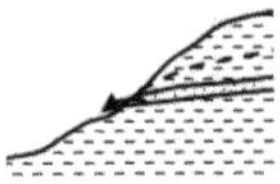
Wedge failure	Failure surface defined by discontinuities	Fissured matrix-dominated tills	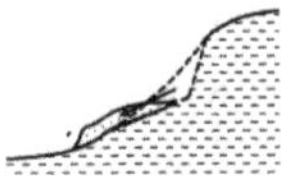
Toppling	Usually associated with rock slopes but occurs with eroding cliffs composed of matrix-dominated tills		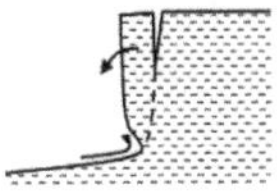

- Local and regional hydrogeological conditions
- Relict failure zones and shear planes

It is prudent in glacial soils to undertake a scenario analyses to investigate combinations of geological, hydrogeological and geotechnical profiles. Analyses of slopes can be based on the following:

- Limit equilibrium methods based on circular and non-circular failure mechanisms
- Numerical methods
- Physical models
- Observational method
- Stability charts

Slopes are designed not to fail and failure is usually the critical limit state since deformation is often not an issue unless the slope is supporting a structure. However, the analysis of a slope has to be placed in context. Stability analyses can be used to check that a slope is safe or, in a forensic analysis, the reason for a slope failure. In the latter case, the most likely strength is used. In the former case, a cautious estimate of the strength is used; the choice of strength depending on whether it is a potential first-time slide (intrinsic strength), it is a slope containing relict slip surfaces (reduced strength to allow for discontinuities) or whether there are sensitive structures above the slope (deformation is critical so a reduced strength is used). The upper and lower bound of unit weight of soil should be used in order to take account of the contribution the soil mass makes to the disturbing and restoring moments. A worst credible view of the groundwater conditions should be assumed unless a detailed investigation has been undertaken to fully determine the hydrogeological conditions.

In homogenous soils, the limit equilibrium method may be adequate because the failure mechanism is likely to be circular, which can be automatically modelled to investigate many possible circular failure mechanisms. If the geological profile is not homogenous, then non-circular analyses may be more appropriate. This is more difficult to analyse with limit equilibrium methods because the slip surface has to be specified. An alternative is to use numerical methods that are able to predict where the failure zone is likely to occur. It is challenging to model spatial variation of glacial soils because slope failure in glacial soils may be triggered by local structural or fabric features, which may not be identified in the ground investigation and can be difficult to model at an appropriate scale. It is possible in glaciolacustrine clays and matrix-dominated tills that the structure and fabric of the soil will produce a wedge-type failure mechanism.

In a slope with no external loads, the disturbing force is the soil weight and the restoring force is the shear stress on the slip surface. The ultimate resistance, R, of a section of a slope is defined by

$$R = W\sin\alpha = (W\cos\alpha - uA)\tan\varphi' + Ac' \tag{5.1}$$

where W is the weight of the soil, α the angle of the slip plane to the horizontal, φ' the mobilised angle of friction, c' the cohesion, u the pore pressure on the slip surface and A the area of the slip surface (the base of the section). Circular and non-circular slips (Figure 5.1a and b) can be analysed using the method of slices such that the global factor of safety, F, given by

$$F = \frac{\sum_1^n [(W\cos\alpha - uA)_i \tan\varphi' + A_i c']}{\sum_1^n (W\sin\alpha)_i} \tag{5.2}$$

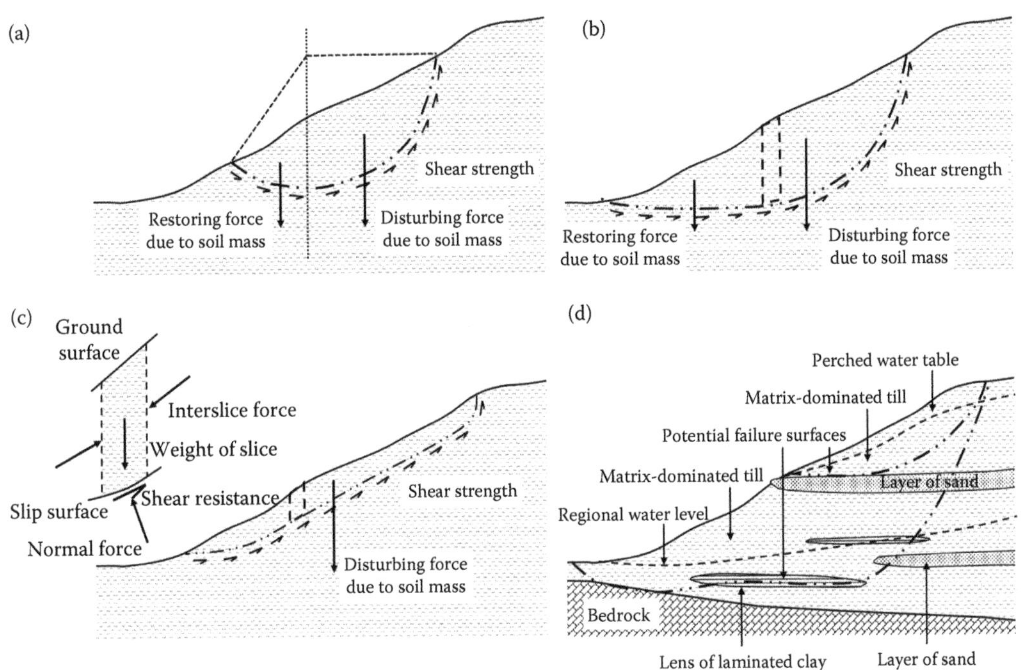

Figure 5.1 Mechanisms for (a) circular, (b) non-circular and (c) translational failures showing the disturbing and restoring forces and (d) how the structure of a glacial soil could govern the failure mechanism.

where W, α and A refer to the slice i. There are a number of methods available to solve this equation, which depend on the assumption for the inter-slice forces (Bromhead, 2012), but the effect on the solution is small and much less important than selecting the correct parameters. The method of slices is the basis of many commercial slope stability packages, which often give the user a choice of methods of analysis such as those listed in Table 5.2. Slope failures are a function of the composition, fabric and structure of the soil, which may be difficult to model because of the difficulty in determining the spatial variation in glacial soils, and are a function of the hydrogeological conditions which can be difficult to assess because of local variations in permeability and, in the case of excavations, the change in permeability due to the excavation. Further, many failures in glacial soils are complex, so limit equilibrium methods are restrictive as they are generally used to analyse a simple failure mechanism. Many slope failures are triggered by structural features such as interbedded layers of weaker and water-bearing soils, which may influence the shape of the slip surface. Detailed modelling of complex soils in which fabric, structure and composition can strongly influence the failure mechanisms is challenging. Hence, a sensitivity study is essential to account for weaker layers of varying thickness and location. Fabric can be taken into account by reducing the strength of the soil. In the case of a zone of weakness parallel to the slope, a translational slip (Figure 5.1c) may take place such that the global factor of safety, F, is

$$F = \frac{c' + (\gamma z - \gamma_w h_w)\cos^2\beta \tan\varphi'}{\gamma z \sin\beta \cos\beta} \tag{5.3}$$

where β is the angle of the slope, z the depth to the slip plane, and h_w the depth of water above the slip plane.

Table 5.2 Advantages and disadvantages of methods of stability analysis

	Reference	*Assumptions*	*Advantages*	*Disadvantages*	*Recommendations*
Circular	Bishop (1955)	Considers force and moment equilibrium for each slice. Rigorous method assumes values for the vertical forces on the sides of each slice until all equations are satisfied. Simplified method assumes the resultant of the vertical forces is zero for each slice	Simplified method compares well with finite element deformation methods (average F within 8%)	Circular arcs do not always fit the failure surface (in the case of failed slopes) or the likely constraints of topography, geological and hydrogeological conditions	Useful where circular failure surfaces can realistically be assumed in homogenous soils
Non-circular	Morgenstern and Price (1965)	Considers forces and moments for each slice; similar to Janbu procedure	Considered more accurate than Janbu	No routine (or simplified) method	Very useful where there are topographic and other constraints on the geometry of the failure surface. Most useful for back analysis of existing landslides
	Janbu (1973)	Generalised procedure considers force and moment equilibrium for each slice. Assumptions on line of action of inter-slice forces must be made. Vertical inter-slice forces not included in routine procedure and calculated F is then corrected to allow for vertical forces	Realistic failure or potential failure surfaces can be adopted	Published correction factors are for homogeneous materials and use of routine procedure can produce large errors in slopes composed of more than one material. Factor of safety is usually underestimated in these cases. Generalised method does not have the same limitations	Very useful where there are topographic and other constraints on geometry of the failure surface. Limitations of the routine method must be considered. Routine method not suitable for embankments where the potential failure surface extends deep into underlying foundation soil
	Sarma (1979)	A modification of Morgenstern and Price method, which reduces the number of iterations necessary by the application of earthquake forces			As for Morgenstern and Price

Source: Based on Geotechnical Engineering Office. *Geotechnical Manual for Slopes*. Geotechnical Engineering Office, Hong Kong, 1984; and Trenter, N.A. *Engineering in Glacial Tills*. CIRIA, London, 1999.

According to Bromhead (2012), errors in stability analyses are due to incorrect assessment of design values for shear strength and the relevant drainage conditions (drained, consolidated drained, or consolidated undrained conditions); incorrect assessment of pore water pressure regime; inadequate assessment of the effect of weak layers in the slope, which may be naturally present in the soil or have been created by construction activities; and a selection of inadequate mobilisation factors, especially where there is uncertainty surrounding the soil parameters and pore water pressures and where there is the risk of progressive failure.

An allowance for uncertainty and risk is made using either global factors or partial factors of safety. Typical global factors of safety are 1.1–1.3 for a temporary situation and 1.3–1.5 for a permanent situation. The soil mass is both a disturbing force and a restoring force depending on its location in the slope. This is an issue when applying partial factors (Table 5.3) because partial factors are applied to the whole of the soil mass, not separately to the disturbing and restoring actions due to the soil mass. In that case, a limit equilibrium analysis must consider upper and lower characteristic values of unit weight.

Limit equilibrium methods are used to assess the ultimate limit state; they cannot be used to assess the serviceability limit state. If slope deformation can cause damage, for example, to structures above a slope, then measures have to be taken to restrict ground movement if the design is based on limit equilibrium methods. Alternatively, BS EN 1997-1:2004+A1:2013 suggests that reducing the shear strength is an option to restrict deformation because a more stable slope is likely to deform less. The alternative is to monitor the slope and take action if necessary. Numerical methods can be used to predict deformation of slopes but, given the complex nature of the ground, the difficulty in correctly modelling the soil and the magnitude of movement necessary to cause damage, they can be used to indicate only where movements may be critical. Hence, it is recommended that critical slopes are monitored.

The fabric, structure and composition of the soils, especially glacial soils, forming the slope often govern the failure mechanism. This includes the presence of weaker and water-bearing lenses and layers in glacial tills, discontinuities in matrix-dominated tills, existing

Table 5.3 Partial factors on actions (γ_F), the effects of actions (γ_E), material properties (γ_m) and resistances (γ_R) for internal failure or excessive deformation of structural elements (STR) or the ground (GEO) and overall stability

	Partial factors on actions (γ_F) or effects of actions (γ_E) (STR and GEO)			
Action	*Symbol*	*A1*		*A2*
Permanent (unfavourable)	γ_G	1.35		1
Permanent (favourable		1		1
Variable (unfavourable	$\gamma_{Q;dst}$	1.5		1.3
Variable (favourable)		0		0
Soil parameter		*M1*		*M2*
Angle of shearing resistance (tan φ′)	$\gamma_{\varphi'}$	1.0		1.25
Effective cohesion (c')	$\gamma_{c'}$	1.0		1.25
Undrained shear strength (c_u)	γ_{cu}	1.0		1.4
Unconfined strength (q_u)	γ_{qu}	1.0		1.4
Unit weight (γ)	γ	1.0		1.0
Resistance		*R1*	*R2*	*R3*
Earth resistance	$\gamma_{R;e}$	1.0	1.1	1.0

Source: After BS EN 1997-1:2004+A1:2013. *Eurocode 7: Geotechnical Design – Part 1: General Rules.* British Standards Institution, London.

Table 5.4 Nominal load due to surcharge at the top of slopes and embankment

Standard load	*Uniformly distributed load (kPa)*	*Typical design cases*
No specified load case	10	Earthworks slopes where maintenance equipment might present an adverse load case
Typical highway loading	10	Common practice is to assume this value. Extreme cases agreed on a site-specific basis
RL loading	30 on area occupied by tracks	Light rail systems
RU loading	50 on area occupied by tracks	All standard rail systems (UK)

Source: After BS 6031:2009. *Code of Practice for Earthworks*. British Standards Institution, London.

shear planes due to existing slips or depositional features. Surcharges at the top of slope contribute to the failure; therefore, it is prudent to assume a minimum surcharge of 10 kPa, for example, to allow for maintenance of cut slopes. Examples of loads on infrastructure embankments due to highway and rail track loading are given in Table 5.4. Other slopes, especially in urban areas, may be developed, so the building loads have to be taken into account if the buildings fall within the zone of influence of a potential slope failure including that due to a complex slide.

Constructed slopes for cutting and embankments can either be designed using an appropriate method of analysis or be specified based on experience. An example of the latter is the specification for slopes produced by the UK Department of Transport for highways, which is often used for slopes other than highways. Slopes in glacial soils are typically 1v:2.5h to 1v:3h. Perry (1989) created a database of failures of motorway embankments and cuttings constructed in the United Kingdom from 1967 to determine whether the specified slopes were stable. The majority of the slopes were less than 5 m high. Slope angles in matrix-dominated tills varied between 1v:2h and 1v:2.5h. The failure rate, shown in Table 5.5, shows that the percentage of failures increases with age. This is attributed to a reduction in strength with time, a feature of over-consolidated sedimented clays but not, according to Skempton (1964), glacial tills. No reference is made to the ground profile, the soil fabric or the groundwater profile in these tills, which are known to influence the failure. In order to have no failures, it is suggested that it would be necessary to reduce the slopes to 1:5, which is neither economic nor consistent with natural slopes in glacial clay tills, which are known

Table 5.5 Performance of British motorways cut slopes from 2.5 to 5 m high

Soil type		*Age*	*Length surveyed*	*Slope angle*	*% Failure*
		3–18	4142	1:2.5	0
		22	204	1:2.5	20.5
Glacial till		7–17	3291	1:2	0
		18	1750	1:2	1
		22	2319	1:2	3.7
		25	463	1:2	5.8
Stiff sedimented clays	London Clay	6	533	1:3	1.9
		10	543	13	3.2
	Gault Clay	10	353	1:2.5	3.8
		22	299	1:2.5	4.4
	Lias Clay	4.5	528	1:2	0
		25	894	1:2	1.4

Source: Adapted from Perry, J. *A Survey of Slope Condition on Motorway Earthworks in England and Wales*. Research Report-Transport and Road Research Laboratory, 199, 1989.

to stand unsupported up to 45°. Table 5.5 also highlights a difference in behaviour between matrix-dominated tills and over-consolidated sedimented clays; that is, failure of slopes in sedimented clays takes place much earlier than those in tills.

5.2.2 Mobilised strength

The uncertainties in stability analysis are the geological profile, the mobilised shear strength and pore pressure, but they are further complicated in glacial soils because of their composite nature and the spatial variation of the composition, fabric and structure. Pore pressure profiles are site specific, but it is possible to use generic strength parameters in homogeneous soils. Most methods of stability analysis assume a simple model for the soil strength, that is, undrained or effective strength. They do not take into account the direction of shear, possible reduction in strength with strain post-peak, fabric or structure. Therefore, conservative values of strength should be used.

BS 6031:2009 suggests that shear box tests on coarse-grained soils and fine-grained soils with a plasticity index less than 25% can be used to determine the post-peak strength effective, which should give conservative values. This applies to glaciofluvial soils and many unweathered tills. The intact peak shear strength can be used for other fine-grained soils though if significant displacements are expected it is prudent to use the angle of friction at constant volume and if relic shear surfaces are known to exist, the residual strength. In brittle soils with a plasticity index greater than 25%, progressive failure is a possibility, which means that the post-peak strength should be used. Matrix-dominated tills generally have a plasticity index less than 25% and do not exhibit brittle behaviour.

Forensic analyses of slips were used by Stark et al. (2005) to produce a database of strength of a variety of soils including glacial soils. BS 8004:2015 suggests that this is a useful source of data on glacial tills for foundation design though, as explained below, the strengths apply to slopes not foundations. Stark et al. (2005) suggested that the residual strength is relevant to slopes that contain relict slip surfaces, which includes historic slips, slopes subject to solifluction, bedding planes in folded strata, sheared joints or faults and foundations of a dam subject to annual cycles of reservoir level. The residual strength is defined by the residual angle friction with no cohesion. A fully softened drained shear strength is applicable to first-time slides, which, according to Skempton (1970), is the strength of the equivalent normally consolidated soil, that is, the angle of friction at constant volume. Skempton (1964) suggested that, in over-consolidated sedimented clays, the cohesion should be set to zero because fissured clays soften with time. Skempton and Brown (1961) suggested that this is not the case with over-consolidated matrix-dominated tills and proposed that representative values of cohesion and the angle of friction are appropriate for analysis of slopes in matrix-dominated tills though allowance must be made for a reduction in the mass strength if the till is fissured. The strength of remoulded matrix-dominated tills at the same density as the *in situ* density is similar to their intact strength since these soils do not exhibit brittle behaviour. This may be a consequence of remoulding that took place during deposition. The variation in the residual angle of friction with clay fraction and liquid limit for fine-grained soils (Figure 5.2) suggests that the residual angle of friction for glacial clay tills will depend only on the clay fraction because the liquid limits are usually low. For example, the residual angle for matrix-dominated tills is likely to be between 25° and 30°, and that for glaciolacustrine clays between 15° and 25°.

Figure 5.3 shows that, for values of liquid limit less than 50% (typical of matrix-dominated tills), the residual angle of friction is about 4–6° less than the fully softened value shown in Figure 5.4. The fully softened angles of friction for matrix-dominated tills are similar to their intact angle though this will depend on the clay fraction; the softening increases with

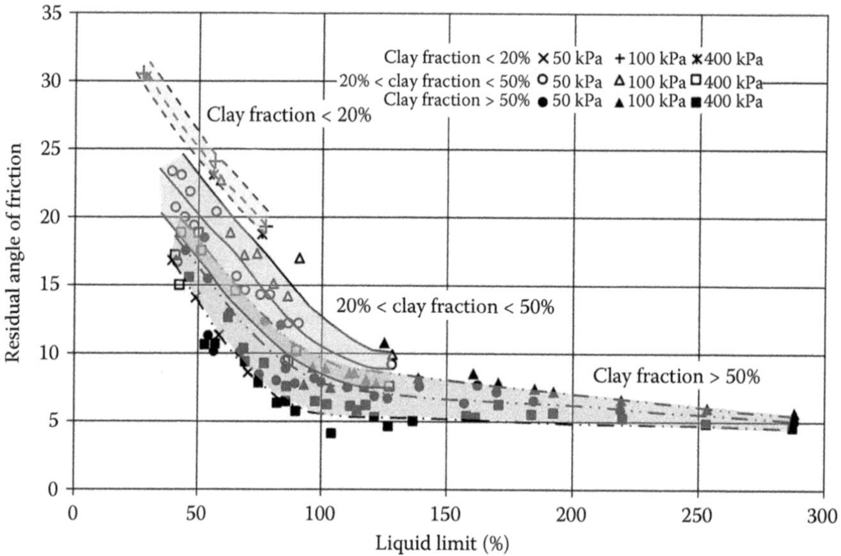

Figure 5.2 Variation in residual angles of friction with clay fraction and liquid limit. (After Stark, T. D. et al. *Journal of Geotechnical and Geoenvironmental Engineering*, 131(5); 2005: 575–588.)

increasing clay content. This is not the case for glaciolacustrine clays because the mobilised strength will depend on the direction of shear due to the anisotropic nature. Figures 5.2 and 5.4 also suggest that the mobilised angle of friction reduces with depth because the confining stress increases. This is consistent with a curved failure envelope unlike the linear envelope normally assumed. There is no reference to cohesion in these studies.

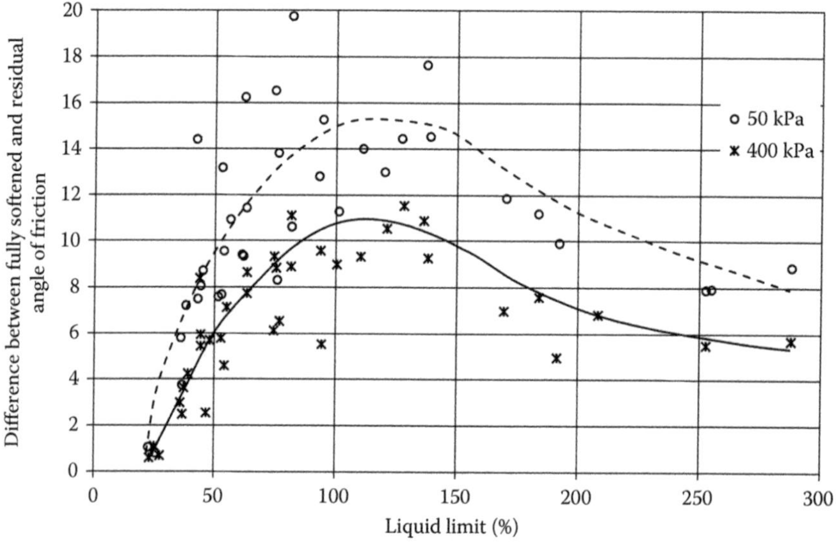

Figure 5.3 Variation in the difference between residual angles of friction and softened angle of friction with liquid limit. (After Stark, T. D. et al. *Journal of Geotechnical and Geoenvironmental Engineering*, 131(5); 2005: 575–588.)

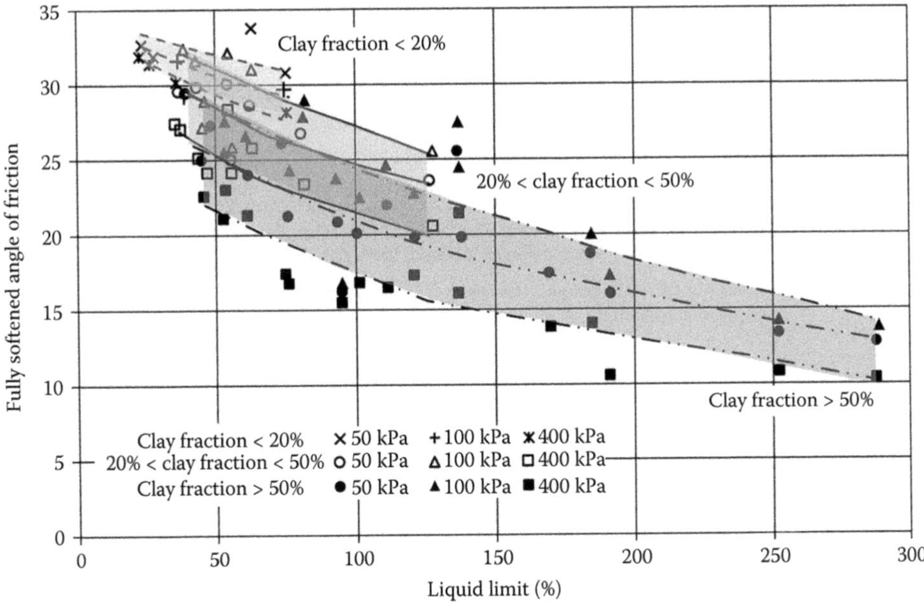

Figure 5.4 Variation in fully softened angles of friction with clay fraction and liquid limit. (After Stark, T. D. et al. *Journal of Geotechnical and Geoenvironmental Engineering*, 131(5); 2005: 575–588.)

Weathering of clay tills increases the liquid limit, so a lower angle of friction than that for the unweathered soil should be assumed. Therefore, it is important to identify the depth of weathering in a glacial clay sequence. This could be based on the profile of consistency limits and, possibly, supported by the colour of the till. The mobilised angle of friction for glaciolacustrine clays will also be less than the intact strength, but given the anisotropic behaviour of these soils, allowance has to be taken of the direction of loading relative to the laminations.

Therefore, it is recommended that, for first-time slides in matrix-dominated tills, the intact effective strength parameters can be used but, for fissured tills, the cohesion should be set to zero and, possibly, the post-peak angle of friction used if brittle behaviour is observed. Glaciolacustrine clays and weathered clay tills have a higher liquid limit and clay fraction, which means that the angle of friction at constant volume or residual angle of friction should be used depending on whether it is a first-time slide or a reactivated dormant slide.

Landslides that are a consequence of engineering or weather-related events may be first-time slides but many slopes are dormant landslides that have taken place following isostatic uplift and development of the current drainage system since the last Ice Age ended. This means that the residual angle should be considered a possibility and a geomorphological investigation is essential.

It is recommended that total stress analysis should not be used to assess the stability of a slope given the difficulty in obtaining representative values of undrained strength and the effects of fabric and structure (Figure 4.22) and sampling (Table 5.6) on the mobilised strength. Skempton (1964) suggested that the mobilised strength in stiff sedimented clays is different from that measured in routine tests due to the reduction in strength because of potential softening along discontinuities. This also applies to glacial tills but, unlike sedimented clays, Skempton and Brown (1961) suggested that there is no loss of cohesion. Henkel (1957), Skempton and Delory (1957), Potts et al. (1997) and others showed that slips in over-consolidated sedimented clays occur sometime after excavation due to a reduction

Table 5.6 Effect of sample disturbance on undrained strength

Material type		*Effect on undrained shear strength*
Soft clay	Low plasticity	Very large decrease
	High plasticity	Large decrease
Stiff clay	Low plasticity	Negligible[a]
	High plasticity	Large increase

Source: After Vaughan, P. R. et al. *Predictive Soil Mechanics, Proceedings of the Wroth Memorial Symposium*, Oxford, 1993: 224–242.

[a] Does not take into account the effect of fabric, structure and composition (e.g. gravel particles).

in cohesion. Vaughan (1994) states that the undrained behaviour of low plasticity stiff clays such as matrix-dominated tills is different to medium to high plasticity clay such as over-consolidated sedimented clays. Matrix-dominated tills are not as brittle as stiff plastic clays, so progressive failure is unlikely to occur; the undrained shear strength is independent of the initial structure – intact or remoulded (Figure 5.5). This means that slope failures are more likely to achieve neutral equilibrium with large displacements rather than collapse that occurs with progressive failure. Skempton and Brown (1961) refer to a forensic analysis of the Selset landslide, N England, which was carried out to establish whether the stability of glacial clay tills reduced with time. The glacial till was a matrix-dominated till with 17% fines such that the residual angle was 2° less than the peak angle, consistent with the findings of Stark et al. (2005) and Vaughan (1994) if the peak angle and post-peak or fully softened values are the same, which is typical of ductile matrix-dominated tills. Skempton and Brown (1961) showed that the peak strength was mobilised in a first-time slide whereas Stark et al. (2005) suggested that it should be the post-peak. Since these matrix-dominated tills do not exhibit brittle behaviour, this is consistent. Skempton and Brown (1961) had observed that glacial clay till slopes showed that the effective cohesion did not reduce with time, unlike slopes in over-consolidated stiff sedimented clays. A forensic analyses with an angle of friction of 32° showed that the mobilised cohesion was 8.6 kPa. Sevaldson (1956) undertook an analysis of a first-time slide in a glacial clay with a cohesion of 12 kPa and the angle of friction of 32°, which showed that cohesion was necessary for the slope to be just stable. The clay fraction was about 32%. This concept of neutral equilibrium accompanied by significant deformation but not failure was observed at Muirhead Dam (Banks, 1948; Vaughan and Hamza, 1977), Cow Green Dam (Vaughan et al., 1975) and Balderhead Dam (Kennard, 1964; Thomas and Ward, 1969).

Skempton (1964) suggested that stiff glacial tills when remoulded due to compaction retain their strength even though the structure is destroyed. This means that the peak strength can be used to determine the stability of compacted embankments constructed of matrix-dominated tills.

Therefore, when assessing the stability of a slope in glacial clay tills,

- A geomorphological investigation is essential to assess whether it is a dormant or first-time slide.
- Effective stress parameters should be used.
- Sufficient representative samples should be taken to ensure that the effect of fabric and composition can be taken into account in the interpretation.
- Tests should be carried out on reconstituted samples consolidated to the *in situ* density if it is not possible to obtain sufficient representative samples.
- The fabric of the till should be taken into account when assessing the mass strength.

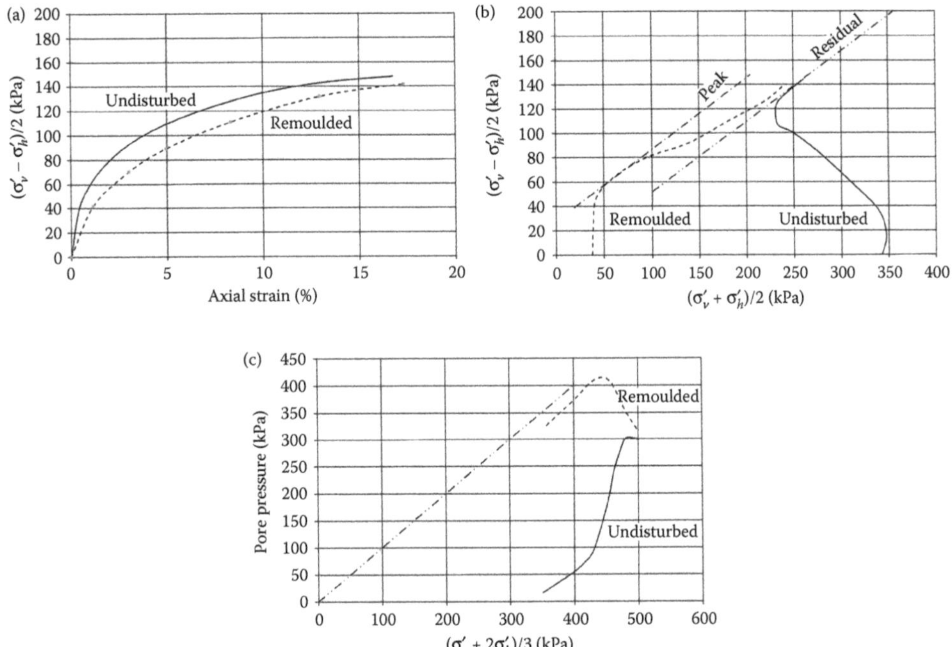

Figure 5.5 Triaxial tests on samples of intact and remoulded matrix-dominated till showing (a) the variation in deviator stress with strain, and (b) the variation in deviator stress and (c) pore pressure with mean stress. They show that stiff, low plasticity clays do not exhibit brittle behaviour and the undrained shear strength is independent of the structure. (After Vaughan, P. R. *Geotechnique*, 44(4); 1994: 573–609.)

- If there is potential for a first-time slide, the angle of friction should be the post-peak or fully softened value, which, in the majority of matrix-dominated tills, is likely to be the lower bound to the peak strength since they do not exhibit brittle behaviour.
- If there is potential for a reactivated dormant slide, the residual angle of friction should be used.

5.2.3 Pore pressures

Pore pressures, like shear strength, are difficult to predict in fine-grained soils and are continually changing due to rising/falling groundwater levels, infiltration and changes in drainage patterns. Many slope failures are attributed to rising groundwater levels though they may also be a result of infiltration leading to a loss of suction. In either case, it is water in a slope that is critical.

There are three ways to allow for pore pressures in an analysis:

- Estimated from the depth below ground level using an r_u value, the ratio of the pore pressure to effective overburden pressure
- Based on ground investigation data including piezometer readings and other observations of water level together with an interpretation of the hydrogeological conditions from boreholes and geological records
- A model of the groundwater conditions to account for seepage and potential changes in flow due to infiltration or rising and falling groundwater levels

Pore pressures will change with time due to seasonal changes, rainfall events, excavation and construction and, in future, climate change. A conservative view of seasonal pore pressure changes means that the groundwater level is at the ground level. However, perched water tables are possible especially if the stratum includes matrix-dominated tills, which can act as aquicludes. Therefore, a detailed hydrogeological study is recommended to avoid overdesign. Pore pressure also changes due to excavation and construction. In the case of excavations in glacial clays (Figure 5.6a), the pore pressure reduces initially due to the unloading of the soil and then increases with time as steady-state conditions are established, which means that the excavated slope becomes more unstable with time. The rate of change depends on the mass permeability, which changes in soils containing discontinuities as they can open up on unloading, and the presence of any more permeable layers. Pore pressure increases if the soil is loaded (Figure 5.6b) and then reduces with time increasing the stability.

Slope failures often take place after rainfall events and are attributed to rising groundwater levels. Skempton et al. (1989) suggested that the ratio of the rise in groundwater level to rainfall intensity was four based on field observations at five sites. This is not the most conservative estimate of groundwater level but may be a more realistic estimate. Matrix-dominated tills may act as aquicludes or may contain more impermeable layers, which means that perched water tables are possible. It is possible to have a regional water level, perched water levels and water-bearing lenses with independent pore pressures in matrix-dominated tills creating complex hydrogeological conditions. Further, it is possible to have glacial clay tills separated by coarse-grained layers, which does reduce the pore pressure in the overlying till (Figure 5.1d). Therefore, a ground investigation should be designed to ascertain the complexity of the hydrogeological conditions.

The pore pressure coefficient, r_u, in the absence of quality site observations or results of seepage studies, can be used to estimate the pore pressure. It was developed for analyses

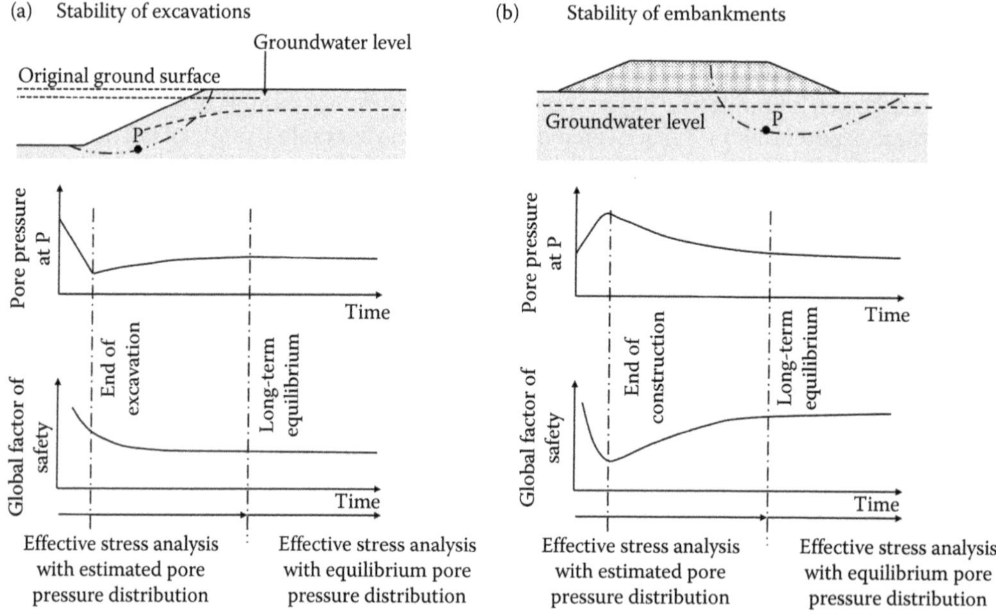

Figure 5.6 Effect of the variation in pore pressure and the impact on the global factor of safety of (a) excavations and (b) embankments showing that the stability changes with time, the time depending on the permeability of the glacial soils.

of cuttings as a simple means of estimating pore pressure based on the total vertical stress. Farrar (1979) suggested that r_u varies between 0.1 and 0.3 for fine-grained soils if the groundwater level is below the base of the excavation.

5.2.4 Fabric, structure and composition

Failures in glacial soils are often triggered by the fabric, structure and composition because of the inhomogeneous nature of these soils. This includes the following:

- Anisotropic strength of glaciolacustrine clays, which can lead to non-circular slides.
- The presence of fissures and discontinuities in matrix-dominated tills, which can lead to non-circular slides and toppling falls on eroding cliffs.
- Weaker layers and lenses in glaciofluvial soils and tills.
- Water-bearing layers in matrix-dominated tills.
- The percentage of coarse-grained particles in matrix-dominated tills can lead to debris flow slides.

Thus, a ground investigation should be designed to locate any structural features and assess the fabric of the soil, and the assessment of the stability should include a sensitivity analysis to explore the variation in fabric and structure.

5.2.5 Methods of analysis

The challenges of analysing natural and cut slopes in glacial soils include the following:

- The influence the fabric and structure have upon the slip surface, both the resistance and orientation of the slip surface
  - Failure in glaciolacustrine clays will be influenced by the laminations leading to non-circular slides.
  - Weaker layers and lenses in glacial tills will influence the location of the slip plane.
  - Discontinuities will reduce the mobilised strength and, possibly, the strength of the discontinuities will reduce with time.
- Whether it is a first-time slide, activation of a dormant landslide or re-activation of a recent landslide
  - The intact strength can be used for a first-time slide in matrix-dominated tills provided the liquid limit is less than 50%.
  - A reduced strength is used for first-time slides in fissured matrix-dominated tills.
  - The post-peak strength is used for weathered clay tills for a first-time slide.
  - Residual strength is used for reactivated slides.
- Selection of appropriate parameters
- An assessment of the pore pressure profile given the inclusion of more permeable layers, perched water levels, possible aquicludes and possible artesian pressures in the underlying rock

The form the landslide will take depends on a combination of these factors, which makes it difficult to predict. Further, slope failures in matrix-dominated tills are often complex.

Embankments are engineered, which means that they are more likely to be more homogeneous than the natural soil from which they were derived. It is possible to incorporate layers of sand in low permeability glacial clay tills to accelerate consolidation. These can act as reinforcement, thus influencing the failure mechanism.

5.3 NATURAL SLOPES

The British Geological Survey has developed classification criteria (Table 5.7) for landslides based on the classification schemes proposed by Varnes (1978) and Cruden and Varnes (1994), which are a function of the type of movement and the material involved. Material type refers to rock, debris, Earth, mud or soil; the type of movement can be a fall, a topple, a slide, a spread or flow. Landslides are further divided into inland and coastal landslides. Trenter (1999) suggested that 20% of the 8835 inland landslides recorded in the UK

Table 5.7 Terms used to classify landslides covering (a) the type of material that is displaced and (b) the kinematic form

(a) Type of material (pre-failure)	*Description*	*Glacial soils*
Soil	An aggregate of solid particles, generally of minerals and rocks, that either was transported or was formed by the weathering of rock in place. Gases or liquids filling the pores of the soil form part of the soil	Glacial soils
Earth	Material in which 80% or more of the particles are smaller than 2 mm, the upper limit of sand-sized particles	Matrix-dominated tills; glaciolacustrine clays
Mud	Material in which 80% or more of the particles are smaller than 0.06 mm, the upper limit of silt-sized particles	Glaciolacustrine clays
Debris	Contains a significant proportion of coarse material; 20%–80% of the particles are larger than 2 mm, and the remainder are less than 2 mm	Glaciofluvial soils Clast-dominated tills Matrix-dominated tills

(b) Kinematic form	*Description*
Fall	A fall starts with the detachment of soil or rock from a steep slope along a surface on which little or no shear displacement takes place. The material then descends largely by falling, bouncing or rolling. This can be a feature of fissured tills, especially when subject to toe erosion
Topple	A topple is the forward rotation, out of the slope, of a mass of soil and rock about a point or axis below the centre of gravity of the displaced mass. This can be a feature of fissured tills, especially when subject to toe erosion
Slide	A slide is the downslope movement of a soil or rock mass occurring dominantly on the surface of rupture or relatively thin zones of intense shear strain (circular, non-circular or translational)
Flow	A flow is a spatially continuous movement in which shear surfaces are short lived, closely spaced and usually not preserved after the event. The distribution of velocities in the displacing mass resembles that in a viscous fluid (common in inland landslides in glacial tills especially on steep slopes)
Spread	A spread is an extension of a cohesive soil or rock mass combined with a general subsidence of the fractured mass of cohesive material into softer underlying material. The rupture surface is not a surface of intense shear. Spreads may result from liquefaction or flow (and extrusion) of the softer material
Complex	These are failures in which one of the five types of movement is followed by another type (or the same type). This is the most common form of landslide in glacial soils

Source: After Varnes, D. J. *Slope Movement Types and Processes*. Transportation Research Board Special Report 176, 1978; Cruden, D. M. and D. J. Varnes. *Landslides: Investigation and Mitigation*. Transportation Research Board National Academy of Sciences, 1994, Chapter 3.

Table 5.8 Frequency and type of landslides in glacial soils

Location	*Type of landslide*	*Frequency (%)*
Inland	Complex	32
	Debris flow	28
	Planar	26
	Rotational	8
	Multiple rotational	4
	Other	2
Coastal	Complex	49
	Debris flow	9
	Planar	5
	Rotational	18
	Multiple rotational	8
	Other	11

Source: After Trenter, N. A. *Engineering in Glacial Tills*. CIRIA, London, 1999 using the database of UK landslides (After Jones, D. K. C. and E. M. Lee. *Landsliding in Great Britain*. Department of the Environment. HMSO, London, 1994.).

landslide database (Jones and Lee, 1994) involved glacial tills taking the forms listed in Table 5.8. Complex landslides, the most common type of landslide in glacial tills, involve more than one mechanism and are difficult to predict.

Hungr et al. (2001) developed further definitions of flow type landslides, which are the second most common types of landslides in glacial soils. The descriptions (Table 5.9) are based on mechanisms, material properties and possible velocity with a number of subclasses to cater for a wide variety of flow or debris slides. Debris flows (Table 5.10) are formed of loose unsorted, low plasticity soils such as those given in Figure 5.7. These examples include glacial tills, which is consistent with the observations of Winter et al. (2013) in their study of the debris flows in Scotland during the winter of 2004.

Table 5.9 Description of landslides of the flow type highlighting the conditions that are necessary to trigger the debris flow slides in Table 5.7

Soil type	*Water content*	*Conditions*	*Velocity*	*Classification*
Silt, sand, gravel, debris	Dry, partially saturated, saturated	No excess pore pressure Limited volume	Various	Non-liquefied flow
Silt, sand, debris	Saturated at rupture surface	Liquefiable material Constant water content	Extremely rapid	Sand flow slide
Sensitive clay	Near liquid limit	Liquefaction *in situ* Constant water content	Extremely rapid	Clay flow slide
Clay	Near plastic limit	Slow movements Sliding	Slow	Earth flow
Debris	Saturated	Established channel	Extremely rapid	Debris flow
Mud	Near liquid limit	Fine-grained debris flow	Extremely rapid	Mud flow
Debris	Free water present	Flood	Extremely rapid	Debris flow
Debris	Partially or fully saturated	No established channel Relatively shallow steep source	Extremely rapid	Debris avalanche

Source: After Hungr, O. et al. *Environmental & Engineering Geoscience*, 7(3); 2001: 221–238.

Table 5.10 Types of materials found in landslides that can flow

Type of soil	*Character*	*Condition*	*Name*
Sorted (e.g. lacustrine)	Non-cohesive $I_P < 5\%$	Dry or saturated	Gravel, sand, silt
	Cohesive $I_P > 5\%$	$I_L < 50\%$ $I_L > 50\%$	Clay Sensitive clay
Unsorted (e.g. glacial tills, glaciofluvial)	Non-cohesive $I_P < 5\%$	Dry or saturated	Debris
	Cohesive $I_P > 5\%$	$I_L < 50\%$ $I_L > 50\%$	Earth Mud

Source: After Hungr, O. et al. *Environmental & Engineering Geoscience*, 7(3); 2001: 221–238.

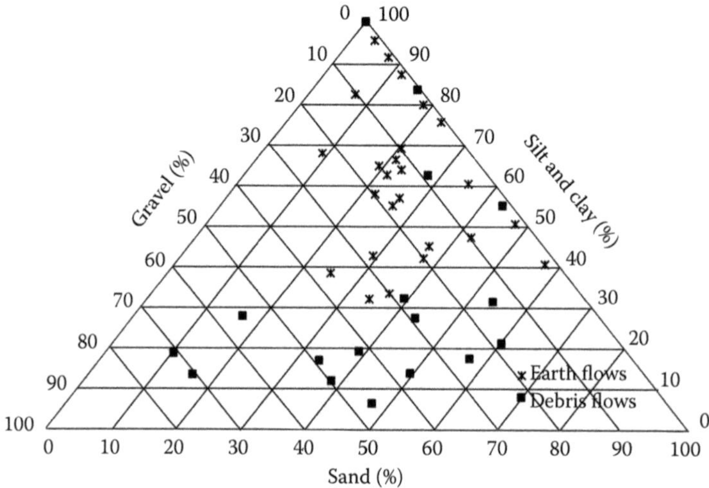

Figure 5.7 Composition of the matrix material in debris flows and earth flows showing that composite soils such as glacial tills are susceptible to flows. (After Hungr, O. et al. *Environmental & Engineering Geoscience*, 7(3); 2001: 221–238.)

5.3.1 Inland slopes

Inland slopes are triggered by rainfall events, construction or erosion depending on their location. Many inland natural slopes are dormant landslides, which have taken place since the Ice Age due to isostatic uplift and the establishment of the current drainage conditions. Failures are likely to be complex landslides or debris flow slides. It is necessary to take into account the spatial variation of the composition, fabric and structure of the glacial soils, the hydrogeological conditions and different failure mechanisms. Limit equilibrium methods are used to back analyse failures, but the complex nature of slopes formed of glacial soils requires more sophisticated analyses including geomorphological studies to appreciate the development of the failure (e.g. Misfeldt et al., 1991; Davies et al., 2014).

Misfeldt et al. (1991) combined limit equilibrium calculations with a seepage analysis to study a dormant landslide at Hepburn, Saskatchewan in order to establish the factor of safety. The original landslide of glacial till created a complex stratigraphy (Figure 5.8) at an average slope of 5°. While this may seem stable, there had been landslides in the region at these slopes. They demonstrated that a retrogressive dormant landslide can be analysed by using a series of stages ensuring that the groundwater conditions are adjusted between

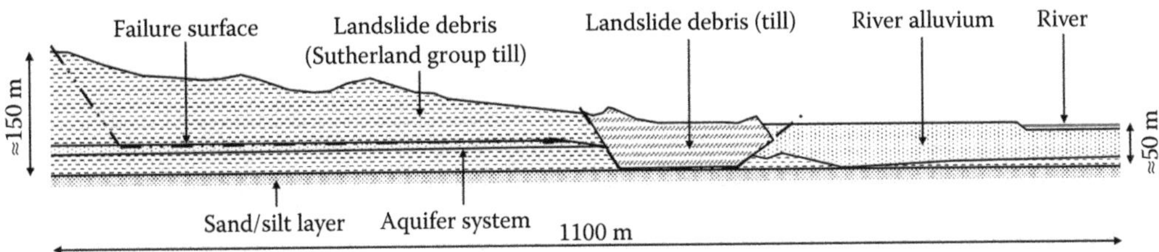

Figure 5.8 Cross section of the dormant landslide at Hepburn, Saskatchewan showing the complex stratigraphy and the influence the aquifer within the Sutherland Group had upon the failure surface. (After Misfeldt, G. A. et al. *Canadian Geotechnical Journal*, 28(4); 1991: 556–573.)

each stage to take into account the modified seepage conditions as a result of the change in stratigraphy.

Davies et al. (2014) found that they could demonstrate that pore water pressure was the most critical factor affecting the stability of a slope on the outskirts of Belfast, Northern Ireland but were unable to relate the changes in pore pressure to rainfall events. They used a coupled hydro-mechanical model to gain a better understanding of the characteristics of the landslide. The cross section of the slope (Figure 5.9) shows the complex nature of the glacial deposits formed of two layers of glacial till separated by medium dense sand, which contained a perched water table. The strength of the lower till was assumed to be residual because of the presence of relic shear surfaces. They used Shetran and FLAC-tp flow to model the pore pressure fluctuations taking into account vegetation, rainfall and movement. They were able to demonstrate using the data in Table 5.11 that elevated pore pressures in the underlying gravels were the main reason for pore pressure changes that triggered movement.

Fish et al. (2006) describe the geomorphology of a complex landslide at Cayton Bay, North Yorkshire. Figure 5.10 shows a possible cross section of the landslide that took the form of multiple notational slides in matrix-dominated tills, which are saturated clay-rich

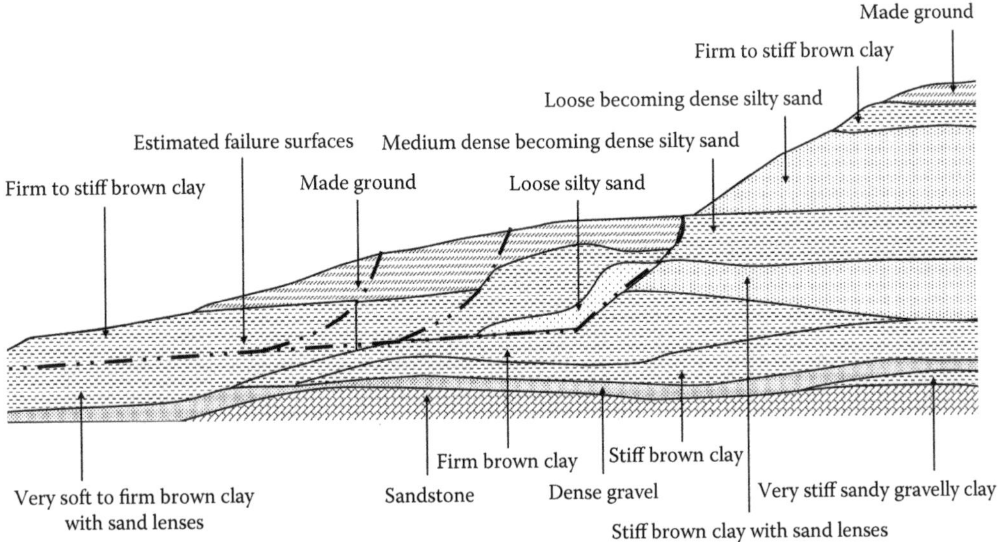

Figure 5.9 Cross section of the failing slope at Belvoir, Northern Ireland showing the complex ground conditions and possible slip surfaces. (After Davies, O. et al. *Engineering Geology*, 178; 2014: 70–81.)

Table 5.11 Properties of soil's (a) mechanical and (b) hydrological behaviour and (c) properties of the vegetation used in the Shetran and FLAC analysis of the Belvoir slope

(a) Material properties

Stratum	*Bulk density (Mg/m³)*	*Undrained shear strength (kPa)*	*Angle of friction*	*Cohesion (kPa)*
Made ground	1.88	11	20[a]	5[a]
Upper boulder clay	2.06	50–100	26[a]	10[a]
Malone sands	2.00	75–128	33[b]	0
Lower boulder clay	2.08	68–125	26	10
Glacial sand	2.00		36[b]	0
Basal gravel	2.31		37[b]	0

(b) FLAC-tp flow analysis

Stratum	*Bulk modulus (kPa)*	*Shear modulus (kPa)*	*Permeability (m/s)*	*Residual saturation*	*Porosity*	*Van genuchten n*	*Van genuchten α*
Made ground	2×10^5	1×10^4	5×10^{-9}	0.115	0.45	3.2	0.048
Upper boulder clay	2×10^3	1×10^3	5×10^{-8}	0.045	0.226	2.7	0.015
Malone sands	1×10^4	6×10^3	5×10^{-7}	0.115	0.45	3.2	0.048
Soft clay	2×10^5	1×10^4	5×10^{-8}	0.055	0.3	2.2	0.03
Lower boulder clay	2×10^5	1×10^4	5×10^{-8}	0.045	0.226	2.7	0.015

(c) Shetran modelling

Canopy storage capacity	m	15×10^{-4}
Fractional rate of change of drainage water storage	l/mm	3.7×10^3
Leaf drainage rate = C_s	mm/s	19×10^{-9}
Canopy resistance factor	s/m	RCF
Actual transpiration as a fraction of potential (number of pairs of fraction:soil moisture tension)	$-150\ m^{-0}$ $-3.33\ m^{-1}$	FET
Leaf area fraction given as a number of pairs (layer thickness:ratio of leaf area to area of element)	$1\ m^{-0}$ $4\ m^{-3}$	LAF

Source: After Davies, O. et al. *Engineering Geology*, 178; 2014: 70–81.

[a] Estimated from comparison with similar materials.
[b] Based on N_{60}.

sediments, leading to debris flow slides. It was slow moving with blocks of intact material 'floating' in the failed debris with the runout resulting in a sea cliff, which is continually triggering further movement, usually accompanied by rising groundwater levels as a result of antecedent effective rainfall. Climate change leading to more intense and frequent rainfall events together with rising sea levels is likely to increase the instability of the slope.

Debris flow slides do occur in mountainous regions in the United Kingdom, but they are on a much smaller scale (e.g. Winter et al., 2013). A review of landslides in 1988 (Jones and Lee, 1994) observed that many of the landslides were due to soil creep or solifluction and deep-seated slips are rare. However, such landslides have occurred in the past changing the

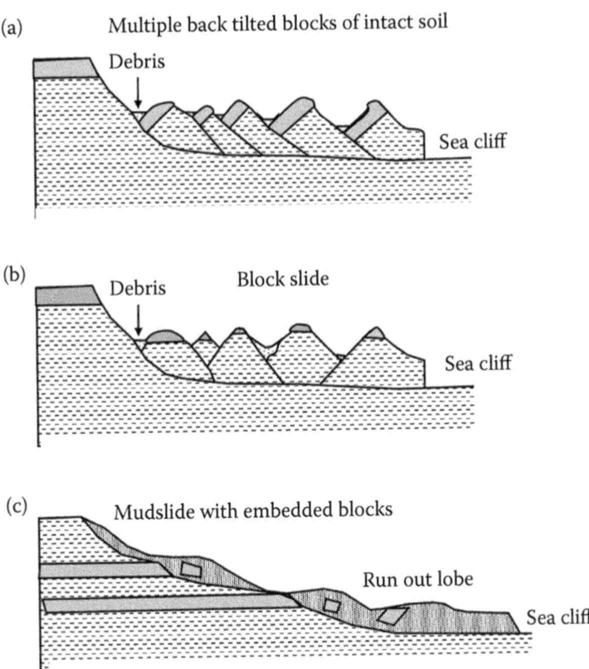

Figure 5.10 Three types of failure at Cayton Bay, Yorkshire involving (a) rotational slips, (b) block sliding or (c) mud slides. (After Fish, P. R. et al. Landslide geomorphology of Cayton Bay, North Yorkshire, In UK. *Proceedings of the Yorkshire Geological and Polytechnic Society*, 56(1); 2006: 5–14 (Geological Society of London).)

topography and hydrogeological conditions. Thus, any engineering project could reactivate these dormant landslides.

There are extensive deposits of glaciolacustrine clays in the French Alps, Scandinavia, Estonia and British Columbia, which have been studied because of the impact slope failures have on infrastructure and urban areas. Fletcher et al. (2002) describe two failures in these soils due to over-steepening, stress release and valley rebound associated with fluvial erosion. The two landslides initially exhibited behaviour typical of over-consolidated clays – slow to rapid intermittent displacements controlled by pore pressure changes. A feature (Figure 5.11) of these landslides is the slip plane associated with the laminations. The Slesse Park landslide, a complex landslide, exhibited features of multiple slip planes. The Attachie landslide was unusual in that a major debris flow slide followed the initial, multiple slides. A forensic analysis using the properties in Table 5.12 suggested that the mobilised angle friction was the residual angle of 14.5° parallel to the laminations and 17° oblique to the laminations with a phreatic surface 13.5 m below the ground surface. A dynamic analysis showed that the mobilised angle of friction, φ_b, given by

$$\tan\varphi_b = (1 - r_u)\tan\varphi_r \tag{5.4}$$

was 8.1°, that is, a reduction in strength triggered the flow slide; φ_b is the residual angle of friction. Fletcher et al. (2002) proposed three possible mechanisms: collapse of metastable structure, reduction in internal shear strength or microscopic brittleness. The first was discounted even though it is the usual cause of failure in other glaciomarine clays (e.g. Hutchinson, 1992; Leroueil et al., 1996). There was no evidence of metastable structures

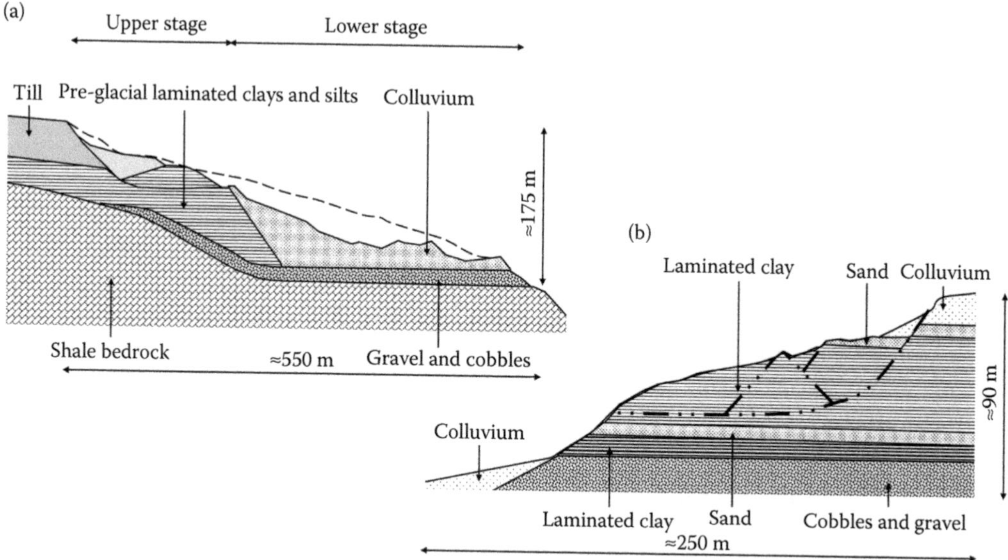

Figure 5.11 Cross section through the central part of the (a) Attachie landslide and (b) the Slesse Park landslide, British Columbia, showing the main stratigraphic units and the approximate shape of the rupture surface, based on an interpretation of surface morphologic features. (After Fletcher, L., O. Hungr, and S. G. Evans. *Canadian Geotechnical Journal*, 39(1); 2002: 46–62.)

Table 5.12 Properties of over-consolidated glaciolacustrine clays from two slides in British Columbia

		Attachie landslide		
Property	*Slesse park landslide*	*Till*	*Lake sediments (clay)*	*Lake sediments (silt)*
% clay	32 (26–65)	31 (19–37)	46 (28–68)	16 (7–27)
% silt	46 (34–74)	46 (25–63)	54 (32–72)	84 (73–91)
% sand	22 (8–32)			
I_L%	48 (10–32)	38 (33–44)	41 (27–59)	30 (NP–34)
I_P%	23 (10–32)	17 (8–22)	18 (8–34)	21 (NP–24)
PI%	24 (10–40)			9 (NP–12)
Activity	0.53 (0.17–0.83)			
Clay mineralogy	Illite, kaolinite, minor smectite			
w%	28 (20–38)	24 (16–35)	31 (20–37)	–
LI%	18 (–6 to 52) (intact) 31 (–38 to 97) (remoulded)	0.27 (–0.01 to 0.57)	0.19 (–3.18 to 0.59)	–
Unconfined compressive strength	490 kPa (intact) (60–200 kPa) (remoulded)			
Vane shear strength	200 kPa (intact) (40–60 kPa) (remoulded) (10–20 kPa) (residual)			

Source: After Fletcher, L., O. Hungr, and S. G. Evans. *Canadian Geotechnical Journal*, 39(1); 2002: 46–62.

within the soils which were deposited in fresh water conditions. Hutchinson (1987) proposed several mechanisms that could trigger rapid movement, which were possible at this location. Two conditions were necessary: a non-circular slip that is a kinematically inadmissible mechanism and a low shear strength slip surface overlain by a high strength brittle mass formed of cemented silt facies within the slope. The third mechanism requires till to be fractured either during deposition or as a result of previous movements. The latter was feasible at the Attachie site. Rain water collects in the fractures resulting in stiff intact fragments of till surrounded by loose material leading to liquefaction of the loose material. This was first noted by Terzaghi (1950) reporting on a landslide in the glacial till. Fletcher et al. (2002) showed that the second mechanism would lead to a 17% reduction in capacity; 33% for the third mechanism. The third mechanism explains why stiff, clast-dominated tills can exhibit both slow intermittent and sudden rapid flow.

A similar landslide to the Attachie landslide occurred in Oso, Washington after three weeks of intense rainfall. It began within a 200-m high slope (Figure 5.12) and led to a debris flow of unconsolidated glaciofluvial deposits and the underlying glaciolacustrine clays that moved up to 1.6 km (Keaton et al., 2014; Iverson et al., 2015). This is an area prone to landslides that had altered the hydrogeological conditions, intact strength and topography. Field stations up to 180 km from Oso were used to study the seismic signature of the landslide. Seismology is being used to establish the dynamics of large mass movements (e.g. Brodsky et al., 2003; Favreau et al., 2010; Schneider et al., 2010; Moretti et al., 2012; Allstadt, 2013; Yamada et al., 2013; Hibert et al., 2015). Field stations up to 180 km from Oso were monitored. Hibert et al. (2015) identified two events at Oso involving 6×10^6 to 7.5×10^6 m^3 of soil in the first event and about 15%–20% of that in the second event, which took place about 3 min later. The seismic analysis of the

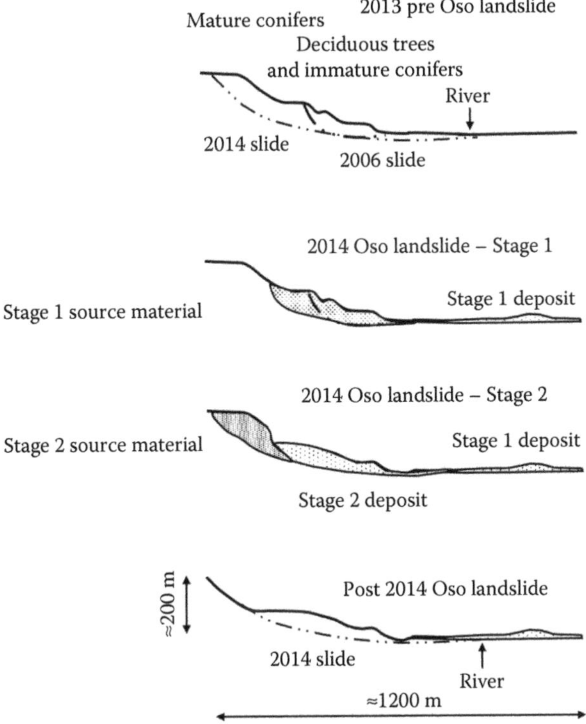

Figure 5.12 Development of the 2014 Oso landslide, Washington.

mass movement was consistent with the field observations and LiDAR mapping undertaken by Iverson et al. (2015).

Cruden et al. (1993) describe a Type 6 (US Geological Survey) landslide, which was triggered by a significant rainfall event which led to a flood eroding the toe of the cross section shown in Figure 5.13 creating a landslide with the slip surface passing beneath the river, thus blocking the river. This was a reactivated, retrogressive, compound landslide formed of about 45 million m^3 of glacial lake clays, till and pre-glacial lake clays. Landslides in this region are mostly translational failures, but in this case, the failure was associated with pre-glacial buried channels leading to a deep-seated slip and upthrust of the river bed creating a 20-m-deep, 3.5-km-long lake.

Landslides in mountainous regions can be affected by the bedrock, drainage conditions and composition in a more dramatic way. For example, Lebourg and Fabre (2000) describe the glacial till instability in the Pyrenees, where these soils are contained in channels within the bedrock. The lateral till covers the bedrock creating slopes between 10° and 50°. The angle of friction of the till varies between 20° and 35° with a cohesion of between 0 and 100 kPa. In this situation, there is a complex subsurface drainage system due to different glaciations fracturing the bedrock and creating braided channels. They used resistivity surveys to identify the channels and thus provide a geomorphological assessment of the risk of failure, which were often triggered by engineering works blocking a natural drainage channel.

Jongmans et al. (2009) also used the Trieves area of the French Alps to apply geophysical techniques to investigate landslides in glaciolacustrine clays, which extend over 300 km^2 and can be up to 200 m deep. Landslides change the geophysical characteristics of a deposit. They used seismic noise measurements, electrical resistivity tomography, P-wave seismic refraction tomography, S-wave seismic tomography and surface wave inversion to show that the S-wave velocity can be inversely correlated to the displacement rates with a distinct difference between the displaced material and the unaffected material, but displacement had little effect on the electrical resistivity.

Glaciolacustrine clays were formed in a fresh water environment in the Alps. The deposits in the Alps vary in thickness over relatively short distances because of the nature of the glacial-modified underlying bedrock. Giraud et al. (1991) studied a number of landslides in the region to the south of Grenoble, France (Trieves) to conclude that surficial slides are triggered by rainfall and melting snow to the extent that construction can take place only in the dry season. They identified two zones. The upper zone is prone to desiccation leading to discontinuities, which subsequently fill with rain water leading to unstable masses that can be

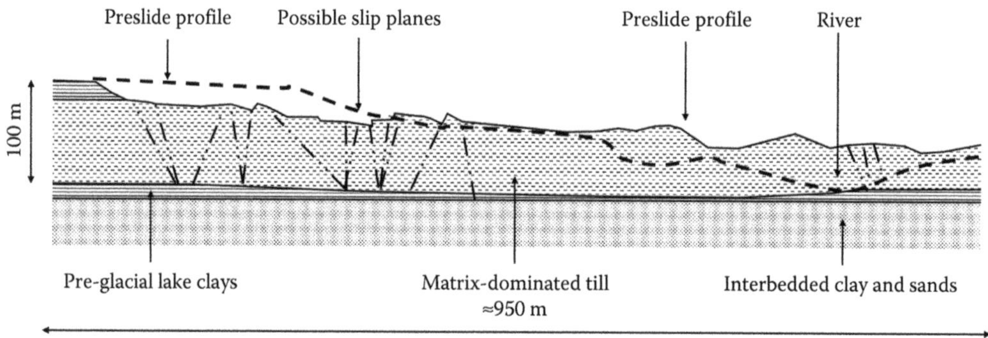

Figure 5.13 Section through the Type 6 Rycroft slide, Alberta. (After Cruden, D. M. et al. *Canadian Geotechnical Journal*, 30(6); 1993: 1003–1015.)

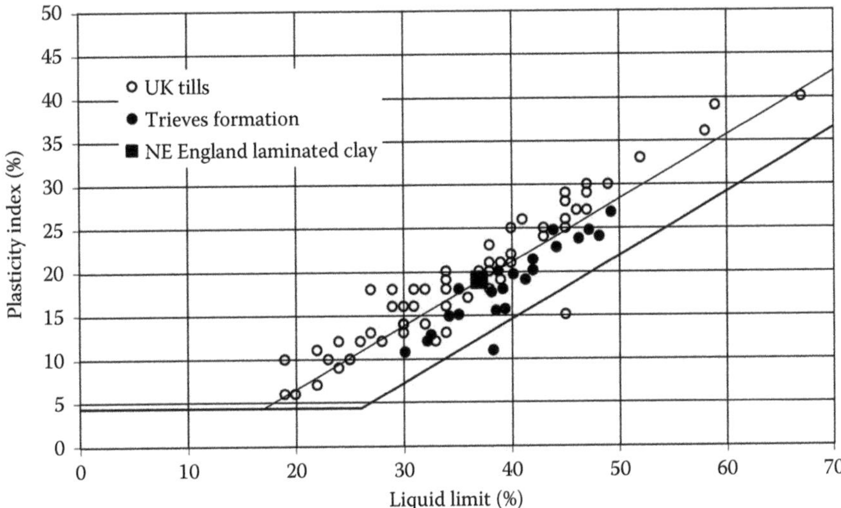

Figure 5.14 Consistency limits of the Trieves clayey formations compared with those of UK tills and the T-line. (After Vuillermet, E. *Les argiles glacio-lacustres du Trièves*. Mém. DEA Univ. de Grenoble, 1989: 55.)

stabilised by appropriate drainage; a more impermeable lower zone can lead to deep-seated failures and is more difficult to stabilise. The consistency limits of the deposits lie about the T-line (Figure 5.14), and since the plasticity index is low, it means that the near-surface layers very quickly reach the liquid limit when subject to seasonal changes in water content. Table 5.13, a summary of the properties, highlights the anisotropic nature. Figure 5.15 is an

Table 5.13 Mechanical properties of glaciolacustrine clays from south of Grenoble, France

Parameter	*Value*	*Test*	*Comment*
Peak angle of friction	23–26	CD triaxial	Across laminae
	20–21	CD triaxial	Along laminae
	22–23	CD direct shear	Along laminae
	20–23	Back analysis	
Residual angle of friction	18–19	CD direct shear	
	17–19	Back analysis	
Peak cohesion (kPa)	13–23	CD triaxial	Across laminae
	1–5	CD triaxial	Along laminae
	1–5	CD direct shear	Along/across laminae
	29–40	Back analysis	
Residual cohesion (kPa)	0	CD direct shear	Along/across laminae
	6–7	Back analysis	
Undrained shear strength (kPa)	46–68	UU triaxial	Across laminae
	30–42	UU triaxial	Along laminae
Dynamic viscosity (kPa s)	2.5×10^8	CD direct shear	Continues creep
	1.4×10^3	CD direct shear	Slip plane
Over-consolidation ratio	13–20	Oedometer	
Elastic modulus (MPa)	1–5	Pressuremeter	Disturbed
	10–60	Pressuremeter	Undisturbed

Source: After Giraud, P.A. et al. *Engineering Geology*, 31(2); 1991: 185–195.

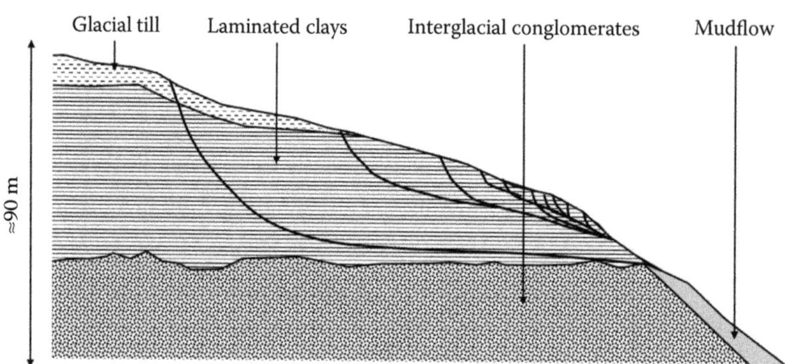

Figure 5.15 Idealised section across a slope in laminated clays in the Trieves clayey formations showing that modes of failure include deep-seated slips, intermediate non-circular slips and shallow slips. (After Giraud, P. A. et al. *Engineering Geology*, 31(2); 1991: 185–195.)

idealised cross section of the types of failure mechanisms, which include shallow, superficial slips, intermediate non-circular slips and deep-seated slips. Giraud et al. (1991) concluded that surface movements (0–5 m) are associated with mudflows because of the low plasticity index; planar movements (5–10 m) are associated with rainfall or meltwater entering shrinkage cracks and penetrating the silty layers.

Kohv et al. (2009) reported that the frequency of landslides in the river valleys cutting through the plains of Western Estonia formed of glaciolacustrine clays was increasing. This is attributed to increased storms, floods and inhabitation. The altitude of the plains varies from 2 to 15 m with 10–15-m-deep river valleys with side slopes up to 30°. A typical geological section, shown in Figure 5.16, is a marine sand overlying glaciolacustrine clay. Table 5.14 lists the slopes investigated, which shows that the majority occurred in the clays (A and C). They found that the thickness of marine sand governed the critical slope angle; 10° if the thickness was less than 3 m and 20° if it exceeded 3 m. Retrogressive failure occurred in the clays as shown in Figure 5.16. These clays are lightly over-consolidated with properties shown in Table 5.14. An analysis of the slope failure shown in Figure 5.16 suggests that the reason for an increase in the frequency of landslides was due to a reduction in groundwater abstraction leading to a rising water table and artesian pressures in the underlying till, also confirmed by Kohv et al. (2010) in their analysis of a landslide in the same region. This was accompanied by a reduction in shear strength to post-peak due to opening of discontinuities and toe erosion. An analysis of the complex slides led to the development of a landslide hazard zone (Table 5.15).

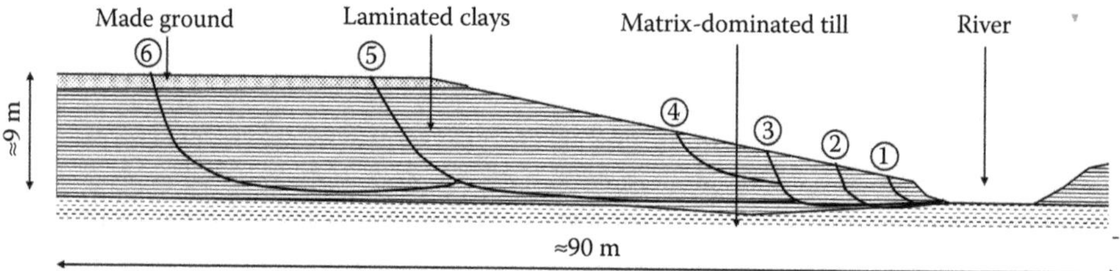

Figure 5.16 Suggested profile of the pre-failure cross section at the location of the Sauga slide showing possible retrogressive slides predicted from numerical analyses. (After Kohv, M. et al. *Geomorphology* 106(3); 2009: 315–323.)

Table 5.14 Location, morphological characteristics and classification of the landslides in glaciolacustrine clays, Western Estonia investigated by Kohv et al. (2009)

Location	*Coordinates*	*Length parallel to river channel (m)*	*Width perpendicular to river channel (m)*	*Scarp height (m)*	*Slope angle*	*Date*	*Group*
Audru-1	N58°25.26′ 24°20.09′	75	36	1.2	10	May 2002	A
Audru-2	N58°25.28′ E24°19.89′	8	4	0.3	?	Spring 2002	C
Audru-3	N58°25.28′ E24°19.89′	16	4	0.4	?	Spring 2002	C
Sauga-1	N58°25.72′ E24°26.41′	13	13	1.4	22	Spring 2002	A
Sauga-2	N58°26.40′ E24°29.28′	137	80	4	11	Dec 2005	A
Parnu-1	N58°22.70′ E24°36.29′	80	42	5.4	20	April 2002	B
Reiu-1	N58°21.60′ E24°36.21′	8	15	1.2	20	2000	A
Reiu-2	N58°21.21′ E24°37.09′	23	16	1.5	25	Feb 2002	B
Reiu-3	N58°19.39′ E24°36.93′	22	10	2.5	30	2000	B

Note: A: slides in glaciolacustrine varved clay covered by <3 m of marine sand/silt; B: slides in marine sand (sand layer >3 m thick); C: small (4–15 m) slides in glaciolacustrine clay directly in the bank of the flow channel.

Table 5.15 Typical properties of the glacial soils at the landslides investigated by Kohv et al. (2010) in Western Estonia

Sediment	*Cohesion (kPa)*	*Angle of friction*	*Unit weight (kN/m³)*	*Hydraulic conductivity (m/s)*	*Groundwater level (m)*
Sand	0	33	20	1×10^{-2} to 1×10^{-3}	−1
Varved clay	0	15	18.5	1×10^{-8} to 5×10^{-10}	–
Till	25	35	22	1×10^{-5} to 1×10^{-7}	1.5 masl

Transport corridors will often follow river valleys in mountainous areas and are therefore vulnerable to ground movements in areas of relict landslides triggered by rainfall. Huntley and Bobrowsky (2014) describe such a case in British Columbia along the Thomson River, the main link between Vancouver and the rest of Canada where landslides have known to occur for over 100 years. There is up to 150 m of glacial till, glaciolacustrine deposits and outwash gravels. Large rotational and retrogressive translational landslides took place as the Thompson River was formed leading to an unstable zone, which is affected by erosion, rainfall events and river levels. They are mostly slow moving complex slides. Reducing risk is either by avoiding the unstable area, by stabilising the area or by monitoring the landslide. The Ripley landslide, a cross section shown in Figure 5.17, was monitored. They concluded that the ground movements were a function of erosion by the Thompson River, the complex geology, fluctuating groundwater conditions and river level and possible anthropogenic activity. Peak movement occurred at the lowest river and groundwater levels in autumn and winter.

Landslide susceptibility assessment (e.g. Guzzetti et al., 2006) is a key tool in asset management in areas prone to instability. These are based on the topography (e.g. Erener et al.,

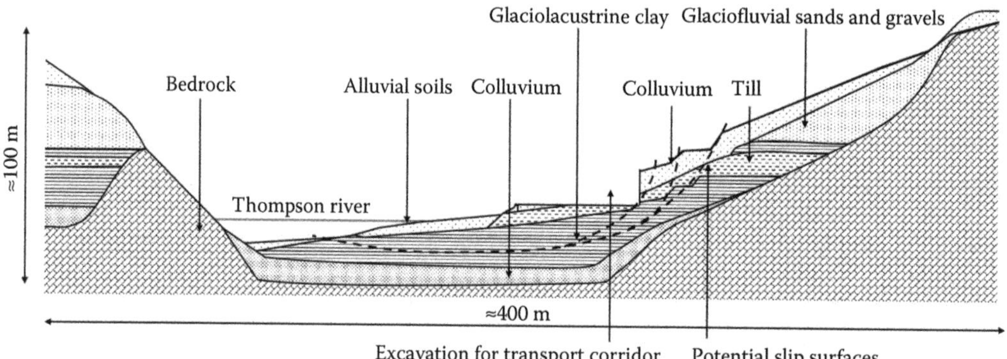

Figure 5.17 Hypothetical geological section across the Ripley Slide, British Columbia, showing potential failure planes influenced by the soil type and transport corridor. (After Huntley, D. and P. Bobrowsky. *Surficial Geology and Monitoring of the Ripley Slide, near Ashcroft.* Geological Survey of Canada, British Columbia, Open File 7531; 2014.)

2007) and soil type and thickness, bedrock type and land use (e.g. Quinn et al., 2009). Melchiorre and Tryggvason (2015) developed an algorithm for assessing the susceptibility of slopes in sensitive clays in Sweden, which often occur without warning and therefore a threat to transport corridors and urban areas. Sensitive clays are those clays with a remoulded strength of less than 0.4 kPa and a sensitivity of at least 50. The characteristics of failures include a flat terrain bounded by a steep slope adjacent to a river or ravine. Mass movements can occur in the flat terrain, so the slope angle is not a critical factor. Berggren et al. (1992) suggested that it is the relative angle of the slope height divided by the distance over which the slope is measured that is more relevant (cf. damage of buildings due to differential settlement) and the critical value was 1:10. They tested their algorithm on the Goba River valley glaciomarine deposits. The area was divided into a series of cells; a decision is taken as to whether the soil is susceptible and whether the relative angle between adjacent cells exceeds the critical value taking into account non-critical features such as ditches. The data included the soil type, depth to bedrock, quick clay susceptibility index (the probability that the soil is a quick clay) and landslide scarps. Persson et al. (2014) used a multi-criteria evaluation to assess the probability of finding quick clay, which was based on stratigraphy, potential for groundwater flux, relative infiltration capacity and geomorphological conditions for high groundwater flux.

Databases of landslides have been developed across Europe and North America. For example, the British Geological Survey (Pennington et al., 2015) has produced a database of 17,000 landslides, which uses social media to build the database providing greater coverage. The database has been used to study the correlation between precipitation and landslide events (Figure 5.18). The spatial extent of the database has been used to create a domain map (Figure 5.19), which relates the types of landslides to the underlying geology and topography. The eight domains (Table 5.16) are further subdivided according to the local geology and geomorphology. The database has been used to develop a landslide susceptibility rating described in Table 5.17, which recommends the level of investigation for planning and engineering based on a geomorphological study of the area. This database and others across Europe have been used to develop an indicative view of periods of landslides in Europe highlighting the extent of landslips since the Ice Age (Figure 5.20). A concern is that further climate change will reactivate these dormant landslides (Cooper, 2007).

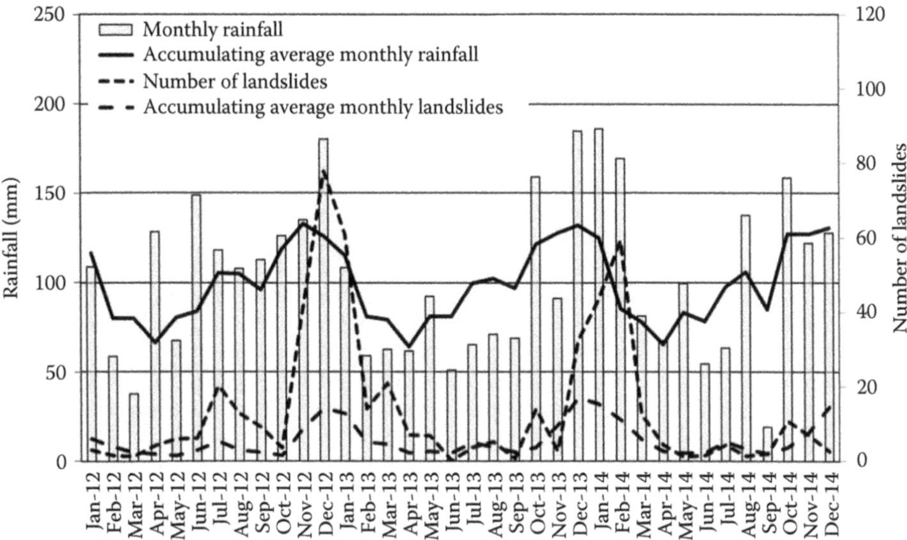

Figure 5.18 Comparison between intensity of landslides and monthly rainfall events in the United Kingdom suggesting there is a possible correlation with the accumulated monthly average. (After Pennington, C. et al. *Geomorphology*, 249; 2015: 44–51.)

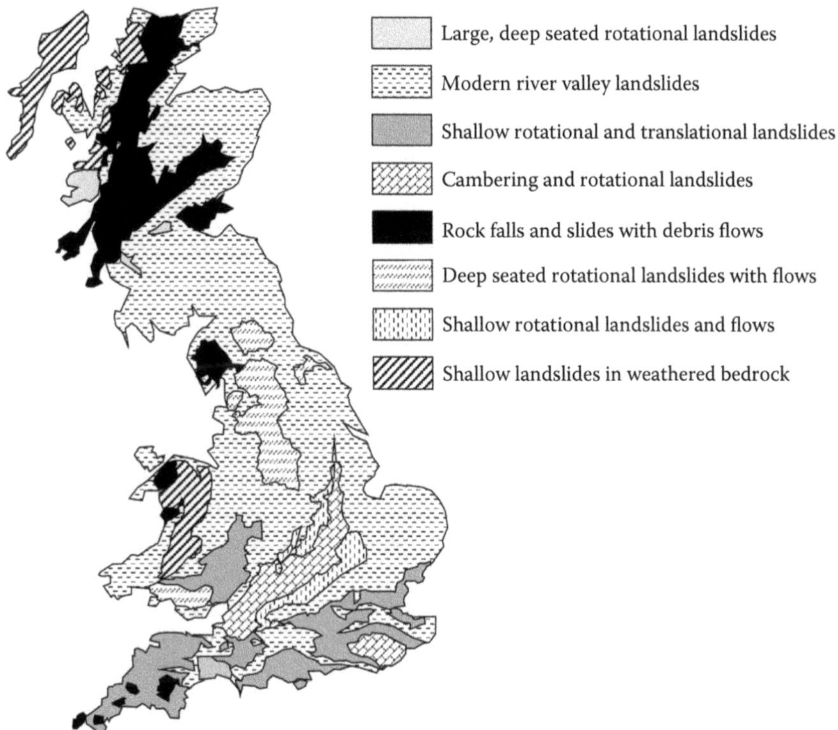

Figure 5.19 UK landslides domain map, which relates the types of landslides to the underlying geology and topography. (After Dashwood, C. et al. *GeoSure Version 7 Methodology: Landslides Slope Instability*. Internal Report, IR/14/014. British Geological Survey, Nottingham, UK, 2014: 31.)

Table 5.16 Landslide domain name and summary which, together with Figure 5.19, shows that the majority of landslides are domain Type 2 and 6

Domain	*Domain name*	*Summary*
1	Large, deep-seated rotational landslides	Large, deep-seated rotational landslides – lithologically and structurally controlled
2	Modern river valley landslides	Landslides predominantly controlled by the presence of modern river valleys. Gently undulating low-relief landscapes, spreads of weathered till, drumlins, dissected by palaeo and modern river valleys with associated valley side rotational and planar landslides
3	Shallow rotational and translational landslides	Shallow rotational and planar landslides controlled by processes in a weathered zone, both in bedrock and in superficial veneer, involving contemporary landsliding as well as more ancient landsliding
4	Cambering and rotational landslides	Cambering and rotational landslides involving clay-rich bedrock leading to, for example, spreads in the Weald and cambering and rotational landslides in the Cotswolds
5	Rock falls and slides with debris flows	Bedrock controlled rock slope failures, including falls, toppling/ spreading, rock creep, translational landslides occurring in harder bedrock with V-shaped valleys and including some large rock slope failures in western Scotland. Large rock slope failures in an eroded, rounded bedrock geomorphology with U-shaped valleys. Flows and landslides in superficial deposits are also present
6	Deep-seated rotational landslides with flows	Deep-seated rotational landslides, often degrading into flows, in plateau and valley landscapes mainly where competent bedrock is overlying incompetent bedrock. Landslides also occur in till and head deposits, mainly along river valleys sides
7	Shallow rotational landslides and flows	Escarpment cap-rock related landslides, mostly involving shallow rotational landslides; occurring in harder or more resistant rock type overlying a weaker or less resistant rock type. The majority of landslides are shallow rotational features with a strong flow element
8	Shallow landslides in weathered bedrock (regolith)	Bedrock controlled landslides in thin regolith in landscapes that are glacially eroded and smoothed with ice scoured hard bedrock, or involve hard bedrock. Landslide occurrence controlled by the presence of weathered bedrock material and thin superficial deposits

Source: After Dashwood, C., D. Diaz Doce, and K.A. Lee. *GeoSure Version 7 Methodology: Landslides Slope Instability*. Internal Report, IR/14/014. British Geological Survey, Nottingham, UK, 2014: 31.

5.3.2 Coastal cliffs

The stability of coastal cliffs is a particular problem because of the effect of toe erosion and infiltration of salt water. Unlike inland landslides, many of these are first-time landslides. The Holderness Coast, East Yorkshire, is the fastest eroding coastline in Europe at between 1.5 and 2 m/year. This is due to a combination of relatively softer glacial tills compared to the chalk headland to the North, the longshore drift which removes the beach sediment and the impact of waves particularly during storms. The 45-km coastline is up to 20 m high and has been extensively studied. Pickwell (1878) gave a detailed account of the state of the coast and the engineering works to protect the various villages, tracing the history of the coast back to about 1100 AD. Groynes were used in the nineteenth century to retain the beaches in front of coastal villages, thus preventing toe erosion and maintaining the load on the toe. The Cowden test bed site, referred to in Chapter 4, is a typical profile of weathered till overlying unweathered till, which becomes stiffer with depth (Figure 5.21a). The clay tills are separated by sand and gravel layers, which affect the pore pressure distribution. Note the stone-free soft brown clay till at the toe of the cliff. The sand and gravel layers act as drains, which means

Table 5.17 Landslide susceptibility rating

Hazard rating	*Planning issues*	*Engineering issues*
No indication of landslides in the area	No constraints to land use	Normal desk study and walk-over study
Slope stability problems unlikely to be present on the site though evidence of stability problems in the area	No constraints to land use but need to assess possible effects of slips	Normal desk study, walk-over study and an assessment of overall stability
Slope stability problems may be present on the site and evidence of stability problems in the area triggered by extreme events	Implications on stability of excavations and changes to drainage conditions	Ground investigation should address stability problems taking into account effects of excavations, drainage conditions and seasonal changes
Slope stability problems are present on the site and evidence of stability problems in the area triggered by extreme events	Implications on stability of excavations and changes to drainage conditions which means mitigation measures should be considered	Ground investigation should address stability problems taking into account effects of excavations, drainage conditions and seasonal changes leading to possible design of mitigation measures
Slope stability problems are present on the site and evidence of stability problems in the area triggered by moderate events and possible erosion	Permission to develop the land may take into account possible stability assessment, mitigation measures and remedial works	Ground investigation should address stability problems taking into account effects of excavations, drainage conditions and seasonal changes leading to design of mitigation measures
Slope stability problems are present on the site and evidence of stability problems in the area triggered by minor events including erosion	Permission to develop the land must take into account possible stability assessment, mitigation measures and remedial works to assess whether the site is suitable	An initial ground investigation is necessary to assess the suitability of the site. Ground investigation should address stability problems taking into account effects of excavations, drainage conditions and seasonal changes leading to design of mitigation measures

Source: After BGS. British Geological Survey, 2015. www.bgs.ac.uk.

that the pore pressure is constant below 3 m. Further inland the pore pressure profile was hydrostatic, varying seasonally by up to 1 m. Butcher (1991) monitored the movement of the cliff using inclinometers, shear tubes and electro-optical distance measuring. Figure 5.21b shows the observed complex mechanism, which was triggered by the pore pressure regime and the removal of the base load and the reduction in strength of the basal clay. A forensic

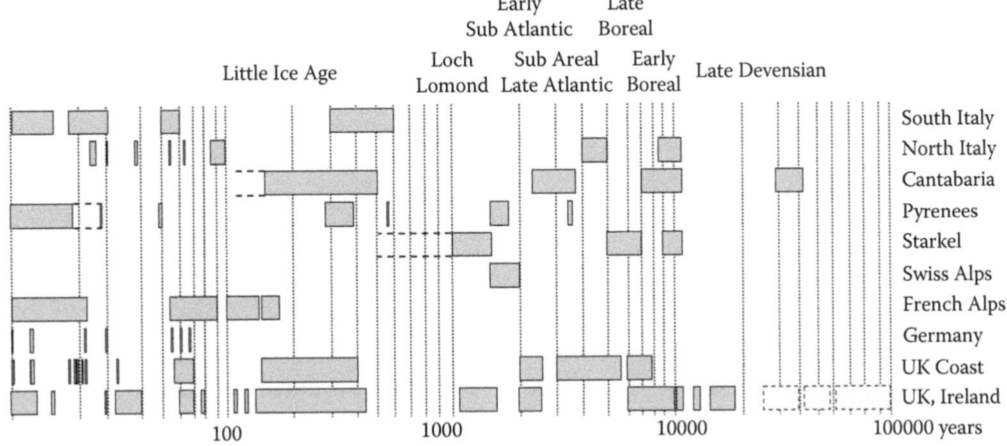

Figure 5.20 Indicative periods of major landslide activity in Europe highlighting the extent of landslips since the Ice Age. (After Brunsden, D. and M.-L. Ibsen. Gustav Fisher Verlag, Stuttgart, 1997: 401–407.)

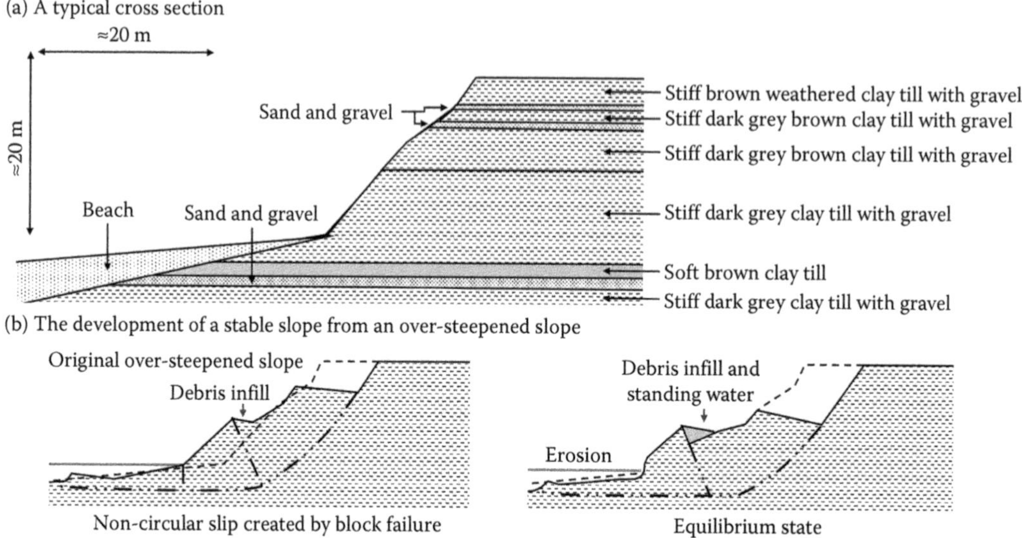

Figure 5.21 Cowden test bed site showing (a) the soil profile at the Cowden cliff site and (b) the failure mechanism derived from instrumentation results. (After Butcher, A. P. 39. *Slope Stability Engineering Developments and Applications: Proceedings of the International Conference on Slope Stability.* Institution of Civil Engineers, Thomas Telford Publishing, 1991: 271–276.)

analysis showed that the mobilised strength in the basal clay was the post-peak strength providing evidence of progressive failure; the peak strength was mobilised in the overlying glacial tills. The analysis suggested that beach erosion was the most critical factor.

The episodic nature of the cliff recession, of the form shown in Figure 5.22, was determined from 114 locations by simply measuring the distance from a datum to the cliff edge, and GPS data to show the average rate of recession (Figure 5.23). Table 5.18 shows that significant erosion takes place once every 11 years and the median extent of erosion takes place every 2 years. The periodic recession appears to comprise four phases: above-average recession preceding a period of high recession followed by another above-average recession and then below-average recession. This behaviour has been attributed to either longshore drift periodically removing the beach or to toe erosion and pore pressure transient conditions. The types of landslides include rotational, wedge and planar failures, falls, topples and mud flows, which are influenced by the fabric and structure of the till, the groundwater and toe erosion. Anthropogenic interventions have a local effect as they can both increase and reduce the rate of recession in adjacent cliff sections depending on their alignment and their relation to the longshore drift.

Appreciating the impact the local geological, geotechnical, geomorphological and hydrogeological conditions have upon regional cliff recession is necessary to gain a better understanding of the periodic cycle of recession (Quinn et al., 2010). The Holderness Coast is formed of three distinct tills: Basement Till, Skipsea Till and Withernsea Till, which have similar characteristics. Quinn et al. (2010) undertook a detailed study of six sites using GPS and TLS surveys. The slopes were modelled using FLAC, a finite difference software, to develop an empirical model, which showed that the landslide-induced recession is a function of the cliff height, the pre-failure slope angle, the till type in the lower half of the slope, the beach level, the level of the phreatic surface and the presence of laminated clay in the toe of base of the cliff. Quinn et al. (2010) found that this empirical model (Figure 5.24) could predict the 0.1-m contour of recession derived from numerical modelling to within 5.5%.

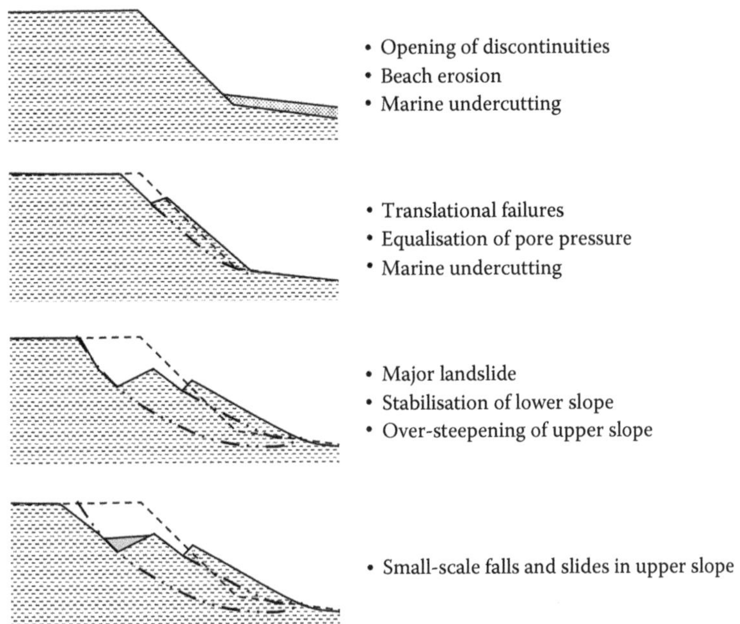

Figure 5.22 Episodic nature of coastal recession using observations at 114 locations along the Holderness Coast showing how beach erosion and marine undercutting leads to translational failures followed by major slips and subsequent planar failures due to over-steepening. (After Quinn, J. D. et al. 2010. Identifying the behavioural characteristics of clay cliffs using intensive monitoring and geotechnical numerical modelling. *Geomorphology*, 120(3); 107–122.)

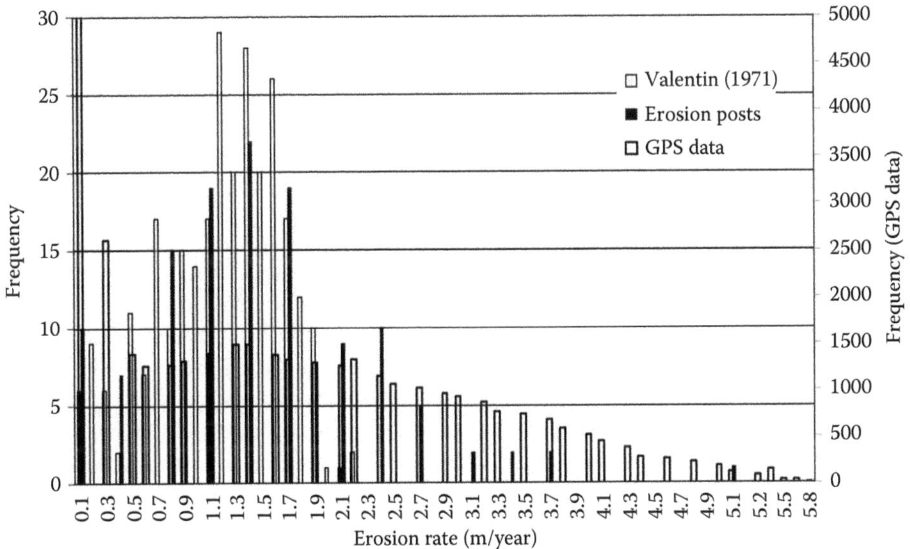

Figure 5.23 Distributions of the average values of recession of the Holderness Coast using data presented. (After Quinn, J. D., L. K. Philip, and W. Murphy. *Quarterly Journal of Engineering Geology and Hydrogeology*, 42(2); 2009: 165–178.)

Table 5.18 Average periodicities of values higher or lower than percentile values for the frequency distributions of annual measurements of erosion for Erosion Post dataset showing that significant erosion takes place once every eleven years and the median extent of erosion takes place every 2 years

Percentile	*Average periodicity of values greater than the percentile (years)*	*Average periodicity of values less than the percentile (years)*
10	1.6	0
15	1.6	0
25	1.6	0
50	2.1	2.0
75	4.1	1.3
85	7.3	1.2
90	11	1.1

Source: After Quinn, J. D. et al. *Quarterly Journal of Engineering Geology and Hydrogeology*, 42(2); 2009: 165–178.

Note: The average 25th percentile value for each dataset is 0 m; therefore, results below this value are identical.

They found that deep-seated failures occurred if the cliff height was more than 15 m and failures on lower cliffs were due to structural weaknesses. Abrasion and erosion were the prime causes of recession for cliffs less than 7 m high. They also found that the relationship between the cliff recession and the beach wedge area was similar to the glacial cliffs along the Norfolk Coast described by Hobbs et al. (2008) though the empirical model is site specific.

Hobbs et al. (2008) undertook terrestrial surveys and numerical studies of the Norfolk Coast at three representative sites to understand the factors influencing coastal recession. They used LiDAR and GPS techniques to create 3D models of the cliffs formed of Lowestoft Till and Cromer Tills. The numerical analyses used FLAC/slope (Itasca, 2016), which does not require a failure mechanism to be identified; hence, the failure mechanism will appear as contours of displacement. The factor of safety is calculated using the principle of strength reduction in which the strength of the soil is gradually reduced until the analysis does not converge to a solution. This means that the slope has failed and the factor of safety, *F*, is defined as

$$\frac{\tau}{F} = \frac{c'}{F} + \frac{\tan\varphi'}{F} \quad (5.5)$$

where τ is the strength of the soil.

The groundwater conditions were estimated and the geotechnical characteristics were taken from tests or databases of geotechnical parameters. They found that the recession in winter was driven by rising groundwater and wave action and the summer recession by wave action alone at Happisburgh (Figure 5.25). Deep-seated rotational and translational slides leading to debris flows (Figure 5.26) occur to the east of Happisburgh where the stratigraphy is more complex. Hobbs et al. (2008) found that numerical models could predict possible failure mechanisms but the predicted cliff recession was greater than that observed, which was assumed to be due to soil suction delaying the recession. However, the quality of the geotechnical data and the inability to model the groundwater conditions, which included perched water tables, also affected the results.

Clark and Fort (2009) undertook a review of techniques used to stabilise soft cliffs around the United Kingdom. This included glacial deposits along the east coast. Stabilisation measures included drainage, soil reinforcement retaining structures and slope support, as shown

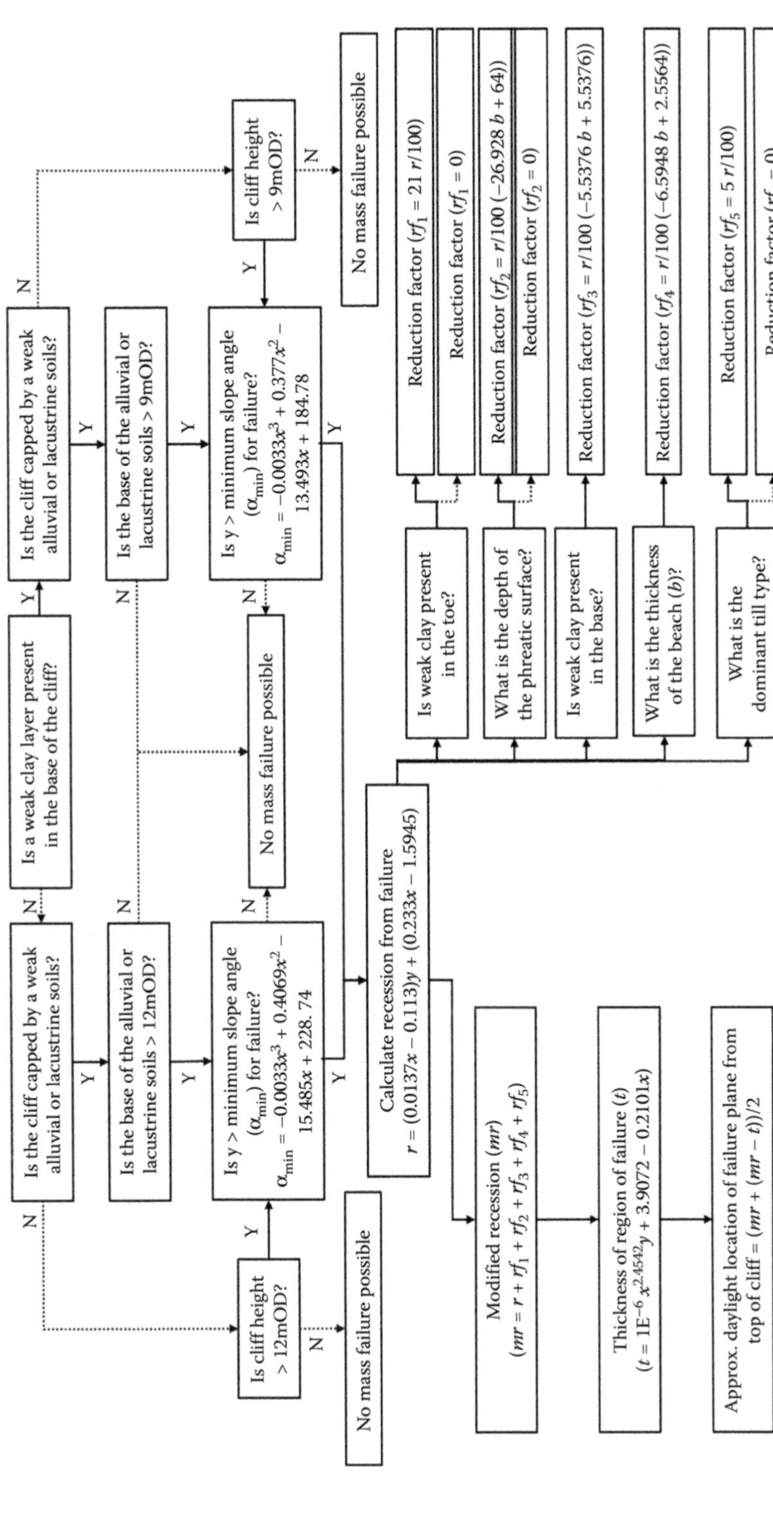

Figure 5.24 Empirical method to predict the recession rate of the Holderness Coast based on numerical and geomorphological studies. (After Quinn, J. D. et al. 2010. *Geomorphology*, 120(3); 107–122.)

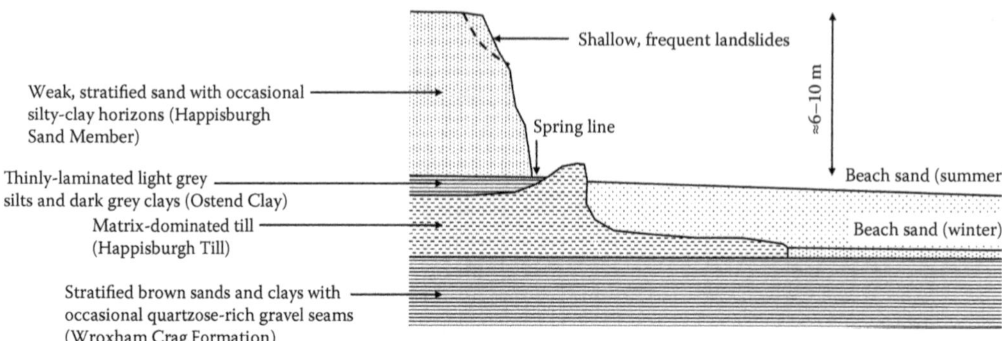

Figure 5.25 Geological profile at Happisburgh, Norfolk showing the importance of the beach in maintaining cliff stability. (After Hobbs, P. R. N. et al. *Slope Dynamics Project Report: Norfolk Coast (2000–2006)*. British Geological Survey Research Report, OR/08/018, 2008: 166.)

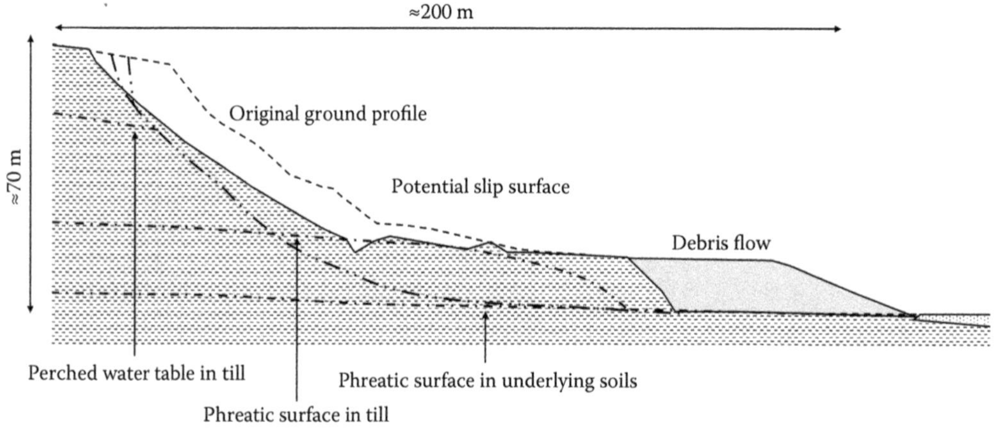

Figure 5.26 Landslide at Cromer, Norfolk showing a potential non-circular slide leading to debris flow and the water levels in different stratum with a possible perched water level near the top of the cliff. (After Hutchinson, J. N. *Coastal Landslides in Cliffs of Pleistocene Deposits between Cromer and Overstrand*, Norfolk, England. Building Research Establishment, Building Research Station, Watford; 1976.)

in Table 5.19. Erosion of coastal cliffs formed of glacial clay tills is prevented by a combination of drainage and slope reinforcement to stabilise the slope and toe protection including beach loading and retaining structures to prevent erosion.

Landslides that be considered inland failures but can be affected by the sea due to their runout can be found at Runswick Bay, North Yorkshire. Booth (2013) provides details of stabilising the village of Runswick Bay on the Yorkshire Coast, which is a known area of instability since 1682 when the original village was lost due to a major landslide. A monitoring programme was instigated including beach profile surveys, topographic surveys, cliff top recession, wave data, sea bed characterisation, aerial photography and walk-over surveys. A risk assessment based on the surveys and geomorphological mapping was used to assess remedial measures. They were unable to link rainfall to instability and recommended reprofiling, deep drainage to lower the water table, a bored pile portal frame and shear keys fixed into the intact rock to provide structural stability while the other remedial measures took effect and rock armour to prevent erosion.

Table 5.19 Techniques used to stabilise coastal cliffs in North Yorkshire, United Kingdom

Category	*Type*	*Robin hoods bay*	*Runswick bay*	*Holbeck, scarborough*	*Whitby*	*Haggerlythe, whitby*
Earthworks	Regrading	√	√	√	√	√
	Buttressing	√		√	√	√
	Toe weighting		√	√		
Walls	Concrete walls	√				
Reinforcement	Shear keys	√		√		
	Piles	√	√			√
	Reinforced soil	√	√	√		√
	Anchors	√				
Drainage	Trench drains	√	√	√	√	√
	Counterforts			√	√	√
	Drainage blankets			√	√	√
	Sub-horizontal drains	√	√			
	Pumped wells	√				
Erosion control	Revetments	√	√	√	√	√
	Beach replenishment					√
Monitoring		√	√	√	√	

Source: After Clark, A. R. and S. Fort. *Proceedings of the Institution of Civil Engineers-Geotechnical Engineering*, 162(1); 2009: 49–58.

Bridges et al. (2015) describe an ecosystem-based approach to mitigate against coastal storms using natural and nature-based features to control erosion. Coasts formed of glacial soils are classed as primary coasts because they are soft relative to rock and produce sediment that provides protection to the toe of the cliff. Longshore drift, such as that found at Holderness, removes the sediment, thus exposing the cliffs to wave action. Other factors

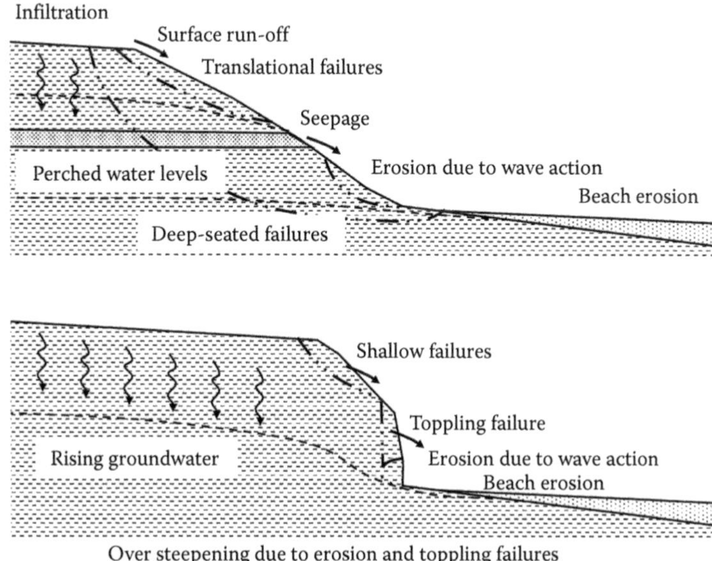

Figure 5.27 Factors that contribute to cliff instability and erosion of glacial till coastal cliffs.

Table 5.20 Relative risk metrics for coastal vulnerability index

Factor	*Very low*	*Low*	*Moderate*	*High*	*Very high*
Mean elevation (m)	>30.0	20.1–30.0	10.1–20.0	5.1–10.0	0.0–5.0
Geology	Igneous rock	Metamorphic rock	Sedimentary rock	Gravel; glacial till	Sands, silts and clays
Geomorphology	Fjords Rock cliffs	Indented coasts Medium cliffs	Low cliffs Salt marsh Coral reefs	Beaches Lagoons Alluvial plains	Barrier beaches Mudflats Deltas
Subsidence trend (mm/year)	<−1.0 (land rising)	−1.0 to 1.0	1.1–2.0	2.1–4.0	>4.0 land sinking
Mean shoreline displacement (m/year)	>2.0 accretion	1.1–2.0	−1.0 to 1.0	−2.0 to −1.1	<−2.0 erosion
Mean tidal range (m)	<1.0 microtidal	1.0–1.9	2.0–4.0	4.1–6.0	>6.0 macrotidal
Maximum significant wave height (m)	0.0–2.9	3.0–4.9	5.0–5.9	6.0–6.9	>6.9

Source: After Gornitz, V. and P. Kanciruk. *Assessment of Global Coastal Hazards from Sea Level Rise*. No. CONF-8907104-1. Oak Ridge National Lab., TN (USA), 1989; Gornitz, V. et al. *Vulnerability of the US to Future Sea Level Rise*. No. CONF-910780-1. Oak Ridge National Lab., TN (USA), 1991; Gornitz, V. M. and T. W. White. *A Coastal Hazards Database for the U.S. East Coast*. ORNL/CDIAC-45, NDP-043 A. Oak Ridge National Laboratory, Oak Ridge, TN, 1992; Bridges, T. S. et al. *Use of Natural and Nature-Based Features (NNBF) for Coastal Resilience*. Engineer Research and Development Center, Vicksburg MS Environmental Lab, Vicksburg, MS, USA, 2015.

that contribute to coastal instability include erosion due to surface run-off, seepage and wind (Figure 5.27). Table 5.20 compares the risk of glacial coasts compared to other types of coasts, showing that glacial coasts are high risk. The rate of recession of the Holderness Coast is consistent with those risk metrics. Glacial coasts can include the following:

- Drowned glacial erosion coast includes partially submerged glacial features in the form of islands and beaches, marshes and scarps.
- Glacial depositional coast includes irregular shorelines, indented river valleys, sand and gravel beaches and barrier spits.

Understanding coastal erosion is key to managing the asset allowing mitigation measures to be targeted. Chase and Kehew (2000) studied six sites on the shores of Lake Michigan formed of glacial soils. They compared results of limit equilibrium analyses with the results of balanced cross sections used in structural analysis of displaced rock bodies to generate the critical failure surface from the observed displacements. Detailed monitoring and analysis of the sites showed that

- Cliffs formed of sand or clay were more stable than those formed of interlayered sand and clay.
- At these sites, the cliffs of saturated clay created translational failures.
- Deep-seated failures occurred in the interlayered deposits.
- Cliff degradation was associated with wave action in the autumn, freezing of the surface preventing natural drainage causing an increase in pore pressure and spring thaw releasing groundwater thus maintaining high pore pressures.

Bridges et al. (2015) suggest a number of metrics to describe coastal characteristics and the external factors that contribute to failure (Table 5.21) that could affect structures, which includes the elevation and distance of the structure relevant to the shoreline, the land mass

Table 5.21 Assessing vulnerability of glacial coasts to erosion

Metric	*Drowned glacial erosional coastal landscape*	*Glacial depositional coastal landscape with bluffs*	*Glacial depositional coastal landscape without bluffs*
Coastal characteristics that affect the vulnerability of an inland structure			
Average elevation at point of interest (m)	Primary driver of coastal vulnerability to storms and should always be included as a metric		
Max elevation between point of interest and nearest shoreline (m)	Considers the presence of protective features such as large rock outcroppings, levees, etc.	Considers the presence of protective features such as hills, levees, etc.	
Shoreline sediment median grain size (mm)	Used as a measure of the erodibility of the coastline and the ability of the shoreline to recover (e.g. gravel vs. sand vs. clay)		
Distance from point of interest to nearest shoreline (m)	Accounts for presence of the landmass, which dissipates wave energy, slows surge propagation and provides a buffer for erosion. Shoreline could be considered at multiple levels such that sub-tidal features could be accounted for		
Land cover type along distance from point of interest to nearest shoreline (Manning n)	The coverage on a landmass also influences wave energy dissipation, surge propagation, and erodibility		
Open-water fetch from nearest shoreline (km)	In the absence of wave and water level data, can be used, along with wind data, as an indicator of the wave energy and storm surges a shoreline may be subject to		
Nearest shoreline change variance (m)	A proxy for measuring the storminess along a sandy coastline, particularly as an indicator of how storminess affects the erosion hazard		
Distance from point of interest to bluff edge (m)		Applicable for locations on profile landward of bluff edge. Locations in proximity to the edge of the bluff are more vulnerable to the erosion hazard than are location more distant	
Long-term nearest shoreline change rate (m)	An eroding shoreline is more vulnerable than an accreting shoreline, and recovery of a beach along a chronically eroding shoreline is less likely		
Average max elevation between nearest shoreline and open coast (m)	Accounts for the presence of a landmass, such as a barrier island, offshore the nearest shoreline		

(*Continued*)

Table 5.21 (Continued) Assessing vulnerability of glacial coasts to erosion

Metric	*Drowned glacial erosional coastal landscape*	*Glacial depositional coastal landscape with bluffs*	*Glacial depositional coastal landscape without bluffs*
Landmass area between nearest shoreline and open coast (km^2)	Accounts for the presence of a landmass, such as a barrier island, offshore the nearest shoreline		
Beach berm width (m)		Beach protects the toe of the bluff from wave impact and erosion. The wider the beach the less vulnerable the toe of the bluff and therefore the less vulnerable locations landward of the bluff edge are to coastal storms	Beach protects the upland from wave impact and erosion. The wider the beach the less vulnerable it is and the less vulnerable the upland is to coastal storms
Beach slope (%)		Influences wave run-up and therefore the likelihood that the toe of a bluff will be subjected to wave run-up impact	Influences wave run-up and therefore the likelihood of flooding from wave run-up
Coastal slope (%)	In the absence of water level data, may be used as an indicator of storm surges that an area may experience during a storm		
Open coast shoreline sediment median grain size (mm)	Used as a measure of the erodibility of the coastline and the ability of the shoreline to recover (e.g. gravel vs. sand vs. clay)		
Triggers that could lead to coastal recession			
Max still-water elevation (m)	Primary driver of coast coastal vulnerability to storms. Application of statistically derived values allows for the consideration of storminess over the temporal reference of interest		
Max wave height (m)	Important driver of coast coastal vulnerability to storms. Application of statistically derived values allows for the consideration of storminess over the temporal reference of interest		
Max wave run-up elevation (m)	Not typically available directly from data, but may be calculated based on other available date (e.g. offshore wave height, period and beach slope). May be the primary source of flooding on some coasts		
Max wind speed (m/s)	Should be considered as a damage driver and can also be used to estimate other metrics (such as wave heights) in the absence of that data		
Relative sea level rise (mm/year)	Important consideration for vulnerability assessments with a long temporal reference		
Tidal range (m)	Shorelines with large tidal ranges typically dissipate more wave energy		

Source: After Bridges, T. S. et al. *Use of Natural and Nature-Based Features (NNBF) for Coastal Resilience*. Engineer Research and Development Center, Vicksburg MS Environmental Lab, Vicksburg, MS, USA, 2015.

and cover, the cliff geometry, the rate of recession, the open-water fetch, the beach geometry and the grain size of the sediment, beach and cliff. The factors that could trigger instability related to the free water include the high water level, the maximum wave height, the maximum wave run-up and elevation, the tidal range and potential sea level rises. The groundwater conditions within the cliff including the presence of perched water levels, more permeable layers, infiltration and the opening of discontinuities due to erosion increasing the mass permeability, also have to be considered.

5.3.3 Recommendations

Natural slopes failures can be divided into those that fail due to erosion, pore pressure changes or a reduction in strength. They can be first-time slides, reactivated slides or reactivated dormant slides. A geomorphological investigation linked to geological and hydrogeological investigations is essential to assess the most likely situation. It may seem obvious that erosion is the prime cause of coastal instability, but erosion also leads to stress relief leading to strength reduction and increased infiltration.

The mobilised strength depends on whether it is a first-time or reactivated slide, the fabric and composition of the glacial soils and possible stress relief due to erosion. A geotechnical investigation must produce a detailed study of the structure of the glacial soils and obtain sufficient samples to determine the fabric and representative strength of the soil. If it is not possible to obtain sufficient quality samples, then *in situ* tests or tests on reconstituted samples will be necessary.

Groundwater conditions in glacial soils can be complex due to the variation in composition, fabric and structure. The intrinsic permeability of glacial soils covers the range for all particle sizes and can change due to stress relief. A glacial soil can contain layers and lenses of soil with significantly different particle sizes creating local perched water levels or acting as aquitards or aquicludes, hence the importance of an adequate hydrogeological investigation.

Failures of glacial soils can take many forms depending on the fabric, composition and structure and the triggering events including pore pressure changes, erosion or strength reduction. Complex and debris flows are the most common making them difficult to predict from simple limit equilibrium analyses or numerical methods. There are guidelines to assess the vulnerability of coastal cliffs and whether a more detailed study is needed based on their susceptibility. A risk assessment of the consequences of failure should be undertaken taking into account the geological, hydrogeological, topographical, geomorphological and geotechnical models.

5.4 ENGINEERED FILL AND EXCAVATIONS

An engineered fill is used to create transport infrastructure, platforms for development (industrial, retail and domestic), defenses (military, flooding and coastal), earth dams and backfill and to reshape the land (landscaping). According to Trenter (2001), current methods of using soil as a construction material can be traced back 2000 years to the time of the Romans when they built some 80,000 km of roads across Europe and North Africa, though fill has always been used as a construction material. It was not until the first UK Industrial Revolution at the end of the eighteenth century with the development of turnpike roads, canals and drainage systems that the modern appreciation of the properties of soil to create engineering structures started to develop and, with the introduction of powered machines in the nineteenth century, a transformational expansion of transport infrastructure, initially focused on the rail network, took place. Table 5.22 is an overview of the use of earthworks

Table 5.22 Historical development of earthworks based on experience in the eighteenth and nineteenth centuries in the United Kingdom primarily because of the increased investment in transport infrastructure

Date	*Earthworks activity in the United Kingdom*				*Related points of interest*
1730–1840	Transport and Industrial Revolutions: Large-scale civil engineering projects, using empirical methods for earthworks design, with most embankments constructed of uncompacted fill material				Earthworks techniques developed through increased practice during the Industrial Revolution, largely by trial and error. Some empirical methods were formalised
1840–1890	Canals: Pre-1740: 'contour canals' – improvements to natural waterways, minimising the need for earthworks 1741: Newry Canal (NI) and 1761: Bridgewater Canal opened as the first truly 'artificial' canals 1770–1830: the 'golden age' of British canal construction with over 3,000 km of canal constructed 1835: completion of the Birmingham to Liverpool Junction Canal marked the end of the canal boom 1887–1893: Manchester Ship Canal construction, largest single earthworks project of the era	Dams: 1748: Grimsthorpe Dam, Lincolnshire by John Grundy, the earliest known dam formed with earthfill shoulders and a central clay core. Design records show the use of experience to select, place and compact the fill (possibly the first use of bulkfill compaction). Various dams were constructed to feed the growing canal network and development of the English landscape gardens	Highways: 1730–1850: Turnpike roads built by rudimentary earthworks methods to link major centres of population. Reliably financed and operated for the benefit of long-distance commercial traffic, rather than to satisfy the limited needs within individual parishes (thus justifying expenditure on earthworks)	Railways: 1758: Middleton railway – the first modern railway, transporting coal to Leeds 1807: Oystermouth railway, Swansea, was the first fare-paying passenger railway in the world 1812: Middleton railway upgraded to become the first commercial railway to use steam locomotives 1829: George Stephenson constructed Liverpool and Manchester Railway across Chat Moss by 'floating' the line on a bed of bound heather and branches topped with tar and covered with rubble 1833–1841: Nine mainline railways constructed, 1,060 km length and 54 million m^3 of excavation 1860: London Underground construction started with extensive earthworks	John Smeaton (1724–1792), first self-proclaimed civil engineer, designed the third Eddystone Lighthouse (1755–1759) and pioneered the use of 'hydraulic lime', which was critical to the rediscovery/development of cement, and led to the invention of Portland cement 1807: Geological Society of London founded 1815: Geological map of Britain published by William Smith, transforming understanding of the nature of natural ground 1818: Institution of Civil Engineers founded, Thomas Telford appointed as the first president in 1820 1862: Molesworth's Pocket Book of Engineering Formulae 1st edn, included standard retaining wall sections, earthwork profiles, cut slope angles, and labour for earthmoving

Source: After Trenter, N.A. *Earthworks: A Guide*. Thomas Telford, London; 2001.

in the eighteenth and the nineteenth centuries in the United Kingdom creating the infrastructure that exists today and is still being maintained. The canal and railway eras led to an appreciation of the use of embankments and cuttings to create a uniform gradient and the use of clay to form an impermeable barrier for canals and earth dams. Embankments were constructed by dumping excavated material directly onto the underlying soil which, if soft, may have been reinforced with timber mattresses (cf. geotextiles). There was limited compaction relying on the self-weight of the soil to compress the fill and underlying foundation soils. Extra fill was used to compensate for settlement during construction. Given the speed of construction, this was perfectly adequate. Subsequent settlement could be accommodated by routine maintenance, for example, use of ballast to maintain the level of rail tracks. The introduction of machines in the nineteenth century and the introduction of standards in the twentieth century led to increasing use of soil as an engineered material; the key developments are shown in Table 5.23.

The design of earthworks depends on whether it is for excavations or fills. Earthworks, in the context of construction, cover the mass movement of soil and rock to create excavations, embankments and backfill. In the case of excavations, the ultimate limit state is assessed using techniques to interpret natural slopes, that is, slope stability analyses based on measured parameters. Fill can be engineered by specifying the material properties to create embankments or backfill. The performance requirements of an engineered fill depend on whether it is being used to support a structure (e.g. infrastructure or platform), load a structure (e.g. backfill behind a retaining wall) or provide a barrier (e.g. flood embankments, dams, defenses).

5.4.1 Excavations

Excavations include excavations for foundations, basements, gravity and water retaining structures, landfill and cuttings. Any excavation will create a slope, which has to be assessed in the short and long term for overall stability and, if necessary, the design of engineered systems to ensure stability is maintained, which includes drainage preventing water entering the excavation or reducing the stability of the excavation and structural supports such as retaining structures or slope stabilisation measures. The stability of an excavation has to be assessed for temporary and permanent conditions.

The nineteenth century was a period when significant construction in glacial soils took place to form the UK rail network. Papers published at that time recorded the construction of the railway network between about 1840 and 1890 highlighting the challenges faced using observational techniques, that is, reacting to what was uncovered during the excavation. For example, Laws (1881) described retrogressive slips in glaciolacustrine clay, which were dealt with by trench drains or by firing the clay to produce a brick-like material that was tipped into place. Technical papers published by the UK Institution of Civil Engineers gave an insight into the challenges faced and how they were overcome. For example, Whitley (1880) described the construction of a railway in the North East of England, which ran through matrix-dominated tills leading to frequent slips in cuttings and embankments. One feature of these stiff low plasticity clays was their tendency to revert to slurry because of their stone content. Frequent, rotational slips took place in excavations. They were stabilised with gravel-filled drains perpendicular to the line of the excavation (counterfort drains) and taken below the failure zone and hence excavation level. An alternative was to excavate the clay, mix it with ash and replace it, a form of ground modification. Whitley (1880) recommended slopes of 1:1.5 in stiff matrix-dominated tills up to 8 m deep and 1:2 above that – much steeper than current standards. The fact that they have remained standing explains why, when the mass properties are the same as the intact properties, slopes in glacial soils

Table 5.23 Development of standards for earthworks in the twentieth century based on scientific research and soil mechanics theory

Date	*Milestone events that shaped earthworks practice in the United Kingdom*	*Extended research and developments in theory*
1890	Mohr–Coulomb theory formalised as a mathematical form for engineering design, providing the opportunity for slope stability analysis to be developed in the twentieth century	1920s to 1940s: Karl Terzaghi (1883–1963) established the fundamentals of soil mechanics enabling a design-based approach to geotechnical engineering. 1936: Terzaghi and Casagrande organised 1st International Conference on Soil Mechanics and Foundation Engineering. Terzaghi published a collation of his work as *Theoretical Soil Mechanics* 1st edn in 1943
1927	Fellenius used limit equilibrium theory to determine the 'ordinary method of slices', providing the first model to enable engineers to calculate the stability of slopes based on circular failure planes	
1933	Proctor developed the concept of optimum water content and the standard compaction test in connection with dam construction. Researchers then studied the compaction process at a laboratory scale. The test was used on major UK earthworks projects	
1936–1937	Chingford dam construction; the first use of mechanised plant to form embankments of compacted fill. The dramatic increase in construction rate led to slope failures. Led to Mowlem establishing the first soils laboratory	
1939–1945	World War II – major advances in earthworks theory and construction practice included Casagrande soil classification system; TRL established the relationship between soil type, water content, compactive effort, fill density and air voids; cement stabilised soils pioneered for airfield runways; towed sheeps foot rollers and small vibrating plate compaction plant developed	
1948	BS 1377 1st edn includes Proctor compaction tests.	1940s to 1980s: Taylor (1948) produced slope stability charts for design. Bishop (1955) developed a method of analysing circular failure surfaces; then extended by Morgenstern and Price (1965) and Janbu (1973) for analysing non-circular failures. Analysis of individual slopes by computer software became routine during the 1980s
1950s	Large self-propelled scrapers become common on major earthworks projects. Development of vibratory rollers	

(*Continued*)

Table 5.23 (Continued) Development of standards for earthworks in the twentieth century based on scientific research and soil mechanics theory

Date	*Milestone events that shaped earthworks practice in the United Kingdom*	*Extended research and developments in theory*
1951	*Specification for Road and Bridge Works* 1st edn (title changed to *Specification for Highways Works* [SHW] in 1986), with compaction control to be achieved by 'end product compaction' criteria	1940s to 1990s: 1946–1990: Comprehensive programme of research by TRL on the compaction of soil and granular materials, by laboratory testing and full-scale field trials to understand earthworks and fill materials and improve compaction plant, which forms the basis of modern earthworks theory in the United Kingdom (the Highways Agency's Design Manual for Roads and Bridges [DMRB] design manual HA 70/94 was prepared based on this work)
1952	*Soil Mechanics for Road Engineers* published: TRL collated World War II advances in earthworks technology for application on civil engineering projects, including a detailed understanding of compaction theory. Compaction plant at this time was limited to dead weight rollers, the most commonly used being 8-ton steam engine rollers	
1951–1954	Jackfield land slip enabled Skempton to clarify the role of relict slip planes and residual angle of shearing resistance	
1951–1968	M1 built, followed soon afterwards by the main routes of the present motorway network	
1959	BS6031 1st edn published (as CP 2003) Code of Practice for Earthworks (largely unchanged in 1981 revision)	
1969 and 1976	SHW 4th and 5th edns replaced end product compaction with 'method compaction' (by 1976 the methods were similar to those of the present). Vibrating rollers were permitted and became commonly used through 1970s/1980s	
1975	BS 1377 updated, formalising the earthworks soil testing methods	
1978	Earthworks conference entirely on 'clayfills' held at the ICE	
1970–1986	M25 orbital motorway constructed in sections (188 km long)	
1986	SHW 6th edn in modern format with: 600 series clauses; engineer to select appropriate fill acceptability properties at Table 6/1 based on relationship testing; MCV test added to improve control on fill material suitability; re-introduction of 'end product compaction' for certain fill classes and updated 'method compaction'. This triggered further developments of compaction plant	
1996–2007	First high-speed railway constructed, 108-km Channel Tunnel Rail Link (HS1)	
2004	Eurocode 7 (EC7) system of common European standards introduced into UK practice	
2009	BS6031 Code of Practice for Earthworks fully updated to reflect modern methods and EC7	

Source: After Trenter, N.A. *Earthworks: A Guide*. Thomas Telford, London; 2001.

can be stable up to 45°. It is the structure, fabric and presence of weaker and water-bearing layers that often trigger failure; many of these were removed when constructing the railways in the nineteenth century.

5.4.2 Cuttings

A cutting is a steepened natural slope or the side slopes of an excavation; therefore, the design is based on the *in situ* properties taking into account the possible strength reduction due to stress relief caused by unloading, the changes in the groundwater profile due to excavation and drainage and the presence of any weak horizons especially in glacial soils.

A number of motorway schemes constructed in the 1960s and 1970s included cuttings in matrix-dominated tills. Parsons and Perry (1985) undertook a survey of 300 km of motorway and found that there were a significant number of failures in both cuttings and embankments. They found that 1% of 45 km of glacial clay till cuttings had failed at slopes of 1:3 after 20 years unlike cuttings in sedimented clays that were more likely to fail within 2–3 years. Embankments were less prone to failure for slopes of 1:3. They recommended that to reduce the failure rate to below 1% in 22 years in cuttings and embankments, the slopes should be 1:4 and 1:3, respectively, for cuttings and embankments formed of glacial clay tills and 1:2 for glacial gravel. Failure in stiff sedimented clays is attributed to a loss of strength with time linked to the fabric of the clay and pore pressure. Skempton and Brown (1961) suggested that there is little loss of cohesion with time in dense clay tills; Stark et al. (2005) suggested that post-peak values of strength should be used for first-time failures; McGown et al. (1977) suggested that the mass strength of glacial clay tills will be less than the intrinsic strength because of the fabric. Failures of cuttings in matrix-dominated tills are a special case if they cross a drumlin field. McGown and Radwan (1975) showed that this was more likely if the cutting was aligned with the axes of the drumlins, which they attributed to the direction of shear during deposition. Hughes et al. (2007) provided a case study of such a failure in Northern Ireland. The 1:2 slope, 19 m high was constructed in 1972 and failed in 1999. The failure was deemed to be a rotational slip due to dissipation of negative pore pressures, a reduction in strength due to progressive failure, strain softening and increase in pore pressure. The failure occurred after several months of above-average rainfall. The glacial till was described as a low plasticity well-graded till with a water content of 12%–15%, liquid limit 38% and plastic limit 18%. The matrix water content was estimated to be 18%–22%. Consolidated undrained triaxial tests gave a cohesion of 4 kPa and the angle of friction of 32°. There is a reference to a highly fissured clay layer in the slip zone, which possibly contributed to the slide since the cutting may have been aligned with the major axis of the drumlin. They recommended positive drainage to lower the water table and prevent water entering the slope.

It is important when excavating spatially variable soils to take action when unsuitable materials are encountered. This observational method, evident in the nineteenth century, reduces the number of failures during construction. Chapman et al. (2004) extended this further because of the lack of recorded knowledge of the behaviour of glacial clay to the north of London. The construction sequence to form 12 m high diaphragm walls for an underpass was undertaken in stages using a temporary berm to support the wall rather than props. This simplified the construction and reduced costs. Table 5.24 shows the range of design parameters and the effect of switching from the most probable (undrained) to worst credible (drained). The difficulty was deciding whether the glacial soils would behave as drained or undrained (i.e. did they behave as coarse- or fine-grained soils) during the temporary works. Figure 5.28 shows that the prediction based on undrained analysis of the temporary works produced similar displacements to those observed.

Table 5.24 The range of design parameters for a glacial till used by Chapman et al. (2004) to model a 12-m high diaphragm wall including its installation showing the parameters to be considered to take account of the short- and long-term conditions

Design case	*Most probable*	*Most unfavourable*	*Worst credible*	*AIP*
Strength	Undrained	Softened	Drained	Drained
Strength parameters	Glacial till and London Clay $c_u = 95 + 7z$ (no softening) Glacial sands and gravels $\varphi' = 36°$	Active side Glacial till – average of relevant drained and undrained strengths London Clay – undrained Passive side Glacial till and London Clay $c_u = 95 + 7z$ (with 25% softening) Sands and gravels $\varphi' = 36°$	Glacial till $\varphi' = 30°$ Sands and gravels $\varphi' = 36°$ London Clay $\varphi' = 25°$	All materials $c' = 6$ kPa; $\varphi' = 26°$
K_o and stiffness	$K_o = 1.0$ $E_u/c_u = 1{,}500$	$K_o = 1.0$ $E_u/c_u = 1{,}200$	$K_o = 1.0$ $E'/c_u = 1{,}200$	$K_o = 1.5$ $E' = 10$–30 MPa
Pore water pressures	Best estimate PWP profile		Hydrostatic on active side; greater than hydrostatic on passive side	
Live load surcharges	Not included	Included	Included	

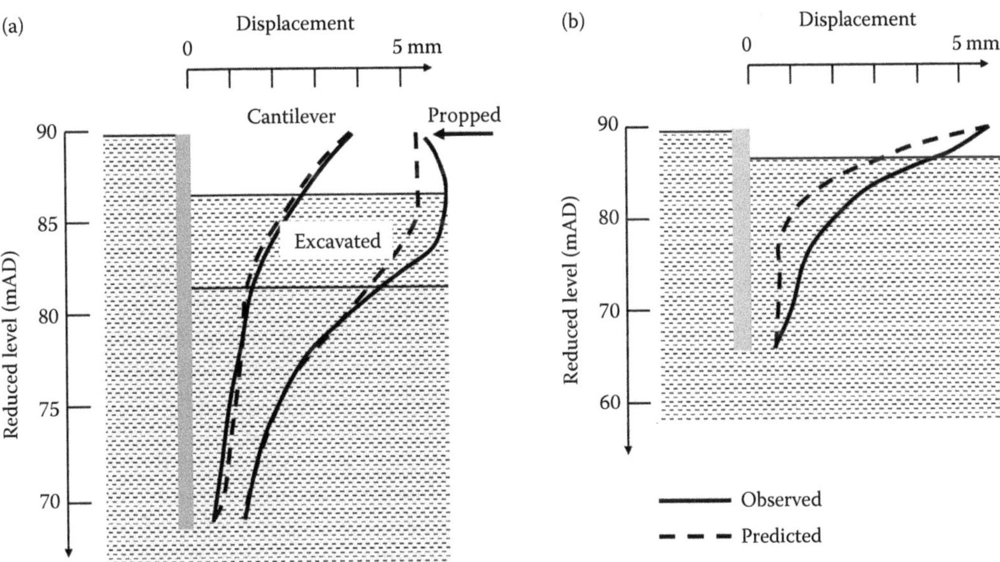

Figure 5.28 Comparison between the observed and predicted displacements for (a) a diaphragm wall and (b) a bored piled wall in matrix-dominated till in North London, United Kingdom. (After Chapman, T. J. P., S. J. Deeble, and D. P. Nicholson. *Advances in Geotechnical Engineering: The Skempton Conference: Proceedings of a Three Day Conference on Advances in Geotechnical Engineering*, organised by the Institution of Civil Engineers and held at the Royal Geographical Society, London, UK, on 29–31 March 2004, pp. 1044–1055. Thomas Telford Publishing, 2004.)

5.4.3 Engineered fill

An engineered fill is natural, modified and manufactured particulate material that is placed in a controlled manner to produce a soil-like material of known properties. Glacial soils are used in their natural and modified mode as an engineered fill. Given the spatial variation in composition, glacial soils may have to be pretreated. For example, very coarse particles are removed because they affect the compaction process; glaciofluvial soils may have to be mixed because of their variable composition; glaciolacustrine soils may have to be treated to avoid the possibility of aligning laminations with a potential slip surface; unacceptable materials such as soft clay may have to be removed from matrix-dominated tills.

5.4.3.1 Soil properties

In the case of excavations, the soil is described using international standards for description of soil for engineering purposes such as BS EN ISO14688-1:2002+A1:2013 and ASTM D2487 – 11. Classifying soil for an engineered fill is different from classifying soils for engineering purposes, as shown in Table 5.25. Further, classification schemes can change with time and country. It is for this reason that any correlations between classification data and mechanical properties are not necessarily universal.

Glacial soils are composite soils. BS 6031:2009 suggests that composite soils are those that contain at least 10% of the secondary fraction, which is consistent with the view that most glacial soils are composite soils. BS EN ISO14688-1:2002+A1:2013 defines a composite fine soil as one in which the fines content determines the engineering behaviour. A composite coarse soil is the one which contains fines but behaves as a coarse-grained soil. In terms of engineered fills (BS6031:2009), soils that contain at least 15% fines are classed as cohesive soils, and for the geotechnical design of cuttings and embankments, cohesive soils are defined as those containing at least 35% fines.

Some composite soils, such as well-graded matrix-dominated tills, are prone to slumping if the water content is too low, which led Jenkins and Kerr (1998) to investigate the relationship between strength and water content. They observed that the saturation water content of the sand and gravel was 5%. The relationship between the undrained shear strength, water content and matrix water content (Figure 5.29) suggests that there may be a more consistent relationship between undrained shear strength and matrix water content rather than the water content based on total solids.

5.4.3.2 Selecting fill materials

Table 5.26 lists the tests to determine the design and construction criteria for an engineered fill showing the test, the material to be tested and the purpose of the test. Table 5.27 is a commentary on the advantages and disadvantages of the criteria for selecting, accepting and controlling fills. There are a number of ways to engineer fill but it is the performance that the user is most interested in. Acceptance criteria can be based on the dry density and water content ranges shown in Figure 5.30. This will produce an engineered fill with acceptable properties for structural support or landfill liners based on experience. If performance criteria are specified, then it is necessary to determine undrained shear strength (structural support), Californian Bearing Ratio (road pavements) or permeability (landfill liners).

Granular fill comprises coarse-grained soils, which, in the United Kingdom, are those soils with less than 15% fines (SHW, 2013). The soil description should include the particle size distribution (particularly the uniformity coefficient), the shape of the gravel particles and the particle mineralogy. Well-graded angular granular fill (uniformity coefficient <10) is possibly the best material for an engineered fill because particle breakage is reduced and

Table 5.25 Comparison of soil definitions in different earthworks circumstances for classification and testing

% passing 63-micron sieve		0	5	10	15	20	25	30	35	40	45	50	55	60	65	70	75	80	85	90	95	*100*
UK standard approach to earthworks material classification by grading (after SHW[a,b])	Fill behaviour	Granular fill				Intermediate fill				Cohesive fill												
UK traditional approach to classification for geotechnical design (after BS 5930:1999+A2[d])	Soil parameters	Coarse-grained						Intermediate[c]								Fine-grained						
BS EN 1997-1:2004 geotechnical design approach (after BS EN ISO 14688-1: 2002[e])	Simplified interpretation for comparison purposes	Coarse soil			Composite coarse soil[f]													Composite fine soil[g]				Fine soil
					BS EN 1997-1:2004 approach does not set any fixed boundary but generally >10% of the secondary fraction is likely to be needed in most soil types to constitute a composite soil																	

Source: From BS 6031:2009. *Code of Practice for Earthworks*. British Standards Institution, London; after BS EN ISO 14688-1:2002+A1:2013. *Geotechnical Investigation and Testing – Identification and Classification of Soil – Part 1: Identification and Description*. British Standards Institution, London; after BS EN 1377-1, *Methods of Test for Soils for Civil Engineering Purposes, General Requirements and Sample Preparation*; British Standards Institution, London; 1990.

[a] SHW sets the granular/intermediate divide at 15% in recognition of pore water pressure in dynamic action of compaction.
[b] The terms 'granular' and 'cohesive' are included here with regard to behaviour, the soil description terms in accordance with BS 5930:1999+A2 are 'coarse' and 'fine'.
[c] The designer has to use judgement of how a soil will behave within the intermediate zone, which is not considered in BS 5930:1999+A2.
[d] Most fills in the United Kingdom that are in the intermediate range are classified as Class 2C. Alternatively, the designer can create a new site-specific class, for example, 'Class 2F, clayey sand'.
[e] The BS EN 1997-2:2007 approach for identification and description of soils is set out within BS EN ISO 14688-1:2002, 4.3, by this system many soils are classified as composite soils and the distinction between soil terms can be summarised as follows.
[f] Composite coarse soil is a soil where the fines content is not sufficient to determine the engineering properties (BS EN ISO 14688-1:2002 should be referred to for the full determination procedures).
[g] Composite fine soil is a soil where the fines content is sufficient to determine the engineering properties.

Particle size (mm)	0.002	0.0063	0.02	0.063	0.2	0.63	2.0	6.3	20	37.5	63	200	630	>630
Soil fraction	Fine soil				Coarse soil						Very coarse soil			
Sub-fraction	Clay	Silt			Sand			Gravel			Cobbles	Boulder	Large boulder	
		Fine	Medium	Coarse	Fine	Medium	Coarse	Fine	Medium	Coarse				
Soil group	Fine-grained soil[a]							Medium-grained soil[b]		Coarse-grained[c]	Soils too coarse to test by standard means			

[a] Soil containing not more than 10% retained on a 2-mm sieve.
[b] Soil containing more than 10% retained on a 2-mm sieve but not more than 10% retained on 20-mm sieve.
[c] Soil containing more than 10% retained on a 20-mm sieve but not more than 10% on a 37.5-mm sieve.

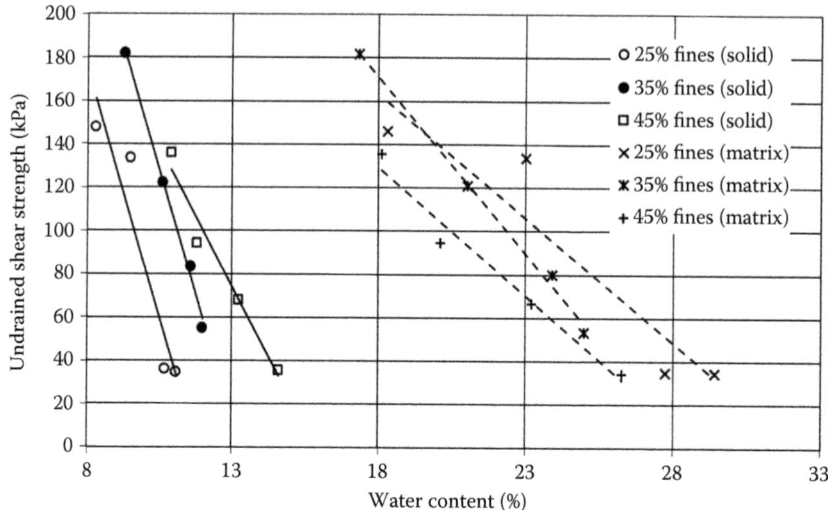

Figure 5.29 Comparison between undrained shear strength and water content using the mass of solids and the mass of fine-grained particles suggesting that there may be a better correlation with the matrix water content. (After Jenkins, P. and I. A. Kerr. *Ground Engineering*, 31(3); 1998.)

the maximum density for granular fill is achieved. Uniformly graded granular fill produces a low density fill. Gap-graded clast-dominated tills do not have the best particle size distribution for compaction because large particles influence the effect of the compactive effort, especially if they exceed 50% of the fill; the fill becomes increasingly difficult to compact if there are limited fines. Ideally, a well-graded fill should be used since it produces the densest fill but it may not be practical to mix soils to achieve an appropriate particle size distribution.

5.4.3.3 Compaction tests

The compaction characteristics of a soil are defined by the relationship between the dry density and water content, the maximum dry density and the optimum water content at that density. They are used to assess whether a soil can be compacted effectively at its *in situ* water content or whether the water content has to be changed.

Compaction tests are carried out on reconstituted samples of soils prepared at different water contents. Coarse particles (>20 mm) are removed before preparing a sample. Thus, it should be possible to test matrix-dominated tills and some glaciofluvial deposits. The Moisture Condition Value (MCV) test is a rapid means of assessing the suitability of a fill that has been successfully used in tills, which suggests that this may be a more appropriate test in clast-dominated tills or matrix-dominated tills with significant stone content.

Given that the properties of an engineered fill can be specified, it is necessary to manage the placement of the fill. This includes the selection of the fill by establishing whether it is suitable and criteria to ensure that it is placed properly to produce a fill with known properties. The selection of a fill is based on its composition and water content; the acceptance is based on the density and water content or air voids; and the performance on strength, stiffness or permeability depends on its use. Compaction tests are used to determine the relationship between dry density and water content for a given compactive effort. They are also used to determine the maximum dry density of coarse-grained soils so that the relative density can be assessed. In some cases, particularly with fine-grained composite soils, it is not possible to adjust the *in*

Table 5.26 Tests to classify, accept, monitor and assess performance of engineering glacial soils

Soil type	*Material type*	*Applicability*			*Applications*	*Comments*
		Design	*Construction control*	*Stabilisation and reuse*		
Water content	F, M, C, R	√	√	√	Classification and compaction control	Used for comparison with laboratory moisture content values determined as part of relationship testing suites
Moisture condition value (MCV)	F, M	√	√			Used for assessing fill material suitability for specification/design and *in situ* monitoring of sources of fill
Particle size distribution	F, M, C, R	√	√	√	Classification	Used for determining fill material grouping and assisting compaction plant selection
Atterberg limits	F, M, C	√		√		Used to derive behavioural characteristics and preliminary engineering properties of cohesive fill
Particle density	F, M, C, R	√	√	√		Used for determining loadings, bulking and compaction control (air voids determination) of fill materials
2.5 kg compaction	F, M	√	√	√	Determination of dry density/moisture relationship	Used to specify moisture content limits for use of material as fill (may be used in conjunction with CBR Test)
4.5 kg compaction	F, M	√	√	√		
Vibrating hammer	F.M	√	√	√		
California bearing ratio (CBR)	F, M	√	√	√	Formation strength determination	Used to determine pavement construction thicknesses and assist *in situ* compaction control during construction
Undrained shear strength parameters	F, M	√		√	Design of earthworks subject to undrained loading conditions	Temporary slope and foundation design for construction purposes; assessment of plant trafficability; permanent works constructed with cohesive soils, subject to rapid loading
Drained shear strength parameters	F, M	√			Design of earthworks subject to drained loading	Slope and foundation design for long-term temporary or permanent works; temporary works constructed with granular soils
Los Angeles abrasion test	M, C	√			Design of permanent works	Used for selected fill materials

(*Continued*)

Table 5.26 (Continued) Tests to classify, accept, monitor and assess performance of engineering glacial soils

Soil type	*Material type*	*Applicability*			*Applications*	*Comments*
		Design	*Construction control*	*Stabilisation and reuse*		
Plate load test	F, M, C	√	√	√	Design and compaction control	Used for assessment of settlement characteristics and bearing capacity at formation level and of compacted fills
Dynamic cone penetrometer	F, M, C	√	√	√		Used for designing foundations (bearing capacity), formations and *in situ* compaction monitoring
MEXE probe	M, C	√	√	√		Used for designing foundations (bearing capacity), formations and *in situ* compaction monitoring
Dynamic plate load test	F, M, C	√	√	√		Used for designing foundations, formations and *in situ* compaction monitoring. The form of test adopted for pavement foundation design is the lightweight falling weight deflectometer
Field density test	F, M	√	√	√		Used to determine *in situ* densities and assist in-compaction control during construction
Clegg impact soil tester	F. M	√	√	√	Formation strength determination	Used to determine pavement construction thicknesses and assist *in situ* compaction control during construction
pH, SO_4, Cl	F, M, C, R	√	√	√	Design and confirmation during construction	Determination of aggressive ground conditions for cementitious products and buried metallic structures and elements
Redox potential/ resistivity	F, M, C, R	√	√	√		Determination of aggressive ground conditions for buried metallic structures and elements
Chemical analysis	F, M, C, R	√	√	√		Determination of toxic elements/compounds for environmental and health and safety control during construction
Waste acceptance criteria	F, M, C, R	√	√	√		Determination of chemical characteristics for disposal offsite to landfill

Source: After BS 6031:2009. *Code of Practice for Earthworks*. British Standards Institution, London.

Note: F = soils not containing more than 10% retained on a 2 mm test sieve; M = soils containing more than 10% retained on a 2-mm test sieve but not containing more than 10% retained on a 20-mm test sieve; C = soils containing more than 10% retained on a 20-mm test sieve but not containing more than 10% retained on a 37.5-mm test sieve; R = zone 'X' material and rock fill B.

Table 5.27 Engineering properties used to select, design, accept and control engineered fills

Property	Design and acceptability		Control	
	For	Against	For	Against
Particle size	Governing factor in performance of fills		Governing factor in performance of fills	Test lengthy to perform and equipment cumbersome to use
Water content with respect to plastic limit	Properties easily measured	Precision of plastic limit test low	Properties easily measured	Precision of plastic limit test low
Optimum water content and maximum dry density	Properties fundamental to engineering performance Particle size up to 20 mm tested	Time taken for compaction test performance (usually in excess of 24 h)	Equipment for control tests accommodated in mobile laboratory	Accuracy of density values from sand replacement tests questioned
Moisture condition value	Particle size up to 20 mm tested	Not readily related to fundamental engineering properties	Equipment accommodated in mobile laboratory	Not readily related to fundamental engineering properties
CBR value	Property frequently used for pavement design Particle size up to 20 mm tested	Not readily related to fundamental engineering properties; use of empirical correlations	Property frequently used for pavement design Particle size up to 20 mm tested	Not readily related to fundamental engineering properties; use of empirical correlations
Undrained triaxial shear strength	Fundamental to trafficability and to bearing capacity of fills Water content and bulk density obtained from test specimen	Not readily related to optimum water content and maximum dry density	Fundamental to trafficability and to bearing capacity of fills Water content and bulk density obtained from test specimen	38-mm test specimens too small for many fills containing gravel 100-mm tests require expensive equipment for testing and sample extraction
Hand vane	Fundamental to trafficability and to bearing capacity of fills Water content and bulk density obtained from test specimen	Not readily related to optimum water content and maximum dry density	Light weight small and easily transportable	Only for clay fill containing little or no coarse sand or gravel
Permeability test	Fundamental to the design of landfill liners, dam cores and other impermeable earth structures; wide variety of test types including *in situ* tests	Wide range of application	Fundamental to the design of landfill liners, dam cores and other impermeable earth structures; wide variety of test types including *in situ* tests	Tests too complex and time consuming for routine site use; better to correlate with results of other test types at the design stage

Source: After Trenter, N.A. *Engineering in Glacial Tills*. CIRIA, London, 1999.

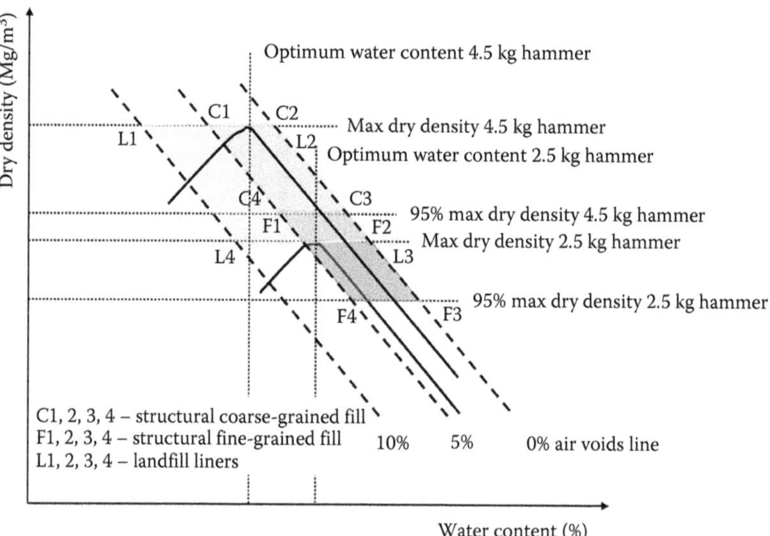

Figure 5.30 Acceptance criteria for an engineered fill based on standard (2.5 kg) and heavy (4.5 kg) compaction tests showing their relationship with air voids content.

situ water content; therefore, compaction tests are used to determine the dry density that can be achieved at the natural water content. Tests include ordinary and heavy compaction tests, compaction using a vibrating hammer and the moisture condition test.

A compaction test (Table 5.28) involves compacting a number of soil water mixes in layers using a standard compactive effort to determine the relationship between dry density and water content from which the maximum dry density and optimum water content can be found. The main difference between the types of test is the amount of compactive effort. In standard compaction tests, particles greater than 20 mm are removed, which is relevant for glacial soils. The density of these particles exceeds the density of the compacted soil; therefore, the density *in situ* exceeds the laboratory derived value for the same compactive effort if the soil contains particles greater than 20 mm. If the gravel content is less than 25% and the particles are distributed throughout the soil, as is the case in many matrix-dominated tills, then it is possible to apply a correction for the stone content described in Table 5.29. If the soil contains a significant number of larger particles, then compaction tests can be carried out with a CBR mould, which allows particles up to 37.5 mm to be included. If there are more than 30% of very coarse particles, then compaction tests may not be feasible. Removing particles greater than 20 or 37.5 mm prior to a compaction test means that the true density of the soil for a given compactive effort is not given. The elimination and

Table 5.28 Compaction test procedures

		Rammer			
Type of test	*Container*	*Mass (kg)*	*Drop (mm)*	*No. of layers*	*Blows/layer*
Standard Compaction	BS mould	2.5	300	3	27
	CBR mould	2.5	300	3	62
Heavy Compaction	BS mould	4.5	450	5	27
	CBR mould	4.5	450	5	62
Vibrating Hammer	CBR mould	32	Vibration	3	1 min

Table 5.29 A procedure to correct for gravel content for soils containing up to 25% of particles exceeding 20 mm which are randomly distributed throughout the specimen

Total mass of sample $= m_m + m_s$

where m_m is the mass of the matrix and m_s the mass of the stones. For a unit volume of sample, the total mass is m (numerically equal to ρ). The mass of the matrix is

$$m_m = (1 - V_s)m$$

where V_s is the volume of stones. The mass of stones is

$$m_s = V_s G_s \rho_w$$

where G_s is the particle-specific gravity. The fraction, F, of the matrix material is

$$F = \frac{m_m}{m_m + m_s}$$

The dry density of the sample (with stones), ρ_d, is

$$\rho_d = \frac{G_s \rho_w}{(1 - F) + F(G_s \rho_w / \rho_{md})}$$

where ρ_{md} is the dry density measured in the compaction test.
The water content of the sample is different from the water content of the matrix. The mass of water in the matrix is

$$w_m m_w = w_m F(m_m + m_s) = w_m F \rho_d$$

where w_m is the water content of the matrix.
The mass of water in the stones is

$$w_s m_s = w_s (1 - F)\rho_d$$

Total mass of water, W, in a unit volume is

$$W = w_m F + w_s (1 - F)\rho_d$$

The water content, w, of the sample is

$$w = F w_m + (1 - F) w_s$$

adjusted maximum dry density methods are used to correct the density to take into account particles >20 mm. In the elimination method, the corrected density, ρ_t, is

$$\rho_t = \frac{\rho_m \rho_w G_s}{\rho_w G_s (1 - F) + F \rho_m} \tag{5.6}$$

where ρ_m is the dry density of the matrix (material smaller than 20 mm) and F is a correction factor equal to the fraction of stones to all particles by dry weight. The corrected density for the adjusted maximum dry density method is

$$\rho_t = \frac{1 - 0.05F}{(F/2.6) + (1 - F/\rho_m)} \tag{5.7}$$

If the control measures are based on water content, then the matrix water content and matrix dry density are used because the stones will be removed from the sample prior to testing.

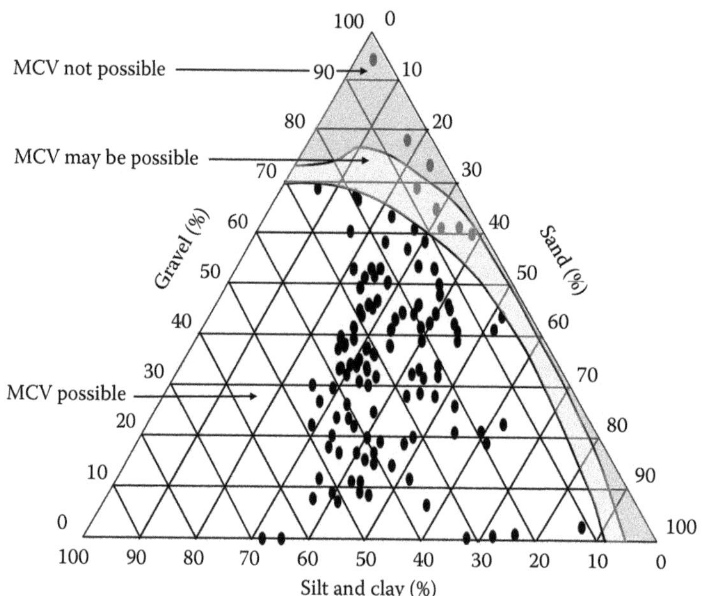

Figure 5.31 Range of soils, which includes glacial soils, for which it is feasible to use the Moisture Condition Apparatus. (After Winter, M. G. and Th. Hólmgeirsdóttir. *Quarterly Journal of Engineering Geology and Hydrogeology* 31(3): 1998: 247–268.)

The MCV test, a form of strength test, is a means of rapidly assessing whether a fill is at a water content suitable for placing. It can be used if the fines content exceeds 18% (e.g. matrix-dominated tills) but cannot be used if the fines content is less than 5% in a sandy soil or 10% in a gravelly soil. Winter and Holmgeristottir (1998) suggested that the MCV test is suitable for soils containing more than 10% fine-grained soil and less than 30% gravel (Figure 5.31). Particles greater than 20 mm are removed prior to testing. It is assumed that the density depends on the compactive effort and water content. A 7-kg rammer is dropped 250 mm onto the sample in a 100-mm-diameter mould and the penetration of the rammer is measured. This is repeated with an increasing number of blows until the penetration between X blows and 4X blows is less than 5 mm. This is assumed to represent the maximum bulk density of the sample.

5.4.3.4 Compaction processes

Compaction reduces the air content leading to an increase in density and therefore an increase in strength and stiffness and a reduction in permeability. Compaction is also used by sedimentologists to describe the processes of gravitational compression, which includes volume changes due to a reduction in air and water content. The initial water content of a partially saturated soil has a significant effect on the performance of an engineered fill following placement. Soils wet of optimum will be difficult to compact and could even be overcompacted creating discontinuities. Clay fills dry of optimum will generate suction pressures, which can lead to a loss in strength with time as the pressures dissipate. Glaciofluvial soils and clast-dominated tills will be generally free draining provided the fines content is less than 15% and are suitable for engineered fill where permeability is not critical.

Winter and Suhardi (1993) suggest that if the percentage of particles greater than 20 mm is less than 50%, the matrix controls the properties. Bolton and Lee (1993) showed that the

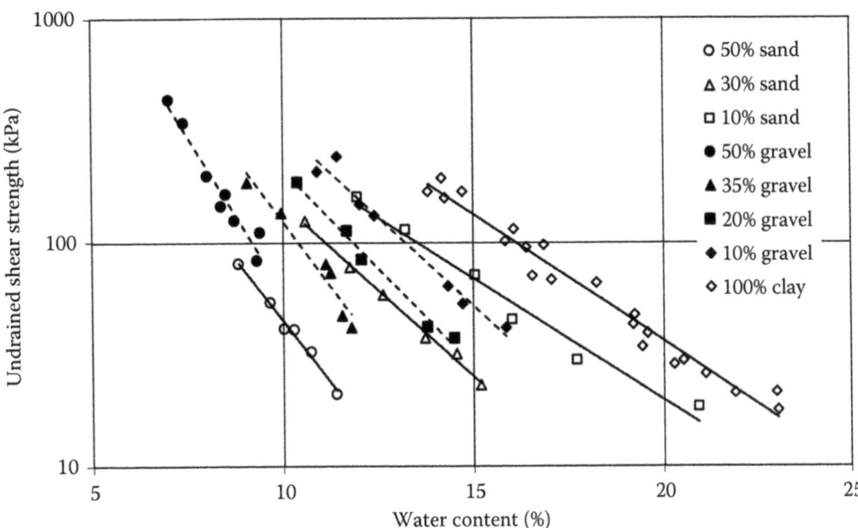

Figure 5.32 Variation of remoulded undrained shear strength with water, sand and gravel content, showing that an increase in coarse-grained content for a given water content reduces the undrained strength of the soil. (After Barnes, G. E. and S. G. Staples. *Ground Engineering*, 21(1); 1988: 22–28.)

angle of friction of a soil increased as the maximum particle size increased provided the proportion of coarse-grained particles exceeded 50% though Winter and Suhardi (1993) showed that the density reduced because the large particles are in contact requiring greater compactive effort to overcome the friction between the particles. Figure 5.32 is an example of the effect of coarse-grained particles on the undrained shear strength of the compacted matrix-dominated soil, which shows that an increase in fine-grained content reduces the undrained strength for a given water content.

There are five factors to consider when using fine-grained soils as engineered soils: trafficability, under-compaction, matteressing, shear surface formation and desiccation. Trafficability refers to bearing failure of the surface due to the construction traffic causing permanent deformation of the surface. Arrowsmith (1979) suggested that the minimum strength of the surface layers should be 35 kPa for tracked vehicles and 50 kPa for rubber-tyred vehicles. Under-compaction occurs when the strength of the intact soil makes it too difficult to compact. This is a problem for dense matrix-dominated tills, which can have a strength in excess of 150 kPa. Mattressing occurs when the water content is too high inducing high pore pressures in the upper layer. Compaction-induced shears can form if a medium to high plasticity clay fill is wet of optimum, which may restrict pore pressure dissipation and possibly create potential failure surfaces.

The fill will compress under its own weight but may collapse on inundation, particularly if it is compacted dry of optimum. The settlement, s, of compacted soils due to their own weight, based on observations of backfill to opencast excavations involving significant depths of fill, is

$$s = 0.5\frac{\gamma H^2}{D} \tag{5.8}$$

where H is the thickness of fill, D the constrained modulus given in Table 5.30 and γ the unit weight of the fill. Infiltration, particularly in the upper layers, can cause the clay to swell though it is less likely with low plasticity clays such as matrix-dominated tills. Figure 5.33

Table 5.30 Suggested values for constrained modulus for engineered fills

	Constrained modulus (MPa)		
Type of fill	*Ht of fill = 10 m*	*Ht of fill = 30 m*	*Ht of fill = 100 m*
Sandy gravel (D_r = 80%) (clast-dominated till, glaciofluvial soils)	50	90	170
Sandy gravel (D_r = 50%) (clast-dominated till, glaciofluvial soils)	30	50	90
Clay (I_p = 15%; LI = 0.1) (matrix-dominated till)	6	10	18

Source: After Charles, J.A. *Building on Fill: Geotechnical Aspects*. Building Research Establishment, Watford, UK; 1993.

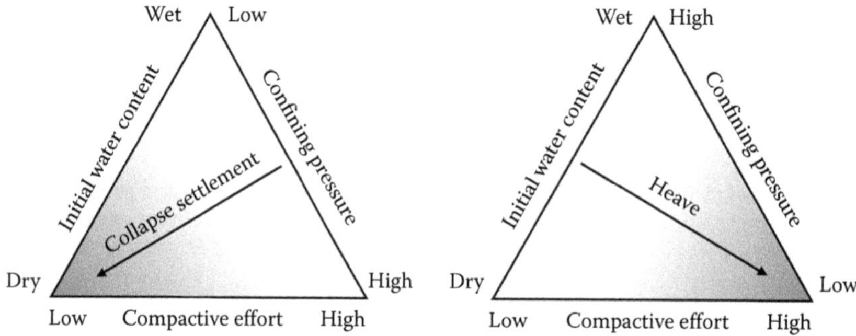

Figure 5.33 Critical factors affecting the performance of engineered clay fill. (After Trenter, N. A. *Earthworks: A Guide. Thomas Telford*, London; 2001.)

Table 5.31 Acceptance criteria for selection of fine-grained soils for a landfill liner

Engineering property	*Reference*	*Criterion*
Plasticity	DETR (1995)	30% > PI > 110%
	Daniel (1993)	PI > 7%–10%
	EA (2013)	LI < 30%; PI < 65%
	Murray et al. (1992)	PI > 12%
	Gordon (1987)	PI > 15%
	Williams (1987)	PI > 15%
% fines	Daniel (1993)	Clay and silt > 20%–30%
	EA (2013)	Clay > 10%
	Gordon (1987)	Clay and silt > 50%
Activity	DETR (1995)	>0.3
% gravel	Daniel (1993)	Gravel (4.8 mm) < 30%
Maximum particle size	Daniel (1993)	<25–30 mm

Source: After Murray, E. J. Properties and testing of clay liners. In *Geotechnical Engineering of Landfills*. Thomas Telford, London, UK, 1988.

is a description of these critical factors showing the effect of the compactive effort, initial water content and confining pressure on the possibility of collapse settlement or heave of clay soils. Soils compacted wet of optimum present problems of consolidation; soils dry of optimum may collapse.

Fine-grained soils can be used to form impermeable liners to landfill. The acceptance criteria are based on the plasticity and fines content (Table 5.31). The control criteria is the permeability that should be less than 10^{-9} m/s according to the US EPA, which can be achieved if the fines content exceeds 20%, the gravel content is less than 30%, the maximum particle size is 50 mm and *PI* exceeds 10%. Murray (1998) suggested that the 4.5-kg hammer should be used for high plasticity clays and the 2.5 kg for low plasticity clays to produce the acceptance criteria. Clay fills are normally compacted in horizontal layers, but in the case of landfill liners, they should be compacted in layers parallel to the slope.

An engineered fill must meet performance criteria, which can be its strength, stiffness or permeability depending on its purpose. There are three types of specification:

- Method specification which covers all aspects of the construction process including the layer thickness, the number of passes and the type of plant
- End product specification which specifies the properties of the compacted fill and is checked by on site testing, for example, MCV, dry density or water content
- Performance specification which sets the limits for the performance of the fill, for example, undrained shear strength, angle of friction, permeability Californian Bearing Ratio or compressibility

Most glacial soils are suitable fill materials and have been extensively used as engineered fills for over 200 years. It may be necessary to remove boulders and zones of weaker and water-bearing materials during excavation. Fill selection will depend on its use; for example, fill for landfill liners must be able to achieve a low permeability when compacted. An engineered fill has to deal with changes in effective stress, water content and erosion during its life. This means that additional measures may be necessary to protect a fill from water. While the target density may be the maximum dry density, an allowance is made for the variation in water content and compacting effort using criteria set out in Figure 5.30. For example, for a fill to be used to support foundations and a fill in front of gravity retaining structures, the air voids content must not exceed 5% and the density must be at least 95% of the maximum dry density based on the heavy compaction method; between 5% and 10% air voids for embankments and other mass fills and a maximum dry density of at least 95% of the maximum dry density from the light compaction test. Figure 5.30 shows the more rigorous criteria for landfill liners where strength and permeability are critical.

Compaction trials are necessary to ensure that a method specification can produce a fill that meets the expected performance, the method of compaction meets the end product specification and the plant type, the number of layers and layer thickness are correct. The number of control tests depends on the volume of the fill, nature of the structure, the uniformity of the fill and the outcome the compaction trials. Figure 5.34 is a suggestion for fills supporting low rise housing, which shows that the number of tests increases with the volume of the fill but the number per 1,000 m^3 reduces. An alternative is to specify tests on two samples per 1,000 m^3 of fill for large projects and five samples per 1,000 m^3 for smaller projects.

Arrowsmith (1979) provided an overview of constructing over 300 miles of motorways which included compacted clay fill embankments, the majority of which were formed with matrix-dominated tills. He emphasised the need to establish the fabric and quantity of boulders from trial pits as boreholes did not provide enough details. Glacial clay tills are insensitive probably because of their method of deposition, which means that there is little difference between the *in situ* strength and strength of the compacted till. A till was deemed acceptable if the water content was within 2% of the plastic limit. However, this included wetter tills, which proved perfectly acceptable, and more stony tills, which were unacceptable. An alternative criterion was developed based on the water content of the matrix and the

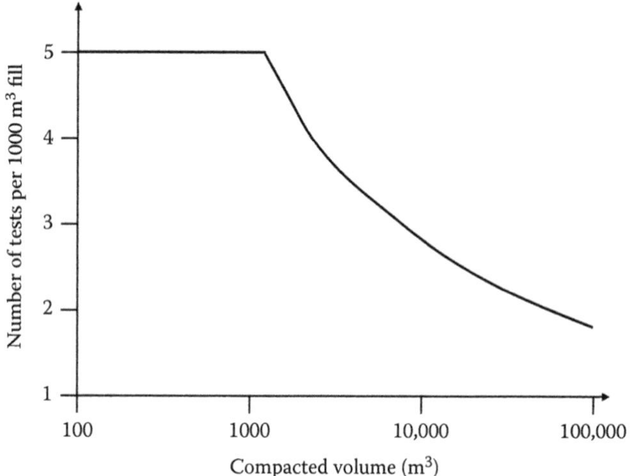

Figure 5.34 Suggested frequency of control tests for fills supporting low rise structures. (After Trenter, N. A. and J. A. Charles. *Proceedings of the Institution of Civil Engineers – Geotechnical Engineering,* 119(4); 1996: 219–230.)

plastic limit; acceptable material had a matrix water content 20% greater than the plastic limit. Arrowsmith (1979) suggested that the water content of the matrix (Table 5.29) could be based on the assumption that the sand content had a natural water content of 9% and larger particles were coated with 0.23-mm-thick water film (Smith, 1952). This compares to Jenkins and Kerr (1998) value of 5%.

Given that representative values of consistency limits of glacial tills are difficult to assess, Arrowsmith (1979) suggested that shear strength should be used as the controlling parameter. The dissipation of pore pressures built up in the compacted till was accelerated with horizontal drainage layers. Glacial tills may contain lenses and layers of weaker and water-bearing soils, which could be deemed unacceptable. Unacceptable wet clay can be treated by drying or lime modification.

Zones of contrasting permeability and interbedded fine-grained and coarse-grained soils can cause instability (Nowak, 2012b). Failure of embankments during construction is mostly due to inappropriate geometry, inadequate foundations, existing shear surfaces in the foundation soils and the variation in embankment materials. Failures during operation are generally translational failures in the slopes or deep-seated failures. Translational failures in the slopes of embankments are rarely more than 1.5 m deep (Perry, 1989) and are often due to a reduction in strength following construction or a change in pore pressure or water content. This could be due to infiltration, poor drainage or poor compaction of the shoulders. Operational failures are mostly a result of seasonal and permanent changes to the water content.

Problems of compacting glacial clays include reduction in grip, softening of acceptable material and rutting, all due to rainfall. Therefore, it is necessary to create drainage paths to remove water from the fill by ensuring appropriate cross falls and longitudinal falls, surface drainage channels and proof rolling to seal the surface. Glacial soils may have a significant percentage of silt so are more prone to slumping. Therefore, slopes should be protected from erosion. Low plasticity clays, such as matrix-dominated tills, can be reduced to a slurry when disturbed, especially if there are water-bearing layers in the till. All very coarse particles have to be removed because the fill is compacted in 250-mm layers. Uniformly graded fine sand is difficult to manage because of its lack of inherent strength and acceptable water content.

5.4.3.5 Embankments

In the nineteenth century, they recognised the issue of internal compression of embankments, which, for railways, could be compensated with ballast. To overcome the effect of rigid structures, such as culverts, they used brushwood as a compressible fill. This changed in the 1920s when mechanised plant and specifications were introduced to produce engineered fills.

There are many specifications for an engineered fill, which covers the selection of suitable materials, method of compaction and method of control. In the United Kingdom, the Highways Agency produced the Specification for Highway Works (SHW, 2013), which is commonly used whenever engineered fill is specified no matter the purpose. Table 5.32, based on SHW (2013), is an example of the criteria used to assess whether a glacial soil is acceptable for a general or selected fill. Table 5.32 shows that the classes of fills are based on their particle distributions, which are listed in Table 5.33. Inspection of this table suggests that all glacial soils could be used as fills. Possible classes of glacial soil fills are listed in Table 5.34 together with their possible application and method of control. Table 5.35 lists the characteristics of engineered fills, including glacial soil fills.

BS 6031:2009 suggests that there are five stages to the lifecycle of an embankment, but as shown in Figure 5.35, adaption should also be included because of changes in technology, use and climate during the life of an embankment, which will lead to changes as is evident from railway embankments constructed in the nineteenth century and still operational today.

Earth embankments include those for infrastructure, flood control, defense, dams and landscaping. Large dams are a special case, which require specialist input. Limit states include loss of overall site stability, internal erosion, surface erosion or scour; deformations in the embankment leading to loss of serviceability; settlements and creep displacements leading to damages or loss of serviceability in nearby structures or utilities; excessive deformation in transition zones; loss of serviceability of traffic areas by climatic influences such as freezing and thawing or extreme drying; and creep in slopes during the freezing and thawing period.It is necessary to consider the effects of construction of the embankment, the effects of adjacent construction or excavation and environmental changes. In the case of embankments used to control floods and defend coasts, the water level on the upstream side and possible drawdown have to be considered.

The underlying soil acts as the foundation to the embankment, which means that the bearing capacity has to be checked. If the foundation soils are not adequate, then stage construction, stabilisation, soil modification, soil replacement, piling and light weight fill are possible solutions. If the foundation soils are glacial in origin, consider the following:

- Matrix-dominated tills are generally dense so stability and settlement of embankments can be dealt with using conventional design methods. The presence of weaker horizons may increase settlement and possibly reduce stability depending on their location, thickness and extent. A drainage layer is necessary at the base of the embankment to allow pore pressures to dissipate, but if the embankment is being used to retain water, then this layer must not be continuous beneath the base of the embankment.
- Glaciolacustrine soils are more compressible and weaker than glacial tills, which means that settlement and stability could be an issue. Stability calculations should include non-linear failure mechanisms due to the layered nature of the foundation soil. Preloading is feasible because the horizontal permeability is greater than the vertical permeability allowing pore pressures to dissipate more rapidly. Differential settlement could generate tensile stresses in the embankment. A working platform may be necessary to construct the embankment.
- Clast-dominated tills and glaciofluvial soils can be considered as coarse-grained soils.

Table 5.32 Acceptability of fine-grained and coarse-grained soils for general and selected fill based on their classification characteristics (after SHW, 2016)

Class	Description	PSD	C_U		w (%)		I_L (%)		I_p	PI (%)	w_{opt}	MCV	c_u	c', φ'	c'_a, δ	k
			min	*max*	*min*	*max*	*min*	*max*								
General fill																
1A	Well-graded granular fill	√	10	–	√							√				
1B	Uniformly graded granular fill	√	–	10	√							√				
1C	Coarse granular material	√	5	–												
2A	Dry cohesive material	√			$I_p - 4$	–			√			√	√			
2B	Dry cohesive material	√			–	$I_p - 4$			√			√	√			
2C	Stony cohesive material	√			√				√				√			
2D	Silty cohesive material	√			√							√	√			
Selected fill																
6A	Well-graded granular material	√	10	–						NP						
6B	Coarse granular material	√			√					NP						
6C	Uniformly graded granular material	√	–	10	√					NP						
6D	Uniformly graded granular material	√	–	10	√		–			NP		√				
6E	Granular material	√			√			45		– 20						
6F/1	Granular material (fine)	√			$w_{opt} - 2\%$	w_{opt}					√					
6F/2	Granular material (coarse)	√			$w_{opt} - 2\%$	w_{opt}					√					
6F/3	Granular material	√			$w_{opt} - 2\%$	w_{opt}					√					
6G	Granular material	√														
6H	Granular material	√			√				NP			√				
6I	Well-graded granular material	√	10	–	√							√		√	√	
6J	Uniformly graded granular material	√	5	10	√							√		√	√	
6K	Granular material	√	5	–	$w_{opt} - 2\%$	$w_{opt} + 1\%$			6		√	√				
6L	Uniformly graded granular material	√														
6M	Granular material	√	5	–	$w_{opt} - 2\%$	$w_{opt} + 1\%$			6		√	√				
6N	Well-graded granular material	√	10	–	√									√	√	

(*Continued*)

Table 5.32 (Continued) Acceptability of fine-grained and coarse-grained soils for general and selected fill based on their classification characteristics

			C_U		w (%)		I_L (%)		I_P	PI (%)		w_{opt}	MCV	c_u	c', φ'	c'_a, δ	k
Class	*Description*	*PSD*	*min*	*max*	*min*	*max*	*min*	*max*									
6P	Granular fill	√	5	–	√								√		√		√
6Q	Granular fill	√			√												
7A	Cohesive material	√			√		–	45		–	25		√		√		
7D	Stony cohesive material	√			√		–	45		–	25		√		√		
7E	Cohesive material	√			√					10	–		√				
7F	Silty cohesive material	√	5	–	√		–	45		–	20		√				
7I	Cohesive material	√			√					10	–		√				

Source: After SHW. *Manual of Contract Documents for Highway Works: Volume 1 Specification for Highway Works: Series 600: Earthworks*. Highways England, London, 2013.

Table 5.33 Particle size distribution of general and selected fill according to SHW (2013)

Class	Particle size (mm)																			
	500	*300*	*125*	*90*	*75*	*37.5*	*28*	*20*	*14*	*10*	*6.3*	*5*	*3.35*	*2*	*1.18*	*0.6*	*0.3*	*0.15*	*0.063*	*0.002*
1A		100	95																<15	
1B			100																<15	
1C	100		10–95													0–25			15	
2A/2B			100											80–100					13–100	
2C			100											15–80					15–80	
2D			100																80–100	0–20
6A	100									0–100		0–85				0–45			0–5	
6B	100		0–10																	
6C			100								0–100		0–35	0–10		0–2				
6D										100		89–100		60–100	30–100	15–80	5–48	0–15		
6E			100	85–100						25–100						10–100				<15
6F1					100	75–100				40–95		30–85				10–50			<15	
6F2			100	80–100	65–100	45–100				15–60		10–45				0–25			0–12	
6F3			100	80–100	65–100	45–100				15–60		10–45				0–25			0–12	
6H								100				60–100			15–45	0–25		0–5		
6I/6J			100		85–100				25–100					15–100		9–100			<15	
6K								100											0–10	
6L										100		89–100		60–100	30–100	15–100	5–70	0–15		
6M					100														0–10	
6N/6P					100														<15	
6S					100									60–100		30–90			0–16	
7A					100													4–45	15–100	
7C			100		85–100				83–100					80–100		60–100			15–45	0–20
7D			100		85–100				40–90					15–79		15–75			15–45	0–20
7E					100		95–100												15–100	
7F			100																15–100	
7I					100		95–100												15–100	

Table 5.34 Possible glacial soil fills based on their particle size distribution, their application and typical method of compaction

Class	*Description*	*Glacial soils*	*Application*	*Compaction*
General fill				
1A	Well-graded granular fill	Glaciofluvial soils	General fill	Tab 6/4 Method 2
1B	Uniformly graded granular fill		General fill	Tab 6/4 Method 3
1C	Coarse granular material		General fill	Tab 6/4 Method 5
2A	Dry cohesive material	Matrix-dominated till	General fill	Tab 6/4 Method 1 (IL < 50%)
2B	Dry cohesive material		General fill	Tab 6/4 Method 2
2C	Stony cohesive material	Clast-dominated till	General fill	Tab 6/4 Method 2
2D	Silty cohesive material	Glaciolacustrine clays	General fill	Tab 6/4 Method 3
Selected fill				
6A	Well-graded granular material	Glaciofluvial soils	Below water	No compaction
6B	Coarse granular material		Starter layer	Tab 6/4 Method 5
6C	Uniformly graded granular material		Starter layer	Tab 6/4 Method 3
6D	Uniformly graded granular material		Starter layer below PFA	Tab 6/4 Method 4
6E	Granular material		Stabilisation to support capping	
6F/1	Granular material (fine-grading)		Capping	Tab 6/4 Method 6
6F/2	Granular material (coarse-grading)		Capping	Tab 6/4 Method 6
6F/3	Granular material		Capping	Tab 6/4 Method 6 capping layer 200 mm
6G	Granular material		Gabion filling	None
6H	Granular material		Drainage layer to reinforced soil and anchored earth structures	Tab 6/4 Method 3
6I	Well-graded granular material		Fill to reinforced soil and anchored Earth structures	Tab 6/4 Method 2
6J	Uniformly graded granular material		Fill to reinforced soil and anchored Earth	Tab 6/4 Method 3
6K	Granular material		Bedding for corrugated steel buried structures	End product 90% γ_{dmax} (vibrating hammer)
6L	Uniformly graded granular material		Upper bedding for corrugated steel buried structures	

(Continued)

Table 5.34 (Continued) Possible glacial soil fills based on their particle size distribution, their application and typical method of compaction

Class	*Description*	*Glacial soils*	*Application*	*Compaction*
6M	Granular material		Surround to corrugated steel buried structures	End product 90% γ_{dmax} (vibrating hammer)
6N	Well-graded granular material		Fill to structures	End product 95% γ_{dmax} (vibrating hammer)
6P	Granular fill		Fill to structures	End product 95% γ_{dmax} (vibrating hammer)
6Q	Granular fill		Stabilisation with lime to form capping	
7A	Cohesive material		Filter to structures	End product 100% γ_{dmax} (2.5-kg hammer) or γ_d corresponding to 5% air voids
7D	Stony cohesive material	Clast-dominated till	Fill to reinforced soil	Tab 6/4 Method 2
7E	Cohesive material	Matrix-dominated till	Stabilisation with lime to form capping	
7F	Silty cohesive material	Glaciolacustrine clays	Stabilisation with cement to form capping	
7I	Cohesive material		Stabilisation with lime and cement to form capping	

Source: After SHW. *Manual of Contract Documents for Highway Works: Volume I Specification for Highway Works: Series 600: Earthworks.* Highways England, London, 2013.

Table 5.35 Typical characteristics of engineered fills

Material type	*Benefits*	*Disbenefits*	*Comments*
Gravel and sand	High permeability; resist development of excess pore water pressures; instantaneous consolidation; instantaneous development of strength	Saturated or loosely packed fine sands can develop 'quick' conditions when subjected to vibration	Uniformly graded fine sands require tight moisture control for compaction Final layers of uniformly graded granular materials may need 'blinding' with appropriate material
Clay	Low permeability (for construction of water retaining structures, etc.)	Excess pore water pressures can develop during construction; suitability for use reliant on natural moisture content; long-term post-construction consolidation	Strength and deformation characteristics for fill and foundations primarily a function of moisture content; foundations influenced by structure and fabric of soil derived from geological history (e.g. over-consolidated clays)
Silt	Intermediate permeability between clay and sand, stability can be maintained by drainage	Strength and deformation behaviour very susceptible to instability caused by disturbance and seepage/high pore water pressures	Cohesive soils with characteristics intermediate between clays and sands
Mixed soils (clay, sand and gravel)	Soil strength can be reasonable if groundwater is managed	Soils of this type (such as glacial till) can be highly moisture susceptible	Properties generally determined by the predominate soil type but consideration must be given to secondary constituents as the soil can behave as either a granular or cohesive soil in different situations (see text above); require careful choice of laboratory testing regime
Interlayered high and low permeability soils	Permeable layers can be utilised as a drainage path if works are designed appropriately	Seepage paths provided which can prove difficult to drain in cuttings; pore water pressures within permeable layers can be unable to dissipate due to the presence of clay layers causing loss of strength; pore water pressures under embankments can be transmitted along silt and sand laminations and beds triggering instability at slope toe	Special attention required to earthworks drainage design

Source: After BS 6031:2009. *Code of Practice for Earthworks*. British Standards Institution, London.

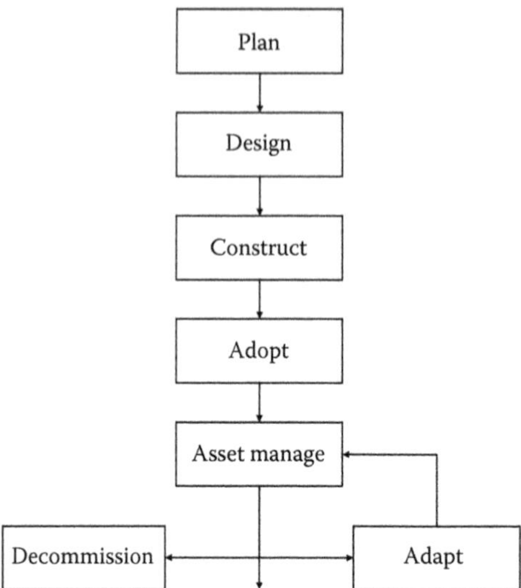

Figure 5.35 Lifecycle of an engineered fill.

Settlement of fill is due to internal compression and compression of the underlying soils and, possibly, due to external loads or changes in groundwater conditions. The compression and settlement of embankments for transport infrastructure, flood control and dams are critical and therefore have to be estimated. The settlement of the surface of an embankment is due to embankment compression because of its self-weight and compression of the underlying soils. Changes to groundwater conditions can cause further movement. Failure to comply with the specification can lead to inundation collapse if compacted dry of optimum or uncompacted. Internal settlement is possible if compacted too wet of optimum. Generally, if an embankment is constructed in line with published guidelines, internal settlement is not usually an issue because of the limited thickness of a fill in an embankment.

The performance criterion for embankments depends on their purpose: rail and road embankments are designed to limit settlement; flood control embankments and earth dams are designed to retain water. In all cases, the embankments have to be checked against overall stability, side slope failure, bearing failure and tensile splitting. It is likely that excavation in glacial soils will yield some unacceptable materials. Provided the amount of such material is limited, it is possible to use glacial soils to construct the embankment as was the case during the canal (eighteenth century), railway (nineteenth century) and motorway (twentieth century) eras in the United Kingdom. Compaction processes do not reinstate the ground as it was *in situ*; the volume of a fill is about 5% greater than the volume of the excavated soil (Nowak, 2012a). The stability of an embankment can be assessed using slope stability analyses though, as embankments are generally formed of homogenous materials, design charts such as those developed by Bond and Harris (2008) can be used. Global factors of safety are given in Table 5.36 and partial factors in Table 5.37.

5.4.3.6 Earth dams

Design and construction of earth dams is a highly technical discipline, which, in the United Kingdom, is regulated by law. Many dams have been built of glacial clays in the United Kingdom for flood control and to provide water for irrigation, canals and drinking over the last 300 years. For example, Kennard and Kennard (1962) report the construction of Selset

Table 5.36 Typical global factors of safety for (a) embankments and cuttings, (b) dam construction, (c) rail embankments and (d) infrastructure embankments

(a) Infrastructure earthworks		*Factors of safety (first-time failure)*	*Factors of safety (reactivation failure)*
Cuttings	Permanent	1.3–1.5	1.1–1.3
	Temporary	1.1–1.3	>1.0–1.2
Embankments	Permanent	1.4–1.6	1.3–1.5
	Temporary	1.2–1.4	1.1–1.3

(b) Earth dams	*Typical factor of safety*
End of construction	1.3–1.5
Steady-state seepage with reservoir full	1.5
Rapid drawdown	1.2

(c) Rail embankments	*Moderately conservative peak parameters*	*Moderately conservative residual parameters*
Affecting trackside and line side services	1.3	1.1
Affecting earthworks	1.2	1.1
Overall stability		1.1

(d) Infrastructure embankments	*Worst credible*	*Moderately conservative*
Shallow failure	1.05	1.15
Deep failure	1.10	1.30

Source: After Trenter, N.A. *Earthworks: A Guide*. Thomas Telford, London; 2001; Johnston, T.A. et al. *An Engineering Guide to the Safety of Embankment Dams in the United Kingdom*. Building Research Establishment, Watford, UK; 1999; Egan, D. *Proceedings of the Conference on Earthworks Stabilisation Techniques and Innovations*, 2005; Perry, J., M. Pedley, and M. Reid. *Infrastructure Embankments: Condition Appraisal and Remedial Treatment*. CIRIA, London, 2003.

Table 5.37 Potential internal erosion in an earth embankment dam with a glacial till core

Max filter size, D_{15} *(mm)*	*Unstable core unstable filter*	*Unstable core + stable filter Stable core + unstable filter*	*Stable core stable filter*
>1.4	High	Increased	Neutral
<1.4 or >0.7	Increased	Neutral	Reduced
<0.7	Neutral	Reduced	Low

Source: After Rönnqvist, H. Predicting surfacing internal erosion in moraine core dams, PhD Thesis, KTH, 2010.

dam in the north of England in which a 39-m-high dam was constructed using over 2 million m^3 of glacial till. At that time, there had been a failure of another dam constructed of glacial till and difficulty in constructing a further, similar dam; therefore, there was a concern about the use of till as a construction material. It was the only suitable available material. The foundation comprised alluvial gravels overlying glacial till. The till was unusual because the upper layers were described as firm (shear strength = 45 kPa), much softer than the lower layers (240 kPa). They installed 3,000, 10-m-long sand drains at 3 m centres to increase the rate of consolidation, achieving an adequate factor of safety with 200 mm of settlement. The nature of the fill and the weather conditions meant that the fill wet of optimum had to be used; they used drainage blankets within the dam to allow pore pressures to dissipate.

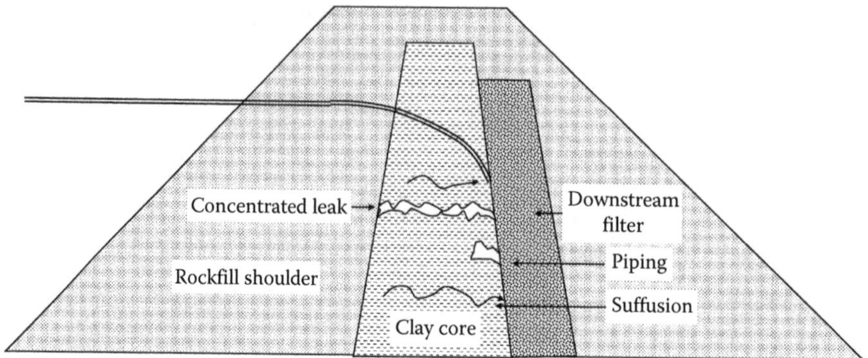

Figure 5.36 Possible internal erosion in a clay core earth dam.

Rodin (1969) and Arrowsmith (1979) described a similar technique for the construction of motorway embankments where it was not feasible to source acceptable fill. They placed layers of unsuitable fill within the embankment and horizontal drainage blankets to allow the soil to consolidate.

Sherard (1979) according to Rönnqvist (2012) recognised that earth dams with glacial till (well-graded) cores were more susceptible to internal erosion (Figure 5.36), leading to more sinkholes than occurred in dams with other types of core materials. Rönnqvist (2010) created a database of 91 earth dams with glacial till cores, which included 21 with internal erosion. He produced three categories of dams:

- *Category 1*, which had suffered internal erosion
- *Category 2*, where internal erosion may be taking place
- *Category 3*, where no internal erosion was obvious

He investigated a number of methods to assess likely filter performance and whether internal erosion would take place. Sherard and Dunnigan (1989) suggested that D_{15} has to be less than 0.7 mm to prevent internal erosion. Foster et al. (2000) suggested that there was a relationship between the fine-medium sand content and the D_{15} filter size that applied to Category 1 dams but did not provide clear guidance. Kenney and Lau (1985) suggested the use of an $H{:}F$ curve in which the ratio of the mass fraction of particle sizes between d and $4d$ (H) and the weight of particles less than d ($=F$) is used to show that instability occurs when $H = 1.3F$. Rönnqvist (2010) produced Table 5.36, which shows the potential for internal erosion.

Rönnqvist and Viklander (2014) applied the Kenney–Lau method to the database to produce Figure 5.37, a guideline to evaluate earth dams with glacial till cores showing that it may be possible to use the method to assess whether internal erosion is likely. This showed a boundary between potentially stable and unstable dams. Ronnqvist (2015) produced a refinement of this method (Figure 5.38) to relate the stability index of dams with glacial soil cores and the susceptibility to internal erosion based on the filter size.

5.4.4 Recommendations

Given the extent and location of glacial soils, they have been a significant resource of construction materials, including aggregates, landfill liners, brick-making earth, non-engineered

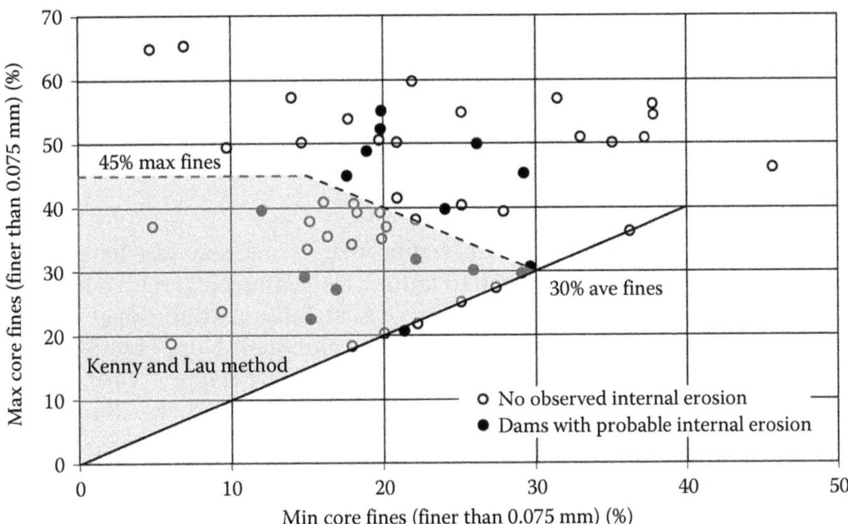

Figure 5.37 Guideline to evaluate earth dams with the glacial till core. (After Rönnqvist, H. and P. Viklander. *Electronic Journal of Geotechnical Engineering*, 19(5); 2014: 6315–6336.)

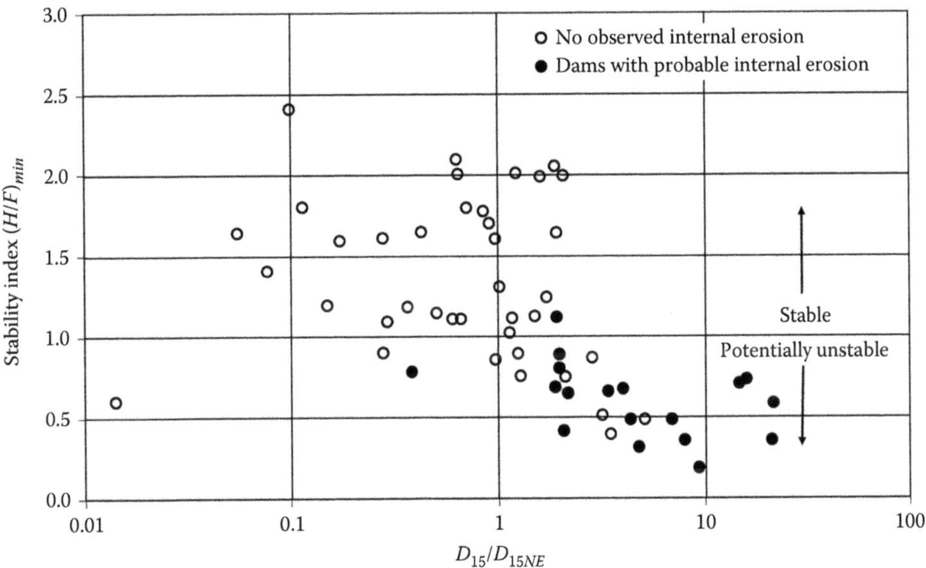

Figure 5.38 Unified approach to evaluate the stability of earth dams with glacial clay cores subject to erosion. (After Rönnqvist, H. *International Conference on Geotechnical Engineering: ICGE2015 Colombo*, Sri Lanka, 10th August and 11th August, 2015.)

fill and engineered fill. Their composition has proved to be an asset because they can be used to form a dense fill, which, depending on the particle size distribution, is relatively incompressible, impermeable and strong. There is a need to be selective when excavating a glacial soil to remove very coarse particles and unsuitable materials. Glacial soils are spatially variable but it is possible to mix the more granular soils prior to placement to create a more homogeneous fill. Some glacial soils can be modified to improve their performance. End product and method specifications can be followed. Given the difficulty in determining the structure of a glacial till, unsuitable material should be expected and mitigation measures, including modification and removal, should be planned during excavation.

5.5 SLOPE STABILISATION

Natural slopes are in equilibrium but in engineering terms may not be considered safe because a change in conditions could lead to failure; for example, pore pressure changes due to infiltration can lead to instability. It is possible to stabilise natural slopes that are potentially unstable using structural and non-structural techniques. Slope stabilisation methods include embedded solutions (Figure 5.39), gravity solutions (Figure 5.40), reinforcement (Figure 5.41), anchors, piling, regrading and drainage (provided the drainage system is maintained). No matter which method is used, it is necessary to control groundwater conditions. The design of embedded and gravity solutions is covered in Chapter 6. Reinforced soil is a combination of reinforcement and engineered fill. The issues of using glacial soil as engineered fill are covered in Section 5.4.3 and the design of reinforced soil is covered in BS 8006-1:2010.

Vegetation can contribute to a stable slope because it reduces erosion, increases the shear strength of the surface layers and removes water by transpiration. These benefits are not usually taken into account in design though Coppin and Richards (1990) suggested that the cohesion of the upper layers could be increased by up to 20 kPa because of vegetation.

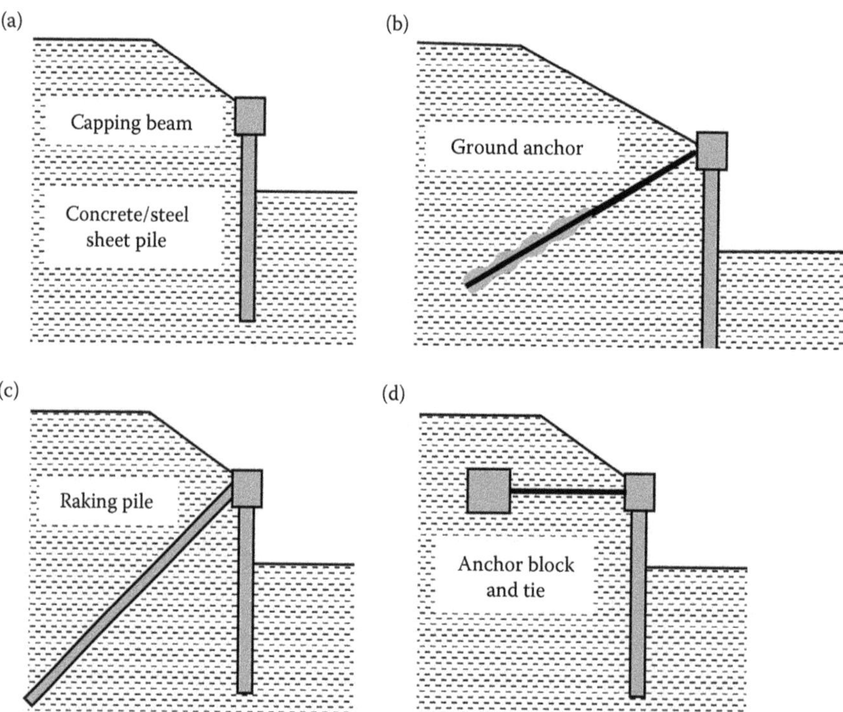

Figure 5.39 Typical embedded means of stabilising slopes: (a) cantilever walls, (b) ground anchors, (c) raking piles and (d) anchor blocks. (After Nowak, P. A. *ICE Manual of Geotechnical Engineering*, Thomas Telford Ltd, London; 2012c: 1087–1091.)

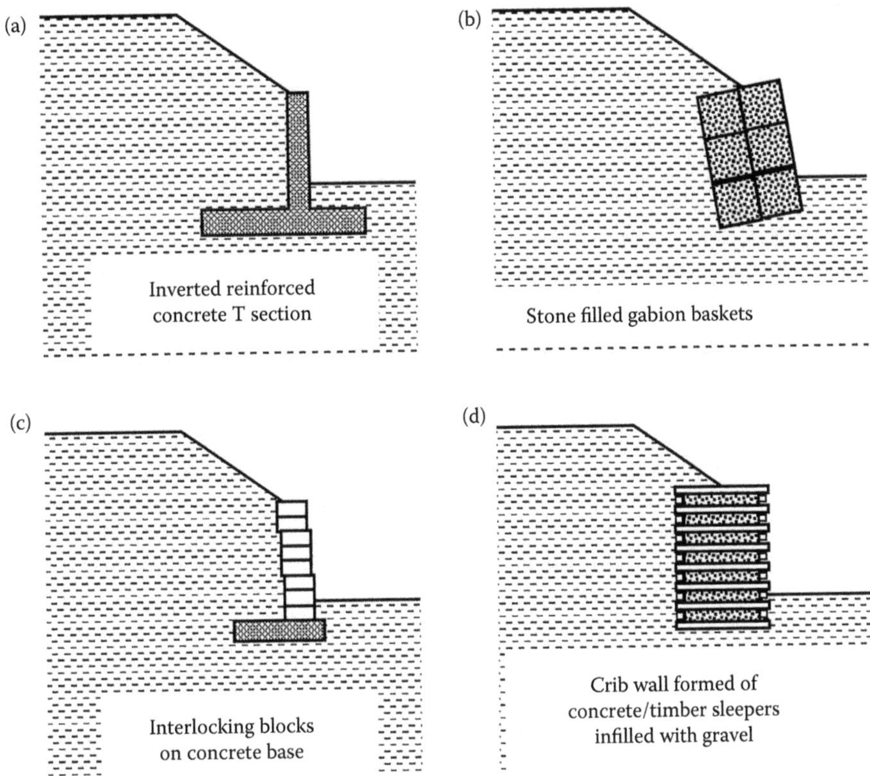

Figure 5.40 Typical gravity means of stabilising slopes: (a) reinforced concrete wall, (b) gabion wall, (c) dry block wall and (d) crib wall. (After Nowak, P. A. *ICE Manual of Geotechnical Engineering*, Thomas Telford Ltd, London; 2012c: 1087–1091.)

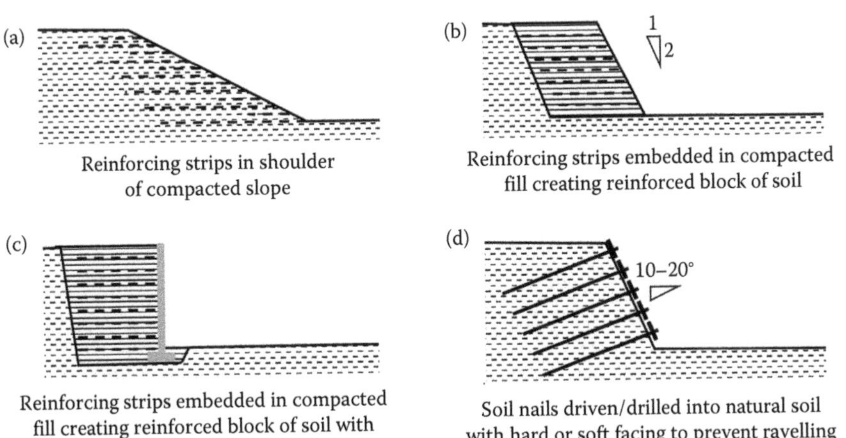

Figure 5.41 Typical reinforced soil solutions: (a) reinforced soil slope, (b) reinforced soil block, (c) reinforced earth slope and (d) soil nail slope. (After Nowak, P. A. *ICE Manual of Geotechnical Engineering*, Thomas Telford Ltd, London; 2012c: 1087–1091.)

5.5.1 Soil nailing

The concept of soil nailing developed from the use of rock bolts in rock excavations and current guidelines in the United Kingdom, United States and Hong Kong are based on the seminal study in France (Clouterre, 1991) in which a number of case studies were analysed in some detail. Soil nails are used to stabilise existing slopes, embankments, cuttings and retaining structures. Failure mechanisms include internal (facing, soil nail, soil nail/grout interface, grout soil interface) and external (rotational, sliding and bearing) failures. There are a number of design guidelines (e.g. BS 8006-1:2010; FHWA, 2003; HK, 2008) based on a soil nail stabilised slope shown in Figure 5.42. Overall stability is normally assessed using slope stability analysis taking into account the contribution the soil nails make to the overall stability; the contribution of the facing may be included.

Installing nails in cuttings is usually a top-down process ensuring stability at all times. This means an observational approach can be used to ensure that enough nails are installed for the ground conditions uncovered. Thus, a design is usually a generic design for the slope based on the ground investigation and the design is adapted to take account of local soil conditions, particularly important in spatially variable glacial soils. BS 8006-1:2010 suggests (Table 5.38) that soil nails can be installed in firm to stiff clays provided the clays are not softened by the installation process. Soil nails cannot be used in loose, coarse-grained soils because they are sensitive to disturbance, unable to stand unsupported as the nails are installed and the uncertainty about the mobilised interface strength because of installation disturbance. Installation on slopes exceeding the angle of friction relies on a combination of soil suction and arching to remain stable during construction.

Soil nailing is based on the assumption that a failing soil mass, the active zone, is nailed back to the underlying soil, the resistance zone. Soil nails can be driven or fired into place or drilled and grouted or placed in predrilled holes. Nails are usually installed at a spacing of 1–2 m and at an angle of 5–10° below horizontal.

Directly installed nails tend to be short and small in diameter making them more vulnerable to damage. Self-drilled nails are also directly installed with the nail acting as the drill rod. The drilling mud used to install the nail is replaced by grout, which has a compressive

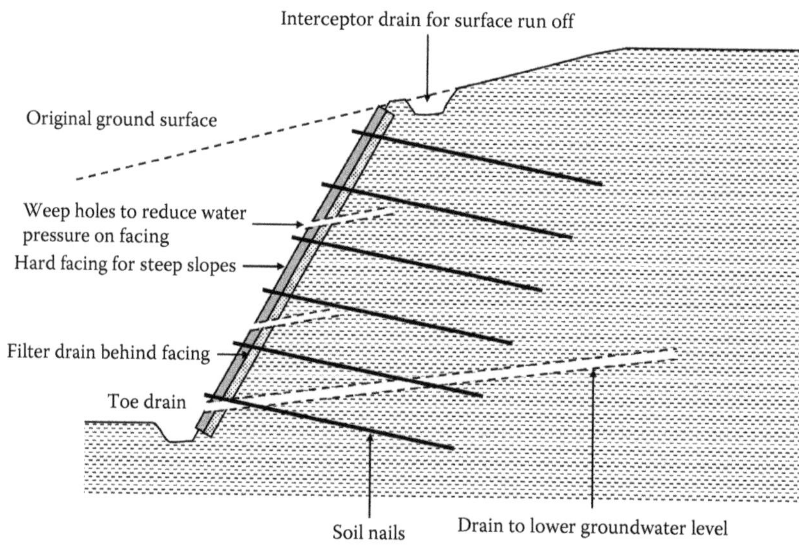

Figure 5.42 Structural and drainage components for a steep slope stabilised by soil nails.

Table 5.38 Summary of ground conditions that are suitable for soil nailing

	Ground conditions best suited for soil nailing	*Ground conditions less suitable for soil nailing*	*Possible measures to improve suitability of ground conditions*
Material to be nailed	• Firm to stiff, low plasticity clays	• Soft cohesive and organic soils prone to creep deformation • High plasticity or highly frost susceptible soil	• None will improve these soils sufficiently for soil nailing • Provide adequate protection against wetting and drying
	• Matrix-dominated tills without cobbles and boulders • Fine to medium sands and silty sands with some apparent cohesion • Medium dense to dense sands and gravels with some apparent cohesion	• Loose, clean sand and gravels with little or no apparent cohesion	• Pre-grouting or ground freezing to improve temporary stability • Limit excavation heights/lengths
	• Engineered fills formed of glacial soils	• Non-engineered fills	• Excavate and replace with suitable material • Use ground improvement to improve non-engineered fills
Groundwater conditions	• Above the groundwater table with a dry excavated face	• Below the water table • Artesian groundwater at depth	• Temporary and permanent dewatering • Allow for in design
	• Perched water or groundwater seepage through coarse-grained soils or pockets of coarse-grained soils		• Temporary and permanent dewatering • Measures to maintain long-term stability
Underlying ground conditions and geological features	• Underlying conditions and geological features that do not compromise the stability and performance of the soil nailed structure	• Adverse underlying ground conditions: • Pre-existing slip surface • Soft compressible soil layers • Voids • Silty soils	• Appropriate measures to deal with the ground conditions

Source: After BS 8006-1:2010. *Code of Practice for Strengthened/Reinforced Soils and Other Fills*. British Standards Institution, London.

strength, typically, of at least 5 N/mm^2 on installation and 28 N/mm^2 at 28 days. It is important that the annulus is completely filled with grout; that is, all the drilling mud is replaced because the design assumes that is the case. It can be difficult to install self-drilled nails in glacial soils containing large and very large particles because the nails may break or bend. In that case, nails should be installed in grout-filled boreholes but note the problems associated with soil nails shown in Figure 5.43 where additional grout may be needed to overcome loss of grout in granular pockets, unstable boreholes collapsing because of weaker or more permeable layers, cobbles or boulders obstructing the nails, reduced nail capacity because of weaker layers and ravelling at the face.

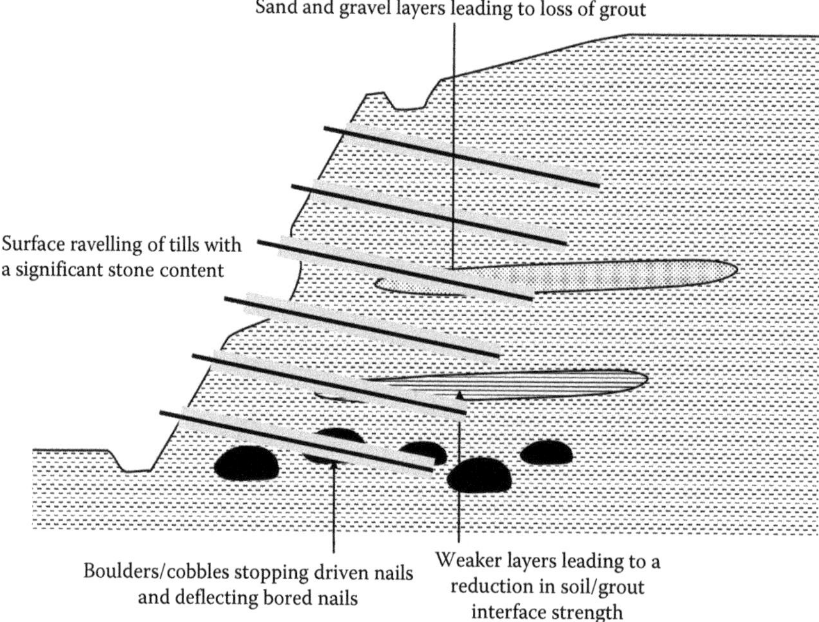

Figure 5.43 Potential problems of installing soil nails in glacial tills.

Drainage is necessary to prevent damage to the excavated face during construction and, if assumed in the design, to prevent water pressure building up on the facing. A rising groundwater level in the slope will reduce the overall stability; increased water pressure on the facing could lead to local failure of the facing. The effect of lenses and layers of water-bearing sands and gravels, the opening of discontinuities due to excavation leading to more permeable till and the effect of matrix-dominated tills as aquicludes giving rise to perched water levels are possible in glacial tills (Figure 5.44).

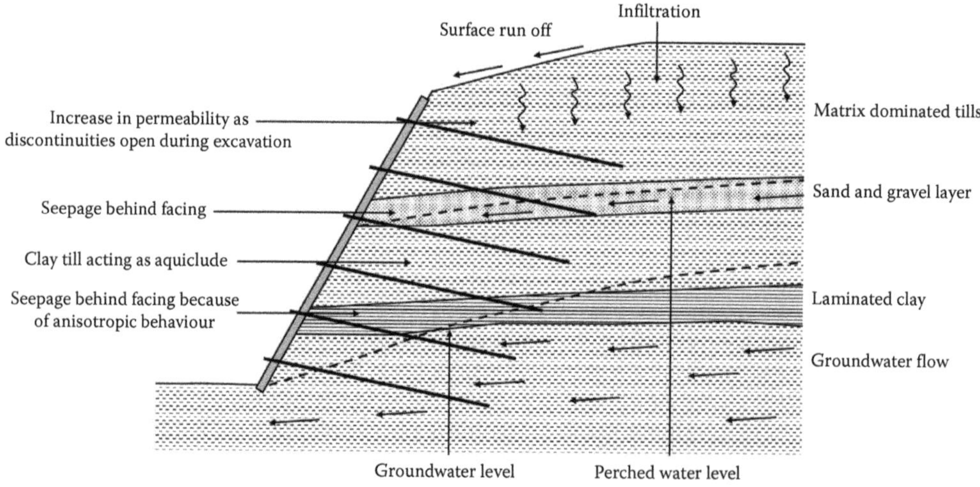

Figure 5.44 Potential impact of hydrogeological conditions on overall stability and the facing of a slope reinforced with soil nails.

Routine ground investigations should provide relevant design information though it is important, in addition, to determine the chemical characteristics including pH of the soil and groundwater, water-soluble sulphate, chloride ion content and soil resistivity because of potential degradation of the nails, especially if there are adverse environmental conditions including partially saturated soils, saline groundwater conditions and fluctuating groundwater levels (BS 8006-1:2010). The category of risk for a range of soil nail systems depends on whether they are for temporary or permanent use and the environmental conditions (Table 5.39). The effect of potential installation problems in glacial soils should be considered in the risk assessment.

Soil nail stabilised slopes are not prescriptive but typical layouts depend on the angle of the finished slope (Figure 5.45). The nails are assumed to act in tension and those tensile forces are applied to the slip surface in limit equilibrium methods.

The capacity of a nail depends on the tensile strength of the nail, the interface friction between the nail and the grout and the interface friction between the grout and the soil. The unit capacity varies along the length of the nail. The contribution a nail makes to the overall stability depends on the length within the resistance zone, which must equal the capacity of the nail in the active zone and the contribution the facing makes which depends on the type of facing and the slope angle. The mobilised capacity depends on the geometry of the nail configuration, the method of installation, the relative stiffness of the nails and

Table 5.39 Categories of risk for different soil nailing systems

	Category of risk								
	Low			*Medium*			*High*		
Type of soil nail	*A*	*B*	*C*	*A*	*B*	*C*	*A*	*B*	*C*
Steel directly in contact with soil	R	R	NR	R	NR	NR	NR	NR	NR
Coated steel directly in contact with soil	R	R	R	R	R	NR	NR	NR	NR
Steel surrounded by cement grout	R	R	R	R	R	NR	R	NR	NR
Self-drilled steel surrounded by cement grout	R	R	R	R	R	NR	R	NR	NR
Coated steel surrounded by cement grout	R	R	R	R	R	NR	R	NR	NR
Self-drilled coated steel surrounded by cement grout	R	R	R	R	R	NR	R	R	NR
Polyester composite surrounded by cement grout	R	R	R	R	NR	NR	R	NR	NR
Vinylester composite surrounded by cement grout	R	R	R	R	R	R	R	R	NR
Stainless surrounded by cement grout	R	R	R	R	R	R	R	R	NR
Self-drilled stainless surrounded by cement grout	R	R	R	R	R	R	R	R	NR
Steel surrounded by grouted impermeable ducting	R	R	R	R	R	R	R	R	R
Coated steel surrounded by grouted impermeable ducting	R	R	R	R	R	R	R	R	R
Stainless steel surrounded by grouted impermeable ducting	R	R	R	R	R	R	R	R	R
Steel surrounded by pre-grouted double impermeable ducting	R	R	R	R	R	R	R	R	R

Source: After BS 8006-1:2010. *Code of Practice for Strengthened/Reinforced Soils and Other Fills*. British Standards Institution, London.

Note: A: temporary (<2 year) or permanent in slightly corrosive environment; B: temporary in highly corrosive environment; C: permanent in highly corrosive environment; R: recommended; NR: not recommended.

Slope angle	Up to 45°	45° to 65°	65° to 90°
Nail length (in terms of slope height)	0.5 to 2.0	0.5 to 1.5	0.5 to 1.2
Nail spacing Vertical	1.5 to 3 m	1.0 to 2.0 m	0.75 to 1.2 m
Horizontal	1.5 to 3 m	1.0 to 2.0 m	0.5 to 2.0 m
Typical facing	Soft, non-structural for erosion control with enlarged nail heads	Flexible facings with a structural role that contribute to stability	Hard facings that provide a structural role that contribute to stability through transfer of forces

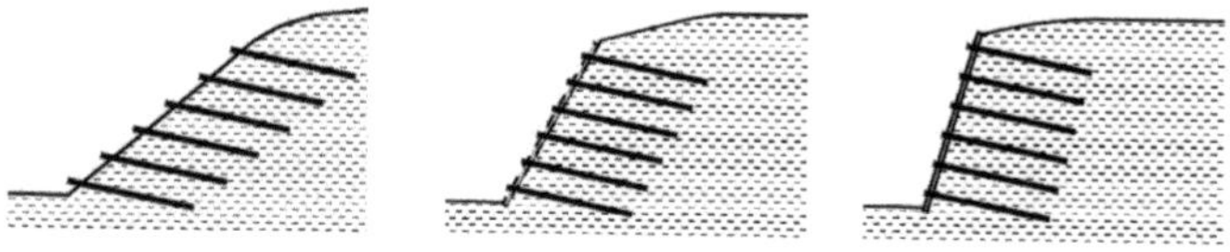

Figure 5.45 Typical dimensions of soil nail slopes showing the type of facing, the vertical and horizontal spacing of the nails and the length of the nails with respect to the height. (After BS 8006-1:2010. *Code of Practice for Strengthened/Reinforced Soils and Other Fills*. British Standards Institution, London.)

the ground and groundwater conditions. The ultimate bond stress, τ_{bu}, based on effective stress, is given by

$$\tau_{bu} = \lambda_f k_f \sigma'_v \tan \Phi'_k \tag{5.9}$$

where k_f is a factor relating the average radial effective stress around the nail to the vertical effective stress, σ'_v, and has a value typically in the range of 0.55–0.9, depending on the relative density of the soil and degree of stress reduction due to slope movements in the active zone of the slope; λ_f is the interface factor which is between 0.7 for smooth interfaces and 1.0 for rough interfaces.

And based on total stress,

$$\tau_{bu} = \alpha c_u \tag{5.10}$$

where α ranges from 0.5 to 0.9 for bond lengths ranging from 3 to 7 m.

Alternatively, the bond stress can be determined by a pullout test using maintained load tests or constant rate of pullout tests. This is recommended in glacial soils because of their spatial variability and difficulty in obtaining representative samples to test. Pullout

Table 5.40 Type of soil nail test

	Minimum frequency of load tests	
Test type	*Sacrificial nail test*	*Production nail test*
Geotechnical Category 1	Optional	Optional
Geotechnical Category 2	Minimum of three nails with at least one nail per soil type; Tests are optional if experience in the soils exists	2% with a minimum of three tests At least one test per soil type and per excavation stage
Geotechnical Category 3	Minimum of five nails with at least two nails per soil type	3% with a minimum of five tests At least one test per soil type and per excavation stage

Source: Adapted from BS EN 14490:2010. *Execution of Special Geotechnical Works. Soil Nailing*. British Standards Institution, London.

tests (Table 5.40) can be used on sacrificial nails when the ultimate capacity of a nail is measured or production nails when the design load is verified. Given the spatial variability of glacial soils, it is recommended that all slopes be treated as Category 3 structures. It is possible with sacrificial nails to determine the overall capacity but unless the failure mechanism is known, not the capacity in the active and resistance zones. The criterion (Table 5.41) for the pullout tests includes the maximum test load, the number of load cycles and the creep rate.

The partial factor for the nails depends on the source of the capacity of the nails (Table 5.42), which includes pullout tests and total and effective stress capacity. Partial factors are shown in Table 5.43. It is recommended that all factors greater than one should be increased by 10% for Category 3 structures, which would be the case for glacial soils.

Table 5.41 Details of tests to check the capacity of soil nails

	Sacrificial nail test	*Production nail test*
Purpose and type of test	To verify the ultimate soil nail to ground bond resistance to be used in the design: • The bond in the passive zone • The bond in the active zone • The bond along the entire length of the nail	To demonstrate satisfactory soil nail performance at a load designated by the designer
Time of testing	Before, during or after production works	During or on completion of production works
Type of nail	Sacrificial	Production
Action for non-compliance	Review method of soil installation and consider alternative nail length and layout	Seek advice
Comments	Test each soil layer	Limit load to prevent overstressing the nail to grout bond or damaging the corrosion prevention Debond the nail in the zone of influence of the facing if hard facing is used
Estimation of maximum test load	The value of the test load shall be based on the design value of the bond resistance, the partial factor and the appropriate value of correlation factor	The value of the test load shall be based on either the design bond resistance multiplied by a proof factor (between 1.1 and 1.5) which is less than the design partial factor to prevent overstressing the nail to grout bond or damaging the corrosion prevention
Number of load cycles	Minimum of two cycles with the bond resistance in the first cycle not exceeding the design value	Single cycle normally satisfactory
Number of load increments	Maximum increment size should be sufficient to define the shape of the load displacement curve and should not exceed 20% of the maximum cycle load	Minimum of five load increments
Interpretation of results	The test result is acceptable provided at the maximum test load the creep rate is less than 2 mm per log cycle of time Maximum extension at the head of nail is not less than the extension of any debonded length of the test nail	Test is acceptable provided that at the maximum proof load the creep rate is less than 2 mm per log cycle of time Maximum extension at the head of nail is not less than the extension of any debonded length of the test nail

Source: After BS EN 14490:2010. *Execution of Special Geotechnical Works. Soil Nailing*. British Standards Institution, London; BS 8006-1:2010. *Code of Practice for Strengthened/Reinforced Soils and Other Fills*. British Standards Institution, London.

Table 5.42 Ultimate limit state approach to determine the bond stress

Method to determine ultimate bond stress (τ_{bu})	*Factors to determine characteristic bond stress from ultimate values* ($\tau_{bk} = \tau_{bu}/\gamma_k$)	*Factors to determine design bond stress from characteristic values for set 1* ($\tau_{bd} = \tau_{bk}/\gamma_{tb}$)	*Factors to determine design bond stress from characteristic values for set 2* ($\tau_{bd} = \tau_{bk}/\gamma_{tb}$)
Empirical pullout test data	$\gamma_k = 1.35–2.0$ Take into account degree of confidence	$\gamma_{tb} = 1.11$	$\gamma_{tb} = 1.50$
Effective stress	$\gamma_k = 1.0–1.35$ Account for dilation and deformation	$\gamma_{tb} = 1.11$	$\gamma_{tb} = 1.50$
Total stress	$\gamma_k = 1.35–2.0$ Account for strain softening, plasticity and shrink swell effects	$\gamma_{tb} = 1.11$	$\gamma_{tb} = 1.50$
Pullout tests	Cautious estimate of test data	$\gamma_{tb} = 1.1–1.3$ for coarse-grained soils $\gamma_{tb} = 1.15–1.7$ for medium and high plasticity soils	$\gamma_{tb} = 1.5–1.7$ for coarse-grained soils $\gamma_{tb} = 2.0–2.25$ for medium and high plasticity soils

Source: After BS 8006-1:2010. *Code of Practice for Strengthened/Reinforced Soils and Other Fills*. British Standards Institution, London.

Table 5.43 Partial factors for soil nail systems

		Symbol		*Set 1*	*Set 2*
Actions	Self-weight of soil (W)	γ_g	Destabilising	1.35	1.0
			Stabilising	1.0	1.0
	Permanent surcharge (q_p)	γ_{qp}	Destabilising	1.35	1.0
			Stabilising	1.0	1.0
	Variable surcharge (q_v)	γ_{qv}	Destabilising	1.5	1.3
			Stabilising	0	0
	Groundwater pressure (u)	γ_u	Destabilising	1.0	1.0
			Stabilising	1.0	1.0
Material properties	Angle of friction ($\tan \phi_k'$)	$\gamma_{\tan\varphi'}$		1.0	1.3
	Cohesion (c_k')	$\gamma_{c'}$		1.0	1.3
	Undrained shear strength (c_{uk})	γ_{cu}		1.0	1.4
	Unit weight (γ_k)	γ_γ		1.0	1.0
Soil nail resistances	Bond stress (τ_{bk})	γ_{rb}	Empirical	1.1	1.5
			Effective stress	1.1	1.5
			Total stress	1.1	1.5
			Pullout tests	1.1–1.7	1.5–2.25
	Tendon strength (T_k)	γ_s		1.0	1.15 (steel)
Model factor	Depends on method of analysis				

Source: After BS 8006-1:2010. *Code of Practice for Strengthened/Reinforced Soils and Other Fills*. British Standards Institution, London.

There are three types of facing:

1. Soft facings are designed for slopes less than 45° to prevent erosion of topsoil while vegetation is established; the vegetation helps reduce infiltration and stabilise the surface of the slope.

2. Flexible facings provide permanent facing by stabilising the soil between the nails and transmitting some of the load on the nails to the soil via the nail plate. They can be used for slopes up to 70° and are usually formed of metallic meshes.
3. Hard facing is formed of sprayed concrete, or cast *in situ* or precast concrete panels. They are designed to resist earth pressures and transfer soil nail load to the soil via the facing.

Details of the design of facing are given in BS 8006-1:2010, FHWA (2003), HK (2008) and Clouterre (1991).

Drainage (Figure 5.42) includes the following:

1. A crest drain above the stabilised slope to remove water flowing down the slope
2. A toe drain to collect water flowing over the facing and water emerging from drains in the slope
3. Weep holes or filter drains behind the facing to reduce the water pressure on the facing
4. Raking drains at 5–10° above horizontal to reduce the water pressure in the soil

Soil nails have been used to stabilise slopes in matrix-dominated tills (e.g. Joy et al., 2010; Lindsay et al., 2015; Smith et al., 2015), which showed that nails should be placed in pre-drilled grouted boreholes. Joy et al. (2010) suggested that the measured capacity of soil nails in glacial soils was almost double that suggested by design guidelines. They studied three sites in Ireland and America and undertook a review of practice through interviews and questionnaires. Figure 5.46 shows the results of pullout tests compared with the values recommended by various design standards and those based on typical α values of 0.45 and 0.72 (Gavin, 2009) for soil nails in Dublin Boulder Clay. They concluded that design guidelines are conservative and recommended ($\alpha = 1.1$).

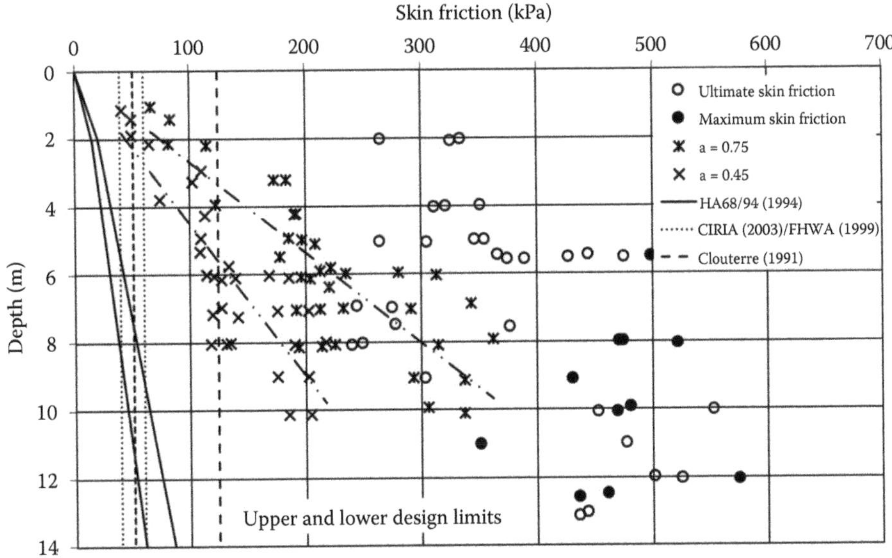

Figure 5.46 Comparison between the measured skin friction based on pullout tests and those predicted from codes of practice using the total stress approach and values back figured from pile tests (Gavin, 2009) for soil nails in Dublin Boulder Clay. (After Joy, J., T. Flahavan, and D. F. Laefer. *Earth Retention Conference* 3; 2010: 252–261.)

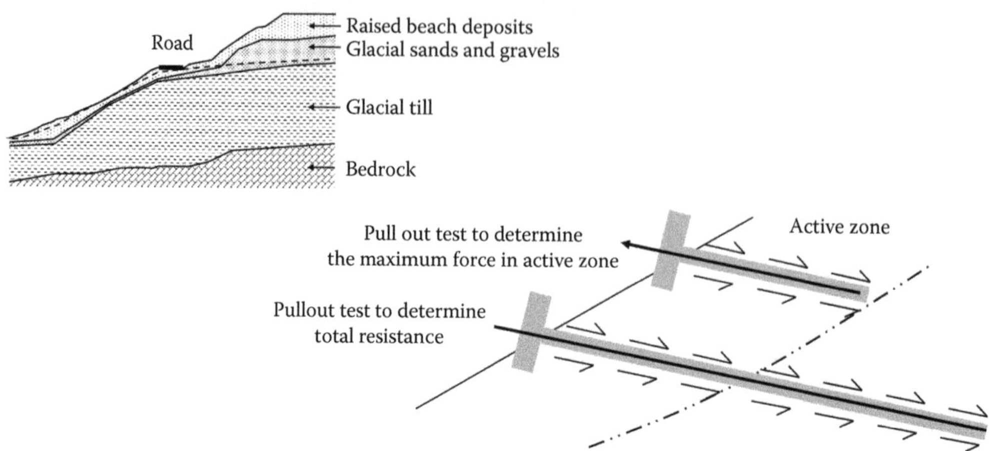

Figure 5.47 Typical ground conditions at a coastal cliff at Stonehaven, Scotland and the type of pullout tests to determine the forces in the active and resistance zones. (After Lindsay, F. M. et al. *Proceedings of the XVI ECSMGE Geotechnical Engineering for Infrastructure and Development*, 2015.)

Lindsay et al. (2015) and Smith et al. (2015) described a project involving 1,500 self-drilled soil nails 7–24 m long to stabilise a 40-m-high coastal slope at Stonehaven, Scotland (Figure 5.47). Hollow steel bars of 38 mm diameter were installed in 100-mm diameter boreholes at 20° to the horizontal. Flexible facing was used with the soil nail heads helping confine the active zone of the soil. The load transfer mechanism is shown in Figure 5.48. The plate spreads the load onto the soil so the bearing capacity has to be checked. They tested 56 sacrificial nails (Figure 5.47) to determine the capacity in the potential active zone and resistance zone. They observed a number of erroneous results; pullout loads for some of the short nails exceeded those for the long nails, the mobilised zone exceeded the drilled zone because grout permeated the surrounding gravel, and potential expansion of borehole diameter due to grout pressure. These observations were consistent with the problems of installing soil nails in glacial soils shown in Figure 5.43. They questioned the use of self-drilled nails in complex ground conditions without control tests and recommended solid nails placed in predrilled grout-filled holes provided the boreholes were of the correct diameter.

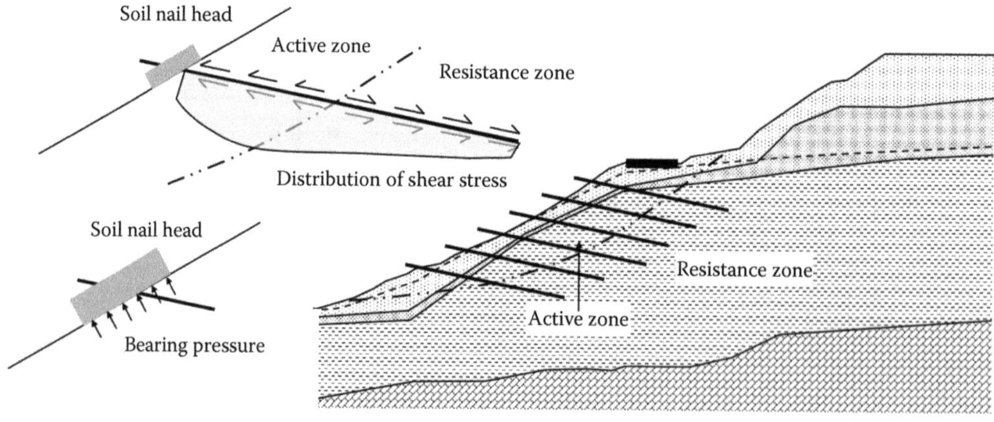

Figure 5.48 Load transfer mechanisms in soil nail stabilised slopes showing the variation in shear stress in the active (failing) and resistance zones and the bearing pressure beneath the soil nail head.

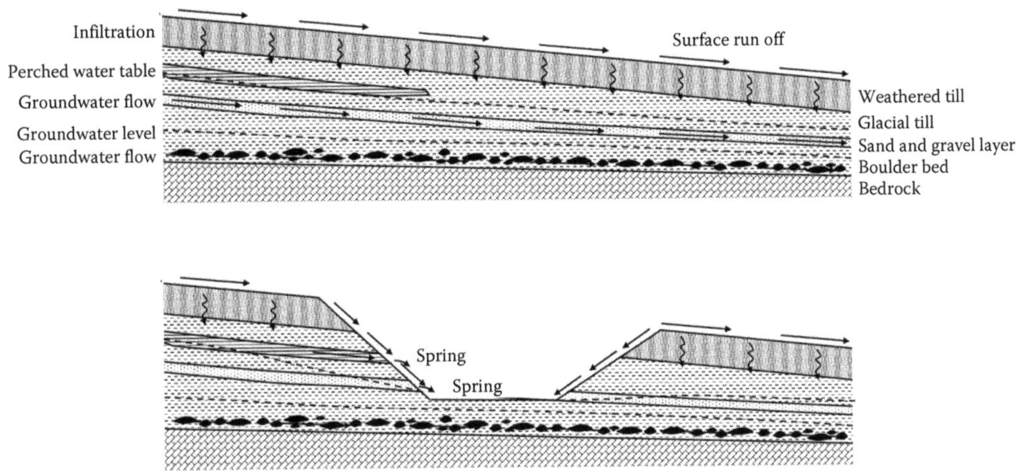

Figure 5.49 Possible flow patterns in glacial soils and the effects that an excavation could have upon those patterns.

5.5.2 Drainage systems

Drainage during and post-construction of temporary and permanent slopes is essential to control the hydrogeological conditions to prevent water entering the active zone of the slope. Patterns of flow (Figure 5.49) affecting an excavation are related to the structure of glacial soils including the presence of more permeable layers and layers with anisotropic permeability. Excavation into these soils intercepts these structural features changing the flow patterns. Perry (1989) suggested that a feature of failed slopes was the lack of drainage.

Drainage systems include cut-off drains, interceptor drains, counterfort and slope drains and herringbone drains. Cut-off drains are designed to prevent water entering a slope by intercepting water flowing through the soil. Water is collected in the gravel drain and taken down to a pipe at the base of the drain, which must be in impermeable soil to prevent water flowing back into the overlying soil. Interceptor drains prevent surface water entering the slope, so they are installed above the slope. Counterfort drains are perpendicular to the slope; thus, the water flows towards the toe of the slope to a drain. These are appropriate for glacial soils as they intercept the structural features. Herringbone drains are shallow gravel-filled trenches to collect water emerging from the soil. It may be necessary to install well points to lower the water level.

Embankments for infrastructure may be built on a working platforms providing access to construction equipment. This may also act as a drainage blanket to allow dissipation of pore pressures in the underlying soils. This would not be appropriate for flood embankments.

5.5.3 Recommendations

The spatial variation in composition, fabric and structure means that any slope in glacial soils can be considered a hazard that could fail and mitigation measures have to take into account that variation if they are going to be successful. Engineering solutions include drainage, reinforcement, embedded retaining structures, gravity structures and reinforced earth structures. Embedded and gravity retaining structures are covered in Chapter 6. Reinforced earth structures are built with engineered fill reinforced with steel and polymer strips and grids. Issues of the selection and placement of engineered fill are covered in this chapter. The design and construction of reinforced soil is covered in BS 8006-1:2010.

The complex hydrogeological conditions that exist in glacial soils mean that drainage solutions have to be designed to meet the local ground conditions; a generic solution may not be appropriate. For example, installing counterfort drains in matrix-dominated tills would not function if they terminate in a sand and gravel layer as this could recharge the drain. Therefore, an appreciation of the local hydrogeological conditions is essential.

Guidance is given on the design of slopes reinforced with soil nails but experience suggests that the observational method is essential. This is especially the case with spatially variable glacial soils where a generic solution may not be appropriate. Experience suggests that the capacity of soil nails installed in matrix-dominated soils exceeds the design capacity, emphasising the importance of testing sacrificial and production nails. It is recommended that rigid nails are installed in grouted boreholes because of the presence of cobbles and boulders making it difficult to install predrilled nails. Care has to be taken of loss of grout in lenses of sands and gravels and a reduction in capacity if weaker layers are present.

5.6 GROUND IMPROVEMENT

Ground improvement is used to densify a soil to increase its strength and stiffness and reduce its permeability or reinforce a soil to increase the strength of the soil mass or alter its characteristics with admixtures. Soil can be modified, stabilised or reinforced using mechanical, chemical and electrical techniques listed in Table 5.44. Improvement techniques may be used in glaciofluvial sands, glaciolacustrine clays and glaciomarine clays to increase their strength and stiffness, in glaciofluvial soils and clast-dominated tills to reduce their permeability and in matrix-dominated tills to create a uniform deposit. The range of suitable soils for densification is shown in Figure 5.50; compression techniques are appropriate for glacial clays and preloading can be used for all glacial soils; admixtures with glacial clays and possibly glaciofluvial sands; permeation grouting can be used in all glacial soils (Figure 5.51).

Improving the properties of glacial soils is normally not necessary. There are exceptions:

- Creating a heterogeneous deposit for low rise structures and pavements especially in matrix-dominated tills
- Reducing the permeability of glaciofluvial soils and clast-dominated tills
- Stabilising slopes in glacial soils
- Densification of glaciofluvial soils
- Compression of glaciolacustrine clays

Lime and cement modification and stabilisation have been used with success to improve road subgrades and foundations for low rise structures. The addition of lime to fine-grained soil absorbs water from the soil, reducing its water content; the lime reacts with clay minerals, the reaction depending on the type of mineral thus modifying the soil provided the plasticity index is at least 10%. This modification process reduces the plasticity. Pozzolanic reactions take place increasing the strength of the soil. Bell (1996) investigated the effect of lime on clay minerals and soils including matrix-dominated tills and glaciolacustrine clays. Bell (1996) showed that 1%–2% of lime will affect the consistency limits of kaolin, montromollinite and fine-grained quartz (i.e. rock flour found in some matrix-dominated tills); the plasticity limit tends to increase but as more lime is added, it reduces. Liquid limit may increase or decrease depending on the composition. The addition of lime to clays increases the optimum water content, reduces the maximum dry density and increases the strength. Tests on glaciolacustrine clays showed that lime reduced its plasticity but had little effect on matrix-dominated tills because the plastic and liquid limits changed by the same amount

Table 5.44 Methods of modifying and stabilising soils (ground improvement)

Principle	*Method*	*Description*	*Glacial soils*
Replacement	Excavate and replace	Natural soil excavated and replaced with an engineered fill	Used to replace weaker layers in till
	Displacement	Fill spread onto the soil and displace the soil	Not appropriate
Densification	Vibro-compaction	Vibrating poker with water flushing	Glaciofluvial soils; clast-dominated tills
	Vibro-stone columns	Vibrating poker to create compacted stone columns	
	Dynamic compaction	Falling surface weight	
	Rapid impact compaction	High-frequency hydraulic hammer	
	Compaction piles	Creating compacted piles with downhole hammer	
	Blasting	Detonation of explosives	Glaciofluvial soils
Compression	Preloading	Surface load applied to consolidate the soil; can be used with vertical drains	Matrix-dominated tills; glaciolacustrine clays; glaciomarine clays
	Vacuum preloading	Application of vacuum at surface to create an atmospheric surcharge	
	Dewatering	Increase in effective stress consolidating the soil	
	Electo-osmosis	Electric potential to reduce pore pressure, increasing the effective stress and consolidating the soil	
Reinforcement	Vibro-stone columns	Vibrating poker to create compacted stone columns	Glaciofluvial soils and clast-dominated tills
	Compaction piles	Creating compacted piles with downhole hammer	
	Soil nails	Nails installed by driving, drilling and grouting or firing	All soils
	Micropiles	Reinforcement inserted into grout-filled boreholes	
Admixtures	Lime columns	Columns of soil mixed with lime	Glaciolacustrine clays; glaciomarine clays
	Deep soil mixing	Columns or blocks of lime or cement soil mixture; mixed *in situ*	
	Subgrade stabilisation	*In situ* mixing of surface layer with lime or cement	Glacial clays
Grouting	Permeation	Replacement of water in voids with grout using low pressures	Glaciofluvial soils and clast-dominated tills
	Hydrofracture	Hydraulic fractures filled with grout	Glaciofluvial sands; glacial clays
	Jet grouting	Grout jetted into soil liquefied by the jetting process creating columns; replacement of soil with grout eroded with water or air jets	
	Compaction grouting	Monitored displacement of ground without fracturing the ground by pumping grout into the ground	

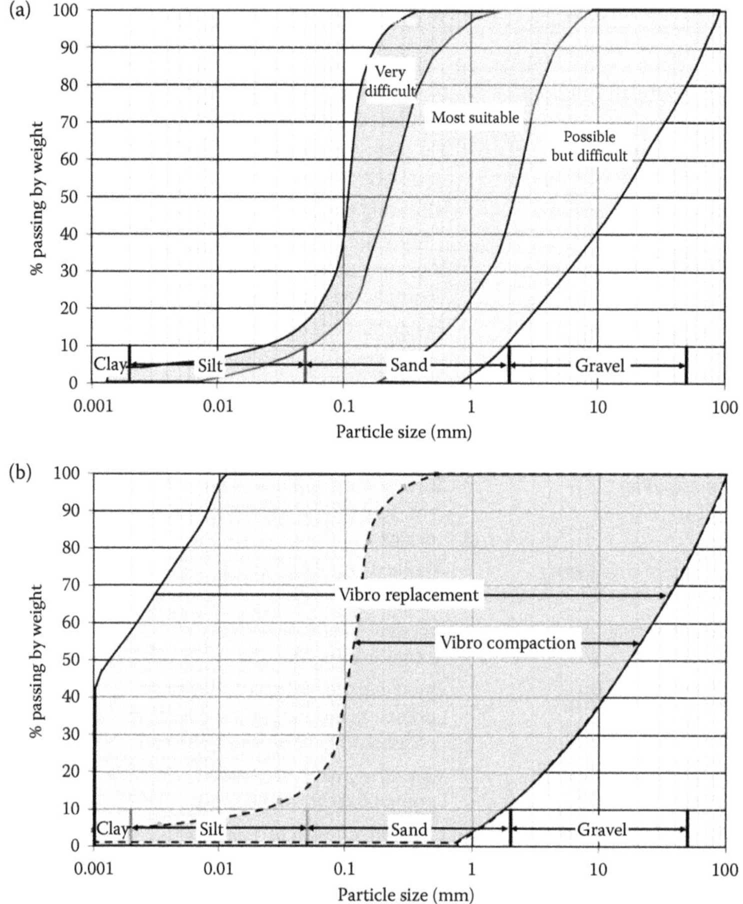

Figure 5.50 Range of soils suitable for densification by (a) vibro-compaction and (b) vibro-replacement. (After Brown, R. E. *Journal of Geotechnical and Geoenvironmental Engineering*, 103, no. ASCE 13415 Proceeding, 1977: 1437–1451; Mitchell, J. M. and F. M. Jardine. *A Guide to Ground Treatment*, Vol. 573. CIRIA, London, 2002.)

(Table 5.45). Adding lime improves the workability of the soil up to the fixation point; thereafter, may increase the strength of the soil.

Heath (1992) describes the use of lime and cement to stabilise subgrade soils at a number of UK airports including Stansted, London, which is founded on glacial clay till. There is 10–30 m of chalky till, which is weathered near the surface. The soil was stabilised with 3% lime and 5% cement following the specification (Figure 5.52) covering the suitability, acceptance and performance tests. The soil had to be scarified to remove the coarse particles and, if the work was carried out in the summer, water had to be added to ensure complete mixing; in winter clay balling took place, which meant that the surface had to be left to dry. Tests for acceptability and acceptance included water content, consistency limits, dry density/water content relationships, compressive strength, CBR, organic content and frost susceptibility. Field trials were undertaken on a minimum 500 m by 100 m section to establish the plant and procedure. Routine testing was undertaken before adding lime, during and completion of the lime stage and during and on completion of the cement stage at a frequency of tests every 1,000 m^2 with a minimum of two per day. Control testing included

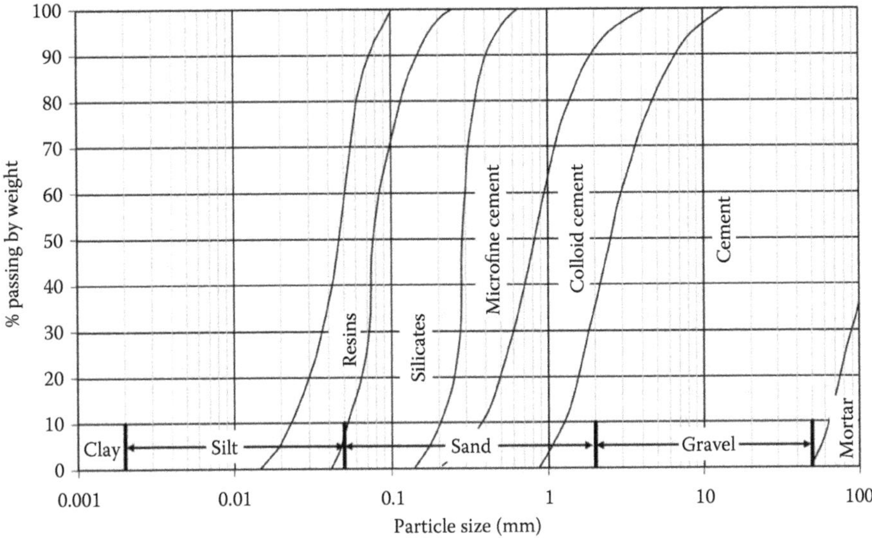

Figure 5.51 Range of soils suitable for permeation grouting. (After Schlosser, F. and I. Juran. *British Geotechnical Society*. 1981; Tausch, N. *Proceeding of the International Symposium on Recent Developments in Grout Improvement Techniques*, 1985: 351–362; Karol, R. H. *Chemical Grouting*. Marcel Dekker, New York; 1990.)

water content, consistency limits, pulverization, dry density at optimum water content, lime and cement content, depth of stabilisation, rate of spread, plate bearing value, *in situ* density and compressive strength.

Trial tests (Table 5.46) were carried out on mixtures of quicklime, hydraulic lime, cement and cement and quicklime to establish the design mix of 3% lime and 5% cement to be mixed in two stages. A summary of the control tests (Table 5.47) showed that they stabilised an area 450,000 m^3 to achieve between 92% and 109% of the specified dry density and the 7-day strength was between 147% and 970% of the strength of the trial specimens.

Quigley (2006) describes the use of lime stabilisation of Dublin Boulder Clays for light industrial units, which proved necessary because of local soft spots and the spatial variation in physical and mechanical characteristics. A key concern is the sulphate content of the soil since sulphate reacts with cement and lime causing volume changes. The variation in CBR (Figure 5.53) with time for soil treated with 1%–2% lime to achieve the specified minimum CBR of 3% showed that the majority of samples exceeded the specified minimum. Quigley (2006) suggested that CBR, MCV and plate bearing tests should be carried out every 1,000 m^2 and pulverisation and sulphate tests every 1,500 m^3 during construction.

Sariosseiri and Muhunthan (2009) tested cement stabilisation of glaciofluvial and glacial tills from the State of Washington, which showed that the soils were more workable and there was a significant increase in strength though the soils became more brittle; therefore, cement treatment has to be treated with caution because of potential progressive failure.

5.7 TUNNELS

The composition, fabric and groundwater conditions are critical in any project but especially in tunnelling because of the health and safety of those involved and damages to adjacent structures due to ground movements. The choice of tunnelling method, which can range

Table 5.45 Physical and mechanical properties of matrix-dominated till and glaciolacustrine clay from Teesside, NE England and the effect of the lime content

Soil	Property	Amount of lime (%) added				
		0	2	4	6	8
Matrix-dominated till	I_P (%)	14	25	23	21	18
	I_L (%)	30	42	40	41	37
	PI (%)	16	17	17	20	19
	I_s (%)	6	2	1	1	1
	c_u (kPa)[a]	270	380	530	800	730
	E_u (MPa)[a]	35	49	56	58	52
	w_{opt} (%)[a]	18			20	
	γ_d (Mg/m^3)[a]	1.81			1.75	
	CBR	9			24	
Laminated clay	I_P	26	36	34	33	31
	I_L	58	57	53	50	49
	PI	32	19	19	17	18
	I_s	10	4	3	2	2
	c_u (kPa)[a]	90	290	445	390	420
	E_u (MPa)[a]	15	21	43	38	40
	w_{opt} (%)[a]	22			25	
	γ_d (Mg/m^3)[a]	1.65			1.60	
	CBR[a]	5			19	

Soil	Lime content (%)	Water content (%)	c_u (kPa) after days			
			0	7	14	286
Matrix-dominated till	2	10	250	300	270	230
		20	350	400	430	480
		30	200	220	280	380
	4	10	300	350	350	300
		20	350	460	630	750
		30	250	310	400	450
	6	10	290	420	430	360
		20	410	600	710	810
		30	300	440	470	560
Laminated clay	2	10	170	180	170	120
		20	200	240	280	300
		30	80	90	80	90
	4	10	240	320	260	230
		20	310	360	400	440
		30	160	180	200	240
	6	10	250	260	160	110
		20	280	330	360	440
		30	160	170	130	180

Source: After Bell, F. G. *Engineering Geology*, 42(4); 1996: 223–237.

[a] 7 days curing at 20°C.

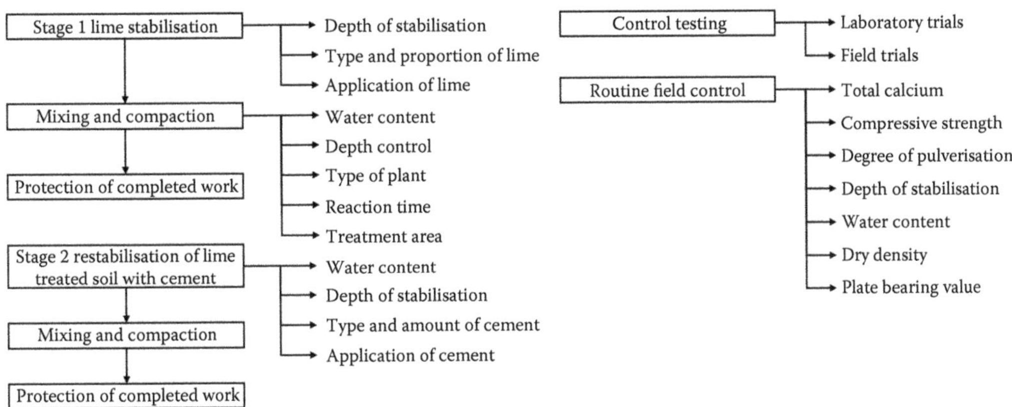

Figure 5.52 Specification followed by Heath (1992) for the selection, acceptance and control of stabilised subgrades for major airports in the United Kingdom including Stansted Airport which is founded on chalky boulder clay.

Table 5.46 Natural properties of the matrix-dominated till and properties of stabilised matrix-dominated till at Stansted Airport, United Kingdom

(a) Natural properties

Property	*MCV*	*w (%)*	I_L *(%)*	I_P *(%)*	*PI (%)*	>425 μ	ρ_d *(Mg/m³)*
Value	12.9–14.5	16–19	19–43	19–38	19–23	56–77	1.67–1.70

(b) Treated soil

Soil type	ρ_d *(Mg/m³)*	*w (%)*	ρ_{dmax} *(Mg/m³)*	*CBR*	*Plate bearing value* K18	*Plate bearing value* K30	*7 day* q_c *(MPa)* 50 mm	*7 day* q_c *(MPa)* 100 mm	*Properties at 1 year* ρ_{dmax} *(Mg/m³)*	*Properties at 1 year* *w (%)*	*Properties at 1 year* q_c *(MPa)*
Natural	1.67	16		11	38	28	0.23–0.31	0.11–0.15			
	1.10	19		9	43	31					
3% quicklime	1.77	13	1.83	48	187	135	0.76–1.03	0.25–0.27			
	1.77	11			168	121					
5% hydraulic lime	1.79	19	1.83	39	153	110	0.70–1.04	0.25–0.26			
	1.84	12			171	134					
5% cement	1.84	15	1.86	60	201	145	0.8–1.51	0.34–0.60	1.62	21	2.4
									1.59	23	2.4
	1.74	15			398	287			1.59	21	2.8
									1.68	17	5.6
3% quicklime/ 5% cement	1.72	12	1.85	52	309	294	1.15–2.27	0.34–0.53	1.55	22	3.3
									1.58	17	3.1
	1.68	20			388	280			1.67	17	3.9
									1.62	17	2.8
									1.63	18	2.9

Source: After Heath, D. C. *Proceedings of the Institution of Civil Engineers:Transport*, 95(1); 1992: 11–50.

Table 5.47 Summary of the test results stabilising the subgrades to the aprons, taxi ways and runways at Stansted Airport, United Kingdom

	Soil mix	*Classification*			*Compaction*			*Strength*		
Location	*3% lime 5% cement*	*w (%)*	I_L *(%)*	I_P *(%)*	ρ_{dmax} *(Mg/m³)*	w_{opt} *(%)*	*In situ* ρ_{dmax} *(Mg/m³)*	*7 day* q_c *(MPa)*	*ρ (Mg/m³)*	*w (%)*
Airside road	15000 m²	12–20	21–36	15–17	1.80	15.6–16.2	1.43–1.64	0.5–2.0	1.82–1.99	13–20
Car parks	Stage 1, 32000 m²	14–28	37–57	16–24	1.79–1.86	13.5–16.5	1.30–1.60	0.6–1.9	1.88–1.96	19.4–23.5
	Stage 2, 62000 m²	18–33	39–57	16–28	1.70–1.74	18.9–20.3	1.36–1.62	0.8–1.4		
	Stage 3, 40000 m²	16–24	41–54	17–22	1.74–1.85	14.8–18.5	1.84–2.06	0.9–2.6	1.84–2.09	24
	Stage 4 17000 m²				1.93–2.01	17.9–19.1	1.83–2.29	0.8–3.3	1.83–2.25	8–25
	Stage 5 18000 m²						1.86–2.06	0.5–1.5	1.80–2.04	15.6–22.1
Aprons/ taxiway	Stage 2, 209000 m²	11–18	36–47	17–20	1.81–1.95	9.4–16.2	1.90–2.12	0.75–2.0	1.88–2.08	10.5–23.8
	Stages 3 and 4000, 93000 m²	13–21	36–48	16–20	1.90–1.95	12.5–14.1	1.92–2.03	0.8–2.3	1.79–2.09	11.9–24

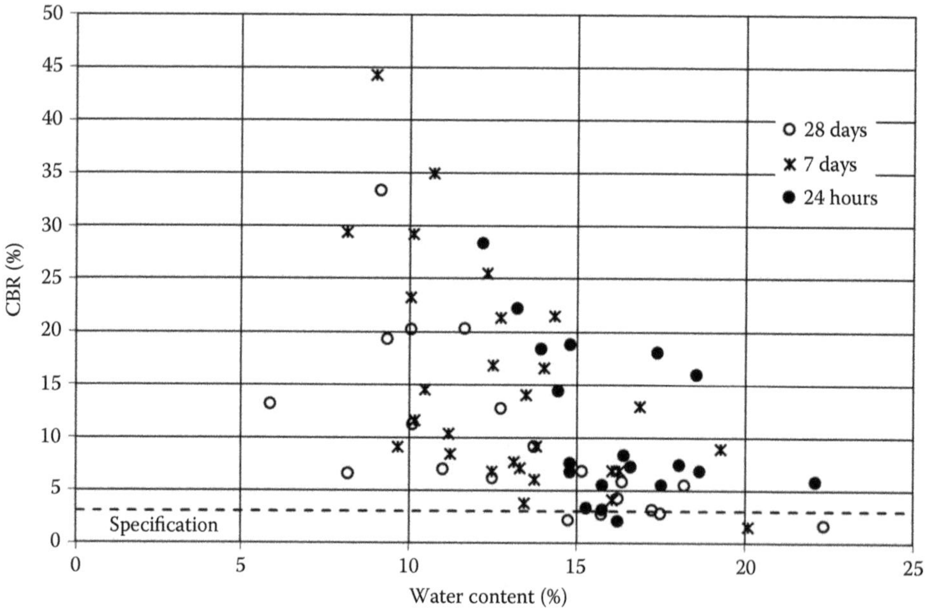

Figure 5.53 Variation in CBR with time for glacial tills treated with 1%–2% lime compared with the specified 3%. (After Quigley, P. *Paper Presented to the Geotechnical Society of Ireland (GSI)*, 16th February 2006. The Institution of Engineers of Ireland.)

from directional drilling to insert small diameter pipes for services to large underground excavations using mobile excavators, depends on the purpose, the diameter and length of the tunnel and the ground conditions. Pressurised closed-face tunnelling systems (TBMs) are used to create infrastructure tunnels varying in diameter from less than 1 m to more than 8 m. There are two types of TBM: slurry machines (STBM) ideal in coarse-grained soils and earth pressure balance machines (EPBMs) for fine-grained soils (BTS, 2005). Both systems maintain the pressure at the face to prevent excessive ground movements. STBMs use pressurised slurry that is mixed with the excavated soil and that mix is pumped to the surface where the soil is separated out so that the slurry can be reused. EPBMs use the excavated soil to maintain the pressure; the soil is excavated and then removed with a screw conveyor. Given that glacial soils are composite soils, the chosen system has to be modified. In the case of STBMs, it is necessary to include means of separating out fine- and coarse-grained soils using screens, hydro-cyclones and centrifuges. Conditioning agents are used in EPBMs to produce a more plastic soil. They both use rotating cutting heads that are designed to deal with a range of particle sizes up to boulders.

A tunnel alignment will be optimised, where possible, to avoid changes in ground conditions. This may not be feasible in glacial soils because of their spatial variability. A particular problem occurs when the alignment is close to rock-head since rock-head in glaciated regions is known to vary because of the nature of the erosive processes during glaciation and pre-glacial drainage systems.

Soil classification for tunnelling is different from the engineering classification of soils, as shown in Table 5.48. Ground investigations should be designed to establish the spatial variability though, given the difficulty in achieving this, a full description of the soils should be provided. The selection of the excavation system will be based on the particle size distribution, plasticity, permeability, pore pressure profile along the tunnel alignment, the settlement limits, the composition of the soil, distribution and size of boulders and the rock-head interface.

The nineteenth century saw significant excavations in glacial soils, so engineers were fully aware of the effects of weaker and water-bearing layers, boulder beds and isolated boulders on the stability of excavations. This was the case in Glasgow, Scotland where substantial excavations were necessary to create the Clyde shipyards. Tunnels were built under the Clyde in 1890 to connect the shipyards. There were three tunnels about 10 m below the river bed level. The tunnels were about 4.9 m diameter with cast iron segments beneath the river and brick arches within the matrix-dominated tills beneath the river banks. The access shaft was sunk under its own weight into the till. Thereafter, underpinning was used because the friction on the shaft wall was too great. Compressed air was used to construct the tunnel because it passed through matrix-dominated tills and sand. Little pressure was necessary in the matrix-dominated tills because of the low mass permeability. However, they did come across a sand lens, which they could cope with because compressed air was available.

Tunnelling in matrix-dominated tills using compressed air proved essential prior to the introduction of TBMs because of the uncertainty of the thickness of the boulder clay and its composition. For example, Haxton and Whyte (1965) and Morgan et al. (1965) describe the construction of twin 9-m diameter tunnels beneath Glasgow. As in 1890, the access shafts were sunk under their own weight into the glacial till and then continued into the glacial till by underpinning but using compressed air. Well points were used to dewater sand lenses in the glacial till; the groundwater level was lowered around the excavations for the portals because of artesian pressures below the till.

Tunnel shields have been in existence since the mid-nineteenth century but the introduction of EPBMs and STBMs transformed tunnelling through glacial soils. STBMs are designed for coarse-grained soils though they can be used in soils with up to 20% fines (Figure 5.54). It is possible to use STBMs with a greater fines content, in particular, in

Table 5.48 Classification of soils for tunnelling

Soil type	*Description*	*Glacial soils*
Firm ground	• Ground in which the tunnel can be advanced safely without providing direct support to the face during the normal excavation cycle and in which ground support or the lining can be installed before problematic ground movement occurs • Where this short-term stability may be attributable to the development of suction in fine-grained soils, significant soil movements and/or ground loading of the tunnel lining may occur later • A closed-face tunnelling machine may not be needed in this ground type	Homogeneous matrix-dominated till with no layers of sands and gravels
Ravelling ground	• Ground characterised by material that tends to deteriorate with time through a process of individual particles or blocks of ground falling from the excavation surface • In this ground, a closed-face tunnelling system may be required to provide immediate support to the ground	Clast-dominated tills; fissured tills; glaciofluvial soils; lenses of sands and gravels in matrix dominated tills
Running or flowing ground	• Ground characterised by material such as sands, silts and gravels in the presence of water, and some highly sensitive clays that tend to flow into an excavation • Above the water table, this may occur in granular materials such as dry sands and gravels. Below the water table, a fluidised mixture of soil and water may flow as a liquid • Such materials can sometimes pass rapidly through small openings and may completely fill a heading in a short period of time • In all running or flowing ground types, there will be considerable potential for rapid over-excavation. A closed-face tunnelling system will be required to support such ground safely unless some other method of stabilisation is used	Clast-dominated tills; glaciofluvial soils; lenses of sands and gravels in matrix dominated tills
Squeezing ground	• Ground in which the excavation-induced stress relief leads to ductile, plastic yield of ground into the tunnel heading • A closed-face machine may be required to provide resistance to squeezing ground, although in some conditions there is also a risk of the TBM shield becoming trapped	Glaciolacustrine clays; glaciomarine clays; matrix-dominated tills
Swelling ground	• Soil characterised by a tendency to increase in volume due to absorption of water • This behaviour is most likely to occur either in highly over-consolidated clay or in clays containing clay minerals naturally prone to significant swelling • A closed-face machine may be useful in providing resistance to swelling ground although, as with squeezing ground, there is a risk of the shield becoming trapped	

(Continued)

Table 5.48 (Continued) Classification of soils for tunnelling

Soil type	*Description*	*Glacial soils*
Mixed ground conditions	• Potentially, the most difficult of situations for a closed-face tunnelling system is that of having to cope with a mixture of different ground types either along the tunnel from zone to zone or sometimes from metre to metre, or within the same tunnel face • Ideally, the vertical alignment would be optimised to avoid, as far as possible, a mixed ground situation; however, in urban locations, the alignment may be constrained by other considerations • For changes in ground types longitudinally, a closed-face machine may have to convert from a closed-face pressurised mode to an open non-pressurised mode when working in harder ground types to avoid overstressing the machine's mechanical functions • Such a change may require some modification of the machine and the reverse once again when the alignment enters a reach of soft, potentially unstable ground • In the case of mixed ground types across the same face, the tunnelling machine will almost certainly have to operate in a compromise configuration. In such cases, great care will be needed to ensure that this provides effective ground control. A common problem, for example, is a face with a hard material in the bottom and running ground at the top. In this situation, the TBM will generally advance slowly while cutting the hard ground but may tend to draw in the less stable material at the top leading to over-excavation of the less stable material and subsequent subsidence or settlement at the surface • Different ground types at levels above the tunnel will also be of significance. For example, in the event that over-excavation occurs, the presence of running or flowing materials at horizons above the tunnel will increase the potential quantity of ground that may be over-excavated and again lead to subsidence or surface settlement • Another potential problem occurs when a more competent layer exists over potentially running ground in which case possible over-excavation would create voids above the tunnel and below the competent material, giving rise to potential longer-term instability problems	Glacial soils which show significant spatial variation, for example, weak and water-bearing layers and lenses in matrix-dominated tills; interface between glacial tills and rock-head; boulder beds

Source: After BTS. *Closed-Face Tunnelling Machines and Ground Stability*. Thomas Telford, London, 2005.

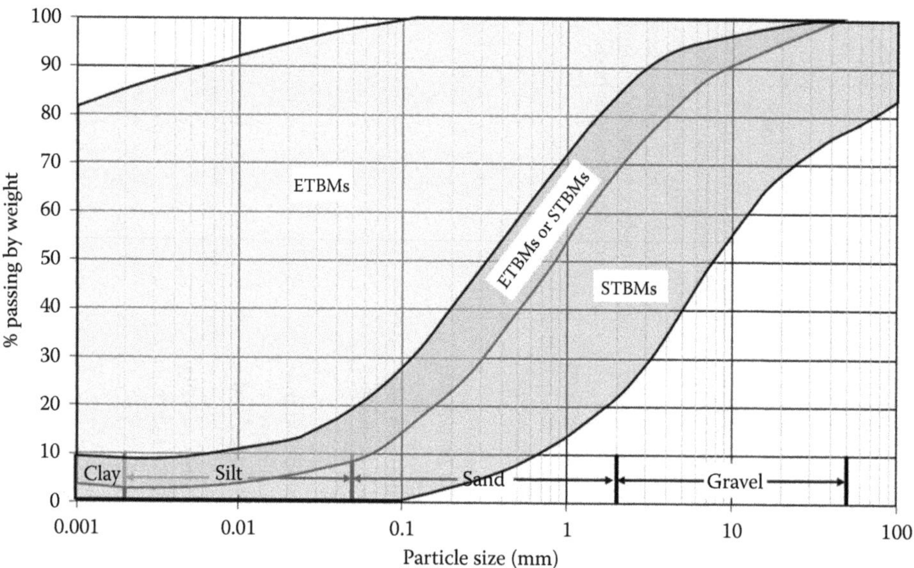

Figure 5.54 Appropriate ground conditions for full face tunnel boring machines. (After BTS. *Closed-Face Tunnelling Machines and Ground Stability*. Thomas Telford, London, 2005.)

micro-tunnels where everything is operated remotely. EPBMs can be used in most soils provided appropriate conditioners are used though, if the fines content is less than 10%, it may be better to use an STBM. Permeability is important because of potential inflow. STBMs can be used in soils with a coefficient of permeability greater than 10^{-5} m/s and EPBMs if it is less than that. It is the mass permeability, including the effect of discontinuities, and the groundwater pressures that are significant, hence, the importance of determining the groundwater conditions.

The importance of a geomorphological study to help establish the spatial variability of the ground was emphasised by Gillarduzzi (2014) who used this technique to investigate property damage along the Dublin Port Tunnel alignment. The tunnel comprises 2.8 km of bored tunnel and 1.9 km of cut and cover and the section of interest was bored using a 11.8-m TBM at 30 m below ground level. The ground conditions comprised up to 3 m made ground overlying up to 20 m of glacial and fluvial glacial deposits. The TBM operated at typically 3–8 m below the soil rock interface. The ground investigation along the alignment identified a shallow valley bed rock profile but a geological study uncovered a number of buried valleys, which included 3.5–17 m of gravel overlain by glacial till. Up to 200 years ago, the area appeared to be undulating with small ponds and lakes, evidence of a drumlin field and moraines. The evolution of the ground model is shown in Figure 5.55. It was assumed that the glacial till acted as an aquiclude creating artesian pressures in the underlying bedrock. The glacial till had a permeability of 10^{-9}–10^{-11} m/s, but braided sand and gravel channels existed at various depths within the till supporting independent aquifers. The permeability of the more gravelly glacial till was as high as 10^{-6} m/s. The surface settlements were attributed to volume loss due to local failure because of unstable rock wedges, dewatering of the glaciofluvial deposits within the glacial till leading to a loss of fines, vibro-densification of the glaciofluvial and alluvial deposits, consolidation of the glacial till and overlying alluvial clays due to the dewatering and the planned reduction in tunnel production to limit noise and vibration leading to an unsupported face. Figure 5.56 is a conceptual model of the dewatering of granular material within the till, which led to the surface settlements.

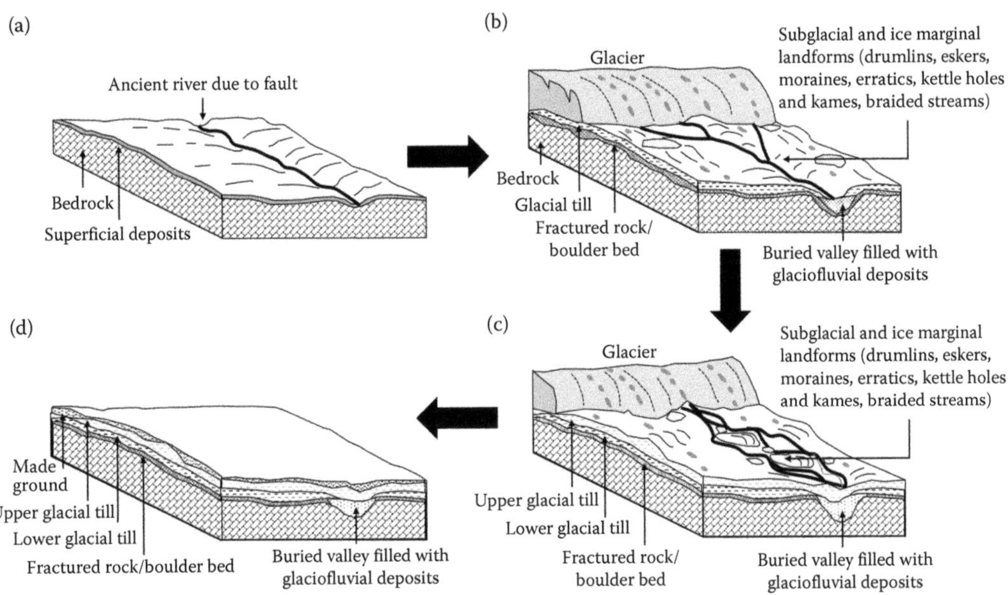

Figure 5.55 Development of the ground model showing the evolutionary phases and how they affected the current situation. (a) Ancient landscape, (b) Irish sea glaciation, (c) Today. (d) Midland glaciation. (After Gillarduzzi, A. *Proceedings of the Institution of Civil Engineers-Forensic Engineering*, 167(3); 2014: 119–130.)

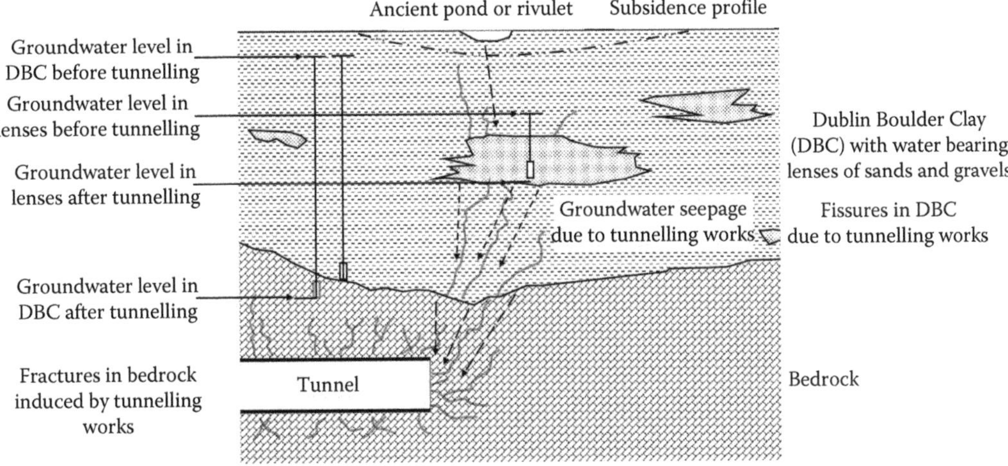

Figure 5.56 Conceptual view of dewatering of water-bearing sand and gravel lenses embedded within matrix-dominated till triggered by tunnelling works for the Dublin Port Tunnel giving rise to subsidence. (After Gillarduzzi, A. *Proceedings of the Institution of Civil Engineers-Forensic Engineering*, 167(3); 2014: 119–130.)

Pressurised systems are used to control ground movements, which means that over-excavation has to be prevented through a combination of factors under the control of the operator. The quantity of excavation allowing for bulking has to balance the volume of excavation based on the speed and diameter of the TBM. This is very dependent on the soil composition. Settlement is limited by the face pressures created by the slurry and cutter head pressures to balance the pore water pressure and the lateral earth pressure and by the slurry radial shield and grout annulus pressures to balance the mean radial earth pressure. The face pressure can be estimated from

$$p_f = K_a\sigma'_v + u + F_s \tag{5.11}$$

where K_a is the coefficient of active earth pressure, σ'_v the effective overburden pressure, u the pore pressure and F_s a nominal safety allowance, typically 20 kPa. In practice, the actual pressure is based on experience. An alternative method is based on limit equilibrium methods such as that proposed by Horn (1961) in which the face pressure has to balance the weight of a soil prism above the face (Figure 5.57). The face pressure, p_f, to maintain stability and, therefore, ground movements are related to the strength of the soil. For drained conditions,

$$p_f = -c'N_c + qN_q + \gamma DN_\gamma \tag{5.12}$$

where γ is the unit weight of the overburden, c' the cohesion, D the diameter of the tunnel, q the surcharge at the surface and N_c, N_q and N_γ are stability numbers. Vermeer et al. (2002) suggested that, if the angle exceeds 20°, stability is independent of overburden pressure. The face pressure, p_f, is then

$$p_f = \gamma DN_\gamma \tag{5.13}$$

$$N_\gamma = \frac{1}{9\sin\Phi'} - 0.05 \tag{5.14}$$

Vermeer et al. (2002) suggested that $N_c = \cot\Phi'$ and $N_q = 0$ if the depth of cover is at least twice the tunnel diameter and the angle of friction is greater than 20°. An increase in stability number means that greater support is required at the tunnel face. An allowance has

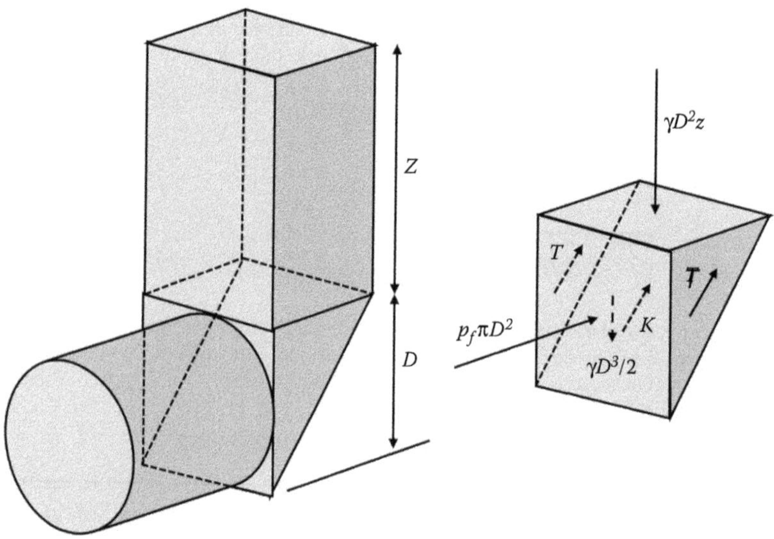

Figure 5.57 Tunnel face pressure based on the assumption that a block of ground the same width as the tunnel diameter is supported by the face pressure.

to be made for the effect of slurry infiltration when passing through coarse-grained soils. These methods produce different results, which means that the experience of the contractor and knowledge of the ground conditions are critical. This demonstrates the difficulty in tunnelling through variable ground such as glacial soils since the stability number will change as the tunnel advances.

Grasmick et al. (2015), Mooney et al. (2014) and Li et al. (2015) highlighted the importance of the face pressures and pressures in the shield and liner annulus on controlling ground movements. They describe the construction of four closely spaced tunnels with a total length of 3.25 km using 6.9 m STBMs in the Queens area of New York. Ground deformation was limited to 10 mm because the tunnels were beneath a live rail yard and mainline rail track. The ground conditions comprise a highly variable glacial till that included lenses of clay, silt, sand and gravel and glaciofluvial deposits. The volume loss at the face was 0.2%, considerably less than the expected 1%, typical of tunnelling operations in 2000 (FHWA, 2009), which was due to better control of the slurry pressures. Mooney et al. (2016) undertook a 3D FE analysis to determine the effect of face and annulus slurry pressures and grout pressure upon surface settlement. A comparison (Figure 5.58), using similar slurry (220 kPa at the springline) and grout pressures (328 kPa) to those used during tunnelling, between the surface settlement and the predicted settlement showed that it was possible to make a reasonable prediction. They undertook a parametric analysis (Figure 5.59) and concluded that the grout pressure was the most critical.

While the rock-head surface is known to be a challenge in glacial soils, Grose and Benton (2005) describe a situation where the interface between the glacial soils and the overlying soils contributed to the failure of a tunnel during construction. A post-failure investigation showed that the soil profile consisted of alluvial sand overlying glaciolacustrine clay, which was underlain by glacial sands and gravel. The laminated clays included bands of fine to medium sand. The failure, as with many geotechnical failures, was complex with no one factor being the cause of the failure.

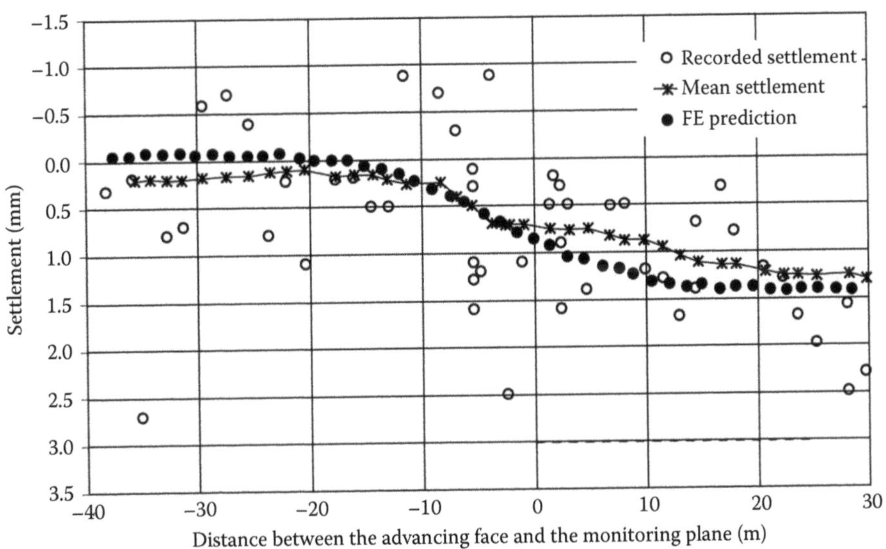

Figure 5.58 Comparison between the mean settlement as the tunnel face advances and that predicted using a FLAC 3D model showing the importance of selecting the correct soil parameters and tunnelling pressures. (After Mooney, M. A. et al. *Tunnelling and Underground Space Technology*, 57; 2016: 257–264.)

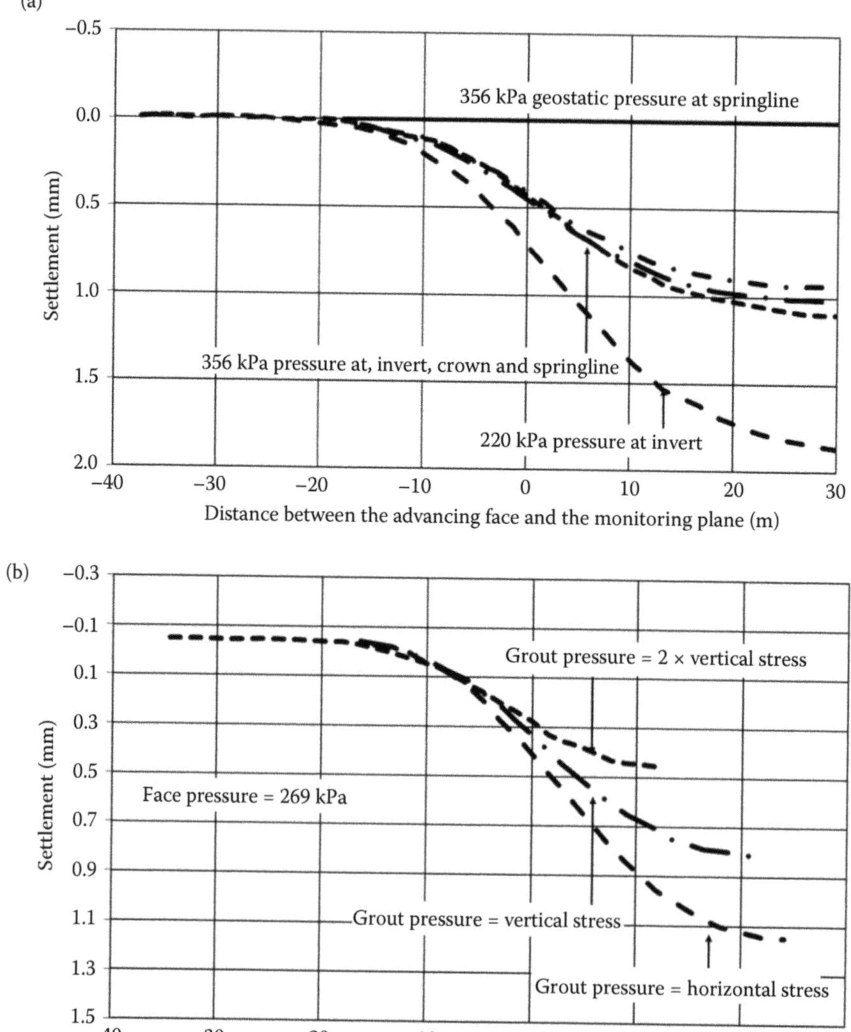

Figure 5.59 Influence of slurry pressure coupling between face and annulus and the effect of grout pressures on the settlement showing that grout pressure are critical. (a) Slurry pressure coupling between face and annulus. (b) Grout pressures. (After Mooney, M. A. et al. *Tunnelling and Underground Space Technology*, 57; 2016: 257–264.)

Benedikt and Beisler (2015) described the use of shotcrete, a sustainable solution for creating underground structures in complex ground conditions, as part of the 120-km-high-speed line between Graz and Klagenfurt. It included a number of tunnels that were too short or in too difficult ground conditions for a TBM. The superficial deposits comprised post-glacial gravel terraces overlying alternate sequences of gravels and lacustrine clays, which were underlain by ground and end moraines. The glaciolacustrine deposits were subdivided into sand- and silt/clay-dominated sequences. Cross sections of two of the tunnels showing how a combination of bored pile walls, jet grouting and shotcrete were used to construct the tunnels are shown in Figure 5.60.

(a)

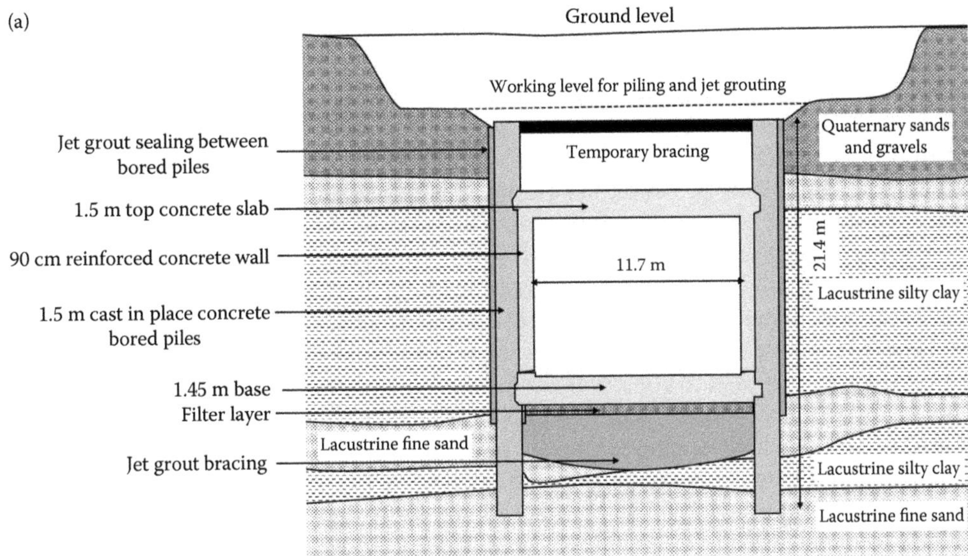

(b)

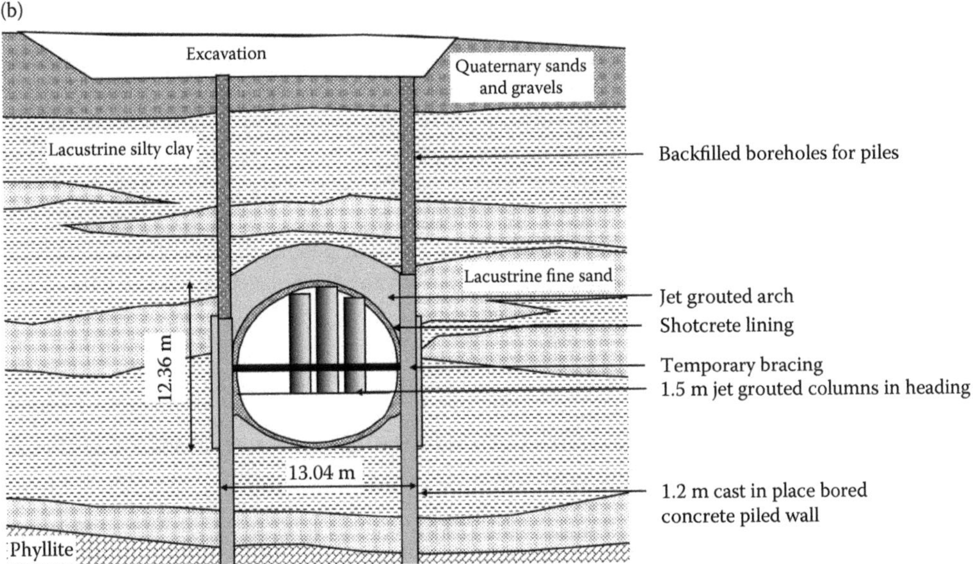

Figure 5.60 Examples of tunnelling through complex glacial soils when TBMs cannot be used as part of the high-speed line between Graz and Klagenfurt showing (a) the Srejach tunnel and (b) the Untersammelsdorf tunnel. (After Benedikt, J. and B. Matthias. Shotcrete – sustainable design for underground structures facing challenging ground conditions. In *Shotcrete for Underground Support XII*, edited by Ming Lu, Oskar Sigl and GuoJun Li, ECI Symposium Series, 2015. http://dc.engconfintl.org/shotcrete_xii/22.)

The challenges of tunnelling through glacial tills are exemplified by Biggart and Sternath (1996) who describe the construction of the undersea 14.82-km-long, 7.7-m-diameter Storebaelt railway tunnel in Denmark. The glacial tills were two well-graded matrix-dominated tills with boulders up to 3 m separated by water-bearing sand lenses, which were found throughout the two tills particularly at the interface between the two tills. Water pressures in those sand lenses were subject to the full hydrostatic pressure of the sea. The properties of the tills were established from a detailed ground investigation including boreholes and geophysical tests. Observations during construction found that piping occurred leading to sea bed depressions 35 m above the tunnel and ravelling of the tunnel face if left unsupported for more than 24 h. An EPBM was used because of the boulders, the range of ground conditions and the quantity of fine-grained particles, which would have been difficult to process using a slurry machine. In a homogenous till, the TBM was used in open mode with no supporting pressure; in variable ground, earth pressure was used to support the face. Dewatering was necessary in places because of the water-bearing lenses and to construct the cross passages. This allowed lower compressed air pressures to be used in the TBM chamber when access was required for maintenance. Vacuum wells were used to reduce local pore water pressures when constructing the cross passages. Freezing was used when sand layers were encountered in the crown of the cross passages.

An EPBM was also used to construct the 1.8-km-long, 8.4-m-diameter St. Clair tunnel at the border between the United States and Canada at Port Huron (Finch, 1996). The tunnel was designed to be wholly within glacial till at the interface between a firm till with cobbles and boulders and a dense clast-dominated till. The clay till was classed as a squeezing clay requiring support at all times. The presence of the soft till proved a challenge when it was necessary to install a temporary shaft to repair the cutter head. The cofferdam was infilled with 30-m-deep secant piles to prevent base heave and retain the soft clay. The only other significant feature of the till was occasional boulders removed by hand.

A number of tunnels have been built in Ireland, significantly around Dublin in the Dublin Boulder Clay where a database of the performance of geotechnical structures has been created. Empirical k values for predicting surface settlement (Mair and Taylor, 1999) assuming a Gaussian distribution curve depend on the dominant particle sizes, fine or coarse-grained. This means that they may not apply to well-graded glacial tills (McCabe et al., 2012). McCabe et al. (2012) studied the Dublin Port Tunnel and 7.50-m-long pipe-jacked 1.5- and 2.1-m-diameter tunnels in Mullingar, both in glacial tills. The Gaussian settlement trough (Figure 5.61) such that the settlement, s_y, at any horizontal distance, y, from the centre line is

$$\frac{s_y}{s_{max}} = e^{-y^2/2i^2} \tag{5.15}$$

where s_{max} is the settlement at the centre line and i the distance to the point of inflection. O'Reilly and New (1982) suggested that

$$i = kz_o \tag{5.16}$$

where z_o is the depth to the tunnel axis and k is an empirical factor depending on soil type. Mair and Taylor (1999) suggested that, for clays, $0.4 < k < 0.6$ and, for sands, $0.25 < k < 0.45$. Based on Equation 5.15, the volume loss, V_l, is

$$V_l = \frac{V_s}{V_t} = \frac{\sqrt{2\pi} i s_{max}}{\pi D^2/4} = 3.192k\frac{z_o}{D}\frac{s_{max}}{D} \tag{5.17}$$

where V_s is the volume of the settlement trough, V_t the volume of the tunnel per metre and D the tunnel diameter. Mair (1996) suggested that for TBMs, V_l of 0.5% can be achieved in sands and 1%–2% in soft clays. Macklin (1999) suggested that for stiff clays

$$V_l = 0.23e^{(4.4^{N/NT})} \tag{5.18}$$

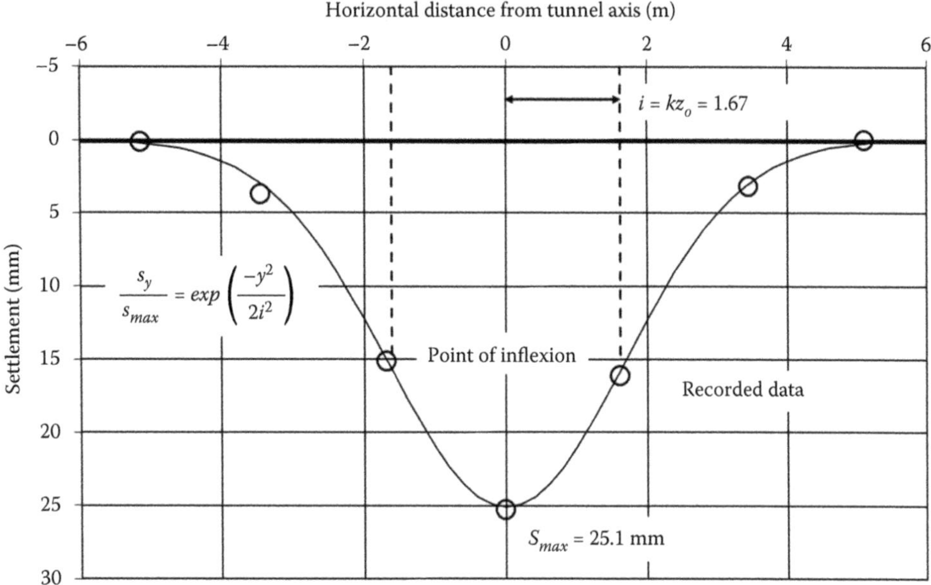

Figure 5.61 Gaussian settlement trough compared with the *in situ* measurements of settlement. (After McCabe, B. A. et al. *Tunnelling and Underground Space Technology*, 27(1); 2012: 1–12.)

where N is the stability number

$$N = \frac{\sigma_v - p_f}{c_u} \tag{5.19}$$

The stability number at collapse, N_T, is an empirical parameter derived from experimental studies (Macklin, 1999; Devriendt, 2010).

Figure 5.62 shows the variation in normalised depth and normalised distance to the point of inflection for three sites in glacial soils showing the variation of k, and Figure 5.63 the impact of the depth on the maximum settlement. McCabe et al. (2012) noted that high values of s_{max}/D were associated with boulders when the rate of tunnelling would reduce in order to remove boulders. They concluded that it is necessary to identify the dominant fraction of glacial soils to make the correct choice of k and, based on these three sites, the values of k suggested by Mair and Taylor (1999) were valid for glacial soils.

Elwood and Martin (2016) investigated ground loss due to the construction of two 6.5-m-wide oval tunnels constructed using a tunnel shield through matrix-dominated tills in Edmonton, Canada. The till was fissured and contained lenses of water-bearing sands. Figure 5.64 shows the variation of shear strength with water content based on tests on various types of sampling devices and Table 5.49 lists the geotechnical characteristics. They used probe holes to identify and drain sand lenses as the tunnel advanced, but there was no other difficulty in excavating the till. An example of the measured and predicted settlement is given in Figure 5.65. The prediction of volume loss for the twin tunnel was based on the principle of superposition (Suwansawat and Einstein, 2007). There was little difference between deep and shallow settlements suggesting that the mechanism was the silo type (Figure 5.57), perhaps because of the fissured nature of the till. Figure 5.66 shows the settlement relative to tunnel depth as a function of the pillar width and tunnel diameter,

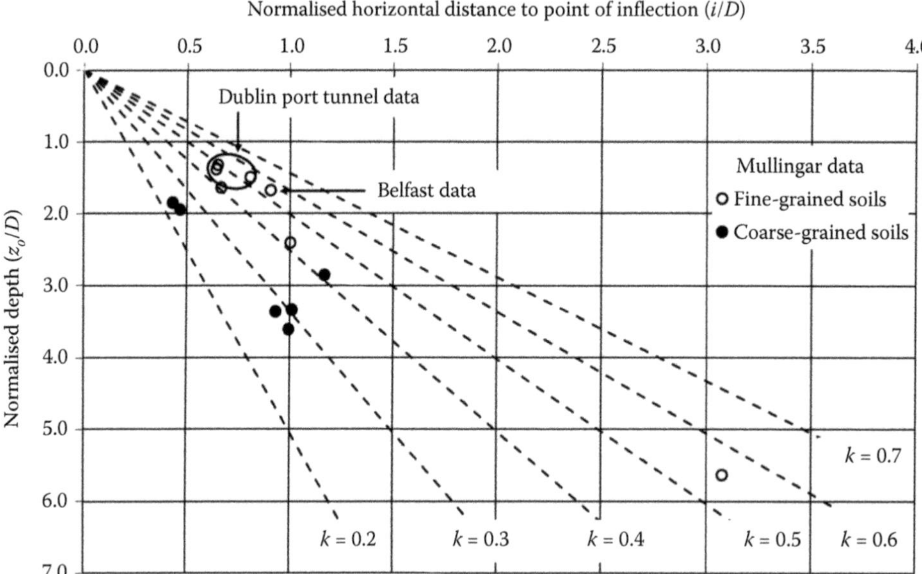

Figure 5.62 Data from tunnelling in glaciofluvial gravels shows that the k values are consistent with the lower bound suggested by Mair and Taylor (1999) and those of the matrix-dominated till are consistent with those suggested by Mair and Taylor (1999). (After McCabe, B. A. et al. *Tunnelling and Underground Space Technology*, 27(1); 2012: 1–12.)

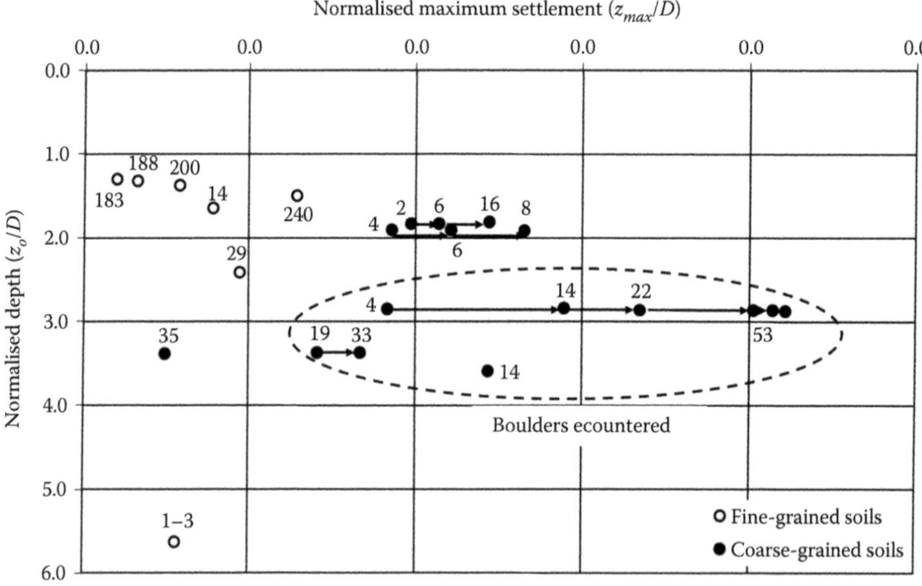

Figure 5.63 Variation in maximum settlement with depth showing the effect of time in days on the amount of settlement highlighting the effect of boulders and the use of large amounts of bentonite slurry in gravels leading to settlement with time. (After McCabe, B. A. et al. *Tunnelling and Underground Space Technology*, 27(1); 2012: 1–12.)

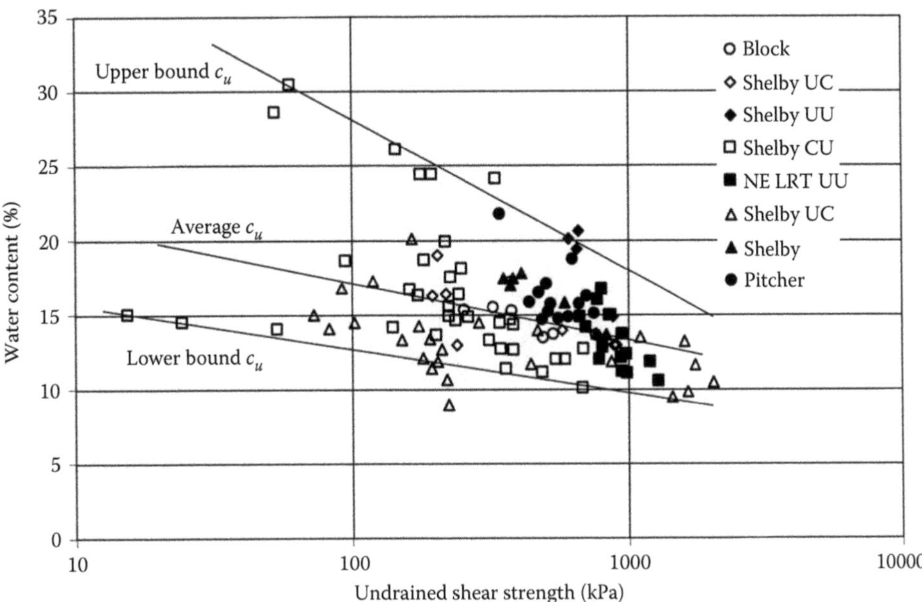

Figure 5.64 Variation of shear strength with water content for Edmonton till based on tests on specimens from various types of sampling devices. (After Elwood, D. E. Y. and C. D. Martin. *Tunnelling and Underground Space Technology*, 51; 2016: 226–237.)

Table 5.49 Geotechnical properties of Edmonton till

Parameter	*Test*	*Number of values*	*Lowest value*	*Mean*	*Highest value*	*Standard deviation*
SPT	*In situ*	173	11	51	100	18
G_{secant} (MPa)	Pressuremeter	15	42.7	83.5	151.7	27.4
c_u (kPa)		15	227	331	425	55
c' (kPa)		10	25	35.5	48	6.9
φ'		10	34	36	39	1.7
K_o		15	0.7	0.85	0.87	–
q_c (kPa)	Laboratory	10	118	236	438	–
c' (kPa)				37.5		
φ'				50		
γ (kN/m^3)			20	20.5	21	
K_o				0.75–0.85		
υ			0.33		0.49	
w (%)		238	5	15	45	3
k (cm/s)			1×10^{-7}	1×10^{-6}	1×10^{-5}	
% sand		9	32	37	42	3
% silt		9	32	45	62	12
% clay		9	16	26	31	6
I_L (%)		15	25	35	42	6
I_P (%)		15	13	15	17	1

Source: After Elwood, D. E.Y. and C. D. Martin. *Tunnelling and Underground Space Technology*, 51; 2016: 226–237.

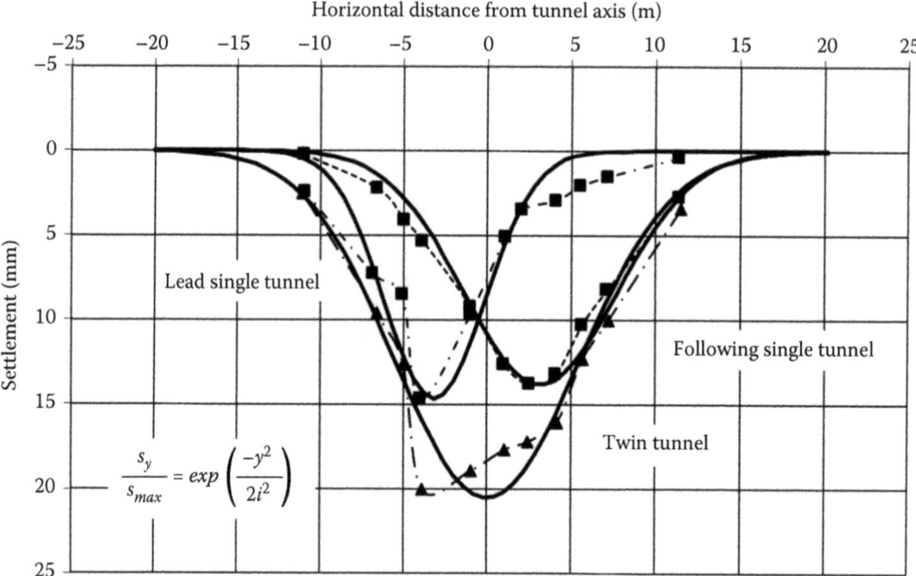

Figure 5.65 Observed settlement profiles above twin tunnels bored through Edmonton till with the derived *k* value for the lead tunnel being 0.2, the following tunnel 0.3 and both tunnels 0.4. (After Elwood, D. E. Y. and C. D. Martin. *Tunnelling and Underground Space Technology*, 51; 2016: 226–237.)

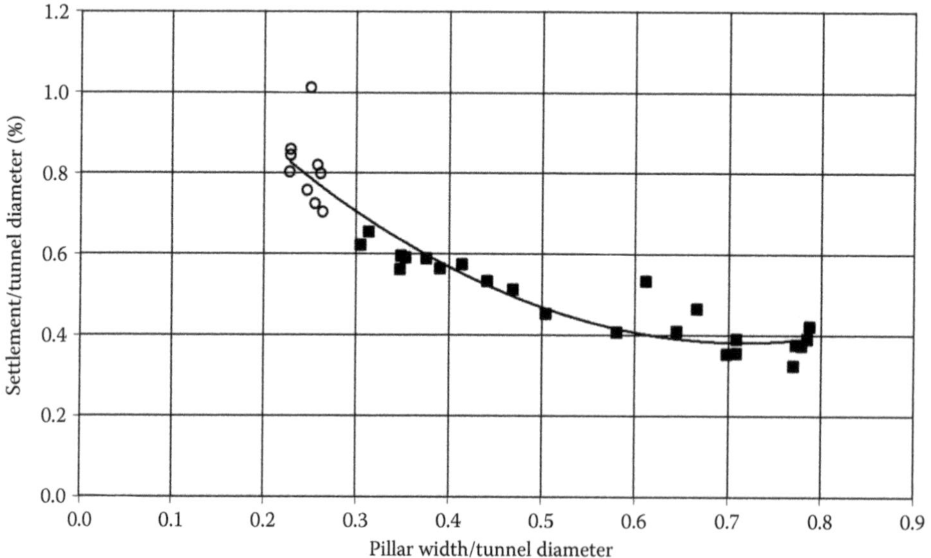

Figure 5.66 Settlement expressed in terms of the tunnel diameter for different pillar widths for twin tunnels bored through Edmonton till showing that at spacing of half the tunnel diameter the settlement is minimised. (After Elwood, D. E. Y. and C. D. Martin. *Tunnelling and Underground Space Technology*, 51; 2016: 226–237.)

suggesting that, for these soils, highly fissured dense tills, a pillar width of $0.5D$ would minimise the interaction of the two tunnels.

5.8 OBSERVATIONS

Earthworks, whether for excavations, tunnels, ground improvement or stabilisation of natural and cut slopes, in glacial soils are influenced by the spatial variability to such an extent that they may dominate the performance:

- Weaker layers in tills and glaciofluvial soils will influence the failure mechanism in slopes and the stability of a tunnel face. Weaker layers in matrix-dominated tills will be unacceptable material for fills.
- Water-bearing layers in matrix-dominated tills can lead to local instability in slopes, create perched water tables in the matrix above the layer and lead to instability of a tunnel face.
- The structure of glaciolacustrine clays will influence a potential failure mechanism within a slope; discontinuities in matrix-dominated tills reduce the mass strength and influence potential failure mechanisms.
- Very coarse particles will impact on any type of excavation.

It is unlikely that a true ground model can be produced, so mitigation measures should be considered. Given the number of glacial periods an area will have experienced, it would be useful to undertake a geomorphological appraisal not only of the current topography but also of historical land surfaces. This would help highlight structural features that could impact on excavations. Geophysical testing will help identify structural features which can be verified by boreholes. Scenario analyses should be undertaken to determine the effect of varying depth, location and thickness of structural features.

The extent of glacial soils means that they are a common source of engineered fills. The particle size distribution of many glacial soils, which are composite soils, means that it is possible to produce a dense fill with high strength and stiffness, and low permeability depending on the percentage of fine-grained particles.

The representative strength of a glacial soil is difficult to determine because of the spatial variability of composition and fabric. In glacial clays, the mobilised strength can vary from the intact to the residual strength depending on the geological history of the clay. It is recommended that

- The intact strength should be used for matrix-dominated tills which show no evidence of discontinuities.
- If a matrix-dominated till does contain fissures, then a reduced strength should be used.
- The mobilised strength of a glaciolacustrine clay will depend on the direction of loading.
- In areas known to have been subject to landslide activity, the residual strength should be used.

A detailed hydrogeological study should be undertaken given the impact groundwater can have on the stability of construction activities and the long-term stability of slopes. The alternative is to consider the worst credible groundwater profile, which could lead to overdesign. Whatever groundwater assumptions are made, it should be assumed that, during construction, groundwater will be encountered. Design should include drainage that will not allow pore pressures to exceed those assumed in the design.

Chapter 6

Geotechnical structures

Spread foundations, piled foundations and retaining structures

6.1 INTRODUCTION

Geotechnical design is based on codes, good practice and experience using a combination of empirical, theoretical and, increasingly, numerical methods. Codes of practice and guidelines can be national, regional, local or company based and can lead to different solutions even if the ground conditions are similar. This is because geotechnical design is based on experience of working with regional soils and rocks, and because construction techniques vary from country to country. However, the *knowledge of the ground conditions depends on the extent and quality of the geotechnical investigations. Such knowledge and the control of workmanship are usually more significant to fulfilling the fundamental requirements than is precision in the calculation models and partial factors* (Eurocode 7). This is particularly the case with glacial soils as the erosion, transport and deposition processes lead to spatially variable composite soils that do not necessarily conform to classic models for soil behaviour based on sedimented soils.

A study of the geological processes that glacial soils undergo has highlighted the spatial variability in composition, structure, fabric and properties such that glacial soils can be considered composite soils. FHWA (2002) suggests that glacial tills are intermediate geomaterials, which are soils that can be considered as intact rock because of their density and strength. The engineering behaviour of composite soils may be different to that based on the dominant soil composition. For example, a glacial till with between 20% and 50% clay size particles will behave as a fine-grained soil because it has such a low permeability yet could be described as a coarse-grained soil and exhibit a higher strength than expected for fine-grained soils. Glaciolacustrine soils are strongly anisotropic; matrix-dominated tills may be fissured such that the intact strength is less than the mass strength. All glacial soils can contain boulders, which may be randomly distributed through the soil or occur as a distinct horizon.

In situ and laboratory test results are strongly influenced by the composition and fabric of glacial soils to the extent that it may not be possible to obtain representative values of the mass properties. Further, the lateral and vertical variation in composition and properties means that a good quality ground investigation requires more boreholes and samples than guidelines suggest. This is to ensure that there are sufficient samples to obtain representative properties and identify features such as lateral variation in glaciofluvial soils, the surface of the underlying rock, lenses and layers of water-bearing sands and gravels and weaker soils. Profiles of soil properties are often variable because of the effects of composition, fabric, structure and sampling, which means that more refined statistical methods are required to obtain design profiles.

This chapter focuses on the design, installation and behaviour of foundations and retaining structures. The principles of design, global and partial factors of safety and design methods are introduced to highlight the properties of glacial soils that influence the design, construction and operation of foundations and retaining walls in glacial soils. Case studies of these structures in glacial soils are used, together with the geological history and geotechnical properties of glacial soils to highlight selection of appropriate design parameters and methods and issues to address during construction.

6.2 DESIGN PHILOSOPHY

Civil engineering structures are large complex systems to the extent that their behaviour is not fully understood. Most civil engineering structures are unique, so it is impossible to refine the design through repeat works. Most civil engineering structures are large, so it is impossible to undertake full-scale tests. However, structural components are small enough to test and can be optimised because they are not unique. The interaction of structural components and the interaction of structural and non-structural components are not fully understood and may not be allowed for in design. Onsite and offsite construction differs due to different levels of quality control and workmanship. Design factors have to comply with international, national, regional and local regulations. New forms of design are being introduced including those based on principles of sustainability, resilience, adaptability, optimisation and probability. The effects of climate change and a sustainable approach to create a resilient future mean a paradigm shift in design philosophy, which is based on a risk-based approach. This is particularly relevant in glacial soils, which are known to be challenging (BS8004:2015).

Failures of civil engineering structures can be catastrophic leading to loss of life and economic loss or result in structures no longer being fit for purpose. The actual capacity of a civil engineering structure is unknown though its performance in service can be measured. The remaining capacity at any time in its life is unknown. Factors of safety are designed to reduce the possibility of catastrophic failure. Factors of safety can be applied to a structural assembly or component, the material or the actions (forces or displacements). Factors can vary with the category of structure, the quality of information available, the design method, the locality and the country. There are two fundamental approaches: a global factor that is applied to the whole of the structure and partial factors applied separately to the material, actions and the components or assembly.

Global factors of safety in geotechnical design vary from 1.3 for stable slopes where deformation may not be so critical to 3 for foundations where deformation may be the governing criteria. Eurocode, a set of harmonised technical rules for the design of construction works, became mandatory in 2010 across Europe introducing partial factors for geotechnical design. There are 10 codes covering structural design including Eurocode 7 for geotechnical design. Their purpose is to ensure compliance of building performance and civil engineering works with mechanical resistance and stability and safety in case of fire, as a means of drawing up specifications for construction and a framework for technical specifications for construction products.

Eurocode 7 covers principles, that is, mandatory statements and definitions, and applications, which are rules that comply with the principles. It does not cover the practice of geotechnical design. This is why a number of codes of practice, such as BS8004:2015, have been introduced. Partial factor design is intended to produce a safe, reliable and durable design for the lifetime of the structure, which for most civil engineering structures will be between 50 and 100 years (Table 6.1). The use of these structures is likely to change during their lifetime because of environmental, technological, social and political changes, which means that most civil engineering structures will have to be adapted at some time as well as undergoing routine maintenance in new and innovative ways.

Table 6.1 Indicative design working life for civil engineering structures

Design working life category	Indicative design working life (years)	Examples
1	0	Temporary structures
2	10–25	Replaceable structural components
3	15–30	Agricultural and similar structures
4	50[a]	Buildings structures and other common structures
5	100[a]	Monumental building structures, bridges and other civil engineering structures

Source: After BS EN 1990:2002+A1:2005. *Eurocode: Basis of Structural Design*. British Standards Institution, London.

[a] The majority of civil engineering structures will fall in these categories.

In geotechnical design, no matter which code is used, there are two limit states to be considered: ultimate limit state (ULS) and serviceability limit state (SLS). ULS considers loss of equilibrium, excessive deformation of the ground, uplift, hydraulic heave, internal erosion and piping. Serviceability is concerned with fit for purpose and is usually associated with deformation. There may also be requirements imposed by national, regional or other local regulations. For example, the minimum depth of a foundation will depend on the local geology, vegetation and temperature and could be specified in the local building regulations. The basis of any design is risk in which hazards are identified and assessments made of likely harm or loss to property, people and the environment. Hazards can vary from frequent to exceptional with the performance level varying from fully operational to near collapse. Figure 6.1 shows the design and performance levels and how the levels can change with time during the life of civil engineering structures. Examples of performance criteria for fully operational strains are given in Table 6.2, which are based on limiting deflections and distortions that are known from experience not to cause excessive structural damage.

Actions or loads can be permanent, transient or accidental where permanent actions are the normal actions observed during the lifetime of a structure; transient actions are temporary conditions that can exist during the lifetime of the structure; and accidental actions can occur at any time. It may be possible to predict the magnitude of accidental actions but not if and when they occur. Given the lifetime of geotechnical structures, it is likely that actions may change. For example, the groundwater pressure profile can change due to changes in groundwater level due to rainfall events, seasonal changes and changes to the hydrogeological environment and, in the long term, climate change.

Geotechnical structures are likely to be modified or adapted during their life because of change of use or change in the environmental conditions or change of loading. For example, consider railway embankments in the United Kingdom, most of which were built in the nineteenth century. They were built using a ‘dig and dump’ approach with little compactive effort. Over time, they have reached a state of equilibrium. The rail track was laid on sleepers supported by a ballast bed, which has been maintained to compensate for movement of the ballast, embankment and underlying soils. The vegetation on the side slopes of the embankment has changed resulting in a change in hydrogeological conditions in the embankment and, in future, climate change could lead to further change with more intense and frequent events. At the same time, the load on the embankment has increased due to changes in train technology. This is an example of difficulties geotechnical engineers face when relying on performance-based criteria, which are often developed from historical data. Engineers have coped with the pace of change that has taken place since the early twentieth century when codes for construction were first introduced but the pace of change is accelerating due to environmental changes and changes in technology and, now, artificial intelligence.

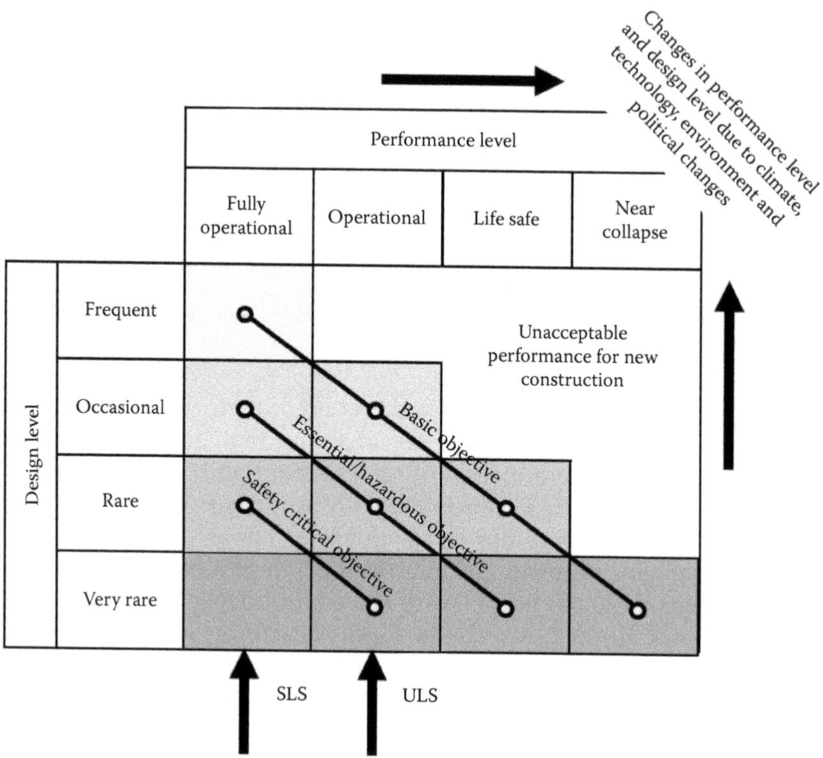

Figure 6.1 Concept of performance-based design taking into account the magnitude and frequency of hazardous events.

Table 6.2 Limiting values of distortion and deflection of structures

		Relative rotation				*Deflection ratio*[a]		
Type of structure	*Type of damage*	*Skempton and MacDonald (1956)*	*Meyerhof (1947)*	*Polshin and Tokar (1957)*	*Bjerrum (1963)*	*Meyerhof (1947)*	*Polshin and Tokar (1957)*	*Burland and Wroth (1974)*
Framed buildings and reinforced load bearing walls	Structural damage	1/150	1/250	1/200	1/150			
	Cracking in walls and partitions	1/300	1/500	1/500	1/500			
Unreinforced load bearing walls	Cracking by sagging					0.4×10^{-3}	$L/H = 3$ 0.3 to 0.4×10^{-3}	$L/H = 1$; 0.4×10^{-3} $L/H = 5$; 0.8×10^{-3}
	Cracking by hogging							$L/H = 1$; 0.2×10^{-3} $L/H = 5$; 0.4×10^{-3}

Source: After Tomlinson, M. J. and R. Boorman. *Foundation Design and Construction*. Pearson Education, Harlow, UK, 2001.

Note: L, width of structural element; H, height of structural element.

[a] Deflection ratio = differential settlement/length over which it is measured.

Codes are updated but typically at 25-year intervals or after a catastrophic event. There is an emerging view that design methodologies will have to change. For example, Eurocode 7 does not address the concept of sustainability, yet this is a requirement of modern design. Other design methodologies being introduced include adaptive, probabilistic, optimised and risk-based design. The embankment is an example of adaption and because it continues to be fit for purpose, it is a sustainable solution but it was not an optimal solution.

Factors of safety are not a replacement for good engineering design and, given the pace of change, there is a need to look at probabilistic methods, which are facilitated by numerical techniques. However, the increased use of numerical techniques has introduced further uncertainty because of the assumptions made in the analysis and the difficulties in obtaining correct parameters required for the constitutive models. Indeed, typical ground investigations do not produce the parameters used to create constitutive models. For example, Eurocode 7 states that *reliable measurements of the stiffness of the ground are often very difficult to obtain from field or laboratory tests. In particular, owing to sample disturbance and other effects, measurements obtained from laboratory specimens often underestimate the in situ stiffness of the soil.* This is particularly the case with glacial soils because of the difficulty in obtaining representative intact samples.

6.3 METHODS OF ANALYSIS

Table 6.3 is a summary of the theoretical and design requirements for a range of methods of analysis. Closed-form solutions, limit equilibrium, stress field and limit analyses and the beam and spring model all have limitations and have to be applied in context using correction factors to take into account observed performance. These correction factors are often based on databases of observations of foundation performance using a best fit line to the data. The data on which the correction factors are based are not necessarily presented. Therefore, these correction factors must be treated with caution unless it can be shown that the original data cover the site-specific ground conditions for the project being considered. Examples of this technique are covered in this chapter. Full numerical analyses are increasingly being used for geotechnical design, which allow more complex ground conditions, soil

Table 6.3 Relevance of theoretical and design requirements for numerical methods

		Theoretical requirements					*Design requirements*		
					Boundary conditions				
Method of analysis		*Equilibrium*	*Compatibility*	*Constitutive behaviour*	*Force*	*Displacement*	*Stability*	*Movements*	*Adjacent structures*
Closed form		√	√	Linear elastic	√	√	X	√	X
Limit equilibrium		√	X	Rigid with failure criterion	√	√	√	X	X
Stress field		√	X	Rigid with failure criterion	√	X	√	X	X
Limit analysis	Lower bound	√	X	Ideal plasticity	√	X	√	X	X
	Upper bound	X	√		X	√	X	?	X
Beam spring approach		√	√	Spring model	√	√	√	X	X
Full numerical analysis		√	√	Any	√	√	√	√	√

Source: After Potts, D. and L. Zdravkovic. *ICE Manual of Geotechnical Engineering*, Thomas Telford Ltd, London; 2012: 35–56.

structure interaction and possible scenarios to be analysed to establish critical conditions. However, these methods depend on full knowledge of the loading conditions, material properties and ground profile. This is particularly important in glacial soils, which are spatially variable both in composition and in properties. The limitations and assumptions of numerical methods and the constitutive models must be fully understood and applied correctly if the results are to have a value.

Other design methods include prescriptive methods, load tests, tests on experimental models and observational methods. The prescriptive method is a conservative approach that is acceptable for light and simple structures, preliminary designs and durability such as depth of foundations to avoid effects of frost, seasonal changes and vegetation. Methods based on load tests, such as pile tests, and experimental models, such as trial embankments, are used to confirm a design or to study the effects of construction on foundation performance. Experience in glacial soils suggests that the results of these tests may be different from those predicted. The reasons may include the following:

- Ground conditions different to those assumed especially in spatially variable glacial soils
- Use of empirical factors derived from tests on fine- and coarse-grained soils not composite soils
- Incorrect parameters, particularly stiffness, used in design
- Duration of a test compared with the time of construction and loading
- Levels of stress mobilised in a test because pilot tests may not replicate those in a full-scale structure
- The effects of installation not taken into account in any prediction resulting from the tests

The potential of the observational method to produce a better engineered solution and inform future designs is increasingly important because of significant improvements in instrumentation, its installation and monitoring and interpretation of the output. If the observational method is going to be used, then it is necessary to agree, in advance, what is to be monitored, the timing of the monitoring in relation to the construction and operation, acceptable limits of behaviour, the actions to be taken should the limits be exceeded and how the data are going to be stored and remain accessible.

6.3.1 Factors of safety

Factors of safety or partial factors are used to deal with uncertainty due to ground conditions, loading combinations and workmanship, none of which are exactly known. Factors of safety usually refer to global factors, which are the ratio of the restoring moment or force to the disturbing moment or force. Global factors of safety depend on the geotechnical structure being analysed. For example, 3 for shallow foundations and 2, 2.5 or 3 for piled foundations.

The alternative approach to deal with uncertainty is to use partial factors, which are applied to the material properties, actions (forces) and resistance. This is the basis of design according to Eurocode.

Eurocode 7 (BS EN 1997-1:2004+A1:2013) for geotechnical design is one of the structural Eurocodes covering the basis of structural design (No. 0 and 1) and the design of structures (2–6, 8 and 9). Eurocode 7 covers geotechnical design, geotechnical data and the design of fill, dewatering, ground improvement and reinforcement, spread foundations, pile foundations, anchorages, retaining structures, hydraulic failure, overall stability and embankments. It gives guidance on design methods, selection of data and partial factors, which may be modified by a National Annex.

The factors to be considered include overall stability, ground movements, nature and size of structure and factors that could affect their design life, ground conditions and the environment. An analysis of the complexity and associated risks leads to light and simple structures and small earthworks (Category 1), which are low risk and can be dealt with by experience and qualitative investigations and other structures that require design calculations. Category 2 includes conventional types of structures and foundations with no exceptional risk or difficult ground or loading conditions. Routine ground investigation is required. Category 3 structures include very large and unusual structures, abnormal risks and difficult ground conditions. BS 8004 (2015) states that glacial soils are most difficult to engineer owing to their variability and, given the points raised in Chapter 4, it is likely that geotechnical structures in glacial soils may be considered Category 3 structures. Certainly, as explained in Chapter 3, investigations in glacial soils are not routine.

6.3.2 Design factors

Short-term and long-term conditions have to be considered but, in composite soils, it may be necessary to consider an intermediate condition. Short-term conditions refer to undrained conditions, which are applicable to fine-grained soils at the end of construction. Long-term conditions, that is, drained conditions, apply to fine- and coarse-grained soils. The ULS of structures on clays can be assessed using undrained shear strength and effective strength parameters; the ULS of structures on sands using effective strength parameters. Serviceability is based on fully drained conditions.

Glacial soils are mostly composite soils, which means that there is likely to be some drainage during construction, which means that partially drained conditions exist. It is necessary to check ULSs for both short- and long-term conditions for foundations on glacial soils.

It is also necessary to consider the following:

- Spatial variability of properties, fabric, structure and composition
- The effect of laminations in glaciolacustrine clays including their alignment with respect to the structure and anisotropic behaviour of those soils
- The variation in the interface between glacial soils and bedrock and the variation in stratum thickness
- The presence of layers of lenses of more permeable/weaker soils within glacial tills
- Possible fissuring in matrix-dominated tills
- Lenses of water-bearing sands and gravels, lenses of weaker soils
- Gradation of glaciofluvial soils

Calculations can either be analytical, semi-empirical or numerical. It is good practice when making initial assessments to use presumed values. There is a general trend to greater use of numerical methods, which allow compatibility of strains between the structure and the ground to be considered and scenario analyses to be undertaken. This is especially useful in glacial soils because the effect of the natural variation in properties and thickness can be studied.

Actions include the weight of soil, rock and water, *in situ* stresses, free water pressures, groundwater pressures, seepage forces, dead and imposed structural loads, surcharges, mooring forces, changes in loads and load combinations including those due to excavation, traffic loads, movements due to underground activity, swelling and shrinkage due to vegetation, climate and water content, soil mass movement, degradation of soils, dynamic loading, pre-stress, downdrag and temperature-dependent loads. Combinations of actions and their duration must be considered. Some actions (e.g. earth pressures) can be both unfavourable (disturbing) and favourable (restoring) actions.

Geotechnical properties can be obtained from a variety of sources, but it must be recognised that they are not necessarily intrinsic properties of the soil and could be different from those mobilised in the field. For example, properties depend on the level of stress and mode of deformation, the presence of discontinuities, softening due to dynamic loading, percolating water especially in fissured tills if the fissures open up an excavation, the construction process and whether a soil behaves as a brittle or ductile material.

Characteristic values of geotechnical properties should be a cautious estimate of the mean value throughout the zone of influence. This depends on the quality of the information, the extent of the investigation, the extent of the zone of influence and the ability of the geotechnical structure to compensate for soil of varying properties and thickness. Statistical methods are often used, which should differentiate between local and regional data making use of *a priori* knowledge. The characteristic value is the value such that the probability of a worse value is not greater than 5%. If using published values, a very cautious value is used. In glacial soils, that can lead to overdesign because the results of laboratory and *in situ* tests are particularly sensitive to disturbance.

6.3.3 Partial factors of safety

Partial factors for the SLS are 1. Serviceability criteria must be less than or equal to the limiting value, C_d; typical values are given in Table 6.2. Soil properties used to assess serviceability may be different to those used to assess ULS. For example, the mobilised secant stiffness at the SLS is greater than that mobilised at the ULS because the level of mobilised strain is less. Brittle soils can exhibit progressive failure, so using peak strength as a limiting value in ultimate state calculations may lead to unsafe design but using the post-peak strength in serviceability calculations may be conservative because the limiting strains may be too small.

The design values of actions, geometrical data and geotechnical properties are factored to take into account safety when assessing the ULS. There are five failure mechanisms: loss of equilibrium (EQU), internal failure or excessive deformation of the structural elements (STR), failure or excessive deformation of the ground (GEO), uplift due to water pressure (UPL) and effects of hydraulic gradients (HYD). The design value of an action (F_d) is

$$F_d = \gamma_F F_{rep} \tag{6.1}$$

where the representative action, F_{rep}, is related to the characteristic value, F_k, given by

$$F_{rep} = \psi F_k \tag{6.2}$$

The partial factor, γ_F, for actions is given in Table 6.4. The factor, ψ, to convert the characteristic value to the representative value normally applies to buildings (BS EN 1990:2002+A1:2005). In geotechnical design, it is assumed to be 1; that is, $F_d = \gamma_F F_k$.

Groundwater pressures are either factored or the characteristic water level is changed to introduce a margin of safety. Groundwater level should be taken as the maximum possible water level, possibly ground surface if it is a disturbing action or the lowest level if it is a restoring action. This can be conservative if a maintained drainage system

Table 6.4 Partial factors (γ_F) on actions for loss of equilibrium (EQU), and internal failure or excessive deformation of structural elements (STR) or the ground (GEO)

	Partial factors on actions (γ_F) (EQU)		*Partial factors on actions (γ_F) or effects of actions (γ_E) (STR and GEO)*		
Action	*Symbol*	*Value*	*Symbol*	*A1*	*A2*
Permanent (unfavourable)	$\gamma_{G;dst}$	1.1	γ_G	1.35	1
Permanent (favourable)	$\gamma_{G;stb}$	0.9		1	1
Variable (unfavourable)	$\gamma_{Q;dst}$	1.5	$\gamma_{Q;dst}$	1.5	1.3
Variable (favourable)	$\gamma_{Q;stb}$	0		0	0

Source: After BS EN 1997-1:2004+A1:2013. *Eurocode 7: Geotechnical Design – Part 1: General Rules*. British Standards Institution, London.

is guaranteed. Groundwater conditions can be complex in glacial soils; therefore, this simple approach may be conservative. For example, a dense matrix-dominated till may act as an aquiclude resulting in a regional water level based on the underlying rock and a perched water table due to infiltration in the overlying soils. Therefore, it is prudent to investigate groundwater conditions in some detail and study the effects of changes in these conditions with time.

The design values (X_d) for geotechnical parameters are derived from the characteristic value (X_k) using the partial factor (γ_m) for material properties

$$X_d = \frac{X_k}{\gamma_m} \tag{6.3}$$

The partial factors are given in Table 6.5.

Eurocode 7 does allow the partial factors to be modified to allow for abnormal loads, temporary works or transient design situations. This would be considered for complex soils such as glacial soils.

The design value of the effect of the actions, E_d, must be less or equal to the design value of the resistance, R_d, when excessive deformation or rupture is considered. The ratio of these two values is sometimes referred to as an overdesign factor, but it is not a global

Table 6.5 Partial factors (γ_m) on material properties for loss of equilibrium (EQU), and internal failure or excessive deformation of structural elements (STR) or the ground (GEO)

			STR and GEO	
Soil parameter	*Factor*	*EQU*	*M1*	*M2*
Angle of shearing resistance (tan φ′)	$\gamma_{\varphi'}$	1.25	1.0	1.25
Effective cohesion (c')	$\gamma_{c'}$	1.25	1.0	1.25
Undrained shear strength (c_u)	γ_{cu}	1.4	1.0	1.4
Unconfined strength (q_u)	γ_{qu}	1.4	1.0	1.4
Unit weight (γ)	γ	1.0	1.0	1.0

Source: After BS EN 1997-1:2004+A1:2013. *Eurocode 7: Geotechnical Design – Part 1: General Rules*. British Standards Institution, London.

factor of safety. If the partial factors are set to 1, the overdesign factor is the same as the global factor of safety. The partial factors on actions may be applied to the actions or their effects.

$$E_d = E\left(\gamma_F F_{rep}; \frac{X_k}{\gamma_M}; a_d\right) \tag{6.4}$$

$$E_d = \gamma_E E\left(\gamma_F F_{rep}; \frac{X_k}{\gamma_M}; a_d\right) \tag{6.5}$$

where a_d is the design value of geometrical data and γ_E is the partial factor for the effect of an action. Partial factors may be applied to ground properties, resistances or both when considering the design resistance, R_d.

$$R_d = R\left(\gamma_F F_{rep}; \frac{X_k}{\gamma_M}; a_d\right) \tag{6.6}$$

$$R_d = \frac{R(\gamma_F F_{rep}; X_k; a_d)}{\gamma_R} \tag{6.7}$$

$$R_d = \frac{R(\gamma_F F_{rep}; (X_k/\gamma_M); a_d)}{\gamma_R} \tag{6.8}$$

where γ_R is the partial factor for resistance given in Table 6.6. There are three Design Approaches (1, 2 and 3). In Design Approach 1, the limit state is assessed for two sets of combinations of partial factors:

$$\text{Combination 1}: A1 + M1 + R1 \tag{6.9}$$

$$\text{Combination 2}: A2 + M2 + R1 \tag{6.10}$$

where $A1$, $A2$, $M1$ and $M2$ refer to the partial factors for actions and materials. These combinations apply to all structures except axial loaded piles and anchors. The combinations for these exceptions are

$$\text{Combination 1}: A1 + M1 + R1 \tag{6.11}$$

$$\text{Combination 2}: A2 + (M1 \text{ or } M2) + R4 \tag{6.12}$$

The combinations for Design Approaches 2 and 3 are

$$\text{Design Approach 2}: A1 + M1 + R2 \tag{6.13}$$

$$\text{Design Approach 3}: A2 + M2 + R3 \tag{6.14}$$

Table 6.6 Partial resistance factors (γ_R) for (a) spread foundations, (b) driven piles, (c) bored piles, (d) CFA piles and (e) retaining structures

Resistance	*Symbol*	*Set*		
		R1	*R2*	*R3*
(a) Spread foundations				
Bearing	$\gamma_{R;v}$	1	1.4	1
Sliding	$\gamma_{R;h}$	1	1.1	1

Resistance	*Symbol*	*Set*			
		R1	*R2*	*R3*	*R4*
(b) Driven piles					
Base	γ_b	1	1.1	1	1.3
Shaft (compression)	γ_s	1	1.1	1	1.3
Total/combined (compression)	γ_t	1	1.1	1	1.3
Shaft (tension)	$\gamma_{s;t}$	1.25	1.15	1.1	1.6
(c) Bored piles					
Base	γ_b	1.25	1.1	1	1.6
Shaft (compression)	γ_s	1	1.1	1	1.3
Total/combined (compression)	γ_t	1.15	1.1	1	1.5
Shaft (tension)	$\gamma_{s;t}$	1.25	1.15	1.1	1.6
(d) CFA piles					
Base	γ_b	1.1	1.1	1	1.45
Shaft (compression)	γ_s	1	1.1	1	1.3
Total/combined (compression)	γ_t	1.1	1.1	1	1.4
Shaft (tension)	$\gamma_{s;t}$	1.25	1.15	1.1	1.6

Resistance	*Symbol*	*Set*		
		R1	*R2*	*R3*
(e) Retaining structures				
Bearing capacity	$\gamma_{R;v}$	1	1.4	1
Sliding resistance	$\gamma_{R;h}$	1	1.4	1
Earth resistance	$\gamma_{R;e}$	1	1.4	1

Source: After BS EN 1997-1:2004+A1:2013. *Eurocode 7: Geotechnical Design – Part 1: General Rules.* British Standards Institution, London.

6.4 GEOTECHNICAL DESIGN REPORT

Any project should conclude with a geotechnical design report, which sets out the assumptions, data, methods of calculation, results of the verification of safety and serviceability and records of any design changes and how they were implemented during construction. The information is important for civil engineering structures because they will be adapted in their lifetime, possibly reused. It is also an opportunity to collate all geotechnical information obtained during the ground investigation and construction. This includes the assumptions and limitations of the design methods.

Eurocode 7 sets outs the information to be included, which, for glacial soils, is as follows:

- *Ground investigation*
  - A description of the site and surroundings
  - A description of the ground conditions
  - Design values of soil properties, including justification, as appropriate
- *Design*
  - A description of the proposed construction, including actions
  - Assumed forces and displacements
  - Statements on the codes and standards applied
  - Geotechnical design calculations and drawings
  - Foundation design recommendations
- *Construction*
  - Statements on the suitability of the site with respect to the proposed construction and the level of acceptable risks
- *Monitoring*
  - A note of items to be checked during construction or requiring maintenance or monitoring
  - The purpose of each set of observations or measurements
  - The parts of the structure, which are to be monitored and the locations at which observations are to be made
  - The frequency with which readings are to be taken
  - The ways in which the results are to be evaluated
  - The range of values within which the results are to be expected
  - The period of time for which monitoring is to continue after the construction is complete
  - The parties responsible for making measurements and observations, for interpreting the results obtained and for maintaining the instruments

The implementation of artificial intelligence in the construction industry and the development of smart buildings and infrastructure mean that continuous records of in-service will be available in future. Therefore, the geotechnical design report will provide useful information for extending the life of a structure and provide a useful reference for future designs.

6.5 SPREAD FOUNDATIONS

Spread foundations include pads, strips, rafts and deep engineering foundations such as caissons, where the width is significant compared to the depth. The overall stability, bearing resistance, punching and squeezing failure, sliding, excessive settlements and excessive heave have to be checked. The effects on the structure and soil structure have also to be considered. The foundation has to be placed on an adequate bearing stratum, taking into account the effects of frost and vegetation, the level of the water table both for excavation level and subsequent seasonal changes, the effects on adjacent structures and changes in geotechnical properties due to groundwater and scour. The overall stability has to be assessed if the foundations are on a slope, adjacent to an excavation or water course or over a buried structure or mine workings.

Figure 6.2 is a guide to foundation selection, which considers the ground conditions, the structure, the site, safety and sustainability. BS 8004 (2015) suggests that spread foundations

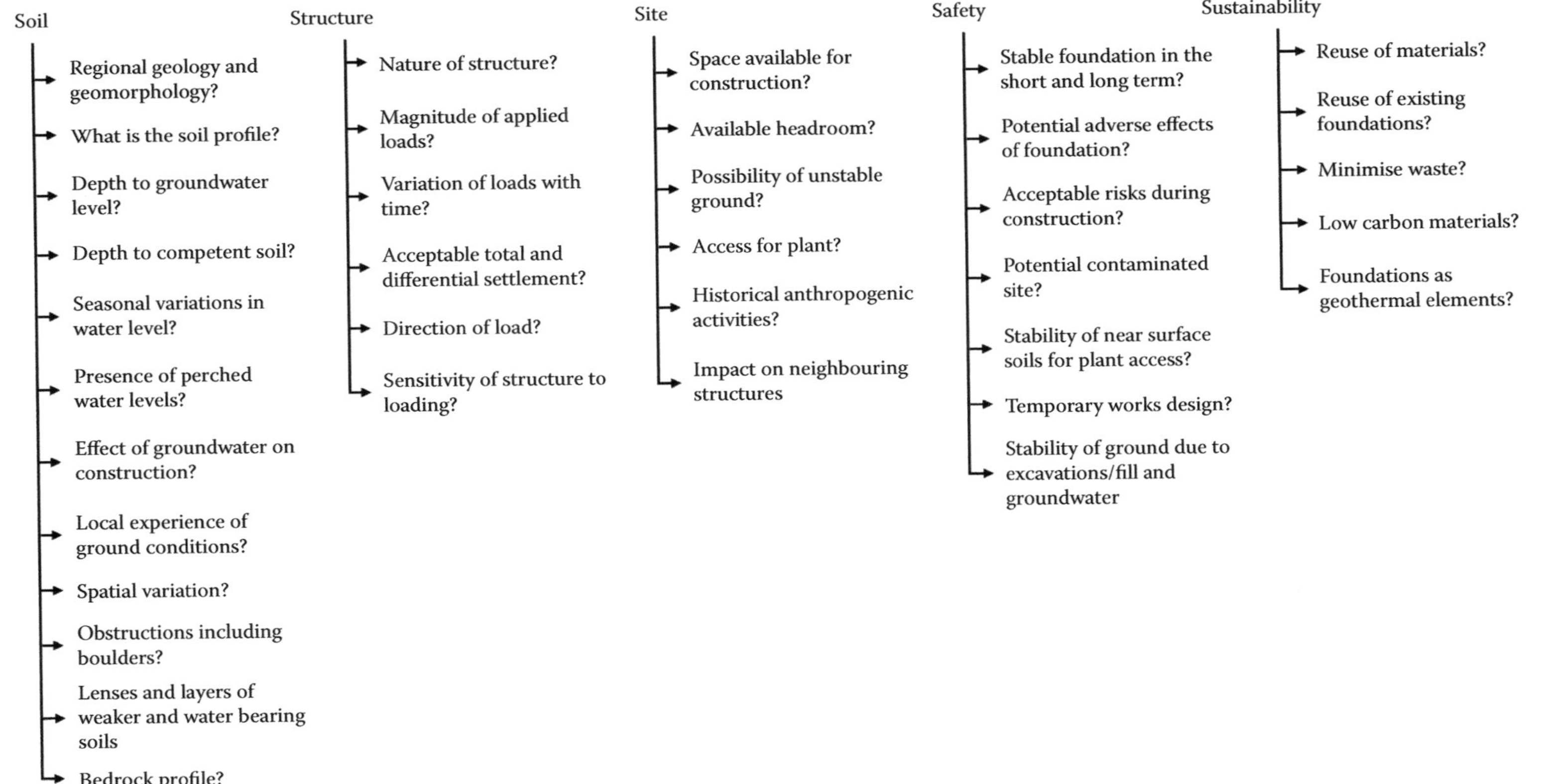

Figure 6.2 Key factors to consider when designing a geotechnical structure. (After O'Brien, A. *CE Manual of Geotechnical Engineering*, Thomas Telford Ltd, 2012: 823–850.)

can be considered where an adequate bearing stratum is found within the top 2 m; on dense coarse-grained soils above the water table; and on medium-strength fine-grained soils.

Presumed bearing resistance, such as those shown in Table 6.7, can be used for Category 1 structures. However, the complexity of the ground should be taken into account when assessing the category. It is possible to have a Category 1 structure sitting on complex ground conditions, which means that the foundations should be designed as if they were Category 2 structures or even Category 3. Glacial soils can be considered complex ground conditions; therefore, presumed bearing pressures should only be used to produce an initial assessment of foundation dimensions. The actual resistance should be assessed.

According to O'Brien and Farooq (2012), the most common cause of excessive movement of shallow foundations is seasonal movement due to trees, particularly for foundations on high plasticity clays. This may be less of a problem in matrix-dominated tills according to NHBC (2003) since unweathered matrix-dominated tills exhibit low plasticity. Groundwater is an issue that influences the depth and type of excavation because of potential flooding, base heave and side instability. A guide to the excavation method to take this into account during construction is given in Figure 6.3. Excavation in matrix-dominated fissured tills can increase the mass permeability and reduce the strength because fissures will open up due to stress relief. This will influence the type of excavation support. Boulders should be expected in glacial tills and glaciofluvial soils. Weaker soils and water-bearing sand and gravels layers and lenses should be expected in glacial tills. The possibility of weaker soil underlying the formation level should be considered using one of the methods shown in Figure 6.4. Thus, the composition, structure and fabric of glacial soils will influence the type of excavation as well as the type of foundation. For example, O'Brien and Farooq (2012) describe a case study which started with the assumption that bored pile foundations would be used. However, bored piles were ruled out because of the possibility of boulders within the glacial tills, which would delay the construction resulting in penalties. An alternative design making use of existing, shallow foundations was proposed because it was known that settlements of foundations on glacial tills are small. It highlights that piles in glacial soils can be considered a risk as the local ground conditions in a spatially variable soil are unknown. The design profile used by O'Brien and Farooq (2012), shown in Figure 6.5, highlights the effect of test and sample type on the profile of strength and the difficulty in selecting design parameters.

A further example of the use of shallow foundations to replace piled foundations in glacial soils was presented by Bentler et al. (2009). The analysis showed that compression of the alluvial

Table 6.7 Presumed bearing resistance for spread foundations on coarse-grained glacial soils (at 0.75 m below ground level) and fine-grained glacial soils (at 1 m below ground level)

			Presumed bearing resistance (kPa)		
Glacial soil type	N_{60}	c_u *(kPa)*	*1 m*	*2 m*	*4 m*
Very dense sands and gravels	>50		800	600	500
Dense gravels	30–50		500–800	400–600	300–500
Medium dense sands and gravels	10–30		150–500	100–400	100–300
Loose sands and gravels	5–10		50–150	50–100	30–100
Hard glacial till		>300	800	600	400
Very stiff glacial till		150–300	400–800	300–500	150–250
Stiff weathered glacial till		75–150	200–400	150–250	75–125
Glaciolacustrine clays		40–75	100–200	75–100	50–75

Source: After Tomlinson, M. J. and R. Boorman. *Foundation Design and Construction*. Pearson Education, Harlow, UK, 2001.

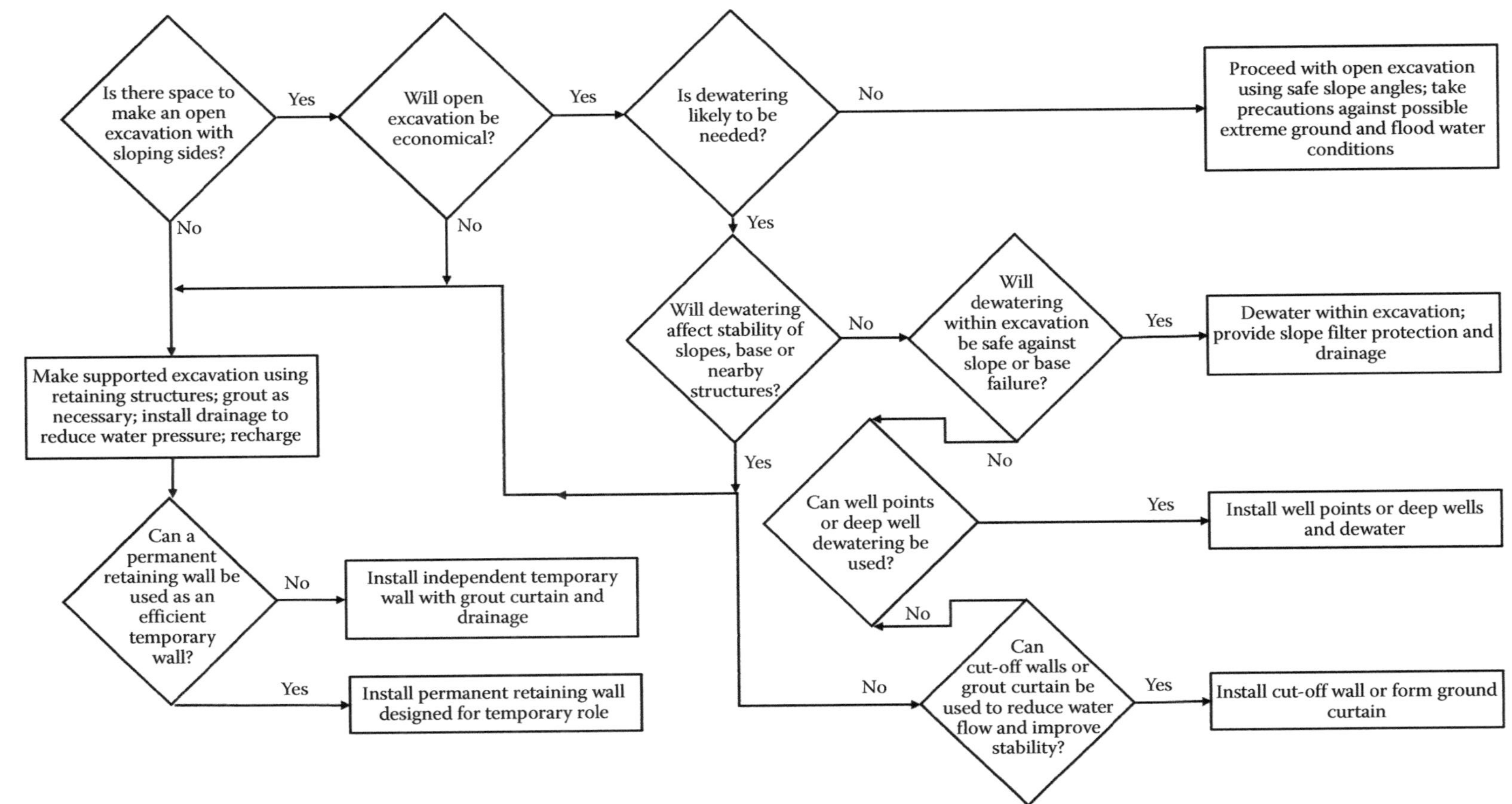

Figure 6.3 Guide to the stability of an excavation for spread foundations. (After Cole, K. W. *Foundations. ICE Works Construction Guides*. Institution of Civil Engineers, London, 1988.)

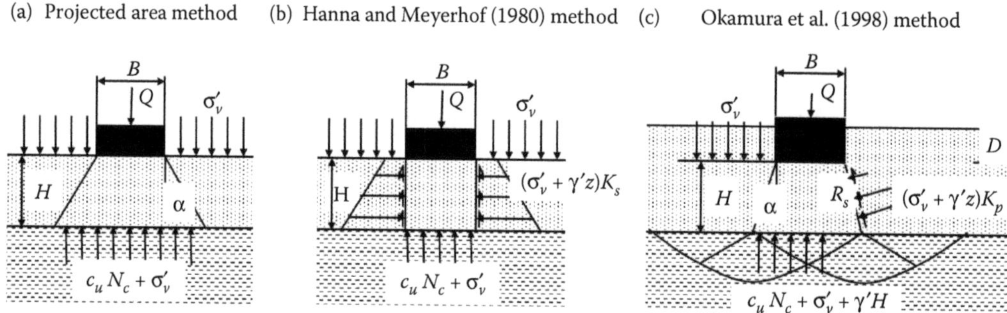

Figure 6.4 Effect of weaker soils on the capacity of spread foundations showing three different mechanisms (a) projected area, (b) punching failure and (c) projected punching failure. (After O'Brien, A. S. and I. Farooq. *ICE Manual of Geotechnical Engineering*, Thomas Telford Ltd, London; 2012: 765–800.)

clay overlying the glacial soils of interbedded matrix-dominated and clast-dominated tills would cause significant downdrag on the piles for a bridge abutment leading to a complex design. An alternative proposal using fill to preload and compress the alluvial clays for 4 months reduced the predicted settlement from 45 to 25 mm using a method developed by Mayne et al. (2001) based on shear modulus derived from shear wave velocity. The abutments were monitored for settlement, lateral movement and rotation to confirm the design assumptions.

6.5.1 Bearing resistance

There are several methods used to determine the bearing resistance of the ground supporting spread foundations (e.g. Hansen, 1970; Meyerhof, 1951; Vesic, 1975). The Hansen

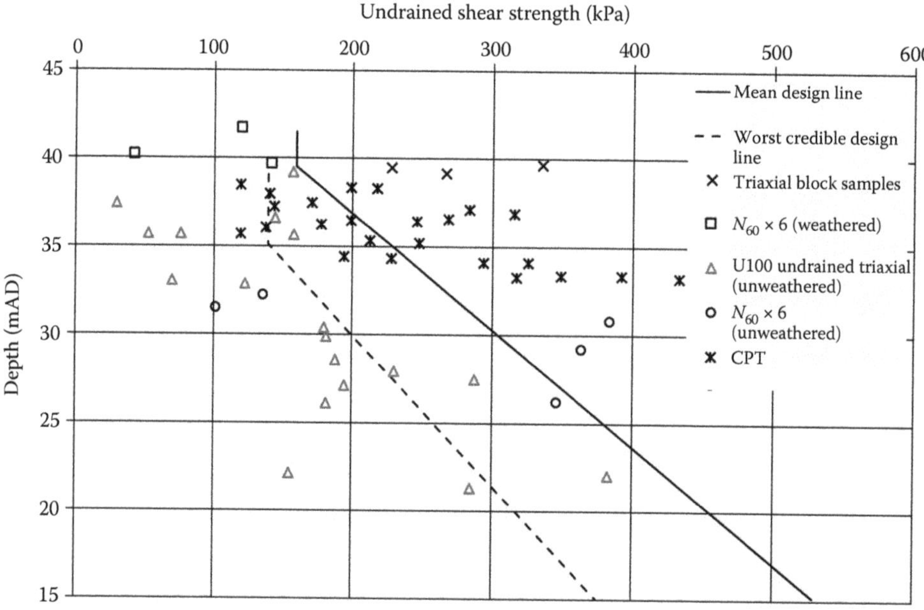

Figure 6.5 Design profile used by O'Brien and Farooq (2012) in opting for spread foundations instead of bored piles for a foundation on glacial till.

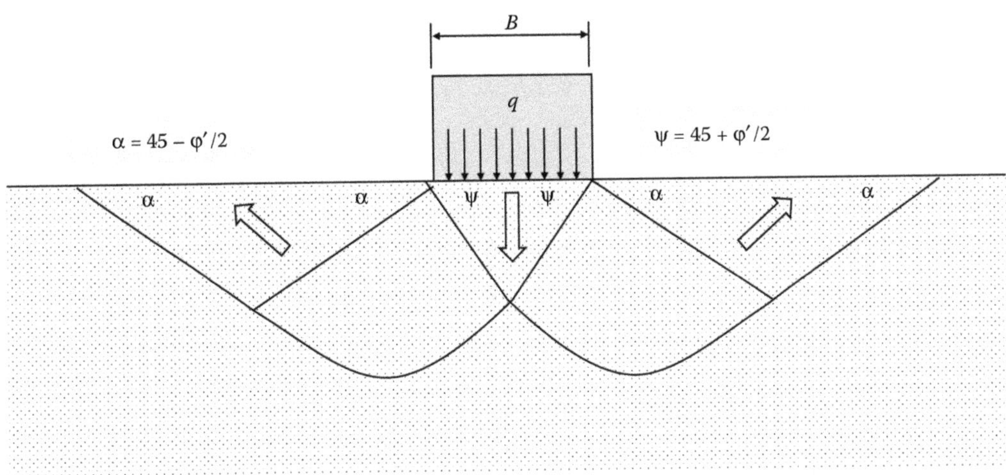

Figure 6.6 **Principle of bearing failure based on Prandtl (1920) analysis of a punch penetrating a softer metal.**

(1970) method is referenced in Eurocode 7; the Meyerhof method in American guidelines. These methods are all based on Prandtl's (1920) method for predicting the indentation load in a metal (Figure 6.6).

Correction factors are applied for the shape and depth of the foundation, the inclination of the load, the ground and the foundation and the rigidity of the foundation.

The effective area, A', of a spread foundation is

$$A' = B' \times L' \tag{6.15}$$

where B' is the effective width and L' the effective length. If a vertical load acts at the centre of the foundation, then the effective width and length are the actual width and length. The effective dimensions are relevant for eccentric loading due to wind loading (if greater than 25% of the vertical load) or walls or columns offset from the centre of the foundation. The design bearing resistance, R, for fine-grained soils is

$$\frac{R}{A'} = (\pi + 2)c_u b_c s_c d_c i_c + \sigma_v \tag{6.16}$$

where b_c is a factor for the inclination of the foundation base, s_c the shape factor, d_c the depth factor, i_c a factor for the inclination of the load due to a horizontal load, H, and σ_v the total overburden pressure at the base of the foundation. The factors are given in Table 6.8.

This applies to undrained conditions which are assumed to exist at the end of construction when building on fine-grained soils. In composite soils, which have characteristics of fine-grained soils (e.g. matrix-dominated tills) and significantly anisotropic soils (e.g. glaciolacustrine soils), the hydraulic conductivity is such that some drainage will take place during construction; therefore, the actual capacity at the end of construction is likely to be greater. Ignoring that effect is a conservative approach. The bearing resistance should be considered for the long-term conditions to check that the resistance increases with time or, at least, is not less than that for the short-term conditions. If it does, it may be possible to produce a more economic design allowing for an increase in strength during construction.

Table 6.8 Ultimate bearing resistance for spread foundations founded on fine-grained soils using undrained strength

The design bearing resistance, R, for fine-grained soils is

$$\frac{R}{A'} = (\pi + 2)c_u b_c s_c d_c i_c + \sigma_v$$

where b_c is a factor for the inclination of the foundation base, s_c the shape factor, d_c the depth factor, i_c a factor for the inclination of the load due to a horizontal load, H and σ_v the total overburden pressure at the base of the foundation.

$$b_c = 1 - \frac{2\alpha}{\pi + 2}$$

$$s_c = 1 + 0.2\left(\frac{B'}{L'}\right) \quad \text{for a rectangular foundation}$$

$$s_c = 1.2 \quad \text{for a square or circular foundation}$$

$$d_c = 1 + 0.27\sqrt{\frac{D}{B'}}$$

$$i_c = \frac{1}{2}\left(1 + \sqrt{1 - \frac{H}{A'c_u}}\right) \quad [H < A'c_u]$$

Source: After BS 8004:2015. *Code of Practice for Foundations*. British Standards Institution, London.

The bearing resistance formula based on effective stress applies to fine-grained soils in the long term and coarse-grained soils in the short and long term but should also be considered for composite soils in the short term. The bearing resistance is given by

$$\frac{R}{A'} = c'N_c b_c s_c i_c d_c g_c r_c + \sigma_v' N_q b_q s_q i_q d_q g_q r_q + 0.5\gamma' B' N_\gamma b_\gamma s_\gamma i_\gamma d_\gamma g_\gamma r_\gamma \tag{6.17}$$

where c' is the soil cohesion, σ_v' the effective overburden pressure at the base of the foundation and γ' the submerged unit weight. The coefficients are given in Table 6.9.

Many foundations are built on level ground with a horizontal base and vertical load, so simplified versions of these equations can be used. However, ignoring the inclination of the ground, base or load is unsafe if they exist.

6.5.2 Settlement

Predicting settlement of structures is difficult but critical. Damage to buildings can be due to differential settlement, settlement due to adjacent structures and differential movement between different parts of a building. Definitions of settlement and distortion are given in Figure 6.7 and the limiting values for various structures are given in Table 6.2. Skempton and MacDonald (1956) suggested that the maximum acceptable settlement for spread foundations on clean sands is about 25 mm; for isolated columns 40 mm; and for rafts 40–65 mm. Skempton and MacDonald (1956) suggested that 40 mm was the limit for differential settlement for foundations on clays; 65 mm total settlement for isolated columns; and 65–100 mm for rafts. These values are based on a limited database of observations and do not take in account differential settlement with adjacent structures or utilities. Further, they do not necessarily apply to composite soils, which include a range of particle sizes.

Table 6.9 Ultimate bearing resistance for spread foundations using effective strength

The bearing resistance is given by

$$\frac{R}{A'} = c'N_c b_c s_c i_c d_c g_c r_c + \Delta\sigma_v' N_q b_q s_q i_q d_q g_q r_q + 0.5\gamma' B' N_\gamma b_\gamma s_\gamma i_\gamma d_\gamma g_\gamma r_\gamma$$

where c' is the soil cohesion, $\Delta\sigma_v'$ the effective overburden pressure at the base of the foundation, γ' the submerged unit weight.

1. Bearing coefficients

$$N_q = e^{\pi \tan\varphi'} \tan^2\left(45 + \frac{\varphi'}{2}\right)$$

$$N_\gamma = a + e^{b\varphi'}$$

$$N_c = (N_q - 1)\cot\varphi$$

where $a = 0.0663$ and $b = 9.3$ for a smooth foundation and $a = 0.1054$ and $b = 9.6$ for a rough foundation.

2. Shape factors

$$s_q = 1 + \frac{B'}{L'}\tan\varphi'$$

$$s_\gamma = 1 - 0.4\frac{B'}{L'}$$

$$s_c = 1 + \frac{B'}{L'}\frac{N_q}{N_c}$$

3. Depth factors

$$d_q = 1 + 2\tan\varphi'(1 - \sin\varphi')^2 \tan^{-1}\left(\frac{D}{B}\right)$$

$$d_\gamma = 1$$

$$d_c = d_q - \left(\frac{1 - d_q}{N_c \tan\varphi'}\right)$$

4. Load inclination factors

$$i_q = \left[1 - \left(\frac{H}{V + A'c'\cot\varphi'}\right)\right]^m$$

$$i_\gamma = \left[1 - \left(\frac{H}{V + A'c'\cot\varphi'}\right)\right]^{m+1}$$

$$i_c = i_q - \left(\frac{1 - i_q}{N_c \tan\varphi'}\right)$$

where H is the horizontal load, V is the vertical load and m is $(2 + B/L)/(1 + B/L)$ if the load is in the direction of B or $(2 + L/B)/(1 + L/B)$ in the direction of L.

(Continued)

Table 6.9 (Continued) Ultimate bearing resistance for spread foundations using effective strength

5. Base inclination factor

$$b_q = b_\gamma = (1 - \alpha \tan\varphi')^2$$

$$b_c = b_q - \left(\frac{1 - b_q}{N_c \tan\varphi'}\right)$$

where α is the angle of the foundation base to the horizontal.

6. Ground inclination factors

$$g_q = g_\gamma = (1 - \tan\omega)^2$$

$$g_c = g_q - \left(\frac{1 - g_q}{N_c \tan\varphi'}\right)$$

where ω is the angle of the ground surface to the horizontal.

7. Rigidity factors

$$r_\gamma = r_q = e^{(-4.4 + 0.6B/L)\tan\varphi' + ((3.07 \sin\varphi' \log_{10} 2I_r)/(1 + \sin\varphi'))}$$

$$r_c = r_q - \left(\frac{1 - r_q}{N_c \tan\varphi'}\right)$$

where $\left[I_r = \frac{G}{(c' + \sigma'_v \tan\varphi')}\right]$

Source: After BS 8004:2015. *Code of Practice for Foundations*. British Standards Institution, London.

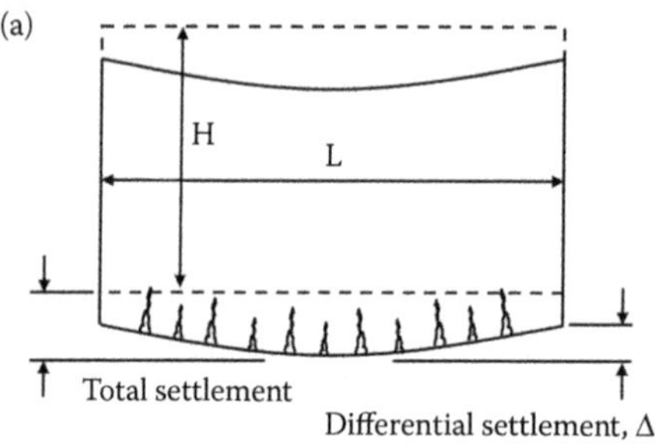

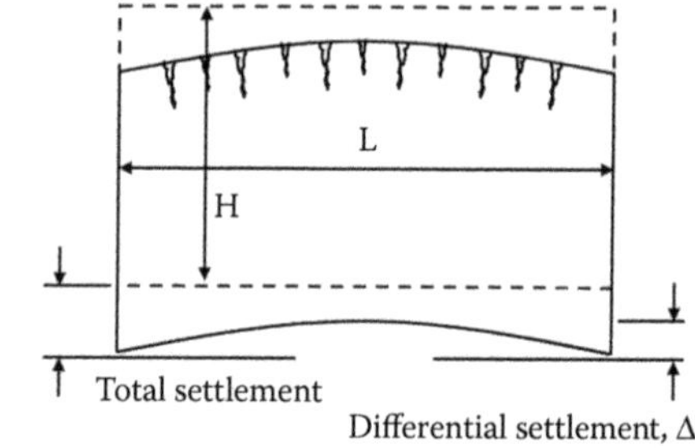

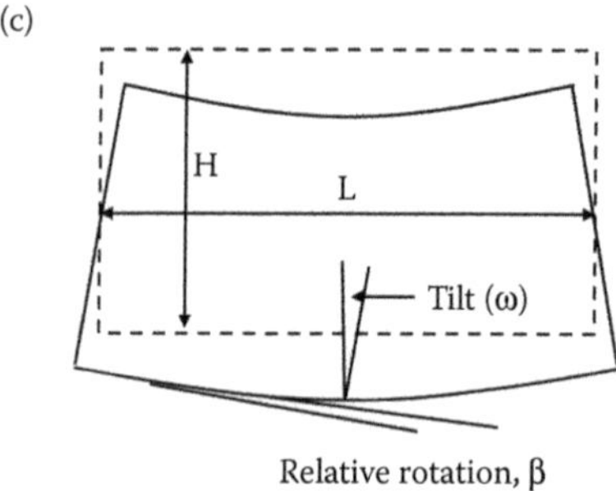

Figure 6.7 Differential settlement and distortion showing (a) settlement due to sagging, (b) settlement due to hogging and (c) tilt. (After Burland, J. B. and C. P. Wroth. *Settlement of Structures, Proceedings of the Conference of the British Geotechnical Society*, Cambridge. Pentech Press, London, UK, 1974: 611–764.)

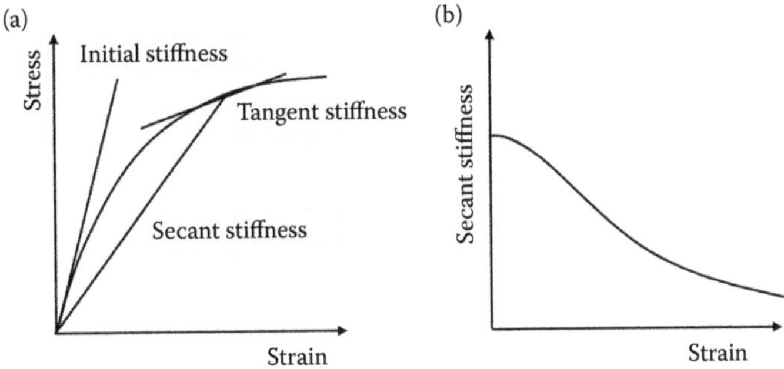

Figure 6.8 Definitions of stiffness and the effect of strain level on the secant stiffness showing (a) the stress strain curve and (b) the stiffness degradation curve.

Settlements can be calculated using the following (BS 8004:2015):

- Theory of elasticity
- 1D consolidation of fine-grained soils
- Empirical methods for coarse-grained soils
- Numerical models

No matter which method is used, the most critical factor is the selection of stiffness, which has to take into account the quality of the investigation, the mobilised strain and the operational stress level (Figures 6.8 and 6.9). Therefore, any prediction of settlement has to be treated with caution and, since many methods are based on field data, a sensitivity

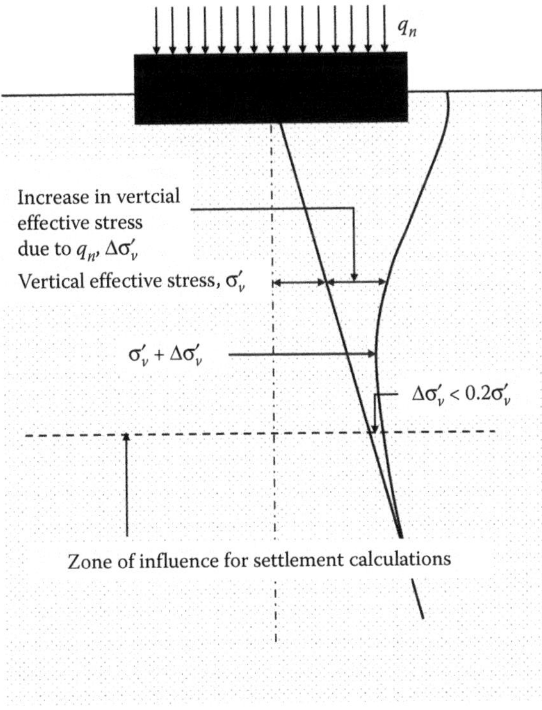

Figure 6.9 Stress distribution beneath a loaded area based on the classic Boussinesq analysis.

analysis should be considered to take into account the range of data on which the empirical methods are based and the variability of the results of the ground investigation. There are numerous papers and text books (e.g. Tomlinson and Boorman, 2001; Bowles, 2001; ICE, 2012) that cover the various methods to predict settlement. A summary of common methods is given here highlighting the relevance to glacial soils.

Burland (2012) suggested that the traditional methods of predicting settlement are perfectly adequate provided the soil stiffness is correctly assessed. The stress distribution beneath a loaded area is calculated using Boussinesq's classic prediction (Figure 6.9) of the increase in stress due to a point load at the surface (Figure 6.10).

The increase in stress beneath a loaded area can be calculated using influence factors developed by Newmark (1942) and Fadum (1948). The increase in stress below a rectangular foundation, according to Newmark (1942), is

$$\Delta\sigma_v = q'\frac{1}{4\pi}\left(\frac{2MN\sqrt{V}}{V+V_1}\frac{V+1}{V} + \tan^{-1}\left(\frac{2MN\sqrt{V}}{V-V_1}\right)\right) \tag{6.18}$$

where q' is the increase in load at formation level and M, N, V and V_1 are given by

$$M = \frac{B'}{z} \tag{6.19}$$

$$N = \frac{L'}{z} \tag{6.20}$$

$$V = M^2 + N^2 + 1 \tag{6.21}$$

$$V_1 = (MN)^2 \tag{6.22}$$

This applies to the corner of a flexible rectangular foundation on the ground surface. This theory is based on elastic isotropic homogeneous medium, which is not typical of

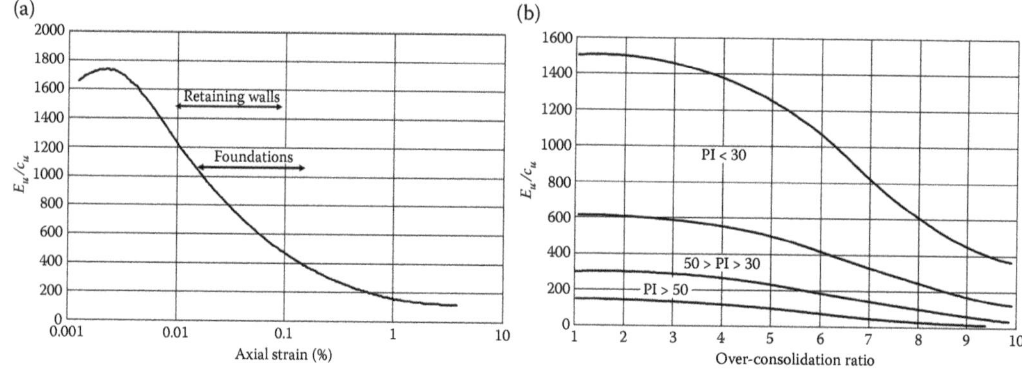

Figure 6.10 Relationship between (a) undrained modulus of elasticity, undrained shear strength and mobilised strain (After Jardine, R. J. et al. *Proceeding of the 11th ICSMFE*, San Francisco, 2; 1985: 511–514) and the relationship between (b) E_u/c_u, plasticity index and over-consolidation ratio. (After Jamiolkowski, M., C. C. Ladd, J. T. Germaine, and R. Lancellotta, New development in field and laboratory testing of soils. In *Proc. 11th ICSMFE*, 1, Balkeema, Holland, 1985: 57–153.)

glacial soils. However, Burland (2012) suggests that this method still gives reasonable predictions of changes in vertical stress except for stiff layers overlying soft layers and for soils that are strongly anisotropic. Eurocode 7 emphasises the need to obtain representative values of stiffness since that is more critical than any other factor in determining settlement. Given the comments on obtaining representative values for glacial soils due to the natural variability of these soils, especially glacial tills and glaciofluvial soils, the difficulty in obtaining Class 1 samples and the difficulty in determining appropriate values from *in situ* and laboratory tests, predicting settlement of structures on glacial soils is challenging. Therefore, it is essential that an assessment of settlement based on a possible range of stiffness be undertaken.

The settlement of the corner of a rectangular foundation on fine-grained soils can be predicted from

$$s = q'B\frac{1-\upsilon^2}{E}I_p \tag{6.23}$$

where B is the width of the foundation, E the soil stiffness, υ Poisson's ratio, I_p an influence factor and q' the net contact stress at formation level. Use of Equation 6.23 is described in Table 6.10 and Figure 6.11.

The undrained stiffness, E_u, is used to predict immediate settlement, the value being dependent on the expected level of strain as soil is non-linear (Figure 6.10). Routine ground investigations generally do not include measurements of E_u, so empirical correlations with shear strength are often used; typical values are given in Table 6.11.

Figure 6.11 is used to determine the influence factor, I_p, based on a maximum layer thickness of $4B$. This is based on the assumption that stiffness is constant with depth. There are techniques that take into account increasing stiffness (e.g. Butler, 1974). Equation 6.23 predicts the settlement of a fully flexible foundation. The settlement of fully rigid foundation is about 80% of that value.

Settlement of foundations on fine-grained soils can be estimated using 1D compression based on results of oedometer tests though appropriate values of stiffness, E_d, can be used. Table 6.12 summarises the steps. The actual settlement is less than the predicted (Skempton and Bjerrum, 1957), so a correction factor is applied (0.7–1.0 for glaciolacustrine clays and 0.2–0.5 for matrix-dominated tills). A further correction is applied for the depth of foundation (Figure 6.12) based on the depth to formation level and the area of foundation.

Table 6.10 Method to predict the settlement beneath the corner of a rectangular foundation

The settlement of the corner of a rectangular foundation on fine-grained soils can be predicted from

$$s = q'B\frac{1-\upsilon^2}{E}I_p$$

where B is the width of the foundation, E the soil stiffness, υ Poisson's ratio, I_p an influence factor and q' the net contact stress at formation level.

- Poisson's ratio for undrained behaviour is 0.5; for drained behaviour 0.1–0.3. Very stiff glacial tills exhibit volume changes so assume $\upsilon = 0.2$.
- H is the stratum thickness or $4B$ whichever is the least.
- For undrained conditions, $I_p = F_1$, where F_1 is taken from Figure 6.10.
- For partially drained or fully drained conditions, $I_p = F_1 + F_2$, where F_2 ranges from 0 to 0.3 depending on υ and L/B and H/B.

Source: After Steinbrenner, W. *Die Strasse*, 1; 1934: 121–124.

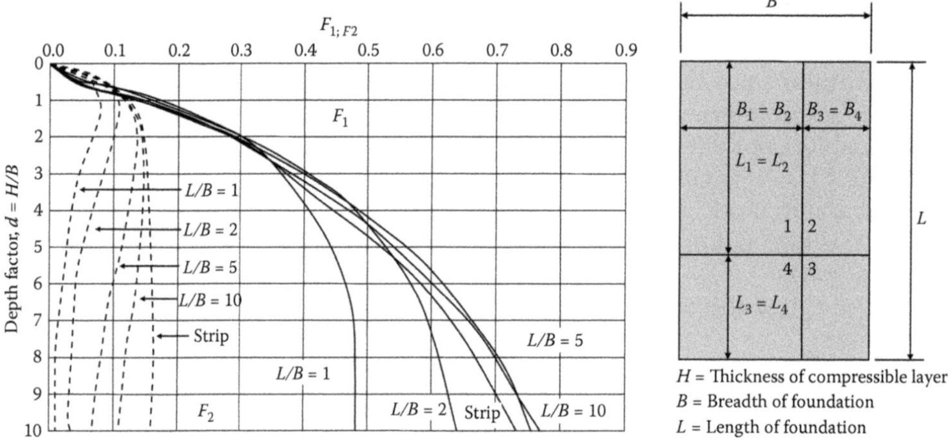

Figure 6.11 Factors used to calculate the settlement beneath the corner of a flexible foundation on fine-grained soil. (After Tomlinson, M. J. and R. Boorman. *Foundation Design and Construction*. Pearson Education, Harlow, UK, 2001.)

Table 6.11 Typical values of stiffness of glacial soils

Glacial soil type	*E based on N_{60}*	*E based on CPT q_c*	*E based on c_u*	m_v
Normally consolidated clay (e.g. glaciolacustrine clay)			100–500 c_u	0.3–1.5 m²/MN
Stiff clays (e.g. weathered matrix-dominated till, supraglacial till)			500–1500 c_u $E = E_{nc}\ OCR^{0.5}$	0.05–0.1 m²/MN
Very stiff clays (e.g. matrix-dominated till)				<0.05 m²/MN
Loose sands	$E = 500\ (N_{60} + 15)$	$E = 4\ q_c\ (q_c < 10$ MPa) $E = (2q_c + 20)$ (10 MPa $< q_c < 50$ MPa) $E = 120\ (q_c > 50$ MPa)		
Dense sands	$E = 18{,}000 + 750\ N_{60}$ $E = E_{nc}\ OCR^{0.5}$	$E = 6$ to $30\ q_c$ $E = 5q_c\ (q_c < 50$ MPa) $E = 250$ MPa ($q_c > 50$ MPa)		
Clayey sands	$E = 320\ (N_{60} + 15)$	$E = 3–6\ q_c$		
Sand and gravel	$E = 1200\ (N_{60} + 6)$ $E = E_{nc}\ OCR^{0.5}$			

Increasingly, settlement predictions are made using numerical analyses, which allow non-homogeneous anisotropic non-linear elastic soils to be analysed. These sophisticated methods are of value only if realistic representative values of stiffness are used. Representative stiffness of glaciolacustrine clay can be obtained from local strain triaxial tests. Representative stiffness of matrix-dominated tills is difficult especially if there is a significant percentage of coarse-grained particles, which may affect sampling, specimen preparation and soil response. Typical values of stiffness, given in Table 6.11, can be used but, as with presumed bearing resistances, should be treated with caution.

Table 6.12 Long-term settlement of spread foundations on fine-grained soils based on the coefficient of volume compressibility, m_v

The compression, s_i, of a clay stratum based on the results of an oedometer test is

$$s_i = m_{vi}\Delta\sigma_{vi}H_i$$

where m_v is the coefficient of volume compressibility, H the layer thickness and $\Delta\sigma_v$ the increase in stress in the layer due to a surface load.

- The stratum is divided into layers and the increase in average stress calculated for each layer. The layer thicknesses can be equal or can allow for a variation in stiffness or composition.
- The zone of influence is either the base of the clay stratum or when $\Delta\sigma_v < 0.1\sigma'_v$.
- The settlement is the sum of the compression of each layer corrected for soil type and depth of foundation

$$s = \mu_g\mu_d\sum m_{vi}\Delta\sigma_{vi}H_i$$

where μ_d is the depth correction factor from Figure 6.12.

- μ_g for matrix-dominated tills is 0.2–0.5, for weathered tills and supraglacial tills 0.5–0.7 and for glaciolacustrine clays 0.7–1.0.

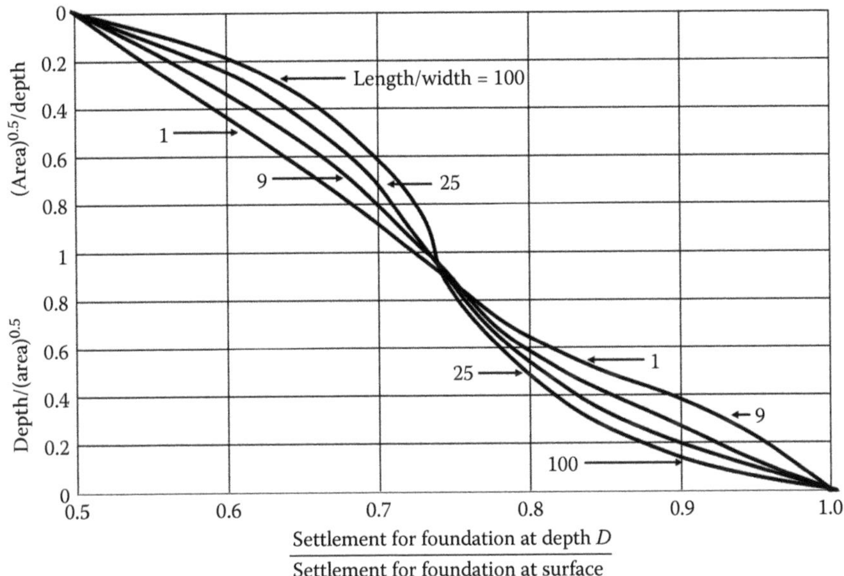

Figure 6.12 Correction for depth of foundation.

Rates of settlement of foundations on fine-grained soils are dependent on the mass permeability, which, in glacial tills, is affected by the presence of continuous layers of sands and gravels and discontinuities and, in glaciolacustrine clays, the anisotropic behaviour. Discontinuities mean that the *in situ* mass permeability is greater than that measured in the laboratory if the effective stress is less than 120 kPa. Therefore, the use of the intact coefficient of hydraulic conductivity is likely to underestimate the rate of settlement. Discontinuities may close because of the increase in stress due to the foundation (see Section 5.1), which means that the conductivity may reduce. Establishing layers of sands and gravels as opposed to pockets should be part of the ground investigation strategy. Glaciolacustrine clays are

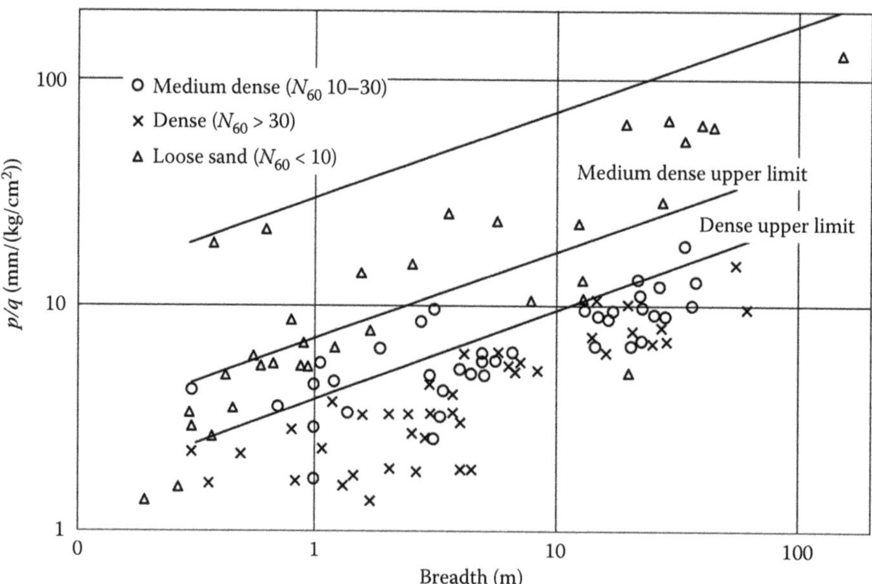

Figure 6.13 Relation between settlement per unit pressure and the breadth of a foundation. (After Burland, J. B., B. B. Broms, and V. F. B. de Mello, Behaviour of foundations and structures. State of the Art Review. *9th International Conference on SMFE*, 2; 1977: 495–546.)

anisotropic, so conventional measurements (perpendicular to the varves) will underestimate the hydraulic conductivity relevant to spread foundations.

There are numerous methods to predict settlement of spread foundations on coarse-grained soils. Simons and Menzies (2000) compared the predictions of six methods to show that they varied by up to a factor of five though this may not be critical because settlement takes place during construction and is small for foundations on medium-dense and dense coarse-grained soils.

Burland et al. (1978) showed that there is a tentative relationship between settlement per unit pressure and foundation breadth for different relative densities (Figure 6.13). Burland and Burbidge (1985) used a database of 200 case studies to produce a method outlined in Table 6.13 (and Figures 6.14 and 6.15), which is based on bearing pressure, foundation width and N_{60}. The settlement, s, of a spread foundation on coarse-grained soils is

$$s = f_s f_l f_t \left(q' - \frac{2}{3}\sigma'_{vmax} \right) B^{0.7} I_c \tag{6.24}$$

Over-consolidated coarse-grained soils are denser and stiffer than normally consolidated coarse-grained soils with a similar composition. Therefore, the net bearing pressure for over-consolidated soils is reduced by $2/3\sigma'_{vmax}$ to allow for this fact. Clast-dominated tills may be over-consolidated and are dense, so a correction should be applied though it may be difficult to determine the maximum overburden pressure. Tomlinson and Boorman (2001) suggest that this method is based on the *in situ* SPT with no correction for effective stress. Terzaghi et al. (1966) recommended that if $N_{60} > 15$ in fine or silty sand, then the N_{60} used in design calculations should be $N_{60} + 0.5(N_{60} - 15)$, and, for gravels or sandy gravels the N_{60} should be increased by 25%.

Table 6.13 Predicting settlement of spread foundations on coarse-grained soils using N_{60}

$$\text{Settlement}, s = f_s f_l f_t \left[\left(q'_n - \frac{2}{3}\sigma'_{vmax} \right) B^{0.7} I_c \right]$$

where q'_n is the net applied pressure, B the breadth (or depth?) of the foundation, σ'_{vmax} the maximum past overburden pressure and I_c the influence factor taken from Figure 6.14.

$$\text{Shape factor}, f_s = \left(\frac{1.25L/B}{(L/B) + 0.25} \right)^2$$

$$\text{Time factor}, f_t = 1 + R_3 + R \log \frac{t}{3}$$

where L is the length and B the breadth of the foundation; t the time from construction (>3 years); R a creep factor (=0.2 for static loads and 0.8 for dynamic loads); and R_3 time depend factor for the first 3 years (=0.3 for static loads and 0.7 for dynamic loads).

$$\text{Correction factor for zone of influence}, f_l = \frac{H}{z_l}\left(2 - \frac{H}{z_l}\right)$$

where H is the thickness of the coarse-grained layer and z_l is the zone of influence given in Figure 6.15.

Source: After Burland, J. B. and M. C. Burbridge. *Proceedings of Institution of Civil Engineers*, Pt 1, Vol. 76, 1985: 1325–1381.

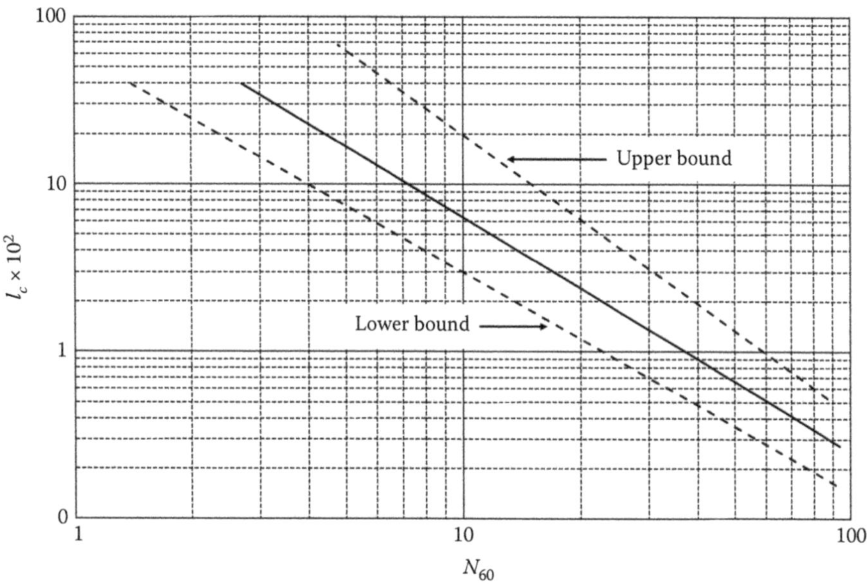

Figure 6.14 Compressibility index for coarse-grained soils. (After Burland, J. B. and M. C. Burbridge. *Proceedings of Institution of Civil Engineers*, Pt 1, 76, 1985: 1325–1381.)

This method can apply to glaciofluvial soils and, possibly, clast-dominated tills. The effect of the composition on the SPT results means that there will be significant scatter in the results, which means that a scenario analysis should be undertaken to investigate the possible variation in stiffness. It is likely that *in situ* tests will be carried out in clast-dominated tills, but there are limited data on the successful use of methods based on sands and gravels

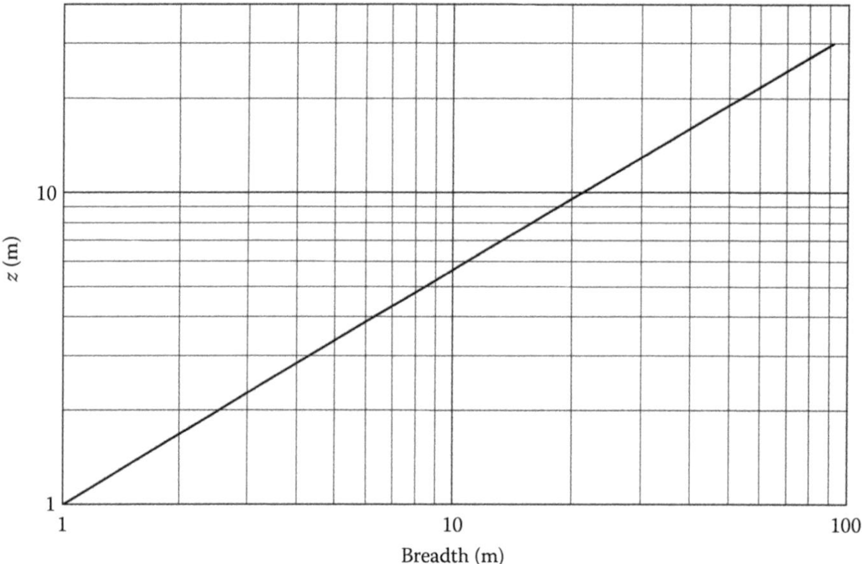

Figure 6.15 Zone of influence for coarse-grained soils. (After Burland, J. B. and M. C. Burbridge. *Proceedings of Institution of Civil Engineers*, Pt 1, 76, 1985: 1325–1381.)

with composite soils. This method can be used with profiles of penetration resistance by converting the cone resistance, q_e, to N_{60} using Figure 6.16.

Settlement in coarse-grained soils can also be predicted directly from cone penetration tests (Schmertmann et al., 1978) using

$$s = C_1 C_2 q' \sum_{0}^{2B} \frac{I_z}{E} h \qquad (6.25)$$

Details are given in Table 6.14 and Figure 6.17.

The deformation modulus, E_s, which corresponds to the secant modulus at 25% of the peak stress, is $2.5q_c$ for square foundations and $3.5q_c$ for rectangular foundations, which is calculated for each layer. Since the soil is divided into layers, it is possible to predict the variation in mobilised stiffness due to an increase in vertical stress within the layer using the method proposed by Tomlinson and Boorman (2001) to determine shear modulus for any increase in stress, $\Delta\sigma_v$, based on estimates of initial tangent modulus:

$$E_d = E_o \sqrt{\frac{\sigma'_v + \Delta\sigma'_v/2}{\sigma'_v}} \qquad (6.26)$$

where E_o is the initial stiffness given in Table 6.11.

6.5.3 Caissons and piers

Spread foundations refer to those foundations that spread the load onto the ground reducing the bearing pressure. Friction on the side of the foundation contributes to the capacity but is ignored if it is small compared to the base capacity and if the soil has been disturbed during

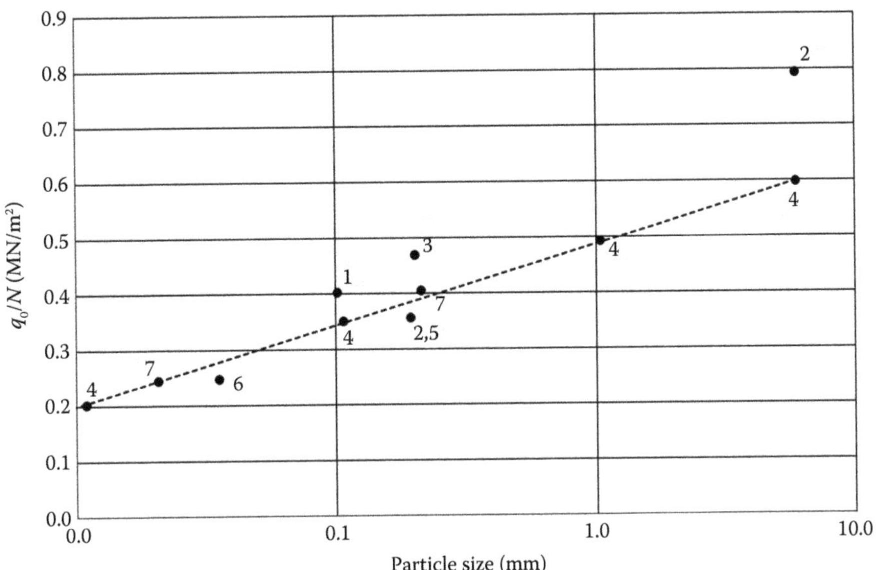

Figure 6.16 Relationship between cone penetration resistance, q_c, and SPTN. (Meigh and Nixon (1), Meyerhof (2), Rodin (3), Schmertmann (4), Shulze and Knausenberger (5), Sutherland and Thorburn (6), McVicar (7)). (After Fleming, K., A. Weltman, M. Randolph, and K. Elson. *Piling Engineering*. CRC Press, Glasgow; 2008.)

Table 6.14 Predicting settlement of spread foundations on coarse-grained soils using *CPT* q_c

$$\text{Settlement, } s = C_1 C_2 q_n' \sum_0^{2B} \left(\frac{I_z}{E_d} \Delta z \right)$$

where q_n' is the net applied pressure, B the breadth of the foundation, E_d the deformation modulus, Δz the thickness of the layer and I_z the influence factor taken from Figure 6.17.

$$\text{Depth correction factor, } C_1 = 1 - 0.5 \frac{\sigma_{vo}'}{q_n'}$$

$$\text{Depth factor, } C_2 = 1 + 0.2 \log \left(\frac{t}{0.1} \right)$$

where σ_{vo}' is the current effective vertical stress at the formation level and t the time since construction. To take into account the variation of q_c with depth, the stratum is divided into layers h thick and $h\ I_z/E_s$ and calculated for each layer.

Source: After Schmertmann, J. H., J. P. Hartman, and P. R. Brown. *Journal of Geotechnical and Geoenvironmental Engineering*, 104, no. Tech Note, 1978.

construction. The exceptions are caissons in which the foundation is pushed into the soil through a combination of weight and excavation. In this case, the bearing resistance, Q, is

$$Q = 2D(B + L)c_{uave} + c_{ubase} f_s N_c BL \tag{6.27}$$

where H is the depth of the caisson, D the diameter, c_{uave} the average undrained shear strength on the shaft, c_{ubase} the undrained shear strength at the base, f_s the shape factor and N_c the bearing resistance factor.

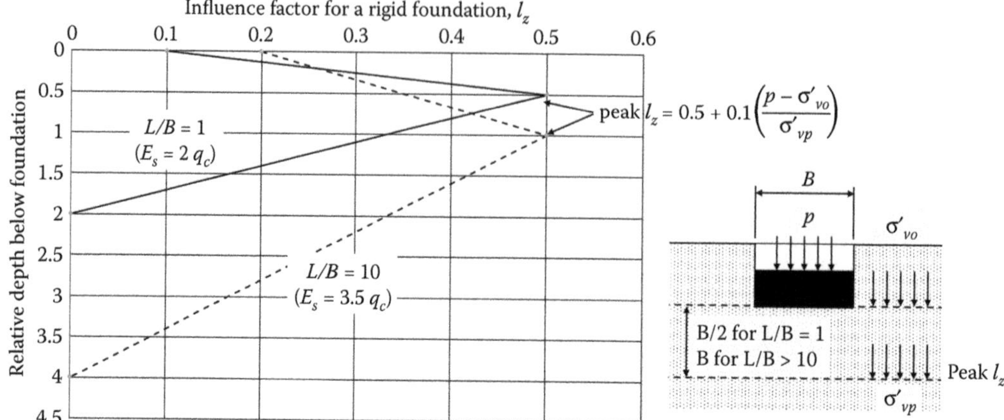

Figure 6.17 Influence factor to predict settlement on coarse-grained soils using the cone penetration resistance. (After Schmertmann, J. H., J. P. Hartman, and P. R. Brown. *Journal of Geotechnical and Geoenvironmental Engineering,* 104, no. Tech Note, 1978.)

The capacity for caissons on sands is given by

$$Q = \left(\sigma'_v N_q s_q d_q + 0.5\gamma' B N_\gamma s_\gamma d_\gamma\right) BL + \sigma'_{vave} K_o \tan\varphi' 2D(B+L) \tag{6.28}$$

6.5.4 Recommendations

Glacial soils are complex soils and, in relation to spread foundations, can vary both horizontally and vertically in properties and thickness. This means a ground investigation has to be designed to fully assess the ground conditions. This may seem obvious and in line with codes of practice and guidelines, but the investigation of glacial soils has to be more thorough than they recommend. Recognised design methods can be applied to glacial soils provided the correct parameters are chosen. Therefore, it is recommended that

- A ground investigation has to establish the geotechnical, geological and hydrogeological conditions for both the short and long term.
    - The spacing and depth of boreholes has to ensure that the spatial variability can be assessed. This includes the variation in thickness and type of strata both vertically and horizontally to ensure the effects of composition and fabric on excavations and foundation performance can be assessed.
    - Sufficient representative *in situ* tests and samples have to be taken to ensure that characteristic properties of stiffness, strength and permeability can be determined. This is especially important in glacial tills and glaciofluvial soils where it may be difficult to develop property profiles due to the natural variability of the soils; scatter in the data is inevitable.
    - Effective and undrained strength of glacial clays is required as the ULS may be governed by the effective strength but the undrained shear strength may be critical in the short term.
    - The fabric of glacial clays must be carefully assessed because it could give an indication of mass behaviour as opposed to the intact behaviour observed in small specimens.

  - Stiffness is required for the SLS. Given the importance of the composition, structure and fabric, enough tests must be undertaken to obtain representative values.
  - Seasonal groundwater conditions should be assessed using regional and site-specific data.
- The ultimate and SLSs can be assessed using a number of semi-theoretical and numerical methods provided the appropriate parameters are selected and the correct adjustment factors are used.
  - The ULS of foundations in fine-grained soils should be assessed using undrained and effective strength parameters. Account must be taken of the fabric when selecting the appropriate values. A cautious estimate should be used to allow for scatter due to sampling and testing unless a regional database is available when more realistic values can be used.
  - The analysis should take account of the possible variation in stratum thickness. This means a scenario analysis should be undertaken.
  - Settlement predictions are unreliable if a single value of stiffness is used because of the natural variability of the soils. Therefore, it is important not only to select the appropriate stiffness model to allow for stress level but to take into account the vertical and horizontal variation in stiffness. This requires a scenario analysis.

6.6 PILED FOUNDATIONS

Piled foundations are those foundations formed of individual piles with a length much greater than its width. These include end bearing, friction, tension and laterally loaded piles installed by driving, jacking, screwing or boring. The limit states to be considered (BS EN 1997-1:2004+A1:2013) include loss of overall stability, bearing failure, tensile failure, lateral failure, excessive settlements, heave or lateral movement. Displacement or load can be considered an action and that load includes eccentric loading due to differential surcharge or excavation on either side of the pile, piles on slopes that are moving, inclined piles in ground that is settling, seismic loading and downdrag due to the compression of the upper layers on one side of the piled foundation (e.g. abutment piles adjacent to an embankment). Consolidating soils due to groundwater lowering or surcharge create negative friction, a disturbing action, on a pile. The upper values of strength of the consolidating layers should be used in the design.

Design can be based on results of static and dynamic tests, observations of piled structures or by empirical or analytical methods. Static tests are often carried out as part of the construction process to validate the capacity of a pile. This has proved to be an issue in matrix-dominated glacial tills because they often show that the capacity of a pile exceeds that based on the undrained shear strength. These soils are often very stiff, which means that some of the load is taken by the soil; that is, the mobilised strength exceeds the undrained shear strength. The mass hydraulic conductivity is such that excess pore pressures dissipate more rapidly if the tills are fissured; that is, the capacity increases after installation in a relatively short time. The undrained shear strength used in the calculations may be lower than that *in situ* strength because of sample disturbance. Thus, piles in matrix-dominated tills often have more capacity than predicted.

The choice of pile depends on the geological and hydrogeological profiles, the method of installation and the effects on adjacent structures, construction constraints, safety, environmental constraints and nature and magnitude of loads and costs. The variation in thickness of glacial soils, the rock head profile, water-bearing and weaker layers and lenses in glacial tills, the boulders within in any glacial soils and the composition of the overlying soils are

Table 6.15 Factors to consider when selecting pile type

Performance	Bearing capacity Uplift capacity Lateral load capacity Durability
Environmental	Noise Vibration Spoil disposal Contamination Carbon efficiency
Site constraints	Restricted access Restricted headroom Restricted working Existing asserts and structures
Safety	Railways Airports Sloping sites Adjacent to sensitive assets
Geotechnical	Very weak strata Deep unstable strata Spatial variation Water-bearing strata[a] Obstructions[b] Lenses and layers[b] Rockhead

Source: After Wade, S., B. Handley, and J. Martin. *ICE Manual of Geotechnical Engineering*, Thomas Telford Ltd, London; 2012: 1191–1223.

[a] Particularly relevant to glacial tills and glaciofluvial soils.
[b] Particularly relevant to all glacial soils (e.g. boulder beds, dropstones).

particularly relevant. Other factors to consider are potential downdrag, heave or lateral loading, obstructions and aggressive ground conditions.

Table 6.15 lists factors that are considered when selecting a pile type, shown in Figure 6.18. Selecting piles for glacial soils is affected by the composition, fabric, structure and strength of the soils, as shown in Figure 6.19. Bored piles include rotary bored and cored piles and continuous flight auger techniques. When boring through glacial soils, it may be necessary

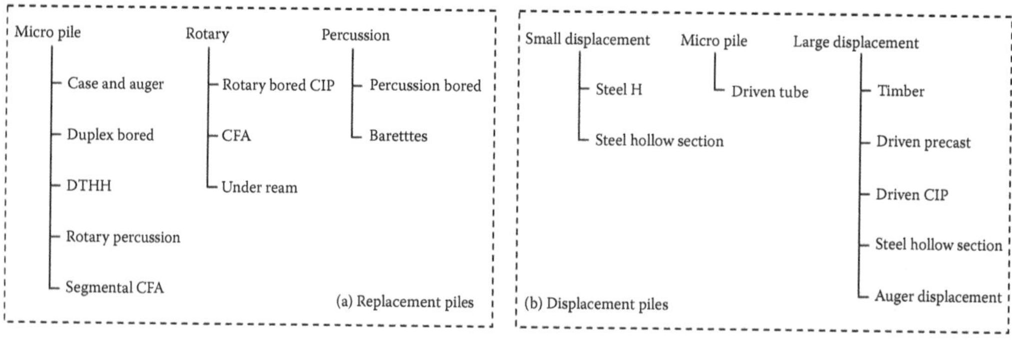

Figure 6.18 Types of bearing piles showing (a) replacement piles and (b) displacement piles. (After Wade, S., B. Handley, and J. Martin. *ICE Manual of Geotechnical Engineering*, Thomas Telford Ltd, London; 2012: 1191–1223.)

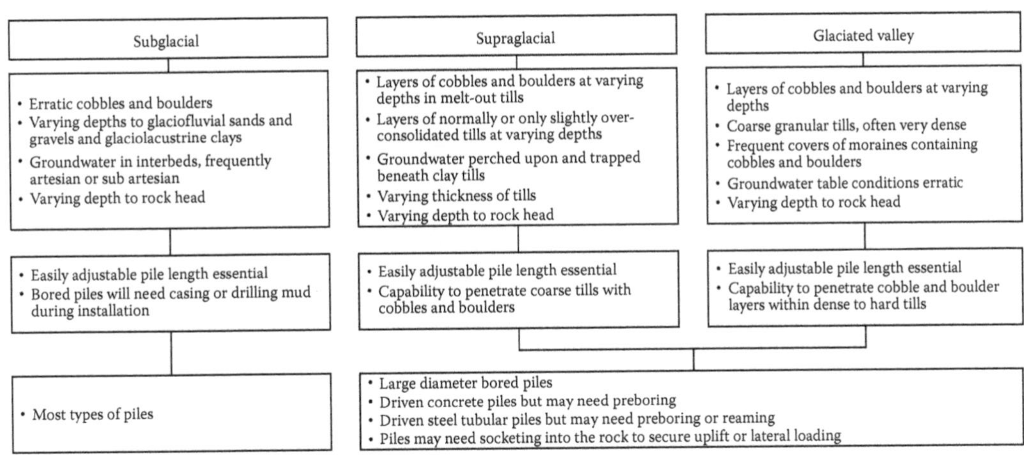

Figure 6.19 Overview of pile selection for compression, tension and laterally loaded piles in tills. (After Trenter, N. A. *Engineering in Glacial Tills*. CIRIA, London, 1999.)

to support the borehole using casing, drilling fluid or soil-filled auger flight (CFA).This is particularly the case with glaciofluvial soils, glaciolacustrine clays (soft), clast-dominated tills and matrix-dominated tills containing weaker and more permeable layers. CFA piles are suitable for coarse-grained soils but not, if they contain boulders, without special equipment, a potential problem in glacial soils especially tills.

Displacement piles are driven piles typically 150- to 400-mm square reinforced concrete piles, which are assembled from segments up to 15 m long. Driven piles also include cast-in-place concrete piles in which a steel or concrete tube is driven into the ground and withdrawn as concrete is cast. Steel *H* piles are small displacement piles as are open-ended tubular steel piles provided they do not plug during driving.

The most important factor in selecting a pile type is knowledge of the ground conditions. However, given the range of pile type, this can still be an issue in glacial soils even with an adequate investigation because of the spatial variation in properties and composition, presence of the boulders and boulder beds, and the interface between glacial soils and the underlying rock, as shown in Figure 6.20. Problems of bored piles include necking of cast concrete in water-bearing soils, suction causing borehole wall instability and over-excavation. Problems of driven piles are over-driving damaging the pile, false sets due to boulders, stiff layers overlying weak layers and temporary delays in driving leading to increased pile capacity due to dissipation of pore pressure. Redriving checks are recommended for matrix-dominated tills. Driven piles in clays consolidate the clay around the pile, which can increase the downdrag.

In glacial tills containing lenses or layers of weaker or coarse-grained soil, care has to be taken to produce a stable hole. Bored piles can penetrate glacial soils of variable composition provided groundwater is cut-off using casing or bentonite. It is necessary to prove that rock is not a large boulder. It may be difficult to penetrate very coarse layers found in glaciofluvial soils and tills, especially near to rock head. Driven piles may not penetrate very-coarse-grained soils or tills containing a significant number of very coarse particles. An alternative is to use driven open-ended steel tube, which can then be left in place or removed. CFA piles can penetrate layers containing particles up to cobble sized though this depends on the spacing of the auger flights. CFA rigs cannot penetrate boulders without additional equipment.

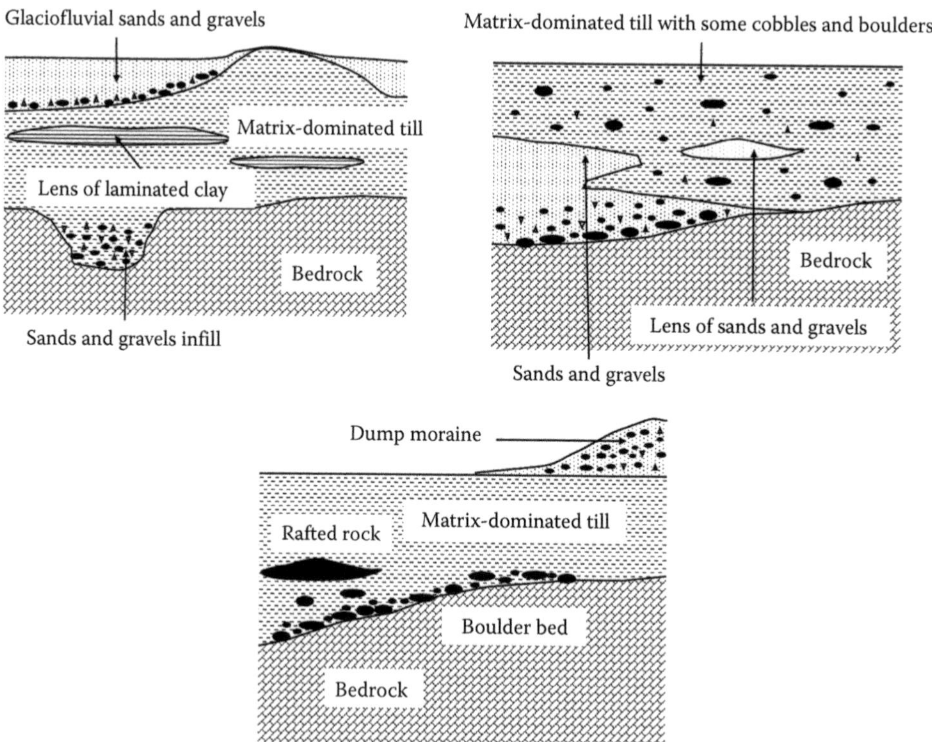

Figure 6.20 Composition and structure of glacial deposits that can impact on pile selection, installation and performance.

Eurocode 7 indicates that pile capacity tests are necessary if there is no experience of piling, if there is a lack of confidence in the capacity or if the pile response is very different to that predicted. Given the variability of glacial tills, it is prudent to tests piles in glacial soils.

If pile tests cannot be carried out, then a cautious estimate of the geotechnical properties should be used. Given the spatial variability of glacial soils, care must be taken to show that the test piles are in ground representative of the site.

Tests can be static, maintained or dynamic load tests or constant rate of penetration, a summary of which is given in Table 6.16. Table 6.17, a suggested strategy for testing piles to reduce risk, shows that the risk of piling in glacial soils is high because the soils are complex. This means that preliminary and working pile tests are essential.

It is also necessary to check the integrity of piles. These include pile driving records and integrity and load tests (Table 6.18), with low strain energy integrity tests the most common. Figure 6.21 is a process to select how many piles to be tested. A more rigorous approach is shown in Figure 6.22, a statistical method based on 90% confidence limits. For example, if 10% of defective piles are permitted and 50 piles are installed, then 20% of the piles have to be tested.

6.6.1 Pile design

The selection of the design method for a pile depends on the quality of the ground investigation, the range of geotechnical data available, the budget and timescale, the scale and sensitivity of the proposed structure, the complexity of the ground conditions and loading

Table 6.16 Advantages and disadvantages of pile capacity tests and their use

Test type	*Advantages*	*Limitations*	*Potential deployment*
Static	Simple test Simple and quick interpretation Well understood and accepted	Slow tests Significant infrastructure (especially as loads increase) High space requirements Safety concerns with increasing loads	Lower loading (<20 MN) Low pile numbers Trial piles/limited number of working piles
Bi-directional	Capable of very high test loads (higher than other techniques) Low infrastructure Low space requirement Cost-effective as load magnitude increases	Pile for testing needs to be preselected Less experience Analysis needs to take account of different surface boundary conditions Specialised analysis and interpretation	Medium to very high loading (2–320 MN) Low pile numbers Trial piles/limited working piles
Dynamic	Low infrastructure Low space requirement Fast tests Quick repeat testing Mature technique	Pile damage may be a concern Perceived reliability: intermediate Tests may be influenced by pile material and geometry Specialised analysis and interpretation	Medium to high loading (1–35 MN) Low to high pile numbers Trial and working piles Where pile driving equipment already on site
Rapid	Low infrastructure Low space requirement Rapid testing Quick repeat testing	Limited experience Analysis techniques under development Limited case study experience Availability of high capacity equipment	Medium to high loading (0.6–40 MN) Medium to high pile numbers Trial and working piles Quality control of working piles Problems with working pile performance

Source: After Brown, M. J. *ICE Manual of Geotechnical Engineering*, Thomas Telford Ltd, London; 2012: 1451–1468.

Table 6.17 Testing strategy to reduce the risk of uncertainty of pile behaviour

Characteristics of the piling works	*Risk level*	*Pile testing strategy*
Complex or unknown ground conditions No previous pile test data New piling technique or very limited relevant experience	High	Both preliminary and working pile tests essential 1 preliminary pile test per 250 piles 1 working pile test per 100 piles
Consistent ground conditions No previous pile test data Limited experience of piling in similar ground	Medium	Pile tests essential Either preliminary and/or working pile tests can be used 1 preliminary pile test per 500 piles 1 working pile test per 100 piles
Previous pile test data available Extensive experience of piling in similar ground	Low	Pile tests not essential If using pile tests either preliminary and/or working pile tests can be used 1 preliminary pile test per 500 piles 1 working pile test per 100 piles

Source: After ICE. *The Specification for Piling and Embedded Retaining Walls*. Thomas Telford, London, 2007.

Table 6.18 Summary of pile integrity testing methods

Test method	*Low strain integrity tests*	*Cross-hole sonic logging*	*Parallel seismic tests*	*High strain integrity tests*
Property measured	Characteristics of the behaviour of acoustic shock waves or stress waves travelling through the pile	Transmission time of an ultrasonic wave through the pile material	Transmission time of acoustic shock waves or stress waves through the pile and intervening soil to a detector	Characteristics of the behaviour of stress waves travelling through the pile from a heavy impact
Pre-planning	None	Access ducts have to be cast into preselected piles	Sinking of measurement bore alongside pile	Not strictly necessary, but access for heavy plant may have to be provided
Time of testing	After concrete has achieved design strength (usually 5–7 days min.)	After concrete has achieved design strength (usually 5–7 days min.)	After construction	After concrete has achieved design strength and typically 7 days min. after construction
Type of pile	All types	Large diameter cast-in-place typically (usually 600 mm diameter or larger)	All types	All types
Relative cost[a]	Low	Low to medium	Medium to very high	Medium to high
Frequency of use (control)[b]	5	4–5	Not applicable	1–2
Frequency of use (retrospective)[b]	3–5	Not applicable	0–1	1–2
Availability	Readily available from specialist testing houses	Available from specialist testing houses	From specialist testing houses	Readily available from piling contractor and specialist test house
Effect of pile length	Yes, signals increasingly attenuated with depth	No	Yes	No, not within normal pile depth
Comments	Very common technique Pile response is investigated in terms of time and/or frequency	Mainly used for large diameter cast-in-place piles, piers and barettes Especially large single piles supporting high column loads Not usually suitable for retrospective investigation because of necessity to install access ducts	Used for retrospective investigation only	Not commonly used in routine testing Typically may be used to investigate a post installation problem, such as pile damage

Source: After French, S. and M. Turner. *ICE Manual of Geotechnical Engineering*, Thomas Telford Ltd, London; 2012: 1419–1448.

[a] Low: <10% of pile cost; medium: 10%–50% of pile cost; high: 50%–100%; very high: >100% of pile cost (excludes mobilisation costs).

[b] 0: very rare; 1: rare; 2: occasional; 3: sometimes; 4: common; 5: very common.

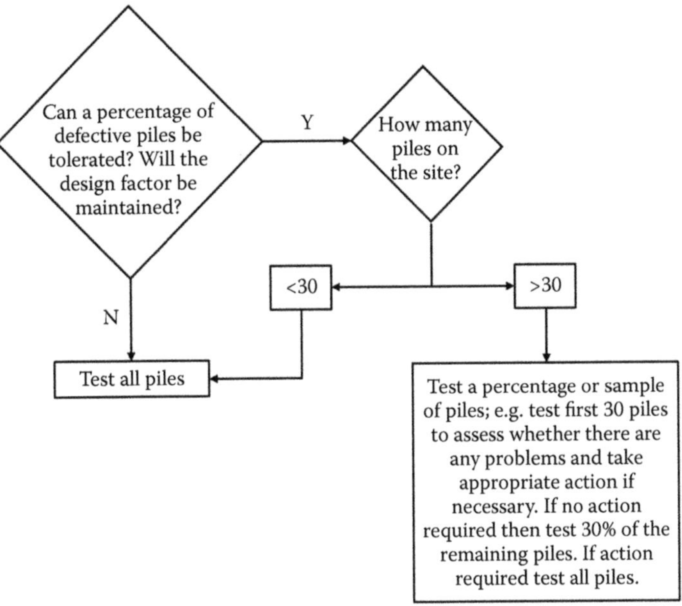

Figure 6.21 Guide to the selection of the number of piles to be tested. (After Williams, H. and R. T. Stain. *Proceedings of the International Conference on Foundations and Tunnels*, London, 1987.)

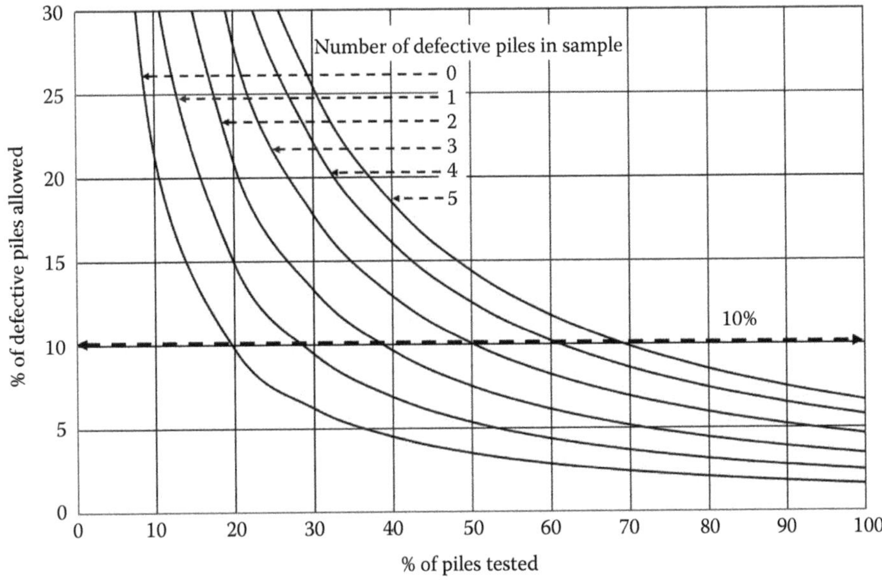

Figure 6.22 Statistical method to select the number of pile tests based on the number of piles tested, the number of defective piles allowed and the number of defective samples found. (Adapted from Cameron, G. and T. Chapman. *Ground Engineering*, 37(2); 2004: 35–40.)

regime (Bell and Robinson, 2012). Care must be taken in applying coefficients derived from tests in clays and sands to the design of piles in glacial soils because the range of particle sizes within glacial soils is likely to be different from that used in assembling the database. The variation in stratum thickness along the length of a pile needs to be considered, especially in glacial soils which contain lenses of different soils.

6.6.2 Axially loaded piles

6.6.2.1 Compressive capacity

The limit states cover the overall stability, the failure of a single pile, the failure of the piled foundation, and excessive settlement or differential settlement of the piled foundation. Failure of compression piles is often defined as settlement equivalent to 10% of the pile base diameter.

The allowable load, Q_a, on a pile is the least of the following if using global factors of safety:

$$Q_a = min\left(\frac{Q_b + Q_s}{2.5}; \frac{Q_b}{3} + \frac{Q_s}{1.5}\right) \tag{6.29}$$

where Q_b is the base capacity and Q_s the shaft capacity.

Glacial soils can contain lenses and layers of different soils and their relation to a pile can impact on the shaft friction or end bearing resistance capacity. Figure 6.23 shows the effect on the base resistance of a dense sand overlying a weak layer and the effect of a weak layer in a stratum on the shaft capacity. This is particularly important in glacial soils where lenses or layers of weaker soils can reduce the end bearing and shaft resistance and increase the settlement. If a weaker layer is within four times the pile base diameter below the base, then punching failure is possible. It is possible in glacial soils to have a combination of fine- and coarse-grained soils supporting a pile; and weak layers below the pile and acting on the

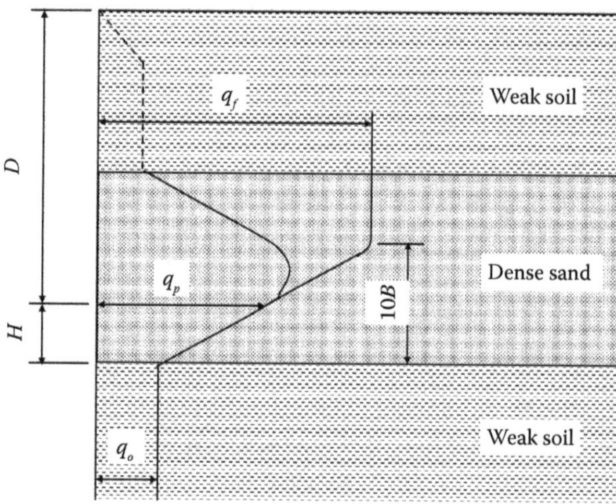

Figure 6.23 Effect on the base capacity of a weak layer underlying a dense sand and the effect of a weak layer on the shaft capacity. (After Meyerhof, G. G. *Journal of the Geotechnical Engineering Division, ASCE*, 102; 1976: 197–228.)

pile shaft. This is why it is important to ensure that the ground investigation is designed to determine the ground conditions in some detail and why test piles are necessary.

The partial factors used to determine resistance depend on whether the design is based on test piles or ground investigation data. The characteristic resistance, $R_{c;k}$, based on the results of static load tests, is given by

$$R_{c;k} = min\left\{\frac{(R_{c;m})_{mean}}{\xi_1};\frac{(R_{c;m})_{min}}{\xi_2}\right\} \tag{6.30}$$

where $R_{c;m}$ is the measured ultimate capacity, R_c, from one or several pile load tests; the factors, ξ_1 and ξ_2, depend on the number of piles tested and are given in Table 6.19. The characteristic resistance, $R_{c;k}$, of the ground is given by

$$R_{c;k} = R_{b;k} + R_{s;k} \tag{6.31}$$

where $R_{b;k}$ is the characteristic base resistance and $R_{s;k}$ the characteristic shaft friction. The design resistance, $R_{c;d}$, is given by

$$R_{c;d} = \frac{R_{c;k}}{\gamma_t} \tag{6.32}$$

or

$$R_{c;d} = \frac{R_{b;k}}{\gamma_b} + \frac{R_{s;k}}{\gamma_s} \tag{6.33}$$

The partial factors γ_t, γ_b and γ_s are set out in Table 6.6.

The design compressive resistance of a pile is given by Equation 6.33 but when using ground investigation data, the characteristic values are given by

$$R_{c;k} = R_{b;k} + R_{s;k} = \frac{R_{b;cal} + R_{s;cal}}{\xi} = \frac{R_{c;cal}}{\xi} = min\left\{\frac{(R_{c;cal})_{mean}}{\xi_3};\frac{(R_{c;cal})_{min}}{\xi_4}\right\} \tag{6.34}$$

Table 6.19 Correlation factors, ξ, to derive characteristic values from *n* static pile load tests and *n* profiles of boreholes/*in situ* test profiles

	Number of static pile tests				
ξ	*1*	2	2	4	≥5
ξ_1	1.4	1.3	1.2	1.1	1.0
ξ_1	1.4	1.2	1.05	1.0	1.0

	Number of borehole/in situ test profiles						
ξ	*1*	2	3	4	5	7	*10*
ξ_3	1.4	1.35	1.33	1.31	1.29	1.27	1.25
ξ_4	1.4	1.27	1.23	1.20	1.15	1.12	1.08

Source: After BS EN 1997-1:2004+A1:2013. *Eurocode 7: Geotechnical Design – Part 1: General Rules*. British Standards Institution, London.

where ξ_3 and ξ_4 are correlation factors (Table 6.19), which depend on the number of profiles of tests, n, applied to the mean values

$$(R_{c;cal})_{mean} = (R_{b;cal} + R_{s;cal})_{mean} = (R_{b;cal})_{mean} + (R_{s;cal})_{mean} \tag{6.35}$$

and to the lowest values

$$(R_{c;cal})_{min} = (R_{b;cal} + R_{s;cal})_{min} \tag{6.36}$$

$R_{c:cal}$, $R_{b:cal}$ and $R_{s:cal}$ are the calculated values of pile capacity, end bearing resistance and shaft resistance, respectively. The characteristic values are related to the unit base resistance, $q_{b;ik}$, and unit skin friction, $q_{s;ik}$ by

$$R_{b;k} = A_b q_{b;k} \tag{6.37}$$

$$R_{s;k} = \sum_{i}^{n} A_{s;i} q_{s;i;k} \tag{6.38}$$

where A_b is the base area and $A_{s;i}$ the shaft area of layer i.

6.6.2.2 Coarse-grained soils

The unit shaft friction in coarse-grained soils is given by

$$q_{s;ik} = K_{s;i} \sigma'_{v;i} \tan \delta_i \tag{6.39}$$

where δ_i is the interface friction, $K_{s;i}$ the earth pressure coefficient and $\sigma'_{v;i}$ the average vertical effective stress of soil layer i. Typical values of $K_{s;i}$ are given in Table 6.20, which

Table 6.20 Values of K_s for piles in coarse-grained (silica) soils

Pile type		*Soil type*	*Typical coefficient (K_s)*
Large displacement	Precast concrete Closed ended tubular steel Timber Driven cast-in-place concrete	All	1.0–1.2
Small displacement	H-section steel bearing piles Open-ended tubular steel Helical steel	All	0.80–0.96
Replacement	Continuous flight auger (CFA)	Clean medium coarse sand	0.9
		Fine sand	0.7–0.8
		Silty sand	0.6–0.7
		Inter-layered silt and sand	0.5–0.6
	Bored cast-in-place concrete Micropiles		0.7

Source: After BS 8004:2015. *Code of Practice for Foundations*. British Standards Institution, London.

Note: K_s values may vary due to installation, soil layering, groundwater pressures and time between installation and testing; K_s values may be adjusted to take account of results of pile tests.

Table 6.21 Values of k_δ for piles installed in coarse-grained soils

Pile type		*Typical coefficient (k_δ)*
Large displacement	Precast concrete Closed ended tubular steel	0.67
	Timber	0.85
	Driven cast-in-place concrete	0.9
Small displacement	H-section steel bearing piles Open-ended tubular steel	0.67
	Helical steel	0.67[a] or 1.0[b]
Replacement	Continuous flight auger (CFA) Bored cast-in-place concrete Micropiles	1.0

Source: After BS 8004:2015. *Code of Practice for Foundations*. British Standards Institution, London.

[a] Soil to steel.
[b] Soil to soil.

shows that K_s can vary from 0.5 to 0.9 for CFA piles in silt and sand layers to coarse sand; 0.7 for bored piles. K_s is 1–2 for full displacement piles. The interface friction is

$$\delta_i = min\left[k_{\delta;i}\varphi'_{pk;i};\varphi'_{cv}\right] \tag{6.40}$$

where $\varphi'_{pk;i}$ is the peak angle of friction, $\varphi'_{cv;I}$ is the constant volume angle of friction and k_δ is given by Table 6.21. Weltman and Healy (1978) suggest that dense granular glacial deposits containing a high proportion of sand and silt are likely to be more disturbed when installing bored piles; so a further reduction factor (Figure 6.24) is necessary than for piles bored into more uniform soils. The unit friction should be restricted to 110 kPa.

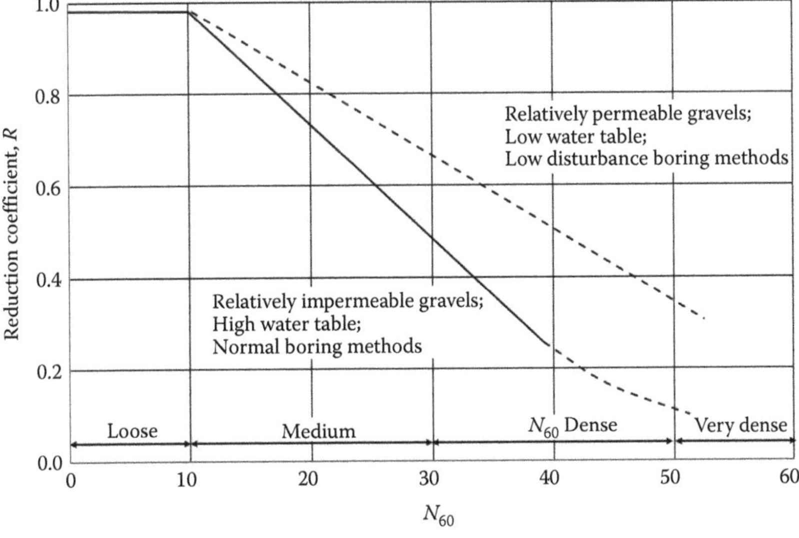

Figure 6.24 Reduction factor to allow for disturbance when installing bored piles in clast-dominated tills; they contain a significant proportion of finer particles. (After Weltman, A. J. and P. R. Healy. *Piling in 'Boulder Clay' and Other Glacial Tills*. CIRIA Report PG5 Monograph, 1978.)

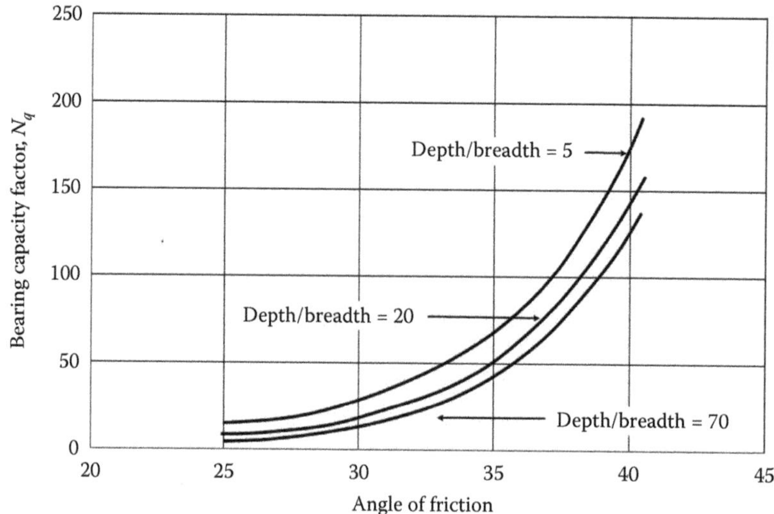

Figure 6.25 Bearing capacity factors for the base capacity of piles. (After Berezantzev, V. G. *Proceeding of the 5th ICSMFE*, Paris, Vol. 2, 1961: 11–12.)

The ultimate base resistance, q_b', for piles founded in coarse-grained soils is

$$q_b' = N_q \sigma_{vb}' \tag{6.41}$$

where σ_{vb}' is the vertical effective stress at the base of the pile and N_q a bearing capacity factor given in Figure 6.25. The maximum unit base resistance should be limited to 10–15 MPa in dense sands and 5–7.5 MPa in medium dense sands if within 10 pile diameters of the top of the stratum.

6.6.2.3 Fine-grained soils

Piles in fine-grained soils generally derive most of their capacity from shaft friction because the pile settlement is insufficient to mobilise the base resistance. Further, the settlement required to fully mobilise the effective strength at the base is unacceptable; therefore, capacity of end bearing piles is normally based on *in situ* undrained shear strength (Bell and Robinson, 2012), which would normally give a lower bound value. The shaft capacity is mobilised at between 0.5% and 2% of the pile diameter, and the base capacity at 10%–20%.

The capacity of piles in fine-grained soils can be assessed using total or effective strength parameters. The unit shaft friction in fine-grained soils based on effective stress is

$$q_{s;ik} = \beta_{j;i} \sigma_{v;i}' \tag{6.42}$$

where β_j is an empirical coefficient given by

$$\beta_j = (1 - \sin\varphi')\tan\varphi' \quad \text{for normally consolidated clays} \tag{6.43}$$

$$\beta_i = 1.5(1 - \sin\varphi')\tan\varphi'\sqrt{OCR} \quad \text{for over-consolidated clays} \tag{6.44}$$

In clays $\delta = \varphi'_{cv}$ for bored piles and the residual angle of friction, φ'_{res}, for driven piles. If the soil's undrained shear strength is known, then

$$q_{s;ik} = \alpha_i c_{u;i} \tag{6.45}$$

where $c_{u;i}$ is the mean undrained shear strength of the layer and α_i an empirical coefficient, which for replacement piles is

$$0.4 \le \alpha_i = k_1\left(1 - k_2 \log\frac{c_u}{100}\right) \le 1 \tag{6.46}$$

And for displacement piles

$$\alpha_i = 0.5\left(\frac{c_u}{\sigma'_v}\right)^{-m} \tag{6.47}$$

where $k_1 = 0.45$ and $k_2 = 1$ for clays and $k_1 = k_2 = 0.75$ for matrix-dominated tills; m is 0.25 if $c_u/\sigma'_v \ge 1$ and 0.5 for $c_u/\sigma'_v < 1$. α also depends on pile type. For bored piles, it varies between 0.3 and 0.6, and in stiff over-consolidated clays 0.7, but note that Figure 6.26 shows values of α analysed from tests on a variety of driven piles in matrix-dominated tills (Weltman and Healy, 1978).

Therefore, it is appropriate to use recommended values of α for piles in glacialacustrine clays; Figure 6.26 shows values of α for piles in clay tills. Table 6.22 lists back-figured values of β and α for a range of pile types.

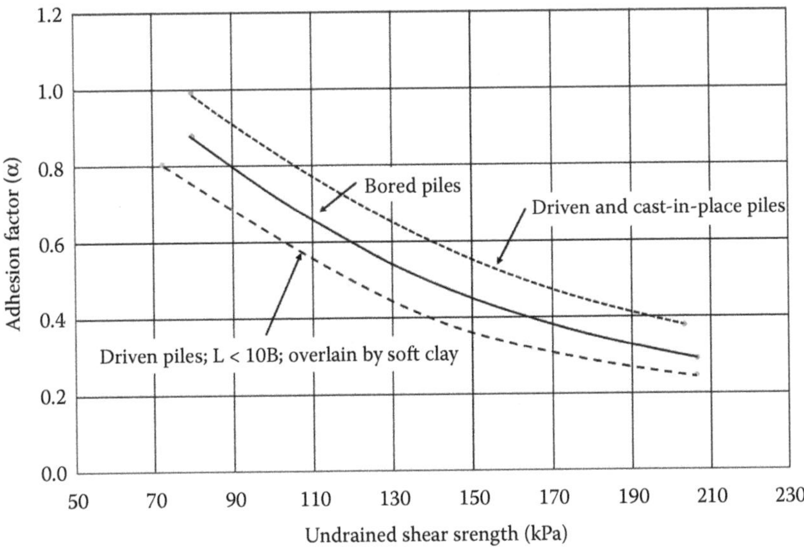

Figure 6.26 Adhesion factors for piles in matrix-dominated tills developed from a database of 73 tests. (After Weltman, A. J. and P. R. Healy. *Piling in 'Boulder Clay' and Other Glacial Tills*. CIRIA Report PG5 Monograph, 1978.)

Table 6.22 Effective stress parameters back-figured from driven pile tests in glacial clay tills showing the factors depend on the pile type and soil strength

Pile type	*Mean undrained shear strength (kPa)*	*Adhesion factor, α*	*Back-analysed factor, β (=K_s tan δ)*	*Assumed K_s*	*Apparent δ*
Bored cast-in-place	126	0.4	0.53	0.75	35°
	125	0.4	0.5	0.75	37.5°
	130	0.65	0.63	1	32°
Driven cast-in-place	120	1	0.71	1	35°
Driven taper piles	80	1	1.3	2.5	27.5°
Driven concrete segmental	220	0.8	1.1	2	29°
Driven precast	107	0.8	0.75	1.5	27°
Timber driven	95	0.54	0.52	2	15°

Source: After Weltman, A. J. and P. R. Healy. *Piling in 'Boulder Clay' and Other Glacial Tills.* CIRIA Report PG5 Monograph, 1978.

The α factor for driven piles is affected by the overlying soil because it is dragged down into the bearing stratum. If the overlying soil is soft clay, the α value is reduced and, if granular soil, increased, as shown in Figure 6.27, the effect of overlying soil layers, pile geometry and soil strength on the adhesion factor in stiff clays. It is not certain that these can be applied to matrix-dominated tills, though published results on Dublin Clay do confirm the values.

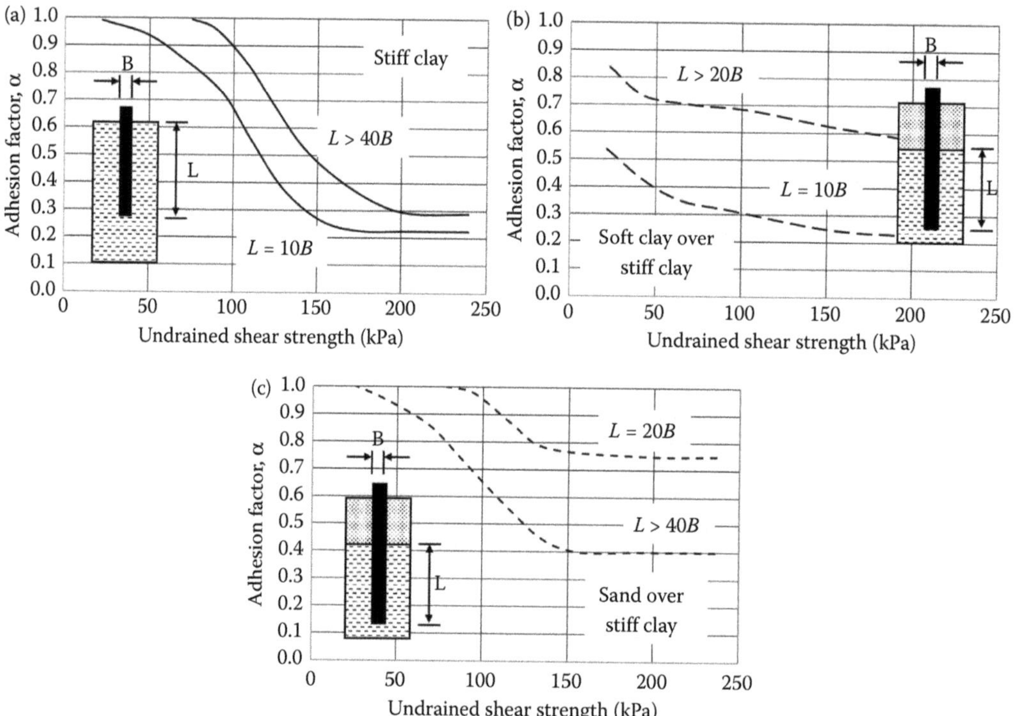

Figure 6.27 Adhesion factors for driven piles in fine-grained soils and the effect of undrained shear strength and overlying soils showing (a) a stiff clay, (b) soft clay overlying a stiff clay and (c) sand overlying a stiff clay. (After Tomlinson, M. J. and R. Boorman. *Foundation Design and Construction.* Pearson Education, Harlow, UK, 2001.)

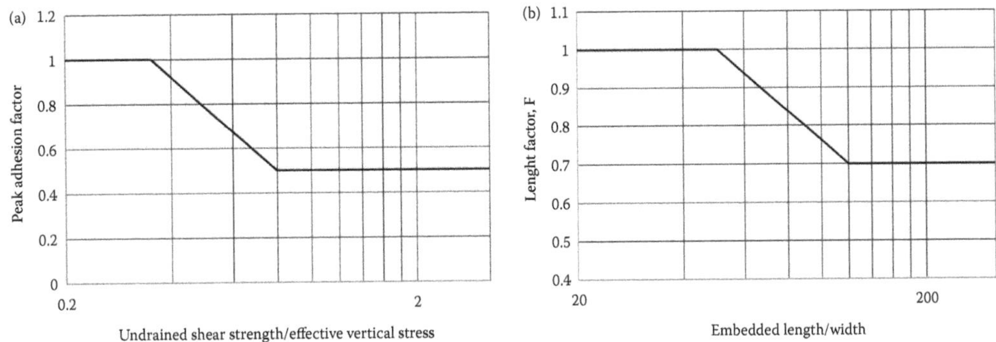

Figure 6.28 Adhesion factors for heavily loaded piles and the effect (a) overburden pressure and (b) pile length. (After Semple, R. M. and W. John Rigden. *Analysis and Design of Pile Foundations, ASCE*, 1984: 59–79.)

Randolph and Wroth (1981) suggested that the adhesion factor could be related to the ratio c_u/σ_v' which is the basis for Figure 6.28, a database of driven steel piles in clay. If $c_u/\sigma_v' > 0.8$, then α is 0.5 and, less than 0.4, α is 1.0. Figure 6.28 also suggests that the α factor should be reduced if the ratio of the embedded length to pile diameter exceeds 50.

The ultimate base resistance, q_b, for piles founded in fine-grained soils based on undrained shear strength is

$$q_b' = N_c c_{ub} \tag{6.48}$$

where c_{ub} is the undrained shear strength at the base of the pile and N_c a bearing capacity factor given by

$$N_c = 9k_1k_2 \tag{6.49}$$

$$k_1 = \frac{2}{3}\left(1 + \frac{L}{6B}\right) \tag{6.50}$$

where B is the width of the pile, L the depth of embedment into the bearing layer and k_2 as given in Table 6.23. N_c is reduced in layered soils if the pile length in the founding strata is less than three diameters.

Table 6.23 Values of k_2 for piles in fine-grained soils

Pile type	*Undrained shear strength (kPa)*	k_2	$9k_2$
Bored, CFA[a]	<25	0.72	6.5
	50	0.89	8
	>100	1.0	9
Driven[b]		1.11	10

[a] Federal Highway Administration (FHWA). *Drilled Shaft Foundations* (Publication No. FHWA-NHI-10016, May 2010).
[b] Salgado (2008).

6.6.2.4 Other design methods

There are a number of methods to predict pile capacity including those based on standard penetration and cone penetration tests, the American Petroleum Institute (API) method for driven piles and the ICP method (Jardine et al., 2005) developed for offshore piles but based on onshore piles.

The unit base resistance of piles in coarse-grained soils based on SPT N_{60} is

$$q_b = KN_{60} \tag{6.51}$$

where K is 0.4 for sands when the penetration into the bearing stratum exceeds six pile diameters and 0.3 for silts. The unit shaft resistance is given by

$$q_s = 2N_{60} \quad \text{for driven piles} \tag{6.52}$$

$$q_s = N_{60} \quad \text{for bored piles} \tag{6.53}$$

The unit base resistance of a pile in fine-grained soil can be estimated from CPT q_c and the unit skin friction from the sleeve resistance q_s. The ultimate base capacity in coarse-grained soils is

$$q_b = 0.9q_{cm} \tag{6.54}$$

where q_{cm} is the weighted average in the zone of influence (8D above the toe to 4D below the toe). The base resistance at an acceptable base settlement is

$$q_b = 0.6q_{cm} \tag{6.55}$$

The API (2000) design method covers piled foundations, particularly driven steel tubular piles for offshore foundations. The unit shaft resistance in fine-grained soils is based on the undrained shear strength using the α factor:

$$\alpha = 0.5\left(\frac{c_u}{\sigma'_v}\right)^{-0.5} \quad \text{for}\left(\frac{c_u}{\sigma'_v} < 1\right) \tag{6.56}$$

$$\alpha = 0.5\left(\frac{c_u}{\sigma'_v}\right)^{-0.25} \quad \text{for}\left(\frac{c_u}{\sigma'_v} > 1\right) \tag{6.57}$$

These values are based on experience with a recommendation that α should be less than 1 and values for $c_u/\sigma'_v > 3$ should be treated with caution. The unit base resistance in fine-grained soils is $9c_u$.

The unit shaft resistance, f_s, in coarse-grained soils is

$$f_s = K\sigma'_v \tan\delta \tag{6.58}$$

K is 0.8 for open end piles and 1 for full displacement piles. The unit base resistance is

$$q_b = \sigma'_v N_q \tag{6.59}$$

where typical factors are given in Table 6.24.

Table 6.24 Design parameters for piles in sands and silts

Density	Soil type	Soil pile interface angle	Maximum unit shaft resistance (kPa)	N_q	Maximum unit end bearing (MPa)
Very loose	Sand	15	48	8	1.9
Loose	Sand–silt				
Medium	Silt				
Loose	Sand	20	67	12	2.9
Medium	Sand–silt				
Dense	Silt				
Medium	Sand	25	81	20	4.8
Dense	Sand–silt				
Dense	Sand	30	96	40	9.6
Very dense	Sand–silt				
Dense	Gravel	35	115	50	12
Very dense	Sand				

Source: After API, RP. 2A-WSD, 2000. *Recommended Practice for Planning, Designing and Constructing Fixed Offshore Platforms—Working Stress Design*, 21st edition. American Petroleum Institute, Washington, DC 2000.

Jardine et al. (2005) describe the ICP method to predict axial capacity of driven piles for offshore use, which was based on field tests and a database of driven piles. An analysis of the data using a number of methods showed that the ICP method proved to be consistent. The shaft resistance is governed by the Coulomb criterion ($\tau_f = \sigma_r' \tan\delta$) in both sands and clays, where δ is the interface shear resistance obtained from ring shear tests. Installation changes the horizontal stress acting on the pile such that the radial effective stress acting on the pile is the sum of the radial stress setup during installation and the dilatant increase in stress during loading. Table 6.25 describes the method used to determine the capacity of driven piles in clay. Table 6.26 shows the method applied to sands.

The method, developed for sands and clays, has been assessed for piles in glacial soils to show that it was feasible but they highlighted the difficulty in obtaining representative samples and the effects of sample disturbance on the parameters used in the design. The information is mostly derived from CPT q_c resistance profiles and can allow for pile installation, loading, cyclic loading and ageing, which may mean converting SPT N_{60} or laboratory test results to cone resistance when it proves too difficult to push a cone into glacial soils.

6.6.3 Vertical displacements

According to Eurocode 7, if a pile is in a medium dense or firm layer overlying rock or very hard soil, the factors applied to assess the ULS are sufficient to take into account settlement, which is typically less than 10% of the pile diameter. Eurocode suggests that for pile diameters up to 600 mm it is unnecessary to predict settlement of isolated piles (Tomlinson and Boorman, 2001) as the partial/global factors are sufficient to restrict the settlement to less than that permitted for piles. For piles in excess of 600 mm, such as monopiles for wind turbines, it is necessary to calculate the settlement. The settlement, s, of an isolated pile can be predicted using formula such as

$$s = \frac{(W_s + 2W_b)L}{2A_s E_{pile}} + \frac{\pi W_b}{4A_b}\frac{B(1-\upsilon^2)I_p}{E_{base}} \tag{6.60}$$

where W_s is the load on the shaft, W_b the load on the base, L the length of the pile, B the pile diameter, E_{pile} the pile stiffness, E_{base} the stiffness of the soil beneath the pile, A_s the area of

Table 6.25 ICP method for clays

The shaft capacity for closed ended piles, Q_s, is

$$Q_s = \pi D \int \tau_f \, dz$$

where D is the pile diameter and τ_f the local shear stress, which is given by

$$\tau_f = \sigma'_{rf} \tan \delta_r$$

where σ'_{rf} is the horizontal stress developed at failure due to the installation and δ_r the angle of interface friction.

This can be expressed in terms of the local radial stress after equalisation, σ'_{rc},

$$\tau_f = \frac{K_f}{K_c} \sigma'_{rc} \tan \delta_r$$

The loading factor, (K_f/K_c), is assumed to be constant, 0.8.

The local radial stress is

$$\sigma'_{rc} = K_c \sigma'_{v0}$$

where K_c is the relevant coefficient of earth pressure:

$$K_c = (2.2 + 0.016YSR - 0.87\Delta I_{vy})YSR^{0.42}\left(\frac{h}{R}\right)^{-0.2}$$

where YSR is the yield stress ratio and (h/R) is the normalised distance from the pile tip limited to a minimum of 8; ΔI_{vy} expressed in terms of the soil sensitivity, S_t, is

$$\Delta I_{vy} = \log_{10} S_t$$

YSR can be estimated from

$$\frac{c_u}{\sigma'_{v0}} = \left(\frac{c_u}{\sigma'_{v0}}\right)_{NC} YSR^{0.85}$$

The shaft capacity of open-ended piles is based on the same method but substituting R^* for R where

$$R^* = \left(R_{outer}^2 - R_{inner}^2\right)^{0.5}$$

The unit base capacity, q_b, of close ended piles is

$$q_b = 0.8q_c \quad \text{for undrained conditions}$$

$$q_b = 1.3q_c \quad \text{for drained conditions}$$

For fully plugged open-ended piles,

$$q_b = 0.4q_c \quad \text{for undrained conditions}$$

$$q_b = 0.65q_c \quad \text{for drained conditions}$$

And for unplugged open-ended piles,

$$q_b = q_c \quad \text{for undrained conditions}$$

$$q_b = 1.6q_c \quad \text{for drained conditions}$$

Table 6.26 ICP method for sands

The shaft capacity for closed ended piles, Q_s, is

$$Q_s = \pi D \int \tau_f \, dz$$

where D is the pile diameter and τ_f the local shear stress, which is given by

$$\tau_f = \sigma'_{rf} \tan \delta_r$$

where σ'_{rf} is the horizontal stress developed at failure due to the installation and δ_r the angle of interface friction.

The local radial stress is

$$\sigma'_{rf} = \left(\sigma'_{rc} + \Delta\sigma'_{rd}\right)$$

where $\Delta\sigma'_{rd}$ is the increase in radial stress:

$$\Delta\sigma'_{rd} = 2G\frac{\Delta r}{R}$$

$$\sigma'_{rc} = 0.029 q_c \left(\frac{\sigma'_{v0}}{100}\right)^{0.13} \left(\frac{h}{R}\right)^{-0.38}$$

$$G = q_c(0.0203 + 0.00125(q_c\sigma'_v)^{-0.5} - 1.2\,1e^{-6} q_c(\sigma'_v)^{-0.5})^{-1}$$

The shaft capacity of open-ended piles is based on the same method but substituting R^* for R where

$$R^* = \left(R^2_{outer} - R^2_{inner}\right)^{0.5}$$

The unit base capacity, q_b, of close ended piles is

$$q_b = q_c\left[1 - 0.5\log\left(\frac{D}{D_{CPT}}\right)\right]$$

For fully plugged open-ended piles,

$$q_b = q_c\left[0.5 - 0.25\log\left(\frac{D}{D_{CPT}}\right)\right]$$

And for unplugged open-ended piles,

$$q_b = q_c$$

the shaft, A_b the area of the base and I_p an influence factor. For $L/B > 5$ and $\upsilon < 0.25$, I_p is 5. This is based on the assumption that the soil at the base governs the settlement.

Figure 6.29 shows the variation of secant modulus derived from pressuremeter tests, plate tests and pile tests in glacial soils with undrained shear strength, which shows, for a given shear strength, the mobilised stiffness from the pile tests is much greater than that from *in situ* and laboratory tests. Figure 6.30 shows the settlement at working load normalised with respect to the pile diameter for a range of pile types provided 50% of the pile is in till, the till is relatively homogenous, the undrained shear strength exceeds 100 kPa, the pile diameter is between 200 and 600 mm in diameter, disturbance during installation is limited and the length of the pile in the till exceeds 10 pile diameters. It shows that there is a trend of an

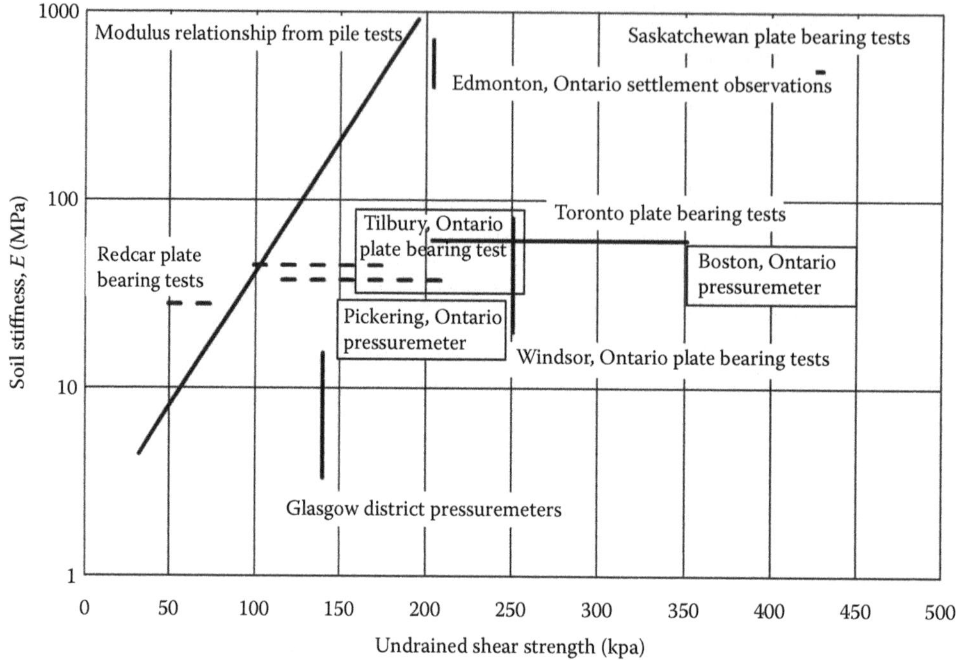

Figure 6.29 Variation of secant modulus derived from pressuremeter tests, plate tests and pile tests in glacial soils with undrained shear strength. (After Weltman, A. J. and P. R. Healy. *Piling in 'Boulder Clay'* and Other Glacial Tills. CIRIA Report PG5 Monograph, 1978.)

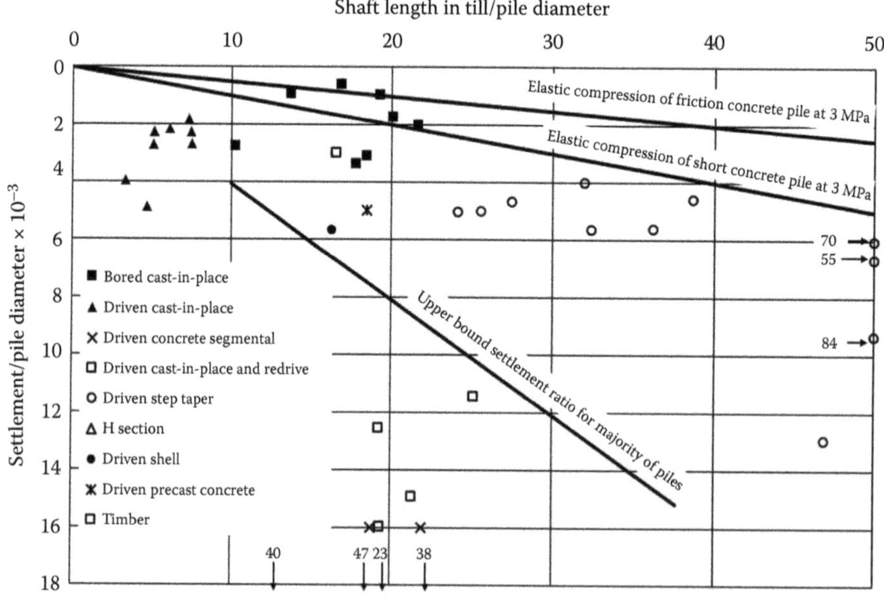

Figure 6.30 Settlement at working load normalised with respect to the pile diameter for a range of pile types (50% of the pile is in till; the length in till greater than 10; pile diameter between 20 and 600 mm; undrained shear strength greater than 100 kPa). (After Weltman, A. J. and P. R. Healy. *Piling in 'Boulder Clay' and Other Glacial Tills*. CIRIA Report PG5 Monograph, 1978.)

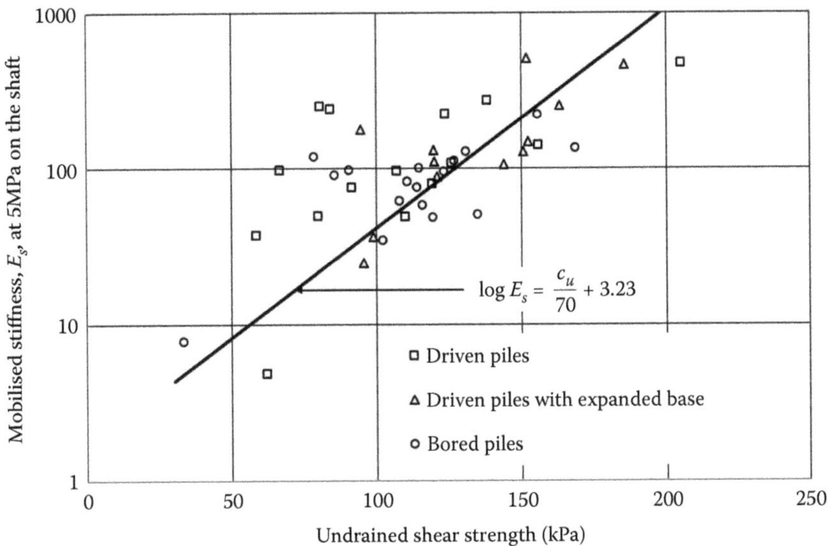

Figure 6.31 Comparison of the pile/till interface modulus with undrained shear strength from tests on driven, bored piles and driven piles with expanded bases. (After Weltman, A. J. and P. R. Healy. *Piling in 'Boulder Clay' and Other Glacial Tills*. CIRIA Report PG5 Monograph, 1978.)

increase in settlement as the length of the pile in the till increases but the settlement is less than 0.5% of the pile diameter. Figure 6.39 shows that the upper bound limit for settlement at the working load is $L_m/2.5$, where L_m is the length of the pile (in m) in till and the settlement is in mm. The estimated settlement for bored cast-in-place piles is $L/4$. Figure 6.31 suggests a pile/till interface modulus with undrained shear strength from a range of pile analyses such that

$$E_s = \frac{c_u}{70} + 3.23 \quad \text{for } c_u < 250\,\text{kPa} \tag{6.61}$$

Since most of the settlement at the working load is due to the movement of the shaft, this figure may provide an estimate of the settlement of a pile in matrix-dominated tills.

There are a number of methods to predict settlement known as t–z methods (e.g. Coyle and Reese, 1966; Coyle and Sulaiman, 1966; Reese and O'Neill, 1971; Vijayvergiya, 1977) in which the mobilised unit shaft resistance, t, is related to the pile deflection, z. Typical curves are shown in Figure 6.32 in which the normalised mobilised unit shaft resistance is plotted against the normalised pile settlement. The maximum value of the interface shear resistance is the unit shaft resistance given by Equation 6.58. The equivalent curve for the base deflection is given in Figure 6.33, where the normalised mobilised base resistance is expressed in terms of the normalised base settlement. These curves can be used to predict settlement by dividing the pile into a number of elements to calculate the compression of each element taking account of the pile/soil interaction.

Fleming (1992) proposed a method based on a hyperbolic relationship following the work of Chin (1970), which is described in Table 6.27 such that the settlement (Figure 6.34) is the sum of

$$s_s = \frac{M_s D_s Q_s}{Q_s - P} \quad \text{or} \quad \frac{0.6 Q_b P_b}{D_b E_b (Q_b - P_b)} \tag{6.62}$$

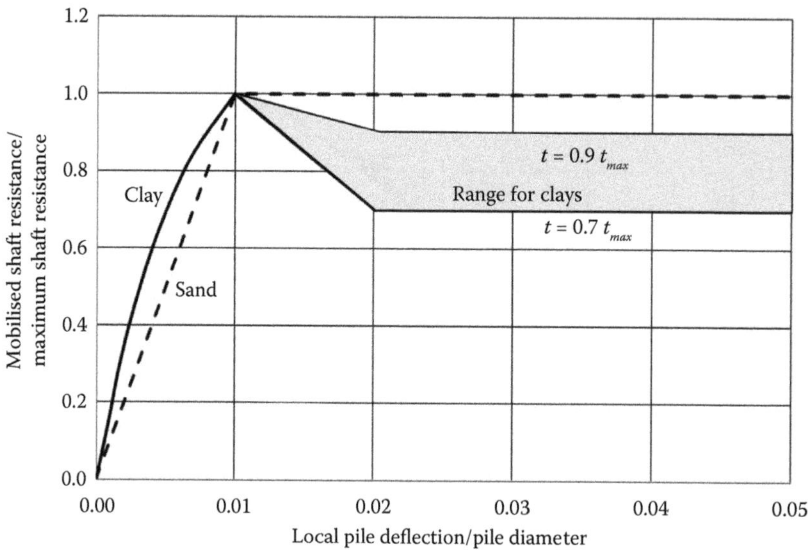

Figure 6.32 Variation in interface shear stress with settlement for axially loaded piles. (After API, RP. 2A-WSD, 2000. *Recommended Practice for Planning, Designing and Constructing Fixed Offshore Platforms—Working Stress Design*, 21st edition. American Petroleum Institute, Washington, DC 2000.)

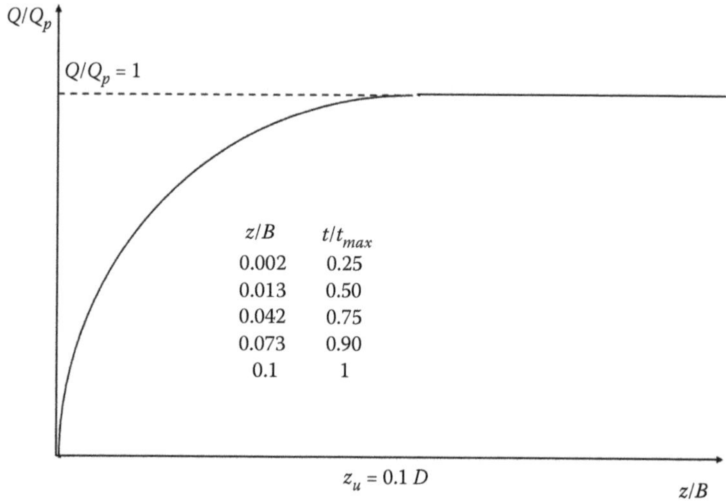

Figure 6.33 Variation in base load, Q, with settlement, z. (After API, RP. 2A-WSD, 2000. *Recommended Practice for Planning, Designing and Constructing Fixed Offshore Platforms—Working Stress Design*, 21st edition. American Petroleum Institute, Washington, DC 2000.)

where s_s is the settlement of the shaft due to the load P, D_s the pile diameter, Q_s the ultimate shaft friction and M_s a factor related to the shaft friction and pile diameter.

$$s_b = \frac{\pi}{4}\frac{q_b}{E_b}D_b(1-\upsilon^2)f_1 \tag{6.63}$$

where s_b is the settlement of the base, D_b the base diameter, q_d the unit base resistance, E_b the stiffness of the soil below the base and f_1 a depth correction factor.

Table 6.27 Settlement and capacity of single isolated piles

The ultimate shaft friction, Q_s, based on Chin's method is

$$Q_s = \frac{s_s}{(s_s/P) - K_s}$$

where s_s is the settlement due to the load P_s and K_s the intercept of the settlement against settlement/load plot (Figure 6.34).

K_s can be expressed in terms of the factor M_s

$$K_s = \frac{M_s D_s}{Q_s}$$

where D_s is the pile diameter. Thus, the settlement due to the shaft load is

$$s_s = \frac{M_s D_s Q_s}{Q_s - P}$$

The base settlement, s_b, is based on the settlement of a circular footing, diameter, D_b:

$$s_b = \frac{\pi}{4}\frac{q_b}{E_b} D_b(1-\upsilon^2) f_l$$

where E_b is the stiffness of the soil below the base, f_l the depth correction factor (Figure 6.12) and q_b the base contact stress.

E_b is assumed to be the mobilised elastic modulus in the pile modulus at 25% of the ultimate stress.

Thus, the total capacity of a pile, Q, is

$$Q = \frac{sQ_s}{M_s D_b + s} + \frac{D_b E_b Q_b s}{0.6Q_b + D_b E_b s}$$

The elastic shortening, s_p, of the pile if most of the capacity of the pile is based on the shaft resistance is given by

$$s_p = \frac{4}{\pi}\frac{Q(L_o + K_e L_f)}{D_s^2 E_c}$$

where L_f is the length of pile over which the capacity of the pile is derived, $(L_o + L_f)$ is the total length of the pile, E_c is the stiffness of the pile and K_e is the coefficient applied to give the effective free length.

In stiff soils, K_e is typically 0.45. The compression of the piles in which some base capacity is mobilised is

$$s_p = \frac{4}{\pi}\frac{1}{D_s^2 E_c}[Q(L_o + L_f) - L_f Q_s(1-K_e)]$$

Source: After Fleming, W. G. K. *Geotechnique*, 42(3); 1992: 411–425.

6.6.4 Pile groups

Small diameter piles (e.g. less than 600 mm) are more likely to be installed in groups. Pile group failure can be due to the sum of the individual capacities, the failure of the block enclosing the piles, a failure of a row of piles or global instability. Table 6.28 is a summary of the factors to be considered.

If the piles are too close, then the resistance of the individual piles is reduced because of the interaction between the piles. The reduction can be limited by specifying a minimum

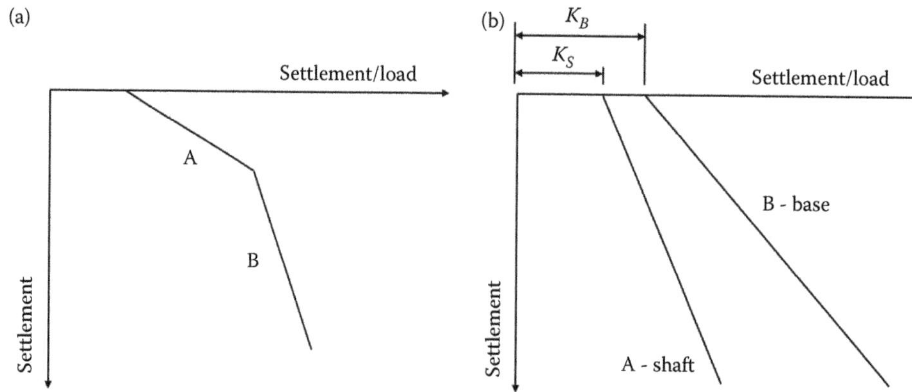

Figure 6.34 Use of a hyperbolic relationship to determine the shaft and end bearing resistance showing (a) the settlement/load vs. settlement for a whole pile, and (b) for the shaft and base load. (After Fleming, W. G. K. *Geotechnique*, 42(3); 1992: 411–425.)

Table 6.28 Factors that influence interaction between piles within a group

	Vertical loading		*Horizontal loading*	
Factor	*Group settlement ratio*	*Non-uniformity of axial loads*	*Group deformation*	*Stresses in piles*
Increase of ground stiffness with depth	↓	↓	↓ (if < L_c)	↓ (if < L_c)
Ground layering, relatively stiff layer at depth below pile toe	↓↓	↓↓	NE	NE
End bearing pile (compared with friction pile)	↓	↓	NE	NE
Ground layering, relatively compressible layer at depth below pile toe	↑↑↑	↑↑↑	↑	↑
Ground, non-linear stress–strain behaviour	↓	↓	↑↑	↑↑
Pile installation	↓↓	↓↓	↑	↑
Pile group layout	NE	NE	↑	↑
Pile spacing (s > 3d)	↓	↓	↓	↓
Pile spacing (s < 3d)	↑	↑	↑	↑
Increasing displacement	↓	↓	↑↑	↑↑
Near-surface soils (<6d) (relatively weak layer)	NE	NE	↑↑↑	↑↑↑
Pile cap stiffness reduces	NE	↓↓	NA	NA
Pile head fixity, fixed to free head condition	NA	NA	↑↑↑	Depends on location in group

Source: After O'Brien, A. *ICE Manual of Geotechnical Engineering*, Thomas Telford Ltd, 2012: 823–850.

Note: ↓ – reduction in interaction effects, for example, smaller group settlement, more uniform axial loads (no of arrows indicates effect is greater); ↑ – increase in interaction effects, for example, larger group deformation, more non-uniform axial loads and stresses in piles; NE – negligible effect; NA – not applicable; L_c – critical pile length.

spacing between the piles. The centre-to-centre spacing for friction piles (BS 8004:2015) to prevent interaction is

$$s \geq 3D \text{ or } P \tag{6.64}$$

where D is the diameter of the largest circular pile and P is the perimeter of the largest non-circular pile; and for end bearing piles,

$$s \geq 2D \text{ or } \frac{2P}{3} \tag{6.65}$$

Figure 6.35 and Table 6.29 summarise the issues that have to be addressed. The zone of influence of pile groups is much greater than that of a single pile, which can be an issue in glacial soils because of the variability in composition (Figure 6.36). The behaviour of a pile group depends on the pile aspect ratio $R = (ns/L)^{0.5}$, where n is the number of piles in the group, s the pile spacing and L the length of pile (Randolph and Clancy, 1993). If the aspect ratio is large, the load is mostly taken by base resistance, there is significant pile interaction and settlement is significantly greater than that for a single pile. If the aspect ratio is small, then most of the load is taken by shaft resistance.

Pile group capacity is often estimated using commercial software such as PIGLET and PGROUP, but, as with any numerical simulation, there are issues related to the numerical model and selection of appropriate parameters. An estimate of the ultimate capacity of a pile group can be obtained by modelling the group as a block, Q, is based on the method used for shallow foundations except that the capacity due to the friction of the side of the block is included. For example, in clays:

$$Q = 2D(B + L)c_{uave} + c_{ub}f_sN_cBL \tag{6.66}$$

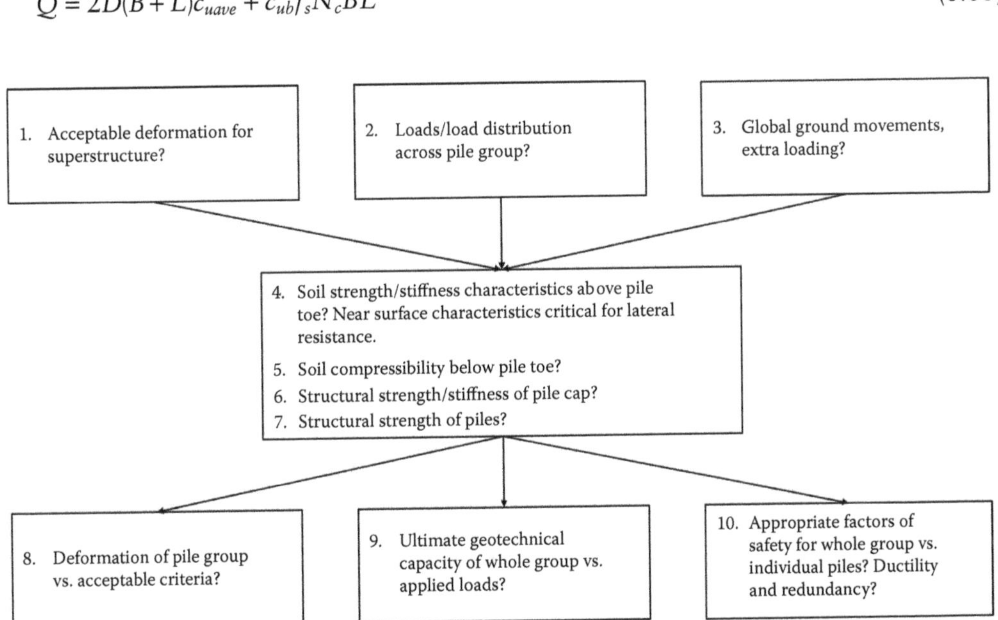

Figure 6.35 Design considerations for pile groups. (After O'Brien, A. *ICE Manual of Geotechnical Engineering*, Thomas Telford Ltd, 2012: 823–850.)

Table 6.29 Factors that have to be considered when designing pile groups

Factor	Comment
Pile cap and substructure stiffness	A stiff pile cap or substructure can redistribute axial loads; hence, the individual pile factor of safety is not significant If the pile cap is flexible, then individual pile factors of safety need to be considered
Number of piles in the group	If more than 5 piles, then there is redundancy and 'failure' of a pile within group does not imply failure of the group For large pile groups, there is considerable redundancy
Code requirements	Many codes do not discuss pile groups in detail and mainly focus on single piles. EC7 provides some guidance AASHTO (and NCHRP Report 507, 2004) gives guidance on reduced risk of failure associated with varying levels of redundancy
Direction of loading	For horizontal and moment loading, carefully check structural strength of piles, pile cap and pile-to-pile cap connections A factor of safety in excess of 1.3 to 1.4 along the perimeter pile row should be met if large long-term moment loading, to avoid excessive group rotation in the long term (creep)
Analysis method	For non-linear methods, reliable calibration of model is important. Use simple methods to check factor of safety against failure Computer-based methods more appropriate for assessing deformation and stresses induced in piles
Reliability and scope of ground investigations	Most important factor to consider Especially important to verify the strength and stiffness of layers below the pile group Near-surface materials important for laterally loaded pile groups The greatest uncertainty lies with establishing the geological model, the idealisation of the ground profile for analysis and the selection of appropriate geotechnical parameters
Nature of loading	Guidance in this chapter is solely for pile groups with predominantly static, monotonic loading Under prolonged cyclic loading, significant degradation of shaft resistance can occur, with associated substantial increases in deformation and reductions in ultimate capacity

Source: After O'Brien, A. *ICE Manual of Geotechnical Engineering*. Thomas Telford Ltd, London; 2012: 823–850.

The average shear strength, c_{uave}, over the block depth, D, is used to calculate the side friction resistance and the average shear strength, c_{ub}, in the zone of influence (0.5 B) of a block of width B and length L, the base resistance. Figure 6.37 shows the bearing capacity and shape factors, N_c and f_s, for pile groups in fine-grained soils.

The capacity of piles in glacial sands based on the group capacity is

$$Q = \left(\sigma'_v N_q s_q d_q + 0.5\gamma' B N_\gamma s_\gamma d_\gamma\right) BL + \sigma'_{vave} K_o \tan\Phi' 2D(B+L) \tag{6.67}$$

The group capacity generally exceeds that of the sum of the single pile's capacity but settlement is often the governing criteria. The outer row of piles in a pile group subject to horizontal loads must be assessed for bearing failure.

O'Brien (2012) suggests that the most common failure of pile groups is due to structural failure because out-of-balance loading applies horizontal load to the piles (e.g. approach embankment to a piled bridge abutment). Failure can also occur if a weaker layer exists within the zone of influence of the base of the piles (Figures 6.38 and 6.39). There are two ground conditions to consider in glacial tills: weaker layers within the glacial till and true

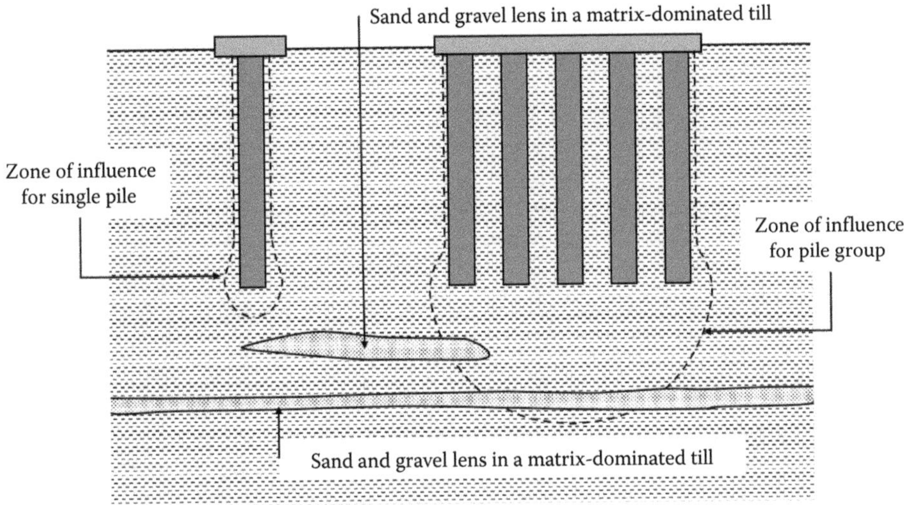

Figure 6.36 Zone of influence of a single piles and pile group to show the effect that a weaker/water-bearing layer could have upon pile group performance and the need to ensure boreholes are deep enough to locate any such layer.

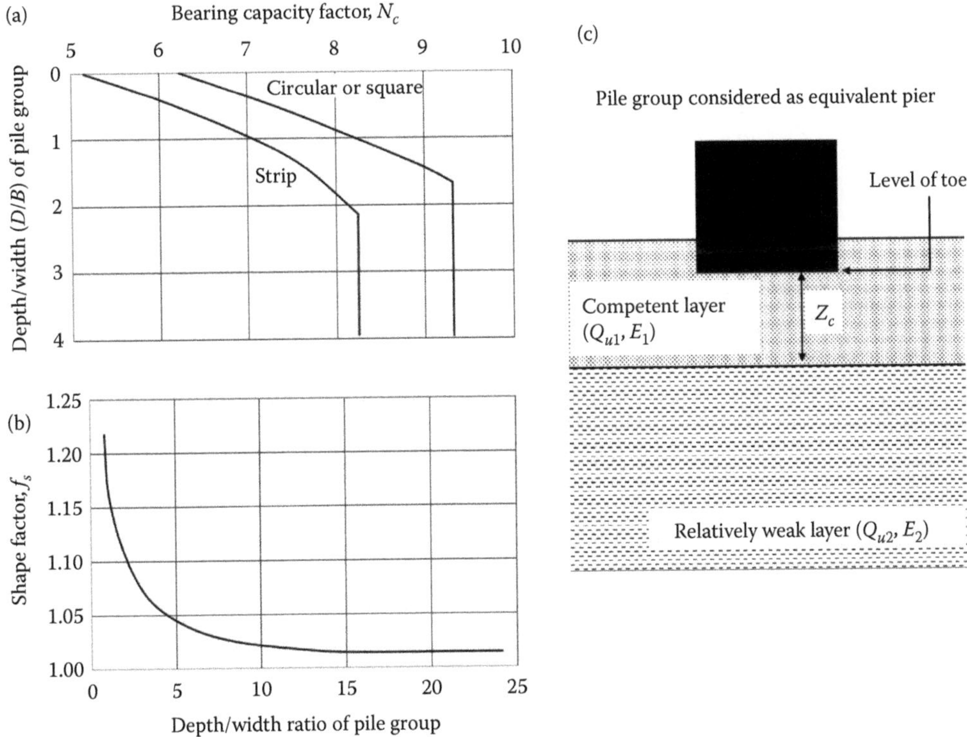

Figure 6.37 Bearing capacity (a) and shape factors (b) for pile groups (after Meyerhof et al.) and the effect of a weaker layer (c) beneath the toe. (After Matsui, T. *Proceedings of the 2nd International Geotechnical Seminar on Deep Foundations on Bored and Auger Piles*, Ghent, Belgium, 1993: 77–102.)

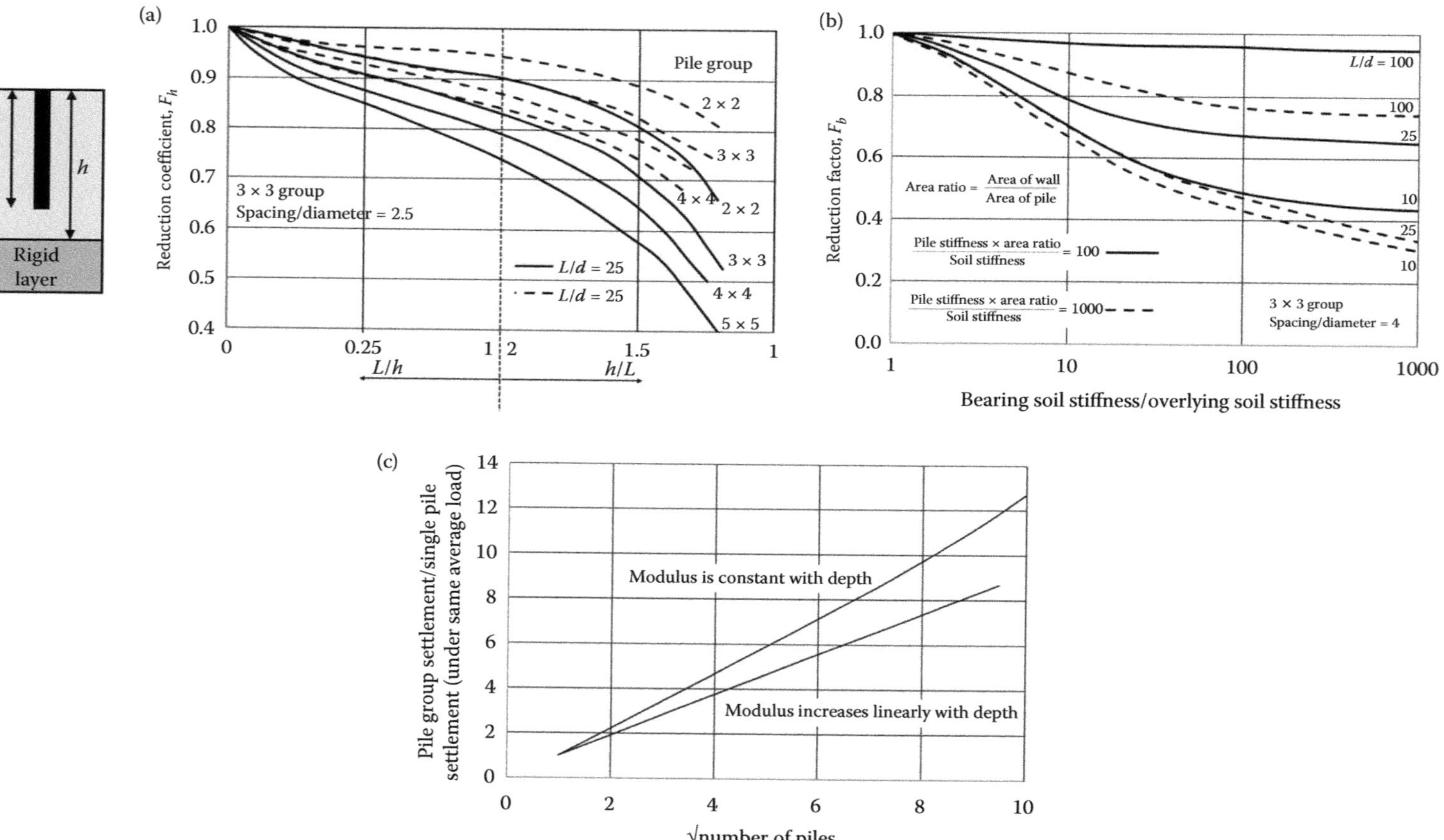

Figure 6.38 Influence of (a) finite layer thickness, (b) bearing stratum stiffness and (c) soil stiffness profile on pile group settlement. (After Fleming, K., A. Weltman, M. Randolph, and K. Elson. *Piling Engineering*. CRC Press, Glasgow; 2008; Poulos, H. G. and E. H. Davis. *Pile Foundation Analysis and Design*. Wiley, New York, 1980.)

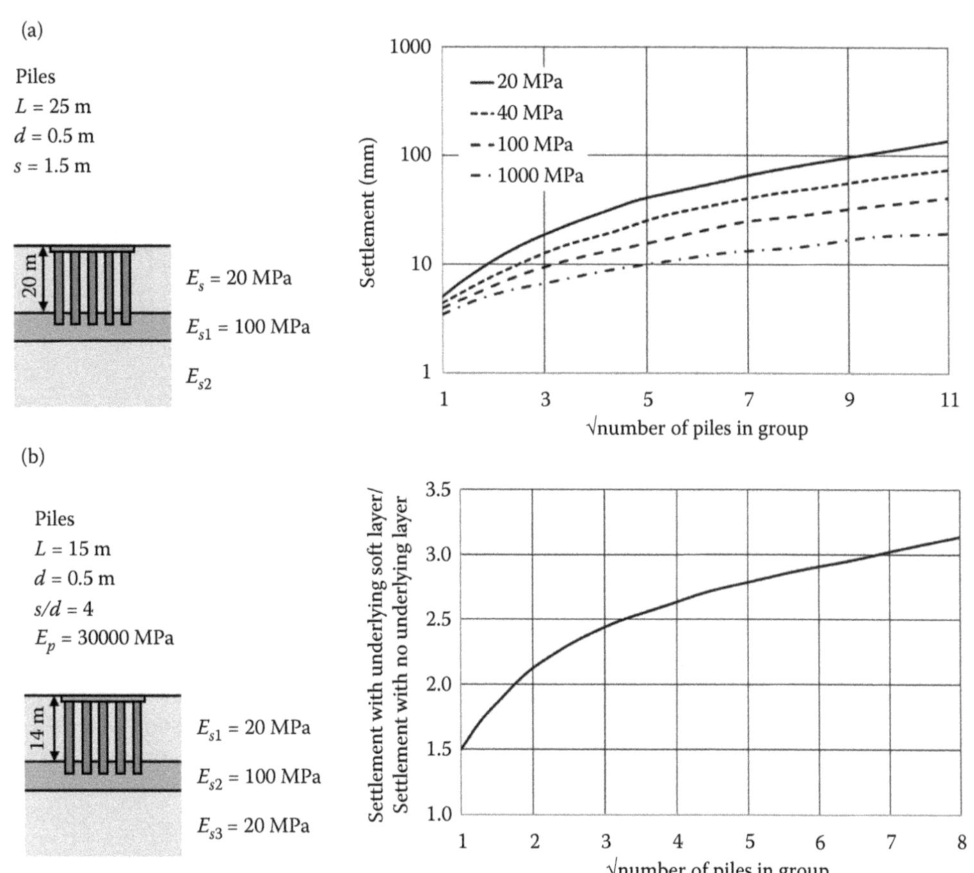

Figure 6.39 Influence of soil layering on pile group settlement showing (a) the variation in settlement with the number of piles in a group, and (b) a comparison between a pile group in a layered soil and one in a homogeneous soil. (After Poulos, H. G. *Journal of Geotechnical and Geoenvironmental Engineering*, 131(5); 2005: 538–563; Poulos, H. G., J. P. Carter, and J. C. Small. *Proceedings of the 15th International Conference on Soil Mechanics and Geotechnical Engineering*, A A Balkema, Istanbul, Vol. 4, 2002: 2527–2606.)

rock head. Matsui (1993) suggested that the capacity of the group should be adjusted to take into account compressible layers within the zone of influence.

$$Q_b = Q_{u2} \quad \text{for } \frac{Z_c}{d_b} < 0.5 \tag{6.68}$$

$$Q_b = Q_{u1} \quad \text{for } \frac{Z_c}{d_b} > 3 \tag{6.69}$$

$$Q_b = Q_{u2} + \left(\frac{0.4Z_c}{d_b} - 0.2\right)(Q_{u1} - Q_{u2}) \tag{6.70}$$

where Q_{ul} is the ultimate base resistance of the upper layer, Q_{u2} the base resistance of the lower layer, Z_c the depth of the underlying layer below the pile group toe level and d_b the equivalent pier diameter.

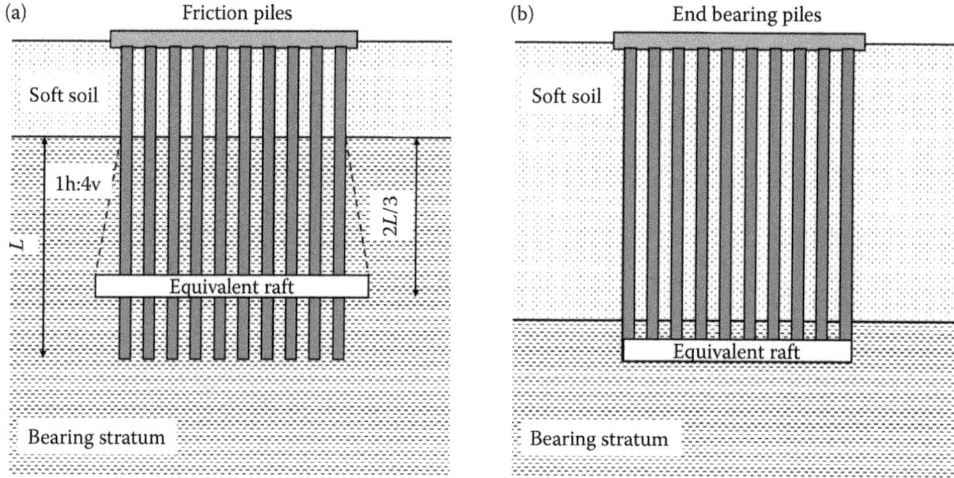

Figure 6.40 Equivalent raft for (a) friction piles and (b) end bearing resistance piles used to estimate the settlement of the pile group.

Increasingly, numerical methods are used to predict pile group settlement but simplified methods do exist including the assumption that a pile group behaves as an equivalent raft or pier (Figure 6.40). Table 6.30 explains the disadvantages and advantages of simple methods to predict pile group settlement, which includes the effects of pile installation, the soil model and the pile interaction. The capacity of a single pile is due to the properties of the soil adjacent to the pile (Figure 6.41) whereas the capacity of a pile group depends on the

Table 6.30 Characteristics of simplified methods to predict pile group settlement

Simplified method	*Advantages*	*Disadvantages*
Empirical settlement ratio	Very quick and simple Best suited for friction piles, in deposits with increasing strength and stiffness with depth Useful as a 'sense' check of more complex methods	The influence of specific geological features, pile or ground properties cannot be assessed Potentially unsafe if weak or compressible strata underlies the bearing stratum Potentially over-conservative, if a relatively stiff layer underlies the bearing stratum
Elastic interaction factors	Influence of varying pile length, diameter and spacing can be quickly checked Best suited for friction piles in deposits with increasing strength and stiffness with depth	Cannot directly check influence of underlying strong or weak layers Care needed in amplifying single pile settlement for large groups, use initial tangent pile stiffness, rather than secant; otherwise can be over-conservative
Equivalent pier	Well suited for pile groups with relatively small aspect ratio, $R < 3.0$ If using elastic solutions for single pile, then it is quick to use Flexible, method can also be used within sophisticated numerical models, in axisymmetric mode	Not appropriate for large pile group aspect ratios, $R > 3.0$ Inappropriate if pile lengths in group vary significantly
Equivalent raft	Most appropriate simplified method for checking influence of strong or weak layers at depth	Over-conservative for pile groups with a small aspect ratio, $R < 3.0$ Significant judgement needed to assess appropriate raft level and dimensions

Source: After O'Brien, A. *ICE Manual of Geotechnical Engineering*, Thomas Telford Ltd, London; 2012: 823–850.

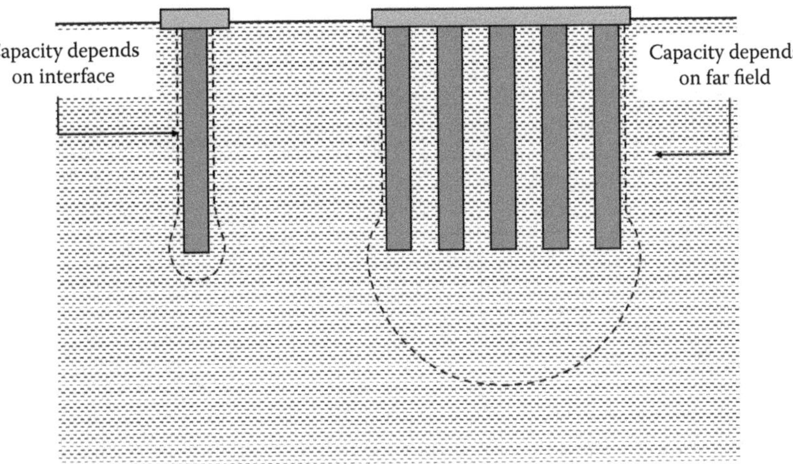

Figure 6.41 Influence of the pile/soil interface on single pile capacity and far-field conditions on pile group settlement. (After Randolph, M. F. *Geotechnique*, 53(10); 2003: 847–876; O'Brien, *ICE Manual of Geotechnical Engineering*. Thomas Telford Ltd, London; 2012: 823–850.)

soil properties within the larger zone of the influence of the group. The mobilised stiffness varies with the direction of the load, the stiffness at the pile/soil interface and the stiffness of the soil remote from the pile or pile group (Figure 6.42).

The group reduction factor (R_g = average group settlement/settlement of a single pile at same total load as a pile in the group) is used to decide whether to model a pile group as a pier or equivalent raft. If $R_g > 3$, then an equivalent raft is used to predict settlement; if $R_g < 3$, then the equivalent pier is more appropriate (O'Brien, 2012). The equivalent raft

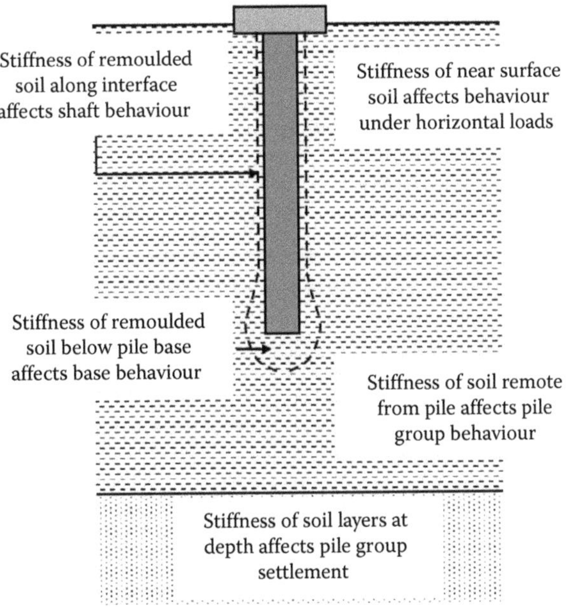

Figure 6.42 Relation between mobilised stiffness and pile. (After O'Brien, A. *ICE Manual of Geotechnical Engineering*. Thomas Telford Ltd, London; 2012: 823–850.)

replaces the pile group with a raft at a depth two-thirds of the length of the pile if the capacity is primarily shaft resistance and at the base of the group for end bearing piles (Figure 6.40).

The settlement, s, is given by

$$s = s_{raft} + s_{pile} \tag{6.71}$$

where s_{raft} is the compression of the soil and s_{pile} is the elastic compression of the piles above the raft.

$$s_{raft} = 0.8 I_d q \sum_{1}^{n} \frac{I_e}{E_s} h_i \tag{6.72}$$

where I_d is the depth factor (Figure 6.12), q the average pressure acting on the raft, I_e an influence factor taken from Figure 6.43, E_s the stiffness of the soil layer i and h_i the thickness of the layer. Unlike the equivalent raft, the equivalent pier includes the effect of shaft resistance. The settlement can be predicted using the method proposed by Fleming (1992) for single piles (Table 6.27). Since the pile is affected by the soil in the zone of influence which extends some distance below the base of the piles, it is necessary to determine the stiffness of the soil within that zone. O'Brien (2012) recommends that boreholes should extend to at least three times the width of the group using the shear modulus ratio given in Table 6.31.

The simplified methods usefully provide an indication of the likely amount of settlement, but they do not take into account of how the stiffness profile and pile soil interaction (Figures 6.36 and 6.37) affect the pile soil system. They are based on a linear elastic assumption, which tends to over-predict settlement. They do not take into account the redistribution of the load due to the pile cap or the effects of installation. Despite these disadvantages, they are useful (cf. presumed bearing capacity) but the limitations (Table 6.30) need to be recognised.

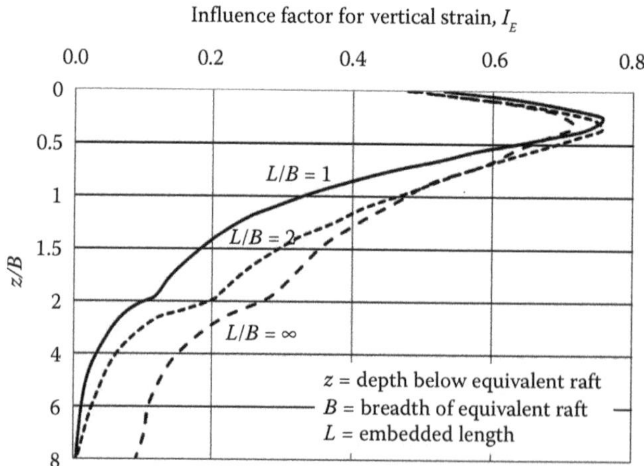

Figure 6.43 Influence factors for the equivalent raft. (After Randolph, M. F. *Proceedings the 13th International Conference on Soil Mechanics and Foundation Engineering*, New Delhi, Balkeema, Rotterdam, 1994: 61–82.)

Table 6.31 Possible values of shear modulus ratio (G/G_{max})

Factor of safety (pile spacing = 3d)	G/G_{max} Single pile	Small- to medium-sized pile group (5–25 piles)	Large pile group (>25 piles)
>5.0	>0.6	0.7–0.8	>0.9
3.0	0.4–0.5	0.6–0.8	0.8–0.9
2.0	0.3–0.45	0.5–0.7	0.7–0.8

Source: After O'Brien, A. *ICE Manual of Geotechnical Engineering*. Thomas Telford Ltd, London; 2012: 823–850.

6.6.5 Tensile capacity

There are two possible failure mechanisms for tensile piles; individual piles can be pulled out of the soil and, the piles together with the block of soil in which the piles sit, are pulled out of the ground. The action $F_{t;d}$ must be less than the design tensile resistance $R_{t;d}$. The tensile capacity can be derived from pile tests or predicted using results of a ground investigation. In the case of pile tests, the design tensile resistance is

$$R_{t;d} = \frac{R_{t;k}}{\gamma_{s;t}} \tag{6.73}$$

using the partial factors given in Table 6.6. The characteristic value of the pile resistance based on these tests is

$$R_{t;k} = min\left\{\frac{(R_{t;m})_{mean}}{\xi_1};\frac{(R_{t;m})_{min}}{\xi_2}\right\} \tag{6.74}$$

using the correlations factors in Table 6.18.

The capacity of a tensile pile is the characteristic shaft resistance due to the shaft friction. Thus, the design value of time resistance is

$$R_{t;d} = \frac{R_{t;k}}{\gamma_{s;t}} \tag{6.75}$$

The characteristic value of tensile resistance based on ground investigation data is

$$R_{t;k} = R_{s;k} = min\left\{\frac{(R_{s;cal})_{mean}}{\xi_3};\frac{(R_{s;cal})_{min}}{\xi_4}\right\} \tag{6.76}$$

$$R_{s;cal} = \sum_{i}^{n} A_{s;i}q_{s;i;k} \tag{6.77}$$

6.6.6 Transverse loaded piles

Transverse piles can fail by rotation or translation (short rigid piles) or bending (long slender piles). Transverse loads can be resisted by raked piles in compression, long slender piles (piles with length to width >10) in bending or rigid piles in translation or rotation. The action must be less than the design resistance, which can be derived from pile load tests or the characteristic properties of the ground and piles. There are a number of commercial software packages used to predict the deformation of transversely loaded piles, which are based on the beam spring model. They produce the bending moments in the pile and the deflected shape of the pile. This means that the stiffness of the ground, the flexural stiffness of the pile and the fixity of the pile are taken into account. The effects of group action and cyclic loading must also be considered.

Hansen (1961) proposed a method to predict the capacity of rigid piles based on the horizontal earth pressure acting on the pile. The unit lateral pressure, p_z, at depth, z, is

$$p_z = \sigma'_v K_{qz} + c' K_{cz} \tag{6.78}$$

where the coefficients K_{qz} and K_{cz} are given in Figure 6.44, σ'_{vz} is the effective vertical stress at z and c' is the cohesion. The pile length, L, is divided into n elements and the lateral pressure calculated for each element such that the force, P_z, on each element is

$$P_z = p_z \frac{L}{n} B \tag{6.79}$$

where L is the length of the pile and B the pile diameter.

The sum of the moments due to the earth pressure acts about the point of rotation, which is defined by

$$\sum M = \sum_{z=0}^{z=x} \left[p_z \frac{L}{n}(e+z)B \right] - \sum_{z=x}^{z=L} \left[p_z \frac{L}{n}(e+z)B \right] \tag{6.80}$$

where x is the depth to the point of rotation. If the transverse load is due to the horizontal force, H, at the top of the pile, then the moment is due to the force (=He), where e is the

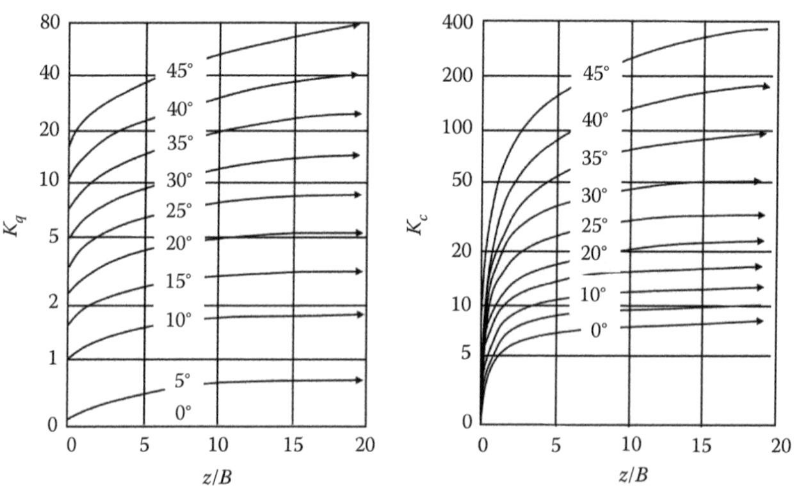

Figure 6.44 Coefficients, K_{qz} and K_{cz}, used in the design of short laterally loaded piles. (After Hansen, B. J. *A General Formula for Bearing Capacity*. Danish Geotechnical Institute, Bulletin No 11, 1961.)

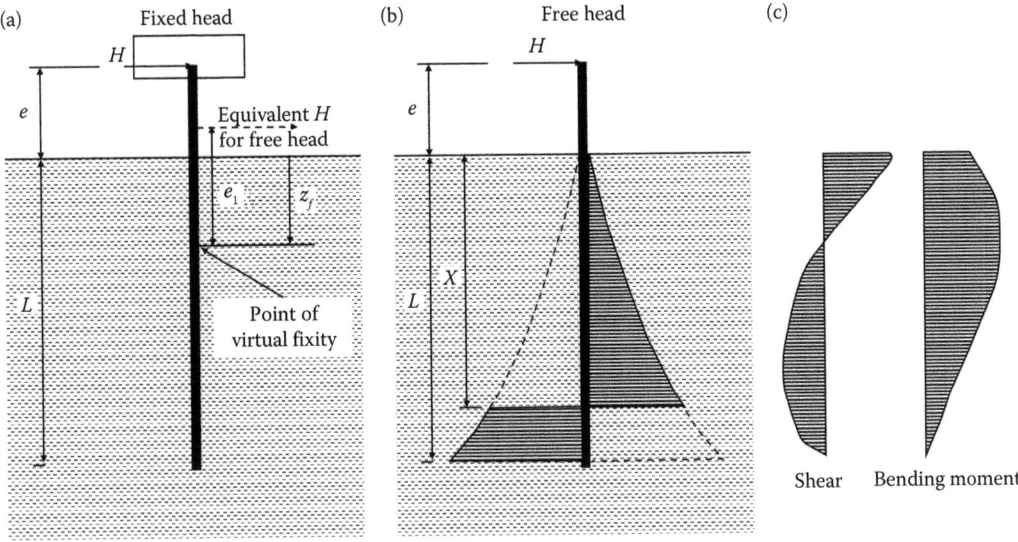

Figure 6.45 Pressure distribution on short laterally loaded piles for (a) fixed head piles, (b) free head piles and (c) the shear force and bending moment diagram. (After Hansen, B. J. *A General Formula for Bearing Capacity*. Danish Geotechnical Institute, Bulletin No 11, 1961.)

distance above the ground surface of the pile head. If the pile head cannot rotate, then the distance to the between the point of load application and the ground surface is $0.5(e + z_f)$, as shown in Figure 6.45.

The ultimate resistance, P, due to a horizontal force, H, is given by

$$H(e+x) = \sum_{0}^{x}\left[p_z \frac{L}{n}(x-z)B\right] - \sum_{x}^{x+L}\left[p_z \frac{L}{n}(z-x)B\right] \tag{6.81}$$

The deflection, d, at the pile head is

$$d = \frac{H(e+z_f)^3}{3EI} \quad \text{for a free headed pile} \tag{6.82}$$

$$d = \frac{H(e+z_f)^3}{12EI} \quad \text{for a fixed headed pile} \tag{6.83}$$

The API (2000) method for predicting lateral capacity is based on p–y curves, where the lateral displacement, y, is related to the unit lateral pressure, p_z. The unit lateral pressure, p_z, in clays at a depth, z, is

$$p_z = 3c_u + \gamma x + J\frac{c_u z}{D} \tag{6.84}$$

where c_u is the undrained shear strength, D the pile diameter, z depth below surface, J a constant ranging between 0.25 and 0.5 and z_R a critical depth depending on the reduced

Table 6.32 p–y curves for clays

p/p_{max}	0	0.50	0.72	1.00	1.0
$y/(2.5\varepsilon_{50}D)$	0	1.0	3.0	8.0	∞

Source: After API, RP. 2A-WSD, 2000. *Recommended Practice for Planning, Designing and Constructing Fixed Offshore Platforms—Working Stress Design*, 21st edition. American Petroleum Institute, Washington, DC 2000.

Note: ε_{50} strain to 50% of peak axial stress in undrained triaxial test.

resistance zone where $p = 9c_u$. The parameters for the p–y curves are given in Table 6.32, where z is normalised with respect to the maximum lateral resistance at that depth and y, the deflection at depth, z, normalised with respect to the pile diameter and the strain to 50% of the peak axial stress. These values are based on static loading but, in practice, offshore piles are subject to cyclic loads, which reduce the lateral capacity for a given displacement. Further, in brittle clays (e.g. stiff over-consolidated clays), account has to be taken of the reduction to post-peak strength if the displacement to peak strength is exceeded. This may not be an issue in matrix-dominated tills as they do not necessarily exhibit brittle behaviour.

The unit lateral resistance for piles in sands is given by

$$p = (C_1H + C_2D)\gamma x \tag{6.85}$$

$$p_{max} = C_3D\gamma x \tag{6.86}$$

where C_1, C_2 and C_3 are derived from Figure 6.46 and p_{max} is the upper limit of p. The lateral resistance/deflection relationship is

$$p = Ap_{max}\tanh\left(\frac{kx}{Ap_{max}}y\right) \tag{6.87}$$

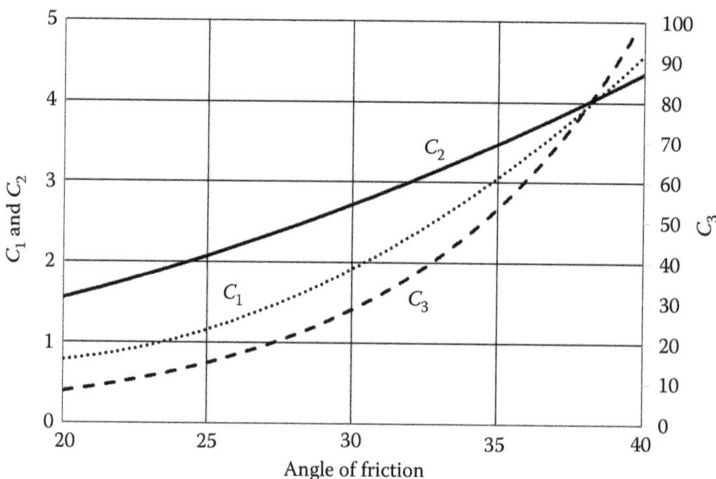

Figure 6.46 Coefficients (C_1, C_2 and C_3) used in the API method to predict the lateral capacity of piles in sands. (After API, RP. 2A-WSD, 2000. *Recommended Practice for Planning, Designing and Constructing Fixed Offshore Platforms—Working Stress Design*, 21st edition. American Petroleum Institute, Washington, DC 2000.)

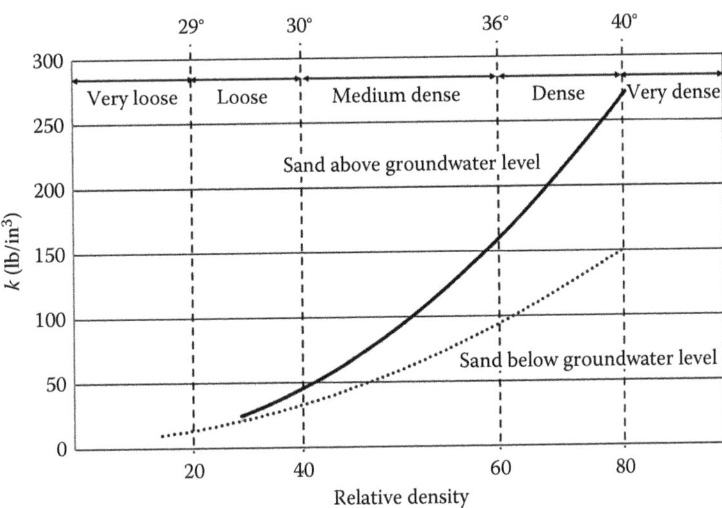

Figure 6.47 Variation of modulus of subgrade reaction with angle of friction to predict the deflection of laterally loaded piles. (After API, RP. 2A-WSD, 2000. *Recommended Practice for Planning, Designing and Constructing Fixed Offshore Platforms—Working Stress Design*, 21st edition. American Petroleum Institute, Washington, DC 2000.)

where A is 0.9 for cyclic loading and 3–0.8 H/D for static loading, and k is the modulus of subgrade reaction from Figure 6.47.

6.6.7 Pile tests

The design capacity of piles can be validated by pile load tests, important when installing piles in glacial soils. There are two categories of pile tests: integrity testing to check on pile quality and estimate capacity; and tests to check capacity. Integrity tests include those in which acoustic or radiometric logging devices are lowered into a tube installed in the pile and those that apply a load or vibrations to the top of a pile. These include stress wave tests in which the top of a pile is subject to a hammer blow and dynamic response to tests in which a vibrator is applied to the top of a pile.

Pile load tests include maintained load tests, constant rate of penetration tests and dynamic tests. Maintained load tests involve load increments up to the design verification load plus 50% of the specified working load (ICE, 2007) with each increment being maintained until the rate of settlement is less than 0.25 mm/h. CRP tests are those that push piles into the ground at 0.5–2 mm/min up to 10% of the pile diameter. Various methods exist to interpret pile tests including Chin (1970), Hansen (1963) and Fleming (1992) (Figure 6.48), which compare the actual pile head deflection with pile load assuming it is a hyperbolic relationship.

Dynamic pile tests are interpreted using wave equation analyses and interpretation of the shear wave data to determine capacity. Commercial systems include pile driving analysis and dynamic load testing with signal matching techniques such as CAPWAP, DLTWAVE and PDPWAVE (TNO). CAPWAP generally produces a conservative estimate of the ultimate capacity for both driven and cast *in situ* piles (Likins et al., 2008). The techniques are subjective since it depends on the assumptions used when analysing the tests. There are a number of wave equation models, including dimensional discrete pile and soil models and dimensional continuous piles and discrete soil model.

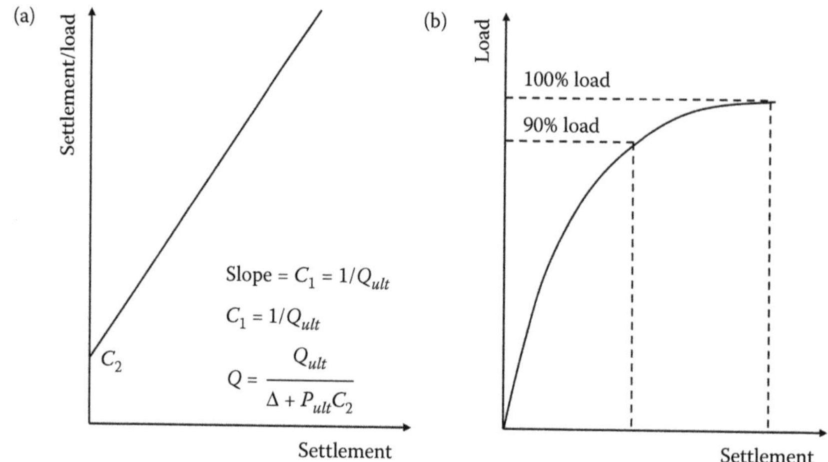

Figure 6.48 Interpretation of pile tests after (a) Chin (1970) and (b) Hansen (1963).

6.6.8 Case studies of piles in glacial soils

There are a number of methods to calculate the capacity of piles and different guidelines on the factors to be used. This means that there is no unique value of pile capacity other than that obtained from a pile test. For example, Karim et al. (2014) undertook tests on CFA and bored piles in glacial tills and compared the results to predictions based on American (FHWA-NHI-10-016, 2010) and Canadian (CFEM, 2006) guidelines. FHWA-NHI-10-016 (2010) recommends that the adhesion factor, α, is

$$\alpha = 0.55 - 0.1\left(\frac{c_u}{100} - 1.5\right) \tag{6.88}$$

for $250 > c_u > 150$ kPa and $\alpha = 0.55$ for $c_u < 150$ kPa. FHWA-NHI-10-016 (2010) recommends β for coarse-grained soils:

$$\beta = (1 - \sin\varphi')\left(\frac{\sigma'_{vmax}}{\sigma'_v}\right)^{\sin\varphi'} \quad \tan\varphi' \leq K_p \tan\varphi' \tag{6.89}$$

where σ'_{vmax} is the effective preconsolidation pressure given by

$$\frac{\sigma'_{vmax}}{100} \approx 0.47(N_{60})^m \tag{6.90}$$

where m is 0.6 for clean sands and 0.8 for silty sands and sandy silts and for gravelly soils

$$\frac{\sigma'_{vmax}}{100} \approx 0.15N_{60} \tag{6.91}$$

Glacial tills are neither coarse-grained soils nor fine-grained soils as assumed in the design guidelines, which is why Karim et al. (2014) undertook these tests to establish values of the soil coefficients of composite soils. The soil profile was glacial till overlain by deposits of sand and silt. Figure 6.49 shows the variation of blow count with depth, not an unusual

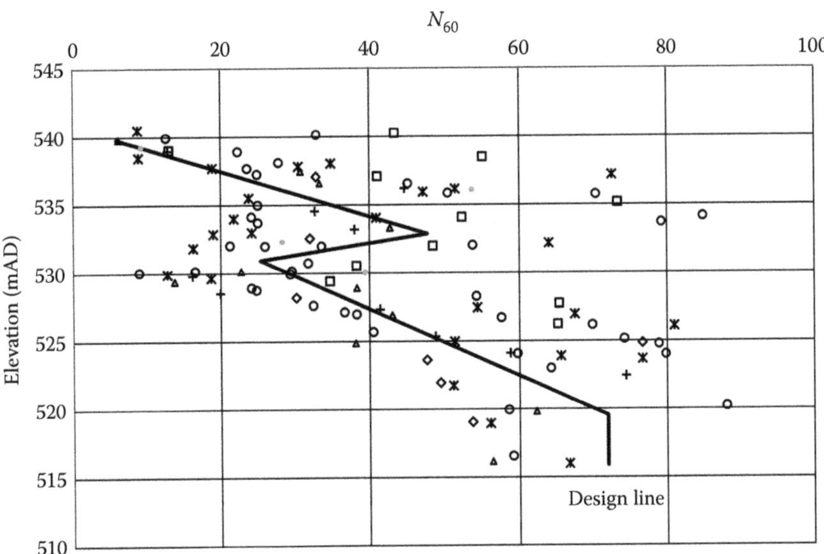

Figure 6.49 Variation of blow count with depth showing the design profile used by Karim et al. (2014) to compare the Canadian, US and UK design methods for pile capacity.

Table 6.33 Estimated ultimate shaft and end bearing resistance using methods proposed by CFEM (β method), FHWA (α method) and α method after Weltman and Healy (1978)

			CFEM (2006)			*FHWA (2010)*			*Weltman and Healy (1978)*	
Depth (m)	σ'_v *(kPa)*	c_u *(kPa)*	f_s *(kPa)*	f_b *(kPa)* $N_t = 7$	f_b *(kPa)* $N_t = 9$	α	f_s *(kPa)*	f_b *(kPa)* $N_t = 9$	α	f_s *(kPa)*
5	51	175	14	357	460	0.53	92	1580	0.40	70
10	102	135	28.5	714	920	0.55	75	1220	0.55	74
15	153	245	43	1071	1380	0.46	112	2210	0.35	85
20	214	355	60	1500	1925	0.35	124	3200	0.35	124
30	336		94	2352	3024					

Source: After Karim, M., C. Zubrowski, and D. C. LePoudre. *Proc Geo Regina*, 2014.

Note: Unit weight 20 kN/m³, 0–15 m; 22 kN/m³, >15 m; groundwater level at surface.

plot for glacial tills, and the assumed design profile. The groundwater level was typically about 2 m below the ground surface. Table 6.33 is a summary of the predicted unit shaft and base resistance, which shows that the UK and USA methods give similar unit shaft resistance, about twice that of the Canadian method. Table 6.34 provides details of the piles and predicted capacities. The measured shaft resistance and base resistance were determined from strain gauge measurements. They observed that all methods underestimated pile capacity and suggested that typical α values quoted for fine-grained soils are not appropriate for piles in matrix-dominated tills.

Wang et al. (2015) also showed that pile capacities exceeded the predictions using the CFEM (2006) design method, which is based on undrained shear strength. They installed instrumented concrete bell piles in matrix-dominated tills at three sites in Edmonton. The till was overlain by glaciolacustrine clay and made ground. The shaft diameter of the

Table 6.34 Comparison between the predicted capacity and measured capacity

	f_s (kPa)						
		Estimated			f_b (kPa)		
Depth (m)	*Measured*	*CFEM (%)*	*FHWA (%)*	*W&H (%)*	*Measured*	*CFEM*	*FHWA*
5	98	−85	−6	−27		460	1580
10	103	−72	−27	−28		920	1220
15	125	−82	−13	−40		1380	2210
20	110	−45	+12	+12	3250	1925	3200

Source: After Karim, M., C. Zubrowski, and D. C. LePoudre. *Proc Geo Regina*, 2014.

Table 6.35 Properties of the glaciolacustrine clay and clay till in Edmonton

Location	*Soil type*	*Depth (m)*	*Water content (%)*	N_{60} *(range and ave)*
Site 1	Glaciolacustrine clay	0.5–5.0	24–40	6–13 (9)
	Matrix-dominated till	5.0–18	9–22	23–42 (32)
Site 2	Glaciolacustrine clay	2.6–10.2	34–37	8–10 (9)
	Matrix-dominated till	10.2–13.2	19–21	16–29 (23)
Site 3	Glaciolacustrine clay	2–6.2	35–40	6–7 (7)
	Matrix-dominated till	6.2–34.3	15–25	14–42 (26)

Source: After Wang, X., R. Tweedie, and R. Clementino. *Proceedings of 68th Canadian Geotechnical Conference*, Canadian Geotechnical Society, Quebec, 2015.

Table 6.36 Ultimate shaft and toe resistance estimated from the loading tests and the back-figured coefficients, α and N_c, based on the undrained strength derived from N_{60}

			Shaft resistance		Base resistance		Coefficients	
Location	*Soil type*	*Depth (m)*	*Resistance (kPa)*	*SPTN (ave)*	*Resistance (kPa)*	N_{60} *(ave)*	α	N_c
Site 1	Glaciolacustrine clay	0.9–3.4	37	9	2730	31	0.69	
	Glaciolacustrine clay Matrix-dominated till	3.4–5.5	123	–			0.55	>14
	Matrix-dominated till	5.5–7.6	90	–				
	Matrix-dominated till	7.6–9.8	122	37				
Site 2	Glaciolacustrine clay	5.3–8.3	38	9	925	23	0.73–1.13	–
	Glaciolacustrine clay	8.3–10.8	68	10				
Site 3	Glaciolacustrine clay	1.3–6.3	20	7	1190	28	0.48	–
	Matrix-dominated till	6.3–9.3	63	21			0.5–0.88	7.1
	Matrix-dominated till	9.3–12.1	152	34				
	Matrix-dominated till	12.1–15.2	121	23				

Source: After Wang, X., R. Tweedie, and R. Clementino. *Proceedings of 68th Canadian Geotechnical Conference*, Canadian Geotechnical Society, Quebec, 2015.

piles ranged between 0.9 and 1.2 m with bell diameters 1.8–2.7 m at between 12 and 17 m below ground level. Static pile tests (ASTM Standard D1143/D1143M-07) were used to check the predicted ultimate load. Tables 6.35 and 6.36 list the soil properties and ultimate shaft resistance and base resistance. The end bearing resistance from SPTs was $36N_{60}$ ($c_u = 6N_{60}$; $N_c = 6$). Back-figured values of N_c (assumed to be 6 in design) varied between 6.7 and >14 and α between 0.5 and 1.13. They concluded that the capacity of these types

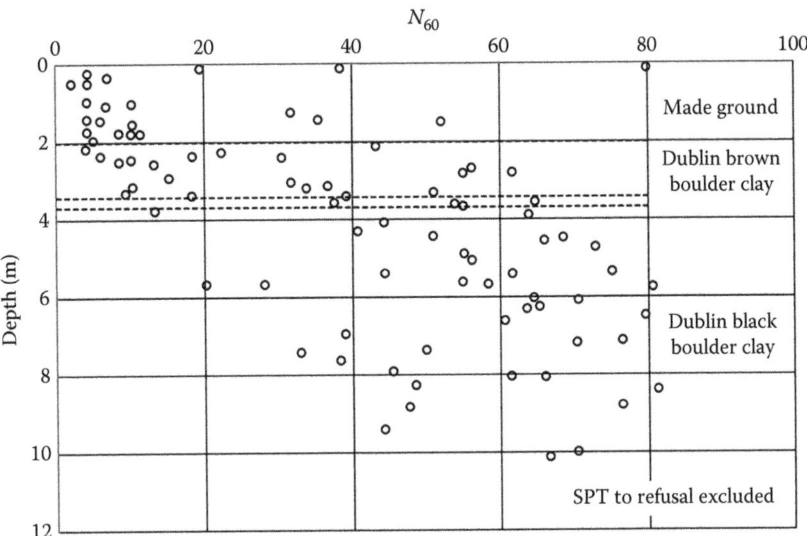

Figure 6.50 Typical profile of Dublin Boulder Clay. (After Farrell, E. R., N. G. Bunni, and J. Mulligan. *Transactions of the Institution of Engineers of Ireland*, 112; 1988: 77–104.)

of piles exceeded that predicted from national design guidelines – a similar conclusion to Karim et al. (2014).

A number of test piles in Dublin Boulder Clay have been reported by Armishaw and Bunni (1993), Farrell and Lawler (2008), Farrell et al. (1998) and Gavin et al. (2008). Figure 6.50 shows a typical profile of SPT tests in Dublin Boulder Clay, which according to Lehane and Simpson (2000) behaves as a low permeability sand. Table 6.37 lists typical properties. In 1993, precast concrete piles were commonly used in Dublin when Armishaw and Bunni (1993) undertook an extensive series of pile tests to understand the relationship between pile set and capacity. Three test piles with predicted capacity 1400 and 1800 kN were installed through the upper brown till into the black till. The test and working piles were 250- and 300-mm concrete square piles driven, typically, to sets of 20 mm for 25 blows using a 4-tonne hammer (Table 6.38). The driving records raised concerns; the driving capacity of some piles increased and some reduced on redriving. CAPWAP analyses estimated that the mobilised total and end bearing resistance of 1300 kN, 840 kN for a settlement of 7.4 mm

Table 6.37 Properties of Dublin Boulder Clay

Property	*Range*
Bulk unit weight (kN/m^3)	21.5 ± 0.5
Water content (%)	11 ± 3
Liquid limit (%)	25 ± 4
Plastic limit (%)	14 ± 2
Clay fraction (%)	$15 + 5$
Gravel fraction (%)	$30 + 5$
Permeability (m/s)	1×10^{-8} to 1×10^{-10}
Initial shear modulus (MPa)	250–350
Peak angle of friction	43
Angle of friction at constant volume	34

Table 6.38 Pile driving records of 300- and 250-mm precast concrete piles driven into Dublin Black Boulder Clay

	First drive		*Sets (mm) for redrives*								*Load test details*			
			1st		*2nd*		*3rd*		*4th*			*Type*		
Pile No	*Depth (m)*	*Set (mm)*	s_i	s_f	s_i	s_f	s_i	s_f	s_i	s_f	*Depth (m)*	*First*	*Last*	*Set (mm)*
1	5.95	20*	25**								5.95	DYN	ML	0
2	5.85	12*									5.85	ML		
3	5.80	11*	15	13	5	5					5.85	MML	CRP	6
4	5.90	6*	7	5	4	3					5.94	MML		3
5	5.40	4*	19*	8*	15	10	20	15			5.76	CRP		5
6	6.10	11*	20*	6*	30*	8*	12*	3*	5	3	9.61	CRP	MML	2
7	6.40	12*	15*	7*	25*	9*	12	10	0		7.05	CRP	MML	10
8	6.35	10***	25***	9***	7	6					6.95	MML	ML	4
9	6.50	11***	35***	8***	20	12	20	10	6	5	9.35	MML	CRP	4
10	5.65	11*	10	8	50						5.75	MML	CRP	
11	8.45	11	7	6	35						8.52	MML	CRP	
12	10.80	15	30	15	12	10					12.53	ML		

Source: After Armishaw, J. W. and N. G. Bunni. *Piling: European Practice and Worldwide Trends, Piling in Difficult Ground and Locations I*, 1993: 272–279.

Note: Test piles: 1 and 2; working piles: 3–12; 300-mm piles: 1–7, and 10–12; 250-mm piles: 8 and 9. 300-mm piles driven with 5-tonne hammer, 0.4-m fall; except *4-tonne hammer, 0.3-m fall; **5-tonne hammer, 0.5-m fall. 250-mm piles driven with 4-tonne hammer, 0.3-m fall; except ***4-tonne hammer, 0.2-m fall. ML, maintained load; MML, modified load; CRP, constant rate of penetration; DYN, dynamic.

for the 300-mm pile; and 1150 kN, 300 kN and 5.3 mm for the 250-mm pile. Maintained load tests on the 250- and 300-mm piles gave 2080 kN for a settlement of 9.5 mm for the 250-mm pile; and 1000 kN for the 300-mm pile at a settlement of 8.8 mm (Table 6.39). This meant that the working piles had to be redriven because the test pile had failed. The criterion was a set of 15 mm for 25 blows after a delay of a week using a 5-tonne hammer. This led to a series of maintained load tests to establish the reasons for the deterioration. Maintained load tests were carried out on four piles using various load configurations; six piles were subject to load increments until the rate of penetration was less than 0.025 mm over 15 min; and CRP tests on seven piles. The installation and test details are given in Table 6.38. They estimated the shaft and base resistance (Table 6.40) using Jain and Kumar's (1963) method with data from maintained load tests. The estimated ultimate capacity was used to back figure the bearing capacity factors (Table 6.41). They concluded that a 57 hammer with a 0.4-m fall for 300-mm piles and 0.3-m fall for 250-mm piles achieved satisfactory capacity if a set of 12 mm for 25 blows achieved. The increase in capacity of some piles was attributed to dissipation of pore pressures but the reduction in capacity of some piles was not explained. The back-figured bearing capacity factors using the effective strength approach gave values of angle of friction greater than typical values for the till (Tables 6.40 and 6.41). They established that a maximum capacity was achieved if a pile was driven at least 2.2 m into the till.

Farrell et al. (1998) undertook static compression and tension tests on instrumented precast concrete piles driven into Dublin Boulder Clay immediately after installation and 1.8, 17 and 24 days after installation. Table 6.42 summarises the results of the tests. Figure 6.51 compares the results with those observed in practice. They observed negative pore pressures during installation, increasing to over 750 kPa and then dissipating after a week. The instrumented pile showed that most of the capacity was derived from the shaft resistance. The pore pressure response is consistent with heavily over-consolidated soils and dense soils; that is,

Table 6.39 Pile test records of 300- and 250-mm precast concrete piles driven into Dublin Black Boulder Clay

	Length		*Sets (mm)*			*Settlement of pile head (mm)*					*Chin*	*Hansen*	
Pile No	*Length (m)*	*Till (m)*	*Before test*	*After test*	*Max load (kN)*	Q_u	*After unload*	Q_w	$1.5Q_w$	$2Q_w$	Q_u *(kN)*	Q_u *(kN)*	S_{max} *(mm)*
1	5.95	3.0	25	0	2080	9.5	4.3	3.8	5.8	8.4	2927	2568	37.5
2	5.85	2.0	12		1750	29.8	23.7	8.8	18	Fail	2238	1966	83.8
3	5.84	2.5	5	6	2000	9.7	3.4	4.1	6.3	9.7			
4	5.94	2.3	3	3	1975	9.1	4.7	4	6.1	9.5			
5	5.66	2.2	15	5	2000	11	4.2	4.6	7	11			
6	9.6	6.0	3	2	2000	11	4.9	4.4	6.6	11			
7	7.05	3.4	0	10	2300	22.2	14.7	4.8	7.9	15	3175	2746	75.6
8	6.95	2.9	6	5	1080	5.1	2.2	2.3	4.6				
9	9.35	5	5	4	1510	6.1	1.2	1.9	3.2	4.2			
10	5.75	2.3	50		2410	34.2	21.4	4.4	8	16.3	3000	2560	68.6
11	8.52	4.7	35		2350	39	30.2	5.4	10.4	21.4	2830	2462	84.2
12	12.5	9.5	10		1480	9.9	2.7	6.5	10				

Source: After Armishaw, J.W. and N. G. Bunni. *Piling: European Practice and Worldwide Trends, Piling in Difficult Ground and Locations I*, 1993: 272–279.

Note: Q_u, ultimate load; Q_w, working load; s_{max}, maximum settlement.

Table 6.40 Pile capacities of 300- and 250-mm precast concrete piles driven into Dublin Black Boulder Clay

	Length (m)			*Jain and Kumar (kN)*			*Ultimate load (kN)*		*Factors of safety*		
Pile No	*Upper till*	*Black till*	*Max load (kN)*	Q_{su}	Q_s	Q_b	Q_u	Q_b	Q_u	Q_s	Q_b
7	3.63	3.42	2300	1450	850	150	2746	1296	2.75	1.73	8.64
10	3.5	2.25	2410	2040	925	75	2560	520	2.56	2.21	6.93
11	3.63	4.65	2350	1068	816	384	2462	1394	2.46	1.73	3.63

Source: After Armishaw, J.W. and N. G. Bunni. *Piling: European Practice and Worldwide Trends, Piling in Difficult Ground and Locations I*, 1993: 272–279.

Table 6.41 Pile test analysis of 300- and 250-mm precast concrete piles driven into Dublin Black Boulder Clay

	Pile shaft												*Pile base*			
	σ'_{vave} *(kPa)*		τ_s *(kPa)*		*K tan* δ		*K*		K/K_{ps}		K/K_{pr}		σ'_b	q'_b		
Pile No	*U*	*L*	*U*	*L*	*U*	*L*	*U*	*L*	*U*	*L*	*U*	*L*	*(kPa)*	*(kPa)*	N_q	Φ'_{base}
7	27	77	51	299	1.89	3.88	4.84	8.47	1.49	2.11	0.81	0.94	100	14.4	144	39.5
10	26	67	109	585	4.2	8.74	10.23	19.03	3.15	4.75	1.71	2.11	82	5.8	70	36.5
11	28	87	23	172	0.83	1.98	2.12	4.28	0.65	1.07	0.35	0.48	118	15.5	131	39

Source: After Armishaw, J. W. and N. G. Bunni. *Piling: European Practice and Worldwide Trends, Piling in Difficult Ground and Locations I*, 1993: 272–279.

Note: σ'_{vave}, average vertical stress over shaft length; τ_s, average shear stress on shaft; K, lateral pressure coefficient on shaft; K_{ps}, lateral pressure coefficient for smooth piles; K_{pr}, lateral pressure coefficient for $\delta = 0.67\Phi'$; σ'_b, vertical stress at the base; q'_b, ultimate base pressure; N_q, bearing capacity factor; U, Upper Dublin Boulder Clay; L, Lower Dublin Boulder Clay.

Table 6.42 Summary of static load tests on instrumented driven precast concrete piles in Dublin Boulder Clay

Test No	*Test type*	*Time delay (days)*	*Max applied load (kN)*	*s at max load (mm)*	*Max shaft load (kN)*	*Base load (kN)*	*Base load as % of max load*	*α (based on c_u = 450 kPa*
1C	Compr	0.1	720	15	279	441	61	0.21
2C	Compr	1.8	944	10	336	608	64	0.25
3C	Compr	17	1350	7.5	546	1450	66	0.55
4T	Tension	24	–450	–14	–450	–	–	0.4

Source: After Farrell, E. R., N. G. Bunni, and J. Mulligan. *Transactions of the Institution of Engineers of Ireland*, 112; 1988: 77–104.

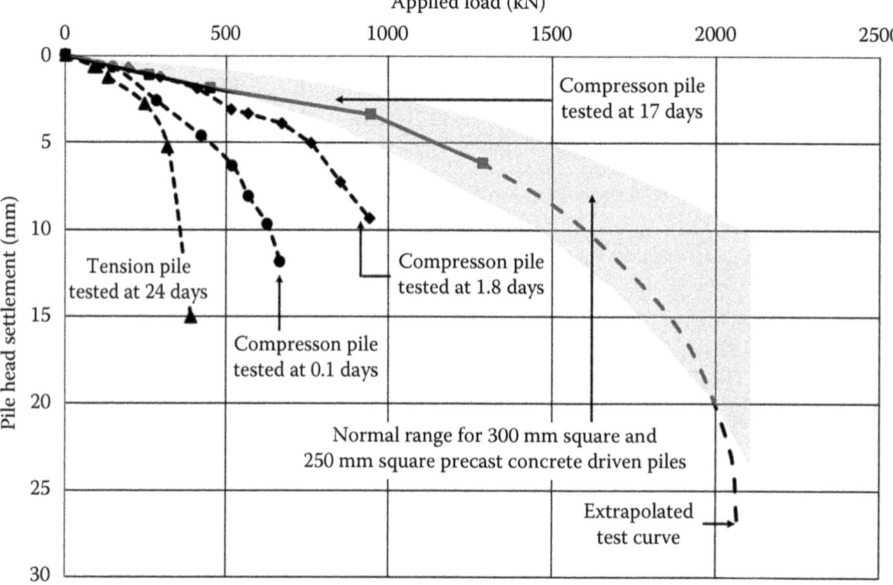

Figure 6.51 Results on an instrumented steel tubular pile driven into Dublin Boulder Clay. (After Farrell, E. R., N. G. Bunni, and J. Mulligan. *Transactions of the Institution of Engineers of Ireland*, 112; 1988: 77–104.)

high pore pressures are generated in the adjacent soil (Bond and Jardine, 1991). The time for dissipation of pore pressure was consistent with typical values of coefficient of consolidation (35 m^2/year). The ultimate capacity of the tension tests was about 80% of that of the compression tests, similar to that observed by Lehane and Jardine (1994) on pile tests in Cowden Till. They suggest that the α method to predict shear resistance is not appropriate because of the increase in resistance with time and it may be more appropriate to use the ICP method (Jardine and Chow, 1996) for driven steel piles.

Farrell and Lawler (2008) installed 450- and 600-mm instrumented CFA piles 12.3 and 11 m into Dublin Boulder Clay and loaded them to 3.15 and 4.5 MN, respectively. The static load tests were taken to 1.5 working load. SPT tests were carried out to give an undrained shear strength profile similar to other sites using 6 N. They used Chin's method to predict the ultimate capacity to derive an N value of 7 for an undrained shear strength of 450 kPa based on 6 N. They estimated α values between 0.65 and 0.75 based on an average shear strength of 350 kPa, which exceeds the 0.45 normally assumed.

Lawler (2003) undertook a numerical study of the same CFA piles and compared it to field observations because experience had shown that CFA piles were overdesigned. The unit shaft resistance of 350 kPa at a load of 4.5 MN on the 600-mm-diameter, 11-m-long pile was in excess of that predicted capacity using factors developed for bored piles. The hardening soil model with a stiffness of 150 MPa was used to simulate the installation process and subsequent loading. It was shown that the installation process had a significant effect on the capacity because the borehole walls are stable and fluid concrete under pressure ensures positive load on the borehole walls. Figure 6.52 shows a comparison between the predicted and measured pile behaviour.

Gavin et al. (2008) reported tests on three instrumented 762-mm-diameter CFA piles in Dublin Boulder Clay to show that the base resistance was less than that of driven piles, but the unit friction resistance was much greater than that for driven piles. Maintained load tests were carried out after 20 days to 250% of their design load. Both compression and tension tests were carried out. In all cases, they found that the majority of the load was taken by shaft resistance. They suggested that effective stress design should be used with the mobilised coefficient of earth pressure, K_m, given by

$$K_m = \frac{q_{max}/\tan\delta}{\sigma_v'} \tag{6.92}$$

where q_{max} is the maximum shear resistance and δ_{cv} the constant volume interface friction. Figure 6.53 shows a comparison of measured and predicted K_m values from CFA piles in Dublin Boulder Clay. They concluded that cyclic loading of driven piles during installation reduces the unit shaft resistance but the driving process increases the base resistance, whereas in CFA piles the unit shaft resistance is increased to such an extent that the base resistance is less than that for driven piles. They suggested that base resistance of CFA piles

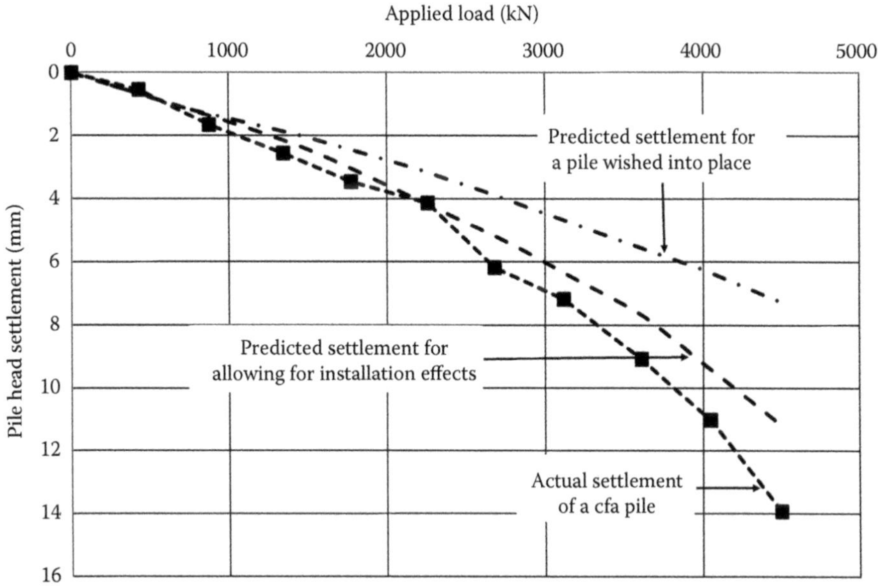

Figure 6.52 Comparison between the predicted and measured behaviour of a 600-mm-diameter, 11-m-long CFA pile in Dublin Boulder Clay. (After Lawler, M. *Electronic Journal of Geotechnical Engineering*, 8; 2003.)

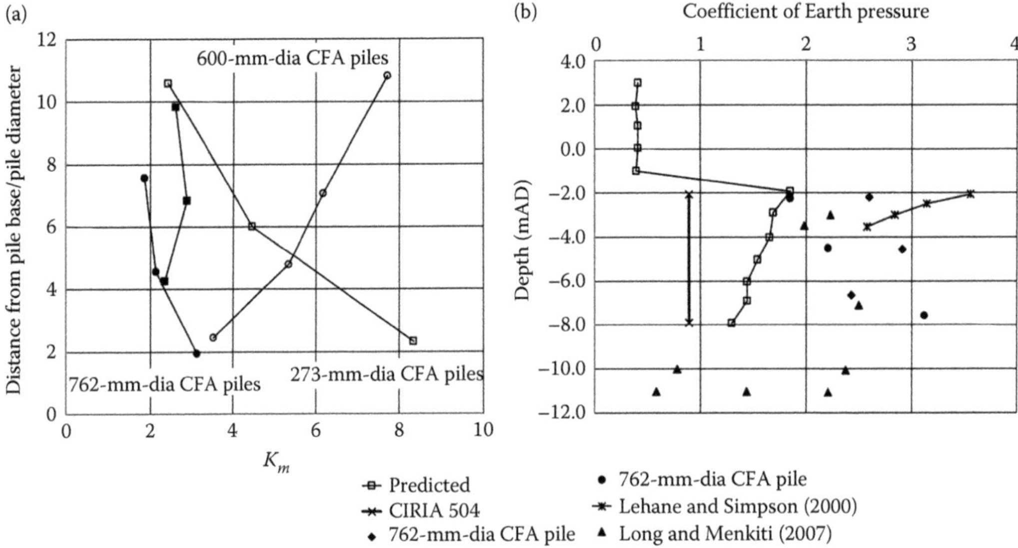

Figure 6.53 Comparison of (a) predicted K_m values from 273-, 600- and 762-mm-diameter CFA piles and (b) measured and predicted K values using various sources for CFA piles in Dublin Boulder Clay. (After Gavin, K., D. Cadogan, and L. Twomey. *Proceedings of the Institution of Civil Engineers-Geotechnical Engineering*, 161(4); 2008: 171–180.)

should be based on N_c of 9 compared with 38 used by the industry for driven piles. This highlights the issue of selecting the appropriate design method for the type of pile. The pile tests suggest that an effective strength approach for base and shaft resistance is more appropriate.

Gavin (2009) used the results of the tests reported by Gavin et al. (2008) to predict the capacity of 900-mm-diameter CFA piles, 20.5 and 21 m long, bored through 5 m of alluvial sands and silts, 5 m of dense gravel into Dublin Boulder Clay. Figure 6.54 shows a comparison between predicted and observed capacity from static load tests on Grimsby Till (Brown et al., 2006) and Dublin Boulder Clay (Gavin et al., 2008). They observed that the factors (skin friction α; end bearing N_c; and parameters f and g used to fit test curves) derived from tests on Dublin Boulder Clay gave a reasonable prediction for piles in Grimsby Till. They observed that the mobilised unit shaft resistance was constant down the length of the pile and some 50% greater than that for driven piles and the back-figured value of α was 0.75. The end bearing capacity should be limited to 15 MPa (cf. piles in coarse-grained soils), and creep factors may become significant if the base resistance exceeds the preconsolidation pressure.

The strength of Dublin Boulder Clay allowed Martin et al. (2007) to use 220-m-diameter minipiles as anchors to resist hydraulic uplift acting on an 8-m-deep basement, 135 m long and 60 m wide. A total of 682 piles were installed. An anchor was formed of 75- or 50-mm-diameter rebar with a yield stress of 500 MPa. The anchor was grouted into 220-mm-diameter hole, at least 9 m into Dublin Boulder Clay. The grout strength was 50–80 MPa after 28 days. A summary of the acceptance tests shown in Table 6.43 gave an average pile head movement of 2.95 mm, which was within the specified 5 mm demonstrating the feasibility of this method.

Doherty et al. (2010) investigated the effect of a soil plug on the base resistance of open-ended piles driven into soft clayey silt by modelling offshore driven piles. They referred to field tests of piles driven into glacial till and observed that the internal unit shaft resistance

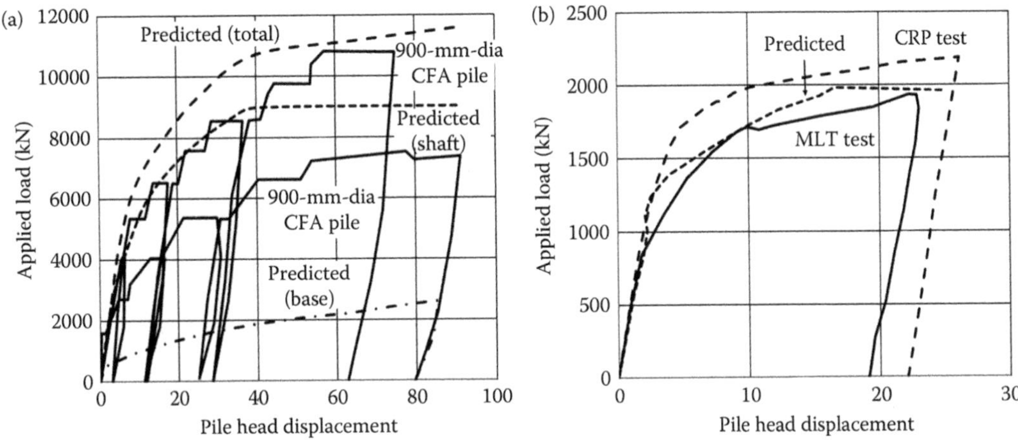

Figure 6.54 Comparison between the predicted and measured settlement of (a) two 21-m-long, 900-mm-diameter CFA piles in 5 m loose sand, 10 m dense sand and gravel overlying Dublin Boulder Clay highlighting the different capacity for similar length piles and the over-prediction of capacity using a non-linear elastic model and (b) comparison between (a) a 600-mm-diameter CFA pile in 2.4 m weathered clay overlying Grimsby Till showing a better fit between the predicted and measured settlement. (Adapted from Gavin, K., D. Cadogan, and L. Twomey. *Proceedings of the Institution of Civil Engineers-Geotechnical Engineering*, 161(4); 2008: 171–180; Brown, M. J., A. F. L. Hyde, and W. F. Anderson. *Geotechnique*, 56(9); 2006: 627–638.)

Table 6.43 Results of tests on anchor minipiles in Dublin Boulder Clay

Minipile	*Elastic movement (mm)*	*Permanent movement (mm)*	*Total head movement (mm)*	*Apparent free Tendon length (mm)*	*Free Tendon length (mm)*	*Tendon bond length (mm)*	*Free length + 50% of tendon bond length (mm)*
1	1.88	0	1.88	1360	1500	9000	6000
2	2.19	0.44	2.63	1581	1500	9500	6000
3	3.70	0.65	4.36	2676	1500	9500	6250
4	3.93	0.74	4.67	2839	1500	9000	6000

Source: After Martin, J. et al. *Proceedings of Conference on Ground Anchorages and Anchored Structures in Service*, Thomas Telford, London, 2007.

was similar to the external unit shaft resistance and CPT q_c and independent of the degree of plugging (Figure 6.55).

The unit plug stress q_{plug} was

$$q_{plug} = q_c(0.8 - 0.6\,IFR) > q_{plugmin} \tag{6.93}$$

$$q_{plugmin} = 0.2q_c \tag{6.94}$$

$$q_{ann} = q_b = q_c \tag{6.95}$$

where IFR is the change in the length of the soil plug, ΔL_p, over a given increment of pile penetration, ΔL, q_c the CPT cone resistance and q_{ann} the annular end bearing pressure.

Glacial soils are complex soils that require more investigation than that specified for homogeneous soils. An example of this requirement is described by Wisniewski et al. (2011)

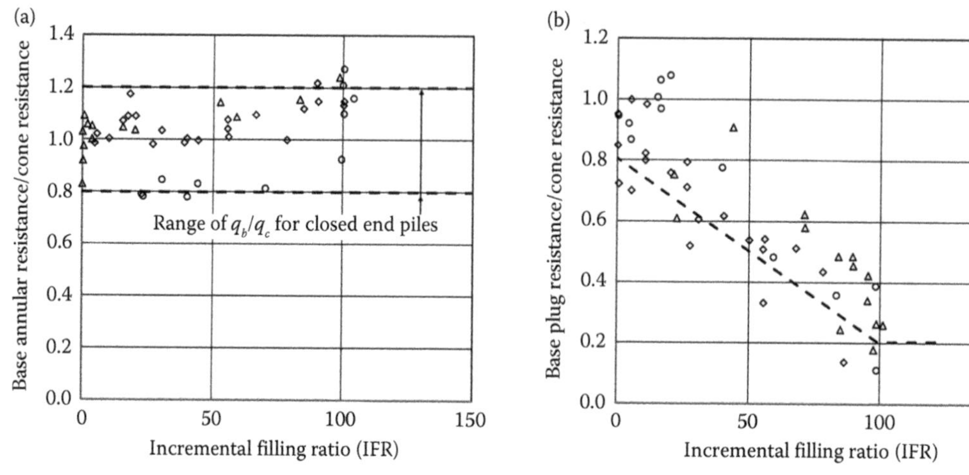

Figure 6.55 Results of an instrumented open-ended pile, 154-mm-diameter, 2-m-long, 9 mm wall thickness pushed into Dublin Boulder Clay showing that (a) the annular resistance is independent of the length of the plug formed in the pile (IFR expressed as a ratio of the change in length of the plug to the length of penetration) and (b) the plug resistance varies with IFR. (After Doherty, P., K. Gavin, and D. Gallagher. *Geotechnical Engineering,* 163GE1; 2010: 13–22.)

who outlined the foundation design for the Yankee Stadium project in New York, which was built over made ground and alluvial soils underlain by glacial deposits of glaciofluvial sands, glaciolacustrine clays and glacial till. The local Building Code required one boring per 150 m^2 for piled structures, but they were permitted to carry out one boring per 370 m^2, a total of 113 borings. 406-mm-diameter concrete pipe piles were driven into sand, till or rock depending on the location to give a working capacity of 1350 kN. A total of 35 static pile tests were taken to twice the design load and held for 96 h to measure the creep load. The pile tests showed that they complied with the local regulations, a limit of 19 mm total settlement and a maximum creep rate of 0.3 mm/48 h. These tests, together with the PDA data, were used to establish the pile driving criteria across the site taking into account the variation in ground conditions.

Fellenius and Ochoa (2009) described field tests to assess the performance of 450- and 600-mm-diameter piles through 3 m of fill, 9 m of sand, 10–15 m of firm lacustrine sandy silty clay, 4 m of stiff clay into matrix-dominated tills at a site in the mid-west United States. Two 457-mm-diameter, 25.6- and 26.2-m-long bored piles were instrumented with an Osterberg cell (O-cell) and strain gauges. The piles were subject to dynamic load tests and O-cell tests in which the pile capacities above and below the O-cell are measured. The field tests were compared with predictions based on CPT and CPTU tests. The aim of the tests was to prove that the predicted pile capacity could be achieved in the till. The load was a combination of the structural load and the downdrag due to the compression of the lacustrine clays. They estimated the shaft resistance to be between 1504 and 1875 kN using methods based on CPT and CPTU tests (Schmertmann et al., 1978; De Kuiter and Beringen, 1979; Bustamante and Gianeselli, 1982; Eslami and Fellenius, 1997). Figure 6.56 shows a reasonable comparison between the shaft resistance based on the results of the CAPWAP interpretation and predicted capacity based on CPT tests and the measured resistance from the O-cells. Following the pile tests, the ground level was raised inducing downdrag in the upper layers. The sustained load was 1300 kN, but the maximum load in the pile including downdrag was 2700 kN. The settlement of the neutral axis governs the settlement of the

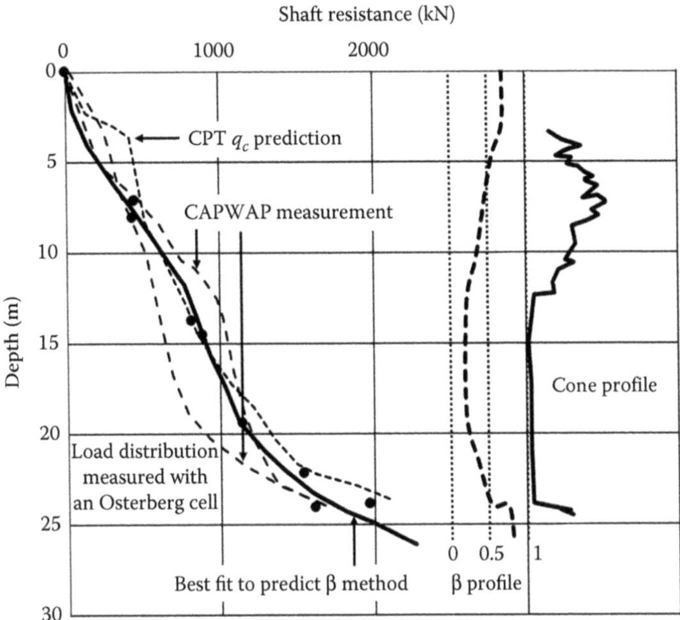

Figure 6.56 Comparison between the predicted shaft resistance of two 457-mm-diameter bored piles using the CPT and CAPWAP methods and the measured capacity based on the results of Osteberg cells at 1.8 m above the pile base and the predicted β value. (After Fellenius, B. H. and M. Ochoa. *Geotechnical Engineering Journal of the Southeast Asian Geotechnical Society*, 40(3); 2009: 129–137.)

pile according to the Unified Design Method, the basis of a number of pile design methods (CFEM, Australian, Hong Kong and FHWA). A 20 mm of settlement was predicted. Further settlement would increase the end bearing load, which would prevent further settlement. Therefore, there is a balance between the location of the neutral plane and end bearing.

Islam and Yang (2002) investigated the use of CPTU to predict pile capacity because of the difficulty in sampling and the fact that the glacial soils are neither coarse- nor fine-grained soils, soil types that are usually used to define design parameters. The 457-mm-diameter close-end steel piles for a bridge abutment in New Jersey were driven through fill, sand, peat and alluvium into glacial till. A review of the many methods (Figure 6.57) of predicting pile capacity from CPT and SPT tests suggested that those due to Bustamante and Gianeselli (1982) and De Kuiter and Beringen (1979) methods were most accurate (Figure 6.58) if site-specific correlations were used. In both cases, the unit skin friction and unit base resistance are related to the cone resistance q_c, using the factors in Table 6.44 where $f_s = q_c > a$ and $q_b = k_c\, q_c$, which produces a range of capacity by up to 40%. They recommended site-specific correlations with soil type to improve the prediction of pile capacity.

These case studies demonstrate that pile designs based on undrained shear strength underestimate the capacity, suggesting that an effective strength approach should be used provided the settlement is checked. Lehane and Jardine (1994) used a 7-m-long, 102-mm-diameter steel pile at the Cowden site to evaluate pile capacity in matrix-dominated tills. This was a fully instrumented pile used to develop the ICP design method. They found that the radial stresses acting on the pile reduce during and immediately after installation and then increased, the rate of increase being dependent on the soil immediately adjacent to the pile and the relative depth of the point being considered to the tip of the pile. They found

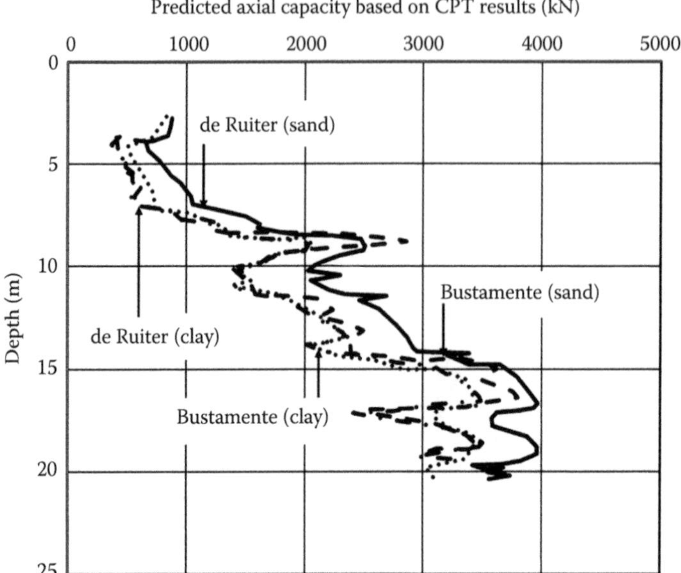

Figure 6.57 Ultimate axial capacity of a 457-mm-diameter concrete-filled closed-end steel pipe driven through alluvium into glaciofluvial sands and gravels using the factors in Table 6.44, which produces a range of capacity by up to 40%. (After Islam, M. Z. and M. Z. Yang. *Deep Foundations 2002: An International Perspective on Theory, Design, Construction, and Performance*, ASCE, 2002: 1247–1260.)

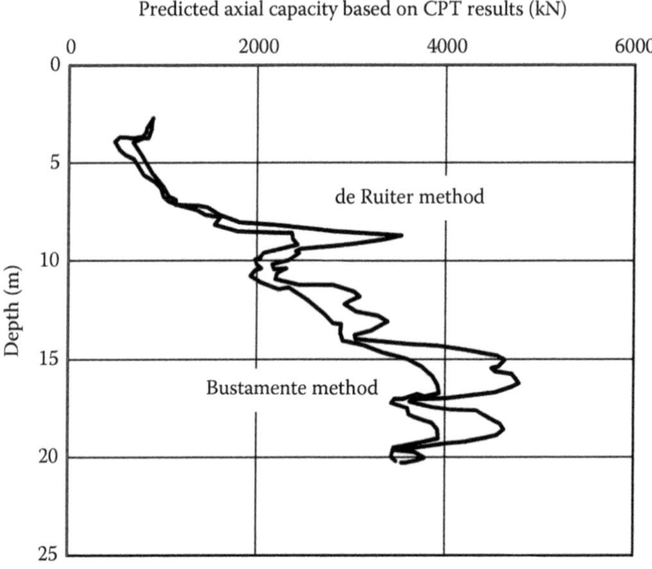

Figure 6.58 Ultimate axial capacity of a 457-mm-diameter concrete-filled closed-end steel pipe driven through alluvium into glaciofluvial sands and gravels using site-specific correlations with soil type to improve the prediction of pile capacity. (After Islam, M. Z. and M. Z. Yang. *Deep Foundations 2002: An International Perspective on Theory, Design, Construction, and Performance*, ASCE, 2002: 1247–1260.)

Table 6.44 Recommended parameter to predict pile capacity from CPT tests according to Bustamante and Gianeselli (1982) and De Kuiter and Beringen (1979)

Soil type	q_c *(MPa)*	α	f_p *(max) (kPa)*	k_c
Soft clay and mud	<1	30	15	0.5
Firm to stiff clay	1–5	80	35	0.45
Stiff to hard clay and hard silt	>5	120	35	0.55
Loose sand	<5	120	35	0.5
Medium dense to dense sand and gravel	5–12	200	80	0.5
Dense to very dense sand and gravel	>12	200	120	0.4

Table 6.45 Back-figured failure parameters from driven pile tests in Cowden Till

Test	f_L	$\delta_p\ (f_s)$	$\delta_p\ (\tau_{rz})$
CW1T	0.98	15.5	16.0
CW2C	0.88	24.0	18.0
CW3T	0.82	17.5	18.5
CW4C	1.30	25.5	22.0
CW4C/S	0.85	19.5	20.0

Source: After Lehane, B. M. and R. J. Jardine. *Canadian Geotechnical Journal*, 31(1); 1994: 79–90.

Table 6.46 Shear resistance during installation and during the load test after equalisation of installation pore pressures

				Installation		*Load testing*		
Test	*Shaft embedment (m)*	*Loading*	*Equalisation period (days)*	τ_{ave}	α	τ_{ave}	α	*Displacement (mm)*
CW1T	0.55–3.55	Tension	4	125	0.83	89	0.60	1.87
CW2C	2.70–6.35	Compression	4	108	0.86	104	0.82	2.42
CW3T	2.46–6.35	Tension	5	118	0.94	72	0.57	4.40
CW4C	2.46–5.97	Compression	0.08	112	0.87	62	0.49	1.55
CW4C/S	2.70–6.38	Compression	4	71	0.56	75	0.60	1.75

Source: After Lehane, B. M. and R. J. Jardine. *Canadian Geotechnical Journal*, 31(1); 1994: 79–90.

that during installation the adhesion factor, α, varied with installation speed; the faster the greater the resistance. The average α values during installation and load testing are given in Table 6.45 with an average value of 0.62, which compares with the Weltman and Healy (1978) value of 0.7. The interface friction angle during installation varied between 12° and 17°, but the mobilised angle was 24° during the CRP compression tests. CRP tension tests gave a value of 17° highlighting that the shaft resistance in tension and compression are different. They observed that the unit shear resistance was given by

$$\tau_f = f_L \sigma'_{r1} \tan\delta_p \tag{6.96}$$

where σ'_{r1} is the preloading radial effective stress (ICP method), f_L a coefficient related to $\sigma'_{rf}/\sigma'_{r1}$ and δ_p the peak mobilised interface friction (Table 6.46), which is based on typical angle of friction for Cowden Till.

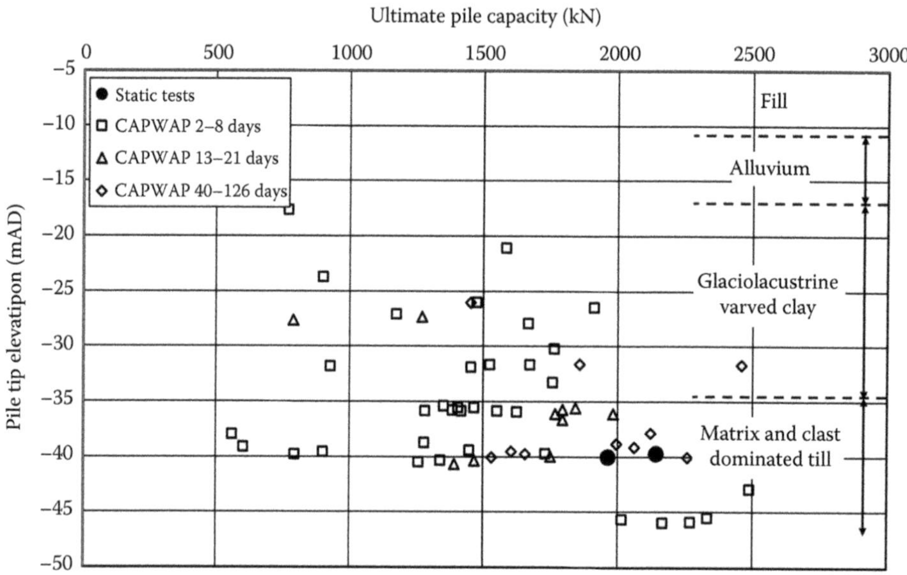

Figure 6.59 CAPWAP and static tests on 610-mm-diameter hollow steel pipes with 13 mm wall thickness driven through alluvium and glaciolacustrine clays into glacial till showing the capacity testing underestimated the static capacity unless the excess pore pressures generated during installation dissipated. (After Morgan, R. et al. *Ports 2013: Success through Diversification*, ASCE, 2013: 1038–1047.)

Morgan et al. (2013) described 610-mm-diameter, 39.5- to 41.3-m-long open-ended steel piles driven into glaciolacustrine deposits overlain by fill, organic silt and sand in Brooklyn, New York (Figure 6.59). Experience had shown that driving piles in glaciolacustrine soils remould the soil adjacent to the pile leading to reduced capacity but that increases with time. Pile testing included dynamic and static tests. The results showed that delaying the pile tests led to an increased capacity, which meant a reduction in pile length. However, this may not be possible because of the construction programme, which means that piles will be overdesigned. Wave equation analysis under-predicted capacity; dynamic testing gave predictions of capacity comparable with the results of static load tests. However, there was a significant variation in the results (Figure 6.60), which suggested that a balance between the number of tests and accepted capacity has to be based on engineering judgement though delaying the test led to more economic designs.

Strandgaard and Vandenbulcke (2002) described the issues of driving 3- and 4.5-m-diameter monopiles 18 and 25 m into glacial till to form foundations for wind turbines. Horizontal loads on wind turbine piled foundations are significant and are often the governing load because the axial capacity is often exceeded in order to provide the necessary lateral resistance. The length of the pile needed to provide that resistance is greater than that required for the axial load. Pile refusal was due to obstructions (boulders), the strength of soil and increase in lateral stress if pile driving was delayed. The pile drivability assessment was based on an evaluation of the resistance to driving (SRD) (Semple and Gemeinhardt, 1981; Stevens et al., 1982), wave equation analysis and the relation between blow count and penetration. Figure 6.61 shows the predicted and back-analysed SRD for piles at two sites in the North Sea justifying the use of the SRD method.

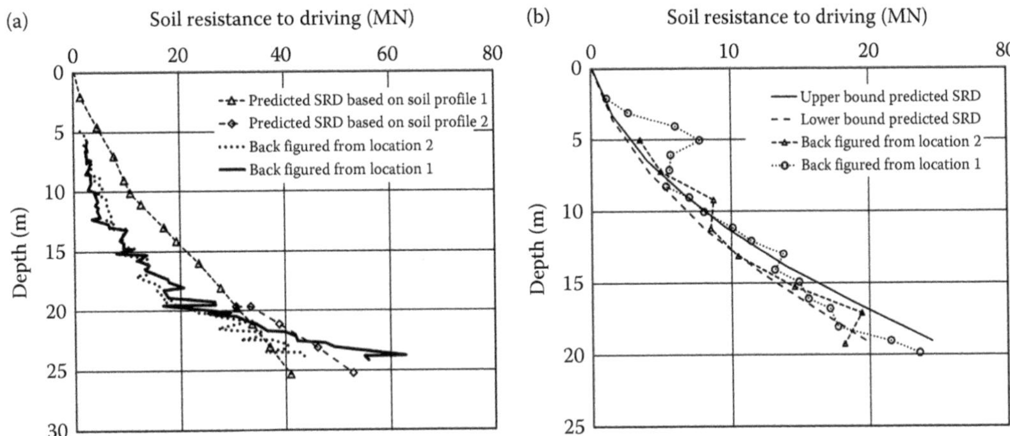

Figure 6.60 Predicted and back-analysed soil resistance to driving (SRD) for (a) Samso: 4.2-m-diameter monopiles with 45–65 mm wall thickness driven to the design depth of 25 m in a matrix-dominated till at Samso, offshore Denmark and (b) Utgrunden: 3-m-diameter monopoles with 45 mm wall thickness driven to the design depth of 19 m in glaciofluvial deposits at Utgrunden, offshore Sweden. (After Strandgaard, T. and L. Vandenbulcke. Driving mono-piles into glacial till. *IBC's Wind Power Europe*, 2002.)

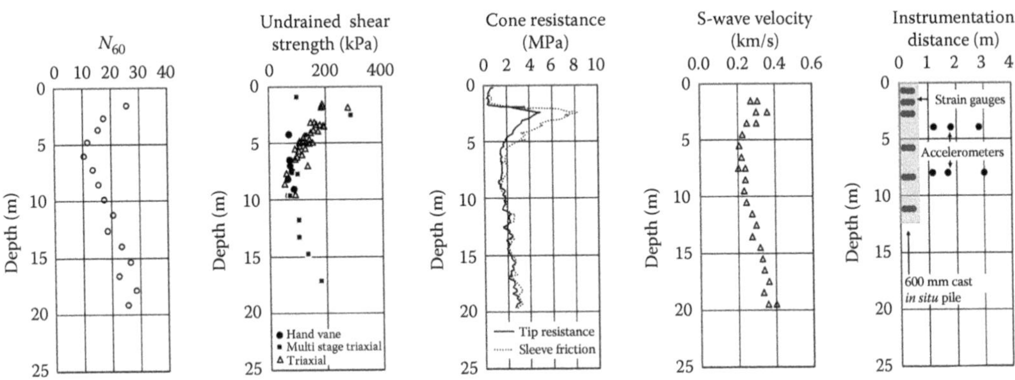

Figure 6.61 Details of an instrumented 600 mm nominal diameter bored cast *in situ* pile and properties of the matrix-dominated till in which it was installed. (After Brown, M. J., A. F. L. Hyde, and W. F. Anderson. *Geotechnique*, 56(9); 2006: 627–638.)

Zdravkovic et al. (2015) and Byrne et al. (2015) describe the analysis, design and testing of large diameter piles for offshore wind applications. Test piles were performed at Cowden. Table 6.47 summarises the Modified Cam Clay parameters, which were developed from the wide range of field and laboratory tests carried out over many years at the Cowden test bed site. They concluded that API *p*–*y* method of design was inappropriate.

Published pile tests in glacial tills show that pile capacities often exceed those predicted from guidelines. The pile tests included integrity and static loading tests, highlighting the benefits of pile testing in complex, composite glacial soils. The benefit of pile testing is an addition to the proof testing recommended in codes of practice, but the scale of the testing requires a more economic method of testing such as rapid pile load tests. Brown et al.

Table 6.47 Parameters for the Modified Cam Clay model of Cowden Till

Component	*Parameters*
Strength	$X = 0.548; Y = 0.698; Z = 0.1$
Hvorslev surface – shape	$\alpha = 0.25; n = 0.40$
Hvorslev surface – plastic potential	$\beta = 0.25; m = 0.40$
Virgin consolidation line	$V_l = 1.757; \lambda = 0.062$
Non-linear elasticity – bulk stiffness	$\kappa = 0.0124$
Non-linear elasticity – shear stiffness	$G_o = 80$ MPa; $p'_{ref} = 100$ kPa
Non-linear elasticity – shear stiffness degradation	$a = 9.78 \times 10^{-5}; b = 0.987; R_{min} = 0.05$

Source: After Zdravkovic, L. et al. *Third International Symposium on Frontiers in Offshore Geotechnics (ISFOG 2015)*, Oslo Norway, 2015.

(2006) used the Statnamic test, which accelerates a mass onto the top of a pile using combustion in a pressure chamber. This reduces the significant costs and time of undertaking static tests. The soil profile consisted of 2–4 m weathered clay overlying a firm to very stiff gravelly clay, shown in Figure 6.61. They measured the response of the ground and the pile using the instrumentation shown in Figure 6.61. Figure 6.62 shows the load settlement curves for rapid and constant rate of penetration tests and static load tests.

The ultimate shaft friction (Randolph and Deeks, 1992) is

$$\tau_d = \tau_s\left[1+\alpha\left(\frac{\Delta v}{v_o}\right)^{\beta} - \alpha\left(\frac{\Delta v_{min}}{v_0}\right)^{\beta}\right] \tag{6.97}$$

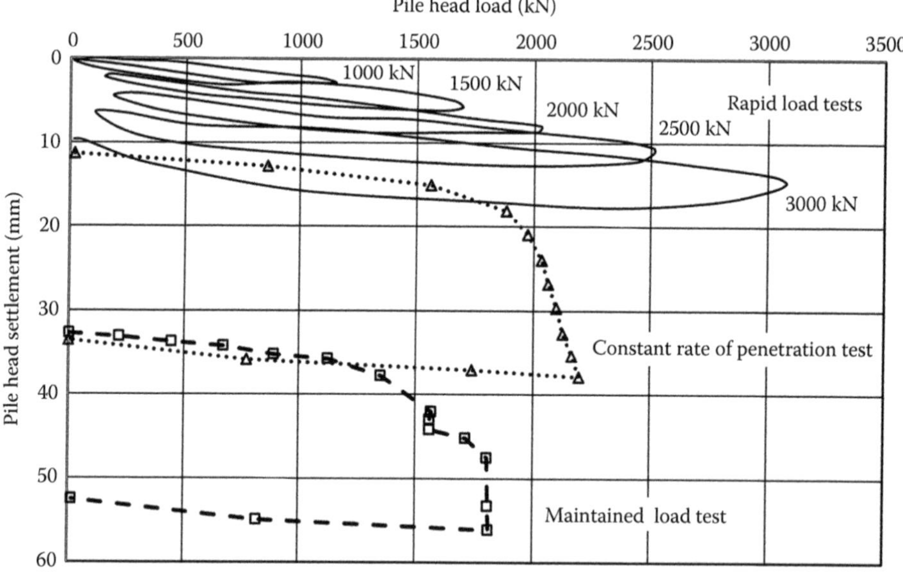

Figure 6.62 Load settlement curves for rapid and static load tests and constant rate of penetration tests for a 600 mm nominal diameter bored cast *in situ* pile in matrix-dominated tills. (After Brown, M. J., A. F. L. Hyde, and W. F. Anderson. *Geotechnique*, 56(9); 2006: 627–638.)

where τ_s is the shaft resistance at low penetration rate (static tests), Δv the relative pile/slip velocity, Δv_{min} the relative pile/slip velocity in a static test, v_o a reference velocity (=1 m/s) and α and β rate parameters. Brown (2004) proposed that the ultimate pile resistance, F_{static}, in fine-grained soils where most of the capacity is derived from shaft friction is given by

$$F_{static} = \frac{F_{STN} - (M\ddot{x})_{pile}}{1+\alpha(\Delta v/v_o)^{\beta} - \alpha(\Delta v_{min}/v_o)^{\beta}} \tag{6.98}$$

where M is the mass of the pile, F_{STN} the measured pile head load and $\ddot{x}$ the pile acceleration. Brown and Powell (2012) undertook rapid loading, maintained load and constant rate of penetration pile tests on CFA piles in glacial till at Cowden. Statnamic tests can be analysed using the ultimate point method (UPM) or non-linear velocity dependent methods. The UPM over-predicts capacity of piles in fine-grained soils but a correction factor based on the liquid limit (Figure 6.63) can be applied. Brown and Powell (2012) investigated two non-linear methods (Brown, 2004; Schmuker, 2005):

$$F_u = \frac{F_{STN} - Ma}{1+(F_{STN}/F_{STNpeak})\alpha(\Delta v/v_o)^{\beta} - (F_{STN}/F_{STNpeak})\alpha(v_{min}/v_o)^{\beta}} \tag{6.99}$$

where F_{STN} is the Statnamic load, Ma the pile inertia, Δv the pile velocity and v_{min} the velocity of the CRP pile test used to define the parameters α and β.

$$F_u = (F_{STN} - Ma)\left(\frac{0.02}{\Delta v}\right)^{I_{vcx}} \tag{6.100}$$

where I_{vcx} is the soil viscosity index (= −7 + 2.55ln(I_L)). The tests allowed revised estimates of μ and α to be made.

$$\mu = -0.0033 I_L + 0.69 \tag{6.101}$$

$$\alpha = 0.027 I_L + 0.25 \tag{6.102}$$

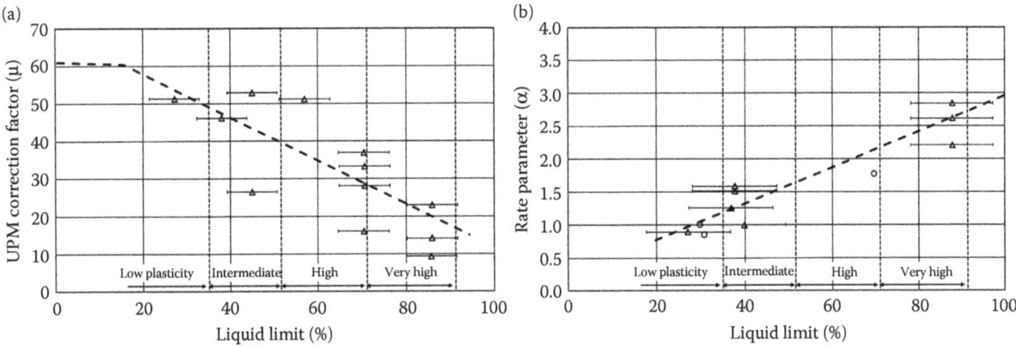

Figure 6.63 Correction factors, (a) μ and (b) α, to predict pile capacity from rapid loading tests. (After Weaver, T. J. and K. M. Rollins. *Journal of Geotechnical and Geoenvironmental Engineering*, 136(4); 2009: 643–646. Brown, M. J. and J. J. M. Powell. *Journal of Geotechnical and Geoenvironmental Engineering*, 139(1); 2012: 152–161.)

$$F_u = (F_{STN} - Ma)\left(\frac{0.02}{\Delta v}\right)^{[I_{v\alpha}(F_{STN}/F_{STNpeak})]} \quad (6.103)$$

Then concluded that it is important to ensure that the test load has enough capacity to ensure sufficient penetration. They recommended that the target load must be 70% greater than the predicted static capacity. They found that the Randolph and Deeks (1992) method predicted a capacity between static and constant rate of penetration tests.

6.6.9 Recommendations

The capacity of a pile is dependent on a soil's mechanical characteristics, the method of installation of the pile and the geometry of the pile. It can be estimated using semi-empirical formula and from dynamic impact tests, pile driving formulae and wave equation analysis or measured directly with load tests. Given the variability of glacial soils, pile tests should be undertaken as a matter of course. Preliminary pile tests can be carried out to verify design assumptions; working pile tests to verify workmanship and materials to reduce risk.

Glacial soils are complex soils and can vary both horizontally and vertically in properties and thickness, which means that the capacity of a test pile may not be the same as the working piles because the spatial variation within a site is considerable. A statistical approach based on the results of the quality and findings of the ground investigation should be used to decide on the number of pile tests.

A ground investigation has to be designed to fully assess the ground conditions. This may seem obvious and in line with codes of practice and guidelines, but the investigation of glacial soils has to be more thorough than they recommend. Therefore, it is recommended that

- A ground investigation has to establish the geotechnical, geological and hydrogeological conditions for both the short and long term.
  - The spacing and depth of boreholes has to ensure that the spatial variability can be assessed. This includes the variation in thickness and type of strata both vertically and horizontally to ensure that the effects of composition and fabric pile performance can be assessed. The depth of the boreholes has to extend below the base of the pile or pile group to pick up any weaker or water-bearing layers, especially in glacial tills. Boulders will influence the choice of pile. The bedrock, if relevant, has to be identified especially as glacial soils may contain rafted rock and the bedrock may not be planar. Water-bearing and weaker layers over the shaft length need to be identified and depth and thickness established across the site.
  - Sufficient representative *in situ* tests and samples have to be taken to ensure that characteristic properties of stiffness, strength and permeability can be determined. This is especially important in glacial tills and glaciofluvial soils where it may be difficult to develop property profiles due to the natural variability of the soils; scatter in the data is inevitable.
  - Effective and undrained strengths of glacial clays are required as the ULS may be governed by the effective strength, but the undrained shear strength may be critical in the short term. The undrained shear strength could affect the choice of pile.
  - The fabric of glacial clays must be carefully assessed because it could give an indication of mass behaviour as opposed to the intact behaviour observed in small specimens.
  - Stiffness is required for the SLS. Given the importance of the composition, structure and fabric, enough tests must be undertaken to obtain representative values.

- Seasonal groundwater conditions should be assessed using regional and site-specific data.
- The ultimate and servicability limit states can be assessed using a number of semi-theoretical and numerical methods provided the appropriate parameters are selected and the correct adjustment factors are used.
  - The ULS of piled foundations in fine-grained soils should be assessed using undrained and effective strength parameters. Account must be taken of the fabric when selecting the appropriate values. A cautious estimate should be used to allow for scatter due to sampling and testing unless a regional database is available when more realistic values can be used.
  - The analysis should take account of the possible variation in stratum thickness. This means that a scenario analysis should be undertaken.
  - Settlement predictions are unreliable if a single value of stiffness is used because of the natural variability of the soils. Therefore, it is important not only to select the appropriate stiffness model to allow for stress level but also to take into account the vertical and horizontal variation in stiffness. This requires a scenario analysis.

6.7 RETAINING STRUCTURES

There are many types of retaining structures (Table 6.48) including gravity wall, embedded walls and composite retaining structures. Gravity walls include precast reinforced concrete stem walls, masonry walls, crib walls and gabion walls. These walls are designed against sliding and overturning due to the active pressures acting on the wall. The active pressures are due to the fill placed behind the wall. Embedded walls include sheet pile walls, king post walls, contiguous bored pile walls, secant piled walls and diaphragm walls.

The issues to address in glacial soils include effects of installation, earth pressures and groundwater pressure:

- Sheet piles cannot be driven into glacial tills or glaciofluvial sands and glaciolacustrine clays containing a significant number of boulders.
- Declutching of sheet piles can occur in coarse-grained soils.
- Contiguous and secant piled walls are created using bored pile or CFA rigs. Therefore, the issues of installation associated with these types of piles apply to these walls.
- Contiguous bored piles do not, on their own, provide water retention because of the gaps between the piles which can be an issue in any type of glacial soil because of layers of water-bearing sands and gravels in matrix-dominated tills, clast-dominated tills, glaciofluvial soils and the horizontal permeability of lacustrine deposits. Secant pile walls can be used provided the female piles are either full strength concrete or a mix of at least 10 N/mm^2. The female piles are the piles that are installed first; the male piles interlock with the female piles creating an impermeable boundary.
- Diaphragm walls can be used in any ground conditions.
- Groundwater conditions in glacial tills are complex because of the effect of water-bearing layers in matrix-dominated tills, the increase in permeability due to unloading by excavation causing discontinuities to open up, and matrix-dominated till acting as an aquiclude.

Sheet piles can be used for temporary and permanent structures, water-retaining structures and to prevent water entering excavations.

Table 6.48 Characteristics of retaining walls

Wall type	*Advantages*	*Disadvantages*
Gravity walls – general	Minimises construction activities on site by using precast or modular systems Cost-effective retaining wall solution for supporting raised ground levels	Requires site dewatering for construction below the water table Requires temporary open-cut excavation for retaining wall construction below existing ground level
Precast reinforced concrete stem walls	Provides a wall with a predictable surface finish	Height of wall may be limited practically by transporting and lifting requirements 1–3 m typical retained height Changes in retained height or plan alignment need to be planned in detail for prefabrication of units Non-draining, appropriate drainage required
Masonry walls	Can provide a wall with a predictable surface finish Can be installed around obstructions at isolated points Can be built to a batter	Construction requires bricklaying skills Foundation slab required Non-draining, appropriate drainage required
Dry-stack masonry walls	Simple manual construction Distinctive course effect May be used with planting Can produce curves in plan	Foundation slab required Non-draining, appropriate drainage required
Crib walls	Simple manual construction Distinctive 'honeycomb' appearance May be used with planting Can produce curves in plan	Foundation slab required Free draining although back drain may also be required Backfill in compacted layers
Gabion walls	Simple construction Foundation slab may not be required Fully draining, no back drain required unless they are filled with non-draining materials Can be used with planting Can produce curves in plan	A flexible system which can result in an undulating wall profile
Embedded retaining walls – general	Minimise volume of excavation Enable deep excavation adjacent to existing structures and utilities Enable deep excavation below the water table (water-retaining embedded walls only)	Specialised construction plant and operation on site Costly in comparison with open-cut construction methods
Sheet pile walls	Provides an economic embedded wall with a predictable surface finish No arisings to be removed Suitable as a water-retaining wall Can be used as both a temporary and a permanent wall	Maximum pile length approximately 30 m Potential declutching in coarse-grained soils
King post walls	Can be installed around obstructions at isolated points	Not suitable for retaining water in the long term Cannot be used for excavation below the groundwater table in coarse-grained soils
Contiguous bored pile walls	The cheapest form of concrete piled wall	Not a water-retaining solution Not a permanent solution in any soil due to the gaps between piles, unless a structural facing is applied

(*Continued*)

Table 6.48 (Continued) Characteristics of retaining walls

Wall type	*Advantages*	*Disadvantages*
Hard/soft secant bored pile walls	Acts as a water-retaining temporary wall The use of soft piles enables hard piles to be formed using lower-torque rigs than for hard/hard secant piles	Not usually a permanent solution for retaining water The soft pile mix is not significantly cheaper than concrete Local concrete plant is often unable to batch the soft material, so site batching is required Depth is limited by the verticality tolerance, which may determine the depth of secanting
Hard/firm secant bored pile walls	Permanent water-retaining wall The firm material for the primary (female) piles is either a standard concrete mix, retarded to reduce its strength when the secondary (male) piles are constructed or a reduced strength concrete mix	Depth is limited by the verticality tolerance, which may determine the depth of secanting
Hard/hard secant bored pile walls	A permanent water-retaining wall Installed using standard piling plant with high-torque rigs	The cutting of the hard primary (female) piles requires high-torque rigs or oscillators Depth is limited by the verticality tolerance, which may determine the depth of secanting
Diaphragm walls	A permanent water-retaining wall Can be installed to great depths provided the verticality tolerances can be accepted In some circumstances the face of the diaphragm wall can form the final finish subject to some surface cleaning and removal of protuberances Fewer joints compared with piled walls	Horizontal continuity is difficult to achieve between panels Cannot follow intricate plan outlines The installation equipment is extensive, requiring a large site area for accommodation of the support fluid plant, reinforcement cages and the excavation plant Disposal of the support fluid is costly
Hybrid walls	Provide a practical retaining wall solution where the site conditions will not allow a gravity wall but do not require an embedded retaining wall	Combination of different wall and foundation elements results in site- and solution-specific wall type

Source: After Gaba AR, B. Simpson, W. Powrie, and D. R. Beadman, Embedded Retaining Walls – Guidance for Economic Design. CIRIA, London, UK, Report No. C580; 2003. Chapman, T., H. Taylor, and D. Nicholson. *Modular Gravity Retaining Walls: Design Guidance*. Publication C516, CIRIA, London, 2000; Anderson, S. *ICE Manual of Geotechnical Engineering*, Thomas Telford Ltd, London; 2012: 959–968.

Potential failure mechanisms (Figure 6.64) include loss of overall stability, failure of a structural element such as a wall, anchor, wale or strut or failure of the connection between such elements, combined failure in the ground and in the structural element or failure by hydraulic heave and piping. Unacceptable movements of the structure, seepage through the structure or loss of soil from behind the structure have to be dealt with. In addition, for gravity walls it is necessary to consider bearing failure, failure by sliding at the base and failure by toppling; and for embedded walls, failure by rotation or translation of the wall and failure by lack of vertical equilibrium.

Retaining structures can be subject to the weight of backfill material, surcharges, weight of water, wave and ice forces, seepage forces, collision forces and temperature effects.

An allowance has to be made for excavation or scour in front of the wall, which, in the case of embedded walls, is 10% of the retained height above excavation level, limited to a maximum of 0.5 m and for a propped or anchored wall, 10% of the distance between the lowest support and the excavation level, limited to a maximum of 0.5 m. The water levels shall be based on the hydraulic and hydrogeological observations at the site and account

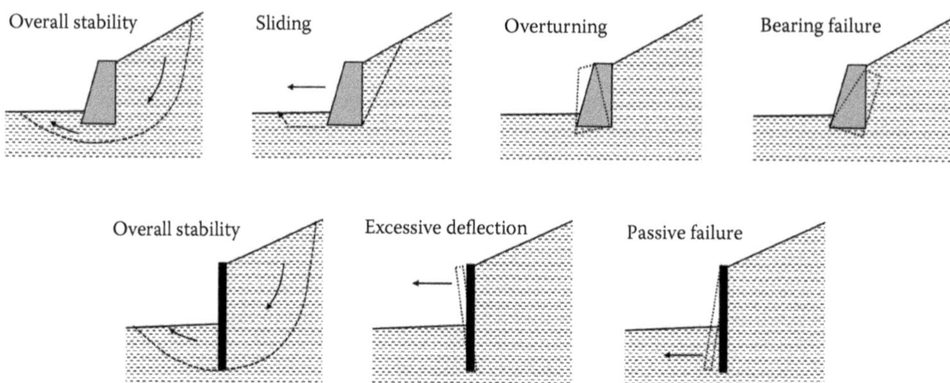

Figure 6.64 Possible geotechnical failure modes for embedded and gravity walls.

shall be taken of the effects of variation in permeability on the groundwater regime and the possibility of adverse water pressures due to the presence of perched or artesian water tables.

The design of retaining structures (e.g. Figure 6.65) shall take account of the effects of constructing the wall, including temporary support, changes of *in situ* stresses and resulting ground movements caused by the excavation for the wall and its construction, disturbance of the ground due to driving or boring operations, the required degree of water tightness of

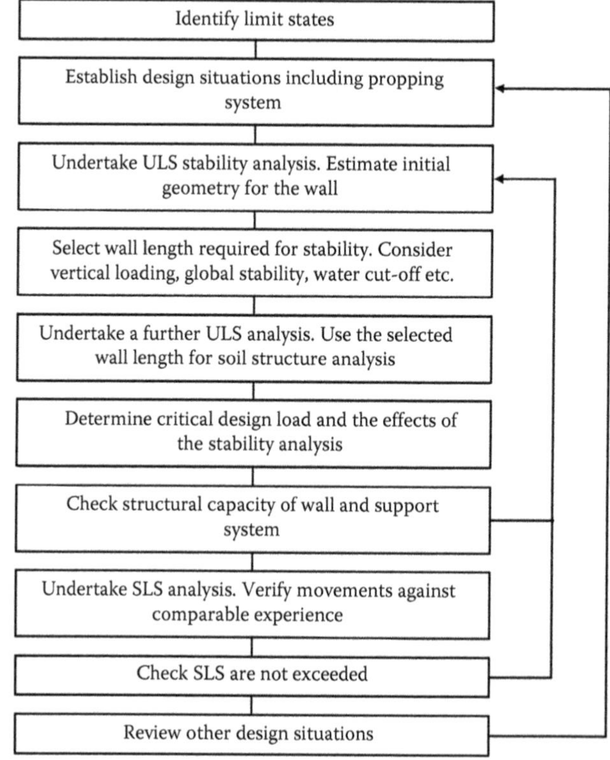

Figure 6.65 Embedded wall design process. (After Pickles, A. *ICE Manual of Geotechnical Engineering*, edited Thomas Telford Ltd, London; 2012: 981–999.)

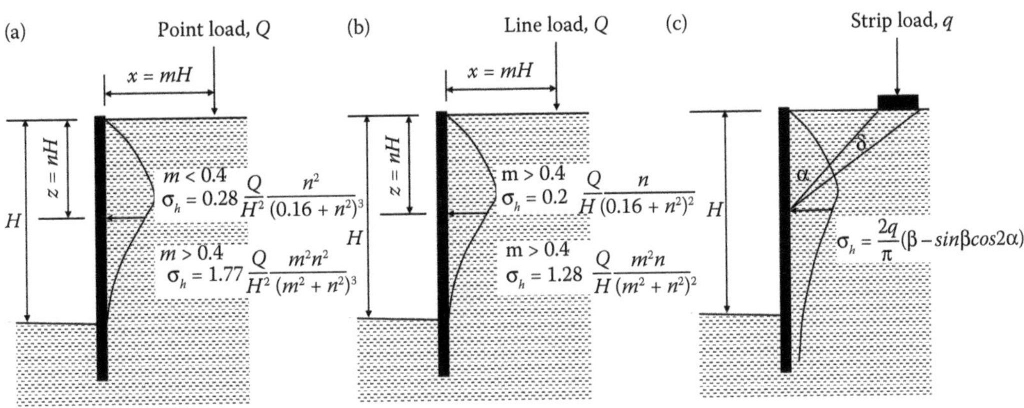

Figure 6.66 Lateral pressure due to (a) a surface point load, (b) a surface line load and (c) a surface strip load on the active side of a retaining wall assuming a uniform linear elastic soil modified by experimental observations.

the finished wall, the practicability of constructing the wall to reach a stratum of low permeability, so forming a water cut-off, the practicability of excavating between any propping of retaining walls, and the ability of the wall to carry vertical load; for sheet piling, the need for a section stiff enough to be driven to the design penetration without loss of interlock, and the stability of borings or slurry trench panels for a diaphragm wall while they are open; and for fill, the nature of materials available and the means used to compact them adjacent to the wall.

The design has to cover earth pressures due to soil and water, the slope of the ground surface, the inclination of the wall to the vertical, water levels and the seepage forces in the ground, the amount and direction of the movement of the wall relative to the ground, the horizontal as well as vertical equilibrium for the entire retaining structure, the shear strength and density of the ground, the rigidity of the wall and its support system and the wall roughness. Loads acting on the active side are included in the analyses using the relationships shown in Figure 6.66. The deflection of the wall creates friction between the soil and the wall such that it increases the passive resistance and reduces the active action. It is a function of the strength of the ground, the wall/soil interface friction, the direction and amount of movement of the wall relative to the ground and the ability of the wall to support any vertical forces resulting from wall friction and adhesion.

6.7.1 Earth pressures

The earth pressures depend on the magnitude of wall movement; they reduce as the wall deflects. Thus, the earth pressures for the SLS are different from those at the ULS. If the wall is very stiff, the earth pressures are a function of the at-rest pressure defined in terms of K_o, which, for a horizontal ground surface, is given by

$$K_o = (1 - \sin\phi)\sqrt{OCR} \qquad (6.104)$$

If the ground slopes upwards from the wall at an angle $\beta \leq \phi'$ to the horizontal, then the horizontal force on the wall is given by

$$K_{o;\beta} = K_o(1 + \sin\beta) \qquad (6.105)$$

The active pressure, $\sigma_{a(z)}$, is

$$\sigma_{a(z)} = \left[K_a \int \gamma\, dz + q - u\right] + u - c' K_{ac} \tag{6.106}$$

where z is the depth below the top of the wall, q is a uniformly distributed surcharge and $K_{ac} = 2\sqrt{[K_a(1+a/c')]}$ $\left(max = 2.56\sqrt{K_a}\right)$. The passive pressure, $\sigma_{p(z)}$,

$$\sigma_{p(z)} = \left[K_p \int \gamma\, dz + q - u\right] + u + c' K_{pc} \tag{6.107}$$

where $K_{pc} = 2\sqrt{[K_p(1+a/c')]}$ $\left(max = 2.56\sqrt{K_p}\right)$. Values of K_a and K_p can be taken from charts published in Eurocode or, for a more complete range, in Caquot and Kérisel (1948).

The earth pressures at the SLS can be estimated from Tables 6.49 and 6.50.

According to Powrie (2012), the earth pressures obtained from limit equilibrium analysis using factored strengths are unlikely to correspond to those *in situ*, which could overestimate the bending moments because

- The relationship between wall movement and shear strain is different on either side of the wall.
- The stress paths followed either side of the wall are different.
- The relative flexibility of the wall and the relative movement of the wall depth varies over the depth of the wall.

This is further complicated if props exist making the use of numerical methods to analyse the soil structure interaction necessary. However, the issues of selecting the correct parameters and profile still apply.

Wall adhesion is mobilised on the wall but it depends on the wall movement. The pressures on a wall are different in service and at failure. The interface friction, δ, varies with the structural material; for concrete or steel sheet pile walls supporting sand or gravel, $\delta_d = k\phi_{cv;d}$

Table 6.49 Ratio of maximum deflection, v_a, at the top of the wall and the height of supported wall, h, at full active conditions for coarse-grained soils; intermediate value can be assessed assuming a linear relation between no movement at K_o conditions

Kind of wall movement		*v_a/h for loose soil (%)*	*v_a/h for dense soil (%)*
Rotation about toe		0.4–0.5	0.1–0.2
Translational		0.2	0.05–0.1
Rotation about top		0.8–1.0	0.2–0.5
Bowing		0.4–0.5	0.1–0.2

Source: After BS EN 1997-1:2004+A1:2013. *Eurocode 7: Geotechnical Design – Part 1: General Rules*. British Standards Institution, London.

Table 6.50 Ratio of maximum deflection, v_p, at the top of the wall and the height of supported wall at full active conditions for coarse-grained soil

Kind of wall movement	v_p/h (v/h for $0.5\sigma_p$) for loose soil (%)	v_p/h (v/h for $0.5\sigma_p$) for dense soil (%)
Rotation about toe	7 (1.5)–25 (4)	5 (1.1)–10 (2.0)
Translational	5 (0.9)–10 (1.5)	3 (0.5)–6 (1.0)
Rotation about top	6 (1.0)–15 (1.5)	5 (0.5)–6 (1.3)

Source: After BS EN 1997-1:2004+A1:2013. *Eurocode 7: Geotechnical Design – Part 1: General Rules*. British Standards Institution, London.

Note: v is the wall displacement; v_p is the displacement to mobilise full passive pressures.

with $k \leq 2/3$; if concrete is cast against soil, $k = 1$; and immediately after driving a steel sheet pile in clay, no adhesive or frictional resistance should be assumed though it will increase with time.

The construction sequence affects the earth and water pressures acting on the wall in the short term and long term. On the active side, the earth pressures can vary from K_o conditions to fully active conditions depending on the wall stiffness, the presence of props or anchors and the depth below the top of the wall. This can be taken into account using numerical methods and stage construction.

Drained conditions should be assumed in coarse-grained soils including glaciofluvial soils and clast-dominated tills and layers of sands and gravels in matrix-dominated tills and possibly glaciolacustrine clays if the mass horizontal coefficient of permeability exceeds 10^{-8} m/s. Matrix-dominated tills often contain discontinuities, so if the mass permeability as opposed to the intact permeability is more than 10^{-8} m/s, then drained conditions apply.

The characteristic design value should be a cautious estimate of the strength of the soil. In matrix-dominated tills, account should be taken of the effect of discontinuities on the strength and the possibility of weaker layers in the till.

Constructing retaining walls leads to ground movement due to movement of the wall, reduction in stress if soil is excavated to create the wall, heave due to excavation in front of the wall and groundwater movement. Ground movement can be estimated using databases of existing structures, pseudo-finite element models (e.g. FREW, WALLAP) and numerical methods (e.g. Plaxis, FLAC, SAFE). Design has to take into account interface friction, passive softening in front of the wall, tension cracks, sloping ground, reverse passive pressures due to props or anchors, thermal effects, surcharges and groundwater.

Eurocode 7 recommends that water pressures should normally be expected in clays and silts unless a reliable drainage system is installed, or infiltration is prevented, or the values of water pressure correspond to a water table at the surface of the retained material. Groundwater pressure should represent the most unfavourable conditions taking into account the hydrogeology, the source of water, the drainage and climate and seasonal changes.

The groundwater conditions that exist depend on the groundwater level(s) and the mass permeability of the ground, the time since construction and the type of wall.

- 'Impermeable' walls including sheet pile walls, secant piles walls, diaphragm walls and reinforced concrete mass gravity walls. These walls set up a flow regime if there is a potential difference between the active and passive side of the wall causing water to flow beneath the wall unless drainage is installed.
- 'Permeable' walls including masonry walls, gabion walls, reinforced earth walls and contiguous bored pile walls. These walls allow water to flow through the wall unless an impermeable facing is placed in front of the wall. In that case, flow will occur beneath the wall in the case of gravity walls and through the contiguous bored piles below excavation level.

In glaciofluvial soils and clast-dominated tills, long-term conditions are established during construction. A flow net can be constructed and pore pressures calculated though a simple model is often used in which the water pressure at the base of the wall is assumed to be balanced or equal.

In matrix-dominated tills and glaciolacustrine clays, steady-state conditions exist sometime after construction so that the simple model applied to coarse-grained soils can be used. In the short term, an out-of-balance water pressure exists at the base of the wall. According to Fleming et al. (2008), the assumption of balanced water pressure at the base of an embedded wall has little effect on the depth of embedment and bending moment if the difference in water levels on the active and passive side of the walls is less than 4 m. If the difference exceeds 4 m, then applying a balanced water pressure leads to a deeper wall and increased bending moment.

Groundwater profiles in glacial soils are complicated because the mass permeability can vary due to the presence of discontinuities and the inclusions of soils of different permeability. Matrix-dominated tills can act as an aquiclude, which means that the regional groundwater level exists in the underlying soil/rock and there is a perched water level above the top of the till due to infiltration. Discontinuities can open up due to wall movement on the active side increasing the permeability. They can also open up on the passive side due to the excavation.

6.7.2 Limit states

The most adverse design values for the strength and resistance of the ground are used. Short- and long-term behaviour are considered for fine-grained soils. If there are differential water pressures, safety against failure due to hydraulic heave and piping shall be checked.

The distortion and displacement of retaining walls, and the effects on supported structures and services, is based on comparable experience, which includes the effects of construction of the wall. More detailed calculations are required if

- The wall is adjacent to structures that are sensitive to movement, which includes utilities.
- There is no or little comparable experience.
- The wall retains more than 6 m of cohesive soil of low plasticity such as matrix-dominated tills.
- The wall retains more than 3 m of soils of high plasticity.

- The wall is supported by soft clay within its height or beneath its base a possibility in matrix-dominated tills.
- The ground and groundwater profile are complex, a likely situation for glacial soils.

Displacement calculations, if they are to be of any use, must take account of the stiffness of the ground and structural elements and the sequence of construction.

Walls formed of 500- to 600-mm contiguous bored piles can be designed as cantilevers if the retained height is less than 5 m, and up to 10 m in stiff/dense soils but with pile diameters of at least 1.2 m (Fleming et al., 2008). Cantilever walls are designed as fixed earth walls in which the depth of embedment is increased beyond the point of fixity, the point about which the wall rotates (Figure 6.67). An embedded wall with a single prop or anchor is designed as a free earth wall with the point of rotation about the prop or anchor (Figure 6.67). Walls with more than one prop or anchor are more complex and can be analysed using numerical methods (e.g. FREW, WALLAP, FLAC, PLAXIS).

There are very few published records of performance of retaining walls in glacial soils. The governing criteria may be wall deformation, assuming that the overall and structural stabilities are satisfactory. Observations of wall deflection may not be made, so the actual performance is unknown. The wall design may be accepted on deflections predicted from a numerical analysis, which is very dependent on using appropriate representative stiffness. The difficulty in obtaining representative samples and carrying out appropriate tests means that stiffness may be selected from published data. Powrie and Li (1991) found, using a numerical analysis with a Cam Clay constitutive model and a Hvorslev surface (Table 6.51), that the deformation of a very stiff cantilever wall was governed by the soil stiffness rather than the wall stiffness and the bending moments are strongly influenced by the *in situ* horizontal stress. The wall, propped at formation level, formed a 9-m-deep cutting in matrix-dominated tills.

A recommended maximum deflection is 0.4% of the retained height (Gaba et al., 2003). Long (2001) analysed a significant number of retaining walls in stiff clay to show that, on average, the maximum deflection was 0.18% of the retained height. Long et al. (2012)

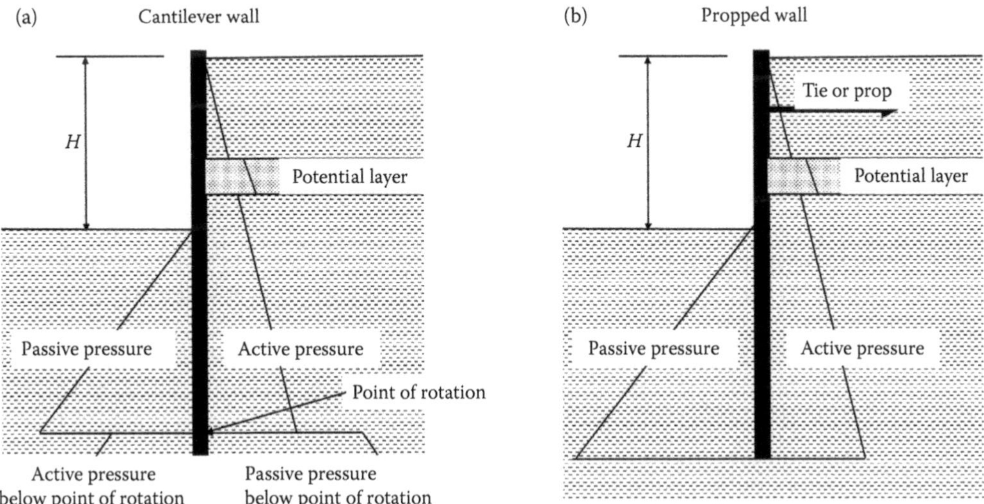

Figure 6.67 Soil pressure profiles for (a) cantilever and (b) propped embedded walls showing the effect of a layer within a glacial till stratum on the pressure distribution and the pressure distribution beneath the point of rotation, which could be affected by possible weaker layers.

Table 6.51 Soil parameters used to model a matrix-dominated till

Slope of one-dimensional compression line in $v - \ln p'$ space	$\lambda = 0.155$
Slope of unload/reload line in $v - \ln p'$ space	$\kappa = 0.016$
Specific volume (and void ratio) on critical state line at $p' = 1$ kPa	$\Gamma = 2.41$ ($e_o = 1.41$)
Slope of critical state line in q–p' space	$M = 1.03$
Poisson's ratio	$\upsilon = 0.2$
Unit weight of water	$\gamma_w = 9.81$ kN/m^3
Bulk unit weight of soil	$\gamma = 22.0$ kN/m^3
Permeability in vertical direction	$k_v = 10^{-10}$ m/s
Permeability in horizontal direction	$k_h = 10^{-10}$ m/s
Angle of Hvorslev surface in τ–σ' space	$\varphi_H = 15.5''$
Slope of no-tension cut-off in q–p' space	$S = 2.0$
Permeability in vertical direction for tensile fracture region	$k_x = 10^{-6}$ m/s
Permeability in horizontal direction for tensile fracture region	$k_y = 10^{-6}$ m/s

Source: After Powrie, W. and E. S. F. Li. *Geotechnique*, 41(4); 1991: 499–51.

found that the average value for retaining structures in Dublin Boulder Clay was 0.08% (Figure 6.68). Cantilever walls up to 7.5 m that retained height in Dublin Boulder Clay, shown in Figure 6.69, were found to have deflected by 0.13%. They observed that the deflection of cantilever walls was dependent on the excavation depth and system stiffness. Many of these walls were part of permanent structures, so the deflections were short term but Long et al. (2003) concluded that the low permeability, high strength and stiffness and the slow dissipation of excess pore pressures led to this very stiff behaviour. O'Leary et al. (2015) took the opportunity to study retaining walls in Dublin Boulder Clay that had stood for some 7 years. They found that movements continued after the end of the construction period which they attributed to dissipation of pore pressures. They found that walls designed as temporary structures were overdesigned for their purpose but may fail to meet the limiting criteria for a long-term structure. Walls designed as permanent structures performed better than predicted leading to the conclusion that retaining walls in Dublin Boulder Clay are overdesigned.

6.7.3 Recommendations

As with all geotechnical projects, the quality of information is critical to the success of the project. In the case of retaining structures, the key information is the following:

- Groundwater is critical because the impact of changes in pore pressure can reduce the effective stress and, therefore, mobilise strength and increase the water pressure acting on the wall.
  - It is important to establish the regional hydrogeological conditions as well as the site-specific groundwater conditions. Care should be taken to measure the pore pressure throughout the profile of the glacial deposit to assess changes in pore pressure and their link to weather-related events.
  - The mass conductivity of soils may not be the same as the intact conductivity; the deposit may be highly anisotropic; there could be water-bearing layers of sands and gravels; and a till may include discontinuities. It is important to determine the vertical and horizontal variation in the hydrological properties and stratum.

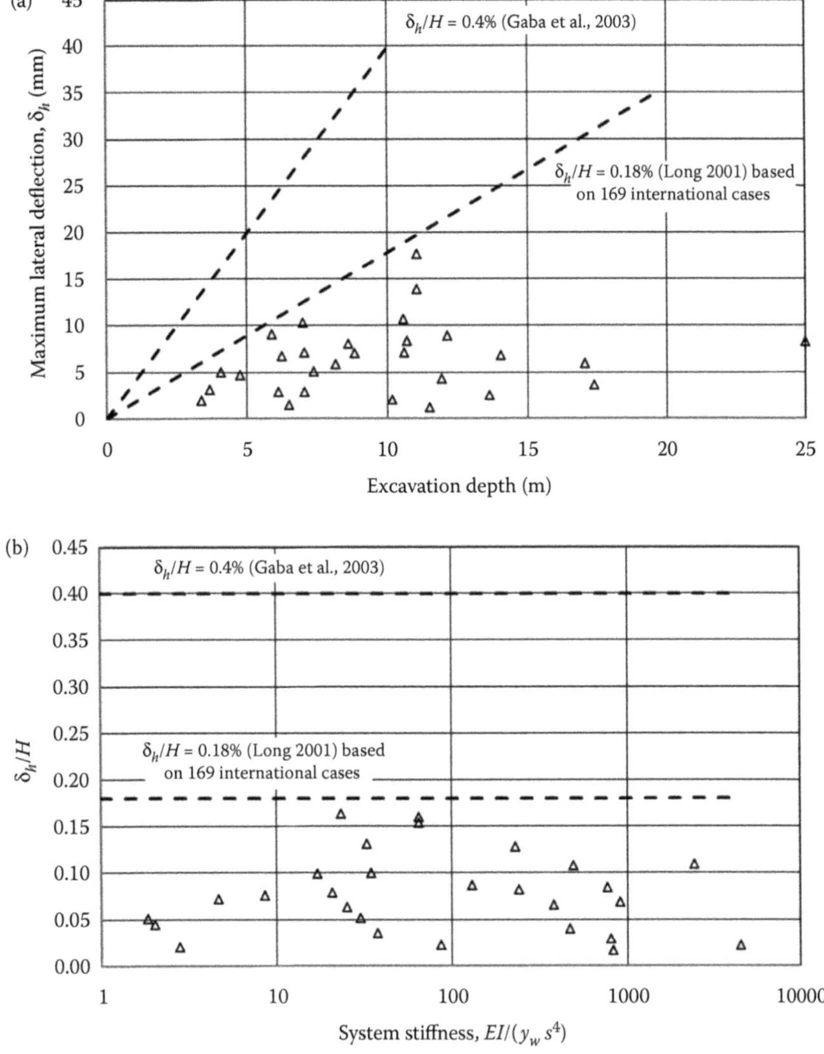

Figure 6.68 Deflection of propped walls in glacial tills showing (a) the maximum lateral movement plotted against excavation depth and (b) the normalised maximum movement plotted against system stiffness comparing the deflection to acceptable values and case studies. (After Long, M. et al. *Proceedings of the Institution of Civil Engineers-Geotechnical Engineering*, 165(4); 2012: 247–266.)

  - Excavations for embedded walls could create negative pore pressures within fine-grained soils, which will dissipate with time leading to increased active pressure. Therefore, it is necessary to consider short- and long-term conditions.
- The choice of embedded wall depends on whether it can be installed and whether it can be structurally capable of resisting the earth pressures.
  - It is important to establish the profile of the bedrock surface especially beneath glacial soils if it exists within the possible embedded depth.
  - Glacial soils can contain boulders which may prevent sheet piles or steel tubular piles being driven to depth or secant or contiguous piles being drilled to depth without specialist equipment.

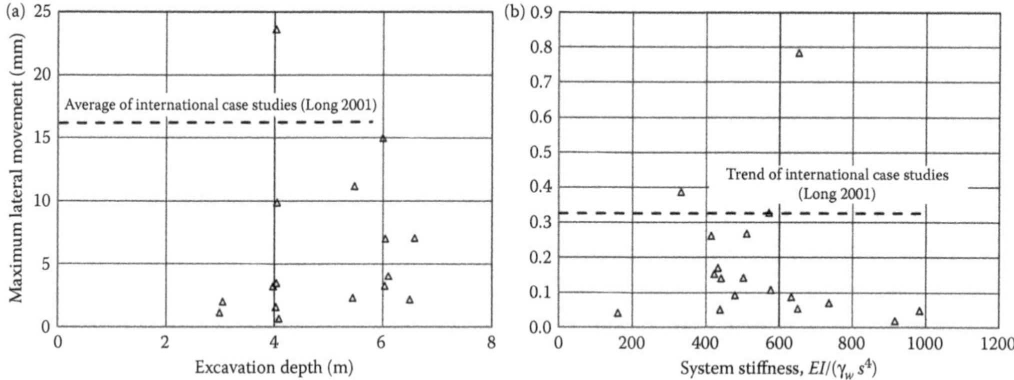

Figure 6.69 Deflection of cantilever walls in glacial tills showing (a) the maximum lateral movement plotted against excavation depth and (b) the normalised maximum movement plotted against system stiffness comparing the deflection to international case studies. (After Long, M. et al. *Proceedings of the Institution of Civil Engineers-Geotechnical Engineering*, 165(4); 2012: 247–266.)

  - The soils, especially subglacial tills, may be so dense that it proves too difficult to drive piles into the ground.
- The failure criterion for embedded walls includes overall stability, structural stability and deformation. Hence, it is necessary to determine the appropriate soil properties, which include the effective strength parameters and stiffness.
  - Sufficient samples have to be taken to determine the representative strength. The effective strength is best obtained from stress path plots combining the results of several tests to ensure that a representative angle of friction is determined. Matrix-dominated tills may exhibit cohesion, a consequence of their density. Ignoring cohesion could lead to overdesign since the cohesion of glacial soils does not necessarily reduce with time.
  - The stiffness of glacial tills and glaciofluvial soils are difficult to assess, but if wall deflections are to be predicted with any accuracy, then every effort must be made to use appropriate techniques to obtain representative values. It is likely that a cautious estimate will be used. The mobilised stiffness for stability calculations will be different from that used for servicability limit states.
  - Analysing embedded walls in glacial tills should take into account the variation in thickness of soil strata; that is, a scenario analysis should be undertaken to investigate the effect of stratum thickness and soil stiffness upon the wall stability and serviceability.

6.8 ANCHORS

Anchors are designed to support retaining structures, ensure stability of slopes, cuts or tunnels, resist uplift forces on a structure, prevent sliding or tilting or restrain tension cables. According to Eurocode 7, the limit states for all anchors are as follows:

- Structural failure of the tendon or anchor head
- Failure of the connection between the tendon and the resisting element in the ground
- Loss of anchor force and excessive displacements of the anchor head due to creep and relaxation

- Failure or excessive deformation of parts of the anchored structure due to the applied anchor force
- Loss of overall stability of the retained ground and the retaining structure
- Limit states in supported or adjacent structures, including those arising from pre-stressing forces
- Instability or excessive deformation of the zone of ground into which tensile forces from a group of anchors are to be transferred
- Failure at the interface between the resisting element and the ground

There are additional limit states for grouted anchors:

- Failure at the interface between the body of grout and the ground
- Failure of the bond at the interfaces of tendon, encapsulation and grout
- Failure of the bond between the tendon and the grout

The ULS resistance, $R_{ULS;d}$, of an anchor is

$$R_{ULS;d} \geq max(F_{ULS;d}F_{Serv;d}) \tag{6.108}$$

where $F_{ULS;d}$ is the design value of the force required to prevent any ULS in the supported structure and $F_{Serv;d}$ is

$$F_{Serv;d} = \gamma_{Serv}F_{Serv;k} \tag{6.109}$$

where the partial factor, γ_{Serv}, is typically 1.35 and $F_{Serv;k}$ the characteristic value of the maximum anchor force.

The SLS of an anchor is given by

$$F_{Serv;k} \leq R_{SLS;d} \tag{6.110}$$

The measured ULS resistance of an anchor, $R_{ULS;m}$, shall be determined by load tests as the lesser of the proof load or the load causing a limiting condition (R_m). The limiting condition depends on the test method and may be the following:

- The asymptote to the creep rate versus load curve
- The load corresponding to a limit value of the creep rate (α_{ULS})
- The load corresponding to a limit value of load loss ($k_{l;ULS}$)

$$R_{ULS;m} \leq min(R_m(\alpha_{ULS} \text{ or } k_{l;ULS}) \text{ and } P_p) \tag{6.111}$$

Recommended values for persistent and transient situations are given in Table 6.52.

The characteristic value of the ULS geotechnical resistance of an anchor, $R_{ULS;k}$, shall be derived from

$$R_{ULS;k} = \frac{R_{ULS;m}}{\xi_{ULS}} \tag{6.112}$$

where ξ_{ULS} is a factor given in Table 6.53.

Table 6.52 Limiting criteria for investigation, suitability and acceptance tests for persistent and transient design situations at the ultimate and serviceability limit states

Test method	*Limiting criterion*	*Investigation and suitability tests*		*Acceptance tests*	
		ULS	*SLS*	*ULS*	*SLS*
1	α_1	2 mm	0.01Δe/NA	2 mm	0.01Δe/NA
2	k_1	2%/log cycle of time	2%/log cycle of time	2%/log cycle of time	2%/log cycle of time
3	α_3	5 mm	P_c	NA	1.5 mm

Source: After BS EN 1997-1:2004+A1:2013. *Eurocode 7: Geotechnical Design – Part 1: General Rules*. British Standards Institution, London.

Note: $\Delta e = \dfrac{F_{serv;k} \times \text{tendon free length}}{\text{area of tendon} \times \text{elastic modulus of tendon}}$

Table 6.53 Design factors for permanent anchors for persistent and transient design situations at the ultimate and serviceability limit states

Symbol	*Test method 1*	*2*	*3*
ξ_{ULS}	1	1	1
$\gamma_{a;SLS}$	NA	1	1.2
n	3	3	2
$\gamma_{a;acc;ULS}$	1.1	1.1	NA
$\gamma_{a;acc;SLS}$	NA	1	1.25

Source: After BS EN 1997-1:2004+A1:2013. *Eurocode 7: Geotechnical Design – Part 1: General Rules*. British Standards Institution, London.

Tests include those carried out as part of the ground investigation and tests as part of the quality control strategy. Investigation tests should normally be loaded to the estimated ultimate resistance of the ground/grout interface and may require tendons and other structural components of greater capacity than used in suitability or acceptance tests.

The design value of the ULS resistance of an anchor shall be derived from

$$R_{ULS;d} = \frac{R_{ULS;k}}{\gamma_{a;ULS}} \tag{6.113}$$

where $\gamma_{a;ULS}$ is a factor given in Table 6.53. The SLS of a test anchor, $R_{SLS;m}$, is the least of the proof load or the load causing a limiting condition. The limiting condition depends on the test method and is the critical creep load (P_c) or the load corresponding to a limit value of the creep rate (α_{SLS}) or load loss ($k_{l;SLS}$), where

$$R_{SULS;m} = min(R_m(\alpha_{ULS} \text{ or } k_{l;ULS} \text{ or } P_c) \text{ and } P_p) \tag{6.114}$$

Recommended values are given in Table 6.52. The design value of the SLS anchor resistance is

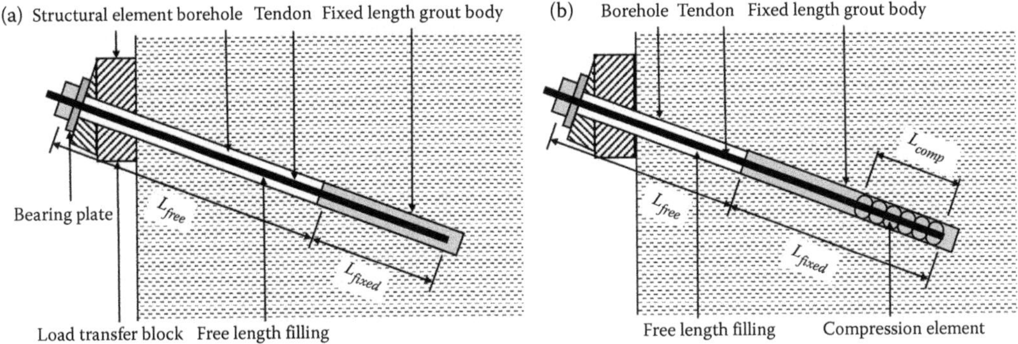

Figure 6.70 Definition of free and fixed length for (a) bond type grouted anchor and (b) compression grouted anchor. (After BS 8081:2015. *Code of Practice for Ground Anchors.* British Standards Institution, London.)

$$R_{SLS;d} = \frac{R_{SLS;k}}{\lambda_{a;SLS}} \tag{6.115}$$

where $\lambda_{a;SLS}$ is a factor given in Table 6.53.

BS8081:2015 was introduced to cover the design, construction, stressing, testing, monitoring and maintenance of grouted anchors, shown in Figure 6.70. Anchors can be vertical, horizontal or inclined, which means that care has to be taken to establish the ground conditions over the length of the anchor, especially important in glacial tills. This means more boreholes. Anchors can fail structurally if the tensile capacity of the tendon is exceeded, or by shear failure between the grout and the soil, and the tendon and the grout. There are four types of anchor (Figure 6.71). Type A anchors are used in rock and possibly in very stiff to hard fine-grained soils. Type B anchors, in which grout permeates the surrounding soil under low

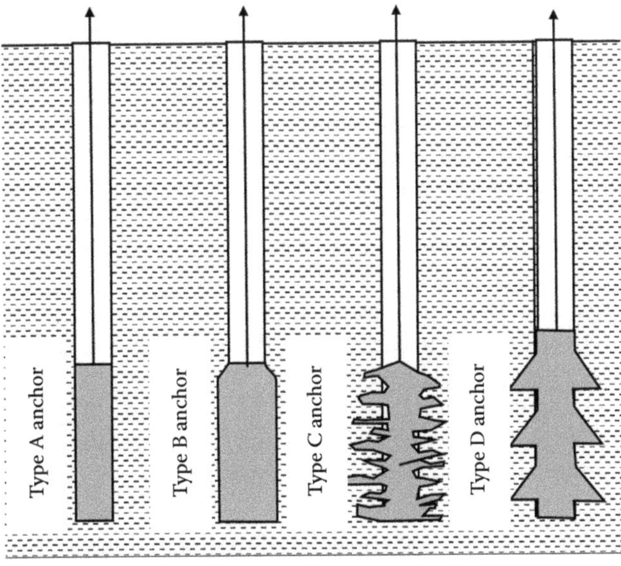

Figure 6.71 Types of cemented grout anchors. (After BS 8081:2015. *Code of Practice for Ground Anchors.* British Standards Institution, London.)

pressure, are used in coarse-grained soils (e.g. glaciofluvial soils and clast-dominated tills) and, possibly, fine-grained soils of low plasticity (e.g. matrix-dominated tills). The grout permeates the voids and increases the density of the surrounding soil. The capacity depends primarily on shear between the grout and the soil, but there is also an end bearing component at the top of the anchor. Type C anchors, in which grout is injected under high pressure, are used in fine-grained soils of low or no plasticity and stiff, fine-grained soils of high plasticity. The capacity is based on the grout/soil interface shear. Type D anchors, in which a series of enlargements are made along the length of the borehole, are used in firm to hard, fine-grained soils (e.g. matrix-dominated tills). The anchor capacity is a combination of end bearing and interface shear.

The ultimate interface shear resistance is not mobilised uniformly along the length of the anchor (BS8081:2015) due to the action of progressive debonding at either or both of the grout/tendon and grout/soil interfaces because of the different elastic properties of the tendon, grout and surrounding soil. Therefore, the ultimate ground/grout interface resistance, $R_{GG;m}$, is less than that assuming a uniform stress distribution along the fixed anchor length (Figure 6.70). A correction factor, f_{eff}, for anchors in matrix-dominated tills is

$$f_{eff} = 1.6L_{fixed}^{-0.57} \tag{6.116}$$

And for fine sands

$$f_{eff} = 0.91^{L_{fixed}\tan\Phi} \tag{6.117}$$

where L_{fixed} is the length of the anchor.

The anchor capacity, $R_{GG;calc}$, for Type B anchors (interface shear and end bearing) in coarse-grained soil is

$$R_{GG;calc} = f_{eff}A\sigma_v'\pi DL_{fixed}\tan\Phi' + B\gamma h\frac{\pi}{4}(D^2 - d^2) \tag{6.118}$$

where A is the ratio of the contact stress to the average effective stress, σ_v' is the average effective stress over the length of the anchor, D is the diameter of the anchor, d the diameter of the free length, B is the bearing capacity factor (=N_q/1.4), where N_q is taken from Figure 6.72, and h is depth to the top of the fixed anchor.

This calculated value depends on knowledge of the anchor geometry which is difficult to assess. Therefore, the capacity can be based on interface shear only assuming either K_o conditions or the grout pressure, p_i.

$$R_{GG;calc} = f_{eff}K\pi DL_{fixed}\,\sigma_v'\tan\varphi' \tag{6.119}$$

where K is the coefficient of earth pressure.

The capacity of Type C anchors is based on field observations using Figure 6.73.

The capacity of anchors in fine-grained soils depends on experimental observations (Type C) or calculated values (Type D). Figure 6.74 are results of load tests on anchors in medium to high plasticity clays which suggests that Type D anchors are more likely to be used in matrix-dominated tills. In that case, the capacity is

$$R_{GG;calc} = (\pi DL_{fixed}c_s)_U + (\pi DL_{fixed}c_a)_{NU} + \frac{\pi}{4}(D^2 - d^2)N_c c_b \tag{6.120}$$

where U refers to the under ream length and NU the length with no under ream.

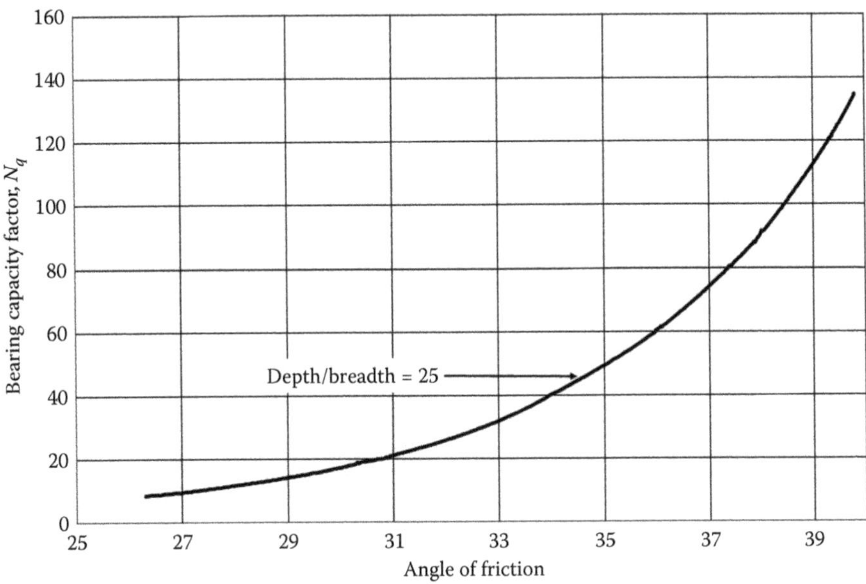

Figure 6.72 Use of bearing capacity factor, N_q (Berezantzev, 1961) in ground anchors. (After BS 8081:2015. *Code of Practice for Ground Anchors*. British Standards Institution, London.)

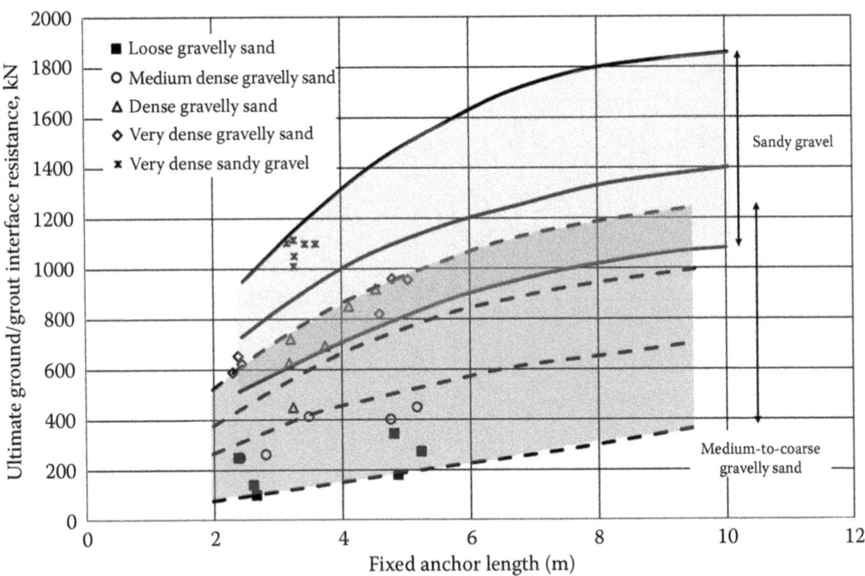

Figure 6.73 Ultimate capacity of anchors in sandy gravels and gravelly sands, showing influence of soil type, density and fixed anchor length for Type C anchors. (After Ostermayer, H. and F. Scheele. *Proceedings of the 9th International Conference on Soil Mechanics and Foundation Engineering*, Tokyo, V97, 1977.)

6.8.1 Recommendations

The capacity of an anchor is dependent on a soil's mechanical characteristics, the method of installation of the anchor and the type of anchor. The type of anchor will depend on how the soil responds to the grout pressure, which in turn depends on the local ground conditions.

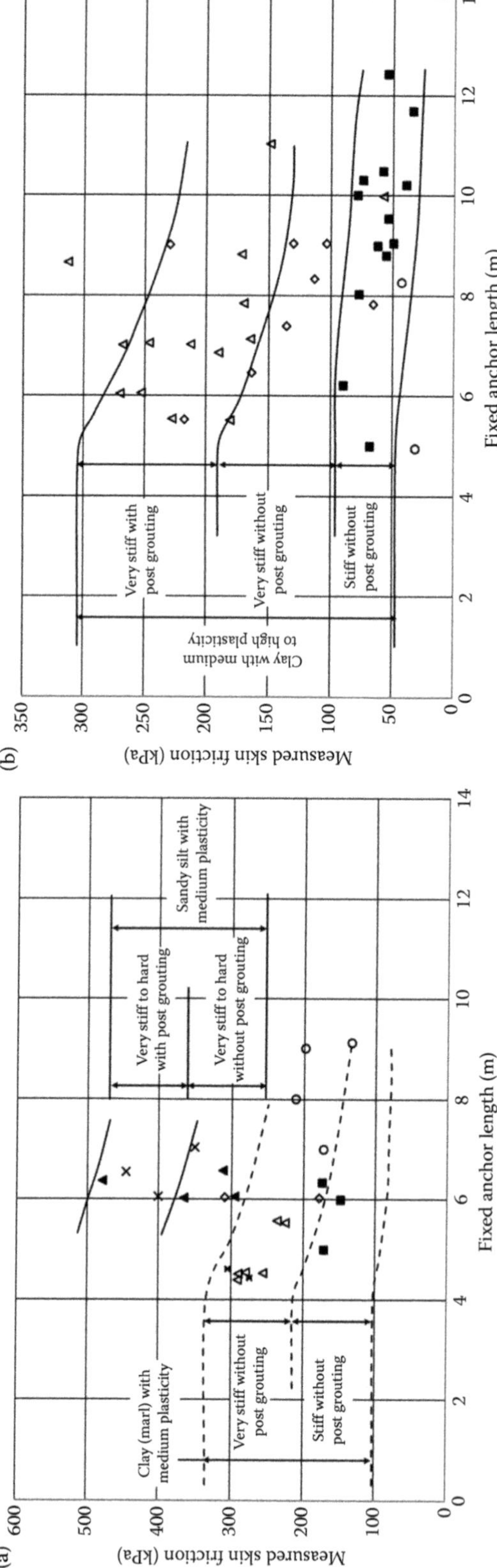

Figure 6.74 Skin friction in fine soils for various fixed anchor lengths, (a) with and (b) without post-grouting. (After Ostermayer, H. *Proceedings of ICE Conference on Diaphragm Walls and Anchorages*, London, 1974: 141–151.)

Glacial soils are complex soils and can vary both horizontally and vertically in properties and thickness, which means that the capacity of a test anchor may not be the same across a project if the spatial variation within a site is considerable. It is prudent to test all anchors.

Anchors are most likely to be used in dense glacial tills. A ground investigation has to be designed to fully assess the ground conditions. This may seem obvious and in line with codes of practice and guidelines but the investigation of glacial soils has to be more thorough than they recommend. Boreholes are normally vertical; anchors inclined. Therefore, it is recommended that

- A ground investigation has to establish the geotechnical, geological and hydrogeological conditions for both the short and long term.
    - The spacing and depth of boreholes has to ensure that the spatial variability can be assessed over the length of the anchor. As this is unlikely, trial anchors should be considered if spatial variation is a concern. The bedrock interface, if relevant, has to be identified especially as glacial soils may contain rafted rock and the bedrock may not be planar.
    - Sufficient representative *in situ* tests and samples have to be taken to ensure that characteristic strength properties can be determined. This is especially important in glacial tills and glaciofluvial soils where it may be difficult to develop property profiles due to the natural variability of the soils; scatter in the data is inevitable.
    - Effective and undrained strengths of glacial clays are required as the ULS may be governed by the effective strength but the undrained shear strength may be critical in the short term.
    - The fabric of glacial clays must be carefully assessed because it could give an indication of mass behaviour as opposed to the intact behaviour observed in small specimens.
    - Seasonal groundwater conditions should be assessed using regional and site-specific data.

6.9 OBSERVATIONS

As with all geotechnical projects, risk is reduced if an adequate ground investigation is undertaken. It is necessary to develop the ground model to take into account the spatial variation of the structure, fabric and composition of glacial soils. It is difficult, especially for glacial tills, to obtain representative values; establish the bedrock interface, if relevant; identify weaker and water-bearing layers and lenses; and locate randomly distributed boulders. Many semi-empirical design methods are based on observations of fine- and coarse-grained soils, which means that the factors may not be appropriate for glacial soils which are generally composite soils. Design codes vary from country to country as do the empirical correlations on which they are based, which means that care has to be taken when using these codes and correlations unless the assumptions and source data are known. There is increasing use of numerical methods but routine ground investigations do not necessarily provide the data needed for the constitutive models. Numerical methods, however, enable scenario analyses to be undertaken, which is extremely beneficial when engineering glacial soils as the variation in structure can be modelled. The key to any design is to obtain representative design parameters. This can be difficult because of the fabric, composition and strength of glacial tills, the composition of glaciofluvial soils and the sensitivity of glaciolacustrine and glaciomarine soils to sampling. The strength and stiffness of matrix-dominated tills and the impact of foundation construction can lead to uneconomic or unsafe designs.

- Preliminary design
  - The preliminary design will be based on a desk study, so the presence of glacial soils will be known. This knowledge should be used to ensure that the ground investigation is properly designed to identify the hazards and obtain representative design parameters.
  - A ground investigation should include two phases to develop the ground model and allow design changes to be made.
  - The spacing and depth of boreholes should be sufficient to pick up the variation in composition, fabric and structure of the glacial soils; locate the bedrock interface, if relevant; and allow for a possible change in design.
  - The ambient and seasonal regional and local groundwater conditions should be established to ensure that the impact on the foundation capacity and construction is understood.
  - Sufficient samples should be taken to obtain representative design parameters.
  - If numerical methods are going to be used, then tests should be designed to obtain the relevant parameters for the constitutive models.
  - Recommendations in codes of practice should be treated with caution because of the composite nature of glacial soils requiring increased investigation.
- Main design
  - The main design is likely to be based on a combination of semi-empirical and numerical methods. The semi-empirical methods are useful in proving that the foundation solution is possible and, for smaller projects, the geometry of the foundation. Numerical methods can provide a more detailed design and, importantly, a study of the impact of the variation within glacial soils upon the foundation. It is important, however, to understand the assumptions and limitation of the numerical methods to ensure that the ground conditions are modelled correctly.
  - The possibility of weaker layers within the zone of influence should be considered.
  - Many semi-empirical methods are based on site-specific correlations mostly of fine-grained or coarse-grained soils. It is important to assess whether these are relevant to glacial soils, especially tills, as their density and composition may be different from the soils used to create the correlations.
  - The stiffness and mass permeability of matrix-dominated tills mean that their capacity in the short term exceeds that based on the undrained strength but less than that mobilised in the long term.
  - Tests should be carried out on anchors and piles to check the design calculations, ideally ahead of the main construction to allow the design to be altered. Tests should be carried out sometime after construction to realise the benefits of the increase in capacity with time.
  - Overall stability should take into account the possible effect of the structure of glacial soils particularly lenses and layers of weaker soils.
  - Temporary work designs should be carried out to ensure that the construction is safe at all stages. This is particularly important when excavating in glacial tills containing water-bearing and weak lenses and layers.
- Foundation construction
  - Glacial soils are likely to contain very coarse particles which will have an impact on the construction of piles, anchors and retaining walls.

- Matrix-dominated tills are likely to contain water-bearing layers of sands and gravels which will affect the construction of any foundation or retaining wall.
- Excavation in matrix-dominated tills could expose lenses and layers of water-bearing and weaker soils.
- Excavations in glaciolacustrine clays could fail unless the anisotropic nature is taken into account.

Chapter 7

Engineering of glacial soils

7.1 INTRODUCTION

Glacial soils are soils that are spatially variable in composition, fabric and structure: they are composite soils. This variation is a function of the geological processes that take place on a macro-, meso- and micro-scales, resulting in soils with physical and mechanical properties that are also spatially variable. They are one of the most diverse of generic soil types with compositions ranging from clay size particles to boulders and may be described as fine- or coarse-grained soils but exhibit features of both fine- and coarse-grained soils. They are complex soils, which are known to be difficult to deal with. Codes of practice and design guidelines may not provide sufficient, relevant information to assess the ground conditions and produce design parameters and, possibly, give wrong information. Design and construction practice based on sound scientific and engineering principles will produce safe, economic, sustainable structures that are fit for purpose. This has been demonstrated using case studies to highlight the characteristics of glacial soils and how they behave. Recent developments in data collection, management, interpretation and analyses have created opportunities to develop regional databases for glacial soil characteristics and use them to improve the selection of site-specific data. The purpose of this chapter is to set out a strategy using the guidelines in Eurocode 7 (BS EN 1997-1:2004+A1:2013, BS EN 1997-2:2007) to clarify what is required when engineering glacial soils.

7.2 THE STRATEGY

Predicting the performance of geotechnical structures and natural slopes is based on appropriate geological, topographical, hydrogeological, geomorphological, geotechnical and structural models and the interaction of those models, as shown in Figure 7.1. Codes of practice, regulations and guidelines are there to support design and construction, but they do not replace good engineering. This is particularly important for glacial soils that are considered a hazard. Over- and under-design do occur leading to cost overruns, excessive movement and possibly failure but, in most cases, can be avoided. The guidelines and codes of practice are based on good practice and have been developed over the years, so, provided they are followed, the risk is reduced but not necessarily eliminated. The expression 'ground is a hazard and you pay for a ground investigation regardless of whether one is carried out' is relevant to all soils, but the risks associated with glacial soils are greater because of their spatial variability. Therefore, it is important to fully investigate a site to reduce risk and cost to ensure the design and construction is fit for purpose. A vast quantity of data on glacial soils has been collected over the years, and advances in data storage, integration, modelling

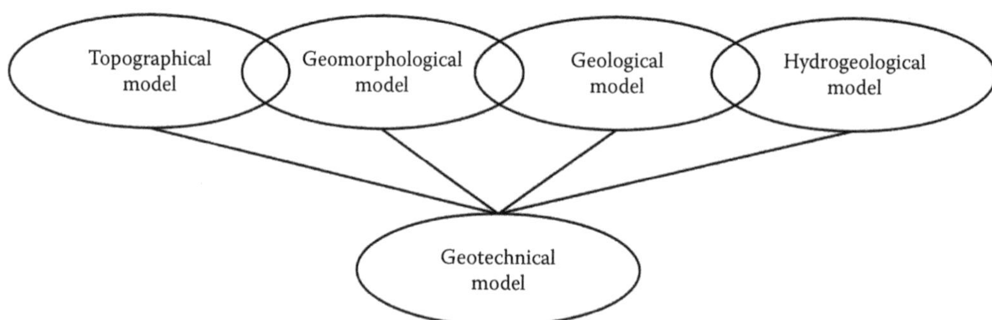

Figure 7.1 Models needed to design geotechnical structures.

and analysis offer the opportunity to develop regional databases to improve the quality of the site-specific data within the region, thus reducing the risk further.

7.2.1 Topographical survey

A topographical survey is of immense value in areas subject to glaciation because it may provide a view of historical land forms, particularly in rural areas where there may be limited anthropogenic alteration of the ground surface. Landforms can be an indicator of types of glacial soils though the last glacial deposit may be overlain by made ground or post-glacial deposits masking the profile immediately following the end of the last glacial period, or altered by the formation of the current drainage system or by ground movements following isostatic uplift. Linear infrastructure projects may cross several landforms suggesting a variation in glacial soils, whereas a development project may be the site of a single landform. Hence, the scale of a landform and its relation to a project footprint are relevant. In the latter case, a study should be undertaken to determine the regional landforms.

It is normally expected that a detailed topographical map is produced indicating the position and level of boreholes. This is essential in glacial soils in order to produce a 3D model of the stratigraphy. The interface between the glacial soils and underlying pre-glacial deposits is unlikely to be planar. Weaker soils and water-bearing soils may be encountered at various depths in glacial tills, which may be horizontal or inclined layers or lenses. Boulders may be randomly distributed or form a boulder bed. Soils of different descriptions, especially in glaciofluvial soils, may have been deposited at the same time but the deposition process naturally produces a graded deposit. Post-glacial soils may lie unconformably on glacial soils covering historical landforms; this is a particular issue in urban areas that could include extensive deposits of artificial soil or made ground. There may have been several periods of glaciation resulting in buried landforms and boundaries of unconformity. These are reasons why it is necessary to use a staged process for the exploratory work and specify more boreholes than would be required for deposits of sedimented soils.

7.2.2 Geomorphological study

Geomorphological studies are essential for linear infrastructure projects that cross glaciated areas as they describe the landforms and, importantly, whether there are signs of ground movement due to isostatic uplift and establishment of the drainage system following the last glacial period. There could be dormant landslides, which may be triggered by construction works or extensive periods of rainfall or intense rainfall events, increasingly likely due to climate change. Geomorphological studies are also useful in development sites, even though the area of a site may be limited, as the regional landforms may help identify possible glacial soils at the site.

While geomorphological studies focus on the land surface, there is benefit in applying the principles to describe historical surfaces overlain by more recent deposits: artificial soil and post-glacial deposits. It is likely that an area will have been subject to several glacial periods, each period leaving a landform that may have been modified. This will be challenging but the benefits will help reduce the risk of encountering problems during construction and operation. This is likely to be carried out at a regional scale, a useful exercise in urban areas provided sufficient data are available.

7.2.3 Geological investigation

An overview of subsurface soils can be obtained from geological maps, existing borehole data, landforms and exposures of glacial soils. However, a detailed geological model requires exploratory work because of the significance of structure, fabric and composition and their spatial variation on the soil behaviour. This detailed exploratory work should be undertaken by someone who is familiar with glacial soils, their mode of deposition and the impact it can have on the structure, fabric and composition of glacial soils. The geological model will be based on topographical, historical and geological maps, natural and excavated exposures, geophysical surveys and borehole logs.

A ground investigation is a staged process, essential when establishing the ground model for a site underlain by glacial soils. The first stage should include a number of strategically placed boreholes and trenches to provide an overview of the ground conditions. The boreholes should be deep enough to extend beyond the zone of influence of the civil engineering structure. This is probably unknown at the time of the investigation because the geometry of the geotechnical structure depends on the results of the ground conditions. It is prudent to prove rock if encountered, and to obtain sufficient samples to produce detailed borehole logs rather than just to rely on the driller's description. Borehole locations should extend beyond the footprint of the structure. Trenches are preferred to trial pits as they can show the lateral variation in near-surface soils, an indicator of the complex structure of the near-surface soils and their fabric. These exposures should be described in detail using knowledge of the formation of glacial soils so that the likely type of glacial soil can be identified.

The second stage provides more detailed knowledge of the ground conditions and geotechnical properties. At this stage, there should be more information on the possible geotechnical structures: their location, geometry and loads. These will be based on preliminary designs using published data and information from the first-stage investigation. Therefore, borehole locations and depths can be strategically chosen to optimise the information producing a cost-effective investigation and reduce the risk. These locations should take into account the location of the geotechnical structures and possible structure of the glacial soils based on the desk study, site reconnaissance and first stage of the investigation. Sampling should take account of the need to obtain sufficient samples to obtain classification data and geotechnical properties. For example, it is likely that the geotechnical properties of matrix-dominated tills will be based on tests on 100 mm diameter samples rather than 38 mm diameter samples because of the effect of fabric and gravel content on the properties. Geotechnical characteristics will be derived from tests on representative samples, which at the time of the investigation are unknown. Hence, the laboratory tests will also be staged to classify the soils and decide which samples are representative of the geological profile. It may be possible to retrieve undisturbed samples provided the number of very coarse particles is limited. However, it is more likely that it will be impossible to obtain undisturbed samples from glacial soils because of their composition, fabric and density. This is especially the case for glacial tills and glaciofluvial soils. Disturbed samples may be reconstituted to the *in situ* density in order to determine the geotechnical characteristics.

In situ tests are valuable means of assessing the variation in composition and density and, provided appropriate correlations, or better still, site-specific correlations, are used, it is possible to assess the spatial variation in geotechnical characteristics. The SPT is the most versatile and can be used in any glacial soil but, like all *in situ* tests, the results are susceptible to the local composition, which means that sufficient tests are required to allow a statistical analysis to be undertaken. The electric cone can be used in glacial clays and sands provided they are not too dense or contain too much gravel and very coarse particles. The piezocone is particularly useful in glaciolacustrine clays to assess the thickness of laminations. More specialist tests, such as pressuremeter tests, can be used provided the composition is appropriate.

Geophysical testing is potentially a useful means of assessing interfaces, boulders and spatial variation in stiffness depending on the technique used. They are of particular value on linear infrastructure projects where longitudinal profiles may be developed.

In many cases, there will be only a single-stage ground investigation, which is a high-risk strategy. It can lead to overdesign specifying geotechnical structures with greater capacity than is needed; or overlook potential hazards that could trigger construction difficulties, excessive ground movement or possibly failure. An interactive staged investigation reduces risk leading to a more economical design that is fit for purpose and that should be specified at the outset.

7.2.4 Hydrogeological model

Groundwater location and fluctuations in groundwater level are critical to any design, cause construction problems and lead to ground movements, possibly failure. Yet establishing the groundwater conditions is not routinely undertaken increasing the risk to a project. The risk can be reduced by assuming the worst possible conditions leading to overdesign. It is prudent to assume that groundwater will be encountered during construction no matter the type of glacial soil.

Ideally, groundwater levels should be monitored over sufficient time to establish seasonal variations, the presence of water-bearing layers, potential aquifers and aquicludes. The mass permeability should be assessed especially if an excavation is necessary when changes in permeability are possible, especially in matrix-dominated tills that contain discontinuities as these discontinuities can open up on excavation.

It is important to establish the local and regional geology and their link to the hydrogeological conditions so that estimates can be made of likely groundwater scenarios. This allows an assessment to be made of changes in groundwater conditions and how they may impact on the stability and movement of any geotechnical structure including natural slopes. The life of any civil engineering project (design life: 40–120 years) means that climate change could have an effect. Therefore, for Category 3 structures, it is essential to model potential groundwater changes in the long term.

7.2.5 Geotechnical model

The geotechnical model, the basis for the selection of design parameters, is based on all the other models, the classification data and the geotechnical characteristics. Glacial soils are composite soils, so sampling, testing, interpreting tests and selecting design parameters are difficult. Failure to take this into account increases the risk of excessive ground movement, possibly failure. The spatial variability of properties means that it can be difficult to produce a representative set of geotechnical properties; therefore, scenario analyses covering a range of parameters are required.

Classification data are useful in that they help in distinguishing between different types of soil but, without a quality description, may be misleading. For example, consistency limits of glacial clay soils lie above the A-line, about the T-line, suggesting that they are clays; yet glaciolacustrine clays are layered silts and clays and matrix-dominated tills have a significant amount of clay-sized rock particles, which may or may not be clay minerals depending on the source rock. This implies that correlations developed between consistency limits and other properties from tests on sedimented soils may not apply to glacial clays. Glacial clays are generally low to medium plasticity, with the plasticity increasing with weathering.

Subglacial tills are very dense, which accounts for their stiffness and strength. Other glacial soils are gravitationally deposited and, therefore, their density can be similar to that of sedimentary soils. Density measurements of any glacial soil containing significant gravel content are difficult because of the difficulty of recovering undisturbed samples or, in the case of *in situ* tests, the effect of the gravel particles on the results. It is inevitable that the recorded density of glacial tills and glaciofluvial soils will show a greater standard deviation than that for glaciolacustrine clays. The variation may be the natural variation or the result of the testing/sampling procedure. Density values are important when assessing the quality of other properties such as strength and can be used to assess whether those other properties are credible. For example, a low strength for a high-density glacial till should be questioned but not necessarily rejected.

Like density, water content is a useful indicator of soil properties and is difficult to determine for the same reasons. Data show a medium to high standard deviation for clay tills, which is due to the effects of sampling and composition. It is useful to quote the water content based on both the whole sample weight and the weight of fine-grained particles when testing clay tills because the matrix water content may be a better indicator of soil behaviour. The water content of a glaciolacustrine clay is more likely to show a low standard deviation but a poor indicator of strength because the water content of the clay layers may be different from the water content of the silt layers.

The most useful indicator of engineering behaviour is the percentage of fine-grained particles. Therefore, particle size distribution tests are invaluable. Fifteen to twenty percent of fine-grained particles suggest that the glacial soil will exhibit characteristics of fine-grained soils. However, this can be modified by the effects of fabric. For example, a glacial till could be classed as a clay because of the fine-grained content but its permeability could be significantly greater than that for a fine-grained soil because of fabric. The shape of a particle size distribution curve can be an indicator of glacial till type and its behaviour.

Hence, classification tests are a useful means of categorising glacial soils and their expected behaviour. Sufficient samples are needed to take into account the natural variation of properties due to the spatial variation in composition and the effects of sampling composite soils. Consistency limits, density, water content and particle size distribution are useful indicators of soil behaviour, means of assessing the quality of other tests, indicators of the spatial variation in composition and useful in developing site-specific correlations.

The behaviour of a glacial soil can be described by its strength, stiffness and permeability, but obtaining representative values is difficult because of the spatial variation in composition, and fabric, and difficulties in sampling and testing and interpreting the test results. The geological descriptions and classification data are used to identify the types of glacial soils. Ideally, this should be completed at the end of the first stage of the exploratory work so that sampling and testing in the second stage can be more strategic though modifications to the planned investigation may be necessary as more information becomes available.

The mechanical characteristics of clast-dominated tills and glaciofluvial soils will primarily be based on *in situ* tests using a range of correlations often developed for a narrower range of particle sizes. Very coarse particles, gap-graded particle size distributions and a small percentage of fine-grained particles may have a significant impact on the results. The mechanical characteristics of soils are dependent on many factors but perhaps the most significant is density; a dense soil is stronger, stiffer and less permeable than a soil of similar composition but with a lower density. Therefore, if it is possible to estimate the density from the field tests, it may be feasible to test specimens reconstituted to the *in situ* density. The type of test and the composition will dictate what is possible. It is not feasible to carry out laboratory tests on samples with a significant number of very coarse particles unless specially designed equipment is used. It will be possible to test sands and gravels though large specimens may be necessary if the gravel content is significant. Shear box texts can be used to determine the strength of the soil if it is consolidated to the *in situ* density. Constant head tests are possible on large enough samples though *in situ* permeability tests would be better. Stiffness is the most difficult parameter to determine; indeed, it may not be possible for composite coarse-grained soils. Representative values will have to be inferred from *in situ* tests taking into account the challenges of obtaining representative test results.

It will be possible to sample glacial clays but the quality of the sample will be a factor in deciding whether the results are representative. The composition and fabric of glacial clays inevitably produce a significant variation in results, making selection of design parameters difficult, less so with glaciolacustrine clays. It is likely that tests will be carried out on large samples (e.g. 100 mm samples) because of the gravel content and fabric. This means that sufficient samples, more than suggested in guidelines, are required. When carrying out strength tests, it is likely that a minimum of three 100 mm samples will be needed to obtain a representative value of strength. Even then, the variation in the results may be unacceptable. As with coarse-grained glacial soils, density is an indicator of soil behaviour. Therefore, tests on reconstituted soils removing particles greater than 10 mm (provided they are less than 10% of the sample) will produce more consistent results, which, if prepared at the *in situ* density, will be a lower bound to the representative strength. A correction may have to be made for fabric since sample preparation will have destroyed the fabric, hence the importance of properly describing the soil.

The stiffness is more difficult to determine though, as density is an indicator of stiffness, it should be possible to obtain a representative value from local strain measurements on reconstituted specimens.

The effect of fabric on the mass permeability means that *in situ* tests are the only reliable way of obtaining representative values. However, it may be possible to obtain representative values of intrinsic permeability of the matrix and apply a correction to allow for fabric. It is important to assess the anisotropic nature of glaciolacustrine clays.

The credibility of the results should be assessed by assessing the quality of the samples and *in situ* tests and the impact that would have upon the results, by ensuring that the physical and mechanical properties are consistent with one another, and by ensuring that the results are consistent with what is expected from the geological model. This can be supported by published data, constitutive models, correlations between results and statistical analyses.

7.3 SELECTION OF DESIGN PARAMETERS

A geotechnical design is based on the worst credible parameters derived from the five models to reduce the risk of failure and most likely to predict deformation with any degree of accuracy. It is generally accepted that the amount spent on ground investigations is less than necessary increasing the likelihood of cost overruns, failures during construction and excessive deformation and possible failure during operation. The exceptions to this are designs for Category 1 structures where the geotechnical solution is straightforward based on a limited investigation and Category 3 structures where the consequences of failure are deemed unacceptable. Category 2 structures should be considered Category 3 in order to reduce the risk because glacial soils are complex soils. The risk can also be reduced by considering the worst possible ground conditions leading to overdesign. This may be acceptable if the cost of the geotechnical structure is a very small percentage of the total project.

In order to reduce the risk, it is necessary to conduct an adequate investigation that will produce more data possibly increasing the difficulty of creating the geotechnical model because of the spatial variability in composition, fabric and structure. To overcome this difficulty, it is necessary to interpret the data within a framework based on sound engineering principles of soil behaviour using statistical analyses and to undertake scenario modelling. Principles of soil behaviour extend from relationships between water content and density to application of constitutive models to check that the data are acceptable. Even if the data do not fit within a framework, it does not mean that they should be rejected without justification. Statistical analyses extend from simple averaging to application of regional datasets to produce design parameters. It is prudent to set upper and lower bounds to the design parameters, which will be set by a statistical analysis and framework.

This approach will produce a geotechnical model, which provides a range of conditions that have to be modelled in a design. For example, it could include varying thicknesses of weaker layers of soil within a glacial till; intrinsic effective strength parameters based on tests on reconstituted soil; upper, lower and median profiles of undrained shear strength with depth; and profiles of stiffness based on the site-specific ground investigation modified by a regional database.

This model can be used to undertake a scenario analysis to investigate the effect of varying the geotechnical model on the performance of the structure. At some point in this process, a decision will have to be taken as to what is the most likely scenario.

7.4 OBSERVATIONS

Codes of practice and design guidelines give useful, generic information, which has to be interpreted for a particular project. Table 7.1 summarises factors to consider when undertaking a ground investigation and deriving design parameters in glacial soils using generic statements from Eurocode 7. Table 7.2 summarises factors to consider when considering the engineering of glacial soils.

Glacial soils are naturally spatially variable complex composite soils. In order to reduce risk,

- Take a strategic approach to undertake an interactive ground investigation
- Apply sound scientific and engineering principles using knowledge of the formation of glacial soils, constitutive models, statistical analyses, regional databases, appropriate correlations and scenario analyses
- Expect the unexpected and take action to reduce the risk

Table 7.1 Ground investigation and selection of design parameters highlighting aspects of glacial soils that are relevant to published guidelines

Guidelines	*Factors to consider*
Ensure that relevant geotechnical information and data are available at the various stages of the project[a]	The information should take into account the design requirements for temporary and permanent works, and construction processes including excavations, piling and engineering fill. Critically, in glacial soils, it is important to consider local variations in composition, fabric and structure as they can have a significant effect on construction
Adequate to manage identified and anticipated project risks[a]	Hazards associated with glacial soils are a function of the spatial variation in the composition, fabric and structure, their strength and the hydrogeological conditions. These hazards are discussed below
The suitability of the site with respect to the proposed construction and the level of acceptable risks[a]	It should be assumed that local features within glacial soils will have impact on their overall performance. Therefore, both mass properties and local variations must be assessed
The field reconnaissance of the site of the project and the surrounding area noting particularly: (a) evidence of groundwater; (b) behaviour of neighbouring structures; (c) exposures in quarries and borrow areas; (d) areas of instability; (e) any exposures of mining activity at the site and in the neighbourhood; (f) difficulties during excavation; (g) history of the site; (h) geology of the site, including faulting; (i) survey data with plans showing the structure and the location of all investigation points; (j) information from aerial photographs; (k) local experience in the area; (l) information about the seismicity of the area[a]	This should include a geomorphological study of the area to establish regional land systems and site-specific landforms
Establish the soil, rock and groundwater conditions, to determine the properties of the soil and rock[a]	Glacial soils are deposited either by ice or by water. In the former case, the process varies from deformation of the pre-glacial deposits to transportation and remoulding of the pre-glacial deposits. In the latter case, glacial soils can be deposited from water or in water creating complex lateral and vertical profiles • The interface between the glacial deposits and the underlying pre-glacial deposits is not necessarily planar creating buried valleys • There may have been several periods of glaciation creating complex deposits with similar classification properties but different geotechnical properties • The structure of glacial soils is variable, which means that the thickness of a deposit can vary over short distances • The composition of a glacial soil is variable and can include lens of different soils of varying volume, and layers of different soils of varying thicknesses • Variations in strength and stiffness of glacial soils may be a result of the deposition processes; that is, they are natural

(Continued)

Table 7.1 (Continued) Ground investigation and selection of design parameters highlighting aspects of glacial soils that are relevant to published guidelines

Guidelines	*Factors to consider*
	• Glacial soils can contain very large particles, which may, incorrectly, be assumed to be bedrock unless proven otherwise • Fine-grained deposits can contain water-bearing lens and layers of coarse-grained soils • Glaciolacustrine soils are anisotropic • Matrix-dominated tills can act as aquicludes resulting in perched water levels • The matrix of glacial tills can be formed of clay minerals or rock flour depending on the source material • Upper layers of glacial soils may be weathered • Stage 1 and 2 investigations are essential to ensure that the spatial variation of glacial soils is assessed. The depth of boreholes, borehole spacing, sample levels and *in situ* test locations should take into account possible variations in the interface between post-glacial soils including artificial soil, glacial soils and pre-glacial soils and rocks. Trenches are more useful than trial pits to obtain an indication of the variation of the near-surface soils. Piezometers should be installed at various depths with the pocket embedded in the layer being investigated. Readings of water levels should be taken to assess the seasonal variations in water level. Geophysical testing should be considered for linear infrastructure projects
Description of the geometry of the strata – detailed descriptions of all strata including their physical properties and their deformation and strength characteristics, referring to the results of the investigations – comments on irregularities such as cavities and zones of discontinuous material[a]	• There should be sufficient information to make an assessment of the spatial variation in composition, fabric and structure • There should be sufficient data on the classification of the soils to take account of the variation in composition and to indicate representative strata as well as local variations so that the local and mass characteristics can be assessed
The documentation of the evaluation should substantiate the following aspect: strata in which ground parameters differ only slightly may be considered as one stratum[a]	• The natural variation in classification and geotechnical characteristics of glacial soils suggests that the description of one stratum has to be based on a broad range of results for any one parameter
A sequence of fine layers with greatly differing composition and/or mechanical properties may be considered as one stratum if the overall behaviour is relevant, and the behaviour can be adequately represented by ground parameters selected for the stratum[a]	• This applies, in particular, to glacial soils

(Continued)

Table 7.1 (Continued) Ground investigation and selection of design parameters highlighting aspects of glacial soils that are relevant to published guidelines

Guidelines	*Factors to consider*
When deriving the boundary between different ground layers and the groundwater level, a linear interpretation may be used between investigation points provided the spacing is sufficiently small and the geological conditions are sufficiently homogeneous. This should be justified and reported[a]	• Account should be taken of local variations in composition as they can impact on construction processes and affect foundation performance and slope stability • Given the possibility of variations in ground structure, it is prudent to undertake scenario analyses to see the effect of a variation in ground structure upon the performance of the geotechnical structure • The interfaces between post-glacial soils, glacial soils and pre-glacial soils and rocks are not necessarily planar, so linear interpolation may not be correct
Groundwater investigations should provide, when appropriate, information on the depth, thickness, extent and permeability of water-bearing strata in the ground, and joint systems in the rock[a]	• This should include an assessment of the fabric of matrix-dominated tills, whether water-bearing soils within matrix-dominated tills form layers or lenses, the mass permeability of glacial soils, and the anisotropic nature of glaciolacustrine clays
The elevation of the groundwater surface or piezometric surface of aquifers and their variation over time and actual groundwater levels including possible extreme levels and their periods of recurrence[a]	• The presence of perched water tables within and artesian pressures below matrix-dominated tills should be considered • Seasonal variations in groundwater levels should be assessed • Infiltration in glacial days containing discontinuities and the possibility of the discontinuities opening up on excavation increasing the permeability should be considered
The information obtained should be sufficient to assess the following aspects, where relevant: the scope for and nature of groundwater-lowering work; possible harmful effects of the groundwater on excavations or on slopes (e.g. risk of hydraulic failure, excessive seepage pressure or erosion); any measures necessary to protect the structure (e.g. waterproofing, drainage and measures against aggressive water); the effects of groundwater lowering, desiccation, impounding etc. on the surroundings; the capacity of the ground to absorb water injected during construction work; whether it is possible to use local groundwater, given its chemical constitution, for construction purposes[a]	• Excavations in matrix-dominated tills containing discontinuities can cause the discontinuities to open reducing mass strength and increasing permeability • Water-bearing lens and layers can trigger local instability • Water-bearing layers may be subject to artesian pressures
The evaluation of the geotechnical information shall be documented and the results of the field investigations and laboratory tests evaluated according to relevant standards	• Results of tests should be in accordance with standards modified to take account of the effect of composition and fabric on the volume of soil tested • It is particularly important to assess the quality of samples to make a judgement of the value of the results • Classification data (particle size distribution, consistency limits, density and water content) are essential in assessing the quality of the results

(*Continued*)

Table 7.1 (Continued) Ground investigation and selection of design parameters highlighting aspects of glacial soils that are relevant to published guidelines

Guidelines	*Factors to consider*
The evaluation of the geotechnical information shall include as appropriate a review of the field and laboratory work. Any limitations in the data (e.g. defective, irrelevant, insufficient or inaccurate) shall be pointed out and commented upon. The sampling and sample transportation and storage procedures shall be considered when interpreting the test results. Any particularly adverse test results shall be considered carefully in order to determine if they are misleading or represent a real phenomenon that must be accounted for in the design; a review of the derived values of geotechnical parameters; any proposals for necessary further field and laboratory work, with comments justifying the need for this extra work. Such proposals shall be accompanied by a detailed programme for the extra investigations to be carried out with specific reference to the questions that have to be answered.[b]	• Stage 1 and Stage 2 investigations are essential to reduce the risk. Stage 1 should provide sufficient information to provide an overview of the ground conditions and enough samples to make a preliminary assessment of the major soil types. Stage 2 should be a strategically designed investigation with the aim of assessing the spatial variation of the glacial soils at local level and providing representative design parameters • It is impossible to retrieve Class 1 samples in many glacial soils because of the composition and fabric. Therefore, it is likely that design parameters will be based on a combination of results of laboratory tests on imperfect samples and reconstituted samples, results of field tests, published data, site-specific correlations within a soil mechanics framework and regional databases • Tests should be carried out on as large samples as possible to reduce the effect of randomly distributed larger particles and take into account the effect of fabric on the soil properties • Site-specific correlations should be used to convert field test data to design parameters. If generic correlations are to be used, then the original source should be checked to ensure that they were developed for similar soils. • Classification data should be used to verify the credibility of geotechnical characteristics • The possibility of additional investigations should never be discounted, especially given the risk of construction in complex soils
Averaging can mask the presence of a weaker zone and should be used with caution. It is important that weak zones are identified. Variations in geotechnical parameters or coefficients can indicate significant variations in site conditions[a]	• The geotechnical properties of glacial soils are naturally variable because of the deposition process • Results of tests in glacial soils may also exhibit scatter because of the spatial variation in properties due to the effects of sampling and testing very dense soils of ranging density containing a range of particle sizes and discontinuities
The documentation should include comparisons of the specific results with experience for each geotechnical parameter, giving special consideration to anomalous results for a given stratum when compared with any results from other types of laboratory and field tests capable of measuring the same geotechnical parameter[a]	• Individual results should be assessed against their classification data to check that they conform to expected behaviour based on principles of soil mechanics and geotechnical practice • Regional databases can create a framework to assess the quality of the results • Classification data are used to help identify stratum and separate out any outliers • Geotechnical characteristics from one stratum should be viewed together rather than reported as individual test results

(Continued)

Table 7.1 (Continued) Ground investigation and selection of design parameters highlighting aspects of glacial soils that are relevant to published guidelines

Guidelines	*Factors to consider*
In addition, the evaluation of the geotechnical data should include the following, if relevant: Tabulation and graphical presentation of the results of the field and laboratory work in relation to the requirements of the project and, if deemed necessary Histograms illustrating the range of values of the most relevant data and their distribution Depth of the groundwater table and its seasonal fluctuations Subsurface profile(s) showing the differentiation of the various formations Detailed descriptions of all formations including their physical properties and their deformation and strength characteristics Comments on irregularities such as pockets and cavities The range and any grouping of derived values of the geotechnical data for each stratum[b]	• Profiles of results should be presented against elevation to assist in the interpolation between boreholes • Data from a stratum should be plotted together to assess the range, standard deviation and median • Groundwater information should include reference to aquicludes, aquifers, discontinuities, artesian pressures and water-bearing lenses, pods and layers • An assessment of the quality of the data based on the difficulty of sampling and testing and the formation of the glacial soils should be included • The spatial variation in composition, fabric and structure of glacial soils should be described

[a] From BS EN 1997-1:2004+A1:2013. *Eurocode 7: Geotechnical Design – Part 1: General Rules*. British Standards Institution, London.

[b] From BS EN 1997-2:2007. *Eurocode 7: Geotechnical Design – Part 2: Ground Investigation and Testing (Incorporating Corrigendum 2010)*. British Standards Institution, London.

Table 7.2 Application of design parameters of glacial soils highlighting aspects of glacial soils that are relevant to published guidelines

When defining the design situations and the limit states, the following factors should be considered: • Site conditions with respect to overall stability and ground movements • Nature and size of the structure and its elements, including any special requirements such as the design life • Conditions with regard to its surroundings (e.g. neighbouring structures, traffic, utilities, vegetation, hazardous chemicals) • Ground conditions • Groundwater conditions • Regional seismicity • Influence of the environment (hydrology, surface water, subsidence, seasonal changes of temperature and moisture)[a]	• Ground and groundwater conditions should be assessed at meso- and macro-scales
The deformation of the ground caused by the structure or resulting from construction works, its spatial distribution and behaviour over time[b]	• The most difficult parameter to assess in any investigation is the stiffness of a soil yet, with increasing use of numerical methods, it is a critical parameter if the results of an analysis are to have any meaning. Given that Class I samples are highly unlikely in most glacial soils, then it is necessary to consider a range of values in any analysis. It is also necessary to consider the local and mass stiffness and how they may vary over the depth and width of a structure
The safety with respect to limit states (e.g. subsidence, ground heave, uplift, slippage of soil and rock masses, buckling of piles)[b]	• In general, primary deposits of glacial soils (i.e. glacial tills deposited by ice) are dense and strong. Failures are often triggered by fabric or local variations in structure • The *in situ* strength of matrix-dominated tills may be less than the intrinsic strength because of fabric • Glaciolacustrine clays are anisotropic, so stability will be a function of the alignment to the layers to the slip surface
The loads transmitted to the structure from the ground (e.g. lateral pressures on piles) and the extent to which they depend on its design and construction[b]	• The spatial variation of glacial soils means that an assessment of lateral loads must take into account the possibility of varying thicknesses of weaker layers • The presence of more permeable layers can create complex groundwater pressure distributions

The foundation methods (e.g. ground improvement, whether it is possible to excavate, driveability of piles, drainage) and the sequence of foundation work[b]	• Glacial soils are complex spatially variable soils that can create construction problems. The problems include
	  • Very dense soils
	  • Pockets of weaker or water-bearing soils
	  • Confined layers of weaker or water-bearing soils
	  • Boulders or boulder beds
	  • Excavation in varved clays

(*Continued*)

Table 7.2 (Continued) Application of design parameters of glacial soils highlighting aspects of glacial soils that are relevant to published guidelines

Geotechnical investigations for use as construction materials shall provide a description of the materials to be used and shall establish their relevant parameters. The information obtained should enable an assessment of the following aspects: the suitability for the intended use; the extent of deposits; whether it is possible to extract and process the materials, and whether and how unsuitable material can be separated and disposed of; the prospective methods to improve soil; the workability of soil during construction and possible changes in their properties during transport, placement and further treatment; the effects of construction traffic and heavy loads on the ground; the prospective methods of dewatering and/or excavation, effects of precipitation, resistance to weathering, and susceptibility to shrinkage, swelling and disintegration	• Glacial soils can be used as engineered fill to form embankments; used as backfill; as landfill liners; and to form road subgrades. Unsuitable materials include boulders, layers and lenses of weaker clays and silts, and laminated clays (unless reworked to the extent that the laminations are no longer an issue). In some cases, unsuitable materials can be imported into engineered fill. The extent of the unsuitable material may not be proven in an investigation but some indication should be given based on the ground investigation; this requires exploratory work that exceeds that normally recommended. Glacial clays can be stabilised/ modified if necessary

[a] From BS EN 1997-1:2004+A1:2013. *Eurocode 7: Geotechnical Design – Part 1: General Rules*. British Standards Institution, London.

[b] From BS EN 1997-2:2007. *Eurocode 7: Geotechnical Design – Part 2: Ground Investigation and Testing (Incorporating Corrigendum 2010)*. British Standards Institution, London.

Symbols

a	Constant
a	Area ratio of electric cone
A	Contact area; ratio of the contact stress to the average effective stress for anchors; area of a slip surface
A, B	Pore pressure coefficients
A'	Effective area of a spread foundation
A_b	Base area of a pile
a_d	Design value of geometrical data
A_s	Area of shaft of a pile; pore pressure coefficient during sampling
A_v	Structural compression factor
b	Constant
B	Plate width; width of a foundation; pile diameter
B'	Effective width of a spread foundation
b_c, b_q, b_γ	Base inclination factors
b_g	Width of base of pile group
C	Shape factor
C_c^*	Intrinsic compression index
C'	Cohesion
c_r'	Residual cohesion
C_1, C_2, C_3	Factors
C_c	Compression index
C_N	Correction factor to *SPTN* for soil type
C_s	Swelling index
C_U	Uniformity coefficient
c_u	Undrained shear strength
c_{uave}	Average undrained shear strength
c_{ubase}	Undrained shear strength at base of pile or caisson
c_{uv}	Vane shear strength
c_v	Coefficient of consolidation
c_{vfield}	*In situ* coefficient of consolidation
d	Particle size
D	Diameter of casing or pile or test pocket; constrained modulus; tunnel diameter

d_c, d_q, d_γ	Depth factors
d_{cap}	Diameter of a capillary
D_F	Pile base diameter
d_o	Reference particle size
D_{p50}	Entrance pore size diameter for 50% of the cumulative pore volume
e	Average penetration of DPT per blow; void ratio; eccentricity
E	Modulus of elasticity (stiffness)
e^*_{100}	Void ratio at a pressure of 100 kPa during one-dimensional compression
e^*_{1000}	Void ratio at a pressure of 1000 kPa
E'	Drained stiffness
E_{base}	Stiffness of the soil beneath the pile
E_d	Design value of the effect of the actions; drained stiffness
E_D	Flat dilatometer modulus
E_m	Pressuremeter modulus
e_{max}	Maximum void ratio
e_{min}	Minimum void ratio
E_o	Initial stiffness
E_{oed}	Stress dependent oedometer modulus
E_{pile}	Pile stiffness
E_r	Energy correction for the standard penetration test
E_s	Secant deformation modulus at 25% of peak stress
E_u	Undrained stiffness
F	Force; shape factor for borehole permeability test; correction factor; global factor of safety
f_1	Depth correction factor
F_d	Design value of an action
f_{eff}	Correction factor
f_I	Factor to correct for zone of influence
F_k	Characteristic value of an action
f_L	Coefficient
F_{rep}	Representative value of an action
f_s	Unit shaft friction on a pile; shaft friction on a cone; shape factor
Fs	Safety allowance
$F_{Serv;d}$	Design value of in service capacity of an anchor
$F_{Serv;k}$	Characteristic value of the maximum anchor force
F_{static}	Ultimate pile resistance from dynamic load tests
F_{STN}	Measured pile head load; statnamic load
f_T	Time factor
$F_{t;d}$	Tensile action
$F_{ULS;d}$	Design value of the force required to prevent any ultimate limit state in the supported structure
G	Shear modulus

G^*	Anisotropic shear modulus
G_0	Threshold shear modulus
G_{0HH}	Small-strain stiffness in the horizontal plane
G_{0VH}	Small-strain stiffness in the vertical plane
g_c, g_q, g_γ	Ground inclination factors
G_{dyn}	Dynamic velocity
G_{max}	Maximium shear modulus
G_s	Specific gravity of soil particles
h	Maximum height of the embankment; depth of cutting; depth of excavation; head of water
H	Height of free fall DPT hammer; horizontal load; thickness of layer; depth of caisson
i	Distance to point of inflection in a settlement trough; hydraulic gradient
i_c	Factor for the inclination of the load on a spread foundation
I_c	Influence factor; consistency index
i_c, i_q, i_γ	Load inclination factors
I_D	Density index; material index from flat dilatometer
I_d	Depth factor
I_e	Influence factor
I_L	Liquid limit
I_p	Plastic limit; influence factor
I_v	Deformability index
I_{vcx}	Soil viscosity index
I_z	Influence factor
J	Coupling parameter linking mean stress with shear strain and deviator stress with volumetric strain; factor
k	Coefficient of hydraulic conductivity (permeability); empirical factor
K	Constant; earth pressure coefficient
k^*_{100}	Coefficient of permeability at an effective stress of 100 kPa
k^*_{1000}	Coefficient of permeability at an effective stress of 1000 kPa
$k^*_{pc/2}$	Coefficient of permeability at an effective stress at OCR of 2
$k^*_{pc/20}$	Coefficient of permeability at an effective stress at OCR of 20
k_1, k_2	Factors
K_a	Coefficient of active earth pressure
k_{cap}	Hydraulic conductivity of a capillary
K_D	Horizontal stress index from flat dilatometer test
K_e	Coefficient to compensate for effective free length of pile
k_f	Average radial effective stress around the nail to the vertical effective stress
k_{field}	Field hydraulic conductivity
$k_{fracture}$	Conductivity of the discontinuities
k_h	Permeability in horizontal direction

k_{intact}	Conductivity of the matrix
$k_{l:uls}$	Limit value of the load loss for anchors
K_m	Coefficient of earth pressure
k_{max}; k_{min}	Maximum and minimum coefficient of hydraulic conductivity
k_n	Intrinsic conductivity for normally consolidated clay; statistical coefficient
K_c	Coefficient of earth pressure for the lateral stress on driven piles
K_{nc}	Coefficient of earth pressure for normally consolidated soils
K_o	Coefficient of earth pressure at rest
K_p	Coefficient of passive earth pressure
K_{qz}, K_{cz}	Coefficients for laterally loaded piles
K_s	Intercept of the settlement against settlement/load; coefficient of earth pressure for shaft friction on piles
k_s	Intrinsic conductivity for over-consolidated clay
$K_{s;i}$	Earth pressure coefficient
k_v	Permeability in vertical direction
k_δ	Factor
L	Length of test pocket; length of pile; length of spread foundation; distance between fractures
L'	Effective length of a spread foundation
L_f	Length of pile over which the capacity is developed
L_{fixed}	Length of the anchor
m	Fractal dimension; factor; dry mass after pretreatment
M	Mass of a pile; mass of DPT hammer; slope of critical state line in q–p′ space
M'	Mass of DPT rods and anvil
m_0	Mass of soil after pretreatment
m_2	Dry mass less than 2 mm
m_3	Mass of riffled soil
M_a	Pile inertia
M_s	Factor
m_v	Coefficient of volume compressibility
m_{vlab}	Coefficient volume compressibility from oedometer tests
n	Number of piles
N	Number of articles; stability number
N_{10}	Number of blows to drive DPT 10 cm
N_{20}	Number of blows to drive DPT 20 cm
N_{60}	Standard penetration test blow count
N_c	Bearing capacity factor for total stress analysis
N_k	Factor to convert cone resistance to undrained shear strength
N_o	Number of particles at reference diameter
N_q	Bearing capacity factor of piles
N_q, N_c, N_γ	Bearing capacity factors for spread foundations using effective strength

N_T	Empirical factor
OCR	Over-consolidation ratio
P	Horizontal pressure on lateral loaded piles; pile perimeter; % by weight of particles smaller than diameter D
p_f	Face pressure
p'_k	Measured effective stress in a specimen
p_{lm}	Pressuremeter limit pressure
p_{max}	Maximum horizontal pressure on lateral loaded piles
p_z	Lateral earth pressure on a pile
P_z	Lateral earth force on a pile
Q	Bearing resistance; function of damping factor; ultimate capacity of a pile group
q	Contact stress; rate of flow of water
q'	Net contact stress at formation level
Q_a	Allowable load on a pile
q_{ann}	Unit capacity of the annulus of a tubular pile
Q_b	Base resistance of a pile
q_b	Ultimate unit base resistance
$q_{b;k}$	Characteristic unit base resistance
q_c	Cone penetration resistance
q_{cm}	Weighted average of cone resistance in zone of influence of the base of a pile
q_{cm}	Weighted average of cone penetration resistance in the zone of influence below the base of a pile
q_d	Dynamic cone resistance
q_{plug}	Unit base capacity of a plugged tubular pile
Q_s	Shaft resistance of a pile
$q_{s;k}$	Characteristic unit shaft resistance
Q_{u1}	Base resistance of upper layer of soil
Q_{u2}	Base resistance of lower layer of soil
R	Particle radius; pile aspect ratio; creep factor; ultimate resistance along a slip surface
R_3	Time factor
R_{ax}	Critical radius controlling regelation and enhanced creep
$R_{b:cal}$	Calculated values of end bearing resistance
$R_{b;k}$	Characteristic base resistance
r_c, r_q, r_γ	Rigidity factors
$R_{c:cal}$	Calculated values of pile capacity
$R_{c;d}$	Design resistance of a pile
$R_{c;k}$	Characteristic resistance of a pile
$R_{c;m}$	Measured ultimate capacity from pile tests
r_d	Unit cone resistance
R_d	Design value of the resistance

R_f	Cone friction ratio
R_g	Group reduction factor
$R_{GG;calc}$	Anchor capacity
$R_{GG;m}$	Ultimate ground/grout interface resistance
R_s	Settlement ratio
$R_{s:cal}$	Calculated values of shaft resistance
$R_{s;k}$	Characteristic shaft friction
$R_{SLS;d}$	Serviceability limit state resistance of an anchor
$R_{t;d}$	Design resistance of a tension pile
$R_{t;k}$	Characteristic resistance of a tension pile
$R_{t;m}$	Measured capacity of tension pile
r_u	Ratio of the pore pressure to effective overburden pressure
$R_{ULS;d}$	Ultimate limit state resistance of an anchor
s	Settlement; proportion of substrate occupied by cavities; pile spacing
s_b	Settlement of a pile due to load on base
s_c	Shape factor for a spread foundation for total stress analysis
S, s_c, s_q, s_γ	Shape factors
s_{max}	Maximum settlement above a tunnel
s_{pile}	Elastic compression of a pile
$SPTN$	Standard penetration test blow count
s_{raft}	Compression of the soil beneath a raft
s_s	Settlement of a pile due to load on shaft
S_v	Structure index
s_y	Settlement of the ground surface above a tunnel
T	Time; time lag in borehole permeability test
t	Embedded length of support to an excavation
t', s'	Stress invariants
T_o	Time of dynamic measurement
u	Pore pressure
u_n	Ice velocity normal to substrate
u_w	Water pressure in cavities at the base of a particle
V	Shear wave velocity
v_a	Deflection of top of retaining wall
V_l	Volume loss above a tunnel
V_l	Volume of a tunnel
v_{min}	Velocity of a CRP pile
V_s	Volume of the settlement trough above a tunnel; volume of stones
V_x	Coefficient of variation
w	Water content
W	Weight of soil above a slip surface
w_1	Stiffness coefficient
w_2	Stiffness exponent
W_b	Load on the base of a pile

W_s	Load on the shaft of a pile
X_d	Design values for geotechnical parameters
X_k	Characteristic values for geotechnical parameters
X_{mean}	Mean value of the parameter
z	Depth
z_I	The zone of influence
z_o	Depth to tunnel axis
α	Coefficient of adhesion on a pile; coefficient relating cone resistance to stiffness; angle of slip plane to horizontal
α, β	Rate parameters; constants
α_{uls}	Limit value of the creep rate for anchors
β	Pile factor; angle of a slope
β_j	Empirical coefficient for pile capacity
γ	Unit weight
Γ	Specific volume (and void ratio) on critical state line at $p' = 1$ kPa
γ'	Submerged unit weight
$\gamma_{c'}$	Partial factor for the effective cohesion
γ_{cu}	Partial factor for the undrained shear strength
γ_E	Partial factor for the effect of an action
γ_F	Partial factor applied to an action
γ_G	Partial factor for a permanent action
$\gamma_{G;dst}$	Partial factor for a permanent destabilising action
$\gamma_{G;stb}$	Partial factor for a permanent stabilising action
γ_m	Partial factor for material properties
γ_Q	Partial factor for a variable action
$\gamma_{Q;dst}$	Partial factor for a variable destabilising action
$\gamma_{Q;stb}$	Partial factor for a variable stabilising action
γ_{qu}	Partial factor for the unconfined strength
γ_R	Partial factor for resistance
$\gamma_{R;e}$	Partial factor for earth resistance for retaining structures
$\gamma_{R;h}$	Partial factor for sliding resistance
$\gamma_{R;b}$	Partial factor for bearing resistance for retaining structures
$\gamma_{R;v}$	Partial factor for bearing resistance for shallow foundations
$\gamma_{s;t}$	Partial factor for tension piles
γ_{Serv}	Partial factor
γ_t, γ_b, γ_s	Partial factors for base resistance, shaft resistance and total resistance of piles
γ_γ	Partial factor for weight density
γ'_ϕ	Partial factor for the angle of shearing resistance (tan ϕ')
δ	Angle of interface friction
δ_{cv}	Constant volume interface friction
δ_i, δ_r	Interface friction
δ_p	Peak mobilised interface friction

Δv	Relative pile/slip velocity; pile velocity
Δv_{min}	Relative pile/slip velocity in a static test
$\Delta\sigma_v$	Increase in vertical stress in the soil due to an external load
ε	Strain
η	Kinematic viscosity
κ	Slope of unload/reload line in v–ln p′ space
μ	Correction factor; viscosity of a pore fluid
μ_d	Depth correction factor
μ_g	Soil type correction factor
ξ_{uls}, ξ_1, ξ_2, ξ_3, ξ_4	Factors
ρ	Settlement; density
ρ_m	Dry density of matrix
ρ_t	Corrected density of soil
σ'_{r1}	Preloading radial effective stress (ICP method)
σ'_v	Effective vertical stress
σ'_{vb}	Effective vertical stress at base of pile
σ'_{vmax}	Preconsolidation pressure
$\sigma'_{a(z)}$	Active earth pressure
σ_h	Total horizontal stress
σ_v	Total vertical stress
σ'_{rf}	Horizontal stress developed at failure due to the installation of a pile
σ'_{rc}	Local radial stress
τ	Shear stress
τ^*	Critical shear stress
τ_{bu}	Ultimate bond stress
τ_d	Ultimate shaft resistance
τ_{max}	Maximum shear resistance
τ_s	Shear resistance from a static test
τ_f	Local shear stress at the interface between a pile and soil
υ	Poisson's ratio
φ'	Angle of friction
φ'_{cv}	Constant volume angle of friction
φ'_{pk}	Peak angle of friction
φ'_{res}	Residual angle of friction
φ^b	Angle of shearing resistance to matric suction
$\Phi_{\varphi H}$	Angle of Hvorslev surface in τ–σ' space
Ψ	Factor to convert the characteristic value to the representative value; the soil suction
λ	Slope of one-dimensional compression line in v–ln p′ space
$\lambda_{a;ULS}$	Factor
λ_f	Interface factor

Glossary

Ablation: All processes by which snow and ice are lost from a glacier, floating ice or snow cover; or the amount which is melted.
Ablation moraines: Formed of supraglacial debris that remains as a glacier retreats.
Abrasion: The mechanical wearing or grinding away of rock surfaces by the friction and impact of rock particles transported by wind, ice, waves, running water or gravity.
Accumulation: All processes that add snow or ice to a glacier or to floating ice or snow cover – snow fall, avalanching, wind transport, refreezing.
Action: Force or displacement creating instability or restoring stability.
Active earth pressure: The horizontal stress exerted by a mass of soil on a retaining wall as the wall moves away from the soil.
Active (Rankine) zone: The area behind a retaining structure that is above the failure plane.
Activity: The ratio of plasticity index to percent by weight of clay.
Adhesion: The shear resistance between soil and a structure.
Air voids ratio: The ratio of the volume of air to the total volume of a mass of soil.
Allowable bearing capacity: The bearing pressure that can be allowed on a foundation soil, usually to limit settlements.
Allowable bearing pressure: The additional pressure above that already existing which can be carried safely by a foundation material.
Alluvial fan: An assemblage of sediments marking place where a stream moves from a steep gradient to a flatter gradient and suddenly loses transporting power.
Alluvial soils: Soils deposited in a valley or slightly graded area by transporting sediments through a mountain river or streams.
Alluvium: Deposit from river water.
Angle of shearing resistance (or friction): For a given soil, the angle on the graph of the shear stress and normal effective stresses at which shear failure occurs.
Angular distortion: The ratio between the relative deflection between two points in a foundation and the distance between them.
Anisotropic: A mass of soil having different properties in different directions.
Aquiclude: An impermeable layer of soil which water cannot flow through because of its low mass permeability.
Aquifer: A stratum of soil with relatively high permeability; a water-bearing stratum of rock or soil.
Aquitard: A material of low permeability that greatly slows the movement of groundwater.
Artesian: A condition that exists when the piezometric surface lies above the ground level; relates to a confined aquifer.
Artificial ground: Ground surface has been significantly modified by human activity (also known as made ground).
Assemblage: The collection of minerals that characterise a facies.

At-rest earth pressure: The horizontal stress developed in a mass of soil loaded in conditions of zero horizontal strain.
Atterberg limits: The water contents of a soil mass corresponding to the transition between a solid, semi-solid, plastic solid or liquid.
Axial strain: Direct strain measured along an axis of a triaxial test sample.
Axial stress: Total or effective stress (both confining and vertical stresses combined) acting along an axis of a triaxial test sample.
Axially loaded compressive piles: Piles that rely on end bearing and shaft resistance to resist compressive load.
Axially loaded tension piles: Piles that rely on shaft resistance to resist tensile load.
Basal shear zone: Basal ice and underlying sediments.
Baseline report: An assessment of risk based on the interpretative report and assessment of the category of the structure.
Bearing capacity: The ability of the underlying soil to support the foundation loads without shear failure.
Bearing capacity factors: Empirically derived factors used in a bearing capacity calculation.
Bearing pressure: The total stress transferred from the structure to the foundation, then to the soil below the foundation.
Bedding: A collective term used to signify the presence of beds, or layers, in sedimentary rocks and deposits.
Bedding plane: Surface separating layers of sedimentary rocks and deposits. Each bedding plane marks termination of one deposit and beginning of another of different character.
Bedrock: Strong rock underlying surface deposits of soil and weathered rock.
Block slide: Translational slide of block of soil or rock.
Bottomset bed: Layer of fine sediment deposited in a body of standing water beyond the edge of a growing delta and which is eventually built over by the advancing delta.
Boudinage: A structure in which brittle beds bounded by more ductile ones have been divided into segments during deformation of subglacial deposits.
Boulder bed: A concentration of boulders within a glacial soil layer.
Boulder clay: Local British term for till, considered inaccurate because neither boulders nor clay is an essential constituent.
Boulder train: Clusters of erratics from same source, with some distinctive characteristic that makes their common source easily recognisable.
Boulders: Soil particles over 200 mm in size.
Boussinesq equation: An equation used to determine the increase in vertical pressure at a particular depth that is caused by an application of a point load at a given surface.
Braided stream: A relatively shallow stream with many branches that migrate across a valley floor.
Brittle: Structural behaviour in which a material deforms permanently by fracturing.
Bulk density: The total mass of water and soil particles contained in a unit volume of soil.
Bulk unit weight: The total weight of water and soil particles contained in a unit volume of soil.
California bearing ratio (CBR): The ratio of the resistance to penetration, by a plunger into the soil being tested, to a standard resistance.
Capillary action: The ability of water to flow through narrow spaces without the assistance of forces such as gravity.
Capillary rise: The rise of pore water due to capillary action.
Circular slide: Movement of a block of soil along a curved failure surface.
Clast-dominated till: Coarse-grained till containing some fine-grained particles.

Clastic: Refers to sediments made up primarily of broken fragments of pre-existing rocks or minerals.
Clay: Soil particles that are finer (smaller) than 0.002 mm in size.
Clay fraction: The fraction by weight of a sediment of size less than 0.002 mm effective spherical diameter.
Clay minerals: A group of alumino-silicate minerals with characteristic sheet structures.
Coarse-grained soils: Soils with more than 50% by weight of grains retained on the #200 sieve (0.075 mm).
Cobbles: Soil particles between 63 and 200 mm in size.
Coefficient of active earth pressure: The ratio of the minimum horizontal effective stress of a soil to the vertical effective stress at a single point in a soil mass retained by a retaining wall as the wall moves away from the soil.
Coefficient of compressibility: The ratio of void ratio difference to the effective pressure difference of two different loadings during primary consolidation.
Coefficient of consolidation: The rate of change of volume during primary consolidation.
Coefficient of curvature: A measure of the shape parameter obtained from a grain size distribution curve.
Coefficient of earth pressure at rest: The ratio of horizontal effective stress of a soil to the vertical effective stress at a specific point in a soil mass in conditions of zero horizontal strain.
Coefficient of permeability: See *Hydraulic conductivity.*
Coefficient of uniformity: A measure of the slope of a grain size distribution curve, and therefore the uniformity of the soil.
Cohesionless soils: Granular soils (sand and gravel type) with values of cohesion close to zero.
Cohesive soils: Fine-grained soils that have an undrained shear strength.
Cold glacier: A glacier in which the bulk of the ice is below the pressure melting point.
Cold ice: Dry ice below the pressure melting point.
Comminution till: Deformation till that is formed entirely of rock flour as a result of abrasion during deformation.
Compaction: Volume change in soils in which air is expelled from the voids, but with the water content remaining constant.
Compaction grouting: Monitored displacement of ground without fracturing the ground by pumping grout into the ground.
Compaction piles: Granular piles created with a downhole hammer.
Compaction test: Test to assess optimum water content and maximum dry density.
Complex failure: Failures in which one of the five types of movement is followed by another type (or the same type).
Compound slide: Combination of rotational and translational slides.
Compressibility: Variation of voids ratio with applied load.
Compression index: The logarithmic slope of the primary consolidation curve.
Cone resistance: The resistance force divided by the end area of the cone tip, measured during the cone penetration test.
Confined aquifer: An aquifer in which the piezometric surface is above the top of the aquifer.
Consolidation test: One-dimensional consolidation test.
Continental ice glacier: An ice sheet that obscures all but the highest peaks of a large part of a continent.
Cross-bedding: See *Inclined bedding.*
Current bedding: Bedding which is formed at an angle to the horizontal by the action of swift local currents of water or air.
Cutting: An excavated slope.

Debris flow: Fast-moving, turbulent mass movement with a high content of both water and rock debris.

Debris slide: Translational slide of debris triggered by rainfall or surface water creating a mantle on the slope.

Deep soil mixing: Columns or blocks of lime or cement soil mixture; mixed *in situ.*

Deformation till: Completely disaggregated and possibly homogenised sediment by shearing in a subglacial deforming layer.

Density index: See *Relative density.*

Depositional environment: The nature of the environment in which sediments are laid down.

Desk study: Studies of topographical, historical and geological maps, aerial photographs, geological memoirs and historical evidence of ground movement.

Dewatering: Increase in effective stress consolidating the soil.

Diamictons: Poorly sorted sediments.

Dilatancy: Tendency of the volume to increase under increasing shear stress.

Direct shear test: A test in which the upper half of the soil is sheared against the lower half of the specimen.

Disturbed samples: Soil samples obtained in a manner which destroys the original orientation and some of the physical and mechanical properties of the natural material.

Drift: Glacial deposits laid down directly by glaciers or laid down in lakes, ocean, or streams as a result of glacial activity.

Dropstone: Clasts released from icebergs to be embedded in water-lain glacial soils.

Drumlin: Streamlined hill, largely of till, with blunt end pointing to the direction from which ice moved; occurs in clusters called drumlin fields.

Ductile: Structural behaviour in which a material deforms permanently without fracturing.

Dump moraine: Debris delivered to a stationary steep ice margin forming a moraine.

Dynamic compaction: Densification by falling surface weight.

Dynamic probe: Low cost, simple, rapid *in situ* test used to obtain profiles of the number of blows every 10–20 cm of a standard weight falling a standard height to drive the cone a certain distance.

Earth pressure balance tunnel boring machines: Use the excavated soil to maintain the pressure; the soil is excavated and then removed with a screw conveyor.

Effective strength parameters: Strength described in terms of effective stress by cohesion and angle of friction.

Effective stress analysis: Stability analysis based on effective stresses and effective strength parameters.

Elastic: Non-permanent structural deformation during which the amount of deformation (strain) is proportional to the stress.

Elastic modulus: The ratio of tensile (or compressive) stress in a material to the corresponding tensile (or compressive) strain.

Elasticity: The tendency for a body to return to its original shape and size when a stress is removed.

Electro-osmosis: Movement of a liquid under an applied electric field through a permeable medium.

Embankment: Usually referred to a 'built-up' section of soil (engineered fill) as for roads or dams.

Embedded walls: Retaining walls that rely on passive earth pressure to resist the active earth pressure.

End bearing capacity: The bearing capacity at the bottom of one member of a deep foundation system.

End bearing pile: A pile that derives the majority of its load bearing ability from the support of the soil layer beneath the tip of the pile.
End moraine: See *Terminal moraine.*
End product specification: A specification for the properties of the outcome of the construction process.
Engineered fill: Soils used as fill, such as retaining wall backfill, foundation support, dams, slopes, etc. placed in accordance to engineered specifications.
Engineering properties: Engineering parameters of a soil such as permeability, shear strength and consolidation; different from index properties.
Englacial: Within a glacier.
Equilibrium line: On a glacier the line separating the zone of accumulation from the zone of ablation.
Equipotential: For a flow net, lines connecting points of equal total head.
Erosion: Erosion is the wearing away of the Earth's surface by the sea, rivers, glaciers and wind.
Erratic: A relatively large rock fragment, lithologically different from its surrounding rock, which has been transported from its place of origin (usually by glacial action).
Esker: A sinuously curving, narrow deposit of coarse gravel that forms along a meltwater stream channel, developing in a tunnel within or beneath the glacier.
Eustatic change in sea level: A worldwide change in sea level, such as caused by melting glaciers.
Excess pore pressure: The increment of pore water pressure greater than the ambient pore pressure at steady state.
Fabric: Size, shape and arrangement of the solid mineral and the associated voids; similar in meaning to soil structure.
Facies: Character of a glacial soil expressed by its formation and composition.
Factor of safety: The ratio of a limiting value of a quantity to the design value of that quantity.
Factual report: Desk study, field work and laboratory tests which presents all relevant topographical, geomorphological, geological, geotechnical and hydrogeological factual data.
Failure envelope: For a given soil, the graph of the shear stress and normal effective stresses at which shear failure occurs.
Fall: When applied to mass movement of material refers to free fall of material moving without contact with the surface.
Fine-grained soils: Soils containing particles smaller than No. 200 sieve or 0.075 mm.
Fines content: Soil grains smaller than No. 200 sieve (0.075 mm).
Fissured clays: Clays which in their natural state show a system of fissures somewhat similar to a jointing system in a hard rock mass, on a reduced scale.
Fjord: Glaciated valleys now flooded by the sea.
Flexible facings: Permanent facing to a slope by stabilising the soil between the nails and transmitting some of the load on the nails to the soil via the nail plate.
Flow line: The path water will follow traveling from high head to low head in a seepage flow analysis.
Flow net: A graphical analysis of seepage flow in a mass of soil to estimate flow quantities and pore pressures.
Flow rate: The ratio of total volume of water flowing to a particular unit of time.
Flow slide: Translational slide in saturated soil due to an increase in water pressure causing the soil to flow as a viscous fluid possibly considerable distances.
Flow velocity: The velocity of water flow through a soil.

Flutes: Surface of ridges typically less than 3 m wide and less than 3 m high.

Footing: An enlargement at the base of a foundation that is designed to transmit forces to the soil.

Foreset bed: Inclined layers of sediment deposited on the advancing margin of a growing delta.

Formation level: The depth below the ground surface where the base of a foundation is located.

Foundation: A component of an engineered structure that transmits the structure's forces into the soil or rock that supports it.

Friction pile: A pile that derives the majority of its load bearing ability from the shaft resistance between the soil and the pile.

Geotextiles: A synthetic fabric used to stabilise soils, retain soils, prevent the mixing of dissimilar soils, provide a filtering function, pavement support, subgrade reinforcement, drainage, erosion control and silt containment.

Glacial deposit: Sediments deposited by ice sheets, glaciers and melt streams.

Glacial drift: The general term for all glacial deposits, both unsorted and sorted (see *Stratified drift*).

Glacial ice: Compacted and inter-grown mass of crystalline ice.

Glaciation: The formation, advance and retreat of glaciers and the results of these activities.

Glacier: A mass of ice and snow which can deform and flow under its own weight.

Glaciofluvial deposits: Terrestrial sediments deposited from flowing water either on (channels), within (tunnels), beneath (tunnels) or beyond the ice margin.

Glaciolacustrine deposits: Formed in a standing body of fresh water such as that found at ice margins.

Glaciomarine deposits: Formed when a glacier terminates in the sea.

Glaciotectonite: Subglacially deformed rock or superficial deposit that retains some of the original structure of the parent material.

Global factor of safety: Ratio of the restoring force to the disturbing force for the whole of the structure.

Graded bedding: Type of bedding sedimentary deposits in which individual beds become finer from bottom to top.

Gravity walls: Retaining walls which depend upon their self-weight to provide stability against overturning and sliding.

Ground anchor: A tie or tendon anchored deep in the ground and stressed to provide a retaining force for a structure at the surface of an excavation.

Ground moraine: Till deposited from main body of glacier during ablation.

Groundwater table: See *Water table.*

Groundwater: Water occupying interstices, fissures and cavities in the ground.

Grout: Often neat cement slurry or a mix of cement and sand or other additives.

Hard facing: Sprayed concrete, or cast *in situ* or precast concrete panels to resist earth pressures and transfer soil nail load to the soil via the facing.

Homogeneous soils: A mass of soil where the soil is of one characteristic having the same engineering and index properties.

Horizontal strain: Strain measured in a horizontal direction.

Horizontal stress: Total or effective stress acting in a horizontal direction.

Hydraulic conductivity: Also, coefficient of permeability; the constant average discharge velocity of water passing through soil when the hydraulic gradient is equal to 1.0.

Hydraulic gradient: Between two points in a hydraulic flow – the difference in total head (piezometric levels) divided by the length of the flow path.

Hydraulic head: The level to which groundwater in the zone of saturation will rise.

Hydrofracture: Hydraulic fractures filled with grout.

Hydrometer test: Laboratory test used to determine the amount and distribution of finer particles of a soil sample.

Hydrostatic pore pressure: Pore water pressures exerted under conditions of no groundwater flow where the magnitude of pore pressures increases linearly with depth below the ground surface.

Ice sheet: A broad, mound-like mass of glacier ice that usually spreads radially outward from a central zone.

Immediate settlement: The settlement of a foundation occurring immediately upon loading.

Inclined bedding (cross-bedding): Bedding laid down at an angle to the horizontal.

Index properties: Attributes of a soil such as moisture content, void ratio, specific gravity, Atterberg limits and grain size distribution.

Infiltration: The downward movement of subsurface water under gravity from the land surface to the water table; that is, it is restricted to the zone of aeration.

***In situ*:** Undisturbed, existing field conditions.

Instrumentation: Geotechnical instruments used to monitor conditions such as deformations, pressures, loads, etc. within the ground.

Interglacial: A phase of relatively warm temperatures between glacial periods.

Internal erosion: Seepage of groundwater along a preferential flow path causing loss of fines or slumping.

Interpretative report: An assessment of the geological, geotechnical and hydrological models.

Isolated footing: A footing designed to support a structural load from a single column; usually a shallow foundation, and square or circular in shape.

Isostatic change in sea level: A sea level change due to change in load on Earth's crust.

Isotropic: A soil mass having essentially the same properties in all directions.

Jet grouting: Grout jetted into soil liquefied by the jetting process creating columns; replacement of soil with grout eroded with water or air jets.

Joint: A surface of fracture in a rock, without displacement parallel to the fracture.

Kame: Stratified drift deposited in depressions and cavities in stagnant ice and left as irregular, steep-sided hills when the ice melts.

Kame delta: A deposit, often triangular, formed where a glacial stream enters into a proglacial lake.

Kame terrace: Stratified drift deposited between wasting glacier and adjacent valley wall.

Kettle: Depression in ground surface formed by the melting of a block of glacier ice buried or partially buried by drift.

Land system: Combination of sediment, landforms and landscapes.

Landform: A singular sculpted feature.

Landslide: Rock or soil displaced downhill by gravity.

Lateral moraine: Moraine formed by valley glaciers along valley sides.

Laterally loaded piles: Piles that rely on lateral support from the soil to resist horizontal load.

Lime columns: Columns of soil mixed with lime.

Limit equilibrium: Failure mechanism in which the restoring force balances the disturbing force.

Liquid limit: The water content above which the soil will flow like a liquid, but below which it will have a plastic consistency.

Liquidity index: A measure of the relationship between the current water content of a soil and its consistency limits.

Lithofacies: Petrological characteristics of a sediment with particular characteristics including colour, clast fabric, clast shape, particle size distribution, composition and sedimentary structures.

Lodgement: Lodgement of glacial debris beneath a glacier by pressure melting or other mechanical processes.
Long-term conditions: Conditions in the ground where full consolidation has taken place and there are no excess pore pressures.
Made ground: See *Artificial ground.*
Maintained load test: Static load test of a pile.
Marchetti dilatometer test: A 250-mm-long, 94-mm-wide and 14-mm-thick blade with a tip angle of 16° with a flat, 60-mm-diameter steel membrane mounted flush on one side used to load the soil.
Matrix-dominated till: Fine-grained till containing some coarse-grained particles.
Matrix water content: The water content expressed in terms of the weight of the matrix rather than the total weight of the solids.
Maximum dry density: Maximum density of a soil for a given compactive effort.
Mean normal stress: The mean value of the three orthogonal stresses.
Medial moraine: Formed by the merging of lateral moraines as two valley glaciers join.
Melt-out till: Glacial debris being deposited from stagnant or slow moving ice without further transport or deformation.
Method specification: A specification for the processes of construction.
Micropile: Piles with relatively small diameters formed of reinforcement inserted into grout-filled boreholes.
Modified Proctor: Laboratory test used to determine maximum dry density and optimum moisture content of soils.
Modulus of subgrade reaction: The unit pressure applied in a plate test divided by the settlement that takes place.
Mohr's circle: A circle constructed in the triaxial test using the principal stresses in order to determine the stresses on the failure plane.
Moisture content: See *Water content*; the ratio between the mass of water and the mass of soil solids.
Moraine: The material eroded by a glacier and carried along by the ice, before being dumped when the glaciers retreat.
Mountain glacier: See *Valley glacier.*
Mudflow: Form of mass movement similar to a debris flow but containing less rock material.
Negative skin friction: Forces induced on deep foundations resulting from downward movement of adjacent soil relative to the foundation element.
Nominal bearing pressure: Allowable bearing pressure for spread foundations on various soil types, derived from experience which provides safety against shear failure or excessive settlements.
Non-circular slide: Movement of a block of soil along a non-circular surface.
Non-conformity: An unconformity that separates profoundly different ground types, such as glacial till and the underlying bedrock.
Normal compression line: The relationship between void ratio and the normal effective stress for soil loaded beyond the current yield stress in an isotropic compression.
Normal force: Force acting normal to the plane of reference.
Normally consolidated soil: Soil having a current state which lies on the normal compression line.
Oedometer test: See *Consolidation test.*
One-dimensional compression: Compression taking place with zero radial and horizontal strain.
One-dimensional modulus: The ratio of the change in vertical effective stress to the change in vertical strain, when there is zero horizontal strain.

Optimum moisture content: The water content at which the maximum dry density of a soil is obtained using a specific effort of compaction.

Outwash: Meltwater-deposited sediment, dominantly sand and gravel, showing increasing rounding and sorting into layers with increasing distance from the ice margin.

Outwash plain: A plain of glaciofluvial deposits of stratified drift from meltwater-fed, braided, and overloaded streams beyond a glacier's morainal deposits.

Overall stability: Mass movement of the ground leading to damage or loss of serviceability of a structure and neighbouring structures, roads or services.

Overburden pressure: The total or effective stress at a given depth due to the weight of overlying soil.

Over-consolidated soil: A soil carrying a higher load in the past.

Over-consolidation ratio: The ratio of maximum past pressure (preconsolidation pressure) to the current effective stress.

Overturning: Overturning failure is a result of excessive lateral earth pressures with relation to retaining wall resistance, thereby causing the retaining wall system to topple or rotate (overturn).

P-wave (primary wave, compressional wave): A seismic body wave that involves particle motion, alternating compression and expansion, in the direction of wave propagation.

Partial factors: Applied separately to the material, actions and the components or assembly.

Particle size distribution: Soil particle sizes that are determined from a representative sample of soil that is passed through a set of sieves of consecutively smaller openings.

Passive earth pressure: The maximum horizontal stress exerted by a mass of soil on a retaining surface as the surface moves towards the soil.

Peak shear strength: The maximum shear strength of a soil at a given normal effective stress and water content.

Perched water table: A water table that develops at a higher elevation than the main water table.

Performance specification: A specification for the performance of the outcome of the construction.

Permeability: A measure of continuous voids in a soil with the capacity of material to transmit water or other fluids.

Permeation grouting: Replacement of water in voids with grout using low pressures.

pH value: A measure of acidity or alkalinity of groundwater or soil water extract based on the hydrogen ion content.

Piezometer: An instrument used to measure *in situ* pore water pressures.

Piezometric surface: An imaginary surface corresponding to the hydrostatic water level of a confined body of groundwater.

Pile: A slender member of a deep foundation system that is driven (hammered), drilled or jetted into the ground.

Pile driving analyser: A method to compute average pile force and velocity by using wave equation analysis with electronic measurements.

Pile spacing: The distance from centre to centre of piles.

Piping: The movement of soil particles as a result of unbalanced seepage forces produced by percolating water.

Planar structure: Term used in geological reconnaissance to describe features, such as bedding, cleavage, schistosity, joints, faults and flow banding, until they can be identified by detailed investigation.

Plane strain: A two-dimensional state of stress, where the out-of-plane strain (i.e. the strain normal to the plane being considered) is zero.

Plastic deformation: The distortion of soil resulting in a permanent and irrecoverable change in shape or volume.
Plastic limit: The water content in which a soil will have a plastic consistency.
Plastic strain: Deformation of soil that is not recovered upon unloading.
Plasticity: The property of a soil which allows it to deform continuously.
Plasticity index: The difference between the water contents of a clay at the plastic and liquid limits, that is, the range of water content for which the clay is plastic.
Plate bearing test: A 300 mm (or larger) diameter rigid metal plate bedded onto the soil in increments of about one-fifth of the design load.
Poisson's ratio: The ratio of the change in strain perpendicular to the direction of loading to the change in strain caused in the same direction.
Polar glacier: A glacier whose temperature throughout is always below freezing.
Pore air pressure: The pressure of air within the void space of a partially saturated soil.
Pore pressure: The pressure exerted by the fluid within the pores or voids in a porous material; in saturated soil, the pore pressure is the pore water pressure.
Pore pressure coefficient: The ratio of the change in pore pressure to the change in deviator stress or change in isotropic stress in undrained loading.
Pore pressure ratio: At a given depth of soil, the ratio of the pore water pressure to the vertical overburden pressure.
Pore water: The water partially or completely occupying pore spaces in soil.
Porosity: The ratio of the volume of voids to the total volume.
Post-peak shear strength: The shear strength of a brittle soil at a given normal effective stress and water content at a strain exceeding the strain at the peak strength.
Potentiometric surface: The level to which water will rise in an artesian system when its confining aquitard is pierced.
Preconsolidation load: The maximum load ever imposed on a particular soil mass in its geological history.
Preloading: Surface load applied to consolidate the soil; can be used with vertical drains.
Pressure head: The height of a column of water required to develop a given pressure at a given point.
Pressure melting: The phenomenon causing increased melting of ice by increase of pressure.
Pressuremeter: Cylindrical flexible membrane inflated within the soil to obtain the stiffness, and in weaker materials the strength, of the ground.
Primary consolidation: The long-term consolidation of a clay from the loss of water from the voids due to a high pressure.
Primary glacial deposits: Glacial soils that are subject to shear as well as compression during deposition.
Principal strains: The strains occurring in the directions of the principal axes of strain.
Principal stresses: Normal stresses acting in the direction of principal axes of stress.
Proctor test: Laboratory test used to determine maximum dry density and optimum moisture content of soils.
Pro-glacial: The area immediately adjacent to a glacier, often affected by outwash and by ice- or moraine-dammed lakes.
Progressive failure: Failure surface develops in brittle soils due to loss of strength post-peak progressively transferring the load along the failure surface.
Push moraines: Moraines formed as the ice advances and bulldozes the pro-glacial sediment.
Quick clays: Clay sensitive to disturbance, whereby shear strength may be substantially reduced.
Radial stress: In a triaxial sample, the total or effective stress acting perpendicular to the longitudinal axis.

Raft: Large piece of rock buried in drift.

Raft foundation: A structural slab utilised as a footing, which usually encompasses the entire building footprint.

Rapid impact compaction: Densification by high-frequency hydraulic hammer.

Recessional moraine: Ridges of glacial till marking halt and slight readvance of glacier during its general retreat.

Relative compaction: A minimum density specification usually designated as a percentage of the maximum dry density.

Relative deflection: The deflection in a foundation due to settlement.

Relative density (density index): The density of a granular soil relative to the minimum and maximum densities achieved for that particular soil.

Residual strength: The strength of a soil at large strains.

Retaining wall: Walls, usually constructed of concrete or rock that provide lateral stability of the Earth, thus preventing the soil from sloughing or slope failure.

Rock flour: Finely divided rock material ground by glacial action and fed by streams fed by melting glaciers.

Rogens: Irregular transverse moraines which are typically 10–20 m high, 50–100 m wide and 1–2 km long.

S-wave (secondary wave, shear wave): A seismic body wave that involves particle motion from side to side, perpendicular to the direction of wave propagation; S-waves are slower than P-waves and cannot travel through a liquid.

Salinisation: A process by which salts accumulate in soil.

Sand: Particles between 0.063 and 2 mm.

Sand fraction: The fraction by weight of a sediment of size range between 0.063 and 2 mm effective spherical diameter.

Saturated density: Density of soil when the voids are filled with water.

Scour: Water flowing across the surface (e.g. run-off, water course) leading to gullies.

Secondary consolidation: The considerably decreased rate of consolidation following the completion of primary consolidation.

Secondary glacial deposits: Glacial soils that are deposited by water or ice within water and only subject to gravitational consolidation.

Sedimentary facies: An accumulation of deposits that exhibits specific characteristics and grades laterally into other sedimentary accumulations that were formed at the same time but exhibit different characteristics.

Sedimentary soils: Soils formed by the deposition of fine-grained soil in water.

Sedimentation compression curve: The relationship between void ratio and overburden pressure for naturally occurring soils.

Seepage: The flow of water through soil.

Seepage force: The force transmitted to a mass of soil due to the seepage of groundwater.

Seepage pressure: The seepage force per unit volume.

Seepage velocity: The average velocity at which groundwater flows through the pores of a soil.

Sensitivity: A measure of the change in ultimate strength of clays between undisturbed and disturbed samples.

Serviceability limit state: Concerned with fit for purpose and is usually associated with deformation.

Settlement: The downward movement of soil, or the downward movement of a foundation.

Shaft resistance: The shear stress on the shaft of a pile, caisson or cone penetrometer.

Shallow foundation: Refers to a foundation system that the depth is much smaller than the foundation width.

Shape factors: Factors used in a general bearing capacity equation which provides an adjustment relating to the footing geometry.

Shear modulus: The ratio of the change in shear stress to the resulting change in shear strain.

Shear strain: The angular distortion or change in shape of a mass of soil.

Shear strength: The maximum shear stress which a soil can sustain under a given set of conditions.

Shear stress: The force per unit area acting tangentially to a given plane or surface.

Sheet pile: Steel section panels that are driven into the ground to provide lateral support.

Shrinkage index: The difference between the plastic and shrinkage limits.

Shrinkage limit: The water content corresponding to the transition between a brittle solid and a semi-solid.

Silt: Particle size between 0.002 and 0.063 mm.

Silt fraction: The fraction by weight of a sediment of size range 0.002–0.063 mm effective spherical diameter.

Site reconnaissance: Observation of regional landforms and exposures of glacial soils in the vicinity of the project.

Slab slide: Translational slide in which the sliding mass remains intact.

Sliding: Sliding failure is a result of excessive lateral earth pressures with relation to retaining wall resistance thereby causing the retaining wall system to move away (slide) from the soil it retains.

Slurry: A thick mixture of soil and water.

Slurry tunnel boring machines: Use pressurised slurry which is mixed with the excavated soil supports the face and that mix is pumped to the surface where the soil is separated out so that the slurry can be reused.

Soft facings: Prevents erosion of topsoil from a slope while vegetation is established.

Soil: All particulate materials above bedrock.

Soil classification: Standardised classification schemes that delineates soil characteristics that are important in determining soil behaviour.

Soil classification system: A system of soil classification based on size, consistency and structure.

Soil horizon: A layer of soil that is distinguishable from adjacent layers by characteristic physical properties such as texture, structure, or colour, or by chemical composition.

Soil moisture: Groundwater in the zone of aeration.

Soil nailing: Slope stabilisation method that involves installing and usually grouting closely spaced metal nails in the soil.

Soil nails: Nails installed by driving, drilling and grouting or firing.

Soil structure: The combination of soil particles into aggregates or clusters which are separated from adjacent aggregates by surfaces of weakness.

Soil suction: See *Capillary rise.*

Soil texture: The physical nature of the soil, according to its relative proportions of sand, clay and silt.

Specific gravity: The density of a substance compared with the density of water at 4°C.

Spread footing: A footing designed to support a structural load from a single column.

Square footing: Isolated/spread footing shaped as a square.

Stabilisation: Process of mechanical and/or chemical treatment of a soil to increase its strength or its other properties of practical importance.

Standard penetration resistance: The number of blows required to drive a split-spoon sampler during a standard penetration test a distance of 0.305 m after the initial penetration of 0.15 m.

Standard penetration test (SPT): A field test that measures resistance of the soil to the penetration of a standard split-spoon sampler that is driven 0.3-m, 63.5-kg hammer dropped from a height of 0.76 m; the N-value is derived from this test.
Standard Proctor: See *Proctor test.*
Static cone penetrometer: Electric cone pushed at 20 ± 5 mm/s into the ground to rapidly give a semi-continuous profile of resistance.
Steady-state pore pressure: The pore water pressure at equilibrium when all excess pore pressures have fully dissipated.
Stiffness: Susceptibility to distortion or volume change under an applied load.
Strain: A measure of the change in size or shape of a mass of soil relative to its original size or shape.
Strain rate: The rate at which a body changes shape or volume as a result of stress.
Stratification: The accumulation of material in layers or beds.
Stratified drift: Sediments deposited by glacial meltwater that are sorted and layered; a major subdivision of glacial drift that includes river, lake and marine deposits.
Strength: The ability to withstand a stress without permanent deformation.
Strength index: Undrained shear strength used for classification.
Stress: The intensity of force per unit area; normal stress is applied perpendicularly to a surface or plane, whereas shear stress is applied tangentially to a surface or plane.
Stress history: The past history of loading and unloading of a soil mass.
Stress path: A path describing the changes in principal stresses due to changes in external load or pore pressure.
Strip footing: A horizontally long footing supporting a wall.
Subglacial: Beneath a glacier.
Subglacial traction tills: Tills deposited by ice and those that undergo deformation.
Subgrade modulus: See *Modulus of subgrade reaction.*
Subgrade stabilisation: *In situ* mixing of surface layer with lime or cement.
Submerged density: Difference between the total density and the density of water.
Substrate: Rock and superficial deposits possibly including remnants of previous glaciations.
Supraglacial: On a glacier.
Surcharge: An additional force applied at the exposed upper surface of a restrained soil.
Swell: Increase in soil volume; volumetric expansion of particular soils due to changes in water content.
Swelling index: The logarithmic slope recompression (reloading) line.
Swelling index: The slope of the swelling (unloading) line.
Temperate glacier: A glacier whose temperature throughout is at, or close to, the pressure point of ice, except in winter when it is frozen for a few metres below the surface.
Tension crack: Cracks appearing at the surface of a soil mass, often adjacent to a retaining wall or top of a failing slope.
Tension crack depth: The depth of a tension crack from the ground surface to a depth at which the horizontal effective stress is zero.
Tension pile: Piles that are designed to resist upward forces.
Terminal moraine (end moraine): Ridge of till marking farthest extent of glacier.
Texture: The general appearance of a soil as shown by the size, shape and arrangement of the materials composing it.
Till: Glacial drift composed of rock fragments that range from clay to boulder size and randomly arranged without bedding.
Tillite: A sedimentary rock composed of till.
Time factor: A dimensionless quantity dependent on the degree of consolidation that is used in primary consolidation analyses.

T-line: The relationship between plasticity index and liquid limit which defines glacial soils.

Toppling: Usually associated with rock slopes but occurs with eroding cliffs composed of matrix-dominated tills.

Topset bed: Layer of sediments deposited over surface of a delta, nearly horizontal and covering the tops of the inclined foreset beds.

Total head: The height of the free water surface above a given datum.

Total stress: The stress acting on or within a soil mass due to self weight and surcharges.

Total stress analysis: Stability analysis based on the total stresses and undrained shear strength.

Translational slide: Movement of a shallow mass of soil along a surface approximately parallel to the surface.

Transported soil: A soil that has been moved from the site of its parent rock.

Triaxial stress test: Laboratory tests such as the consolidated-drained (CD) test, consolidated-undrained (CU) test and unconsolidated-undrained (UU) test that are used to determine the soils' strength characteristics such as cohesion and angle of internal friction.

Turbidite: Sedimentary deposit settled out of turbid water carrying particles of widely varying grade size; characteristically displays graded bedding.

Ultimate bearing capacity: The bearing stress which would cause shear failure in the soil below a foundation; dependent upon the shear strength of the soil, applied loads and on the shape and depth of the foundation.

Ultimate limit state: Loss of equilibrium, excessive deformation of the ground, uplift, hydraulic heave, internal erosion and piping.

Unconformity: A buried erosion surface separating two rock masses.

Unconsolidated: A sediment is unconsolidated if the particles are not in equilibrium with the internal stresses.

Undisturbed samples: Samples of soil from a borehole or trial pit which have been disturbed so little that they can be reliably used for laboratory measurements of their strength and stifffness.

Undrained shear strength: The shear strength of a saturated soil at a given water content (or voids ratio, or specific volume) under loading conditions where no drainage of pore water can take place.

Unfissured clays: Clays which in their natural state do not contain a system of fissures.

Uniformity coefficient: See *Coefficient of uniformity.*

Unit weight: The ratio of the total weight of soil to the total volume of a unit of soil.

Unit weight of water: The weight of a unit volume of water.

Unloading: The release of confining pressure associated with the removal of overlying material (e.g. excavation).

Unsaturated strength: Strength of partially saturated soil.

Vacuum preloading: Application of vacuum at surface to create an atmospheric surcharge.

Valley glacier (alpine glacier, mountain glacier): Streams of ice that flow down valleys in mountainous areas.

Vane: Cruciform vane mounted on a solid rod pushed into the soil, a torque applied to the vane and the rotation and torque measured.

Vane shear test: A field test used to measure the shear strength of a soil that is low-strength, homogeneous and cohesive.

Varve: A pair of sedimentary units, one coarse-grained, the other fine-grained, interpreted as representing one year of sedimentation.

Varved clays: Clays that are layered with fine and coarse varieties.

Vertical stress: The total or effective stress acting vertically in a soil mass at a given depth caused by the soil's own weight.

Vibro-stone columns: Vibrating poker to create compacted stone columns.
Vibro-compaction: Densification by vibrating poker with water flushing.
Void index: The void ratio compared to that at 100 and 1000 kPa during one-dimensional compression.
Void ratio: The ratio of the volume of voids to the volume of solids (soil grains).
Volume of solids: Volume of soil grains in a total soil volume.
Volume of voids: Volume of air space in a total soil volume.
Volume of water: Volume of water in a total soil volume.
Volumetric strain: The ratio of the change in soil volume to the original soil volume.
Warm glacier: A glacier at a temperature of the pressure melting point throughout.
Warm ice: Ice at the melting point regardless of pressure.
Water content: The ratio between the mass of water and the mass of soil solids.
Water table: The level below which the pore spaces of the soil or rocks are completely saturated with water and the hydrostatic water pressure is zero; the surface between the zone of saturation and the zone of aeration.
Weathering: The process by which Earth materials change when exposed to conditions at or near the Earth's surface and different from the ones under which they formed.
Wedge failure: Failure surface defined by discontinuities.
Weight of soil grains: The dry weight of soil grains in a mass of soil.
Weight of water: The weight of water contained in the void space of a body of soil.
Yield point: The point at which the soil loading behaviour changes from elastic to inelastic.
Yield stress: The stress at which yielding takes place in soils; the stress at which the swelling-recompression line joins the normal compression line.
Zero air voids curve: The curve created by plotting dry densities of soils corresponding to saturation at each water content.
Zone of ablation: The area of wastage in a glacier.
Zone of accumulation: The area in which ice accumulates in a glacier.
Zone of aeration: That part of the ground in which the voids are not continuously saturated.
Zone of saturation: The zone below the zone of aeration in which all pore spaces are filled with water.

References

Aas, G., S. Lacasse, T. Lunne, and K. Hoeg. *Use of In Situ Tests in Geotechnical Engineering Proceedings of In Situ '86, a Specialty Conference*. American Society of Civil Engineers. Blacksburg, VA, 1986: 1–30.

Abbiss, C. P. Shear wave measurements of the elasticity of the ground. *Geotechnique*, 31(1); 1981: 91–104.

Abbiss, C. P. Calculation of elasticities and settlements for long periods of time and high strains from seismic measurements. *Geotechnique*, 33(4); 1983: 397–405.

Alley, R. B. Water-pressure coupling of sliding and bed deformation: I. Water system. *Journal of Glaciology*, 35(119); 1989a: 108–118.

Alley, R. B. Water-pressure coupling of sliding and bed deformation: II. Velocity-depth profiles. *Journal of Glaciology*, 35(119); 1989b: 119–129.

Alley, R. B. Deforming-bed origin for southern Laurentide till sheets? *Journal of Glaciology*, 37(125); 1991: 67–76.

Alley, R. B. In search of ice-stream sticky spots. *Journal of Glaciology*, 39(133); 1993: 447–454.

Alley, R. B., D. D. Blankenship, C. R. Bentley, and S. T. Rooney. Deformation of till beneath ice stream B, West Antarctica. *Nature*, 322(6074); 1986: 57–59.

Alley, R. B., D. D. Blankenship, C. R. Bentley, and S. T. Rooney. Till beneath ice stream B: 3. Till deformation: Evidence and implications. *Journal of Geophysical Research: Solid Earth*, 92(B9); 1987: 8921–8929.

Allred, B. J. Survey of fractured glacial till geotechnical characteristics: Hydraulic conductivity, consolidation, and shear strength. *Ohio Journal of Science*, 100(3/4); 2000: 63–72.

Allstadt, K. Extracting source characteristics and dynamics of the August 2010 Mount Meager landslide from broadband seismograms. *Journal of Geophysical Research: Earth Surface*, 118(3); 2013: 1472–1490.

Altmann, J. G. *Versuch einer historischen und physischen Beschreibung der helvetischen Eisbergen* (An attempt at a historical and physical description of the Swiss icebergs). In *Landmarks of Science*, edited by Hartley, H. and Duane, H. D. Roller, Readex Microprint, Heidegger, 1973.

Anandakrishnan, S., R. B. Alley, and C. Ice Stream. Antarctica, sticky spots detected by microearthquake monitoring. *Annals of Glaciology*, 20(1); 1994: 183–186.

Anderson, W. F. The use of multi-stage triaxial tests to find the undrained strength parameters of stony boulder clay. *Proceedings of Institution of Civil Engineers*, 57(2); 1974: 367–372.

Anderson, S. Types of retaining walls. In *ICE Manual of Geotechnical Engineering*, edited by Burland, J., Chapman, T., Skinner, H., and Brown, M., Thomas Telford Ltd, London, 2012: 959–968, Chapter 62.

Andrews, J. T. Techniques of till fabric analysis. No. 1-10. (c/o University of East Anglia, Earlham Hall, Norwich NOR 88C)]. *Geo Abstracts for the British Geomorphological Research Group*, 1971.

Ansted, D. T. Chapter XVIII: River beds, river banks and results of river action. In *Water and Water Supply: Chiefly in Reference to the British Isles*, edited by Allen, W., London, 1888.

API, R. P. 2A-WSD, 2000. *Recommended Practice for Planning, Designing and Constructing Fixed Offshore Platforms—Working Stress Design*, 21st ed., American Petroleum Institute, Washington, DC, 2000.

Araruna, Jr, J. T., A. H. Harwood, and B. G. Clarke. A precise, practical and economical volume change measurement device. *Geotechnique*, 35(3); 1995: 541–544.

Armishaw, J. W. and N. G. Bunni. Driven precast concrete piles in the Dublin black boulder clay. In *Piling: European Practice and Worldwide Trends, Piling in Difficult Ground and Locations I*, Thomas Telford Publishing, London, 1993: 272–279.

Arrowsmith, E. J. Roadwork fills-a materials engineer's viewpoint. In *Proceedings Conference on Clay Fills*, Institution of Civil Engineers, London, 1979; 25–36.

Arrowsmith, E. J. Case histories of slope stability: Failure and remedial measures in NW England. In *Proceedings Conference Glacial Tills 85*, Edinburgh, 1985: 113–120.

ASTM D1194-72. *Standard Test Method for Bearing Capacity of Soil for Static Load and Spread Footings (Withdrawn 2003)*. ASTM International, West Conshohocken, PA, 1972.

ASTM D1140 – 14. *Standard Test Methods for Determining the Amount of Material Finer than 75-μm (No. 200) Sieve in Soils by Washing*. ASTM International, West Conshohocken, PA.

ASTM D1557 – 12e1. *Standard Test Methods for Laboratory Compaction Characteristics of Soil Using Modified Effort (56,000 ft-lbf/ft³ (2,700 kN-m/m³))*. ASTM International, West Conshohocken, PA.

ASTM D1883 – 16. *Standard Test Method for California Bearing Ratio (CBR) of Laboratory-Compacted Soils*. ASTM International, West Conshohocken, PA.

ASTM D2166/D2166M – 13. *Standard Test Method for Unconfined Compressive Strength of Cohesive Soil*. ASTM International, West Conshohocken, PA.

ASTM D2216 – 10. *Standard Test Methods for Laboratory Determination of Water (Moisture) Content of Soil and Rock by Mass*. ASTM International, West Conshohocken, PA.

ASTM D2435/D2435M – 11. *Standard Test Methods for One-Dimensional Consolidation Properties of Soils Using Incremental Loading*. ASTM International, West Conshohocken, PA.

ASTM D2487 – 11. *Standard Practice for Classification of Soils for Engineering Purposes (Unified Soil Classification System)*. ASTM International, West Conshohocken, PA.

ASTM D2850 – 15. *Standard Test Method for Unconsolidated-Undrained Triaxial Compression Test on Cohesive Soils*. ASTM International, West Conshohocken, PA.

ASTM D3080/D3080M – 11. *Standard Test Method for Direct Shear Test of Soils Under Consolidated Drained Conditions*. ASTM International, West Conshohocken, PA.

ASTM D4186/D4186M – 12e1. *Standard Test Method for One-Dimensional Consolidation Properties of Saturated Cohesive Soils Using Controlled-Strain Loading*. ASTM International, West Conshohocken, PA.

ASTM D4254 – 16. *Standard Test Methods for Minimum Index Density and Unit Weight of Soils and Calculation of Relative Density*. ASTM International, West Conshohocken, PA.

ASTM D4318 – 10e1. *Standard Test Methods for Liquid Limit, Plastic Limit, and Plasticity Index of Soil*. ASTM International, West Conshohocken, PA.

ASTM D4546 – 14. *Standard Test Methods for One-Dimensional Swell or Collapse of Soils*. ASTM International, West Conshohocken, PA.

ASTM D4648/D4648M – 16. *Standard Test Methods for Laboratory Miniature Vane Shear Test for Saturated Fine-Grained Clayey Soil*. ASTM International, West Conshohocken, PA.

ASTM D4718/D4718M – 15. *Standard Practice for Correction of Unit Weight and Water Content for Soils Containing Oversize Particles*. ASTM International, West Conshohocken, PA.

ASTM D4767 – 11. *Standard Test Method for Consolidated Undrained Triaxial Compression Test for Cohesive Soils*. ASTM International, West Conshohocken, PA.

ASTM D4829 – 11. *Standard Test Method for Expansion Index of Soils*. ASTM International, West Conshohocken, PA.

ASTM D4943 – 08. *Standard Test Method for Shrinkage Factors of Soils by the Wax Method*. ASTM International, West Conshohocken, PA.

ASTM D6467 – 13. *Standard Test Method for Torsional Ring Shear Test to Determine Drained Residual Shear Strength of Cohesive Soils*. ASTM International, West Conshohocken, PA.

ASTM D6635-15, *Standard Test Method for Performing the Flat Plate Dilatometer*, ASTM International, West Conshohocken, PA, 2015.

ASTM D6836-02, 2008.e2. *Standard Test Methods for Determination of the Soil Water Characteristic Curve for Desorption Using Hanging Column, Pressure Extractor, Chilled Mirror Hygrometer, or Centrifuge*. ASTM International, West Conshohocken, PA.

ASTM D6913-04, 2009.e1. *Standard Test Methods for Particle-Size Distribution (Gradation) of Soils Using Sieve Analysis*. ASTM International, West Conshohocken, PA.

ASTM D698 – 12e2. *Standard Test Methods for Laboratory Compaction Characteristics of Soil Using Standard Effort (12,400 ft-lbf/ft^3 (600 kN-m/m^3))*. ASTM International, West Conshohocken, PA.

ASTM D7181 – 11. *Standard Test Method for Consolidated Drained Triaxial Compression Test for Soil*. ASTM International, West Conshohocken, PA.

ASTM D7263 – 09. *Standard Test Methods for Laboratory Determination of Density (Unit Weight) of Soil Specimens*. ASTM International, West Conshohocken, PA.

ASTM D7382 – 08. *Standard Test Methods for Determination of Maximum Dry Unit Weight and Water Content Range for Effective Compaction of Granular Soils Using a Vibrating Hammer*. ASTM International, West Conshohocken, PA.

ASTM D7608 – 10. *Standard Test Method for Torsional Ring Shear Test to Determine Drained Fully Softened Shear Strength and Nonlinear Strength Envelope of Cohesive Soils (Using Normally Consolidated Specimen) for Slopes with No Preexisting Shear Surfaces*. ASTM International, West Conshohocken, PA.

ASTM D854 – 14. *Standard Test Methods for Specific Gravity of Soil Solids by Water Pycnometer*. ASTM International, West Conshohocken, PA.

ASTM Standard D1143/D1143M-07. *Standard Test Methods for Deep Foundations under Static Axial Compressive Load*. ASTM International, West Conshohocken, PA.

Atkinson, J. H., P. I. Lewin, and C. L. Ng. Undrained strength and overconsolidation of a clay till. In *Proceeding of the International Conference on Construction in Glacial Tills and Boulder Clays*, Edinburgh, 1985: 49–54.

Atkinson, J. H. and J. A. Little. Undrained triaxial strength and stress-strain characteristics of a glacial till soil. *Canadian Geotechnical Journal*, 25(3); 1988: 428–439.

Baguelin, F., J. F. Jezequel, and D. H. Shields. *The Pressuremeter and Foundation Engineering*. Trans Tech Publications, Clausthal, Germany, 1978.

Bahr, D. B. and J. B. Rundle. Stick-slip statistical mechanics at the bed of a glacier. *Geophysical Research Letters*, 23(16); 1996: 2073–2076.

Baker, P. J. and R. Gardener. Penetration testing in glacial till. In *Penetration Testing in the UK, Geotechnology Conference*, Birmingham, United Kingdom, 1989: 223–226.

Ballantyne, C. K. and D. I. Benn. Paraglacial slope adjustment during recent deglaciation and its implications for slope evolution in formerly glaciated environments. *Advances in Hllslope Processes* 2; 1996: 1173–1195.

Banham, P. H. Glacitectonites in till stratigraphy. *Boreas*, 6(2); 1977: 101–105.

Banks, J. A. Construction of Muirhead Reservoir, Scotland. In *Proceedings of the 2nd International Conference on Soil Mechanics and Foundation Engineering*, Rotterdam, Vol. 2, 1948: 24–31.

Baranski, M. P. The mechanical behaviour of normally consolidated tills with reference to their structure. In *Advances in Geotechnical Engineering: The Skempton Conference*, Thomas Telford Publishing, London, UK, 2004: 357–368.

Barnes, G. E. The moisture condition value and compaction of stony clays. In *Compaction Technology, Conference*, London, United Kingdom, 1988: 79–90.

Barnes, G. E. and S. G. Staples. Acceptability of clay fill as affected by stone content. *Ground Engng*, 21(1); 1988: 22–28.

Baumberger, T., F. Heslot, and B. Perrin. Crossover from creep to inertial motion in friction dynamics. *Nature*, 367(6463); 1994: 544–546.

Beeman, M., W. B. Durham, and S. H. Kirby. Friction of ice. *Journal of Geophysical Research: Solid Earth*, 93(B7); 1988: 7625–7633.

Bell, F. G. Lime stabilization of clay minerals and soils. *Engineering Geology*, 42(4); 1996: 223–237.

Bell, F. G. The geotechnical properties of some till deposits occurring along the coastal areas of eastern England. *Engineering Geology*, 63(1); 2002: 49–68.

Bell, F. G. and M. G. Culshaw. Problems soils: A review from a British Perspective. In *Problematic Soils: Proceedings of the Symposium held at the Nottingham Trent University on 8 November 2001*. Thomas Telford Publishing, 2001: 1–35.

Bell, F. G. and A. Forster. The geotechnical characteristics of the till deposits of Holderness. *Geological Society, London, Engineering Geology Special Publications*, 7(1); 1991: 111–118.

Bell, A. and C. Robinson. Single piles. In *ICE Manual of Geotechnical Engineering*, edited by Burland, J., Chapman, T., Skinner, H., and Brown, M. Thomas Telford Ltd, London, 2012: 803–820, 959–968, Chapter 54.

Benedikt, J. and Beisler, M. Shotcrete – sustainable design for underground structures facing challenging ground conditions. In *Shotcrete for Underground Support XII*, edited by Lu, M., Sigl, O., and Li, G. J. ECI Symposium Series, 2015.

Benn, D. I. Fabric signature of subglacial till deformation, Breidamerkurjökull, Iceland. *Sedimentology*, 42(5); 1995: 735–747.

Benn, D. I. Fluted moraine formation and till genesis below a temperate valley glacier: Slettmarkbreen, Jotunheimen, southern Norway. *Sedimentology*, 41(2); 1994: 279–292.

Benn, D. I. and D. J. A. Evans. The interpretation and classification of subglacially-deformed materials. *Quaternary Science Reviews*, 15(1); 1996: 23–52.

Benn, D. and D. J. A. Evans. *Glaciers and Glaciation*. Routledge, London, 2010.

Bennett, M. R. Ice streams as the arteries of an ice sheet: Their mechanics, stability and significance. *Earth-Science Reviews*, 61(3); 2003: 309–339.

Bennett, M. R. and N. F. Glasser. *Glacial Geology: Ice Sheets and Landforms*. John Wiley, London, 1996: 364pp.

Bennett, M. M. and N. F. Glasser eds. *Glacial Geology: Ice Sheets and Landforms*. John Wiley & Sons, Chichester, 2011.

Benson, C. H., H. Zhai, and X. Wang. Estimating hydraulic conductivity of compacted clay liners. *Journal of Geotechnical Engineering* 120(2), 1994: 366–387.

Benson, C. H. and J. M. Trast. Hydraulic conductivity of thirteen compacted clays. *Clays and Clay Minerals*, 43(6); 1995: 669–681.

Bentler, J. G., D. Dasenbrock, and M. J. L. Hoppe. Analysis and performance monitoring of a Spread Footing Bridge Foundation. In *Contemporary Topics in Ground Modification, Problem Soils, and Geo-Support*, edited by Iskander, M., Laefer, D. F., and Hussein, M. H., ASCE, Orlando, 2009: 473–480.

Berezantzev, V. G. Load bearing capacity and deformation of piled foundations. In *Proceeding of the 5th ICSMFE*, Paris, Vol. 2, 1961: 11–12.

Bergdahl, U., E. Ottosson, and B. S. Malmborg. *Plattgrundläggning*. Svensk Byggtjänst, Stockholm. ISBN91-7332-662-3, 1993.

Berggren, B., J. Fallsvik, and L. Viberg. Mapping and evaluation of landslide risk in Sweden. In *Proceedings of the Sixth International Symposium on Landslides*, Christchurch, 1992: 873–878.

Bernard, H. A theoretical model of glacial abrasion. *Journal of Glaciology*, 23(89); 1979: 39–50.

BGS. British Geological Survey, 2015. www.bgs.ac.uk.

Bjerrum, L., Discussion on compressibility of soils. In *Proceedings of the European Conference on Soil Mechanics and Foundation Engineering*, Wiesbaden, 2; 1963: 16–17.

Biegel, R. L., C. G. Sammis, and J. H. Dieterich. The frictional properties of a simulated gouge having a fractal particle distribution. *Journal of Structural Geology*, 11(7); 1989: 827–846.

Biggart, A. R. and R. Sternath. Storebaelt eastern railway tunnel: Construction. *Proceedings of the Institution of Civil Engineers-Civil Engineering*, 114(5); 1996: 20–39.

Bishop, A. W. The use of the slip circle in the stability analysis of slopes. *Geotechnique*, 5(1); 1955: 7–17.

Bishop, A. W. and P. R. Vaughan. Selset Reservoir: Design and performance of the embankment. *Proceedings of the Institution of Civil Engineers*, 21(2); 1962: 305–346.

Black, D. K. and K. L. Lee. Saturating laboratory samples by back pressure. *Journal of Soil Mechanics & Foundations Div* 99(SM1); 1973: 75–93.

Blake, E., G. K. C. Clarke, and M. C. Gérin. Tools for examining subglacial bed deformation. *Journal of Glaciology*, 38, 1992: 388–396.

Blake, W., U. H. Fischer, C. R. Bentley, and G. K. G. Clarke. Instruments and methods: Direct measurement of sliding at the glacier bed. *Journal of Glaciology*, 40(136); 1994: 595–599.

Blankenship, D. D., C. R. Bentley, S. T. Rooney, and R. B. Alley. Seismic measurements reveal a saturated porous layer beneath an active Antarctic ice stream. *Nature*, 322(6074); 1986: 54–57.

Blankenship, D. D., C. R. Bentley, S. T. Rooney, and R. B. Alley. Till beneath Ice Stream B: 1. Properties derived from seismic travel times. *Journal of Geophysical Research: Solid Earth*, 92(B9); 1987: 8903–8911.

Boliiton, G. S. The origin of glacially fluted surfaces-observations and theory. *Journal of Glaciology*, 17(76); 1976: 287–309.

Bolton, M. D. and D. A. Lee. Back-analysis of a pilot scale shear test on coarse granular fill. In *Proceedings Engineered Fills Conference*, Newcastle upon Tyne, 1993: 214–225.

Bölviken, B., G. Kullerud, and R. R. Loucks. Geochemical and metallogenic provinces: A discussion initiated by results from geochemical mapping across northern Fennoscandia. *Journal of Geochemical Exploration*, 39(1); 1990: 49–90.

Bond, A. J. and R. J. Jardine. Effects of installing displacement piles in a high o. c. r. clay. *Geotechnique*, 41(3); 1991: 341–363.

Bond, A. and A. Harris. Decoding Eurocode 7, Taylor and Francis, Abingdon, UK, 2008.

Booth, L. *Runswick Bay Slope Stability*. Halcrow Group Ltd, London, 2013.

Bordier, L. C. *Voyage pitoresque aux glacieres de Savoye, fait en 1772*. Par B. Caille, 1773.

Boston, C. M., D. J. A. Evans, and C. Ó. Cofaigh. Styles of till deposition at the margin of the Last Glacial Maximum North Sea lobe of the British–Irish Ice Sheet: An assessment based on geochemical properties of glacigenic deposits in eastern England. *Quaternary Science Reviews*, 29(23); 2010: 3184–3211.

Boulton, G. Glaciers and glaciation; In *Holmes' Principles of Physical Geology*, edited by Duff, P. M. D. and D. Duff, Taylor & Francis, 1993: 401–438.

Boulton, G. S. Processes and patterns of glacial erosion. In *Glacial Geomorphology*, Springer Netherlands, 1982: 41–87.

Boulton, G. S. Processes and patterns of subglacial sedimentation: A theoretical approach. In *Ice-Ages: Ancient and Modern*, edited by Wright and Moseley, Seel House Press, Liverpool; 1975: 7–42.

Boulton, G. S. Processes of glacier erosion on different substrata. *Journal of Glaciology*, 23(89); 1979: 15–38.

Boulton, G. S. Push-moraines and glacier-contact fans in marine and terrestrial environments. *Sedimentology*, 33(5); 1986: 677–698.

Boulton, G. S. Subglacial processes and the development of glacial bedforms. In Research in Glacial, Glacio-Fluvial and Glacio-Lacustrine Systems. *Proceedings of the 6th Guelph Symposium on Geomorphology*, Vol. 1980, 1982: 1–31.

Boulton, G. S. A theory of drumlin formation by subglacial sediment deformation. In Drumlin Symposium, edited by Menzies, J. and Rose, J., Balkeema, Rotterdam, 1987: 25–80.

Boulton, G. S. Sedimentary and sea level changes during glacial cycles and their control on glacimarine facies architecture. *Geological Society, London, Special Publications*, 53(1); 1990: 15–52.

Boulton, G. S. The origin of till sequences by subglacial sediment deformation beneath mid-latitude ice sheets. *Annals of Glaciology*, 22(1); 1996a: 75–84.

Boulton, G. S. Theory of glacial erosion, transport and deposition as a consequence of subglacial sediment deformation. *Journal of Glaciology*, 42(140); 1996b: 43–62.

Boulton, G. S., D. L. Dent, and E. M. Morris. Subglacial shearing and crushing, and the role of water pressures in tills from south-east Iceland. *Geografiska Annaler. Series A. Physical Geography*, 56(3/4), 1974: 135–145.

Boulton, G. S. and K. E. Dobbie. Consolidation of sediments by glaciers: Relations between sediment geotechnics, soft-bed glacier dynamics and subglacial ground-water flow. *Journal of Glaciology*, 39(131); 1993: 26–44.

Boulton, G. S., K. E. Dobbie, and S. Zatsepin. Sediment deformation beneath glaciers and its coupling to the subglacial hydraulic system. *Quaternary International*, 86(1); 2001: 3–28.

Boulton, G. S. and N. Eyles. Sedimentation by valley glaciers: A model and genetic classification. In *Moraines and Varves*, edited by Schluchter, C., AA Balkema, Rotterdam, 1979: 11–23.

Boulton, G. S. and R. C. A. Hindmarsh. Sediment deformation beneath glaciers: Rheology and geological consequences. *Journal of Geophysical Research: Solid Earth*, 92(B9); 1987: 9059–9082.

Boulton, G. S. and A. S. Jones. Stability of temperate ice caps and ice sheets resting on beds of deformable sediment. *Journal of Glaciology*, 24(90); 1979: 29–43.

Boulton, G. S. and M. A. Paul. The influence of genetic processes on some geotechnical properties of glacial tills. *Quarterly Journal of Engineering Geology*, 9(3); 1976: 159–194.

Boulton, G. S., T. Slot, K. Blessing, P. Glasbergen, T. Leijnse, and K. Van Gijssel. Deep circulation of groundwater in overpressured subglacial aquifers and its geological consequences. *Quaternary Science Reviews*, 12(9); 1993: 739–745.

Bowles, J. E. *Foundation Analysis and Design*. McGraw Hill, Singapore; 2001.

Brabham, P. J. and N. R. Goulty. Seismic refraction profiling of rockhead in the Coal Measures of northern England. *Quarterly Journal of Engineering Geology and Hydrogeology*, 21(2); 1988: 201–206.

Bradwell, T., M. S. Stoker, N. R. Golledge, C. K. Wilson, J. W. Merritt, D. Long, J. D. Everest et al. The northern sector of the last British Ice Sheet: Maximum extent and demise. *Earth-Science Reviews*, 88(3); 2008: 207–226.

Bridges, T. S., P. W. Wagner, K. A. Burks-Copes, M. E. Bates, Z. A. Collier, C. J. Fischenich, J. Z. Gailani et al. *Use of Natural and Nature-Based Features (NNBF) for Coastal Resilience*. Engineer Research and Development Center, Vicksburg MS Environmental Lab, Vicksburg, MS, USA, 2015.

Brodsky, E. E., E. Gordeev, and H. Kanamori. Landslide basal friction as measured by seismic waves. *Geophysical Research Letters*, 30(24); 2003: 2236.

Brodzikowski, K. and A. J. Van Loon. Review of glaciogenic sediments. *Development in Sedimentology*, 49(5); 1991: 688.

Bromhead, E. Slope stability. In *ICE Manual of Geotechnical Engineering*, edited by Burland, J., Chapman, T., Skinner, H., and Brown, M., Thomas Telford Ltd, London, 2012: 247–257, Chapter 23.

Brown, R. E. Vibroflotation compaction of cohesionless soils. *Journal of Geotechnical and Geoenvironmental Engineering*, 103, no. *ASCE 13415 Proceeding*, 1977: 1437–1451.

Brown, M. J. The rapid load testing of piles in fine grained soils. PhD dissertation, University of Sheffield, 2004.

Brown, M. J. Pile capacity testing. In *ICE Manual of Geotechnical Engineering*, edited by Burland, J., Chapman, T., Skinner, H., and Brown, M., Thomas Telford Ltd, London, 2012: 1451–1468, Chapter 98.

Brown, M. J., A. F. L. Hyde, and W. F. Anderson. Analysis of a rapid load test on an instrumented bored pile in clay. *Geotechnique*, 56(9); 2006: 627–638.

Brown, M. J. and J. J. M. Powell. Comparison of rapid load test analysis techniques in clay soils. *Journal of Geotechnical and Geoenvironmental Engineering* 139(1); 2012: 152–161.

Brown, N. E., B. Hallet, and D. B. Booth. Quaternary research center and department of geological sciences university of Washington, Seattle. *Journal of Geophysical Research*, 92(B9); 1987: 8985–8997.

Brunsden, D. and M.-L. Ibsen. *The Temporal Occurrence and Forecasting of Landslides in the European Community: Summary of Relevant Results of the European Community EPOCH Programme*. Rapid Mass Movement as a Source of Climatic Evidence for Holocene. Gustav Fisher Verlag, Stuttgart, 1997: 401–407.

BS EN 1377-1. *Methods of Test for Soils for Civil Engineering Purposes, General Requirements and Sample Preparation*. British Standards Institution, London, 1990.

BS 5930:1999+A2:2010. *Code of Practice for Site Investigation*. British Standards Institution, London.

BS 6031:2009. *Code of Practice for Earthworks*. British Standards Institution, London.

BS 8002:2015. *Code of Practice for Earth Retaining Structures*. British Standards Institution, London.

BS 8004:2015. *Code of Practice for Foundations*. British Standards Institution, London.

BS 8006-1:2010. *Code of Practice for Strengthened/Reinforced Soils and Other Fills*. British Standards Institution, London.

BS 8081:2015. *Code of Practice for Ground Anchors*. British Standards Institution, London.

BS EN 12699:2015. *Execution of Special Geotechnical Work – Displacement Piles*. British Standards Institution, London.

BS EN 12715:2000. *Execution of Special Geotechnical Works – Grouting*. British Standards Institution, London.

BS EN 12716:2001. *Execution of Special Geotechnical Works – Jet Grouting*. British Standards Institution, London.

BS EN 14199:2015. *Execution of Special Geotechnical Works – Micropiles*. British Standards Institution, London.

BS EN 14490:2010. *Execution of Special Geotechnical Works. Soil nailing*. British Standards Institution, London.

BS EN 14679:2005. *Execution of Special Geotechnical Works – Deep Mixing*. British Standards Institution, London.

BS EN 14731:2005. *Execution of Special Geotechnical Works – Ground Treatment by Deep Vibration*. British Standards Institution, London.

BS EN 15237:2007. *Execution of Special Geotechnical Works – Vertical Drainage*. British Standards Institution, London.

BS EN 1536:2010+A1:2015. *Execution of Special Geotechnical Works – Bored Piles*. British Standards Institution, London.

BS EN 1990:2002+A1:2005. *Eurocode: Basis of Structural Design*. British Standards Institution, London.

BS EN 1991-1-7:2006+A1:2014. *Eurocode 1. Actions on Structures. General Actions*. British Standards Institution, London.

BS EN 1997-1:2004+A1:2013. *Eurocode 7: Geotechnical Design – Part 1: General Rules*. British Standards Institution, London.

BS EN 1997-2:2007. *Eurocode 7: Geotechnical Design – Part 2: Ground Investigation and Testing (Incorporating Corrigendum 2010)*. British Standards Institution, London.

BS EN ISO 14688-1:2002+A1:2013. *Geotechnical Investigation and Testing – Identification and Classification of Soil – Part 1: Identification and Description*. British Standards Institution, London.

BS EN ISO 14688-2:2004+A1: 2013. *Geotechnical Investigation and Testing – Identification and Classification of Soil – Part 2: Principles for a Classification*. British Standards Institution, London.

BS EN ISO 14689-1:2003. *Geotechnical Investigation and Testing – Identification and Classification of Rock – Part 1: Identification and Classification*. British Standards Institution, London.

BS EN ISO 17892-1:2014. *Geotechnical Investigation and Testing. Laboratory Resting of Soil. Determination of Water Content*. British Standards Institution, London.

BS EN ISO 17892-10:2004 Ed 1. *Geotechnical Investigation and Testing. Laboratory Testing of Soil. Direct Shear Tests*. British Standards Institution, London.

BS EN ISO 17892-11:2004 Ed 1. *Geotechnical Investigation and Testing. Laboratory Testing of Soil. Determination of Permeability by Constant and Falling Head*. British Standards Institution, London.

BS EN ISO 17892-12:2004 Ed 1. *Geotechnical Investigation and Testing. Laboratory Testing of Soil. Determination of Atterberg Limits*. British Standards Institution, London.

BS EN ISO 17892-2:2014. *Geotechnical Investigation and Testing. Laboratory Testing of Soil. Determination of Bulk Density*. British Standards Institution, London.

BS EN ISO 17892-3:2015. *Geotechnical Investigation and Testing. Laboratory Testing of Soil. Determination of Particle Density*. British Standards Institution, London.

BS EN ISO 17892-4:2014. *Geotechnical Investigation and Testing. Laboratory Testing of Soil. Part 4. Determination of Particle Size Distribution*. British Standards Institution, London.

BS EN ISO 17892-5:2004 Ed 1. *Geotechnical Investigation and Testing. Laboratory Testing of Soil. Incremental Loading Oedometer Test*. British Standards Institution, London.

BS EN ISO 17892-5:2014. *Geotechnical Investigation and Testing. Laboratory Testing of Soil. Part 5. Incremental Loading Oedometer Test*. British Standards Institution, London.

BS EN ISO 17892-6:2014. *Geotechnical Investigation and Testing. Laboratory Testing of Soil. Part 6. Fall Cone Test.* British Standards Institution, London.

BS EN ISO 17892-7:2004 Ed 1. *Geotechnical Investigation and Testing. Laboratory Testing of Soil. Unconfined Compression Test on Fine Grained Soil.* British Standards Institution, London.

BS EN ISO 17892-8:2004 Ed 1. *Geotechnical Investigation and Testing. Laboratory Testing of Soil. Unconsolidated Undrained Triaxial Test.* British Standards Institution, London.

BS EN ISO 17892-9:2004 Ed 1. *Geotechnical Investigation and Testing. Laboratory Testing of Soil. Consolidated Triaxial Compression Tests on Water-Saturated Soil.* British Standards Institution, London.

BS EN ISO 22475-1:2006. *Geotechnical Investigation and Testing. Sampling Methods and Groundwater Measurements. Technical Principles for Execution.* British Standards Institution, London.

BS EN ISO 22476-1:2012. *Geotechnical Investigation and Testing. Field Testing. Electrical Cone and Piezocone Penetration Test.* British Standards Institution, London.

BS EN ISO 22476-11:2006. *Geotechnical Investigation and Testing. Field Testing. Part 11. Flat Dilatometer Test.* British Standards Institution, London.

BS EN ISO 22476-12:2009. *Geotechnical Investigation and Testing. Field Testing. Mechanical Cone Penetration Test (CPTM).* British Standards Institution, London.

BS EN ISO 22476-13:2009. *Geotechnical Investigation and Testing – Field Testing – Part 13: Plate Loading Test.* British Standards Institution, London.

BS EN ISO 22476-15:2014. *Geotechnical Investigation and Testing. Field Testing. Part 15. Measuring While Drilling.* British Standards Institution, London.

BS EN ISO 22476-2:2005+A1:2011. *Geotechnical Investigation and Testing. Field Testing. Dynamic Probing.* British Standards Institution, London.

BS EN ISO 22476-3:2005+A1:2011. *Geotechnical Investigation and Testing. Field Testing. Standard Penetration Test.* British Standards Institution, London.

BS EN ISO 22476-4:2012. *Geotechnical Investigation and Testing. Field Testing. Ménard Pressuremeter Test.* British Standards Institution, London.

BS EN ISO 22476-5:2012. *Geotechnical Investigation and Testing. Field Testing. Flexible Dilatometer Test.* British Standards Institution, London.

BS EN ISO 22476-7:2012. *Geotechnical Investigation and Testing. Field Testing. Borehole Jack Test.* British Standards Institution, London.

BS EN ISO 22476-9:2014. *Ground Investigation and Testing. Field Testing. Part 9. Field Vane Test.* British Standards Institution, London.

BS EN ISO 22477-10:2014. *Geotechnical Investigation and Testing. Testing of Geotechnical Structures. Part 10. Testing of Piles: Rapid Load Testing.* British Standards Institution, London.

BTS. *Closed-Face Tunnelling Machines and Ground Stability.* Thomas Telford, London, 2005.

Burland, J. B. On the compressibility and shear strength of natural clays. *Géotechnique*, 40(3); 1990: 329–378.

Burland, J. B. Settlement and stress distributions. In *ICE Manual of Geotechnical Engineering*, edited by Burland, J., Chapman, T., Skinner, H., and Brown, M., Thomas Telford Ltd, London, 2012: 207–220, Chapter 19.

Burland, J. B., B. B. Broms, and V. F. B. de Mello. Behaviour of foundations and structures. In *Proceedings of the 9th ICSMFE*, Tokyo, ICSMFE; Vol. 1, 1978: 495–546.

Burland, J. B., B. B. Broms, and V. F. B. de Mello, Behaviour of foundations and structures. State of the Art Review. *9th International Conference on SMFE*, 2; 1977: 495–546.

Burland, J. B. and M. C. Burbridge. Settlement of foundations on sand and gravel. In *Proceedings of Institution of Civil Engineers*, Pt 1, vol. 76, 1985: 1325–1381.

Burland, J. B. and C. P. Wroth. Allowable and differential settlement of structures, including damage and soil structure interaction. In *Settlement of Structures, Proceedings of the Conference of the British Geotechnical Society,* Cambridge, Pentech Press, London, UK, 1974: 611–764.

Bustamante, M. and L. Gianeselli. Pile bearing capacity prediction by means of static penetrometer CPT. In *Proceedings of the 2nd European Symposium on Penetration Testing*, edited by Verruijt, A., F. L. Beringen, and E. H. de Leeuw, AA Balkeema, Rotterdam, 1982: 493–500.

Butcher, A. P. 39. The observation and analysis of a failure in a cliff of glacial clay till at Cowden, Holderness. In Slope Stability Engineering Developments and Applications: *Proceedings of the International Conference on Slope Stability*. Institution of Civil Engineers, Thomas Telford Publishing, London, 1991: 271–276.

Butler, F. G. Heavily over-consolidated clays. In *Proceedings British Geotechnical Society Conference on Settlement of Structures,* Cambridge, Pentech Press Ltd, London, 1974: 531–578.

Byrne, B. W., R. A. McAdam, H. J. Burd, G. T. Houlsby, C. M. Martin, K. Gavin, P. Doherty et al. Field testing of large diameter piles under lateral loading for offshore wind applications. In *Proceedings of the 16th European Conference on Soil Mechanics and Geotechnical Engineering*, Edinburgh, UK. 2015.

Cabarkapa, Z., G. W. E. Milligan, C. O. Menkiti, J. Murphy, and D. M. Potts. Design and performance of a large diameter shaft in Dublin Boulder Clay. In *BGA International Conference on Foundations: Innovations, Observations, Design and Practice: Proceedings of the International Conference Organised by British Geotechnical Association*, Thomas Telford Publishing, London, 2003: 175–185.

Cameron, D. Early discoverers XXII, Goethe-Discoverer of the ice age. *Journal of glaciology*, 5(41); 1964: 751–754.

Cameron, G. and T. Chapman. Quality assurance of bored pile foundations. *Ground Engineering*, 37(2); 2004: 35–40.

Caquot, A. I. and J. K. Kérisel. *Tables for the Calculation of Passive Pressure, Active Pressure and Bearing Capacity of Foundations*. Gauthier-Villars, Paris, 1948.

Carman, P. C. Permeability of saturated sands, soils and clays. *The Journal of Agricultural Science*, 29(02); 1939: 262–273.

Carr, S. The micromorphology of last glacial maximum sediments in the southern North Sea. *Catena*, 35(2); 1999: 123–145.

Carr, S. J., H. Haflidason, and H. P. Sejrup. Micromorphological evidence supporting Late Weichselian glaciation of the northern North Sea. *Boreas*, 29(4); 2000: 315–328.

Carr, S. J. and J. Rose. Till fabric patterns and significance: Particle response to subglacial stress. *Quaternary Science Reviews*, 22(14); 2003: 1415–1426.

CFEM. *Canadian Foundation Engineering Manual (CFEM)*, 4th Edition. Canadian Geotechnical Society, c\o BiTech Publisher Ltd, Richmond, BC, 2006.

Chamberlin, T. C. Recent glacial studies in Greenland. *Geological Society of America Bulletin*, 6(1); 1894: 199–220.

Chapman, T., H. Taylor, and D. Nicholson. *Modular Gravity Retaining Walls: Design Guidance*. Publication C516, CIRIA, London, 2000.

Chapman, T. J. P., S. J. Deeble, and D. P. Nicholson. Use of the observational method for the construction of a road cutting in Glacial Till. In *Advances in geotechnical engineering: The Skempton conference: Proceedings of a Three Day Conference on Advances in Geotechnical Engineering, organised by the Institution of Civil Engineers and held at the Royal Geographical Society*. Thomas Telford Publishing, London, UK, 29–31 March 2004: 1044–1055.

Chapuis, R. P. Similarity of internal stability criteria for granular soils. *Canadian Geotechnical Journal*, 29(4); 1992: 711–713.

Charles, J. A. *Building on Fill: Geotechnical Aspects*. Building Research Establishment, Watford, UK, 1993.

Chase, R. B. and A. E. Kehew. *Slope Stability Analysis and Ground-Water Hydrology in Heterogeneous Glacial Material: Elements for Prediction of Bluff Erosion*. Western Michigan University, Michigan, 2000.

Chegini, A. and N. A. Trenter. The shear strength and deformation behaviour of a glacial till. In *Proceedings of Conference on Advances in Site Investigation Practice*, Thomas Telford, London, 1996: 851–866.

Chegini, A. and N. A. Trenter. The shear strength and deformation behaviour of a glacial till. In *Advances in Geotechnical Engineering: The Skempton Conference*, London, UK, 2004: 851–866.

Chin, F. K. Estimation of the ultimate load of piles not carried to failure. In *Proceedings of the 2nd Southeast Asian Conference on Soil Engineering*, Singapore, South East Asian Geotechnical Society, 1970: 81–90.

Cho, W. and R. J. Finno. Stress-strain responses of block samples of compressible Chicago glacial clays. *Journal of Geotechnical and Geoenvironmental Engineering*, 136(1); 2009: 178–188.

Chorley, R. J., S. A. Schumm, and D. E. Sugden. *Geomorphology*. Methuen, New York, 1984.

Christensen, C. W., A. A. Pfaffhuber, H. Anschütz, and T. F. Smaavik. Combining airborne electromagnetic and geotechnical data for automated depth to bedrock tracking. *Journal of Applied Geophysics*, 119; 2015: 178–191.

Church, M. and J. M. Ryder. Paraglacial sedimentation: A consideration of fluvial processes conditioned by glaciation. *Geological Society of America Bulletin*, 83(10); 1972: 3059–3072.

Church, M. and O. Slaymaker. Disequilibrium of Holocene sediment yield in glaciated British Columbia. *Nature*, 337(6206); 1989: 452–454.

Clark, P. U. Unstable behaviour of the Laurentide Ice Sheet over deforming sediment and its implications for climate change. *Quaternary Research*, 41(1); 1994: 19–25.

Clark, C. D. Reconstructing the evolutionary dynamics of former ice sheets using multi-temporal evidence, remote sensing and GIS. *Quaternary Science Reviews*, 16(9); 1997: 1067–1092.

Clark, A. R. and S. Fort. Recent UK experience of coastal cliff stabilisation. *Proceedings of the Institution of Civil Engineers-Geotechnical Engineering*, 162(1); 2009: 49–58.

Clark, C. D. and R. T. Meehan. Subglacial bedform geomorphology of the Irish Ice Sheet reveals major configuration changes during growth and decay. *Journal of Quaternary Science*, 16(5); 2001: 483–496.

Clark, P. U. and J. S. Walder. Subglacial drainage, eskers, and deforming beds beneath the Laurentide and Eurasian ice sheets. *Geological Society of America Bulletin*, 106(2); 1994: 304–314.

Clarke, G. K. C. Subglacial till: A physical framework for its properties and processes. *Journal of Geophysical Research: Solid Earth*, 92(B9); 1987: 9023–9036.

Clarke, B. G. *Pressuremeters in Geotechnical Design*. CRC Press, Glasgow, 1994.

Clarke, B. G., E. Aflaki, and D. Hughes. A framework for characterization of glacial tills. In *Proceedings of the International Conference on Soil Mechanics and Foundation Engineering*, Vol. 1, AA Balkema, 1997a: 263–266.

Clarke, B. G. and C.-C. Chen. Intrinsic properties of permeability. In *Proceedings of the International Conference on Soil Mechanics and Foundation Engineering*, vol. 1, AA Balkema, 1997b: 259–262.

Clarke, B. G., C.-C. Chen, and E. Aflaki. Intrinsic compression and swelling properties of a glacial till. *Quarterly Journal of Engineering Geology and Hydrogeology*, 31(3); 1998: 235–246.

Clarke, G. K. C., S. G. Collins, and D. E. Thompson. Flow, thermal structure, and subglacial conditions of a surge-type glacier. *Canadian Journal of Earth Sciences*, 21(2); 1984: 232–240.

Clarke, B. G., D. B. Hughes, and S. Hashemi. Characteristic parameters of tills in relation to earthworks. In *Proceedings of the Seminar Earthworks in Transportation*, Dublin, Ireland, 2001.

Clarke, B. G., D. B. Hughes, and S. Hashemi. Physical characteristics of subglacial tills. *Géotechnique*, 58(1); 2008: 67–76.

Clayton, C. R. I., M. C. Matthews, and N. E. Simons. *Site Investigation*. Blackwell Scientific, Oxford, 1995.

Clouterre. Recommendations. Soil Nailing Recommendations, 1991 (English translation by Federal Highway Administration), Report No. FHWA-SA-93-093, 1991.

Cole, K. W. *Foundations. ICE Works Construction Guides*. Institution of Civil Engineers, London, 1988.

Coop, M. R., J. H. Atkinson, and R. N. Taylor. Strength, yielding and stiffness of structured and unstructured soils. In *Proceedings of the 11th European Conference on Soil Mechanics and Foundation Engineering*, Copenhagen, Denmark, Vol. 28, 1995: 55–62.

Cooper, R. G. *Mass Movements in Great Britain*. Geological Conservation Review Series No. 26. Joint Nature Conservation Committee, London, 2007.

Coppin, N. J. and I. G. Richards eds. *Use of Vegetation in Civil Engineering*. Construction Industry Research and Information Association, London, 1990.

Coyle, H. M. and L. C. Reese. Load transfer for axially loaded piles in clay. *Journal of Soil Mechanics & Foundations Div* 92, no. SM2, Proc Paper 4702, 1966: 1–26.

Coyle, H. M. and I. H. Sulaiman. Skin friction for steel piles in sand. *Journal of Soil Mechanics & Foundations Div* 92, no. SM5, Proc Paper 490, 1966: 261–278.

Cruden, D. M., T. R. Keegan, and S. Thomson. The landslide dam on the Saddle River near Rycroft, Alberta. *Canadian Geotechnical Journal*, 30(6); 1993: 1003–1015.

Cruden, D. M. and D. J. Varnes. Landslides: Investigation and mitigation. In *Landslides: Investigation and Mitigation*, edited by Turner A. K. and Shuster R. L., Transp Res Board, Spec Rep 247, Washington, 1994, 36–75, Chapter 3.

Cubrinovski, M. and K. Ishihara. Correlation between penetration resistance and relative density of sandy soils. In *15th International Conference on Soil Mechanics and Geotechnical Engineering*, University of Canterbury, Istanbul, Turkey, 2001: 393–396.

Cuffey, K. and R. B. Alley. Is erosion by deforming subglacial sediments significant? (Toward till continuity). *Annals of Glaciology*, 22(1); 1996: 17–24.

Cuffey, K. M. and W. S. B. Paterson. *The Physics of Glaciers*. 4th Ed; Academic Press, Waltham, MA, 2010.

Damsgaard, A., D. L. Egholm, J. A. Piotrowski, S. Tulaczyk, N. K. Larsen, and K. Tylmann. Discrete element modeling of subglacial sediment deformation. *Journal of Geophysical Research: Earth Surface*, 118(4); 2013: 2230–2242.

Daniel, D. E. *Geotechnical Practice for Waste Disposal*. Chapman and Hall, London, UK, 1993.

Dashwood, C., D. Diaz Doce, and K. A. Lee. *GeoSure Version 7 Methodology: Landslides Slope Instability*. Internal Report, IR/14/014. British Geological Survey, Nottingham, UK, 2014: 31.

Davies, O., M. Rouainia, S. Glendinning, M. Cash, and V. Trento. Investigation of a pore pressure driven slope failure using a coupled hydro-mechanical model. *Engineering Geology*, 178; August 2014: 70–81.

De Kuiter, J. and F. L. Beringen. Pile foundations for large North Sea structures. *Marine Georesources & Geotechnology*, 3(3); 1979: 267–314.

Deeley, R. M. VI. – The viscous flow of glacier-ice. *Geological Magazine (Decade IV)*, 2(9); 1895: 408–415.

Demorest, M. Glacier flow and its bearing on the classification of glaciers. *Geological Society of America Bulletin*, 52(12 Pt 2); 1941: 2024–2025.

Derbyshire, E., N. J. Edge, and M. A. Love. Soil fabric variability in some glacial soils. In *Proceeding of the International Conference on Construction in Glacial Tills and Boulder Clays*, Edinburgh, 1985: 169–176.

Desai, C. S., S. Sane, and J. Jenson. Constitutive modeling including creep-and rate-dependent behavior and testing of glacial tills for prediction of motion of glaciers. *International Journal of Geomechanics*, 11(6); 2010: 465–476.

DETR (Department of the Environment, Transport and the Regions) Waste Management Paper 26B – Landfill Design, Construction and Operational Practice. Her Majesty's Stationery Office, London, UK, 1995.

Devriendt, M. Risk analysis for tunnelling ground movement assessments. *Proceedings of the Institution of Civil Engineers-Geotechnical Engineering*, 163(3); 2010: 109–118.

Dobie, M. J. The use of cone penetration tests in glacial till. In *Penetration Testing in the UK, Geotechnology Conference*, Birmingham, United Kingdom, 1989: 212–222.

Doherty, P., K. Gavin, and D. Gallagher. Field investigation of base resistance of pipe piles in clay. *Geotechnical Engineering*, 163GE1; 2010: 13–22.

Dolgoushin, L. D. and G. B. Osipova. *Regime of a Surging Glacier Between Advances*. IAHS Publisher, Renne, France, Vol. 107; 1973: 1150–1159.

Donohue, S., M. Long, and P. O'Connor. Multi-method geophysical mapping of quick clay. *Near Surface Geophysics*, 10(3); 2012: 207–219.

Doran, I. G., V. Sivakumar, J. Graham, and A. Johnson. Estimation of *in situ* stresses using anisotropic elasticity and suction measurements. *Géotechnique*, 50(2); 2000: 189–196.

Dowdeswell, J. A. Processes of glacimarine sedimentation. *Progress in Physical Geography* 11(1); 1987: 52–90.

Dowdeswell, J. A., M. J. Hambrey, and R. Wu. A comparison of clast fabric and shape in Late Precambrian and modern glacigenic sediments. *Journal of Sedimentary Research*, 55(5); 1985: 691–704.

Dowdeswell, J. A. and M. J. Sharp. Characterization of pebble fabrics in modern terrestrial glacigenic sediments. *Sedimentology*, 33(5); 1986: 699–710.

Dreimanis, A. Commission on genesis and lithology of Quaternary deposits (INQUA). *Boreas*, 8(2); 1979: 254–254.

Dreimanis, A. Lithofacies types and vertical profile models, an alternative approach to the description and environmental interpretation of glacial diamict and diamictite sequences. *Discussion Sedimentology*, V31(6); 1984: 885–886.

Dreimanis, A. Tills: Their genetic terminology and classification. In *Genetic Classification of Glacigenic Deposits, edited by* Goldthwait, R. F. and Matsch, C. L., Balkema, Rotterdam, The Netherlands, 1989: 17–83.

Dumbleton, M. J. and G. West. *The Suction and Strength of Remoulded Soils as Affected by Composition*. Report Road Research Laboratory, Ministry of Transport, LR306, 1970.

Duncan, J. M. and P. Dunlop. Slopes in stiff-fissured clays and shales. *Journal of the Soil Mechanics and Foundations Division*, 95(2); 1968: 467–492.

Dunlop, P. and C. D. Clark. The morphological characteristics of ribbed moraine. *Quaternary Science Reviews*, 25(13); 2006: 1668–1691.

Dyke, A. S. and T. F. Morris. Drumlin fields, dispersal trains, and ice streams in Arctic Canada. *The Canadian Geographer/Le Géographe Canadien*, 32(1); 1988: 86–90.

EA (Environment Agency) Groundwater Protection: Principles and Practice. Document GP3, August, Version 1.1. EA, Bristol, UK; 2013.

Edil, T. B. and D. M. Mickelson. Overconsolidated tills in eastern Wisconsin. *National Research Council*, Transportation Res. Bd., Record no. 1479, 1995: 99–106.

Egan, D. Earthworks management – have we got our designs right? In *Proceedings of the Conference on Earthworks Stabilisation Techniques and Innovations*, 2005.

Ehlers, J. and P. L. Gibbard. *Quaternary and Glacial Geology*. J. Wiley & Sons, Chichester, NY, 1996.

Elson, J. A. *The Geology of Tills*. Associate Committee on Soil and Snow Mechanics, National Research Council of Canada, Ottawa, 1961.

Elson, J. A. Comment on glacitectonite, deformation till, and commination till. In *Genetic Classification of Glacigenic Deposits*, edited by Goldthwait, R. E. and Matsch, C. L., Balkema, Rotterdam, 1988: 85–88.

Elwood, D. E. Y. and C. Derek Martin. Ground response of closely spaced twin tunnels constructed in heavily overconsolidated soils. *Tunnelling and Underground Space Technology*, 51; January 2016: 226–237.

Engelhardt, H., N. Humphrey, B. Kamb, and M. Fahnestock. Physical conditions at the base of a fast moving Antarctic ice stream. *Science*, 248(4951); 1990: 57–59.

Engelhardt, H. and B. Kamb. Basal sliding of ice stream B, West Antarctica. *Journal of Glaciology*, 44(147); 1998: 223–230.

Erener, A., S. Lacasse, and A. M. Kaynia. Landslide hazard mapping by using GIS in the Lilla Edet province of Sweden. In *28th Asian Conference on Remote Sensing ACRS2007*, Kuala Lumpur, 2007.

Eslami, A. and B. H. Fellenius. Pile capacity by direct CPT and CPTu methods applied to 102 case histories. *Canadian Geotechnical Journal*, 34(6); 1997: 886–904.

Essex, R. J. Geotechnical baseline reports for underground construction. In *Construction Congress V: Managing Engineered Construction in Expanding Global Markets*, ASCE, Virginia, 1997: 219–225.

Evans, D. J. A. Glacial erratics and till dispersal indicators. *Encyclopedia of Quaternary Science*. Elsevier, Oxford, 2007: 975–978.

Evans, D. *Glacial Landsystems*. Routledge, London, 2014.

Evans, D. J. A., E. R. Phillips, J. F. Hiemstra, and C. A. Auton. Subglacial till: Formation, sedimentary characteristics and classification. *Earth-Science Reviews*, 78(1); 2006: 115–176.

Eyles, N. Glacial geology: A landsystems approach. In *Glacial Geology*, edited by Eyles, N., Pergamon, Oxford, 1983: 1–18.

Eyles, N. ed. *Glacial Geology: An Introduction for Engineers and Earth Scientists.* Elsevier, Philadelphia, 2013.

Eyles, N. and W. R. Dearman. A glacial terrain map of Britain for engineering purposes. *Bulletin of the International Association of Engineering Geology-Bulletin de l'Association Internationale de Géologie de l'Ingénieur,* 24(1); 1981: 173–184.

Eyles, N. and C. H. Eyles. Glacial depositional systems. In Walker, R. G., James, N. P. Facies models: Response to sea level changes. *Geological Association of Canada,* 1992: 73–100.

Eyles, N., C. H. Eyles, and A. M. McCabe. Sedimentation in an ice-contact subaqueous setting: The Mid-Pleistocene 'North Sea Drifts' of Norfolk, UK. *Quaternary Science Reviews,* 8(1); 1989: 57–74.

Eyles, N., C. H. Eyles, and A. D. Miall. Lithofacies types and vertical profile models; an alternative approach to the description and environmental interpretation of glacial diamict and diamictite sequences. *Sedimentology,* 30(3); 1983: 393–410.

Eyles, N., A. M. McCabe, and D. Q. Bowen. The stratigraphic and sedimentological significance of Late Devensian ice sheet surging in Holderness, Yorkshire, UK. *Quaternary Science Reviews,* 13(8); 1994: 727–759.

Eyles, N. and J. A. Sladen. Stratigraphy and geotechnical properties of weathered lodgement till in Northumberland, England. *Quarterly Journal of Engineering Geology and Hydrogeology,* 14(2); 1981: 129–141.

Fadum, R. E. Influence values for estimating stresses in elastic foundations. In *Proceedings of the Second International Conference on Soil Mechanics and Foundation Engineering,* Rotterdam, Vol. 2, 1948.

Farrar, D. M. Settlement and pore-water pressure dissipation within an embankment built of London Clay. In *Proceedings of the Conference on Clay Fills, Institution of Civil Engineers,* London, 1979: 101–106.

Farrell, E. R., N. G. Bunni, and J. Mulligan. The bearing capacity of Dublin black boulder clay. *Transactions of the Institution of Engineers of Ireland,* 112, 1987; 1988: 77–104.

Farrell, E., B. Lehane, and M. Looby. An instrumented driven pile in Dublin boulder clay. *Proceedings of the Institution of Civil Engineers-Geotechnical Engineering,* 131(4); 1998: 233–241.

Farrell, E. R. and M. L. Lawler. CFA pile behaviour in very stiff lodgement till. *Proceedings of the Institution of Civil Engineers-Geotechnical Engineering,* 161(1); 2008: 49–57.

Favreau, P., A. Mangeney, A. Lucas, G. Crosta, and F. Bouchut. Numerical modeling of landquakes. *Geophysical Research Letters,* 37(15); 2010.

Feeser, V. On the mechanics of glaciotectonic contortion of clays. In *Glaciotectonics: Forms and Processes,* edited by Croot, D. G., Balkema, Rotterdam, 1988: 63–76.

Fellenius, B. H. Negative skin friction and settlement of piles. In *Proceedings of the Second International Seminar, Pile Foundations,* Nanyang Technological Institute, Singapore, 1984: 1–12.

Fellenius, B. H. and M. Ochoa. Testing and design of a piled foundation project. A case history. *Geotechnical Engineering Journal of the Southeast Asian Geotechnical Society,* 40(3); 2009: 129–137.

FHWA, Manual for Design & Construction of Soil Nail Walls, FHWA-SA-96-069R, US Federation of Highway Administration, Washington, 1999.

FHWA-SA-02-054, Federal Highway Administration. Shallow Foundations, 2002: 310p.

FHWA-NHI-10-016, Federal Highway Administration. *Drilled Shaft Foundations* (Publication No. FHWA-NHI-10016, May 2010).

FHWA0-IF-03-017, Federal Highway Administration, Geotechnical Engineering Circular No. 7 Soil Nail Walls, (Publication No FHWA0-IF-03-017), 2003.

Finch, A. P. The new St Clair River tunnel between Canada and the USA. *Proceedings of the Institution of Civil Engineers. Civil Engineering,* 114(4); 1996: 150–160, Telford.

Finlayson, A. G. and T. Bradwell. Morphological characteristics, formation and glaciological significance of Rogen moraine in northern Scotland. *Geomorphology,* 101(4); 2008: 607–617.

Finno, R. J. and W. Cho. Recent stress-history effects on compressible Chicago glacial clays. *Journal of Geotechnical and Geoenvironmental Engineering,* 137(3); 2010: 197–207.

Finno, R. J. and C.-K. Chung. Stress-strain-strength responses of compressible Chicago glacial clays. *Journal of Geotechnical Engineering*, 118(10); 1992: 1607–1625.

Fischer, U. R. S. and G. K. C. Clarke. Ploughing of subglacial sediment. *Journal of Glaciology*, 40(134); 1994: 97–106.

Fischer, U. H. and G. K. C. Clarke. Stick slip sliding behaviour at the base of a glacier. *Annals of Glaciology*, 24; 1997: 390–396.

Fischer, U. H. and G. K. C. Clarke. Review of subglacial hydro-mechanical coupling: Trapridge glacier, Yukon Territory, Canada. *Quaternary International*, 86(1); 2001: 29–43.

Fischer, U. H., G. K. C. Clarke, and H. Blatter. Evidence for temporally varying 'sticky spots' at the base of Trapridge Glacier, Yukon Territory, Canada. *Journal of Glaciology*, 45(150); 1999: 352–360.

Fish, P. R., R. Moore, and J. M. Carey. Landslide geomorphology of Cayton Bay, North Yorkshire, UK. *Proceedings of the Yorkshire Geological and Polytechnic Society*, 56(1); 2006: 5–14 (Geological Society of London).

Fleming, W. G. K. A new method for single pile settlement prediction and analysis. *Geotechnique*, 42(3); 1992: 411–425.

Fleming, K., A. Weltman, M. Randolph, and K. Elson. *Piling Engineering.* CRC Press, Glasgow, 2008.

Fletcher, L., O. Hungr, and S. G. Evans. Contrasting failure behaviour of two large landslides in clay and silt. *Canadian Geotechnical Journal*, 39(1); 2002: 46–62.

Flint, R. F. *Glacial and Quaternary Geology.* Wiley, Cambridge, 1971.

Forbes, J. D. *Travels through the Alps of Savoy and Other Parts of the Pennine Chain: With Observations on the Phenomena of Glaciers.* A. and C. Black, Edinburgh, 1845.

Foster, M., R. Fell, and M. Spannagle. The statistics of embankment dam failures and accidents. *Canadian Geotechnical Journal*, 37(5); 2000: 1000–1024.

Fowler, A. C. and C. Johnson. Hydraulic run-away: A mechanism for thermally regulated surges of ice sheets. *Journal of Glaciology*, 41(139); 1995: 454–461.

Frank, R., C. Bauduin, R. Driscoll, M., Kavvadas, N. Krebs Ovesen, T. Orr, and B. Schuppener. *Designer's Guide to EN 1997-1 Eurocode 7: Geotechnical Design-General Rules.* Thomas Telford, London, 2004.

Fredlund, D. G., N. R. Morgenstern, and R. A. Widger. The shear strength of unsaturated soils. *Canadian Geotechnical Journal*, 15(3); 1978: 313–321.

Fredlund, D. G., S. K. Vanapalli, A. Xing, and D. E. Pufahl. Predicting the shear strength function for Unsaturated Soils using the soil-water characteristic curve. In *First International Conference on Unsaturated Soils*, Paris, France, 1995: 6–8.

Fredlund, D. G. and A. Xing. Equations for the soil-water characteristic curve. *Canadian Geotechnical Journal*, 31(4); 1994: 521–532.

French, S. and M. Turner. Pile integrity testing. In *ICE Manual of Geotechnical Engineering*, edited by Burland, J., Chapman, T., Skinner, H., and Brown, M., Thomas Telford Ltd, London, 2012: 1419–1448, Chapter 97.

Gaba, A. R., B. Simpson, W. Powrie, and D. R. Beadman. Embedded Retaining Walls – Guidance for Economic Design. CIRIA, London, UK, Report No. C580, 2003.

Garcia-Bengochea, I., A. G. Altschaeffl, and C. W. Lovell. Pore distribution and permeability of silty clays. *Journal of the Geotechnical Engineering Division*, 105(7); 1979: 839–856.

Gareau, L. F., F. Molenkamp, J. Sharma, C. A. Bregje, and M. H. Hegtermans. Engineering geology of glaciated soils. In *Advances in Geotechnical Engineering: The Skempton Conference*, London, UK, 2004.

Garga, V. K. Effect of sample size on consolidation of a fissured clay. *Canadian Geotechnical Journal*, 34(1); 1988: 76–84.

Gavin, K. Development of design practice for piles in stiff glacial till. *DFI Journal-The Journal of the Deep Foundations Institute*, 3(1); 2009: 57–66.

Gavin, K., D. Cadogan, and L. Twomey. Axial resistance of CFA piles in Dublin Boulder Clay. *Proceedings of the Institution of Civil Engineers-Geotechnical Engineering*, 161(4); 2008: 171–180.

Geikie, A. *On the Phenomena of the Glacial Drift of Scotland.* Published for the Geological Society of Glasgow J. Gray, Glasgow, 1863.

Gens, A. and D. W. Hight. The laboratory measurement of design parameters for a glacial till. In *Proceedings of the 7th European Conference on Soil Mechanics and Foundation Engineering*, Brighton, UK, Vol. 2, 1979: 57–65.

Geotechnical Engineering Office. *Geotechnical Manual for Slopes*. Geotechnical Engineering Office, Hong Kong, 1984.

Ghibaudo, G. Subaqueous sediment gravity flow deposits: Practical criteria for their field description and classification. *Sedimentology*, 39(3); 1992: 423–454.

Gibson, P. J., S. Caloca Casado, X. Pellicer, and D. Jiménez-Martín. Mapping the internal structure of Bull Island, eastern Ireland with ground penetrating radar (GPR) and electrical resistivity tomography (ERT). In *The 46th Conference of Irish Geography*, UCD, Dublin, 2014.

Gillarduzzi, A. Investigating property damage along Dublin Port Tunnel alignment. *Proceedings of the Institution of Civil Engineers-Forensic Engineering*, 167(3); 2014: 119–130.

Giraud, P. A., Th W. J. Van Asch, and J. D. Nieuwenhuis. Geotechnical problems caused by glaciolacustrine clays in the French Alps. *Engineering Geology*, 31(2); 1991: 185–195.

Glen, J. W. Experiments on the deformation of ice. *Journal of Glaciology*, 2(12); 1952: 111–114.

Glen, J. W., J. J. Donner, and R. G. West. On the mechanism by which stones in till become oriented. *American Journal of Science*, 255(3); 1957: 194–205.

Gordon, M. E. Design and performance monitoring of clay-lined landfills. Geotechnical Practice for Waste Disposal '87. In *Geotechnical Special Publication 13*, edited by Woods, R. D., ASCE, New York, USA, 1987: 500–514.

Gordon, M. E. and P. M. Huebner. Evaluation of the performance of zone of saturation landfills in Wisconsin. In *Proceedings of the Sixth Annual Madison Conference*, Madison, 1983: 23–53.

Gornitz, V. M., T. W. Beaty, and R. C. Daniels. *A Coastal Hazards Data Base for the US West Coast*. Oak Ridge National Laboratory, Oak Ridge, 1997.

Gornitz, V. and P. Kanciruk. *Assessment of Global Coastal Hazards from Sea Level Rise*. No. CONF-8907104-1. Oak Ridge National Lab., TN (USA), 1989.

Gornitz, V., T. W. White, and R. M. Cushman. *Vulnerability of the US to Future Sea Level Rise*. No. CONF-910780-1. Oak Ridge National Lab., TN (USA), 1991.

Gornitz, V. M. and T. W. White. *A Coastal Hazards Database for the U.S. East Coast*. ORNL/CDIAC-45, NDP-043 A. Oak Ridge National Laboratory (USA), Oak Ridge, TN, 1992.

Graham, J. and G. T. Houlsby. Anisotropic elasticity of a natural clay. *Géotechnique*, 33(2); 1983: 165–180.

Grasmick, J., B. Rysdahl, M. Mooney, B. Robinson, E. Prantil, and A. Thompson. Evaluation of slurry TBM design support pressures using East side access queens bored tunnels data. In *Rapid Excavation and Tunneling Conference (RETC)*, New Orleans, 2015.

Gray, J. M. Quaternary geology and waste disposal in south Norfolk, England. *Quaternary Science Reviews*, 12(10); 1993: 899–912.

Greenwood, S. L. and C. D. Clark. Subglacial bedforms of the Irish ice sheet. *Journal of Maps*, 4(1); 2008: 332–357.

Greenwood, S. L. and C. D. Clark. Reconstructing the last Irish Ice Sheet 1: Changing flow geometries and ice flow dynamics deciphered from the glacial landform record. *Quaternary Science Reviews*, 28(27); 2009a: 3085–3100.

Greenwood, S. L. and C. D. Clark. Reconstructing the last Irish Ice Sheet 2: A geomorphologically-driven model of ice sheet growth, retreat and dynamics. *Quaternary Science Reviews*, 28(27); 2009b: 3101–3123.

Grisak, G. E. and J. A. Cherry. Hydrologic characteristics and response of fractured till and clay confining a shallow aquifer. *Canadian Geotechnical Journal* 12(1); 1975: 23–43.

Grisak, G. E., J. Ar Cherry, J. A. Vonhof, and J. P. Blumele. *Hydrogeologic and Hydrochemical Properties of Fractured Till in the Interior Plains Region*. University of Toronto Press, Toronto, 1976.

Grose, W. J. and L. Benton. Hull wastewater flow transfer tunnel: Tunnel collapse and causation investigation. *Proceedings of the Institution of Civil Engineers-Geotechnical Engineering*, 158(4); 2005: 179–185.

Gruner, G. S. *Die Eisgebirge des Schweizerlandes* (The ice mountains of the Swiss country), Vol. 1, Verlag der neuen Buchhandlung, 1760.

Guzzetti, F., P. Reichenbach, F. Ardizzone, M. Cardinali, and M. Galli. Estimating the quality of landslide susceptibility models. *Geomorphology*, 81(1); 2006: 166–184.

HA68/94, Design methods for the reinforcement of highway slopes by reinforced soil and soil nailing techniques, *Design Manual for Roads and Bridges*, 4: Part 4, Department of Transport, UK; 1994.

Hallet, B. A theoretical model of glacial abrasion. *Journal of Glaciology 23*, no. 89; 1979: 39–50.

Hallet, B. and R. S. Anderson. Detailed glacial geomorphology of a proglacial bedrock area at Castleguard Glacier, Alberta, Canada. *Zeitschrift fur Gletscherkunde und Glazialgeologie*, 16(2); 1981: 171–184.

Hambrey, M. J. *Glacial Environments*. UCL Press, London, 1994.

Hanna, A. and Meyerhof, G. G. Design charts for ultimate bearing capacity of sand overlying soft clay. *Canadian Geotechnical Journal*; 1980: 17.

Hansbo, S. *A New Approach to the Determination of the Shear Strength of Clay by the Fall-Cone Test*. Royal Swedish Geotechnical Institute, Linkoping, Sweden, 1957.

Hansen, B. J. *A General Formula for Bearing Capacity*. Danish Geotechechnical Insistute, Copenhagen, Bulletin No 11, 1961.

Hansen B. J. Discussion on Hyperbolic stress-strain response of cohesive soils. *Journal of Soil Mechanics and Foundation Engineering, ASCE*, 89(2); 1963: 241–242.

Hansen, J. B. *A Revised and Extended Formula for Bearing Capacity*. Danish Geotechnical Institute, Copenhagen, Bull No 28, 1970: 5–1.

Harbor, J., M. Sharp, L. Copland, B. Hubbard, P. Nienow, and D. S. Mair. Influence of subglacial drainage conditions on the velocity distribution within a glacier cross section. *Geology*, 25(8); 1997: 739–742.

Harrison, W. D., M. Truffer, K. A. Echelmeyer, D. A. Pomraning, K. A. Abnett, and R. H. Ruhkick. Probing the till beneath black rapids glacier, Alaska, USA. *Journal of Glaciology*, 50(171); 2004: 608–614.

Hart, J. K. Proglacial glaciotectonic deformation at Melabakkar–Ásbakkar, West Iceland. *Boreas*, 23(2); 1994: 112–121.

Hart, J. K. Subglacial erosion, deposition and deformation associated with deformable beds. *Progress in Physical Geography*, 19(2); 1995: 173–191.

Hart, J. K. Athabasca Glacier, Canada – a field example of subglacial ice and till erosion? *Earth Surface Processes and Landforms*, 31(1); 2006: 65–80.

Hart, J. K. An investigation of subglacial shear zone processes from Weybourne, Norfolk, UK. *Quaternary Science Reviews*, 26(19); 2007: 2354–2374.

Hart, J. K. and G. S. Boulton. The interrelation of glaciotectonic and glaciodepositional processes within the glacial environment. *Quaternary Science Reviews* 10(4); 1991: 335–350.

Hart, J., R. W. Baker, R. L. Hooke, B. Hanson, and P. Jansson. Coupling between a glacier and soft bed: 1. a relation between effective pressure and local shear stress determined from till elasticity. *Journal of Glaciology*, 45(149); 1999: 31–40.

Hart, J. K., A. Khatwa, and P. Sammonds. The effect of grain texture on the occurrence of microstructural properties in subglacial till. *Quaternary Science Reviews*, 23(23); 2004: 2501–2512.

Hart, J. K. and K. Martinez. Environmental sensor networks: A revolution in the earth system science? *Earth-Science Reviews*, 78(3); 2006: 177–191.

Hart, J. K., K. Martinez, R. Ong, A. Riddoch, K. C. Rose, and P. Padhy. A wireless multi-sensor subglacial probe: Design and preliminary results. *Journal of Glaciology*, 52(178); 2006: 389–397.

Hart, J. K. and D. H. Roberts. Criteria to distinguish between subglacial glaciotectonic and glaciomarine sedimentation, I. Deformation styles and sedimentology. *Sedimentary Geology*, 91(1); 1994: 191–213.

Hart, J. and J. Rose. Approaches to the study of glacier bed deformation. *Quaternary International*, 86(1); 2001: 45–58.

Hart, J. K., K. C. Rose, K. Martinez, and R. Ong. Subglacial clast behaviour and its implication for till fabric development: New results derived from wireless subglacial probe experiments. *Quaternary Science Reviews*, 28(7); 2009: 597–607.

Hashemi, S., D. B. Hughes, and B. G. Clarke. The characteristics of glacial tills from Northern England derived from a relational database. *Geotechnical & Geological Engineering*, 24(4); 2006: 973–984.

Hättestrand, C. Ribbed moraines in Sweden – distribution pattern and palaeoglaciological implications. *Sedimentary Geology*, 111(1); 1997: 41–56.

Hättestrand, C. and J. Kleman. Ribbed moraine formation. *Quaternary Science Reviews*, 18(1); 1999: 43–61.

Haxton, A. F. and H. E. Whyte. Clyde tunnel: Construction problems. *Proceedings of the Institution of Civil Engineers*, 30(2); 1965: 323–346.

Hazen, A. *Some Physical Properties of Sands and Gravels: With Special Reference to Their Use in Filtration*. Publisher not Identified, Lawrence, MA, 1892.

Head, K. H. *Manual of Soil Laboratory Testing, Vol. 1, Soil Classification and Compaction Tests*. Pentech, London, 1984.

Head, K. H. *Manual of Soil Laboratory Testing, Vol. 2, Permeability, Shear Strength and Compressibility Tests*. Pentech, London, 1988a.

Head, K. H. *Manual of Soil Laboratory Testing, Vol. 3, Effective Stress Tests*. Pentech, London, 1988b.

Heath, D. C. The application of lime and cement soil stabilization at BAA airports. *Proceedings of the Institution of Civil Engineers: Transport*, 95(1); 1992: 11–50.

Hendry, M. J. Hydraulic conductivity of a glacial till in Alberta. *Ground Water*, 20(2); 1982: 162–169.

Henkel, D. J. Investigations of two long-term failures in London Clay slopes at Wood Green and Northolt. In *Proceedings of 4th International Conference on Soil Mechanics and Foundation Engineering*, Butterworths Scientific Publishers, London, 1957: 315–320.

Herzog, B. L. and W. J. Morse. Hydraulic conductivity at a hazardous waste disposal site: Comparison of laboratory and field-determined values. *Waste Management & Research* 4(2); 1986: 177–187.

Hibert, C., C. P. Stark, and G. Ekström. Dynamics of the Oso-Steelhead landslide from broadband seismic analysis. *Natural Hazards and Earth System Sciences*, 15(6); 2015: 1265–1273.

Hicock, S. R. Lobal interactions and rheologic superposition in subglacial till near Bradtville, Ontario, Canada. *Boreas*, 21(1); 1992: 73–88.

Hicock, S. R., J. R. Goff, O. B. Lian, and E. C. Little. On the interpretation of subglacial till fabric. *Journal of Sedimentary Research*, 66(5); 1996: 928–934.

Hiemstra, J. F. and K. F. Rijsdijk. Observing artificially induced strain: Implications for subglacial deformation. *Journal of Quaternary Science*, 18(5); 2003: 373–383.

Hight, D. W., R. Böese, A. P. Butcher, C. R. I. Clayton, and P. R. Smith. Disturbance of the Bothkennar clay prior to laboratory testing. *Géotechnique*, 42(2); 1992: 199–217.

Hight, D. W. and S. Lerouiel. Characterisation of soils for engineering purposes. In *Characterisation and Engineering Properties of Natural Soils*, edited by Tan T. S. Swets and Zeitlinger, Lisse, The Netherlands 1; 2003: 255–362.

Hindmarsh, R. C. A. Drumlinization and drumlin-forming instabilities: Viscous till mechanisms. *Journal of Glaciology*, 44(147); 1998: 293–314.

Hindmarsh, R. C. A. Coupled ice–till dynamics and the seeding of drumlins and bedrock forms. *Annals of Glaciology*, 28(1); 1999: 221–230.

Hird, C. C., J. J. M. Powell, and P. C. Y. Yung. Investigations of the stiffness of a glacial clay till. In *Proceeding of the 10th European Conference on Soil Mechanics and Foundation Engineering*, Florence, 1991: 107–110.

Hird, C. C. and S. M. Springman. Comparative performance of 5 and 10 cm^2 piezocones in a lacustrine clay. *Geotechnique*, 56(6); 2006: 427–438.

HK Guide No 7, Guide to soil nail design and construction, Geotechnical Engineering Office, Hong Kong, 2008.

Ho, D. Y. F., D. G. Fredlund, and H. Rahardjo. Volume change indices during loading and unloading of an unsaturated soil. *Canadian Geotechnical Journal*, 29(2); 1992: 195–207.

Hobbs, P. R. N., C. V. L. Pennington, S. G. Pearson, L. D. Jones, C. Foster, J. R. Lee, J. R. and A. Gibson. *Slope Dynamics Project Report: Norfolk Coast (2000–2006)*. British Geological Survey Research Report, OR/08/018, 2008: 166.

Hodge, S. M. Direct measurement of basal water pressures: Progress and problems. *Journal of Glaciology*, 23(89); 1979: 309–319.

Hooke, R. LeB. and N. R. Iverson. Grain-size distribution in deforming subglacial tills: Role of grain fracture. *Geology*, 23(1); 1995: 57–60.

Hooke, R. LeB., N. R. Iverson, B. Hanson, P. Jansson, and U. H. Fischer. Rheology of till beneath Storglaciären, Sweden. *Journal of Glaciology*, 43(143); 1997: 172–179.

Hooyer, T. S. and N. R. Iverson. Diffusive mixing between shearing granular layers: Constraints on bed deformation from till contacts. *Journal of Glaciology*, 46(155); 2000: 641–651.

Hooyer, T. S., N. R. Iverson, F. Lagroix, and J. F. Thomason. Magnetic fabric of sheared till: A strain indicator for evaluating the bed deformation model of glacier flow. *Journal of Geophysical Research: Earth Surface*, 113(F2); 2008.

Horn, N. Horizontaler erddruck auf senkrechte abschlussflächen von tunnelröhren. *Landeskonferenz der Ungarischen Tiefbauindustrie*, 1961: 7–16.

Hossain, D. and D. G. McKinlay. The influence of fissures on the consolidation of a glacial till. Geological Society, London, Engineering Geology Special Publications 7(1); 1991: 143–149.

Hubbard, B. and M. Sharp. Basal ice facies and their formation in the western Alps. *Arctic and Alpine Research*, 22(4); 1995: 301–310.

Hughes, A. L. C., C. D. Clark, and C. J. Jordan. Subglacial bedforms of the last British Ice Sheet. *Journal of Maps*, 6(1); 2010: 543–563.

Hughes, A. L. C., C. D. Clark, and C. J. Jordan. Flow-pattern evolution of the last British Ice Sheet. *Quaternary Science Reviews*, 89; April 2014: 148–168.

Hughes, D., V. Sivakumar, D. Glynn, and G. Clarke. A case study: Delayed failure of a deep cutting in lodgement till. *Proceedings of the Institution of Civil Engineers-Geotechnical Engineering*, 160(4); 2007: 193–202.

Humphrey, N., B. Kamb, M. Fahnestock, and H. Engelhardt. Characteristics of the bed of the lower Columbia Glacier, Alaska. *Journal of Geophysical Research: Solid Earth*, 98(B1); 1993: 837–846.

Hungr, O., S. G. Evans, M. J. Bovis, and J. N. Hutchinson. A review of the classification of landslides of the flow type. *Environmental & Engineering Geoscience*, 7(3); 2001: 221–238.

Huntley, D. and P. Bobrowsky. *Surficial Geology and Monitoring of the Ripley Slide, Near Ashcroft.* British Columbia, Geological Survey of Canada, Open File 7531, 2014.

Huntley, D., P. Bobrowsky, Z. Qing, W. Sladen, C. Bunce, T. Edwards, M. Hendry, D. Martin et al. Fiber optic strain monitoring and evaluation of a slow-moving landslide near Ashcroft, British Columbia, Canada. In *Landslide Science for a Safer Geoenvironment.* Springer International Publishing, 2014: 415–421.

Hutchinson, J. N. Coastal landslides in cliffs of Pleistocene deposits between Cromer and Overstrand, Norfolk, England. Building Research Establishment, Building Research Station, Watford; 1976.

Hutchinson, J. N. Mechanisms producing large displacements in landslides on pre-existing shears. *Memoir of the Geological Society of China*, 9; 1987: 175–200.

Hutchinson, J. N. Flow slides from natural slopes and waste tips. In *Proceedings of the 3rd National Symposium on Slopes and Landslides*, La Coruna, Spain, 1992: 827–841.

ICE. *The Specification for Piling and Embedded Retaining Walls*. Thomas Telford, London, 2007.

ICE. *ICE Ground Engineering Manual*. Thomas Telford, London, 2012.

Iken, A. and R. A. Bindschadler. Combined measurements of subglacial water pressure and surface velocity of Findelengletscher, Switzerland: Conclusions about drainage system and sliding mechanism. *Journal of Glaciology*, 32(110); 1986: 101–119.

Islam, M. Z. and M. Z. Yang. Pile capacity prediction in glacial soils using piezocone. In Deep Foundations 2002: An International Perspective on Theory, Design, Construction, and Performance, Orlando, Michael O'Neill and Frank Townsend, ASCE, 2002: 1247–1260.

ISSMGE, IRTP. *International Reference Test Procedure for the Cone Penetration Test (CPT) and the Cone Penetration Test with Pore Pressure (CPTU).* Report of the ISSMGE Technical Committee 16, 1999.

Itasca. http://www.itascacg.com/software, 2016.

Iverson, N. R. Coupling between a glacier and a soft bed: II Model results. *Journal of Glaciology*, 45(149); 1999: 41–53.

Iverson, N. R. Shear resistance and continuity of subglacial till: Hydrology rules. *Journal of Glaciology*, 56.200; 2010: 1104–1114.

Iverson, N. R. A theory of glacial quarrying for landscape evolution models. *Geology*, 40(8); 2012: 679–682.

Iverson, N. R., R. W. Baker, and T. S. Hooyer. A ring-shear device for the study of till deformation: Tests on tills with contrasting clay contents. *Quaternary Science Reviews*, 16(9); 1997: 1057–1066.

Iverson, N. R., R. W. Baker, R. Leb. Hooke, B. Hanson, and P. Jansson. Coupling between a glacier and a soft bed: I. A relation between effective pressure and local shear stress determined from till elasticity. *Journal of Glaciology*, 45(149); 1999: 31–40.

Iverson, N. R., D. Cohen, T. S. Hooyer, U. H. Fischer, M. Jackson, P. L. Moore, G. Lappegard, and J. Kohler. Effects of basal debris on glacier flow. *Science*, 301(5629); 2003: 81–84.

Iverson, R. M., D. L. George, K. Allstadt, M. E. Reid, B. D. Collins, J. W. Vallance, S. P. Schilling et al. Landslide mobility and hazards: Implications of the 2014 Oso disaster. *Earth and Planetary Science Letters*, 412; 2015: 197–208.

Iverson, N. R., B. Hanson, R. LeB. Hooke, and P. Jansson. Flow mechanism of glaciers on soft beds. *Science*, 267(5194); 1995: 80.

Iverson, N. R., T. S. Hooyer, and R. V. V. Baker. Ring-shear studies of till deformation: Coulomb-plastic behavior and distributed strain in glacier beds. *Journal of Glaciology*, 44(148); 1998: 634–642.

Iverson, N. R., T. S. Hooyer, U. H. Fischer, D. Cohen, P. L. Moore, M. Jackson, G. Lappegard et al. Soft-bed experiments beneath Engabreen, Norway: Regelation infiltration, basal slip and bed deformation. *Journal of Glaciology*, 53(182); 2007: 323–340.

Iverson, N. R., T. S. Hooyer, and R. L. Hooke. A laboratory study of sediment deformation: Stress heterogeneity and grain-size evolution. *Annals of Glaciology*, 22(1); 1996: 167–175.

Iverson, N. R., T. S. Hooyer, J. F. Thomason, M. Graesch, and J. R. Shumway. The experimental basis for interpreting particle and magnetic fabrics of sheared till. *Earth Surface Processes and Landforms*, 33(4); 2008: 627–645.

Iverson, N. R. and B. B. Petersen. A new laboratory device for study of subglacial processes: First results on ice-bed separation during sliding. *Journal of Glaciology*, 57(206); 2011: 1135–1146.

Iverson, N. R. and L. K. Zoet. Experiments on the dynamics and sedimentary products of glacier slip. *Geomorphology*, 244; 2015: 121–134.

Jain, G. S. and V. Kumar. Calculations for separating skin friction and point bearing in piles. *Materials Research and Standards, ASTM*, 3(4); 1963: 290–293.

Jamieson, T. F. On the history of the last geological changes in Scotland. *Quarterly Journal of the Geological Society*, 21(1–2); 1865: 161–204.

Jamiolkowski, M., C. C. Ladd, J. T. Germaine and R. Lancellotta, New development in field and laboratory testing of soils. In *Proc. 11th ICSMFE*, 1, Balkeema, Holland, 1985: 57–153.

Jamiolkowski, M., R. Lancellotta, S. Marchetti, R. Nova, and E. Pasqualini. Design parameters for soft clays. In *Proceedings of the 7th European Conference on Soil Mechanics and Foundation Engineering*, Brighton, UK, 1979: 10–13.

Janbu, N. Slope stability computations. In *Embankment-Dam Engineering*, edited by Hirschfeld, R. C. and S. J. Poulos. Wiley and Sons, New York, 1973: 40p.

Jardine, R. J. and F. C. Chow. *New Design Methods for Offshore Piles*. Marine Technology Directorate, London, 1996.

Jardine, R. J., A. Fourie, J. Maswoswe, and J. B. Burland. Field and laboratory measurements of soil stiffness. In *Proceeding of the 11th ICSMFE*, San Francisco, Vol. 2; 1985: 511–514.

Jardine, R. J., D. M. Potts, K. G. Higgins, T. J. Chapman, S. J. Deeble, and D. P. Nicholson. Use of the observational method for the construction of a road cutting in Glacial Till. In *Advances in Geotechnical Engineering: The Skempton Conference*, London, UK, 2004: 1044–1055.

Jardine, R., F. Chow, R. Overy, and J. Standing. *ICP Design Methods for Driven Piles in Sands and Clays*. London, Thomas Telford, 2005.

Jeffery, G. B. The motion of ellipsoidal particles immersed in a viscous fluid. *Proceedings of the Royal Society of London A: Mathematical, Physical and Engineering Sciences*, 102(715); 1922: 161–179 (The Royal Society).

Jenkins, P. and I. A. Kerr. The strength of well graded cohesive fills. *Ground Engineering*, 31(3); 1998: 38–41.

Jones, D. K. C. and E. M. Lee. *Landsliding in Great Britain*. Department of the Environment. HMSO, London, 1994.

Johnston, T. A., J. P. Millmore, J. A. Charles, and P. Tedd. *An Engineering Guide to the Safety of Embankment Dams in the United Kingdom*. Watford, UK, Building Research Establishment, 1999.

Jongmans, D., G. Bievre, F. Renalier, S. Schwartz, N. Beaurez, and Y. Orengo. Geophysical investigation of a large landslide in glaciolacustrine clays in the Trièves area (French Alps). *Engineering Geology*, 109(1); 2009: 45–56.

Joy, J., T. Flahavan, and D. F. Laefer. Soil nailing in glacial till: A design guide evaluation based on Irish and American field sites. In *Earth Retention Conference* 3; 2010: 252–261.

Jung, Y.-H., R. J. Finno, and W. Cho. Stress–strain responses of reconstituted and natural compressible Chicago glacial clay. *Engineering Geology*, 129; 2012: 9–19.

Kamb, B. Sliding motion of glaciers: Theory and observation. *Reviews of Geophysics*, 8(4); 1970: 673–728.

Kamb, B. Glacier surge mechanism based on linked cavity configuration of the basal water conduit system. *Journal of Geophysical Research: Solid Earth*, 92(B9); 1987: 9083–9100.

Kamb, B. Rheological nonlinearity and flow instability in the deforming bed mechanism of ice stream motion. *Journal of Geophysical Research: Solid Earth*, 96(B10); 1991: 16585–16595.

Karim, M., C. Zubrowski and D. C. LePoudre. Drilled shaft capacity in compression – comparison of prediction methods. *Proc Geo Regina*, 2014.

Karlsrud, K. Prøveforstyrrlse–siltig leire. *Proceedings Norwegian Geotechnical Society Geoteknikkdagen*, 1995.

Karol, R. H. *Chemical Grouting*. Marcel Dekker, New York, 1990.

Karrow, P. F. Lithofacies types and vertical profile models; an alternative approach to the description and environmental interpretation of glacial diamict and diamictite sequences. *Sedimentology*, 31(6); 1984: 883–884.

Kavanaugh, J. L. and G. K. C. Clarke. Discrimination of the flow law for subglacial sediment using *in situ* measurements and an interpretation model. *Journal of Geophysical Research: Earth Surface*, 111(F1); 2006.

Keaton, J. R., J. Wartman, S. Anderson, J. Benoît, J. deLaChapelle, R. Gilbert, and D. R. Montgomery. *The 22 March 2014 Oso Landslide*, Snohomish County, Washington, GEER report, NSF Geotechnical Extreme Events Reconnaissance, http://www.geerassociation.org/GEER_Post_EQ_Reports/Oso_WA_2014/

Keller, C. K., G. van der Kamp, and J. A. Cherry. Fracture permeability and groundwater flow in clayey till near Saskatoon, Saskatchewan. *Canadian Geotechnical Journal*, 23(2); 1986: 229–240.

Kemmis, T. J. and G. R. Hallberg. Lithofacies types and vertical profile models; an alternative approach to the description and environmental interpretation of glacial diamict and diamictite sequences. *Sedimentology*, 31(6); 1984: 886–890.

Kennard, J. and M. F. Kennard. Selset reservoir: Design and construction. *Proceedings of the Institution of Civil Engineers*, 21(2); 1962: 277–304.

Kennard, M. F. *The Construction of Balderhead Dam*. Civ Eng and Publ Wks Rev, ICOLD supplement, 1964: 35–39.

Kenney, T. C. Multiple-stage triaxial test for determining c' and φ' of saturated soils. In Technical Memorandum, Division of Building Research, National Research Council Canada, 72-2; 1961: 1–5.

Kenney, T. C. and D. Lau. Internal stability of granular filters: Reply. *Canadian Geotechnical Journal*, 23(3); 1986: 420–423.

Kettles, I. M. and W. W. Shilts. Geochemistry of drift over the Precambrian Grenville Province, south eastern Ontario and southwestern Quebec. *Drift Prospecting*, 89(20); 1989: 97.

Khatwa, A. and S. Tulaczyk. Microstructural interpretations of modern and Pleistocene subglacially deformed sediments: The relative role of parent material and subglacial processes. *Journal of Quaternary Science*, 16(6); 2001: 507–517.

Kilfeather, A. A. and J. J. M. Van der Meer. Pore size, shape and connectivity in tills and their relationship to deformation processes. *Quaternary Science Reviews*, 27(3); 2008: 250–266.

Kirkaldie, L. and J. R. Talbot. The effects of soil joints on soil mass properties. *Bulletin of the Association of Engineering Geologists*, 29(4); 1992: 415–430.

Kjær, K. H., J. Krüger, and J. J. M. van der Meer. What causes till thickness to change over distance? Answers from Mýrdalsjökull, Iceland. *Quaternary Science Reviews*, 22(15); 2003: 1687–1700.

Klassen, R. A. The application of glacial dispersal models to the interpretation of till geochemistry in Labrador, Canada. *Journal of Geochemical Exploration*, 67(1); 1999: 245–269.

Klassen, R. A. and F. J. Thompson. *Glacial History, Drift Composition, and Mineral Exploration, Central Labrador*. Energy, Mines and Resources Canada, Geological Survey of Canada, 1993.

Kleven, A., S. Lacasse, and K. H. Anderson. *Foundation Engineering Criteria for Gravity Platforms: Soil Parameters for Offshore Foundation Design*, Part II, Norwegian Geotechnical Institute, Oslo, 1986, NGI Report 40013-34.

Knight, J. Glacial sedimentary evidence supporting stick-slip basal ice flow. *Quaternary Science Reviews*, 21(8); 2002: 975–983.

Knight, J. and A. Marshall McCabe. Identification and significance of ice-flow-transverse subglacial ridges(Rogen moraines) in northern central Ireland. *Journal of Quaternary Science*, 12(6); 1997: 519–524.

Kohv, M., P. Talviste, T. Hang, V. Kalm, and A. Rosentau. Slope stability and landslides in proglacial varved clays of western Estonia. *Geomorphology*, 106(3); 2009: 315–323.

Kohv, M., P. Talviste, T. Hang, and V. Kalm. Retrogressive slope failure in glaciolacustrine clays: Sauga landslide, western Estonia. *Geomorphology*, 124(3); 2010: 229–237.

Ku, T. and P. W. Mayne. Evaluating the *in situ* lateral stress coefficient (K 0) of soils via paired shear wave velocity modes. *Journal of Geotechnical and Geoenvironmental Engineering*, 139(5); 2012: 775–787.

Lacasse, S., T. Berre, and G. Lefebvre. Block sampling of sensitive clays. *Proceedings of the 11th International Conference on Soil Mechanics and Foundation Engineering*, San Francisco. Balkema, Rotterdam, 1985: 887–892.

La Rochelle, P. and G. Lefebvre. Sampling disturbance in Champlain clays. In *Sampling of Soil and Rock*. ASTM International, 1971: 143–163.

Lachniet, M. S., G. J. Larson, D. E. Lawson, E. B. Evenson, and R. B. Alley. Microstructures of sediment flow deposits and subglacial sediments: A comparison, *Boreas*, 30(254262); 2001: 2006a.

Laefer, D. F., T. Flahavan, and J. Joy. Soil nailing in glacial till: A design guide evaluation based on Irish and American field sites. In 2010 Earth Retention Conference, American Society of Civil Engineers, Bellevue, Washington, 2010: 252–261.

Lambe, T. W. and R. V. Whitman. *Soil Mechanics SI Version*. John Wiley & Sons, New York, 2008.

Larsen, N. K., J. A. Piotrowski, and F. Christiansen. Microstructures and microshears as proxy for strain in subglacial diamicts: Implications for basal till formation. *Geology*, 34(10); 2006: 889–892.

Lawler, M. Predicting the behaviour of CFA piles in boulder clay. *Electronic Journal of Geotechnical Engineering*, 8; 2003.

Laws, W. G. Earthwork slips. *Proceedings of the Institution of Civil Engineers*, 66(4); 1881: 263–265.

Lawson, D. E. Sedimentological characteristics and classification of depositional processes and deposits in the glacial environment. No. CRREL-81-27. *Cold Regions Research and Engineering Lab*, HANOVER NH, 1981.

Lebourg, T. and R. Fabre. Glacial tills instability on mountains sides, influence of the geomorphological inheritance and the heterogeneity, for forecasting the behaviour of slopes movements. In *International Symposium on Landslides, Landslides in Research, Theory and Practice*, edited by Cardiff, E. B., N. Dixon, and M.-L. Ibsen, Thomas Telford, London, Vol. 2, 2000: 26–30.

Lefebvre, G. and C. Poulin. A new method of sampling in sensitive clay. *Canadian Geotechnical Journal*, 16(1); 1979: 226–233.

Lehane, B. and A. Faulkner. Stiffness and strength characteristics of a hard lodgement till. In *The Geotechnics of Hard Soil and Soft Rocks, Proceedings of the Second International Symposium on Hard Soils, Soft Rocks*, edited by Evangelista, A. and L. Picarelli, Naples, Italy, 1998: 637–646.

Lehane, B. M. and R. J. Jardine. Displacement pile behaviour in glacial clay. *Canadian Geotechnical Journal*, 31(1); 1994: 79–90.

Lehane, B. M. and B. Simpson. Modelling glacial till under triaxial conditions using a BRICK soil model. *Canadian Geotechnical Journal*, 37(5); 2000: 1078–1088.

Leroueil, S., J. Locat, J. Vaunat, L. Picarelli, H. Lee, and R. Faure. Geotechnical characterization of slope movements. In *Proceedings of the 7th International Symposium on Landslides*, Trondheim, Norway, Vol. 1, 1996: 53–74.

Lewin, P. I. and J. J. M. Powell. Patterns of stress strain behaviour for a clay till. In *Proceeding of the 11th International Conference on Soil Mechanics and Foundation Engineering*, San Francisco, V1, 1985: 553–556.

Li, Z., J. Grasmick, and M. Mooney. Influence of slurry TBM parameters on ground deformation. In *ITA WTC 2015 Congress and 41st General Assembly*, Croatia, 2015.

Likins, G., G. Piscsalko, F. Rausche, and S. Roppel. PDA testing, 2008 state of the art. In *Proceeding of the 8th International Conference on the Application of Stress Wave Theory to Piles*, Lisbon, Portugal, 2008.

Lindsay, F. M., S. B. Mickovski, and M. J. Smith. Testing of self-drilled hollow bar soil nails. In *Proceedings of the XVI ECSMGE Geotechnical Engineering for Infrastructure and Development*, Edinburgh, 2015.

Littlejohn, G. S., N. R. Arber, C. Craig, and M. C. Forde. *Inadequate Site Investigation*. Institution of Civil Engineers, 1991: 1.

Little, J. A. Engineering properties of glacial tills in the Vale of St. Albans. PhD dissertation, City University, 1984.

Liu, M. D. and J. P. Carter. Virgin compression of structured soils. *Géotechnique*, 49(1); 1999: 43–57.

Lo, K. Y. The operational strength of fissured clays. *Geotechnique*, 20(1); 1970: 57–74.

Long, M. Sample disturbance effects in very soft clays. PhD Thesis, University College Dublin, 2000.

Long, M. Database for retaining wall and ground movements due to deep excavations. *Journal of Geotechnical and Geoenvironmental Engineering*, 127(3); 2001: 203–224.

Long, M. Sampling disturbance effects in soft laminated clays. *Proceedings of the Institution of Civil Engineers—Geotechnical Engineering*, 156(4); 2003: 213–224.

Long, M. Sample disturbance effects on medium plasticity clay/silt. *Proceedings of the Institution of Civil Engineers-Geotechnical Engineering*, 159(2); 2006: 99–111.

Long, M. M. and N. J. O'Riordan. Field behaviour of very soft clays at the Athlone embankments. *Geotechnique*, 51(4); 2001: 293–309.

Long, M, Menkiti, C. O., N. Kovacevic et al. An observational approach to the design of steep sided excavations in Dublin glacial till. *Proceedings of Underground Construction 2003 – UC2003, ExCeL*. London, Hemming-Group Ltd, 2003: 443–454.

Long, M. and G. Gudjonsson. T-bar testing in Irish soft soils. *Proceedings of the 2nd International Conference on Geotechnical Site Characterisation, Porto*. Millpress, Rotterdam, 1; 2004: 719–726.

Long, M. and C. O. Mentiki. Geotechnical properties of Dublin Boulder Clay. *Geotechnique*, 57(7); 2007: 595–611.

Long, M., C. Brangan, C. O. Menkiti, M. Looby, and P. Casey. Retaining walls in Dublin boulder clay, Ireland. *Proceedings of the Institution of Civil Engineers-Geotechnical Engineering*, 165(4); (2012): 247–266.

Lory, D. and F. Anthony. Long term stability of slopes in over-consolidated clays. PhD dissertation, Imperial College London (University of London), 1957.

Lumb, P. Multi-stage triaxial tests on undisturbed soils. *Civil Engineering and Public Works Review*, 59; May 1964: 591–595.

Lunne, T., T. Berre, and S. Strandvik. Sample disturbance effects in soft low plastic Norwegian clay. In *Symposium on Recent Developments in Soil and Pavement Mechanics*, 1997: 81–92.

Lunne, T., P. K. Robertson, and J. J. M. Powell. *Cone Penetration Testing in Geotechnical Practice*, Chapman and Hall, London, 1997.

Lunne, T., M. Long, and C. F. Forsberg. Characterisation and engineering properties of Onsøy clay. *Characterisation and Engineering Properties of Natural Soils*, 1; 2003: 395–428.

Lupini, J. F., A. E. Skinner, and P. R. Vaughan. The drained residual strength of cohesive soils. *Geotechnique*, 31(2); 1981: 181–213.

Lyell, C. *Principles of Geology: Being an Inquiry How Far the Former Changes of the Earth's Surface are Referable to Causes Now in Operation*. J. Murray, London, 1837.

Macayeal, D. R., R. A. Bindschadler, and T. A. Scambos. Basal friction of ice stream E, West Antarctica. *Journal of Glaciology*, 41(138); 1995: 247–262.

Macklin, S. R. The prediction of volume loss due to tunnelling in overconsolidated clay based on heading geometry and stability number. *Ground Engineering*, 32(4); 1999: 30–33.

Madgett, P. A. and J. A. Catt. Petrography, stratigraphy and weathering of Late Pleistocene tills in East Yorkshire, Lincolnshire and north Norfolk. *Proceedings of the Yorkshire Geological and Polytechnic Society, Geological Society of London*, 42(1); 1978: 55–108.

Mair, R. J. Settlement effects of bored tunnels. In *Proceedings of the International Symposium on Geotechnical Aspects of Underground Construction in Soft Ground*, London, 1996: 43–53.

Mair, R. J. and R. N. Taylor. Theme lecture: Bored tunnelling in the urban environment. In *Proceedings of the Fourteenth International Conference on Soil Mechanics and Foundation Engineering*, Balkema, Hamburg, 1999: 2353–2385.

Mair, R. J. and D. M. Wood. *In-Situ Pressuremeter Testing: Methods Testing and Interpretation*. CIRIA Ground Engineering Report, 1987.

March, A. Mathematische Theorie der Regelung nach der Korngestah bei affiner Deformation. *Zeitschrift für Kristallographie-Crystalline Materials*, 81(1–6); 1932: 285–297.

Marshall, T. J. A relation between permeability and size distribution of pores. *Journal of Soil Science*, 9(1); 1958: 1–8.

Marsland, A. *In situ* and laboratory tests on Boulder clay at Redcar. In *Midland Soil Mechanics and Foundation Engineering Society Symposium on Engineering Behaviour of Glacial Materials*, Birmingham, 1975: 7–17.

Marsland, A. The evaluation of the engineering design parameters for glacial clays. *Quarterly Journal of Engineering Geology and Hydrogeology*, 10(1); 1977: 1–26.

Marsland, A. and J. J. M. Powell. Field and laboratory investigations of the clay tills at the building research establishment test site at Cowden, Holderness. In *Proceedings of the International Conference on Construction in Glacial Tills and Boulder Clays*, Edinburgh, 1985: 147–168.

Marsland, A. and J. J. M. Powell. Field and laboratory investigations of the clay tills at the test bed site at the Building Research Establishment, Garston, Hertfordshire. *Geological Society, London, Engineering Geology Special Publications*, 7(1); 1991: 229–238.

Martin, J., P. Daynes, C. McDonnell, and M. J. Pedley. The design, installation & monitoring of high capacity antiflotation bar anchors to restrain deep basements in Dublin. In *Proceedings of Conference on Ground Anchorages and Anchored Structures in Service*, Thomas Telford, London, 2007: 438–449.

Martinez, K., J. K. Hart, and R. Ong. Environmental sensor networks. *Computer*, 37(8); 2004: 50–56.

Martini, I. P. and M. E. Brookfield. Sequence analysis of upper Pleistocene (Wisconsinan) glaciolacustrine deposits of the north-shore bluffs of Lake Ontario, Canada. *Journal of Sedimentary Research*, 65(3); 1995.

Matsui, T. Case studies on cast-in-place bored piles and some considerations for design. In *Proceedings of the 2nd International Geotechnical Seminar on Deep Foundations on Bored and Auger Piles*, Ghent, Belgium, 1993: 77–102.

Mayne, P. W., B. R. Christopher, and J. DeJong. *Manual on Subsurface Investigations*. Nat. Highway Inst. Sp. Pub. FHWA NHI-01–031. Fed. Highway Administ, Washington, DC, 2001.

McCabe, A., J. K. Marshall, and S. G. McCarron. Ice-flow stages and glacial bedforms in north central Ireland: A record of rapid environmental change during the last glacial termination. *Journal of the Geological Society*, 156(1); 1999: 63–72.

McCabe, B. A., T. L. L. Orr, C. C. Reilly, and B. G. Curran. Settlement trough parameters for tunnels in Irish glacial tills. *Tunnelling and Underground Space Technology*, 27(1); 2012: 1–12.

McGown, A. Genetic influences on the nature and properties of basal melt out tills. PhD dissertation, University of Strathclyde, 1975.

McGown, A. and E. Derbyshire. Genetic influences on the properties of tills. *Quarterly Journal of Engineering Geology and Hydrogeology*, 10(4); 1977: 389–410.

McGown, A. and A. M. Radwan. The presence and influence of fissures in the boulder clays of west central Scotland. *Canadian Geotechnical Journal*, 12(1); 1975: 84–97.

McGown, A., A. M. Radwan, and A. W. A. Gabr. Laboratory testing of fissured and laminated clays. In *Proceeding of the 9th International Conference on Soil Mechanics and Foundation Engineering*, Tokyo, V1, 1977: 205–210.

McKay, L., J. A. Cherry, and R. W. Gillham. Field experiments in a fractured clay till: 1. Hydraulic conductivity and fracture aperture. *Water Resources Research*, 29(4); 1993: 1149–1162.

McKinlay, D. G., M. J. Tomlinson, and W. F. Anderson. Observations on the undrained strength of a glacial till. *Geotechnique*, 24(4); 1974: 503–516.

Meigh, A. C. *Cone Penetration Testing: Methods and Interpretation*. Elsevier, Philadelphia, 2013.

Meigh, A. C. and I. K. Nixon. Comparison of *in situ* tests for granular soils. In *Proceedings Fifth International Conference on Soil Mechanics and Foundation Engineering*, Vol. 1, 1961: 499.

Melchiorre, C. and A. Tryggvason. Assessing landslide-prone areas in sensitive clay. *Nat. Hazards Earth Syst. Sci.*, 15; 2015: 2703–2713.

Menzies, J. Investigations into the Quaternary deposits and bedrock topography of central Glasgow. *Scottish Journal of Geology*, 17(3); 1981: 155–168.

Menzies, J. Towards a general hypothesis on the formation of drumlins. In *Drumlin Symposium*, edited by Menzies, J. and J. Rose, A.A. Balkema, Rotterdam, 1987: 9–24.

Menzies, J. Subglacial hydraulic conditions and their possible impact upon subglacial bed formation. *Sedimentary Geology*, 62(2); 1989: 125–150.

Menzies, J. Micromorphological analyses of microfabrics and microstructures indicative of deformation processes in glacial sediments. *Geological Society, London, Special Publications*, 176(1); 2000: 245–257.

Menzies, J. and A. J. Maltman. Microstructures in diamictons – evidence of subglacial bed conditions. *Geomorphology*, 6(1); 1992: 27–40.

Menzies, J. and W. W. Shilts. Subglacial environments. In *Modern and Past Glacial Environments: A Revised Student Edition*, edited by Menzies, J., Butterworth-Heineman, Oxford, 2002: 183–278.

Menzies, J. and J. J. M. Van Der Meer. Sedimentological and micromorphological examination of a Late Devensian multiple diamicton sequence near Moneydie, Perthshire, east-central Scotland. *Scottish Journal of Geology*, 34(1); 1998: 15–21.

Menzies, J., J. J. M. van der Meer, and J. Rose. Till – as a glacial 'tectomict', its internal architecture, and the development of a 'typing' method for till differentiation. *Geomorphology*, 75(1); 2006: 172–200.

Meyerhof, G. G. The settlement analysis of building frames. *The Structural Engineer*, 25(9); 1947: 369.

Meyerhof, G. G. The ultimate bearing capacity of foundations. *Geotechnique*, 2(4); 1951: 301–332.

Meyerhof, G. G. The ultimate bearing capacity of foundations on slopes. In *Proceedings of the 4th International Conference on Soil Mechanics and Foundation Engineering*, Butterworths, London, vol. 1, 1957: 384–386.

Meyerhof, G. G. Bearing capacity and settlement of pile foundations. 11th Terzaghi Lecture. 10. *Journal of the Geotechnical Engineering Division, ASCE*, 102; 1976: 197–228.

Miall, A. D. Architectural-element analysis: A new method of facies analysis applied to fluvial deposits. *Earth Science Reviews*, 22(4); 1985: 261–308.

Millmore, J. P. and R. McNicol. Geotechnical aspects of the Kielder Dam. *Proceedings of the Institution of Civil Engineers*, 74(4); 1983: 805–836.

Misfeldt, G. A., E. Karl Sauer, and E. A. Christiansen. The Hepburn landslide: An interactive slope-stability and seepage analysis. *Canadian Geotechnical Journal*, 28(4); 1991: 556–573.

Mitchell, J. M. and F. M. Jardine. *A Guide to Ground Treatment*, Vol. 573. CIRIA, London, 2002.

Moncrieff, A. C. M. Classification of poorly-sorted sedimentary rocks. *Sedimentary Geology*, 65(1–2); 1989: 191–194.

Mooney, M., J. Grasmick, A. Clemmensen, A. Thompson, E. Prantil, and B. Robinson. Ground deformation from multiple tunnel openings: Analysis of Queens Bored Tunnels. In *Proceeding of Conference on North American Tunneling*, Los Angeles, 2014.

Mooney, M. A., J. Grasmick, B. Kenneally, and Y. Fang. The role of slurry TBM parameters on ground deformation: Field results and computational modelling. *Tunnelling and Underground Space Technology*, 57; August 2016: 257–264.

Moore, P. L. and N. R. Iverson. Slow episodic shear of granular materials regulated by dilatant strengthening. *Geology*, 30(9); 2002: 843–846.

Moretti, L., A. Mangeney, Y. Capdeville, E. Stutzmann, C. Huggel, D. Schneider, and F. Bouchut. Numerical modeling of the Mount Steller landslide flow history and of the generated long period seismic waves. *Geophysical Research Letters*, 39(16); 2012.

Morgan, H. D., C. K. Haswell, and E. S. Pirie. Clyde Tunnel – design, construction and tunnel services. *Proceedings of the Institution of Civil Engineers, London*, 30(2); 1965: 291–322.

Morgan, R., P. Pizzimenti, K. Walsh, and G. Margeson. Pile capacity setup in fine-grained glacial deposits at the South Brooklyn Marine Terminal. In *Ports 2013: Success through Diversification*. ASCE, 2013: 1038–1047.

Morgenstern, N. R. and V. E. Price. The analysis of the stability of general slip surfaces. *Geotechnique*, 15(1); 1965: 79–93.

Murray, E. J. Properties and testing of clay liners. In *Geotechnical Engineering of Landfills*. Thomas Telford, London, UK, 1988.

Murray, T. and G. K. C. Clarke. Black-box modeling of the subglacial water system. *Journal of Geophysical Research: Solid Earth*, 100(B6); 1995: 10231–10245.

Murray, E. J., N. Dixon, and D. R. V. Jones. Properties and testing of clay liners. In *Proceedings of the Symposium on Geotechnical Engineering of Landfills*, Thomas Telford Services Limited, Nottingham, 1998: 37.

Murray, T. and J. A. Dowdeswell. Water throughflow and the physical effects of deformation on sedimentary glacier beds. *Journal of Geophysical Research: Solid Earth*, 97(B6); 1992: 8993–9002.

Murray, T. and P. R. Porter. Basal conditions beneath a soft-bedded polythermal surge-type glacier: Bakaninbreen, Svalbard. *Quaternary International*, 86(1); 2001: 103–116.

Nash, D. F. T., J. J. M. Powell, and I. M. Lloyd. Initial investigations of the soft clay test site at Bothkennar. *Geotechnique* 42(2); 1992: 163–181.

Newmark, N. M. Numerical procedure for computing deflections, moments, and buckling loads. In Selected Papers By Nathan M. Newmark: Civil Engineering Classics, ASCE, 1942: 197–270.

N.H.B.C. *Building Near Trees*. National House Building Council, York, UK; 2003.

Nowak, P. A. Design of new earthworks. In *ICE Manual of Geotechnical Engineering*, edited by Burland, J., Chapman, T., Skinner, H., and Brown, M., Thomas Telford Ltd, London, 2012a: 1043–1046, Chapter 69.

Nowak, P. A. Design of new earthworks. In *ICE Manual of Geotechnical Engineering*, edited by Burland, J., Chapman, T., Skinner, H., and Brown, M., Thomas Telford Ltd, London, 2012b: 1047–1063, Chapter 70.

Nowak, P. A. Slope stabilisation methods. In *ICE Manual of Geotechnical Engineering*, edited by Burland, J., Chapman, T., Skinner, H., and Brown, M., Thomas Telford Ltd, London, 2012c: 1087–1091, Chapter 72.

Nye, J. F. The flow of glaciers and ice-sheets as a problem in plasticity. *Proceedings of the Royal Society of London A: Mathematical, Physical and Engineering Sciences*, 207(1091); 1951: 554–572 (The Royal Society).

Nye, J. F. Glacier sliding without cavitation in a linear viscous approximation. *Proceedings of the Royal Society of London A: Mathematical, Physical and Engineering Sciences*, 315(1522); 1970: 381–403 (The Royal Society).

Nye, J. F. Water at the bed of a glacier. *International Association of Hydrology Publication*, 95; *Symposium at Cambridge 1969—Hydrology of Glaciers*; 1973: 189–194.

O'Brien, A. Pile-group design. In *ICE Manual of Geotechnical Engineering*, edited by Burland, J., Chapman, T., Skinner, H., and Brown, M., Thomas Telford Ltd, London, 2012: 823–850, Chapter 55.

O'Brien, A. S. and I. Farooq. Shallow foundations. In *In ICE Manual of Geotechnical Engineering*, edited by Burland, J., Chapman, T., Skinner, H., and Brown, M., Thomas Telford Ltd, London, 2012: 765–800, Chapter 53.

Okamura, M., Takemura, J., and Kimura, T. Bearing capacity predictions of sand overlying clay based on limit equilibrium methods. *Soils and Foundations*, 38; 1988: 181–194.

O'Leary, F., M. Long, and M. Ryan. The long-term behaviour of retaining walls in Dublin. *Proceedings of the Institution of Civil Engineers-Geotechnical Engineering*, 169(2); 2015: 99–109.

O'Reilly, M. P. and B. M. New. Settlements above tunnels in the United Kingdom-their magnitude and prediction. In *Tunnelling 82 Proceedings of the 3rd International Symposium*, Institution of Mining & Metallurgy, 1982: 173–181.

Ostermayer, H. Construction, carrying behaviour and creep characteristics of ground anchors. In *Proceedings of ICE Conference on Diaphragm Walls and Anchorages*, London, 1974: 141–151.

Ostermayer, H. and F. Scheele. Research on ground anchors in non-cohesive soils. In *Proceedings of the 9th International Conference on Soil Mechanics and Foundation Engineering*, Tokyo, V97, 1977.

Otto, G. H. An interpretation of the glacial stratigraphy of the city of Chicago. PhD dissertation, Univ. of Chicago, 1942.

Parsons, A. W. and J. Perry. Slope stability problems in ageing highway earthworks. In *Proceedings of the Symposium on Failures in Earthworks*, Institution of Civil Engineers, London, 1985: 63–78.

Pedersen, S. A. S. *Glacitectonite: Brecciated Sediments and Cataclastic Sedimentary Rocks Formed Subglacially*, Vol. 43. *Geological Survey of Denmark*, Copenhagen, 1988.

Pennington, C., K. Freeborough, C. Dashwood, T. Dijkstra, and K. Lawrie. The National Landslide Database of Great Britain: Acquisition, communication and the role of social media. *Geomorphology*, 249; November 2015: 44–51.

Perry, J. *A Survey of Slope Condition on Motorway Earthworks in England and Wales*. Research Report-Transport and Road Research Laboratory, Vol. 199, 1989.

Perry, J., M. Pedley, and M. Reid. *Infrastructure Embankments: Condition Appraisal and Remedial Treatment*. CIRIA, London, 2003.

Persson, M. A., R. L. Stevens, and Å. Lemoine. Spatial quick-clay predictions using multi-criteria evaluation in SW Sweden. *Landslides*, 11(2); 2014: 263–279.

Phillips, E. R. and C. A. Auton. Micromorphological evidence for polyphase deformation of glaciolacustrine sediments from Strathspey, Scotland. *Geological Society, London, Special Publications*, 176(1); 2000: 279–292.

Pickles, A. Geotechnical design of retaining walls. In *ICE Manual of Geotechnical Engineering*, edited by Burland, J., Chapman, T., Skinner, H., and Brown, M., Thomas Telford Ltd, London, 2012: 981–999, Chapter 64.

Pickwell, R. The encroachments of the sea from Spurn Point to Flamborough Head and the works executed to prevent the loss of land. In *Minutes of the Proceedings of the Institution of Civil Engineers*, Thomas Telford, Vol. 51, 1878: 191–212.

Piotrowski, J. A. Genesis of the Woodstock drumlin field, southern Ontario, Canada. *Boreas*, 16(3); 1987: 249–265.

Piotrowski, J. A. and A. M. Kraus. Response of sediment to ice-sheet loading in northwestern Germany: Effective stresses and glacier-bed stability. *Journal of Glaciology*, 43(145); 1997: 489–502.

Piotrowski, J. A., N. K. Larsen, and F. W. Junge. Reflections on soft subglacial beds as a mosaic of deforming and stable spots. *Quaternary Science Reviews*, 23(9); 2004: 993–1000.

Piotrowski, J. A., N. K. Larsen, J. Menzies, and W. Wysota. Formation of subglacial till under transient bed conditions: Deposition, deformation, and basal decoupling under a Weichselian ice sheet lobe, central Poland. *Sedimentology*, 53(1); 2006: 83–106.

Piotrowski, J. A. and S. Tulaczyk. Subglacial conditions under the last ice sheet in northwest Germany: Ice-bed separation and enhanced basal sliding? *Quaternary Science Reviews*, 18(6); 1999: 737–751.

Plaxis. http://www.plaxis.nl/, 2016.

Polshin, D. E. and R. A. Tokar. Maximum allowable non-uniform settlement of structures. In *Proceedings Fourth International Conference on Soil Mechanics and Foundation Engineering*, 1; 1957: 402–405.

Potts, D. M., N. Kovacevic, and P. R. Vaughan. Delayed collapse of cut slopes in stiff clay. *Géotechnique*, 47(5); 1997: 953–982.

Potts, D. and L. Zdravkovic. Computer analysis principles in geotechnical engineering. In *ICE Manual of Geotechnical Engineering*, edited by Burland, J., Chapman, T., Skinner, H., and Brown, M., Thomas Telford Ltd, London, 2012: 35–56, Chapter 6.

Poulos, H. G. Pile behaviour – consequences of geological and construction imperfections 1. *Journal of Geotechnical and Geoenvironmental Engineering*, 131(5); 2005: 538–563.

Poulos, H. G., J. P. Carter, and J. C. Small. Foundations and retaining structures-research and practice. In *Proceedings of the 15th International Conference on Soil Mechanics and Geotechnical Engineering*, A A Balkema, Istanbul, Vol. 4, 2002: 2527–2606.

Poulos, H. G. and E. H. Davis. *Pile Foundation Analysis and Design*. Wiley, New York, 1980.

Powell, J. J. M. and C. R. I. Clayton. Field geotechnical testing. In *ICE Manual of Geotechnical Engineering*, edited by Burland, J., Chapman, T., Skinner, H., and Brown, M., Thomas Telford Ltd, London, 2012: 629–650, Chapter 47.

Powrie, W. Earth Pressure Theory. In *ICE Manual of Geotechnical Engineering*, edited by Burland, J., T. Chapman, H. Skinner, and M. Brown, Thomas Telford Ltd, London, 2012: 221–226, Chapter 20.

Powrie, W. and E. S. F. Li. Finite element analyses of an *in situ* wall propped at formation level. *Geotechnique*, 41(4); 1991: 499–514.

Prandtl, L. Über die härte plastischer körper. *Nachrichten von der Gesellschaft der Wissenschaften zu Göttingen, Mathematisch-Physikalische Klasse*, 1920; 1920: 74–85.

Prudic, D. E. Hydraulic conductivity of a fine-grained Till, Cattaraugus County, New York. *Ground Water* 20(2); 1982: 194–204.

Quigley, P. Modification/stabilisation of low strength cohesive soils under foundations and floor slabs. In Paper Presented to the Geotechnical Society of Ireland (GSI), 16th February 2006. The Institution of Engineers of Ireland.

Quinn, J. D., L. K. Philip, and W. Murphy. Understanding the recession of the Holderness Coast, east Yorkshire, UK: A new presentation of temporal and spatial patterns. *Quarterly Journal of Engineering Geology and Hydrogeology*, 42(2); 2009: 165–178.

Quinn, J. D., N. J. Rosser, W. Murphy, and J. A. Lawrence. Identifying the behavioural characteristics of clay cliffs using intensive monitoring and geotechnical numerical modelling. *Geomorphology*, 120(3); 2010: 107–122.

Ramberg, W. and W. R. Osgood. *Description of Stress-Strain Curves by Three Parameters*. NAC-TN-902, NASA, Washington, DC, 1943.

Randolph, M. F. Design methods for pile groups and piled rafts. In *Proceedings the 13th International Conference on Soil Mechanics and Foundation Engineering*, New Delhi, Balkeema, Rotterdam, 1994: 61–82.

Randolph, M. F. Science and empiricism in pile foundation design. *Geotechnique*, 53(10); 2003: 847–876.

Randolph, M. F. and P. Clancy. Efficient design of piled rafts. In *Proceedings 2nd International Seminar Deep Foundation*, Ghent, Belgium, 1993: 119–130.

Randolph, M. F. and A. J. Deeks. Dynamic and static soil models for axial pile response. In *Proceedings of the 4th International Conference on the Application of Stress Wave Theory to Piles*, The Hague, The Netherlands, 1992: 21–24.

Randolph, M. F. and C. P. Wroth. An analysis of the vertical deformation of pile groups. *Geotechnique*, 29(4); 1979: 423–439.

Randolph, M. F. and C. P. Wroth. Application of the failure state in undrained simple shear to the shaft capacity of driven piles. *Geotechnique*, 31(1); 1981: 143–157.

Reese, L. C. and M. W. O'Neill. *Criteria for the Design of Axially Loaded Drilled Shafts*. Final Rpt. Center for Highway Research, University of Texas at Austin, 1971.

Reynolds, J. M. *An Introduction to Applied and Environmental Geophysics*. John Wiley & Sons, Chichester, UK, 2011.

Reynolds, J. M. Geophysical exploration and remote sensing. In *ICE Manual of Geotechnical Engineering*, edited by Burland, J., Chapman, T., Skinner, H., and Brown, M., Thomas Telford Ltd, London, 2012: 601–618, Chapter 45.

Reynolds, J. M. Some basic guidelines for the procurement and interpretation of geophysical surveys in environmental investigations. In *Proceedings of the 4th International Conference, Construction on Polluted and Marginal Land*, edited by Forde, M. C., Brunel University, London, 1996: 57–64.

Robertson, T. L., B. G. Clarke, and D. B. Hughes. Geotechnical properties of Northumberland Till. *Ground Engineering*, 27(10); 1994: 29–34.

Rodin, S. Using wetter fills in motorway embankments. *Ground Engineering*, 4(1); 1969: 19–20.

Rönnqvist, H. Predicting surfacing internal erosion in moraine core dams. PhD Thesis, KTH, 2010.

Rönnqvist, H. Applying available internal erosion criteria to dams with cores of glacial till-a reassessment of a 1980s sinkhole. In *Proceedings 17th Biennial Conference of the British Dam Society, Dams: Engineering in a Social and Environmental Context, Institution of Civil Engineers*, 2012: 131–144.

Rönnqvist, H. A tale of two dams: A comparative study of performance and internal erosion. In *International Conference on Geotechnical Engineering: ICGE2015 Colombo*, 10–11 August 2015.

Rönnqvist, H. and P. Viklander. Laboratory testing of internal stability of glacial Tills, a Review. *Electronic Journal of Geotechnical Engineering*, 19(5); 2014: 6315–6336.

Rose, J. Status of the Wolstonian glaciation in the British Quaternary. *Quaternary Newsletter*, 53(9); 1987.

Rose, J. Castle Bytham. In Quaternary Research Association Field Meeting Guide – West Midlands, Quaternary Research Association, Cambridge, 1989a: 117–122.

Rose, J. Glacier stress patterns and sediment transfer associated with the formation of superimposed flutes. *Sedimentary Geology*, 62(2); 1989b: 151–176.

Rose, J. and J. M. Letzer. Superimposed Drumlinsm. *Journal of Glaciology*, 18(80); 1977: 471–480.

Röthlisberger, H. Water pressure in intra-and subglacial channels. *Journal of Glaciology*, 11(62); 1972: 177–203.

Rowe, P. W. The relevance of soil fabric to site investigation practice. *Geotechnique*, 22(2); 1972: 195–300.

RP2A-WSD, A. P. I. *Recommended Practice for Planning, Designing and Constructing Fixed Offshore Platforms–Working Stress Design*. American Petroleum Institute, Washington, 2002.

Saarnisto, M. An outline of glacial indicator tracing. In *Glacial Indicator Tracing*, edited by Kujansuu, R. and Saarnisto, M., A. A. Balkema, Rotterdam, 1990: 1–15.

Salgado, R. *The Engineering of Foundations*. New York: McGraw-Hill, 2008, ISBN 978-007-125940-9.

Sarala, P., J. Räisänen, P. Johansson, and K. O. Eskola. Aerial LiDAR analysis in geomorphological mapping and geochronological determination of surficial deposits in the Sodankylä region, northern Finland. *GFF*, 137(4); 2015: 293–303.

Sariosseiri, F. and B. Muhunthan. Effect of cement treatment on geotechnical properties of some Washington State soils. *Engineering Geology*, 104(1); 2009: 119–125.

Sarma, S. K. Stability analysis of embankments and slopes. *Journal of Geotechnical and Geoenvironmental Engineering* 105(12); 1979: 1511–1524.

Sauer, E. K., A. K. Egeland, and E. A. Christiansen. Preconsolidation of tills and intertill clays by glacial loading in southern Saskatchewan, Canada. *Canadian Journal of Earth Sciences*, 30(3); 1993: 420–433.

Scheuchzer, J. J. *Johannis Jacobi Scheuchzeri ... Herbarium diluvianum*. Sumptibus Petri Vander Aa, 1723.

Schlosser, F. and I. Juran. Design parameters for artificially improved soils. In *British Geotechnical Society Conference*. 1981.

Schmertmann, J. H., J. P. Hartman, and P. R. Brown. Improved strain influence factor diagrams. *Journal of Geotechnical and Geoenvironmental Engineering*, 104, no. Tech Note, 1978.

Schmuker, C. Comparison of static load tests and Statnamic load test. German: Vergleich statischer und statnamischer Pfahlprobebelastungen). *MSc Thesis*, Biberach University, Germany, 2005.

Schneider, H. R. Determination of characteristic soil properties Determination des valeurs caracteristiques. In *Geotechnical Engineering for Transportation Infrastructure: Theory and Practice, Planning and Design, Construction and Maintenance: Proceedings of the 12th European Conference on Soil Mechanics and Geotechnical Engineering*, Taylor & Francis, Amsterdam, Netherlands, 1999: 273.

Schneider, D., P. Bartelt, J. Caplan-Auerbach, M. Christen, C. Huggel, and B. W. McArdell. Insights into rock-ice avalanche dynamics by combined analysis of seismic recordings and a numerical avalanche model. *Journal of Geophysical Research: Earth Surface*, 115(F4); 2010: 1–20.

Schweizer, J. and A. Iken. The role of bed separation and friction in sliding over an undeformable bed. *Journal of Glaciology*, 38(128); 1992: 77–92.

Seed, H. B. and I. M. Idriss. *Soil Modulus and Damping Factors for Dynamic Response Analysis.* University of California Report GRC, 1970.

Seierstad, H.-H. Prøveforstyrrelese i leire vurdering av Ø75 mm stempelprøvetaker. Diploma Thesis, NTNU Trondheim, 2000.

Semple, R. M. and J. P. Gemeinhardt. Stress history approach to analysis of soil resistance to pile driving. In *Offshore Technology Conference*, Housto, TX, 1981.

Semple, R. M. and W. John Rigden. Shaft capacity of driven pipe piles in clay. In *Analysis and Design of Pile Foundations*, edited by Meyer, J. R., ASCE, New York, 1984: 59–79.

Sevaldson, R. A. The slide in Lodalen, October 6th, 1954. *Geotechnique*, 6(4); 1956: 167–182.

Sharp, J. M. Hydrogeologic Characteristics of Shallow Glacial Drift Aquifers in Dissected Till Plains (North-Central Missouri). *Ground Water* 22(6); 1984: 683–689.

Sharp, M. Surging glaciers behaviour and mechanisms. *Progress in Physical Geography*, 12(3); 1988: 349–370.

Sharp, M., J. Campbell Gemmell, and J.-L. Tison. Structure and stability of the former subglacial drainage system of the Glacier de Tsanfleuron, Switzerland. *Earth Surface Processes and Landforms*, 14(2); 1989: 119–134.

Shaw, J. Drumlin formation related to inverted melt-water erosional marks. *Journal of Glaciology*, 29(103); 1983: 461–479.

Sherard, J. L., Sinkholes in dams of coarse, broadly graded soils. In *13th ICOLD Congress*, India, 2; 1979: 25–35.

Sherard, J. L. and L. P. Dunnigan. Critical filters for impervious soils. *Journal of Geotechnical Engineering*, 115(7); 1989: 927–947.

Shilts, W. W. Geological Survey of Canada's contributions to understanding the composition of glacial sediments. *Canadian Journal of Earth Sciences*, 30(2); 1993: 333–353.

Shilts, W. W., C. M. Cunningham, and C. A. Kaszycki. Keewatin Ice Sheet – Re-evaluation of the traditional concept of the Laurentide Ice Sheet. *Geology*, 7(11); 1979: 537–541.

Shilts, W. W. and S. L. Smith. Drift prospecting in the Appalachians of Estrie-Beauce, Québec. In *Drift Prospecting*, edited by DiLabio, R. N. W. and Coker, W. B. Geological Survey of Canada, Paper 89–20, 1989: 41–59.

Shreve, R. L. Movement of water in glaciers. *Journal of Glaciology*, 11(62); 1972: 205–214.

Shumway, J. R. and N. R. Iverson. Magnetic fabrics of the Douglas Till of the Superior lobe: Exploring bed-deformation kinematics. *Quaternary Science Reviews*, 28(1); 2009: 107–119.

SHW. *Manual of Contract Documents for Highway Works: Volume 1 Specification for Highway Works: Series 600: Earthworks.* Highways England, London, 2013.

Simons, N. E. and Menzies, B. K. *A Short Course in Foundation Engineering*, Vol. 5. Thomas Telford, London, 2000.

Sims, J. E., D. Elsworth, and J. A. Cherry. Stress-dependent flow through fractured clay till: A laboratory study. *Canadian Geotechnical Journal*, 33(3); 1996: 449–457.

Sissons, J. B. The geomorphology of central Edinburgh. *Scottish Geographical Magazine*, 87(3); 1971: 185–196.

Skempton, A. W. The colloidal activity of clays. In *Selected Papers on Soil Mechanics*. Thomas Telford, London, 1953: 106–118.

Skempton, A. W. Horizontal stresses in an over-consolidated Eocene clay. In *Proceedings of the 5th International Conference on Soil Mechanics and Foundation Engineering*, Paris, France, V1, 1961: 351–357.

Skempton, A. W. Long-term stability of slopes. *Geotechnique*, 14(2); 1964: 75–102.

Skempton, A. W. The consolidation of clays by gravitational compaction. *Quarterly Journal of the Geological Society*, 125(1–4); 1969: 373–411.

Skempton, A. W. First-time slides in over-consolidated clays. *Geotechnique*, 20(3); 1970: 320–324.

Skempton, A. W. Slope stability of cuttings in brown London clay. In *Selected Papers on Soil Mechanics*. Thomas Telford Publishing, London, 1984: 241–250.

Skempton, A. W. Standard penetration test procedures and the effects in sands of overburden pressure, relative density, particle size, ageing and overconsolidation. *Geotechnique*, 36(3); 1986: 425–447.

Skempton, A. W. and O. T. Jones. Notes on the compressibility of clays. *Quarterly Journal of the Geological Society*, 100(1–4); 1944: 119–135.

Skempton, A. W. and A. W. Bishop. Soils. In *Chapter 10 Building Materials*, edited by Reiner, M., Amsterdam, 1954: 417–482.

Skempton, A. W. and L. Bjerrum. A contribution to the settlement analysis of foundations on clay. *Geotechnique*, 7(4); 1957: 168–178.

Skempton, A.W. and Delory, F.A. Stability of natural slopes in London Clay. In *Proceedings Fourth Conference on Soil Mechanics London*, 2; 1957: 378–381.

Skempton, A. W. and J. D. Brown. A landslide in boulder clay at Selset, Yorkshire. *Geotechnique*, 11(4); 1961: 280–293.

Skempton, A. W. and O. T. Jones. Notes on the compressibility of clays. *Quarterly Journal of the Geological Society*, 100(1–4); 1944: 119–135.

Skempton, A. W., A. D. Leadbeater, and R. J. Chandler. The Mam Tor landslide, North Derbyshire. *Philosophical Transactions of the Royal Society of London A: Mathematical, Physical and Engineering Sciences*, 329(1607); 1989: 503–547.

Skempton, A. W. and D. H. MacDonald. The allowable settlements of buildings. *Proceedings of the Institution of Civil Engineers*, 5(6); 1956: 727–768.

Skempton, A. W. and V. A. Sowa. The behaviour of saturated clays during sampling and testing. *Geotechnique*, 13(4); 1963: 269–290.

Sladen, J. A. and W. Wrigley. Geotechnical properties of lodgement till – a review. In *Glacial Geology: An Introduction for Engineers and Earth Scientists*, edited by Eyles, N., Pergamon Press, Oxford, 1983: 184–212.

Small, R. J. Englacial and supraglacial sediment: Transport and deposition. *Glacio-Fluvial Sediment Transfer: An Alpine Perspective*. John Wiley and Sons, New York, 1987: 111–145.

Smart, C. C. Some observations on subglacial ground-water flow. *Journal of Glaciology*, 32(111); 1986: 232–234.

Smith, H. H. Gravel compaction and testing and concrete mix design at London Airport. *Proceedings of the Institution of Civil Engineers*, 1(1); 1952: 1–54.

Smith, N. D. and G. M. Ashley. Proglacial lacustrine environment. In *Glacial Sedimentary Environments*, edited by Ashley, G. M., J. Shaw, and N. D. Smith, No. 16. Society of Economic Paleontologists and Mineralogists, Tulsa, 1985, Chapter 4.

Smith, M. J., J. D. Black, F. M. Lindsay, and S. B. Mickovski. Soil nailing the green way: Sustainable stabilisation of a failing slope using innovative soil nail head design to give a fully vegetated green slope finish. In *Proceedings of the XVI European Conference on Soil Mechanics and Geotechnical Engineering*, Thomas Telford, Edinburgh, London, 2015: 1885–1890.

Smith, M. J. and C. D. Clark. Methods for the visualization of digital elevation models for landform mapping. *Earth Surface Processes and Landforms*, 30(7); 2005: 885–900.

SSIG. *Site Investigation Steering Group. Specification for Ground Investigation. No. 3.* Thomas Telford, London, 1993.

Stark, T. D., H. Choi, and S. McCone. Drained shear strength parameters for analysis of landslides. *Journal of Geotechnical and Geoenvironmental Engineering*, 131(5); 2005: 575–588.
Steinbrenner, W. Tafeln zur Setzungsberechnung. *Die Strasse*, V1, 1934: 121–124.
Stephenson, D. A., A. H. Fleming, and D. M. Mickelson. Glacial deposits. In *Vol O2 Hydrology: The Geological Society of America*, edited by Back, W., J. S. Rosenshein, and P. R. Seaber, The Geology of North America; 1988: Chapter 5.
Stevens, R. S., E. A. Wiltsie, and T. H. Turton. Evaluating Drivability for Hard Clay, Very Dense Sand, and Rock. In *Offshore Technology Conference*, Houston, TX, 1982.
Stokes, C. R., C. D. Clark, O. B. Lian, and S. Tulaczyk. Ice stream sticky spots: A review of their identification and influence beneath contemporary and palaeo-ice streams. *Earth-Science Reviews*, 81(3); 2007: 217–249.
Strandgaard, T. and L. Vandenbulcke. Driving mono-piles into glacial till. *IBC's Wind Power Europe*, 2002.
Stroud, M. A. and F. G. Butler. The standard penetration test and the engineering properties of glacial materials. In *Symposium on Engineering Properties of Glacial Materials, Midland Geotechnical Society*, 1975.
Suwansawat, S. and H. H. Einstein. Describing settlement troughs over twin tunnels using a superposition technique. *Journal of Geotechnical and Geoenvironmental Engineering*, 133(4); 2007: 445–468.
Tanaka, H. Pore size distribution and hydraulic conductivity characteristics of marine clays. In *2nd International Symposium on Contaminated Sediments*, Quebec, 2003: 151–157.
Tanaka, H., P. Sharma, T. Tsuchida, and M. Tanaka. Comparative study on sample quality using several types of samplers. *Soils and Foundations* 36(2); 1996: 57–68.
Tausch, N. A special grouting method to construct horizontal membranes. In *Proceeding of the International Symposium on Recent Developments in Grout Improvement Techniques*, 1985: 351–362.
Taylor, D. W. *Fundamentals of Soil Mechanics*. John Wiley and Sons, New York; 1948: 700p.
Taylor, J., A. Reid, R. Simpson, R. Cameron, P. McLaughlin, P. Bryson, R. Scott, and P. Morgan. Ground investigation and Eurocode 7: A Scottish perspective. *Ground Engineering*, 44(7); 2011: 24–29.
Terzaghi, K. Discussion of settlement of structures. In *Proceedings of the 1st International Conference on Soil Mechanics and Foundation Engineering*, Cambridge, UK, Vol. 3, 1936: 79–87.
Terzaghi, K. Undisturbed clay samples and undisturbed clays. *Journal of the Boston Society of Civil Engineers*, 28(3); 1941: 211–231.
Terzaghi, K. Geologic aspects of soft-ground tunnelling. In *Applied Sedimentation*, edited by Trask, P. D., Wiley, New York, 1950: 193–209.
Terzaghi, K., R. B. Peck, and G. Mesri. *Soil Mechanics in Engineering Practice*. John Wiley & Sons, New York, 1996.
Thabet, K. M. A. Geotechnical Properties and Sedimentation Characteristics of Tills in S.E. Northumberland, PhD Thesis, Newcastle University, UK, 1973.
Thomas, H. S. H. and W. H. Ward. The design, construction and performance of a vibrating-wire earth pressure cell. *Geotechnique*, 19(1); 1969: 39–51.
Thomason, J. F. and N. R. Iverson. Microfabric and microshear evolution in deformed till. *Quaternary Science Reviews*, 25(9); 2006: 1027–1038.
Thomason, J. F. and N. R. Iverson. A laboratory study of particle ploughing and pore-pressure feedback: A velocity-weakening mechanism for soft glacier beds. *Journal of Glaciology*, 54(184); 2008: 169–181.
Tomlinson, M. J. and R. Boorman. *Foundation Design and Construction*. Pearson Education, Harlow, UK, 2001.
Trenter, N. A. *Engineering in Glacial Tills*. CIRIA, London, 1999.
Trenter, N. A. *Earthworks: A Guide*. Thomas Telford, London, 2001.
Trenter, N. A. and J. A. Charles. A model specification for engineered fills for building purposes. *Proceedings of the Institution of Civil Engineers – Geotechnical Engineering*, 119(4); 1996: 219–230.

Truffer, M., K. A. Echelmeyer, and W. D. Harrison. Implications of till deformation on glacier dynamics. *Journal of Glaciology*, 47(156); 2001: 123–134.

Truffer, M. and W. D. Harrison. *In situ* measurements of till deformation and water pressure. *Journal of Glaciology*, 52(177); 2006: 175–182.

Tulaczyk, S. Structure of subglacial sediments from beneath Ice Stream B, West Antarctica, and their possible modes of deformation. In *Program and Abstracts of the American Quaternary Association Conference*, Vol. 13; 1994: 171.

Tulaczyk, S. Ice sliding over weak fine-grained soils: Dependence of ice-till interactions on till granulometry. In *Glacial Processes Past and Present.* Special papers, No. 337. Geological Society of America, Boulder, 337; 1999: 159.

Tulaczyk, S., W. Barclay Kamb, and H. F. Engelhardt. Basal mechanics of ice stream B, West Antarctica: 1. Till mechanics. *Journal of Geophysical Research: Solid Earth*, 105(B1); 2000: 463–481.

Tulaczyk, S., B. Kamb, R. P. Scherer, and H. F. Engelhardt. Sedimentary processes at the base of a West Antarctic ice stream: Constraints from textural and compositional properties of subglacial debris. *Journal of Sedimentary Research*, 68(3); 1998.

Tyndall, J. *The Forms of Water in Clouds and Rivers, Ice and Glaciers.* With 35 Illustrations drawn and engraved under the Direction of the Author. Henry S. King & Company, 1873.

Valentin, H. Land loss at holderness. In *Applied Coastal Geomorphology*, edited by Steers, J. A. MacMillan, London; 1971: 116–137.

van der Meer, J. J. M. Micromorphology of glacial sediments as a tool in distinguishing genetic varieties of till. *Geological Survey of Finland, Special Paper*, 3; January 1987: 77–89.

van der Meer, J. J. M. Microscopic evidence of subglacial deformation. *Quaternary Science Reviews*, 12(7); 1993: 553–587.

van der Meer, J. J. M. Micromorphology. In *Glacial Environments, Past Glacial Environments – Processes, Sediments and Landforms*, edited by Menzies, J., Vol. 2. Butterworth and Heinemann, Oxford, 1996: 335–355.

van der Meer, J. J. M. Particle and aggregate mobility in till: Microscopic evidence of subglacial processes. *Quaternary Science Reviews*, 16(8); 1997: 827–831.

van der Meer, J. J. M., J. Menzies, and J. Rose. Subglacial till: The deforming glacier bed. *Quaternary Science Reviews*, 22(15); 2003: 1659–1685.

Van der Wateren, F. M., S. J. Kluiving, and L. R. Bartek. Kinematic indicators of subglacial shearing. *Geological Society, London, Special Publications*, 176(1); 2000: 259–278.

Van Landeghem, K. J. J., A. J. Wheeler, and N. C. Mitchell. Seafloor evidence for palaeo-ice streaming and calving of the grounded Irish Sea Ice Stream: Implications for the interpretation of its final deglaciation phase. *Boreas*, 38(1); 2009: 119–131.

Varnes, D. J. *Slope Movement Types and Processes.* Transportation Research Board Special Report 176, 1978.

Vaughan, P. R. Assumption, prediction and reality in geotechnical engineering. *Geotechnique*, 44(4); 1994: 573–609.

Vaughan, P. R., R. J. Chandler, J. P. Apted, W. M. Maguire, and S. S. Sandroni. Sampling disturbance-with particular reference to its effect on stiff clays. In *Predictive Soil Mechanics, Proceedings of the Wroth Memorial Symposium*, Oxford, 1993: 224–242.

Vaughan, P. R. and M. M. A. F. Hamza. Clay embankments and foundations: Monitoring stability by measuring deformations. In *Proc. Proceedings of the Ninth International Conference on Soil Mechanics and Foundation Engineering*, Tokyo, Speciality Session, no. 8, 1977: 37–48.

Vaughan, P. R., D. W. Hight, V. G. Sodha, and H. J. Walbancke. Factors controlling the stability of clay fills in Britain. In *Proceedings of a Conference on Clay Fills*, Institution of Civil Engineers, Thomas Telford, London, 1978: 203–217.

Vaughan, P. R., H. T. Lovenbury, and P. Horswill. The design, construction and performance of Cow Green embankment dam. *Geotechnique*, 25(3); 1975: 555–580.

Vaughan, P. R. and H. J. Walbancke. The stability of cut and fill slopes in boulder clay. In *Proceedings of the Symposium on the Engineering Behaviour of Glacial Materials*, Birmingham, 1975: 209–219.

Vermeer, P. A., N. Ruse, and T. Marcher. Tunnel heading stability in drained ground. *Felsbau*, 20(6); 2002: 8–18.

Vesic, A. S. Bearing capacity of shallow foundations. In *Foundation Engineering Handbook*. Van Nostrand Reinhold Book Co, New York, 1975: 121–147, Chapter 3.

Vijayvergiya, V. N. Load-movement characteristics of piles. In Ports'77. 4th Annual Symposium of the American Society of Civil Engineers, Waterway, Port, Coastal and Ocean Division, Long Beach, California, Vol. 2, 1977: 269–284.

Vuillermet, E. *Les argiles glacio-lacustres du Trièves*. Mém. DEA Univ. de Grenoble, 1989: 55.

Waddington, E. D. Life, death and afterlife of the extrusion flow theory. *Journal of Glaciology*, 56(200); 2010: 973–996.

Wade, S., B. Handley, and J. Martin. Types of bearing piles. In *ICE Manual of Geotechnical Engineering*, edited by Burland, J., Chapman, T., Skinner, H., and Brown, M., Thomas Telford Ltd, London, 2012: 1191–1223.

Walder, J. Geometry of former subglacial water channels and cavities. *Journal of Glaciology*, 23(89); 1979: 335–346.

Walder, J. S. and A. Fowler. Channelized subglacial drainage over a deformable bed. *Journal of Glaciology*, 40(134); 1994: 3–15.

Walker, R. G. Facies, facies models and modern stratigraphic concepts. In *Facies Models: Response to Sea Level Change*, edited by Walker, R. G. and James, N. Geological Association of Canada, 1992: 1–14.

Wang, X., R. Tweedie, and R. Clementino. Full-scale pile loading tests on instrumented concrete piles in clay till, in Edmonton, Alberta. In *Proceedings 68th Canadian Geotechnical Conference*, Quebec, Canadian Geotechnical Society, 2015.

Warren, W. P. and G. M. Ashley. Origins of the ice-contact stratified ridges (eskers) of Ireland. *Journal of Sedimentary Research*, 64(3); 1994: 433–449.

Wartman, J., J. R. Keaton, A. Scott, J. Benoit, J. delaChapelle, R. Gilbert, and D. R. Montgomery. The 22 March 2014 Oso Landslide, Snohomish County, Washington: Findings of the GEER Reconnaissance Investigation. In *AGU Fall Meeting Abstracts*, vol. 1, 2014: 04.

Watabe, Y., J-P. LeBihan, and S. Leroueil. Probabilistic modelling of saturated/unsaturated hydraulic conductivity for compacted glacial tills. *Géotechnique*, 56(4); 2006: 273–284.

Weaver, T. J. and K. M. Rollins. Reduction factor for the unloading point method at clay soil sites. *Journal of Geotechnical and Geoenvironmental Engineering*, 136(4); 2009: 643–646.

Weertman, J. On the sliding of glaciers. *Journal of Glaciology*, 3(21); 1957: 33–38.

Weertman, J. General theory of water flow at the base of a glacier or ice sheet. *Reviews of Geophysics*, 10(1); 1972: 287–333.

Weltman, A. J. and P. R. Healy. *Piling in 'Boulder Clay' and Other Glacial Tills*. CIRIA Report PG5 Monograph, 1978.

Whitley, H. M. Earthwork slips on the Castle Eden and Stockton Railway. *Minutes of the Proceedings of the Institution of Civil Engineers*, V62(1880); 1880: 280–284 (Thomas Telford-ICE Virtual Library).

Wiens, D. A., S. Anandakrishnan, J. Paul Winberry, and M. A. King. Simultaneous teleseismic and geodetic observations of the stick-slip motion of an Antarctic ice stream. *Nature*, 453(7196); 2008: 770–774.

Williams, H. and R. T. Stain. Pile integrity testing – Horses for courses. In *Proceedings of the International Conference on Foundations and Tunnels*, London, 1987.

Winberry, J. P., S. Anandakrishnan, R. B. Alley, R. A. Bindschadler, and M. A. King. Basal mechanics of ice streams: Insights from the stick-slip motion of Whillans Ice Stream, West Antarctica. *Journal of Geophysical Research: Earth Surface*, 114(F1); 2009.

Winter, M. G. and S. Suhardi. The effect of stone content on the determination of acceptability for earthworking. In *Proceedings Engineered Fills Conference*, Newcastle upon Tyne, Thomas Telford, London, 1993: 312–319.

Winter, M. G. and S. Suhardi. The effect of stone content on the determination of acceptability for earthworking. In *Engineered Fill*, edited by Clarke, B. G., C. J. F. P. Jones, and A. I. B. Moffat, Thomas Telford, London. 1993: 312–319.

Winter, M. G. and Th Hólmgeirsdóttir. The effect of large particles on acceptability determination for earthworks compaction. *Quarterly Journal of Engineering Geology and Hydrogeology* 31(3); 1998: 247–268.

Winter, M. G., M. Harrison, F. Macgregor, and L. Shackman. Landslide hazard and risk assessment on the Scottish road network. *Proceedings of the Institution of Civil Engineers-Geotechnical Engineering*, 166(6); 2013: 522–539.

Wisniewski, R. T., M. Weckler, and A. H. Brand. Subsurface conditions and foundation solutions for the New Yankee Stadium. In *Geo-Frontiers 2011: Advances in Geotechnical Engineering*, edited by Han, J. and D. E. Alzamora. ASCE, Washington, 2011: 85–93.

WGMS: Global Glacier Change Bulletin No. 1 (2012–2013). Edited by Zemp, M., Gärtner-Roer, I., S. U. Nussbaumer, F. Hüsler, H. Machguth, N. Mölg, F. Paul, and M. Hoelzle. ICSU(WDS) / IUGG(IACS)/ UNEP/UNESCO/WMO, World Glacier Monitoring Service, Zurich, Switzerland, 2015: 230 pp.

Wroth, C. P. and D. M. Wood. The correlation of index properties with some basic engineering properties of soils. *Canadian Geotechnical Journal*, 15(2); 1978: 137–145.

Yamada, M., H. Kumagai, Y. Matsushi, and T. Matsuzawa. Dynamic landslide processes revealed by broadband seismic records. *Geophysical Research Letters*, 40(12); 2013: 2998–3002.

Yassir, N. A. The undrained shear behaviour of fine-grained sediments. *Geological Society, London, Special Publications*, 54(1); 1990: 399–404.

Zdravkovic, L., D. M. G. Taborda, D. M. Potts, R. J. Jardine, M. Sideri, F. C. Schroeder, B. W. Byrne et al. Numerical modelling of large diameter piles under lateral loading for offshore wind applications. *Third International Symposium on Frontiers in Offshore Geotechnics (ISFOG 2015)*, Oslo Norway, 2015.

Zoet, L. K. and N. R. Iverson. Experimental determination of a double-valued drag relationship for glacier sliding. *Journal of Glaciology*, 61(225); 2015: 1–7.

Index

Milton Keynes UK
Ingram Content Group UK Ltd.
UKHW052026071024
449327UK00027B/2443